FUNDAMENTALS OF ELECTRICAL ENGINEERING

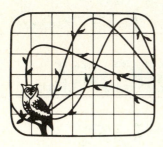

**HRW
Series in
Electrical and
Computer Engineering**

M. E. Van Valkenburg, Series Editor

FUNDAMENTALS OF ELECTRICAL ENGINEERING

Leonard S. Bobrow

UNIVERSITY OF MASSACHUSETTS, AMHERST

Holt, Rinehart and Winston, Inc.
New York   Chicago   San Francisco   Philadelphia
Montreal   Toronto   London   Sydney   Tokyo
Mexico City   Rio de Janeiro   Madrid

Acquisitions Editor: Deborah L. Moore
Production Manager: Paul Nardi
Project Editor: Bob Hilbert
Interior Design: Rita Naughton
Cover Illustration: Marc Cohen
Cover Design: Gail Lovig
Illustrations: J&R Services

Library of Congress Cataloging in Publication Data

Bobrow, Leonard S.
  Fundamentals of electrical engineering.

  Includes index.
  1. Electric engineering.   I. Title.
TK146.B693   1984   621.3   84-15829

ISBN 0-03-059571-1

Printed in the United States of America
Published simultaneously in Canada
5 6 7   016   9 8 7 6 5 4 3 2

CBS College Publishing
Holt, Rinehart and Winston
The Dryden Press
Saunders College Publishing

In memory of my sister Adrienne
and my brother-in-law Jay,
who introduced me to electrical engineering

CONTENTS

PREFACE

This book is intended to be the text for a first course in electrical engineering. It is partitioned into four parts: circuits, electronics, digital systems, and electromechanics. Although many topics are covered in each of these parts, this book is more than just a survey of the basics of electrical engineering. Even at this introductory level, material is usually covered in sufficient detail such that the reader may gain a good understanding of the fundamental principles on which modern electrical engineering is based.

The circuits portion of this book includes the traditional topics of Ohm's law, Kirchhoff's laws, resistive analysis techniques, various circuit theorems and principles, time-domain and frequency-domain analysis procedures, power, three-phase circuits, resonance, frequency response, and elementary system concepts. For this section of the book, it is assumed that the reader has had a course in college freshman-level (or higher) physics in which the concepts of the physics of electricity and magnetism (e.g., electric charge, electric potential, and magnetic fields) were introduced and studied. In addition, it is assumed that the reader has had one or more courses in elementary calculus in which the topics of differentiation and integration were covered.

The electronics portion of this book deals with both theory and applications of the major semiconductor devices: diodes and transistors—bipolar junction transistors (BJTs) and field-effect transistors (FETs)—in both discrete and integrated-circuit (IC) form. In addition to the coverage of the application of semiconductor devices to digital logic circuits, established analog topics such as small-signal, operational, and power amplifiers are included. Also covered are the concepts of feedback, oscillations, and elementary notions of communications via modulation. For a thorough understanding of the electronics section of the book, knowledge of a number of the principles in the circuits section is necessary. When studying particular electronics topics, it should be obvious which circuits material is a prerequisite.

In the digital systems portion of this book, basic digital logic elements and logic design in both discrete and IC forms are covered. Included is sequential as well as combinational logic. In addition to the design of logic circuits which perform various types of operations, digital devices which count and store information are discussed. The digital systems section requires a minimal background from the circuits portion of the book. Familiarity with the electronics material, while illuminating as to the internal behavior of digital logic elements, is not a prerequisite for this part of the book.

The electromechanics portion of this book covers topics such as magnetic circuits, magnetic induction, and transformers on an elementary level. In addition, the basic principles of electromechanical devices such as transducers, meters, generators, and motors are studied. This section of the book employs numerous concepts from the circuits part of the book.

At the end of each chapter there is a set of problems. Answers to selected problems are located in the back of the book. Students may use these problems as drill exercises.

In the study of electrical engineering I feel that it is imperative that a student solve problems related to the material covered. This not only reinforces the understanding of the subject matter, but in some cases, it allows for the extension of various concepts discussed in the text. In certain situations, even new material is introduced via the problem sets.

This book can be used as the text for various types of courses. The entire book may be employed as the text for a two-semester or three-quarter sequence which is an introduction to electrical engineering. The circuits section (Chapters 1–5) and the electromechanics section (Chapters 14 and 15) can be utilized as the text for a course on introductory circuits. Portions of the circuits section and the electronics section (Chapters 6–10) can be used as the text for a first electronics course. Furthermore, a part (Chapters 6–8) of the electronics section along with the digital-systems section (Chapters 11–13) can serve as the text for a course on digital electronics and logic design.

I would like to thank the reviewers who read and commented on portions of the manuscript. I received many excellent comments and suggestions. The reviewers were:

Professor Edwin J. Allen, University of Florida
Professor David Besch, University of Nevada at Reno
Professor John C. Clegg, Brigham Young University
Professor Fawzi Emad, University of Maryland
Professor Ross Johnson, Michigan Technological University
Professor Allen Nussbaum, University of Minnesota
Professor William J. Ross, Pennsylvania State University
Professor Andrew Sage, George Mason University
Professor Donald Thorn, University of Akron
Professor David Wade, University of Lowell
Professor Richard Williams, University of Akron
Professor Willard Worley, Texas A & M University

I would also like to thank Mr. Larry Leshan for his graphic-art work and Mr. Bob Hilbert for his contribution to the production of this book.

LEONARD S. BOBROW

FUNDAMENTALS
OF ELECTRICAL
ENGINEERING

PART I
Circuits

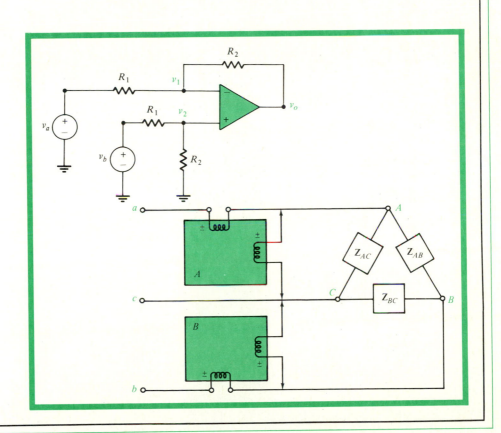

Chapter 1

Basic Elements and Laws

INTRODUCTION

The study of electric circuits is fundamental in electrical engineering education and can be quite valuable in other disciplines as well. The skills acquired not only are useful in such electrical engineering areas as electronics, communications, microwaves, electromechanics, and control and power systems, but also can be employed in other, seemingly different fields (e.g., acoustics and mechanics).

By an **electric circuit** or **network,** we mean a collection of electric devices (e.g., voltage and current sources, resistors, inductors, capacitors, transformers, amplifiers, and transistors) that are interconnected in some manner. Our prime interest in this section will be in the process of determining the behavior of a given circuit; this is referred to as **analysis.**

We begin our study by discussing some basic electric elements and the laws that describe them. It is assumed that the reader has been introduced to the concepts of electric charge, potential, and current in various science and physics courses in high school and college.

1.1 IDEAL SOURCES

Electric charge* is measured in **coulombs** (abbreviated C) in honor of the French scientist Charles de Coulomb (1736–1806); the unit of work or energy—the **joule**

---

*An electron has a charge of 1.6×10^{-19} C.

(J)—is named for the British physicist James P. Joule (1818–1889). Although for energy expended on electric charge the unit is J/C, we give it the special name **volt** (V) in honor of the Italian physicist Alessandro Volta (1745–1827), and we say that it is a measure of **electric potential difference** or **voltage.** These units are part of the **Système International d'Unités** (International System of Units). Units of this system are referred to as **SI** units. Unless indicated to the contrary, SI units are the units used in this book.

An **ideal voltage source,** which is represented in Fig. 1.1, is a device that produces a voltage or potential difference of v volts across its terminals *regardless of what is connected to it.*

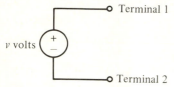

Fig. 1.1 Ideal voltage source.

For the device shown in Fig. 1.1, terminal 1 is marked plus (+) and terminal 2 is marked minus (−). This denotes that terminal 1 is at an electric potential that is v volts higher than that of terminal 2. (Alternatively, the electric potential of terminal 2 is v volts lower than that of terminal 1.)

The quantity v can be either a positive or a negative number. For the latter case, it is possible to obtain an equivalent source with a positive value as demonstrated by the following example.

Example 1.1

Suppose that $v = -5$ V for the voltage source shown in Fig. 1.1. Then the potential at terminal 1 is -5 V higher than that of terminal 2. However, this is equivalent to saying that terminal 1 is at a potential of $+5$ V lower than terminal 2. Consequently, the two ideal voltage sources shown in Fig. 1.2 are equivalent.

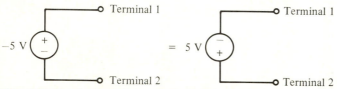

Fig. 1.2 Equivalent ideal voltage sources.

In the discussion above, we may have implied that the value of an ideal voltage source is constant, that is, it does not change with time. Such a situation is plotted in Fig. 1.3. For occasions such as this, an ideal voltage source is commonly represented by the equivalent notation shown in Fig. 1.4. We refer to such a device as an ideal **battery.** Although an actual battery is not ideal, there are many circumstances under which an ideal battery is a very good approximation. One such example is the 9-V battery that you use for your portable transistor radio—or you may have the type that uses four or six C or D $1\frac{1}{2}$-V batteries. A 12-V automobile storage battery is another case in point. In general, however, the voltage produced by an ideal voltage source will be a function of time. A few of the multitude of possible voltage waveforms are shown in Fig. 1.5. The waveform in Fig. 1.5(a) is a typical AM (amplitude modulation) signal while that in Fig. 1.5(b) is an FM (frequency modulation) signal. Both types of signals are used in consumer radio communications. The sinusoid shown in Fig. 1.5(c) has a wide variety of uses— this is the shape of ordinary household voltage. The frequency of such a waveform is often used as a reference signal, such as for color detection in television. Also, the sawtooth waveshape shown in Fig. 1.5(d) is used to sweep the electron beam in a television set (and an oscilloscope, as well). A pulse train such as that in Fig. 1.5(e) is part of a TV video signal—it contains the horizontal and vertical synchronization information. The waveforms shown in Fig. 1.5(f) and 1.5(g) are obtained from Fig. 1.5(e). That in Fig. 1.5(f) synchronizes the horizontal oscillator, and the one in Fig. 1.5(g) the vertical oscillator.

Fig. 1.3 Constant voltage.

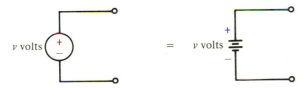

Fig. 1.4 Battery symbols.

Since the voltage produced by a source is, in general, a function of time, say $v(t)$, then the most general representation of an ideal voltage source is that shown in Fig. 1.6. There should be no confusion if the units "volts" are not included in the representation of the source. Thus, the ideal voltage source in Fig. 1.7 is identical to the one in Fig. 1.6 with "volts" being understood.

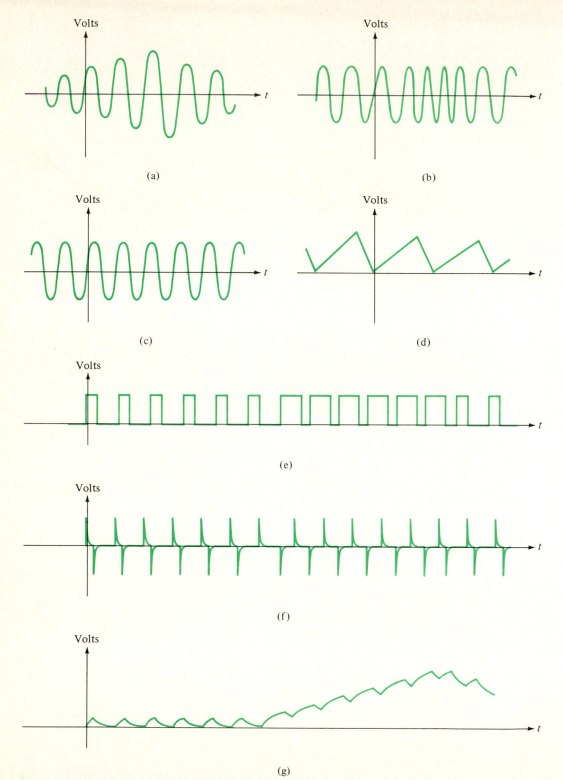

Fig. 1.5 Typical voltage waveforms.

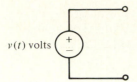

Fig. 1.6 Generalized ideal voltage source.

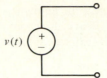

Fig. 1.7 Equivalent generalized ideal voltage source.

Placing an electric potential difference (voltage) across some material generally results in a flow of electric charge. Negative charge (in the form of electrons) flows from a given electric potential to a higher potential. Conversely, positive charge tends to flow from a given potential to a lower potential. Charge is usually denoted by q, and since this quantity is generally time dependent, the total amount of charge that is present in a given region is designated by $q(t)$.

We define **current**, denoted $i(t)$, to be the flow rate of the charge; that is,

$$i(t) = \frac{dq(t)}{dt}$$

is the current in the region containing $q(t)$. The units of current (coulombs per second, or C/s) are referred to as **amperes** (A) or **amps** for short, in honor of the french physicist André Ampère (1775–1836). Following the convention of Benjamin Franklin (a positive thinker), the direction of current has been chosen to be that direction in which positive charge would flow.

An **ideal current source**, represented as shown in Fig. 1.8, is a device which, *when connected to anything*, will always move I amperes in the direction indicated by the arrow.

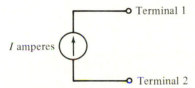

○ Terminal 1

I amperes

○ Terminal 2

Fig. 1.8 Ideal current source.

As a consequence of the definition, it should be quite clear that the ideal current sources in Fig. 1.9 are equivalent.

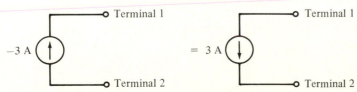

−3 A   Terminal 1   = 3 A   Terminal 1

Terminal 2   Terminal 2

Fig. 1.9 Equivalent ideal current sources.

Again, in general, the amount of current produced by an ideal source will be a function of time. Thus, the general representation of an ideal current source is shown in Fig. 1.10, where the units "amperes" are understood.

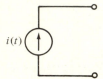

Fig. 1.10 Generalized ideal current source.

1.2 RESISTORS AND OHM'S LAW

Suppose that some material is connected to the terminals of an ideal voltage source $v(t)$ as shown in Fig. 1.11. Suppose that $v(t) = 1$ V. Then the electric potential at the top of the material is 1 V above the potential at the bottom. Since an electron has a negative charge, electrons in the material will tend to flow from bottom to top. Therefore, we say that current tends to go from top to bottom. Hence, for the given polarity, when $v(t)$ is a positive number, $i(t)$ will be a positive number with the direction indicated. If $v(t) = 2$ V, again the potential at the top is greater than at the bottom, so $i(t)$ will again be positive. However, because the potential is now twice as large as before, the current will be greater. (If the material is a "linear" element, the current will be twice as great.) Suppose now that $v(t) = 0$ V. Then the potentials at the top and the bottom of the material are the same. The result is no flow of electrons and, hence, no current. In this case, $i(t) = 0$ A. But suppose that $v(t) = -2$ V. Then the top of the material will be at a potential lower than at the bottom of the material. A current from bottom to top will result, and $i(t)$ will be a negative number. Note that the current $i(t)$ through the material must also go through the voltage source as there is nowhere else for it to go.

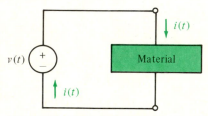

Fig. 1.11 Material with an applied voltage.

If in Fig. 1.11 the resulting current $i(t)$ is always directly proportional to the voltage for any function $v(t)$, then the material is called a **linear resistor**, or **resistor** for short.

Since voltage and current are directly proportional for a resistor, there exists a proportionality constant R, called **resistance**, such that

$$v(t) = Ri(t)$$

In dividing both sides of this equation by $i(t)$, we obtain

$$R = \frac{v(t)}{i(t)}$$

The units of resistance (volts per ampere) are referred to as **ohms** * and are denoted by the capital Greek letter omega, Ω. The accepted schematic representation of a resistor whose resistance is R ohms is shown in Fig. 1.12. A plot of voltage versus current for a (linear) resistor is given in Fig. 1.13.

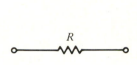

Fig. 1.12 Circuit symbol for a resistor.

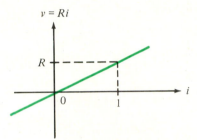

Fig. 1.13 Plot of voltage versus current for a resistor.

It was Ohm who discovered that if a resistor R has a voltage $v(t)$ *across* it and a current $i(t)$ *through* it, then if one is the cause, the other is the effect. Furthermore, if the polarity of the voltage and the direction of the current are as shown in Fig. 1.14, then it is true that

$$v(t) = Ri(t)$$

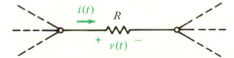

Fig. 1.14 Current and voltage convention for Ohm's law.

This equation is often called **Ohm's law**. From it, we may immediately deduce that

---

*Named for the German physicist Georg Ohm (1787–1854).

$$R = \frac{v(t)}{i(t)}$$

and

$$i(t) = \frac{v(t)}{R}$$

These last two equations are also referred to as Ohm's law.

It should be pointed out here that directions of currents and polarities of voltages are crucial when writing Ohm's law—the accepted convention is given in Fig. 1.14. For example, in the situation depicted in Fig. 1.15, before writing Ohm's law, note that the current's direction is opposite to that dictated by convention. However, this difficulty is easily remedied by redrawing the figure in the equivalent form given in Fig. 1.16. (A current of 5 A going through the resistor to the left is the same as a current of −5 A going to the right.) Since the direction of the current and the polarity of the voltage now conform to convention, we use Ohm's law to write

$$v_1(t) = R[-i_1(t)]$$

or

$$v_1(t) = -Ri_1(t)$$

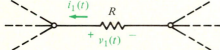

Fig. 1.15 Situation requiring negative sign for Ohm's law.

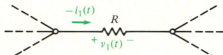

Fig. 1.16 Equivalent of resistor in Fig. 1.15.

Alternatively, we could have redrawn Fig. 1.15 in another equivalent form as shown in Fig. 1.17. Again we use Ohm's law to write

$$-v_1(t) = Ri_1(t)$$

from which we get

$$v_1(t) = -Ri_1(t)$$

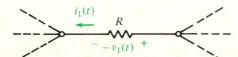

Fig. 1.17 Another equivalent of resistor in Fig. 1.15.

Example 1.2

Consider the circuit shown in Fig. 1.18. We use the letter "k" to represent the prefix "kilo," which indicates a value of 10^3. Following is a table of the more common symbols.

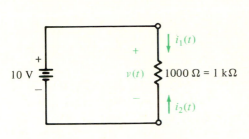

Value	Prefix	Symbol
10^{-12}	pico	p
10^{-9}	nano	n
10^{-6}	micro	μ
10^{-3}	milli	m
10^{3}	kilo	k
10^{6}	mega	M
10^{9}	giga	G

Fig. 1.18 Circuit for Example 1.2.

For the circuit in Fig. 1.18, the voltage across the 1-kΩ resistor is, by the definition of an ideal voltage source, $v(t) = 10$ V. Thus, by Ohm's law, we get

$$i_1(t) = v(t)/R = 10/1000 = 1/100 = 0.01 \text{ A} = 10 \text{ mA}$$

and

$$i_2(t) = -v(t)/R = -10/1000 = -1/100 = -0.01 \text{ A} = -10 \text{ mA}$$

Note that $i_2(t) = -i_1(t)$ as expected.

For the circuit shown in Fig. 1.19, by the definition of an ideal current source, $i(t) = 25 \ \mu\text{A} = 25 \times 10^{-6}$ A. By Ohm's law, we have that

$$v(t) = -R\, i(t) = -(2 \times 10^6)(25 \times 10^{-6}) = -50 \text{ V}$$

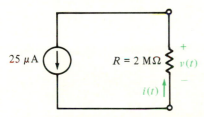

Fig. 1.19 Another circuit for Example 1.2.

Given a resistor R connected to an ideal voltage source $v(t)$ as shown in Fig. 1.20, we conclude the following. Since $i(t) = v(t)/R$ for any particular ideal source $v(t)$, the amount of current $i(t)$ that results can be made to be any finite value

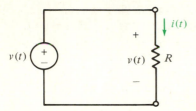

Fig. 1.20 Current through a voltage source.

by choosing the appropriate value for R (e.g., to make $i(t)$ large, make R small). Thus, we see that an ideal voltage source is capable of supplying *any* amount of current, and that amount depends on what is connected to the source—only the voltage is constrained to be $v(t)$ volts at the terminals.

When a resistor R is connected to an ideal current source as in Fig. 1.21, we know that $v(t) = Ri(t)$. Therefore, for a given current source $i(t)$, the voltage $v(t)$ that results can be made to be any finite value by appropriately choosing R (e.g., to make $v(t)$ large, make R large). Hence, we conclude that an ideal current source is capable of producing *any* amount of voltage across its terminals, and that amount depends on what is connected to the source—only the current is constrained to be $i(t)$ amperes through the source.

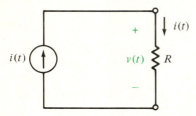

Fig. 1.21 Voltage across a current source.

Physical (nonideal) sources do not have the ability to produce unlimited currents and voltages. As a matter of fact, an actual source may approximate an ideal source only for a limited range of values.

Now consider the two ideal voltage sources whose terminals are connected as shown in Fig. 1.22. By definition of an ideal 3-V source, $v(t)$ must be 3 V. However, by the definition of a 5-V ideal voltage source, $v(t) = 5$ V. Clearly, both conditions cannot be satisfied simultaneously. (Note that connecting a resistor, or

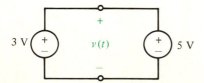

Fig. 1.22 Example of nonallowable connection of ideal voltage sources.

anything else, for that matter, to the terminals will not alleviate this problem.) Therefore, to avoid this paradoxical situation, we will insist that two ideal voltage sources never have their terminals connected together as in Fig. 1.22; the only exception is two sources with the same value and the same polarity.

Now consider a resistor whose value is zero ohms. An equivalent representation, called a **short circuit**, of such a resistance is given in Fig. 1.23. By Ohm's law, we have that

$$v(t) = Ri(t)$$
$$= 0i(t)$$
$$= 0 \text{ V}$$

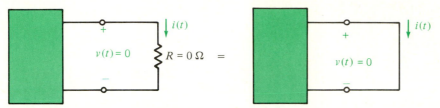

Fig. 1.23 Short-circuit equivalents.

Thus, no matter what finite value $i(t)$ has, $v(t)$ will be zero. Hence, we see that *a zero-ohm resistor is equivalent to an ideal voltage source whose value is zero volts*, provided that the current through it is finite. Therefore, for a zero resistance to be synonymous with a constraint of zero volts (and to avoid the unpleasantness of infinite currents), we will insist that we never be allowed to place a short circuit directly across a voltage source. In actuality, the reader will be spared a lot of grief by never attempting this in a laboratory or field situation.

Next consider a resistor having infinite resistance. An equivalent representation, called an **open circuit**, of such a situation is depicted in Fig. 1.24. By Ohm's law,

$$i(t) = v(t)/R$$
$$i(t) = 0 \text{ A}$$

as long as $v(t)$ has a finite value. Thus, we may conclude that *an infinite resistance*

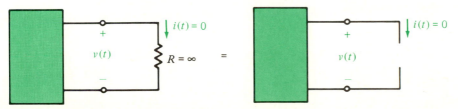

Fig. 1.24 Open-circuit equivalents.

is equivalent to an ideal current source whose value is zero amperes. Futhermore, we will always assume that an ideal current source has something connected to its terminals.

1.3 KIRCHHOFF'S CURRENT LAW (KCL)

It is a consequence of the work of the German physicist Gustav Kirchhoff (1824–1887) that enables us to analyze an interconnection of any number of elements (voltage sources, current sources, and resistors, as well as elements not yet discussed). We will refer to any such interconnection as a **circuit** or a **network**.

For a given circuit, a connection of two or more elements shall be called a **node**. An example of a node is depicted in the partial circuit shown in Fig. 1.25. In addition to using a solid dot, we may also indicate a node by a hollow dot, as was done for a terminal. Conversely, we may use a solid dot for the terminal of a device.

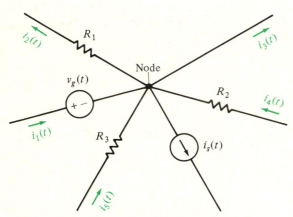

Fig. 1.25 Portion of a circuit.

We now present the first of Kirchhoff's two laws, his current law (KCL), which is essentially the law of conservation of electric charge.

> **KCL: At any node of a circuit, at every instant of time, the sum of the currents into the node is equal to the sum of the currents out of the node.**

Specifically for the portion of the network shown in Fig. 1.25, by applying KCL we obtain the equation

$$i_1(t) + i_4(t) + i_5(t) = i_2(t) + i_3(t) + i_s(t)$$

Note that one of the elements [the one in which $i_3(t)$ flows] is a short circuit—KCL holds regardless of the kinds of elements in the circuit.

An alternative, but equivalent, form of KCL can be obtained by considering currents directed into a node to be positive in sense and currents directed out of a node to be negative in sense (or vice versa). Under this circumstance, the alternative form of KCL can be stated as follows:

> **KCL: At any node of a circuit, the currents algebraically sum to zero.**

Applying this form of KCL to the node in Fig. 1.25 and considering currents directed in to be positive in sense, we get

$$i_1(t) - i_2(t) - i_3(t) + i_4(t) - i_s(t) + i_5(t) = 0$$

A close inspection of the last two equations, however, reveals that they are the same!

From this point on, we will simplify our notation somewhat by often abbreviating functions of time t such as $v(t)$ and $i(t)$ as v and i respectively. For instance, we may rewrite the last two equations, respectively, as

$$i_1 + i_4 + i_5 = i_2 + i_3 + i_s$$

and

$$i_1 - i_2 - i_3 + i_4 - i_s + i_5 = 0$$

It should always be understood, however, that lowercase letters such as v and i, in general, represent time-varying quantities.*

Example 1.3

Let us find the voltage v in the two-node circuit given in Fig. 1.26 in which the directions of i_1, i_2, and i_3 and the polarity of v were chosen arbitrarily. (The directions of the 2-A and 13-A sources are given.)

By KCL (at either of the two nodes), we have

$$13 - i_1 + i_2 - 2 - i_3 = 0 \tag{1.1}$$

From this we can write

$$i_1 - i_2 + i_3 = 11$$

By Ohm's law,

$$i_1 = v/1, \qquad i_2 = -v/2, \qquad i_3 = v/3$$

---

*A constant is a special case of a function of time.

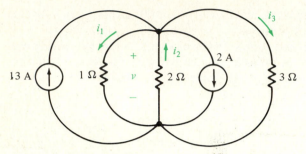

Fig. 1.26 Circuit for Example 1.3.

Substituting these into the preceding equation yields

$$(v/1) - (-v/2) + (v/3) = 11$$

$$v + v/2 + v/3 = 11$$

$$\frac{6v + 3v + 2v}{6} = 11$$

$$11v/6 = 11$$

$$v = 6 \text{ V}$$

Having solved for v, we now find that

$$i_1 = v/1 = 6/1 = 6 \text{ A}$$

$$i_2 = -v/2 = -6/2 = -3 \text{ A}$$

$$i_3 = v/3 = 6/3 = 2 \text{ A}$$

Note that a reordering of the circuit elements, as shown in Fig. 1.27, will result in the same equation (1.1) when KCL is applied. Since Ohm's law remains unchanged, the same answers are obtained.

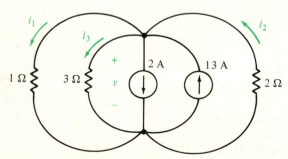

Fig. 1.27 Circuit equivalent to Fig. 1.26.

Just as KCL applies to any node of a circuit (i.e., to satisfy the physical law of conservation of charge, the current going in must equal the current coming out), so must KCL hold for any closed region.

For the circuit shown in Fig. 1.28, three regions have been arbitrarily identified. Applying KCL to Region 1, we get

$$i_1 + i_4 + i_5 = i_2$$

Applying KCL to Region 2, we obtain

$$i_5 + i_6 + i_7 = 0$$

and by applying KCL to Region 3, we get

$$i_2 + i_7 = i_4 + i_g$$

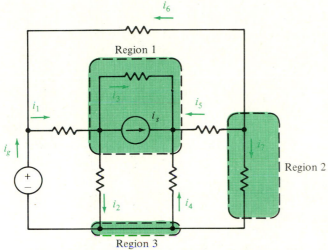

Fig. 1.28 Circuit with three arbitrarily selected regions.

Note that Region 3 apparently contains two nodes. However, since they are connected by a short circuit, there is no difference in voltage between these two points. Therefore, we can shrink the short circuit so as to coalesce the two points into a single node without affecting the operation of the circuit. Applying KCL at the resulting node again yields $i_2 + i_7 = i_4 + i_g$.

The converse process of expanding a node into apparently different nodes interconnected by short circuits also does not affect the operation of a circuit. For example, the portion of a circuit shown in Fig. 1.25 has the equivalent form shown in Fig. 1.29. Applying KCL to the region indicated results in the same equation as is obtained when KCL is applied to the node shown in Fig. 1.25. Thus, although it may appear that there are four distinct nodes in the region depicted in Fig. 1.29, they actually constitute a single node.

It is because of the foregoing that we can redraw the circuit given in Fig. 1.26 in the equivalent form shown in Fig. 1.30(a). Even though it may not appear so at

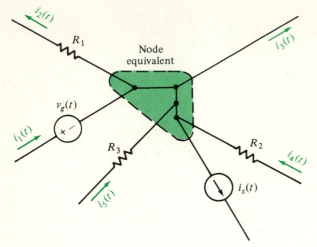

Fig. 1.29 Equivalent of circuit portion in Fig. 1.25.

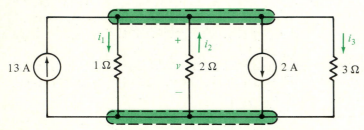

Fig. 1.30(a) Equivalent form of circuit in Fig. 1.26.

first glance, the circuit shown in Fig. 1.30(a) has two distinct nodes—they are indicated by the depicted regions. With this fact in mind, from this point on we shall draw such a circuit without the indication of regions, as is shown in Fig. 1.30(b).

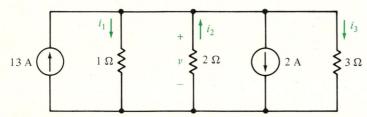

Fig. 1.30(b) Circuit in Fig. 1.30(a) with region indications omitted.

Current Division

We now consider the circuit given in Fig. 1.31. For this circuit, let us express the current i in terms of R_1, R_2, and v. Clearly, the voltage across each resistor is v.

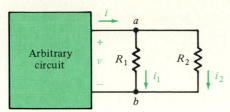

Fig. 1.31 Parallel connection of resistors.

By Ohm's law, we can write

$$i_1 = v/R_1 \qquad \text{and} \qquad i_2 = v/R_2$$

Applying KCL (at either node), we get

$$i = i_1 + i_2 = v/R_1 + v/R_2 = v(1/R_1 + 1/R_2) \tag{1.2}$$

Note that resistors R_1 and R_2 are connected to the same pair of nodes (nodes a and b). We call this a **parallel connection.** Specifically, we say that two elements are **connected in parallel** if they are connected to the same pair of nodes, regardless of what else is connected to those two nodes. This definition holds not only for two elements, but for three or four, or more, as well. As a consequence of the definition, we see that elements connected in parallel all have the same voltage across them.

As far as the arbitrary circuit in Fig. 1.31 is concerned, there is a resistance R between nodes a and b. But how is this value related to R_1 and R_2? From Equation (1.2), by Ohm's law, we have

$$\frac{i}{v} = \frac{1}{R} = \frac{1}{R_1} + \frac{1}{R_2}$$

Thus, we see that the parallel connection of resistors R_1 and R_2 is equivalent to a single resistor R, provided that

$$\boxed{\frac{1}{R} = \frac{1}{R_1} + \frac{1}{R_2}}$$

From this equation,

$$\frac{1}{R} = \frac{R_1 + R_2}{R_1 R_2}$$

or

$$\boxed{R = \frac{R_1 R_2}{R_1 + R_2}}$$

Using reasoning as in the above discussion, we deduce that m resistors R_1, R_2, R_3, . . . , R_m connected in parallel are equivalent to a single resistor R, provided that

$$\frac{1}{R} = \frac{1}{R_1} + \frac{1}{R_2} + \cdots + \frac{1}{R_m}$$

In the discussion above, the reciprocal of a resistance appeared a number of times. It is because of the frequent occurrence of the quantity $1/R$ that we denote it by a separate symbol. Given an R-ohm resistor, we define its **conductance** G to be

$$G = \frac{1}{R}$$

Furthermore, since the units of R are ohms, denoted Ω, the units* of G are often called **mhos** and are denoted by the symbol $\mho$. Figure 1.32 shows two equivalent ways of representing the same element. The equivalent forms of Ohm's law are

$$v = Ri = \frac{1}{G}i$$

$$i = \frac{1}{R}v = Gv$$

$$R = \frac{v}{i} = \frac{1}{G}$$

$$G = \frac{i}{v} = \frac{1}{R}$$

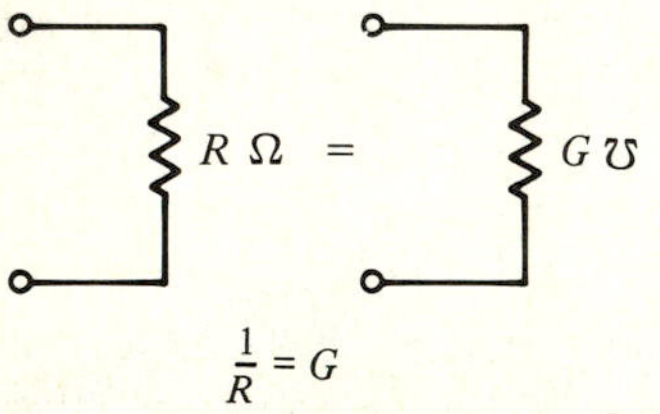

$R\ \Omega$ = $G\ \mho$

$\frac{1}{R} = G$

Fig. 1.32 Resistance and conductance.

---

*In terms of SI units, conductance is measured in **siemens** (S) in honor of the British inventor William Siemens (1823–1883). However, because it is in widespread use, we will use mhos.

Note that when combining resistors in parallel, to obtain an equivalent resistor, we add conductances. This fact is demonstrated in Fig. 1.33.

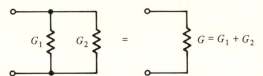

Fig. 1.33 Combining conductances in parallel.

Now consider the circuit given in Fig. 1.34. We ask the question that has been haunting scientists for a long time: "If you were an ampere, where would you go?" To answer this question, we simply replace the parallel connection of R_1 and R_2 by its equivalent resistance as shown in Fig. 1.35. We then have

$$v = Ri = \frac{R_1 R_2}{R_1 + R_2} i$$

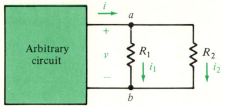

Fig. 1.34 Current division.

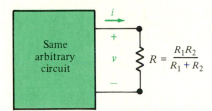

Fig. 1.35 Equivalent resistance of parallel connection.

Since this is the same v that appears in Fig. 1.34, we conclude that

$$i_1 = v/R_1 \qquad \text{and} \qquad i_2 = v/R_2$$

Thus,

$$i_1 = \frac{R_2}{R_1 + R_2} i \qquad \text{and} \qquad i_2 = \frac{R_1}{R_1 + R_2} i$$

Since, by KCL, $i_1 + i_2 = i$, the above two boxed formulas describe how the current i is divided by the resistors. For this reason, a pair of resistors in parallel is often referred to as a **current divider.** Note that if R_1 and R_2 are both positive, and if R_1 is greater than R_2, then i_2 is greater than i_1. In other words, a larger amount of current will go through the smaller resistor—amperes tend to take the path of least resistance!

For the circuit shown in Fig. 1.36, since $G_1 = 1/R_1$ and $G_2 = 1/R_2$, then by the current-divider formula for i_1, we have

$$i_1 = \frac{R_2}{R_1 + R_2}i = \frac{1/G_2}{(1/G_1) + (1/G_2)}i = \frac{1/G_2}{(G_1 + G_2)/G_1 G_2}i = \frac{G_1}{G_1 + G_2}i$$

Similarly,

$$i_2 = \frac{G_2}{G_1 + G_2}i$$

Thus, we also have current-divider formulas in terms of conductances.

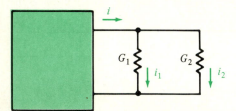

Fig. 1.36 Conductances in parallel.

Example 1.4

For the circuit given in Fig. 1.37, let us determine the currents indicated by using current division.

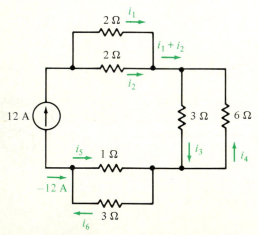

Fig. 1.37 Example of current division.

The current that is to be divided by the two 2-Ω resistors is 12 A. By the current-division formulas,

$$i_1 = i_2 = \frac{2(12)}{2 + 2} = 6 \text{ A}$$

The current that is to be divided by the 3-Ω and 6-Ω resistors is $i_1 + i_2 = 12$ A. By the current-division formulas, we have

$$i_3 = \frac{6(12)}{3 + 6} = 8 \text{ A}$$

and*

$$-i_4 = \frac{3(12)}{3 + 6} = 4 \text{ A} \quad \Rightarrow \quad i_4 = -4 \text{ A}$$

Finally, the current that is to be divided by the 1-Ω and 3-Ω resistors is -12 A. Thus, by current division,

$$i_5 = \frac{3(-12)}{1 + 3} = -9 \text{ A}$$

and

$$-i_6 = \frac{1(-12)}{1 + 3} = -3 \text{ A} \quad \Rightarrow \quad i_6 = 3 \text{ A}$$

1.4 KIRCHHOFF'S VOLTAGE LAW (KVL)

We now present the second of Kirchhoff's laws—the voltage law. To do this, we must introduce the concept of a "loop."

Starting at any node n in a circuit, we form a **loop** by traversing through elements (open circuits included!) and returning to the starting node n, and never encountering any other node more than once. As an example, consider the partial circuit shown in Fig. 1.38. In this circuit, the 1-Ω, 2-Ω, 3-Ω, and 6-Ω resistors, along with the 2-A current source, constitute the loop a, b, c, e, f, a. A few (but not all) of the other loops are: (1) a, b, c, d, e, f, a; (2) c, d, e, c; (3) d, e, f, d; (4) a, b, c, e, d, f, a; (5) a, b, e, f, a; (6) b, c, e, b.

---

*The symbol $\Rightarrow$ means "implies."

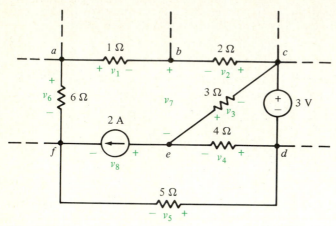

Fig. 1.38 Portion of a circuit.

We now can present Kirchhoff's voltage law (KVL):

> **KVL: In traversing any loop in any circuit, at every instant of time, the sum of the voltages having one polarity equals the sum of the voltages having the opposite polarity.***

For the partial circuit shown in Fig. 1.38, suppose that the voltages across the elements are as shown. Then by KVL around loop a, b, c, e, f, a, we have

$$v_1 + v_8 = v_2 + v_3 + v_6$$

and around loop b, c, d, e, b, we have

$$v_2 + v_7 = 3 + v_4$$

In this last loop, one of the elements traversed (the element between nodes b and e) is an open circuit—KVL holds regardless of the nature of the elements in the circuit.

Since voltage is energy (or work) per unit charge, then KVL is another way of stating the physical law of the conservation of energy.

An alternative statement of KVL can be obtained by considering voltages across elements that are traversed from plus to minus to be positive in sense and voltages across elements that are traversed from minus to plus to be negative in sense (or vice versa). Under this circumstance, KVL has the following alternative form.

---

*Another form of this statement is: The sum of the voltage rises equals the sum of the voltage drops, where traversing a voltage from minus to plus is a voltage rise and from plus to minus is a voltage drop.

> **KVL: Around any loop in a circuit, the voltages algebraically sum to zero.**

By applying this form of KVL to the partial circuit shown in Fig. 1.38, and selecting a traversal from plus to minus to be positive in sense, around loop a, b, c, e, f, a, we get

$$v_1 - v_2 - v_3 + v_8 - v_6 = 0$$

and around loop b, c, d, e, b, we get

$$-v_2 + 3 + v_4 - v_7 = 0$$

These two equations are the same, respectively, as the preceding two equations.

Voltage Division

Now consider the circuit given in Fig. 1.39. For this circuit, let us express the voltage v in terms of R_1, R_2, and i. By Ohm's law, we can write

$$v_1 = R_1 i \qquad \text{and} \qquad v_2 = R_2 i$$

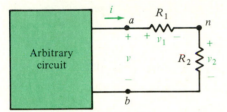

Fig. 1.39 Series connection of resistors.

Applying KVL, we get

$$v = v_1 + v_2 = R_1 i + R_2 i = (R_1 + R_2)i \qquad (1.3)$$

Note that resistors R_1 and R_2 have a node in common (node n), and no other element is connected to this node. This is known as a **series connection**. Specifically, we say that two elements are **connected in series** if they have a node in common and no other element is connected to this common node. In general, if m elements are connected together such that each resulting node joins no more than two of the elements, then these elements are connected in series. As a consequence of the definition, we see that elements connected in series all have the same current through them.

As far as the arbitrary circuit in Fig. 1.39 is concerned, there is a resistance R between nodes a and b. From Equation (1.3), by Ohm's law, we have

$$\frac{v}{i} = R = R_1 + R_2$$

Thus, we see that the series connection of resistors R_1 and R_2 is equivalent to a single resistor R, provided that

$$R = R_1 + R_2$$

Using reasoning as above, we deduce that m resistors R_1, R_2, R_3, . . . , R_m connected in series are equivalent to a single resistor R, provided that

$$R = R_1 + R_2 + R_3 + \cdots + R_m$$

Example 1.5

Let us find i in the circuit shown in Fig. 1.40. To find i, we can replace series and parallel connections of resistors by their equivalent resistances. We begin by noting that the 1-Ω and 3-Ω resistors are in series. Upon combining them, we get the circuit in Fig. 1.41(a).

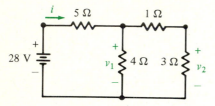

Fig. 1.40 Series-parallel circuit.

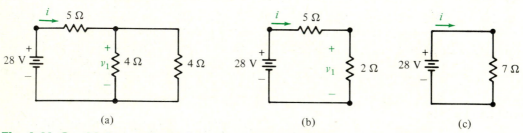

(a)   (b)   (c)

Fig. 1.41 Combining resistors in series and in parallel.

Note that it is not possible to display the voltage v_2 given in Fig. 1.40 in the circuit in Fig. 1.41(a). Since the two 4-Ω resistors are connected in parallel, we can further simplify the circuit as shown in Fig. 1.41(b). Here the 5-Ω and 2-Ω resistors are in series, so we may combine them and obtain the circuit in Fig.

1.41(c). Hence, by Ohm's law, we have

$$i = \tfrac{28}{7} = 4 \text{ A}$$

Now consider the circuit given in Fig. 1.42. To determine how the voltage v is divided between R_1 and R_2, we express i in terms of R_1, R_2, and v by replacing the series connection of the resistors by its equivalent resistance as shown in Fig. 1.43. By Ohm's law,

$$i = v/R = v/(R_1 + R_2)$$

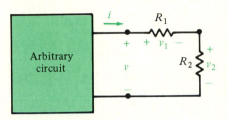

Fig. 1.42 Voltage division.   **Fig. 1.43 Equivalent resistance of series connection.**

Thus, for the circuit in Fig. 1.42, we have

$$v_1 = R_1 i \qquad \text{and} \qquad v_2 = R_2 i$$

so

$$v_1 = \frac{R_1}{R_1 + R_2} v \qquad \text{and} \qquad v_2 = \frac{R_2}{R_1 + R_2} v$$

and the two boxed formulas describe how the voltage v is divided by the resistors. Because of this, a pair of resistors in series is often called a **voltage divider.** Note that for positive-valued resistors, if R_1 is greater than R_2, then v_1 is greater than v_2. In other words, a larger voltage will be maintained across the larger resistor.

Example 1.6

Let us find the voltage v_2 in the circuit given in Fig. 1.40. Combining the series connection of the 1-Ω and 3-Ω resistors, we obtain the circuit shown in Fig. 1.41(a). The resulting pair of 4-Ω resistors in parallel can be combined as shown in Fig. 1.41(b). By voltage division, we have

$$v_1 = \frac{2(28)}{2 + 5} = \frac{56}{7} = 8 \text{ V}$$

Returning to the original circuit (Fig. 1.40) and applying voltage division again yields

$$v_2 = \frac{3v_1}{3 + 1} = \frac{3(8)}{4} = 6 \text{ V}$$

◼

For a given circuit, there is generally more than one way to analyze it; as long as the different techniques are employed correctly, of course, the same results will be obtained. To demonstrate this, let us again consider the circuit given in Fig. 1.40 and take a slightly different approach.

Example 1.7

Consider the circuit shown in Fig. 1.44. By Ohm's law,

$$i_2 = v_2/3$$

By KCL,

$$i_3 = i_2 = v_2/3$$

By Ohm's law,

$$v_3 = 1i_3 = v_2/3$$

By KVL,

$$v_1 = v_2 + v_3 = v_2 + v_2/3 = \tfrac{4}{3}v_2$$

By Ohm's law,

$$i_1 = v_1/4 = v_2/3$$

By KCL,

$$i = i_1 + i_3 = (v_2/3) + (v_2/3) = \tfrac{2}{3}v_2$$

By Ohm's law,

$$v_4 = 5i = \tfrac{10}{3}v_2$$

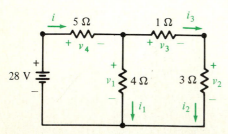

Fig. 1.44 Series-parallel circuit given in Fig. 1.40.

By KVL,

$$28 = v_1 + v_4 = \tfrac{4}{3}v_2 + \tfrac{10}{3}v_2 = \tfrac{14}{3}v_2$$

Hence,

$$v_2 = 28(\tfrac{3}{14}) = 6 \text{ V}$$

Combining Sources

We mentioned earlier that we will not be allowed to connect ideal voltage sources in parallel. However, consider the series connection of two ideal voltage sources as shown in Fig. 1.45. By KVL, we have that $v = v_1 + v_2$, and by the definition of an ideal voltage source, this must always be the voltage between nodes a and b, regardless of what is connected to them. Thus, a series connection of two voltage sources as shown in Fig. 1.45 (remember, no other element can be connected to node c!) is equivalent to the ideal voltage source shown in Fig. 1.46. Clearly, the obvious generalization to m voltage sources in series holds.

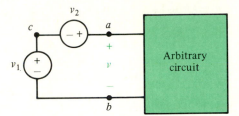

Fig. 1.45 Series connection of voltage sources.

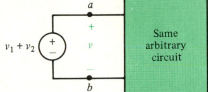

Fig. 1.46 Equivalent of voltage sources in series.

In a dual manner, consider a parallel connection of ideal current sources as shown in Fig. 1.47. By KCL, we have that $i = i_1 + i_2$, and by the definition of an ideal current source, this must always be the current into the arbitrary circuit. Therefore, a parallel connection of two current sources as shown in Fig. 1.47 (remember, other elements *can* be connected to nodes a and b!) is equivalent to the ideal current source shown in Fig. 1.48. Again, this result can be generalized to the case of m current sources in parallel.

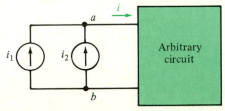

Fig. 1.47 Parallel connection of current sources.

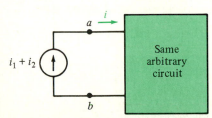

Fig. 1.48 Equivalent of current sources in parallel.

Example 1.8

Reconsider the circuit shown in Fig. 1.30. Combining the current sources in parallel, we obtain the circuit given in Fig. 1.49. Since the equivalent resistance of the three resistors in parallel is R, where

$$\frac{1}{R} = \frac{1}{1} + \frac{1}{2} + \frac{1}{3} = \frac{6 + 3 + 2}{6} = \frac{11}{6}$$

then

$$R = \tfrac{6}{11} \ \Omega$$

Thus, by Ohm's law we have that

$$v = (\tfrac{6}{11})(11) = 6 \text{ V}$$

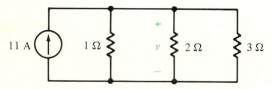

Fig. 1.49 Equivalent of circuit in Fig. 1.30.

1.5 DEPENDENT SOURCES

Up to this point, we have been considering voltage and current sources whose values are, in general, time dependent. However, these values were given to be independent of the behavior of the circuits to which the sources belonged. For this reason, we say that such sources are **independent.**

We now consider an ideal source, either voltage or current, whose value depends upon some variable (usually a voltage or current) in the circuit to which the source belongs. We call such an ideal source a **dependent** (or **controlled**) **source,** and represent it as shown in Fig. 1.50. Note that a dependent source is represented by a diamond-shaped symbol so as not to confuse it with an independent source.

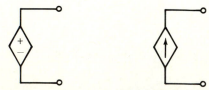

Dependent voltage source   Dependent current source

Fig. 1.50 Circuit symbols for dependent sources.

Example 1.9

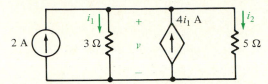

Fig. 1.51 Circuit with current-dependent current source.

Consider the circuit shown in Fig. 1.51 in which the value of the dependent current source depends on the current i_1 in the 3-Ω resistor. Thus, we say that it is a **current-dependent current source.** Specifically, the value of the dependent source is $4i_1$, and the units are amperes, of course. (This implies that the constant 4 is dimensionless.) Hence, if the current through the 3-Ω resistor is $i_1 = 2$ A, then the dependent current source has a value of $4i_1 = 8$ A; if the current is $i_1 = -3$ A, then the dependent current source has a value of $4i_1 = -12$ A; and so on. To find the actual quantity i_1, we proceed as follows:

Applying KCL (at either node), we can write

$$i_1 + i_2 = 4i_1 + 2$$

Simplifying this equation, we have

$$-3i_1 + i_2 = 2$$

By Ohm's law,

$$i_1 = v/3 \qquad \text{and} \qquad i_2 = v/5$$

Therefore,

$$-3(v/3) + v/5 = 2$$

$$-4v/5 = 2$$

$$v = -\tfrac{10}{4} = -\tfrac{5}{2} \text{ V}$$

Hence, we see that in actuality

$$i_1 = v/3 = -\tfrac{5}{6} \text{ A}$$

so that the value of the dependent current source is

$$4i_1 = -\tfrac{20}{6} = -\tfrac{10}{3} \text{ A}$$

Remember, if the independent source has a different value, then the quantity $4i_1$ will in general be different from $-\tfrac{10}{3}$ A.

Example 1.10

Let us now consider the circuit given in Fig. 1.52. In this circuit, the value of the dependent current source is specified by a voltage! In other words, the value of the source is $4v$ amperes, where v is the *amount* of voltage across the 3-Ω resistor (also the 5-Ω resistor in this case). This implies that the constant 4 has as units A/V, or mhos—so it is a conductance. Such a device is a **voltage-dependent current source.**

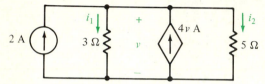

Fig. 1.52 Circuit with voltage-dependent current source.

To solve for v, we apply KCL, and obtain

$$i_1 + i_2 = 4v + 2$$

Thus,

$$v/3 + v/5 = 4v + 2$$

$$-4v + 8v/15 = 2$$

$$-52v/15 = 2$$

$$v = -\tfrac{30}{52} = -\tfrac{15}{26} \text{ V}$$

Consequently, the numerical value of the dependent current source is

$$4v = -\tfrac{30}{13} \text{ A}$$

Again, we will point out that if the independent source has a value other than 2 A, then the dependent source's value of $4v$ amperes will be something other than $-\tfrac{30}{13}$ A.

The other variables in this circuit are

$$i_1 = v/3 = -\tfrac{5}{26} \text{ A} \qquad \text{and} \qquad i_2 = v/5 = -\tfrac{3}{26} \text{ A}$$

Example 1.11

The circuit shown in Fig. 1.53 contains a **current-dependent voltage source.** The value of this dependent voltage source is determined by the loop current i. In this case, the constant 4 has as its units V/A, or ohms—so it is a resistance.

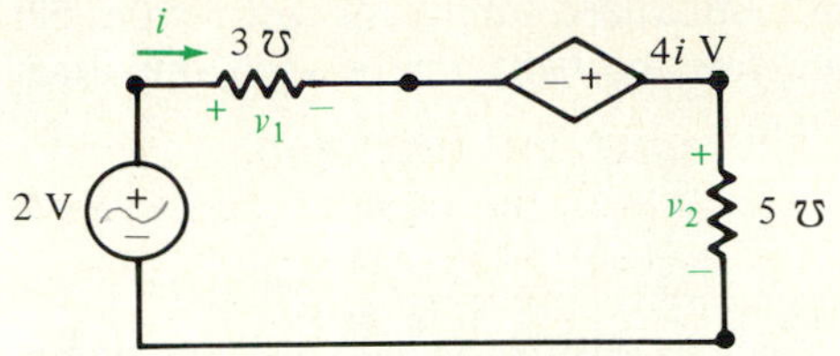

Fig. 1.53 Circuit with current-dependent voltage source.

Before we analyze this circuit, let us compare it with the circuit given in Fig. 1.52. What is a parallel connection there is a series connection here; what is a voltage (v) there, is a current (i) here; what is a current there (2 A, i_1, i_2, $4v$ amperes) is a voltage here (2 V, v_1, v_2, $4i$ volts, respectively), and what is a resistance there (3 Ω, 5 Ω) is a conductance here (3 ℧, 5 ℧, respectively). For this reason, we say that this circuit is the **dual** of the previous circuit (and vice versa). Once a circuit has been analyzed, its dual is analyzed automatically. A more formal discussion of the concept of duality is given in Chapter 3. However, for the time being, let us verify it for this particular case.

By KVL,

$$v_1 - 4i + v_2 = 2$$

By Ohm's law,

$$v_1 = i/3 \quad \text{and} \quad v_2 = i/5$$

Thus,

$$\frac{i}{3} - 4i + \frac{i}{5} = 2$$

from which

$$\frac{5 - 60 + 3}{15} i = 2$$

or

$$i = -\tfrac{30}{52} = -\tfrac{15}{26} \text{ A}$$

which is the numerical value for v in volts in the dual circuit (Fig. 1.52). Therefore, the value of the dependent source is

$$4i = 4(-\tfrac{15}{26}) = -\tfrac{30}{13} \text{ V}$$

Furthermore,

$$v_1 = i/3 = -\tfrac{5}{26} \text{ V} \quad \text{and} \quad v_2 = i/5 = -\tfrac{3}{26} \text{ V}$$

which are the respective values for i_1 and i_2 in the dual circuit.

The concept of dependent sources is not introduced simply for abstraction, but it is precisely this type of source that models the behavior of such important electronic devices as transistors and amplifiers.

Comments about Circuit Elements

Certain types of nonideal circuit elements, such as positive-valued resistors, come in a wide variety of values and are readily available in the form of discrete components. For many situations, the assumption that an actual resistor behaves as an ideal resistor is a reasonable one.

With a few exceptions, however, this is not the case with independent voltage and current sources. Although an actual battery often can be thought of as an ideal voltage source, other nonideal independent sources are approximated by a combination of circuit elements (some of which will be discussed later). Among these elements are the dependent sources, which are not discrete components as are many resistors and batteries, but are in a sense part of electronic devices such as transistors and operational amplifiers (which consist of numerous transistors and resistors). But don't try to peel open a transistor's metal can (for those that are constructed that way) so that you can see a little diamond-shaped object. The dependent source is a theoretical element that is used to help describe or model the behavior of various electric devices.

In summary, although ideal circuit elements are not off-the-shelf circuit components, their importance lies in the fact that they can be interconnected (on paper or on a computer) to approximate actual circuits that are comprised of nonideal elements and assorted electrical components—thus allowing for the analysis of such circuits.

1.6 POWER

Recall that electrons flow through a resistor from a given potential to a higher potential, and hence (positive-valued) current goes from a given potential to a lower one. Potential difference or voltage is a measure of work per unit charge (i.e., joules per coulomb). To obtain a current through an element, as shown in Fig. 1.54, it takes a certain amount of work or energy—and we say that this energy is absorbed by the element. By taking the product of voltage (energy per unit charge) and current (charge per unit time), we get a quantity that measures energy per unit time. Such a term is known as **power**. It is for this reason that we define $p(t)$, the **instantaneous power absorbed** by an element as in Fig. 1.54, as the product of

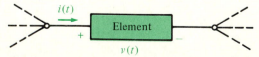

Fig. 1.54 Arbitrary element with conventional current direction and voltage polarity.

. voltage and current. That is,

$$p(t) = v(t)i(t)$$

The unit of power (joules per second) is called the **watt*** (W). Given an expression for instantaneous power absorbed $p(t)$, to determine the power absorbed at time t_o, simply substitute the number t_o into the expression.

As with Ohm's law, when using the formula for power absorbed, we must always be conscious of both the voltage polarity and the current direction. For example, for the situation shown in Fig. 1.55, going back to the definition for instantaneous power absorbed, we have in this case that

$$p_1(t) = -v_1(t)i_1(t)$$

Fig. 1.55 Nonconventional current direction with respect to voltage polarity.

Since power absorbed in a given element can be either a positive or a negative quantity (depending on the relationship between voltage and current for the element), we can say that the element absorbs x watts, or equivalently, that it **supplies** or **delivers** $-x$ watts.

Example 1.12

Consider the circuit in Fig. 1.56. Let us determine the power absorbed by each of the elements for the case that $I = 10$ A. Note that the voltage across each of the elements is 6 V since all the elements are in parallel. Therefore, by Ohm's law,

$$i_1 = 6/1 = 6 \text{ A}, \qquad i_2 = 6/2 = 3 \text{ A}, \qquad i_3 = 6/3 = 2 \text{ A}$$

and the power absorbed by the 1-Ω, 2-Ω, and 3-Ω resistors is

$$p_1 = 6i_1 = 36 \text{ W}, \qquad p_2 = 6i_2 = 18 \text{ W}, \qquad p_3 = 6i_3 = 12 \text{ W}$$

respectively, or a total of

$$p_R = p_1 + p_2 + p_3 = 36 + 18 + 12 = 66 \text{ W}$$

---

*Named for the Scottish inventor James Watt (1736–1819).

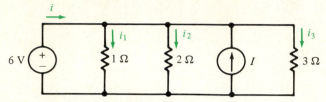

Fig. 1.56 Circuit for Example 1.12.

absorbed by the resistors. By KCL,

$$i + 10 = i_1 + i_2 + i_3 = 6 + 3 + 2 = 11$$

or

$$i = 11 - 10 = 1 \text{ A}$$

Thus, the power absorbed by the voltage source is

$$p_V = -6i = -6 \text{ W}$$

Also note that the power absorbed by the current source is

$$p_I = -6(10) = -60 \text{ W}$$

Hence, the total power absorbed is

$$p_R + p_V + p_I = 66 - 6 - 60 = 0 \text{ W}$$

Recalling from elementary physics that power is work (energy) per unit time, we see that the fact that the total power absorbed is zero is equivalent to saying that the principle of the conservation of energy is satisfied in this circuit (as it is in any circuit). In this case, the sources supplied all the power and the resistors absorbed it all.

Now, however, consider the situation in which $I = 12$ A. In this case,

$$i_1 = 6 \text{ A}, \qquad i_2 = 3 \text{ A}, \qquad i_3 = 2 \text{ A}$$

and

$$p_1 = 36 \text{ W}, \qquad p_2 = 18 \text{ W}, \qquad p_3 = 12 \text{ W} \qquad \Rightarrow \qquad p_R = p_1 + p_2 + p_3 = 66 \text{ W}$$

as before. But by KCL,

$$i + 12 = i_1 + i_2 + i_3 \qquad \Rightarrow \qquad i = 11 - 12 = -1 \text{ A}$$

Therefore,

$$p_V = 6(1) = 6 \text{ W} \qquad \text{and} \qquad p_I = -6(12) = -72 \text{ W}$$

so that the total power absorbed is

$$p_R + p_V + p_I = 66 + 6 - 72 = 0 \text{ W}$$

and again energy (power) is conserved. However, in this case, not only do the resistors absorb power, but so does the voltage source. It is the current source that supplies all the power absorbed in the circuit.

■

In all the examples worked so far, the reader may have noted that the power absorbed in every resistor was a nonnegative* number. As we now shall see, this is always the case; consider Fig. 1.57. By definition, the power absorbed by the resistor is $p = vi$. But by Ohm's law, $v = Ri$. Thus, $p = (Ri)i$, or

$$p = Ri^2$$

Also, $i = v/R$, so that $p = v(v/R)$, or

$$p = \frac{v^2}{R}$$

and both formulas for calculating the power absorbed by a resistor R demonstrate that p is always a nonnegative number when R is positive.

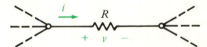

Fig. 1.57 Resistor with conventional current direction and voltage polarity.

A resistor always absorbs power. In a physical resistor, this power is dissipated as heat. In some types of resistors (e.g., an incandescent lamp or bulb, a toaster, or an electric space heater), this property is desirable in that the net result may be light or warmth. In other types of resistors, such as those in electronic circuits, the heat dissipated by a resistor may affect the operation of its circuit. In this case, heat dissipation cannot be ignored.

The common carbon resistor comes in values that range from less than 10 Ω to more than 10 MΩ. The physical size of such resistors will determine the amount of power that they can safely dissipate. These amounts are referred to as **power ratings.** Typical power ratings of electronic-circuit carbon resistors are $\frac{1}{8}$, $\frac{1}{4}$, $\frac{1}{2}$, 1, and 2 W. The dissipation of power that exceeds the rating of a resistor can damage the resistor physically. When an application requires the use of a resistor that must dissipate more than 2 W, another type—the wire-wound resistor—is often used.

---

*A **positive number** is a number that is greater than zero; a **nonnegative number** is a number that is greater than or equal to zero.

Metal-film resistors, although more expensive to construct, can have higher power ratings than carbon resistors and are more reliable and more stable. In addition to the lumped circuit elements such as carbon, wire-wound, and metal-film resistors, there are resistors in integrated-circuit (IC) form. These types of resistors, which dissipate only small amounts of power, are part of the countless ICs employed in present-day electronics.

SUMMARY

1. An ideal voltage source is a device that produces a specific (not necessarily constant with time) electric potential difference across its terminals regardless of what is connected to it.

2. An ideal current source is a device that produces a specific (not necessarily constant with time) current through it regardless of what is connected to it.

3. A (linear) resistor is a device in which the voltage across it is directly proportional to the current through it. If we write the equation (Ohm's law) describing this as $v = Ri$, then the constant of proportionality R is called resistance. If we write $i = Gv$, then G is known as conductance. Thus, $R = 1/G$.

4. An open circuit is an infinite resistance and a short circuit is a zero resistance. The former is equivalent to an ideal current source whose value is zero; the latter is equivalent to an ideal voltage source whose value is zero.

5. At any node of a circuit, the currents algebraically sum to zero—Kirchhoff's current law (KCL).

6. Around any loop in a circuit, the voltages algebraically sum to zero—Kirchhoff's voltage law (KVL).

7. Current going into a parallel connection of resistors divides among them—the smallest resistance has the most current through it.

8. Voltage across a series connection of resistors divides among them—the largest resistance has the most voltage across it.

9. Resistances in series behave as a single resistance whose value equals the sum of the individual resistances.

10. Conductances in parallel behave as a single conductance whose value equals the sum of the individual conductances.

11. Voltage sources in series can be combined into a single voltage source and current sources in parallel can be combined into a single current source.

12. Voltage and current sources can be dependent as well as independent.

13. A dependent source is a voltage or current source whose value depends on some other circuit variable.

14. The power absorbed by an element is the product of the current through it and the voltage across it.

Problems for Section 1.1

1.1 An ideal voltage source has a value of $v(t) = 10e^{-t}$ V. Give the value of this voltage source when (a) $t = 0$ seconds; (b) $t = 1$ second; (c) $t = 2$ seconds; (d) $t = 3$ seconds; (e) $t = 4$ seconds.

1.2 Repeat Problem 1.1 for the case $v(t) = 5 \sin(\pi/2)t$ volts.

1.3 Repeat Problem 1.1 for the case $v(t) = 3 \cos(\pi/2)t$ volts.

1.4 The total charge in some region is described by the function $q(t) = 4e^{-2t}$ coulombs. Find the current in this region.

1.5 Repeat Problem 1.4 for the case that $q(t) = 3 \sin \pi t$ coulombs.

1.6 Repeat Problem 1.4 for the case that $q(t) = 6 \cos 2\pi t$ coulombs.

1.7 Repeat Problem 1.1 for the case that $v(t)$ is described by the function shown in Fig. P1.7.

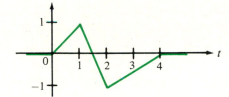

Fig. P1.7

1.8 The total charge $q(t)$ in some region is described by the function given in Problem 1.7. Sketch the current $i(t)$ in this region.

Problems for Section 1.2

1.9 Consider the circuit shown in Fig. P1.9.
   (a) Given $i_1 = 4$ A, find v_1.

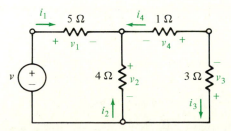

Fig. P1.9

(b) Given $i_2 = -2$ A, find v_2.

(c) Given $i_3 = 2$ A, find v_3.

(d) Given $i_4 = -2$ A, find v_4.

1.10 Consider the circuit in Fig. P1.10.

(a) Given $v_1 = -\frac{2}{3}$ V, find i_1.

(b) Given $v_2 = \frac{8}{3}$ V, find i_2.

(c) Given $v_3 = \frac{11}{9}$ V, find i_3.

(d) Given $v_4 = \frac{7}{3}$ V, find i_4.

(e) Given $v_5 = -\frac{14}{9}$ V, find i_5.

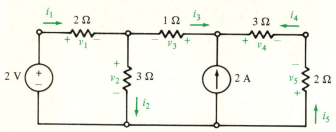

Fig. P1.10

1.11 Consider the circuit shown in Fig. P1.11.

(a) Given $i_1 = -2$ A, find v_1.

(b) Given $v_2 = -\frac{11}{7}$ V, find i_2.

(c) Given $i_3 = \frac{30}{7}$ A, find v_3.

(d) Given $v_4 = 2$ V, find i_4.

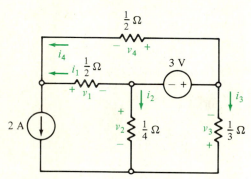

Fig. P1.11

1.12 Consider the circuit given in Fig. P1.9.

(a) Find $i_1(t)$ for the case $v_1(t) = 20$ V.

(b) Find $i_2(t)$ for the case $v_2(t) = 3e^{-2t}$ V.

(c) Find $v_3(t)$ for the case $i_3(t) = 6 \sin 2t$ A.

(d) Find $v_4(t)$ for the case $i_4(t) = -e^{-t} \cos 5t$ A.

1.13 Consider the circuit in Fig. P1.13.

(a) Given $i_1 = 2$ A and $v_1 = 4$ V, find R_1.

(b) Given $i_2 = 2$ A and $v_2 = -12$ V, find R_2.

(c) Given $i_3 = -\frac{4}{3}$ A and $v_3 = -4$ V, find R_3.

(d) Given $i_4 = -\frac{2}{3}$ A and $v_3 = 4$ V, find R_4.

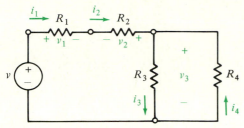

Fig. P1.13

1.14 Consider the circuit shown in Fig. P1.13.

(a) Given $i_1(t) = -e^{-2t}$ and $v_1(t) = -2e^{-2t}$, find R_1.

(b) Given $i_2(t) = -e^{-2t}$ and $v_2(t) = 6e^{-2t}$, find R_2.

(c) Given $i_3(t) = \frac{2}{3}e^{-2t}$ and $v_3(t) = 2e^{-2t}$, find R_3.

(d) Given $i_4(t) = \frac{1}{3}e^{-2t}$ and $v_3(t) = -2e^{-2t}$, find R_4.

Problems for Section 1.3

1.15 For the circuit shown in Fig. P1.15, find v for the case that

(a) $i = 1$ A,

(b) $i = 2$ A,

(c) $i = 3$ A.

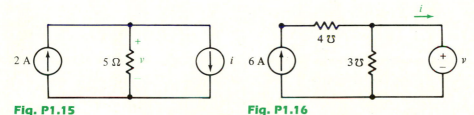

Fig. P1.15   **Fig. P1.16**

1.16 For the circuit given in Fig. P1.16, find i for the case that

(a) $v = 1$ V,

(b) $v = 2$ V,

(c) $v = 3$ V.

1.17 For the circuit given in Fig. P1.17, find v_1 for the case that

(a) $v = 2$ V,

(b) $v = 4$ V,

(c) $v = 6$ V.

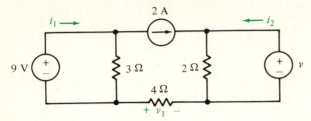

Fig. P1.17

1.18 For the circuit shown in Fig. P1.18, assuming $i_1 = 5$ A, determine i_2, i_3, i_4, and i_5.

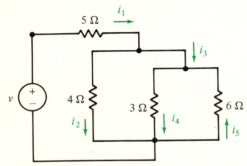

Fig. P1.18

1.19 For the circuit given in Fig. P1.18, suppose that $i_4 = 4$ A. Find i_1, i_2, i_3, and i_5.

1.20 For the circuit given in Fig. P1.18, suppose that $i_5 = 4$ A. Find i_1, i_2, i_3, and i_4.

1.21 For the circuit given in Fig. P1.18, suppose that $i_2(t) = 6 \cos 5t$ A. Find $i_1(t)$, $i_3(t)$, $i_4(t)$, and $i_5(t)$.

1.22 Given the circuit shown in Fig. P1.22, suppose that $i_1 = 2$ A. Find v for the case that
(a) $i_2 = 1$ A,
(b) $i_2 = 2$ A,
(c) $i_2 = 3$ A.

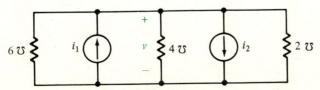

Fig. P1.22

1.23 For the circuit given in Fig. P1.22, suppose that $i_1(t) = 3 \cos 2t$ A. Find $v(t)$ for the case that $i_2(t) = 5 \cos 2t$ A.

1.24 Repeat Problem 1.23 for the case that $i_2(t) = 5 \sin 2t$ A.

Problems for Section 1.4

1.25 For the circuits given in Fig. P1.25, what values of v are permissible? Is the circuit of part (b) a series or parallel connection?

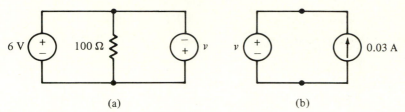

(a)   (b)

Fig. P1.25

1.26 For the circuits shown in Fig. P1.26, what values of i are permissible?

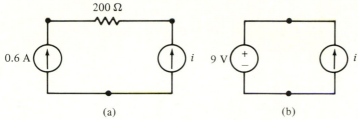

(a)   (b)

Fig. P1.26

1.27 For the circuits given in Fig. P1.27, find the variables indicated.

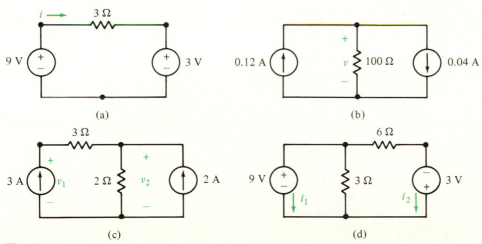

(a)   (b)

(c)   (d)

Fig. P1.27

1.28 For the circuits shown in Fig. P1.28, find the variables indicated.

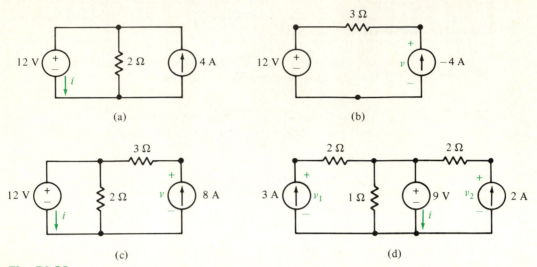

Fig. P1.28

1.29 For the circuits given in Fig. P1.29, find the variables indicated.

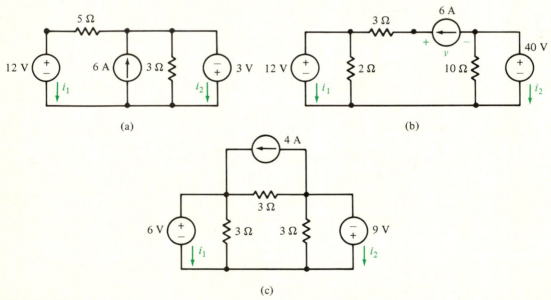

Fig. P1.29

1.30 For the circuit shown in Fig. P1.30, find the variables indicated for:
    (a) $R = 2 \ \Omega$
    (b) $R = 4 \ \Omega$

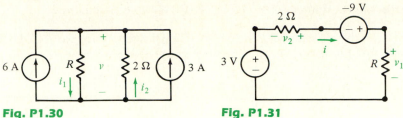

Fig. P1.30   **Fig. P1.31**

1.31 Repeat Problem 1.30 for the circuit shown in Fig. P1.31.
1.32 For the series-parallel circuit given in Fig. P1.32, find v and i.

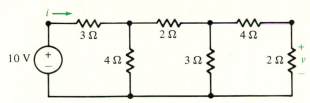

Fig. P1.32

1.33 Repeat Problem 1.32 for the circuit shown in Fig. P1.33.

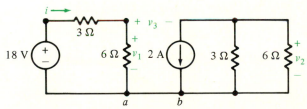

Fig. P1.33

1.34 Given the circuit shown in Fig. P1.34:
 (a) Find i, v_1, v_2, and v_3.
 (b) Remove the short circuit between a and b (erase it). Find i, v_1, and v_2. (Don't try to find v_3—it can't be done!)

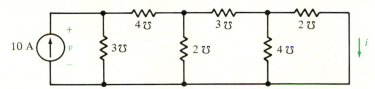

Fig. P1.34

1.35 Given the series-parallel circuit shown in Fig. P1.35:
 (a) If $v_1 = 2$ V, what is v?

(b) If $i_3 = 3$ A, what is v?

(c) If $i_5 = 4$ A, what is v?

(d) What is the resistance $R_{eq} = v/i$ "seen" by the battery for part (a)? For part (b)? For part (c)?

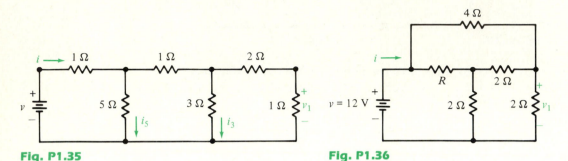

Fig. P1.35

Fig. P1.36

1.36 Given the non-series-parallel circuit shown in Fig. P1.36:

(a) When $R = 1/2 \ \Omega$, then $v_1 = 6$ V. What is the resistance $R_{eq} = v/i$ seen by the battery?

(b) When $R = 4 \ \Omega$, then $v_1 = 4$ V. What is R_{eq}?

(c) When $v_1 = 3$ V, what is R and R_{eq}?

1.37 The non-series-parallel circuit shown in Fig. P1.37 is known as a **twin-T network.**

(a) Suppose that $R_1 = 1 \ \Omega$, $R_2 = 3 \ \Omega$, and $v_2 = 3$ V. Find the resistance $R_{eq} = v/i$ seen by the battery.

(b) Suppose that $R_2 = \frac{2}{7} \ \Omega$ and $v_2 = 1$ V. Find R_1 and R_{eq}.

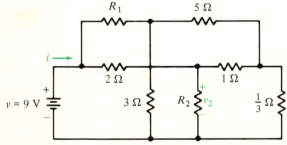

Fig. P1.37

1.38 Shown in Fig. P1.38 is a non-series-parallel connection known as a **bridge circuit.**

(a) Given that $R_1 = 6 \ \Omega$, $R_2 = 3 \ \Omega$, and $v_1 = 7$ V, find v_2, i, v_3, and the resistance $R_{eq} = v_g/i_g$ seen by the voltage source.

(b) Repeat part (a) for the case $R_1 = 3 \ \Omega$, $R_2 = 6 \ \Omega$, and $v_1 = 4$ V.

(c) When the current $i = 0$, we say that the bridge is **balanced.** Under what condition (find an expression relating R_1 and R_2) will this bridge be balanced?

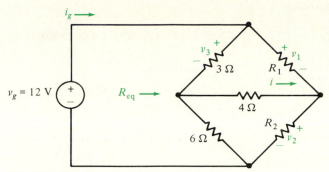

Fig. P1.38

Problems for Section 1.5

1.39 Given the circuit shown in Fig. P1.39, find i_1 for
(a) $K = 2$,
(b) $K = 3$,
(c) $K = 4$.

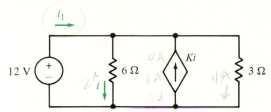

Fig. P1.39

1.40 The circuit given in Fig. P1.40 contains a **voltage-dependent voltage source** as well as a current-dependent current source. Repeat Problem 1.39 for this circuit.

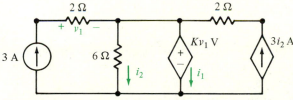

Fig. P1.40

1.41 Consider the circuit shown in Fig. P1.41. Find v for
(a) $K = 2$,
(b) $K = 4$.

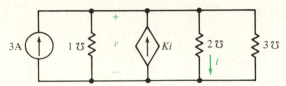

Fig. P1.41

1.42 Repeat Problem 1.41 for the circuit given in Fig. P1.42.

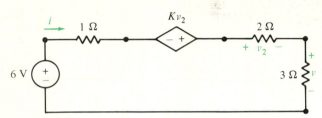

Fig. P1.42

1.43 For the circuit shown in Fig. P1.43, find the output voltage v_2 in terms of the input voltage v_1. Also find $R_{eq} = v_1/i_1$.

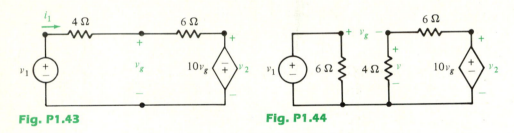

Fig. P1.43   **Fig. P1.44**

1.44 Consider the circuit given in Fig. P1.44.
   (a) Use voltage division to find v in terms of v_g.
   (b) Find the output voltage v_2 in terms of the input voltage v_1.
1.45 The circuit shown in Fig. P1.45 is a simple field-effect transistor (FET)

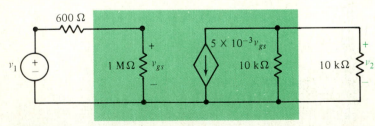

Fig. P1.45

amplifier in which the input is v_1 and the output is v_2. The portion of the circuit in the shaded box is an approximate model of an FET.

(a) Find v_{gs} in terms of v_1.

(b) Find v_2 in terms of v_1.

(c) Find v_2 given that $v_1 = 0.1 \cos 2000\pi t$ volts.

1.46 The circuit shown in Fig. P1.46 is a simple bipolar junction transistor (BJT) amplifier. The portion of the circuit in the shaded box is an approximate model of a BJT in the common-emitter configuration.

(a) Find i_b in terms of the input voltage v_1.

(b) Find the output voltage v_2 in terms of v_1.

(c) Find v_2 given that $v_1 = 0.1 \cos 2000\pi t$ volts.

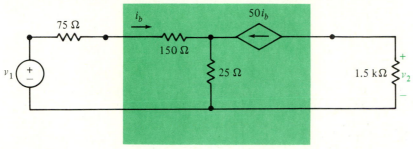

Fig. P1.46

1.47 The circuit shown in Fig. P1.47 is also a simple BJT amplifier. The portion in the shaded box is an approximate model of a BJT in the common-base configuration.

(a) Find i_e in terms of the input voltage v_1.

(b) Find the output voltage v_2 in terms of v_1.

(c) Find v_2 given that $v_1 = 0.1 \cos 2000\pi t$ volts.

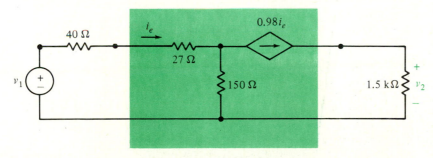

Fig. P1.47

Problems for Section 1.6

1.48 A flashlight incandescent lamp is rated at $\frac{1}{2}$ W when used with a (series) connection of two $1\frac{1}{2}$-V batteries. Assuming ideal batteries, what current is drawn by the lamp?

1.49 For the circuit given in Fig. 1.51, find the power absorbed by each element.

1.50 For the circuit given in Fig. 1.52, find the power absorbed by each element.

1.51 For the circuit given in Fig. 1.53, find the power absorbed by each element.

1.52 For the circuit given in Fig. P1.41, find the power absorbed by each element for

   (a) $K = 2$,
   (b) $K = 4$.

1.53 For the circuit given in Fig. P1.42, find the power absorbed by each element for

   (a) $K = 2$,
   (b) $K = 4$.

$P = vi$

$2A$ 3Ω $4i_1 A$ 5Ω

$i_1 + i_2 = 4i_1 + 2 \Rightarrow -3i_1 + i_2 = 2$

$i_1 = v/3$ $-3(v/3) + v/5 = 2$

$i_2 = v/5$ $v = -5/2 \ V$

$i_1 = -5/6 \ A$ $i_3 = -20/6 = -10/3 \ A$

$i_2 = -1/2 \ A$

$P = vi$

$P_1 = -5W$ $P_2 = 25/12 \ W$ $P_3 = 50/6 = 25/3 W$ $P_4 = 5/4 \ W$

Chapter 2
Circuit Analysis Principles

The process by which we determine a variable (either a voltage or a current) of a circuit is called **analysis.** Up to this point, we have been dealing with circuits that are not very complicated. Don't be fooled, though, some very simple circuits can be quite useful. As part of the set of problems, we have already come across some simple single-stage amplifier circuits whose analysis was accomplished by applying the basic principles covered to date—Ohm's law, KCL, and KVL. Nonetheless, we will want to analyze more complicated circuits—circuits for which it is simply not possible conveniently to write and solve a single equation that has only one unknown. Instead, we will have to write and solve a set of linear algebraic equations. To obtain such equations, we again utilize Ohm's law, KCL, and KVL.

There are two distinct approaches that we will take. In one, we will write a set of simultaneous equations in which the variables are voltages; this is known as **nodal analysis.** In the other, we will write a set of simultaneous equations in which the variables are currents; this is known as **mesh analysis** (also called **loop analysis**). Although nodal analysis can be used for any circuit, mesh analysis is valid only for **planar networks**—that is, circuits that can be drawn in a two-dimensional plane in such a way that no element crosses over another.

An increasingly important circuit element is the operational amplifier. Because of the evolution of integrated-circuit (IC) technology, the operational amplifier (or op amp) is both small in size and inexpensive. Its versatility and usefulness have made it extremely popular. In this chapter, we study the operational amplifier from the point of view of an ideal circuit element, and a number of applications are presented. A discussion of the operational amplifier from the electronics point of view is given in Section 9.4.

A very important circuit concept is Thévenin's theorem. This result says, in essence, that an arbitrary circuit behaves as an appropriately valued voltage source in series with an appropriately valued resistance to the outside world. A major consequence of this fact is the determination of the maximum power that can be delivered to a load and the condition for which this occurs.

When a circuit contains more than one independent source, a response (a voltage or a current) of the circuit can be obtained by finding the response to each individual independent source, and then summing these individual responses. This notion, known as the principle of superposition, is another important circuit concept that is frequently used.

Although in this chapter we consider **resistive circuits**—that is, circuits that contain only resistors and sources (both independent and dependent)—we will see later that the same techniques are applicable to networks that contain other types of elements as well.

2.1 NODAL ANALYSIS

Given a circuit to be analyzed, the first step in employing nodal analysis is the arbitrary choice of one of the nodes of the circuit as the **reference** (or **datum**) **node.** Although this node can be indicated by any of the three symbols shown in Fig. 2.1, we will use the symbol depicted in Fig. 2.1(a) exclusively.

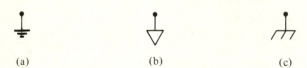

(a)   (b)   (c)

Fig. 2.1 Symbols used to indicate a reference node.

We can label any node in a circuit with its potential with respect to the reference node in the following manner.

Suppose that the voltages between nodes a, b, c, and the reference node are as shown in Fig. 2.2. Then, Fig. 2.3 indicates how the nodes are labeled. Of course, since the potential difference between the reference node and itself must be zero volts, we can mark the reference node "0V" if we wish—but this is redundant.

It is possible, and easy, to express the voltage between any pair of nodes in terms of the potentials (with respect to the reference node) of those two nodes. Suppose that we wish to determine the voltage v_{ac} between nodes a and c, with the plus at node a and the minus at node c, in terms of v_a, and v_c. This situation is depicted in Fig. 2.4. Since the labeling in Fig. 2.5 is equivalent to Fig. 2.4, by KVL, we have that $v_{ac} = v_a - v_c$. Similarly, the voltage v_{ca} between nodes a and c, with the plus at node c and the minus at node a, is $v_{ca} = v_c - v_a$. With reference

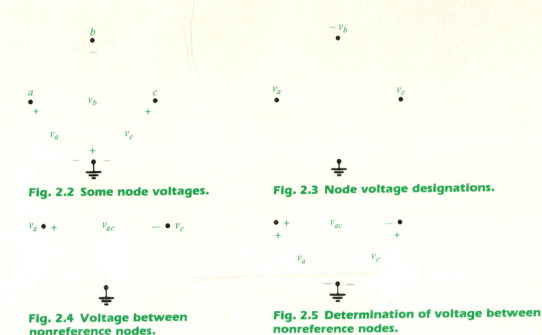

Fig. 2.2 Some node voltages.

Fig. 2.3 Node voltage designations.

Fig. 2.4 Voltage between nonreference nodes.

Fig. 2.5 Determination of voltage between nonreference nodes.

to Fig. 2.3, we see that the voltage v_{ab} between nodes a and b (plus at node a and minus at node b) is $v_{ab} = v_a - (-v_b) = v_a + v_b$. Also, v_{bc}, the voltage between nodes b and c (plus at node b and minus at node c), is $v_{bc} = -v_b - v_c$.

With these ideas in mind, let us see how to employ nodal analysis for the circuit shown in Fig. 2.6. After choosing the reference node arbitrarily, the two non-reference nodes are labeled v_1 and v_2. The next step is to indicate currents (the names and directions are arbitrary) through elements that do not already have currents indicated through them. For this example, the values and directions of the current sources are given whereas i_1, i_2 and i_3 were chosen arbitrarily.

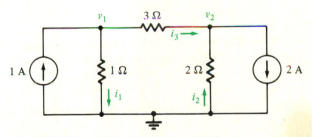

Fig. 2.6 Example of nodal analysis.

Now apply KCL at the node labeled v_1. The result is

$$i_1 + i_3 = 1 \tag{2.1}$$

By Ohm's law,

$$i_1 = v_1/1 \qquad \text{and} \qquad i_3 = (v_1 - v_2)/3$$

Substituting these into Equation (2.1) yields

$$\frac{v_1}{1} + \frac{v_1 - v_2}{3} = 1$$

and simplifying this we get

$$4v_1 - v_2 = 3 \tag{2.2}$$

When we apply KCL at node v_2 (i.e., the node labeled v_2), we obtain

$$i_2 + i_3 = 2$$

and by Ohm's law this equation becomes

$$-\frac{v_2}{2} + \frac{v_1 - v_2}{3} = 2$$

from which

$$2v_1 - 5v_2 = 12 \tag{2.3}$$

Equations (2.2) and (2.3) are simultaneous, linear equations in the unknowns v_1 and v_2. Multiplying Equation (2.3) by 2 and subtracting the result fom Equation (2.2), we get

$$9v_2 = -21 \quad \Rightarrow \quad v_2 = -\tfrac{7}{3}\text{V}$$

Substituting this value of v_2 in Equation (2.2), we obtain

$$4v_1 - (-\tfrac{7}{3}) = 3 \quad \Rightarrow \quad v_1 = \tfrac{1}{6}\text{V}$$

Having determined the values of v_1 and v_2, it is a simple matter to determine i_1, i_2, and i_3. By Ohm's law,

$$i_1 = \frac{v_1}{1} = \frac{1}{6}\text{A}, \quad i_2 = \frac{-v_2}{2} = \frac{7}{6}\text{A}, \quad i_3 = \frac{v_1 - v_2}{3} = \frac{(1/6) + (7/3)}{3} = \frac{5}{6}\text{A}$$

Now that we have seen an example of the application of nodal analysis, we can state the rules for nodal analysis for circuits that do not contain voltage sources. We will discuss the case of circuits with voltage sources immediately thereafter.

Nodal Analysis with No Voltage Sources

Given a circuit with n nodes and no voltage sources, proceed as follows:

1. Select any node as the reference node.

2. Label the remaining $n - 1$ nodes (e.g., $v_1, v_2, \ldots, v_{n-1}$).

3. **Arbitrarily assign currents to the elements in which no current is designated.**

4. **Apply KCL at each nonreference node.**

5. **Use Ohm's law to express the currents through resistors in terms of the node voltages, and substitute these expressions into the current equations obtained in Step 4.**

6. **Solve the resulting set of $n - 1$ simultaneous equations for the node voltages.**

Having seen an example of nodal analysis for a circuit without a voltage source, let us now consider a circuit in which a voltage source is present.

Example 2.1

Figure 2.7 shows a circuit that contains a 3-V independent voltage source, as well as a dependent current source whose value depends on the voltage across the $\frac{1}{2}$-Ω resistor which is drawn vertically. This circuit contains four nodes. One is the reference node, and the other three are labeled v_1, v_2, and v_3. The directions of the currents i_1, i_2, i_3, and i_4 through the resistors were chosen arbitrarily.

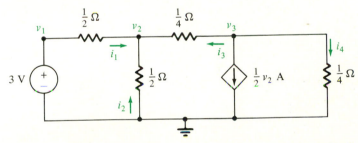

Fig. 2.7 Nodal analysis of a circuit with an independent voltage source.

Because of the voltage source between node v_1 and the reference node, we do not have to apply KCL at node v_1. Actually, we see by inspection that $v_1 =$ 3 V—and this is the equation obtained at node v_1. Thus, essentially, there are only two unknowns, v_2 and v_3. To determine these, we proceed as discussed previously.

At node v_2, by KCL,

$$i_1 + i_2 + i_3 = 0$$

By Ohm's law, this equation becomes

$$\frac{v_1 - v_2}{1/2} + \frac{-v_2}{1/2} + \frac{v_3 - v_2}{1/4} = 0$$

from which

$$-8v_2 + 4v_3 = -6 \tag{2.4}$$

At node v_3, by KCL,

$$\tfrac{1}{2}v_2 + i_3 + i_4 = 0$$

and so by Ohm's law,

$$\frac{1}{2}v_2 + \frac{v_3 - v_2}{1/4} + \frac{v_3}{1/4} = 0$$

from which

$$-7v_2 + 16v_3 = 0 \tag{2.5}$$

From Equation (2.5),

$$v_3 = \tfrac{7}{16}v_2 \tag{2.6}$$

Substituting this into Equation (2.4), we get

$$-8v_2 + 4(\tfrac{7}{16}v_2) = -6 \quad \Rightarrow \quad v_2 = \tfrac{24}{25} = 0.96 \text{ V}$$

and from Equation (2.6),

$$v_3 = \tfrac{7}{16}\left(\tfrac{24}{25}\right) = \tfrac{21}{50} = 0.42 \text{ V}$$

By Ohm's law,

$$i_1 = 2(v_1 - v_2) = \tfrac{102}{25} = 4.08 \text{ A}, \qquad i_2 = -2v_2 = -\tfrac{48}{25} = -1.92 \text{ A}$$

$$i_3 = 4(v_3 - v_2) = -\tfrac{54}{25} = -2.16 \text{ A}, \qquad i_4 = 4v_3 = \tfrac{42}{25} = 1.68 \text{ A}$$

For the last two circuits, we had the situation in which we were required to solve two equations in two unknowns. Although the approaches taken in the two examples were slightly different, there is yet another technique, known as **Cramer's rule,** that can be employed. To see how this is done, first consider the pair of simultaneous equations,

$$a_1x_1 + a_2x_2 = d_1$$
$$b_1x_1 + b_2x_2 = d_2$$

where the coefficients a_1, a_2, b_1, b_2 are given, as are d_1, d_2, and the unknowns x_1, x_2, are to be determined.

The **determinant** D of the coefficients a_1, a_2, b_1, b_2, is defined by

$$D = \begin{vmatrix} a_1 & a_2 \\ b_1 & b_2 \end{vmatrix} = a_1b_2 - b_1a_2$$

To use Cramer's rule, two more determinants D_1 and D_2 must be calculated. To find

D_1, replace coefficients a_1 and b_1 by d_1 and d_2, respectively, that is,

$$D_1 = \begin{vmatrix} d_1 & a_2 \\ d_2 & b_2 \end{vmatrix} = d_1b_2 - d_2a_2$$

To find D_2, replace the coefficients a_2 and b_2 by d_1 and d_2 respectively. That is,

$$D_2 = \begin{vmatrix} a_1 & d_1 \\ b_1 & d_2 \end{vmatrix} = a_1d_2 - b_1d_1$$

By Cramer's rule,

$$x_1 = \frac{D_1}{D} \quad \text{and} \quad x_2 = \frac{D_2}{D}$$

Example 2.2

Consider the pair of equations (2.2) and (2.3), which are now repeated.

$$4v_1 - v_2 = 3 \tag{2.2}$$

$$2v_1 - 5v_2 = 12 \tag{2.3}$$

Forming determinants, we have

$$D = \begin{vmatrix} 4 & -1 \\ 2 & -5 \end{vmatrix} = (4)(-5) - (2)(-1) = -18$$

$$D_1 = \begin{vmatrix} 3 & -1 \\ 12 & -5 \end{vmatrix} = (3)(-5) - (12)(-1) = -3$$

$$D_2 = \begin{vmatrix} 4 & 3 \\ 2 & 12 \end{vmatrix} = (4)(12) - (2)(3) = 42$$

and by Cramer's rule,

$$v_1 = \frac{D_1}{D} = \frac{-3}{-18} = \frac{1}{6} \text{ V} \quad \text{and} \quad v_2 = \frac{D_2}{D} = \frac{42}{-18} = -\frac{7}{3} \text{ V}$$

For the case of three equations and three unknowns

$$a_1x_1 + a_2x_2 + a_3x_3 = d_1$$

$$b_1x_1 + b_2x_2 + b_3x_3 = d_2$$

$$c_1x_1 + c_2x_2 + c_3x_3 = d_3$$

the determinant D is given by

$$D = \begin{vmatrix} a_1 & a_2 & a_3 \\ b_1 & b_2 & b_3 \\ c_1 & c_2 & c_3 \end{vmatrix} = a_1b_2c_3 + a_2b_3c_1 + a_3b_1c_2 - a_3b_2c_1 - a_1b_3c_2 - a_2b_1c_3$$

A simple way to remember this formula is to repeat the first two columns and then put in solid and dashed lines as follows:

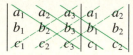

The solid lines indicate the positive products in D and the dashed lines indicate the negative products.

To obtain D_1, replace the column consisting of a_1, b_1, c_1 with the column consisting of d_1, d_2, d_3 and calculate the resulting determinant. Similarly, D_2 is obtained by replacing a_2, b_2, c_2 with d_1, d_2, d_3; and D_3 is obtained by replacing a_3, b_3, c_3 with d_1, d_2, d_3. Again, by Cramer's rule,

$$x_1 = \frac{D_1}{D}, \qquad x_2 = \frac{D_2}{D}, \qquad x_3 = \frac{D_3}{D}$$

Just as Cramer's rule can be applied to the cases of two simultaneous equations and three simultaneous equations, so can it be employed for the case of four or more simultaneous equations. However, this subject is beyond the scope of this book.

Example 2.3

Consider the circuit shown in Fig. 2.8, which contains two voltage sources—one independent and one dependent. In this circuit, the reference node and the node voltages have been labeled. Even though currents have not been explicitly indicated, let us go ahead and employ nodal analysis.

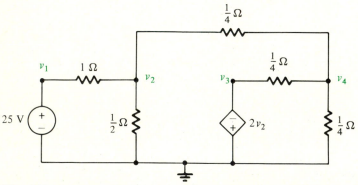

Fig. 2.8 Nodal analysis of a circuit with two voltage sources.

Since there is a voltage source between node v_1 and the reference node, by inspection we have

$$v_1 = 25 \text{ V}$$

Summing the currents out of node v_2 to zero, by KCL,

$$\frac{v_2 - 25}{1} + \frac{v_2}{1/2} + \frac{v_2 - v_4}{1/4} = 0$$

from which

$$7v_2 - 4v_4 = 25 \tag{2.7}$$

Since there is a voltage source (even though it is a dependent source) between node v_3 and the reference node, we again obtain, by inspection,

$$v_3 = -2v_2$$

or

$$2v_2 + v_3 = 0 \tag{2.8}$$

Summing the currents into node v_4 to zero, we get

$$\frac{v_2 - v_4}{1/4} + \frac{v_3 - v_4}{1/4} + \frac{-v_4}{1/4} = 0$$

from which

$$v_2 + v_3 - 3v_4 = 0 \tag{2.9}$$

The three equations in three unknowns that must be solved are given in Equations (2.7), (2.8), and (2.9). Again, these equations are

$$7v_2 + 0v_3 - 4v_4 = 25$$

$$2v_2 + v_3 + 0v_4 = 0$$

$$v_2 + v_3 - 3v_4 = 0$$

Calculating determinants:

$$D = \begin{vmatrix} 7 & 0 & -4 \\ 2 & 1 & 0 \\ 1 & 1 & -3 \end{vmatrix} = \begin{aligned} &(7)(1)(-3) + (0)(0)(1) + (-4)(2)(1) \\ &-(-4)(1)(1) - (7)(0)(1) - (0)(2)(-3) = -25 \end{aligned}$$

$$D_2 = \begin{vmatrix} 25 & 0 & -4 \\ 0 & 1 & 0 \\ 0 & 1 & -3 \end{vmatrix} = \begin{aligned} &(25)(1)(-3) + (0)(0)(0) + (-4)(0)(1) \\ &-(-4)(1)(0) - (25)(0)(1) - (0)(0)(-3) = -75 \end{aligned}$$

$$D_3 = \begin{vmatrix} 7 & 25 & -4 \\ 2 & 0 & 0 \\ 1 & 0 & -3 \end{vmatrix} = \begin{aligned} &(7)(0)(-3) + (25)(0)(1) + (-4)(2)(0) \\ &-(-4)(0)(1) - (7)(0)(0) - (25)(2)(-3) = 150 \end{aligned}$$

$$D_4 = \begin{vmatrix} 7 & 0 & 25 \\ 2 & 1 & 0 \\ 1 & 1 & 0 \end{vmatrix} = \begin{aligned} &(7)(1)(0) + (0)(0)(1) + (25)(2)(1) \\ &-(25)(1)(1) - (7)(0)(1) - (0)(2)(0) = 25 \end{aligned}$$

Thus, by Cramer's rule,

$$v_2 = \frac{D_2}{D} = \frac{-75}{-25} = 3 \text{ V}, \qquad v_3 = \frac{D_3}{D} = \frac{150}{-25} = -6 \text{ V}, \qquad v_4 = \frac{D_4}{D} = \frac{25}{-25} = -1 \text{ V}$$

Up to this point, when we employed nodal analysis, a voltage source in a circuit was connected between a nonreference node and the reference node. For the case that a voltage source is connected between two nonreference nodes, say v_a and v_b, and no other voltage source is connected to either of these two nodes, assign a current, say i, through the voltage source in question. Apply KCL at nodes v_a and v_b, and then combine the two equations to eliminate i. Thus, instead of getting two equations from nodes v_a and v_b, only one is obtained. However, we can get the other equation from KVL by expressing the value of the voltage source in terms of v_a and v_b.

Circuits with more than one voltage source connected between nonreference nodes are treated in a manner similar to that discussed above.

2.2 MESH ANALYSIS

As mentioned previously, mesh analysis can be used only for planar networks, that is, circuits that can be drawn in the plane in such a way that elements do not cross. As a result of its definition, we deduce that a planar network necessarily partitions the plane into regions or "windows" called **meshes**. One of the meshes is infinite; this is the mesh that surrounds the circuit. The remaining meshes are finite.

The first step in employing mesh analysis is to visualize a current, called a **mesh current**, around each finite mesh (whether clockwise or counterclockwise is arbitrary). To be specific, consider the circuit shown in Fig. 2.9. The (finite) mesh bounded by the 5-V source, the leftmost 1-Ω resistor, and the 2-Ω resistor has the clockwise mesh current i_1. The other finite mesh (the mesh bounded by the 10-V source, the 2-Ω resistor, and the rightmost 1-Ω resistor) has the clockwise mesh current i_2.

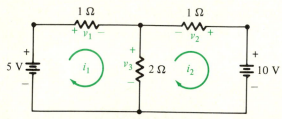

Fig. 2.9 Example of mesh analysis.

Although mesh currents are an invention, we can express the actual current through any element in a circuit in terms of the mesh currents. For the circuit in Fig.

2.9, the current going to the right through the leftmost 1-Ω resistor is i_1, and that going to the right through the other 1-Ω resistor is i_2. The current going down through the 2-Ω resistor is $i_1 - i_2$. What is the current through the 5-V source? What is it through the 10-V source?

Having selected mesh currents for the circuit in Fig. 2.9, we now apply KVL to each mesh. For mesh i_1 (i.e., the mesh labeled with the mesh current i_1), by KVL,

$$v_1 + v_3 = 5$$

By Ohm's law,

$$1i_1 + 2(i_1 - i_2) = 5$$

from which

$$3i_1 - 2i_2 = 5 \tag{2.10}$$

For mesh i_2, by KVL,

$$v_2 + v_3 = 10$$

By Ohm's law,

$$-1i_2 + 2(i_1 - i_2) = 10$$

from which

$$2i_1 - 3i_2 = 10 \tag{2.11}$$

In solving Equations (2.10) and (2.11), we get the mesh currents

$$i_1 = -1 \text{ A} \qquad \text{and} \qquad i_2 = -4 \text{ A}$$

Once we know the mesh currents, we can find any current or voltage in the circuit. For exmple, the current going down through the 2-Ω resistor is $i_1 - i_2 = 3$A. The voltages labeled in the circuit are

$$v_1 = 1i_1 = -1 \text{ V}, \qquad v_2 = -1i_2 = 4 \text{ V}, \qquad v_3 = 2(i_i - i_2) = 6 \text{ V}$$

Now that we have seen an example of the application of mesh analysis, we can state the rules for mesh analysis for planar circuits that do not contain current sources. We will discuss the case of circuits with current sources immediately thereafter.

Mesh Analysis with No Current Sources

Given a planar circuit with m meshes and no current sources, proceed as follows:

1. Place mesh currents (e.g., $i_1, i_2, \ldots, i_m$) in the m (finite) meshes.

2. Arbitrarily assign voltages to the elements for which no voltage is designated.

3. Apply KVL to each of the *m* meshes.

4. Use Ohm's law to express the voltages across resistors in terms of the mesh currents, and substitute these expressions into the voltage equations obtained in Step 3.

5. Solve the resulting set of *m* simultaneous equations for the mesh currents.

Having seen an example of mesh analysis for a circuit without a current source, let us now consider a circuit in which a current is present.

Example 2.4

Figure 2.10 shows a circuit that contains a 3-A independent current source, as well as a dependent voltage source. The mesh currents i_1 and i_2 were arbitrarily chosen to be clockwise and i_3 to be counterclockwise. Since the only mesh current that goes through the current source is i_1, we have by inspection that $i_1 = 3$ A—and this is the equation obtained for mesh i_1. Thus, essentially, there are only two unknowns, i_2 and i_3. To determine these, we proceed as discussed previously.

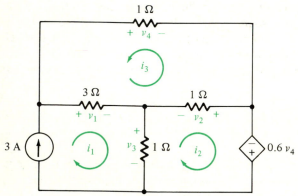

Fig. 2.10 Mesh analysis of a circuit with a current source.

For mesh i_2, by KVL,

$$v_2 + v_3 + 0.6v_4 = 0$$

By Ohm's law, this equation becomes

$$1(-i_2 - i_3) + 1(i_1 - i_2) + 0.6(-1i_3) = 0$$

from which

$$10i_2 + 8i_3 = 5i_1 = 15 \qquad (2.12)$$

For mesh i_3, by KVL,

$$v_1 - v_2 - v_4 = 0$$

By Ohm's law, this equation becomes

$$3(i_1 + i_3) - 1(-i_2 - i_3) - (-1i_3) = 0$$

from which

$$i_2 + 5i_3 = -3i_1 = -9 \qquad (2.13)$$

Solving Equations (2.12) and (2.13), we obtain

$$i_2 = 3.5 \text{ A} \qquad \text{and} \qquad i_3 = -2.5 \text{ A}$$

By Ohm's law, since $i_1 = 3$ A,

$$v_1 = 3(i_1 + i_3) = 1.5 \text{ V}, \qquad v_2 = 1(-i_2 - i_3) = -1 \text{ V}$$

$$v_3 = 1(i_1 - i_2) = -0.5 \text{ V}, \qquad v_4 = -1i_3 = 2.5 \text{ V}$$

In the preceding example, the circuit contained a current source that was common to a finite mesh and the infinite mesh. In such a case, we can express the finite mesh current in terms of the value of the current source by inspection. For the case that a current source is common to two finite meshes, say i_a and i_b, assign a voltage, say v, across the current source in question. Apply KVL around meshes i_a and i_b, and then combine the two equations to eliminate v. Thus, instead of getting two equations from meshes i_a and i_b, only one is obtained. However, we can get the other equation by expressing the value of the current source in terms of the mesh currents i_a and i_b.

2.3 THE OPERATIONAL AMPLIFIER

Of fundamental importance in the study of electric circuits is the **ideal voltage amplifier**. Such a device, in general, has two inputs, v_1 and v_2, and one output, v_o. The relationship between the output and the inputs is given by $v_o = A(v_2 - v_1)$, where A is called the **gain** of the amplifier. The ideal amplifier is modeled by the circuit shown in Fig. 2.11(a), which contains a dependent voltage source. Note that since the input resistance $R = \infty$, when such an amplifier is connected to any circuit, no current will go into the input terminals. Also, since the output v_o is the voltage across an ideal source, we have that $v_o = A(v_2 - v_1)$ regardless of what is connected to the output. For the sake of simplicity, the ideal amplifier having gain A is often represented as shown in Fig. 2.11(b). (Note that the reference node is not

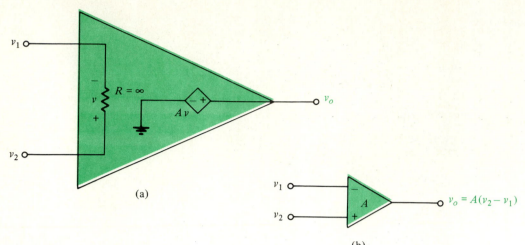

Fig. 2.11 Model and circuit symbol of an ideal amplifier having gain A.

explicitly displayed in this figure.) We refer to the input terminal labeled "−" as the **inverting input** and the input terminal labeled "+" as the **noninverting input**.

Example 2.5

In the circuit given in Fig. 2.12, the noninverting input is at the reference potential, that is, $v_2 = 0$. Furthermore, since node v_s and node v_o are constrained by voltage sources (independent and dependent, respectively), in using nodal analysis we need only sum currents at node v_1. Since the amplifier inputs draw no current, by KCL,

$$i_1 + i_2 = 0$$

or

$$\frac{v_s - v_1}{R_1} + \frac{v_o - v_1}{R_2} = 0$$

from which

$$R_2 v_s - R_2 v_1 + R_1 v_o - R_1 v_1 = 0$$

$$R_2 v_s = (R_1 + R_2) v_1 - R_1 v_o \qquad (2.14)$$

But, due to the amplifier,

$$v_o = A(v_2 - v_1) = -A v_1$$

so

$$v_1 = -\frac{v_o}{A} \qquad (2.15)$$

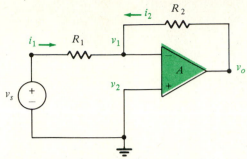

Fig. 2.12 Ideal-amplifier circuit.

and substituting this into Equation (2.14), we get

$$R_2 v_s = (R_1 + R_2)(-v_o/A) - R_1 v_o = -[(1/A)(R_1 + R_2) + R_1] v_o$$

Thus,

$$v_o = \frac{-R_2 v_s}{R_1 + (1/A)(R_1 + R_2)} \qquad (2.16)$$

For a circuit such as that in Fig. 2.12, let us consider the case that the gain A becomes arbitrarily large. We can rewrite Equation (2.16) as

$$v_o = \frac{-R_2/R_1}{1 + (1/A)(1 + R_2/R_1)} v_s$$

Therefore, when $A \to \infty$,

$$v_o = -\frac{R_2}{R_1} v_s$$

We see that although the gain of the amplifier is infinite, for a finite input voltage v_s, the output voltage v_o is finite (provided, of course, that $R_1 \neq 0$). Inspection of Equation (2.15) indicates why the output voltage remains finite—as $A \to \infty$, then $v_1 = -v_o/A \to 0$. This result occurs because there is a resistor connected between the output and the negative input. Such a connection is called **negative feedback.**

An ideal amplifier having gain $A = \infty$ is known as an **operational amplifier**, or **op amp**. In an op-amp circuit, because of the infinite gain property, we must have a feedback resistor and must not connect a voltage source directly between the amplifier's input terminals. For the circuit given in Fig. 2.12, the corresponding op-amp circuit is usually drawn as shown in Fig. 2.13. Using the fact that $v_1 = 0$, and summing the currents at node v_1, we have

$$\frac{v_s}{R_1} = -\frac{v_o}{R_2}$$

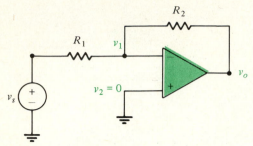

Fig. 2.13 Op-amp circuit—an inverting amplifier.

from which

$$v_o = -\frac{R_2}{R_1} v_s$$

and the gain of the overall circuit is

$$\frac{v_o}{v_s} = -\frac{R_2}{R_1}$$

This circuit is called an **inverting amplifier**.

Notice how simple the analysis of the op-amp circuit in Fig. 2.13 is when we use the fact that $v_1 = 0$. Although this result was originally deduced from Equation (2.16), the combination of infinite gain and feedback constrains the voltage applied to the op-amp (between terminals v_1 and v_2) to be zero. In other words, we must have that $v_1 = v_2$.

Example 2.6

Consider the op-amp circuit with feedback in Fig. 2.14(a). Again, the inputs of the amplifier draw no current, and so in applying KCL at node v_1, we have

$$i_1 + i_2 = 0$$

so

$$\frac{v_1}{R_1} + \frac{v_1 - v_o}{R_2} = 0$$

Since the input voltage to the amplifier is zero, or equivalently, since both input terminals must be at the same potential, then $v_1 = v_2$, and by KCL,

$$\frac{v_2}{R_1} + \frac{v_2 - v_o}{R_2} = 0$$

from which

$$R_2 v_2 + R_1 v_2 - R_1 v_o = 0$$

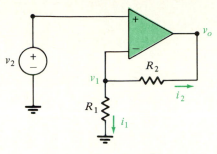

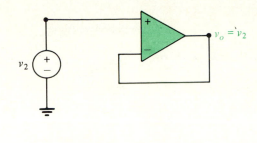

(a) Noninverting amplifier

(b) Voltage follower

Fig. 2.14 Op-amp circuits.

or

$$(R_1 + R_2)\, v_2 = R_1 v_o$$

Hence,

$$v_o = \frac{R_1 + R_2}{R_1} v_2 = \left(1 + \frac{R_2}{R_1}\right) v_2$$

The overall gain of this circuit is, therefore,

$$\frac{v_o}{v_2} = \frac{R_1 + R_2}{R_1} = 1 + \frac{R_2}{R_1}$$

This circuit is called a **noninverting amplifier**.

Note that if $R_2 = 0$, then $v_o = v_2$. Under this circumstance, R_1 is superfluous and may be removed. The resulting op-amp circuit shown in Fig. 2.14(b) is known as a **voltage follower**. This configuration is used to isolate or buffer one circuit from another.

■

Example 2.7

The op-amp circuit given in Fig. 2.15 can be used as an electronic ohmmeter, where R is the resistance whose value is to be determined. The device connected between the output and the reference potential is a dc voltmeter. Since the positive terminal of the voltmeter is connected to the reference, the voltmeter displays $-v_o$. By KCL,

$$\frac{12}{12{,}000} + \frac{v_o}{R} = 0 \qquad \Rightarrow \qquad \frac{v_o}{R} = -\frac{1}{1000}$$

Thus,

$$R = -1000 v_o$$

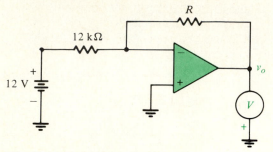

Fig. 2.15 Electronic ohmmeter.

Suppose that the voltmeter "reads" 10 V. Then $R = 10 \text{ k}\Omega$. If the voltmeter "reads" 27 V, then $R = 27 \text{ k}\Omega$.

Operational-amplifier circuits can be quite useful in some situations in which there is more than a single input.

Example 2.8

For the op-amp circuit shown in Fig. 2.16, since $v_2 = 0$, then $v_1 = 0$. Thus from

$$i_1 + i_2 = i_3$$

we get

$$\frac{v_a}{R_1} + \frac{v_b}{R_1} = -\frac{v_o}{R_2}$$

or

$$v_o = -\frac{R_2}{R_1}(v_a + v_b)$$

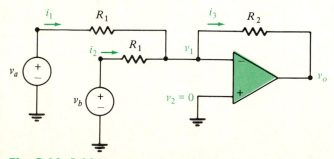

Fig. 2.16 Adder.

Since the output v_o is the sum of the voltages v_a and v_b (multiplied by the constant $-R_2/R_1$), this circuit is called an **adder** (or **summer**). A common application of such a circuit is as an "audio mixer," where, for example, the inputs are voltages attributable to the two separate microphones and the output voltage is their sum.

Example 2.9

Another op-amp circuit with two inputs is given in Fig. 2.17. Using the fact that $v_1 = v_2 = v$, by KCL at the inverting input

$$\frac{v_a - v}{R_1} = \frac{v - v_o}{R_2}$$

from which

$$R_1 v_o + R_2 v_a = (R_1 + R_2)\, v \qquad (2.17)$$

Applying KCL at the noninverting input, we get

$$\frac{v_b - v}{R_1} = \frac{v}{R_2}$$

from which

$$R_2 v_b = (R_1 + R_2)\, v \qquad (2.18)$$

Combining Equations (2.17) and (2.18) results in

$$R_1 v_o + R_2 v_a = R_2 v_b$$

and hence,

$$v_o = \frac{R_2}{R_1}(v_b - v_a)$$

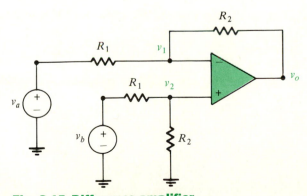

Fig. 2.17 Difference amplifier.

Since the output is the difference of the inputs (multiplied by the constant R_2/R_1), such a circuit is called a **difference amplifier** or **differential amplifier.**

For a physical operational amplifier, the gain and the input resistance are large, but not infinite. A more practical model of an actual op amp is to have a resistance R_o (called the **output resistance**) in series with the dependent voltage source. Yet, assuming that an op amp is ideal often yields a simple analysis with very accurate results.

Although its usefulness is great, the physical size of an op amp typically is small. This is a result of the fact that op amps are commonly available on IC chips, which normally contain between one and four of them. In our discussions above, we depicted the op amp as a three-terminal device. In actuality, a package containing one or more op amps typically has between 8 and 14 terminals. As we have seen, three of the terminals are the inverting input, the noninverting input, and the output. However, there are also terminals for applying dc voltages (power supplies) to "bias" the numerous transistors comprising an op amp, for frequency compensation, and for other details to be discussed later. One of the important practical electronic considerations for proper op-amp operation is that when feedback is between the output and only one input terminal, that input terminal should be the inverting input.

Example 2.10

Shown in Fig. 2.18 is a simple practical amplifier that uses the popular 741 op amp. For this circuit, terminal (pin) 4 is the inverting input, terminal 5 is the noninverting input, and terminal 10 is the output. Furthermore, pin 6 has a dc voltage of -15 V applied to it and pin 11 a dc voltage of $+15$ V. The 741 op amp typically has a gain of $A = 200,000$, an input resistance of $R = 2$ MΩ, and an output resistance of $R_o = 75$ Ω.

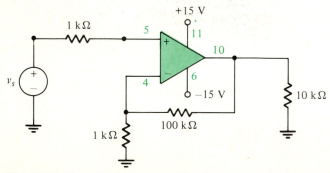

Fig. 2.18 Practical op-amp circuit.

Suppose that v_j is the voltage at terminal j, where $j = 1, 2, 3, \ldots$. Assuming that the op amp is ideal, then $v_5 = v_s$ since the (ideal) op amp inputs draw no current. Also, due to feedback, $v_4 = v_5 = v_s$, and by KCL at terminal 4,

$$\frac{v_{10} - v_s}{100{,}000} = \frac{v_s}{1000}$$

from which

$$v_{10} - v_s = 100 v_s$$

Thus,

$$v_{10} = 101 v_s$$

We can get a more accurate result by not assuming that the op amp is ideal. To do this, which requires a much greater analysis effort (try it and see), yields

$$v_{10} \approx 100 v_s$$

which is not significantly different from the simple approach taken above.

2.4 THÉVENIN'S THEOREM

Suppose that a resistor R_L, called a **load resistor**, is connected to an arbitrary (in the sense that it contains only elements discussed previously) circuit as shown in Fig. 2.19. What value of the load R_L will absorb the maximum amount of power? Knowing the particular circuit, we can use nodal or mesh analysis to obtain an expression for the power absorbed by R_L, then take the derivative of this expression to determine what value of R_L results in maximum power. The effort required for such an approach can be quite great. Fortunately, though, a remarkable and important circuit theory concept states that as far as R_L is concerned, the arbitrary circuit shown in Fig. 2.19 behaves as though it is a single independent voltage source in series with a single resistance. Once we determine the values of this source and this resistance, we simply apply the results on maximum power transfer, which follows shortly, to find the appropriate value of R_L and the resulting maximum power.

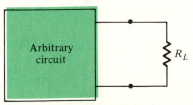

Fig. 2.19 Arbitrary circuit and its associated load.

Suppose we are given an arbitrary circuit containing any or all of the following elements: resistors, voltage sources, current sources. (The sources can be dependent as well as independent.) Let us identify a pair of nodes, say node a and node b, such that the circuit can be partitioned into two parts as shown in Fig. 2.20. Furthermore, suppose that circuit A contains no dependent source that is dependent on a variable in circuit B, and vice versa. Then we can replace circuit A by an appropriate independent voltage source, call it v_{oc}, in series with an appropriate resistance, call it R_o, and the effect on circuit B is the same as that produced by circuit A. This voltage source and resistance series combination is called the **Thévenin equivalent** of circuit A. In other words, circuit A in Fig. 2.20 and the circuit in the shaded box in Fig. 2.21 have the same effect on circuit B. This result is known as **Thévenin's theorem,*** and is one of the more useful and significant concepts in circuit theory.

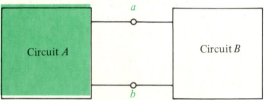

Fig. 2.20 Circuit partitioned into two parts.

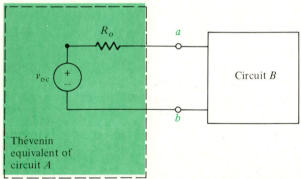

Fig. 2.21 Application of Thévenin's theorem.

To obtain the voltage v_{oc}—called the **open-circuit voltage**—remove circuit B from circuit A, and determine the voltage between nodes a and b. This voltage, as shown in Fig. 2.22, is v_{oc}.

To obtain the resistance R_o—called the **Thévenin-equivalent resistance** or the **output resistance** of circuit A—again remove circuit B from circuit A. Next, set all independent sources in circuit A to zero. Leave the dependent sources as is! (A

---

*Named for the French engineer M. L. Thévenin (1857–1926).

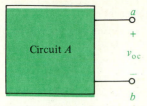

Fig. 2.22 Determination of open-circuit voltage.

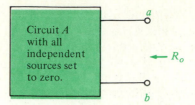

Fig. 2.23 Determination of Thévenin-equivalent (output) resistance.

zero voltage source is equivalent to a short circuit, and a zero current source is equivalent to an open circuit.) Now determine the resistance between nodes a and b—this is R_o—as shown in Fig. 2.23.

If circuit A contains no dependent sources, when all independent sources are set to zero, the result may be simply a series-parallel resistive network. In this case, R_o can be found by appropriately combining resistors in series and parallel. In general, however, R_o can be found by applying an independent source between nodes a and b and then by taking the ratio of voltage to current. This procedure is depicted in Fig. 2.24.

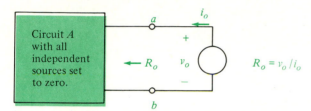

Fig. 2.24 Determination of output resistance.

In applying Thévenin's theorem, circuit B (which is often called the **load**) may consist of many circuit elements, a single element (e.g., a load resistor), or no elements (that is, circuit B already may be an open circuit).

Example 2.11

Let us find the voltage v_L across the 3-Ω load resistor for the circuit shown in Fig. 2.25 by replacing circuit A by its Thévenin equivalent. First, we remove the 3-Ω load resistor and calculate v_{oc} from the circuit shown in Fig. 2.26. Summing the currents out of node v, by KCL, we get

$$\frac{v - 20}{1} + \frac{v - v_{oc}}{2} + 15 + 15 = 0$$

Summing the currents into node v_{oc}, by KCL, we get

$$\frac{v - v_{oc}}{2} + \frac{20 - v_{oc}}{6} + 15 = 0$$

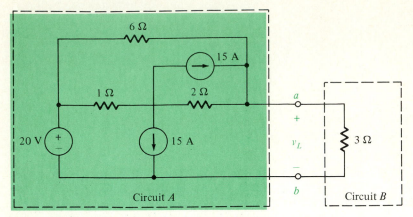

Fig. 2.25 Example of Thévenin's theorem.

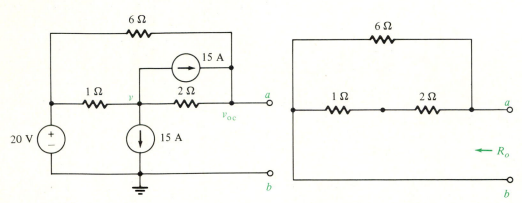

Fig. 2.26 Determination of open-circuit voltage.

Fig. 2.27 Determination of output resistance.

Solving these equations for v_{oc} yields $v_{oc} = 30$ V.

To find the Thévenin equivalent resistance R_o, set the independent sources to zero. The resulting circuit is given in Fig. 2.27. Since the 1-Ω and 2-Ω resistors are in series, their combined resistance is 3 Ω. But this resulting 3-Ω resistance is in parallel with 6 Ω. Thus, $R_o = (3)(6)/(3 + 6) = 2\ \Omega$.

In replacing circuit A by its Thévenin equivalent, we get the circuit shown in Fig. 2.28. By voltage division, we have that

$$v_L = \frac{3(30)}{3 + 2} = 18 \text{ V}$$

Direct analysis of the circuit given in Fig. 2.25, of course, yields the same result for v_L (see Problem 2.5).

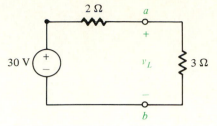

Fig. 2.28 Use of the Thévenin-equivalent circuit.

Example 2.12

Let us now find the Thévenin equivalent of the op-amp circuit (with no load) shown in Fig. 2.29. We have already seen (Fig. 2.13) that

$$v_{oc} = -\frac{R_2}{R_1}v_s$$

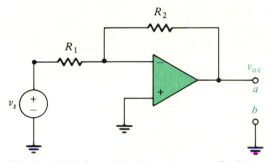

Fig. 2.29 Thévenin's theorem applied to an op-amp circuit.

To find the output resistance R_o, we set v_s to zero and take the ratio $R_o = v_o/i_o$ as depicted in Fig. 2.30. By KCL,

$$i_1 + i_2 = 0$$

Thus,

$$\frac{0}{R_1} + \frac{v_o}{R_2} = 0$$

from which

$$v_o = 0$$

Hence,

$$R_o = v_o/i_o = 0\ \Omega$$

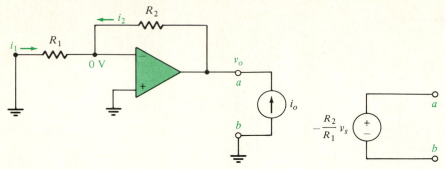

Fig. 2.30 Determination of output resistance. **Fig. 2.31 Thévenin-equivalent circuit.**

and the Thévenin equivalent of the op-amp circuit given in Fig. 2.29 is the (ideal) voltage source shown in Fig. 2.31.

Now let us find the Thévenin equivalent of the simple circuit shown in Fig. 2.32(a). Clearly, $v_{oc} = Ri$ and $R_o = R$. Thus, the Thévenin equivalent of the circuit given in Fig. 2.32(a) is as shown in Fig. 2.32(b). Furthermore, $i = v_{oc}/R = v_{oc}/R_o$. But placing a short circuit between nodes a and b results in a current $i_{sc} = v_{oc}/R_o = i$ through the short circuit directed from node a to node b. Since the circuits in Figs. 2.32(a) and 2.32(b) are equivalent, we deduce that the circuits shown in Figs. 2.33(a) and 2.33(b) are equivalent. The relationships between v_{oc}, i_{sc}, and R_o are

$$v_{oc} = R_o i_{sc}, \qquad i_{sc} = v_{oc}/R_o, \qquad R_o = v_{oc}/i_{sc}$$

If the circuit given in Fig. 2.33(a) is the Thévenin equivalent of circuit A, then the circuit shown in Fig. 2.33(b) is known as the **Norton*equivalent**.

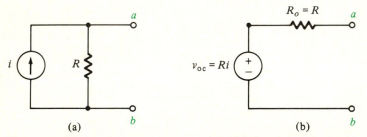

(a)   (b)

Fig. 2.32 Thévenin equivalent of a simple parallel circuit.

*Named for the American engineer Edward L. Norton (1898–   ).

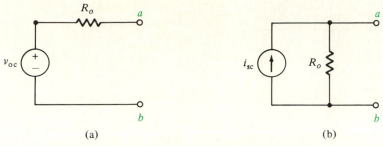

(a)   (b)

Fig. 2.33 Thévenin- and Norton-equivalent circuits.

To obtain the Norton equivalent of circuit A without first finding the Thévenin equivalent, determine the short-circuit current i_{sc} by placing a short circuit between nodes a and b and then calculate the resulting current through it. The output resistance R_o can be found as was done for the Thévenin-equivalent circuit. Alternatively, v_{oc} can be determined as was done for the Thévenin-equivalent circuit, and then the formula $R_o = v_{oc}/i_{sc}$ employed.

Example 2.13

For the circuit shown in Fig. 2.34, let us find the voltage across the $\frac{1}{4}$-Ω resistor by replacing the remainder of the circuit by its Norton equivalent.

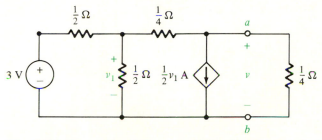

Fig. 2.34 Example of the use of the Norton-equivalent circuit.

First, replace the $\frac{1}{4}$-Ω resistor between nodes a and b with a short circuit, and calculate i_{sc} as shown in Fig. 2.35. Summing the currents out of node v_1, by KCL we obtain

$$\frac{v_1 - 3}{1/2} + \frac{v_1}{1/2} + \frac{v_1}{1/4} = 0$$

Solving this equation for v_1 yields $v_1 = \frac{3}{4}$ V. By KCL,

$$i_{sc} = \frac{v_1}{1/4} - \frac{v_1}{2} = \frac{7}{2}v_1 = \frac{21}{8} \text{ A}$$

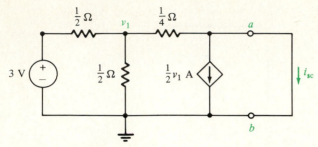

Fig. 2.35 Determination of short-circuit current.

To find R_o, we remove the $\frac{1}{4}$-Ω load resistor, set the independent source to zero, and calculate $R_o = v_o/i_o$ between nodes a and b as depicted in Fig. 2.36. (Be careful—do not use the value of v_1 calculated for the circuit in Fig. 2.35; the circuit in Fig. 2.36 is a different circuit.) Summing the currents out of node v_1, we get

$$\frac{v_1}{1/2} + \frac{v_1}{1/2} + \frac{v_1 - v_o}{1/4} = 0$$

from which $v_1 = v_o/2$. By KCL,

$$i_o = \frac{1}{2}v_1 + \frac{v_o - v_1}{1/4} = \frac{1}{2}\left(\frac{v_o}{2}\right) + 4\left(v_o - \frac{v_o}{2}\right) = \frac{9}{4}v_o$$

Hence,

$$R_o = \frac{v_o}{i_o} = \frac{4}{9}\ \Omega$$

An alternative procedure for finding R_o is to remove the $\frac{1}{4}$-Ω resistor from the circuit in Fig. 2.34, and calculate the open-circuit voltage v_{oc} between nodes a and b. Doing so results in $v_{oc} = \frac{7}{6}$ V. Thus, we also have that

$$R_o = \frac{v_{oc}}{i_{sc}} = \frac{7/6}{21/8} = \frac{4}{9}\ \Omega$$

as was determined above.

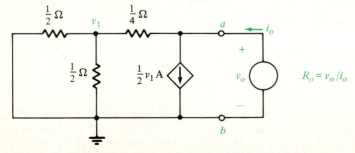

Fig. 2.36 Determination of output resistance.

To determine the voltage v indicated in Fig. 2.34, we may replace the portion of the circuit to the left of nodes a and b by its Norton equivalent. This is illustrated by Fig. 2.37. Using the current-divider formula, we get

$$i = \frac{4/9}{4/9 + 1/4}\left(\frac{21}{8}\right) = \frac{42}{25} \text{ A}$$

and thus,

$$v = \tfrac{1}{4}\left(\tfrac{42}{25}\right) = 0.42 \text{ V}$$

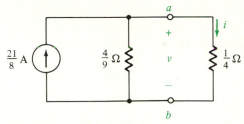

Fig. 2.37 Use of the Norton-equivalent circuit.

Compare this result with that obtained in Example 2.1.

In the preceding example, we mentioned that R_o can be determined by finding v_{oc} and i_{sc} and using $R_o = v_{oc}/i_{sc}$ rather than taking the direct approach shown in Fig. 2.36. However, $R_o = v_{oc}/i_{sc}$ cannot be employed if neither v_{oc} nor i_{sc} exists. For instance, the op-amp circuit given in Fig. 2.29 is equivalent to an ideal voltage source (Fig. 2.31). This means that i_{sc} does not exist for this circuit. In other words, an ideal voltage source cannot be equivalent to a current source in parallel with a resistance. Nor can an ideal current source be equivalent to a voltage source in series with a resistance. This means that for a circuit that is equivalent to an ideal current source, v_{oc} does not exist.

There are circuits whose Thévenin and Norton equivalents are identical. For example, in general, a circuit that contains no independent sources will have $v_{oc} = 0$ and $i_{sc} = 0$. This means that the Thévenin and Norton equivalents simply consist of the output resistance R_o (e.g., see Problem 2.33). Of course, in such a case, the formula $R_o = v_{oc}/i_{sc}$ cannot be used to find R_o. However, there are the exceptional circuits with nonzero independent sources that also have Thévenin and Norton equivalent circuits comprised only of a resistance (e.g., see Problem 2.40).

Maximum Power Transfer

We may now answer the question that was posed at the beginning of this section. For the situation depicted in Fig. 2.19, we wish to determine the value of the load R_L that will absorb the maximum amount of power. By Thévenin's theorem, we can replace the arbitrary circuit by its Thévenin equivalent and the effect on R_L will be

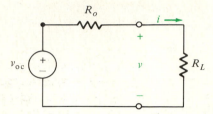

Fig. 2.38 Determination of maximum power transfer.

the same. Doing so, we obtain Fig. 2.38. Since the power absorbed by the load is $p = i^2 R_L$, it may be tempting to believe that to increase the power absorbed, we simply increase the load resistance R_L. This reasoning, however, is invalid since an increase in R_L will result in a decrease in i. To find the value for which maximum power is delivered to the load, we proceed as follows:

By Ohm's law,

$$i = \frac{v_{oc}}{R_o + R_L}$$

The (instantaneous) power absorbed by the load is

$$p = i^2 R_L = (\frac{v_{oc}}{R_o + R_L})^2 R_L = \frac{v_{oc}^2 R_L}{(R_o + R_L)^2} \qquad \textbf{(2.19)}$$

Thus, we have an expression for the power in terms of the variable R_L. For what value of R_L is the power a maximum? From introductory calculus we know that if we have a function $f(x)$ of the real variable x, then maxima occur for the values of x where the derivative of $f(x)$ is equal to zero. Thus, let us take the derivative of p with respect to the variable R_L. Using the fact that

$$\frac{d}{dx}\left[\frac{f(x)}{g(x)}\right] = \frac{g(x)\, df(x)/dx - f(x)\, dg(x)/dx}{[g(x)]^2}$$

we obtain

$$\frac{dp}{dR_L} = \frac{(R_o + R_L)^2\, d(v_{oc}^2 R_L)/dR_L - v_{oc}^2 R_L\, d(R_o + R_L)^2/dR_L}{[(R_o + R_L)^2]^2}$$

Simplifying this expression yields

$$\frac{dp}{dR_L} = \frac{(R_o - R_L)v_{oc}^2}{(R_o + R_L)^3}$$

Clearly, $dp/dR_L = 0$ when the numerator $(R_o - R_L)v_{oc}^2 = 0$. Furthermore, given that v_{oc} is a nonzero source, then $dp/dR_L = 0$ when

$$R_o - R_L = 0 \qquad \text{or} \qquad R_o = R_L$$

Therefore, when $R_o = R_L$, the power absorbed by the load [Equation (2.19)] becomes

$$p_m = \frac{v_{oc}^2}{4R_o} = \frac{v_{oc}^2}{4R_L} \qquad \text{(2.20)}$$

and this quantity is either a maximum or a minimum. However, since we can make the expression for power absorbed by the load arbitrarily small by making R_L arbitrarily small, the quantity $v_{oc}^2/4R_o$ cannot be a minimum. We may also verify that $p_m = v_{oc}^2/4R_o$ is maximum by taking the second derivative of p, setting $R_o = R_L$, and obtaining a negative quantity.

In summary, for the situation depicted in Fig. 2.19 (or Fig. 2.38), the maximum power that can be delivered to a resistive load R_L is obtained when $R_L = R_o$, and that power is given by Equation (2.20).

2.5 THE PRINCIPLE OF SUPERPOSITION

Given a circuit that contains two or more independent sources (voltage or current or both), as we have seen, one way to determine a specific variable (either voltage or current) is by the direct use of nodal or mesh analysis. An alternative, however, is to find that portion of the variable attributable to each independent source, and then sum these up. This concept is known as the **principle of superposition.**

The justification for the principle of superposition is based on the concept of linearity. Specifically, let us first consider the relationship between voltage and current for a resistor (i.e., Ohm's law). Suppose that a current i_1 (the **excitation** or **input**) is applied to a resistor, R. Then the resulting voltage v_1 (the **response** or **output**) is $v_1 = Ri_1$. Similarly, if i_2 is applied to R, then $v_2 = Ri_2$ results. But if $i = i_1 + i_2$ is applied, the response is

$$v = Ri = R(i_1 + i_2) = Ri_1 + Ri_2 = v_1 + v_2$$

In other words, the response to a sum of inputs is equal to the sum of the individual responses (Condition I).

In addition, if v is the response to i (i.e., $v = Ri$), then the response to Ki is

$$R(Ki) = K(Ri) = Kv$$

In other words, if the input is scaled by the constant K, then the response is also scaled by K, (Condition II).

Because Conditions I and II are satisfied, we say that the relationship between current (input) and voltage (ouptut) is **linear** for a resistor. Similarly, by using the alternative form of Ohm's law $i = v/R$, we can show that the relationship between voltage (input) and current (output) is also linear for a resistor.

Although the relationships between voltage and current for a resistor are linear, the power relationships $p = Ri^2$ and $p = v^2/R$ are not. For instance, if the current through a resistor is i_1, then the power absorbed by the resistor R is $p_1 = Ri_1^2$, while if the current is i_2, then the power absorbed is $p_2 = Ri_2^2$. However, the power absorbed due to the current $i_1 + i_2$ is

$$P_3 = R(i_1 + i_2)^2 = Ri_1^2 + Ri_2^2 + 2Ri_1i_2$$

so

$$p_3 \neq p_1 + p_2$$

Hence, the relationship $p = Ri^2$ is **nonlinear**.

Since the relationships between voltage and current are linear for resistors, we say that a resistor is a **linear element**. A dependent source (either current or voltage) whose value is directly proportional to some voltage or current is also a linear element. Because of this, we say that a circuit consisting of independent sources, resistors, and linear dependent sources (as well as other linear elements to be introduced later) is a **linear circuit**. All of the circuits dealt with so far have been linear circuits.

We now return to the simple case of a resistor R connected to a voltage source v. If the voltage is v_1, then the current is $i_1 = v_1/R$, while if v_2 is applied, then $i_2 = v_2/R$ results. By linearity, we have that for the input $v_1 + v_2$, the response is $i_1 + i_2$. However, we can represent this situation as shown in Fig. 2.39. But each component (i_1 and i_2) of the response can be determined as shown in Fig. 2.40. This is a very special case showing how it is possible to obtain the response due to two independent voltage sources by calculating the response to each one separately and then summing the responses. However, for any linear circuit, we can take the same approach: find the response to each independent source (both voltage and current) separately, and then sum the responses. A formal general proof is much more involved, but utilizes analogous ideas.

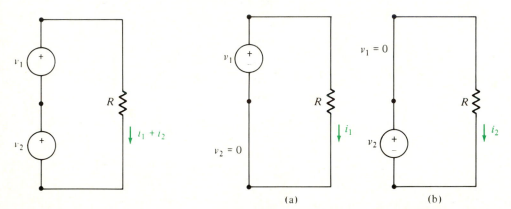

Fig. 2.39 Circuit with two independent voltage sources.

Fig. 2.40 Treating each independent source separately.

For demonstration purposes, consider the linear circuit given in Fig. 2.41, which contains three independent sources. By KCL, at node v,

$$\frac{v}{6} + \frac{v - v_g}{3} + i_y - i_x = 0$$

from which

$$3v = 2v_g + 6i_x - 6i_y \tag{2.21}$$

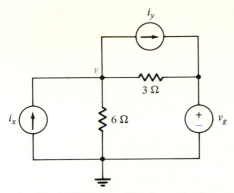

Fig. 2.41 Circuit with three independent voltage sources.

Suppose that when $i_x = i_y = 0$, the resulting node voltage is v_1. Thus, we have the equality

$$3v_1 = 2v_g + 6(0) - 6(0) \qquad (2.22)$$

For the case that $v_g = 0$ and $i_y = 0$, let the resulting node voltage be v_2. This yields the equality

$$3v_2 = 2(0) + 6i_x - 6(0) \qquad (2.23)$$

Finally, let v_3 be the node voltage that results when $v_g = 0$ and $i_x = 0$. This gives us the equality

$$3v_3 = 2(0) + 6(0) - 6i_y \qquad (2.24)$$

Adding the three equalities (2.22), (2.23), and (2.24) produces the equality

$$3(v_1 + v_2 + v_3) = 2v_g + 6i_x - 6i_y$$

which means that $v_1 + v_2 + v_3$ is the solution to Equation (2.21).

We now formally state the principle of superposition:

> **Given a linear circuit with independent sources $s_1, s_2, \ldots, s_n$, let r be the response (either voltage or current) of this circuit. If r_i is the response of the circuit to source s_i with all other independent sources set to zero (dependent sources are left as is), then**
> **$r = r_1 + r_2 + \cdots + r_n$.**

Example 2.14

The circuit given in Fig. 2.42 contains three independent sources. To find the voltage v by using the principle of superposition, we first find that portion of v, call it

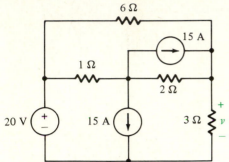

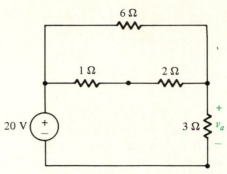

Fig. 2.42 Example of the principle of superposition.

Fig. 2.43 Determination of response due to 20-V source.

v_a, due to the independent voltage source. Thus, we set the two independent current sources to zero as shown in Fig. 2.43. Note that the 1-Ω and 2-Ω resistors are in series, and the resulting equivalent 3-Ω resistance is in parallel with the 6-Ω resistor. These three resistors, therefore, are equivalent to a 2-Ω resistance. By voltage division,

$$v_a = \frac{3}{3 + 2} (20) = 12 \text{ V}$$

To find the voltage v_b due solely to the lower 15-A current source, set the other current source and the voltage source to zero as shown in Fig. 2.44. In this circuit, the 6-Ω resistor is in parallel with the 3-Ω resistor. We can redraw the circuit as shown in Fig. 2.45. The 3-Ω and 6-Ω resistors in parallel are equivalent to a 2-Ω resistor—and this is in series with a 2-Ω resistor. The combination is effectively a 4-Ω resistance that is in parallel with the 1-Ω resistor. By current division,

$$i_2 = \frac{1}{1 + 4} (-15) = -3 \text{ A}$$

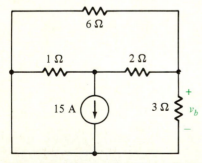

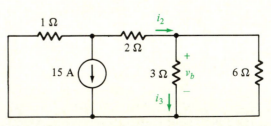

Fig. 2.44 Determination of response due to one 15-A source.

Fig. 2.45 Equivalent circuit for calculating v_b.

Having determined i_2, we can use current division again to get

$$i_3 = \frac{6}{6+3} i_2 = -2 \text{ A}$$

Thus,

$$v_b = 3i_3 = -6 \text{ V}$$

To find the voltage v_c due to the upper 15-A current source, set the other two independent sources to zero. The resulting circuit is shown in Fig. 2.46. Since the 6-Ω and 3-Ω resistors are in parallel, we can redraw the circuit as in Fig. 2.47. Suppose for this circuit that we apply Thévenin's theorem to the 15-A source in parallel with the 2-Ω resistor. The result is a $15(2) = 30$-V source (with the plus toward node n) in series with a 2-Ω resistor. Since the 3-Ω and 6-Ω resistors in parallel are an equivalent 2-Ω resistance, by voltage division we have that

$$v_c = \frac{2}{2+1+2} (30) = 12 \text{ V}$$

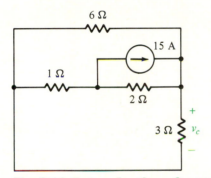

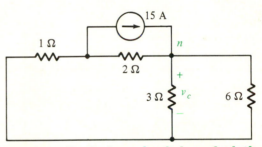

Fig. 2.46 Determination of response due to the other 15-A source.   **Fig. 2.47 Equivalent circuit for calculating v_c.**

By the principle of superposition, we have

$$v = v_a + v_b + v_c = 12 - 6 + 12 = 18 \text{ V}$$

In the example just completed, all the sources in the circuit were independent sources. For the case in which dependent sources are also included, they are left as is; only independent sources are set to zero when applying the principle of superposition.

SUMMARY

1. In nodal analysis, the circuit variables that are determined are the node voltages (with respect to the reference node).

2. In mesh analysis, the circuit variables that are determined are the mesh currents. Mesh analysis can be employed only for planar networks.

3. An operational amplifier is an ideal amplifier with infinite gain. When used in conjunction with feedback, the input voltage to an op amp is constrained to be zero volts. The input terminals of an op amp draw no current.

4. The effect of an arbitrary circuit on a load is equivalent to an appropriate voltage source in series with an appropriate resistance (Thévenin's theorem) or an appropriate current source in parallel with that resistance (Norton's theorem).

5. An arbitrary circuit delivers maximum power to a resistive load R_L when $R_L = R_o$, where R_o is the Thévenin-equivalent (output) resistance of the arbitrary circuit.

6. The response of a circuit having n independent sources equals the sum of the n responses to each individual independent source (the principle of superposition).

Problems for Section 2.1

2.1 For the circuit given in Fig. P2.1, select node d as the reference node and use nodal analysis to determine i_1, i_2, i_3, and i_4.

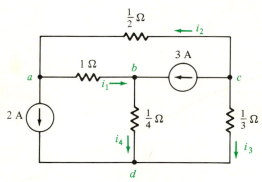

Fig. P2.1

2.2 Repeat Problem 2.1 using node c as the reference node.
2.3 Repeat Problem 2.1 using node b as the reference node.
2.4 For the circuit shown in Fig. P2.4, find the node voltages.

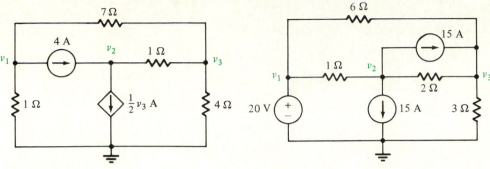

Fig. P2.4   **Fig. P2.5**

2.5 For the circuit given in Fig. P2.5, find the node voltages.
2.6 For the circuit shown in Fig. P2.6, find the node voltages.

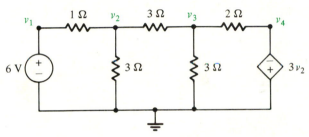

Fig. P2.6

2.7 For the circuit shown in Fig. P2.7, find the node voltages.

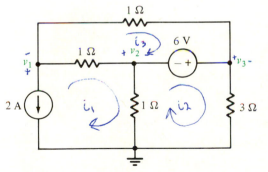

Fig. P2.7

2.8 Fig. P2.8 shows a simple transistor amplifier circuit; the portion in the shaded box is the **hybrid-** or **h-parameter model** of a bipolar junction transistor. Note that h_i is a resistance and h_o is a conductance. Use nodal analysis to find the voltage gain v_2/v_1 of this amplifier.

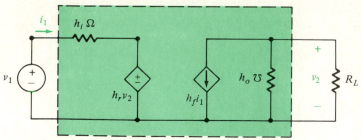

Fig. P2.8

2.9 The circuit shown in Fig. P2.9 is a simple bipolar junction transistor amplifier with "feedback." The portion of the circuit in the shaded box is an approximate T-model of a transistor in the common-emitter configuration. Use nodal analysis to find the voltage gain v_2/v_1 of the amplifier.

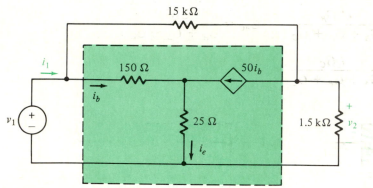

Fig. P2.9

2.10 The circuit shown in Fig. P2.10 is a simple bipolar junction transistor amplifier. The portion of the circuit in the shaded box is a T-model of a transistor in the common-base configuration. Use nodal analysis to find the voltage gain v_2/v_1 of the amplifier.

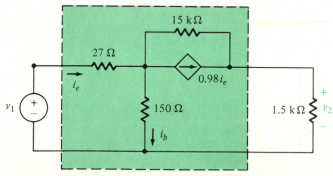

Fig. P2.10

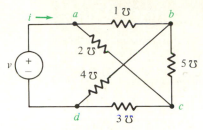

Fig. P2.11

2.11 For the circuit given in Fig. P2.11, use nodal analysis to determine $G = i/v$.

2.12 Suppose that a circuit contains the shaded box on the left side of Fig. P2.12(a). Without affecting the remainder of the circuit, the box on the left side of Fig. P2.12(a) can be replaced by the box on the right side, provided that

$$G_{AB} = \frac{G_A G_B}{G_A + G_B + G_C}, \qquad G_{AC} = \frac{G_A G_C}{G_A + G_B + G_C},$$

$$G_{BC} = \frac{G_B G_C}{G_A + G_B + G_C}$$

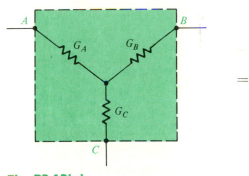

Fig. P2.12(a)

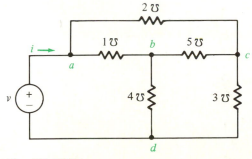

Fig. P2.12(b)

Such a process is called a **Y-Δ (wye-delta) transformation.** The circuit in Fig. P2.12(b) is identical to the circuit given in Problem 2.11. Use a Y-Δ transformation on the 1-℧, 4-℧, and 5-℧ conductances, and then combine elements in series and parallel to determine $G = i/v$.

2.13 With reference to Problem 2.12, the formulas for a **Δ-Y (delta-wye) transformation** are

$$R_A = \frac{R_{AB} R_{AC}}{R_{AB} + R_{AC} + R_{BC}}, \qquad R_B = \frac{R_{AB} R_{BC}}{R_{AB} + R_{AC} + R_{BC}},$$

$$R_C = \frac{R_{AC} R_{BC}}{R_{AB} + R_{AC} + R_{BC}}$$

where $R = 1/G$.

The circuit shown in Fig. P2.13 is identical to the circuit given in Problem 2.11. Use a Δ-Y transformation on the 1-℧, 2-℧, and 5-℧ conductances, and then combine elements in series and parallel to determine $G = i/v$.

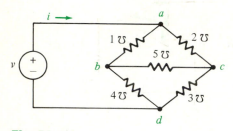

Fig. P2.13

Problems for Section 2.2

2.14 Use mesh analysis to find the node voltages for the circuit shown in Fig. 2.8.

2.15 Use mesh analysis to find the node voltages for the circuit given in Fig. P2.6.

2.16 Use mesh analysis to find the resistance $R = v/i$ for the circuit given in Fig. P2.12.

2.17 Use mesh analysis to find the node voltages for the circuit given in Fig. P2.7.

2.18 Use mesh analysis to find the node voltages for the circuit shown in Fig. 2.7.

Problems for Section 2.3

2.19 For the op-amp circuit given in Fig. P2.19, find v_o.

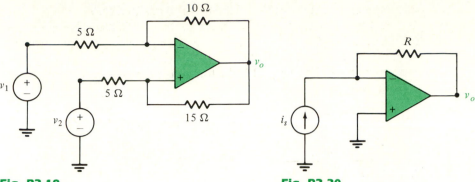

Fig. P2.19   **Fig. P2.20**

2.20 Given the op-amp circuit shown in Fig. P2.20, find the ratio v_o/i_s.

2.21 For the op-amp circuit given in Fig. P2.21, (a) find the ratio v_o/i_s; (b) find the current gain i_2/i_s.

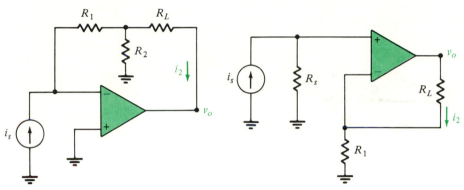

Fig. P2.21   **Fig. P2.22**

2.22 Given the op-amp circuit shown in Fig. P2.22, find v_o/i_s.

2.23 For the op-amp circuit given in Problem 2.22, find the current gain i_2/i_s.

2.24 Given the op-amp circuit shown in Fig. P2.24, find (a) the voltage v; (b) the current gain i_2/i_1.

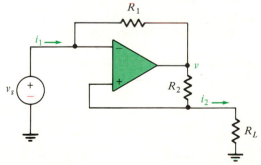

Fig. P2.24

2.25 For the op-amp circuit given in Fig. P2.25, find the voltage gain v_o/v_s.

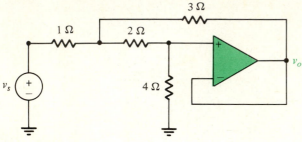

Fig. P2.25

2.26 Find the voltage gain v_o/v_s for the op-amp circuit shown in Fig. P2.26.

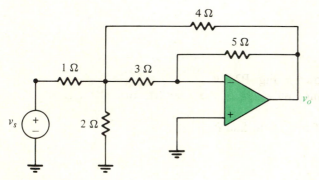

Fig. P2.26

2.27 For the op-amp circuit given in Fig. P2.27, find the voltage gain v_o/v_s.

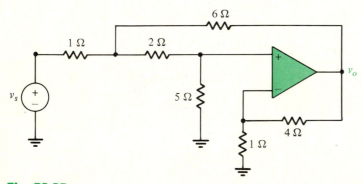

Fig. P2.27

2.28 For the op-amp circuit given in Fig. P2.28, find the voltage gain v_o/v_s.

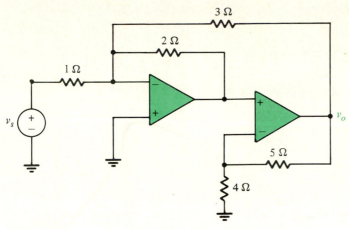

Fig. P2.28

Problems for Section 2.4

2.29 Given the circuit shown in Fig. P2.29:
(a) Find the Thévenin equivalent of the circuit to the left of terminals a and b.
(b) Use the result of part (a) to find i.

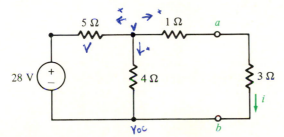

Fig. P2.29

2.30 Repeat Problem 2.29 for the circuit given in Fig. P2.30.

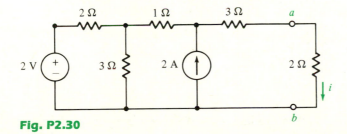

Fig. P2.30

2.31 Given the circuit shown in Fig. P2.31:

(a) Find the Thévenin equivalent of the circuit to the left of terminals a and b.

(b) Find v when $R_L = \frac{1}{3}\,\Omega$.

(c) What value of R_L absorbs the maximum power?

(d) For the value of R_L determined in part (c), find the power absorbed by R_L.

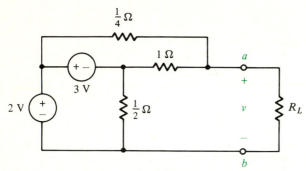

Fig. P2.31

2.32 Find the Norton equivalent of the circuit to the left of terminals a and b for the circuit shown in Fig. P2.32. Use this result to find i.

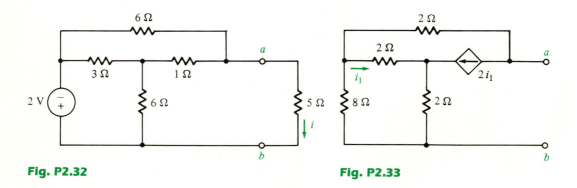

Fig. P2.32

Fig. P2.33

2.33 For the circuit given in Fig. P2.33, find:

(a) The Thévenin-equivalent circuit.

(b) The Norton-equivalent circuit.

2.34 For the circuit given in Fig. P2.34, what value of R_L will absorb the maximum power?

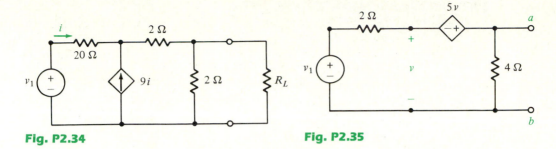

Fig. P2.34

Fig. P2.35

2.35 Find the Thévenin equivalent of the circuit shown in Fig. P2.35.

2.36 Find the Thévenin equivalent of the op-amp circuit shown in Fig. P2.36. (*Hint*: To find R_o, apply a current source i_o and calculate the resulting voltage v_o.)

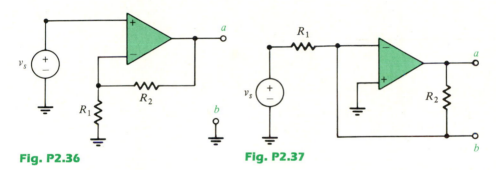

Fig. P2.36

Fig. P2.37

2.37 Find the Thévenin equivalent of the op-amp circuit shown in Fig. P2.37.

2.38 Find the Norton equivalent of the op-amp circuit shown in Fig. P2.37.

2.39 Show that the Norton equivalent of the circuit given in Fig. P2.39 is an ideal current source. (*Hint*: In determining R_o, apply a voltage source v_o and calculate the resulting current i_o.)

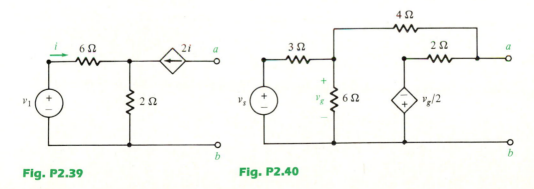

Fig. P2.39

Fig. P2.40

2.40 For the circuit shown in Fig. P2.40, find:
(a) The Norton-equivalent circuit.
(b) The Thévenin-equivalent circuit.

2.41 Figure P2.41 demonstrates the concept of a **source transformation**. Specifically, the voltage source v_s in series with a resistance R_s is equivalent to a current source v_s/R_s in parallel with R_s. Without using Thévenin's or Norton's theorem, confirm the equivalence of Fig. P2.41(a) and (b) by writing expressions relating i and v.

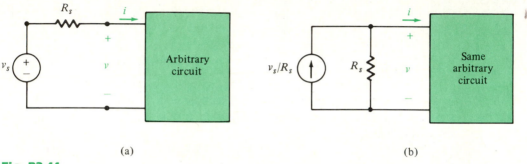

(a)   (b)

Fig. P2.41

2.42 Repeat Problem 2.41 for the source transformation depicted in Fig. P2.42.

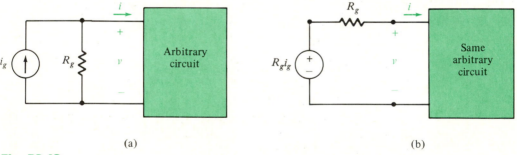

(a)   (b)

Fig. P2.42

2.43 Use source transformations as described in Problems 2.41 and 2.42 to reduce the circuit given in Fig. P2.30 to a circuit with one independent source. Calculate i from this reduced circuit.

2.44 Confirm that the source transformations described in Problems 2.41 and 2.42 can be applied to dependent sources, as well as independent sources, by reducing the circuit given in Fig. P2.6 to a circuit with one independent and one dependent current source, and then determining v_2.

Problems for Section 2.5

2.45 Use the principle of superposition to find v in the circuit shown in Fig. P2.45.

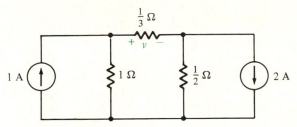

Fig. P2.45

2.46 For the circuit in Fig. P2.45, find the power absorbed by the $\frac{1}{3}$-Ω resistor. Show that the principle of superposition does not hold for power.

2.47 For the circuit shown in Fig. P2.47, use the principle of superposition to find i_1, i_2, and i_3.

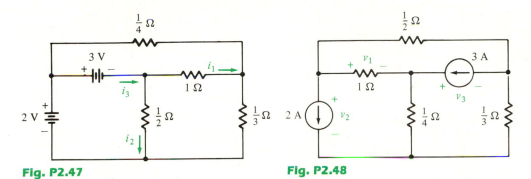

Fig. P2.47   **Fig. P2.48**

2.48 Use the principle of superposition to find v_1, v_2, and v_3 for the circuit given in Fig. P2.48.

2.49 For the circuit shown in Fig. P2.49, use the principle of superposition to find i and v.

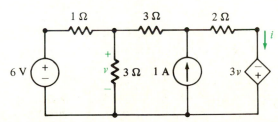

Fig. P2.49

2.50 Repeat Problem 2.49 for the circuit given in Fig. P2.50.

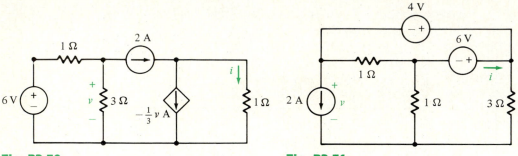

Fig. P2.50　　　　　　　　　　　　　　　　　　　**Fig. P2.51**

2.51 Repeat Problem 2.49 for the circuit shown in Fig. P2.51.
2.52 Repeat Problem 2.49 for the circuit shown in Fig. P2.52 when (a) $i_s = 4$ A; (b) $i_s = 12$ A.

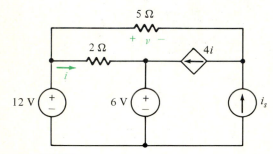

Fig. P2.52

Chapter 3

Circuits with Inductors and Capacitors

INTRODUCTION

The applications of electric circuits consisting solely of sources and resistors are quite limited. Furthermore, sources that produce voltages or currents that vary with time can be extremely useful. Because of this, in the present chapter, we introduce two circuit elements, the inductor and the capacitor, which have voltage–current relationships that are not simple direct proportionalities. Instead, the behavior of each element can be described by either a differential or an integral relationship.

Resistors dissipate power (or energy), but inductors and capacitors store energy. The relationships that describe this property are derived in this chapter. We shall also see how to write node and mesh equations for circuits containing inductors and capacitors. By using the concept of duality, the analysis of a planar circuit results in the automatic analysis of a second circuit.

To analyze a resistive circuit, we can write a set of node or mesh equations—which are algebraic equations—and solve them. However, for circuits with inductors and/or capacitors, not all the equations will be algebraic. To see how to analyze such circuits, we begin by considering simple circuits, which, in addition to resistors, contain a single inductor or capacitor. Such a network, called a **first-order circuit,** is described by a first-order, linear differential equation. We first consider the situation in which an initial condition is present but an independent source is not (the natural response), and then the situation in which an independent source is present (the complete response). The analysis of such circuits requires the solution of "simple" differential equations (a topic that is part of our study). Once a simple circuit has been analyzed, we will see that the use of various principles such as linearity and time-invariance will enable us to handle more complicated situations without having to start from scratch.

After dealing with circuits containing either a single inductor or capacitor, we consider the case of two energy-storage elements in a circuit. Most such networks, called **second-order circuits,** are described by second-order, linear differential equations, the solution of which takes three forms. As with first-order circuits, we start by considering natural responses, and follow that with the study of complete responses. We shall see that the complexity of analysis increases significantly as compared with that encountered with first-order circuits. It is because of this fact that we do not extend such an approach to higher-order circuits, but take a different approach to circuit analysis in the following chapters.

3.1 THE INDUCTOR

In our high school science courses (or earlier), we learned that a current going through a conductor such as a wire produces a magnetic field around that conductor. Furthermore, winding such a conductor into a coil strengthens the magnetic field. For the resulting element, known as an **inductor,** the voltage across it is directly proportional to the time rate of change of the current through it. This fact is credited to the independent work of the American inventor Joseph Henry (1797–1878) and the English physicist Michael Faraday (1791–1867).

To signify a coil of wire, we represent an **ideal inductor** as shown in Fig. 3.1. The relationship between voltage and current (for the polarity and direction, respectively indicated) for this element is given by

$$v = L\frac{di}{dt}$$

(3.1)

where L is the **inductance** of the element, and its unit is the **henry** (abbreviated H) in honor of Joseph Henry. Remember that in the differential relationship just given, v and i represent the functions of time $v(t)$ and $i(t)$ respectively.

Fig. 3.1 Inductor.

Example 3.1

Consider the circuit given in Fig. 3.2 in which the ideal source produces a constant current, called a **direct current** (abbreviated dc). From Equation (3.1), we have that

$$v = L\frac{di}{dt} = 2\frac{d}{dt}[5] = 0 \text{ V}$$

This example is an illustration of the fact that an (ideal) inductor behaves as a short circuit to dc.

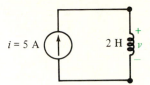

Fig. 3.2 Simple inductor circuit.

In the example just presented, the dc circuit analyzed was quite simple—so let us consider a more complicated situation.

Example 3.2

Let us determine the current i in the circuit shown in Fig. 3.3. This circuit has one independent voltage source whose value is constant. For a resistive circuit, we would naturally anticipate that all voltages and currents are constant. However,

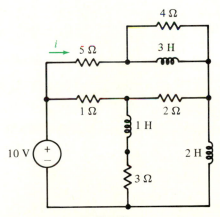

Fig. 3.3 A dc circuit.

this is not a resistive circuit. Yet our intuition suggests that the constant-valued voltage source (even if there were additional constant-valued independent sources, and dependent sources, too) produces constant-valued responses. This fact will be confirmed more rigorously later. In the meantime, we will use the result that a circuit containing only constant-valued sources is a dc circuit.

For the dc circuit given in Fig. 3.3, the inductors behave as short circuits. Thus, as far as the resistors are concerned, we have the circuit shown in Fig. 3.4.

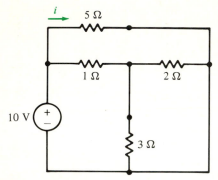

Fig. 3.4 Equivalent form of preceding dc circuit.

Note that in this circuit, the 4-Ω resistor does not appear. The reason is that a 4-Ω resistor in parallel with a short circuit (zero resistance) is equivalent to a short circuit. By KVL, the voltage across the 5-Ω resistor is 10 V (the plus is on the left). Therefore,

$$i = \frac{10}{5} = 2 \text{ A}$$

■

For the circuit given in Fig. 3.3, because of the dc source, we were able to simplify the analysis problem to that of analyzing a resistive circuit. But don't get the impression that the analysis of a circuit containing an inductor or inductors is a snap. Generally, circuits containing inductors have sources that are not simply constant valued.

Example 3.3

For the simple circuit shown in Fig. 3.2, let the current $i(t)$, which is produced by the source, be described by the function of time given in Fig. 3.5. Since

$$v = L\frac{di}{dt} = 2\frac{di}{dt}$$

then a sketch of $v(t)$ is as shown in Fig. 3.6.

$$v(t) = L \frac{di(t)}{dt} = 2 \frac{di(t)}{dt}$$

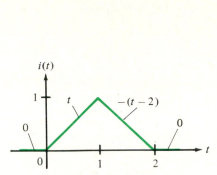

Fig. 3.5 Nonconstant current.

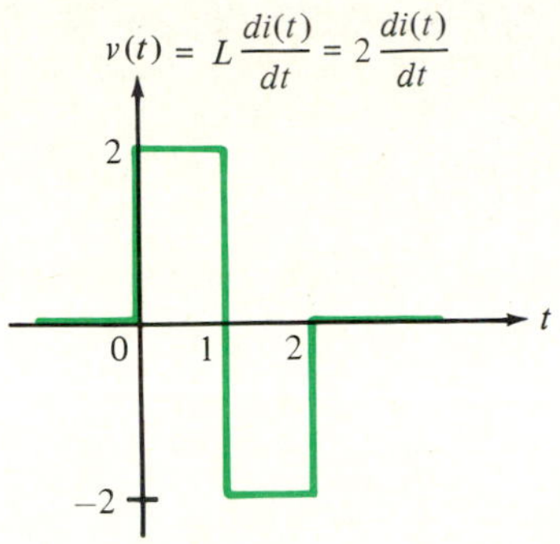

Fig. 3.6 Resulting inductor voltage.

We know that a resistor always absorbs power and that the energy absorbed is dissipated as heat. But how about an inductor? For the inductor in question, the instantaneous power $p(t) = v(t)i(t)$ absorbed by the inductor is as shown in Fig. 3.7. We now see that the power absorbed by the inductor is zero for $-\infty < t \le 0$ and $2 \le t < \infty$. For $0 < t < 1$ s, since $p(t)$ is a positive quantity, the inductor absorbs power (which is supplied by the source). However, for $1 < t < 2$ s, since $p(t)$ is a negative quantity, the inductor actually supplies power (to the source)!

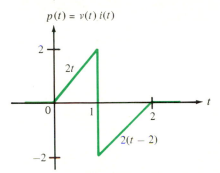

Fig. 3.7 Power absorbed by the inductor.

To get the energy absorbed by the inductor, we simply integrate the power absorbed over time. For this example, the energy absorbed increases from 0 to $\frac{1}{2}(1)(2) = 1$ J as time goes from $t = 0$ to $t = 1$ s. However, from $t = 1$ s to $t = 2$ s, the inductor supplies energy, such that at time $t = 2$ s, and thereafter, the energy absorbed by the inductor is zero. Since the energy absorbed by the inductor is not dissipated, but is eventually returned, we say that the inductor **stores** energy. The energy is stored in the magnetic field that surrounds the inductor.

Given an expression for $p(t)$, the instantaneous power absorbed by an inductor, we can obtain an expression for $w_L(t)$, the energy stored in the inductor—we just sum up the power over time. In other words, the energy stored in an inductor at time t is

$$w_L(t) = \int_{-\infty}^{t} p(t) \, dt$$

However, rather than find the particular expression $w_L(t)$ for a specific example, let us derive a simple formula for the energy stored in an inductor. We proceed as follows:

For an inductor (see Fig. 3.1), since

$$p(t) = i(t)v(t) \qquad \text{and} \qquad v(t) = L\frac{di(t)}{dt}$$

then

$$p(t) = i(t)\left[L\frac{di(t)}{dt}\right] = Li(t)\frac{di(t)}{dt}$$

Thus, the energy absorbed by the inductor at time t is $w_L(t)$, where

$$w_L(t) = \int_{-\infty}^{t} p(t) \, dt = \int_{-\infty}^{t} Li(t)\frac{di(t)}{dt} dt$$

By using the chain rule of calculus, the variable of integration can be changed from t to $i(t)$. Changing the limits of integration appropriately results in

$$w_L(t) = \int_{i(-\infty)}^{i(t)} Li(t) \, di(t) = L\frac{i^2(t)}{2} \Big|_{i(-\infty)}^{i(t)}$$

$$= \frac{1}{2}L[i^2(t) - i^2(-\infty)]$$

Assuming that you can go back far enough in time, you will eventually return to that time when there was no current in the inductor. Thus, we adopt the convention that $i(-\infty) = 0$. The result is the following expression for the energy absorbed at time t by an inductor whose value is L henries*:

$$\boxed{w_L(t) = \tfrac{1}{2}Li^2(t)}$$

The energy stored in the inductor given in Example 3.3 is shown in Fig. 3.8.

Unlike an ideal inductor, an actual inductor (an automobile ignition coil is an example) has some resistance associated with it. This arises from the fact that a physical inductor is constructed from real wire that may have a very small, but still nonzero resistance. For this reason, we model a physical inductor as an ideal

---

*The plural of the unit "henry" is spelled either "henries" or "henrys".

$$w_L(t) = \frac{1}{2} Li^2(t) = i^2(t)$$

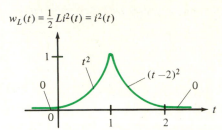

Fig. 3.8 Energy stored in the inductor.

inductor in series with a resistance. Practical inductors (also called "coils" and "chokes") range in value from about 10^{-6} H $= 1\ \mu$H all the way up to around 100 H. To construct inductors having large values requires many turns of wire and iron cores, and consequently results in larger-value series resistances. The series resistance is typically in the range from a fraction of an ohm to several hundred ohms.

3.2 THE CAPACITOR

Another extremely important circuit element is obtained when two conducting surfaces (called **plates**) are placed in proximity to one another, and between them is a nonconducting material called a **dielectric.** For the resulting element, known as a **capacitor** (formerly **condenser**), a voltage across the plates results in an electric field between them and the current through the capacitor is directly proportional to the time rate of change of the voltage across it. The **ideal capacitor** is depicted in Fig. 3.9 and the relationship between current and voltage (for the direction and polarity respectively indicated) is given by*

$$i = C\frac{dv}{dt} \tag{3.2}$$

where C is the **capacitance** of the element, and its unit is the **farad** (abbreviated F) in honor of Michael Faraday.

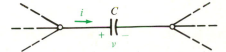

Fig. 3.9 Capacitor.

---

*For those who remember from elementary physics that $q = Cv$, where q is charge, C is capacitance, and v is voltage, by taking the derivative, we get

$$C\frac{dv}{dt} = \frac{dq}{dt} = i$$

Example 3.4

Consider the simple circuit shown in Fig. 3.10. For the case that $v = 5$ V, the circuit in Fig. 3.10 is a dc circuit. Suppose that $C = 2$ F. From the differential relationship for the capacitor, we have

$$i = C\frac{dv}{dt} = 2\frac{d}{dt}[5] = 0 \text{ A}$$

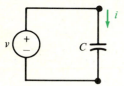

Fig. 3.10 Simple capacitor circuit.

■

We deduce, therefore, that a capacitor (ideal) is an open circuit to dc. A more elaborate dc circuit is presented in the following example.

Example 3.5

Suppose that we wish to find i in the dc circuit shown in Fig. 3.11. Using the fact that a capacitor behaves as an open circuit to dc—as well as the fact that an inductor behaves as a short circuit—we can determine the current i from the resistive circuit given in Fig. 3.12. By current division, we have that

$$i = \frac{4}{4 + 2}(6) = 4 \text{ A}$$

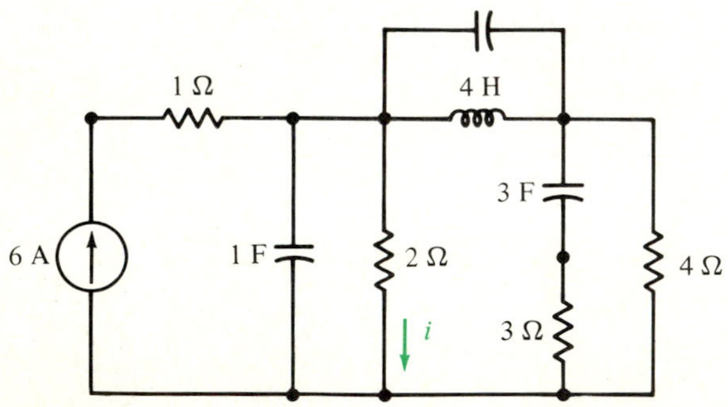

Fig. 3.11 A dc circuit.

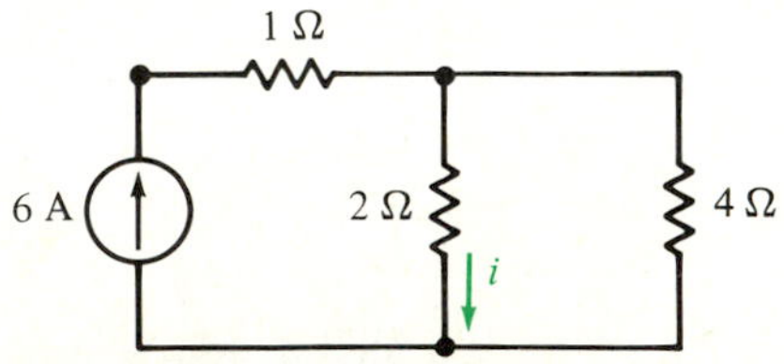

Fig. 3.12 Equivalent form of preceding dc circuit.

■

Perhaps the fact that a capacitor acts as an open circuit to dc does not seem surprising since such an element has a nonconducting dielectric (material that will not allow the flow of charge) between its (conducting) plates. How, then, can there ever be any current through a capacitor? To answer this question, consider the simple circuit shown in Fig. 3.10. For the case that v is a constant, then indeed $i = 0$ (i.e., there is no current through the capacitor). However, there is a voltage v across the capacitor and a charge q on the plates. From elementary physics, we know that $q = Cv$. For the case that the voltage is not constant with time [let's explicitly write $v(t)$], the charge will vary with time, that is, $q(t) = Cv(t)$. Since the net charge on the plates fluctuates, and no charge crosses the dielectric, there must be a transfer of charge throughout the remaining portion of the circuit. Hence, when $v(t)$ is nonconstant, there is a nonzero current $i(t) = C\,dv(t)/dt$. In summary, even though there is no flow of charge across the dielectric of a capacitor, the effect of alternate charging and discharging due to a nonconstant voltage results in an actual current outside of the dielectric.

Example 3.6

For the circuit given in Fig. 3.10, suppose that $C = 2$ F and the voltage v is described by the function given in Fig. 3.13. Then the current through the capacitor is as given in Fig. 3.14. At this point, compare these results with those given in Figs. 3.2, 3.5, and 3.6. Calculating the instantaneous power absorbed by the capacitor in this example results in the same function $p(t)$ that was obtained in Fig. 3.7. As a consequence, a similar subsequent discussion leads us to the conclusion that a capacitor, like an inductor, is an energy-storage element.

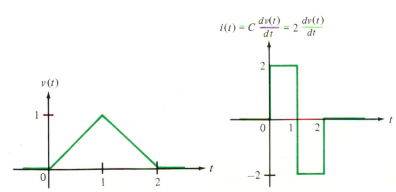

$$i(t) = C\frac{dv(t)}{dt} = 2\frac{dv(t)}{dt}$$

Fig. 3.13 Nonconstant voltage.   Fig. 3.14 Resulting capacitor current.

For a capacitor, energy is stored in the electric field that exists between its plates. To obtain an expression for the energy stored in a capacitor, we proceed in a manner analogous to the case of the inductor. The result is that the energy absorbed by a capacitor of C farads at time t is

$$w_C(t) = \tfrac{1}{2} C v^2(t)$$

Example 3.7

For the op-amp circuit shown in Fig. 3.15, because the inverting input is at a potential of zero volts, then $v_C = v$. Thus,

$$i = C\frac{dv_C}{dt} = 2\frac{dv}{dt}$$

Since the inputs of the op amp draw no current, $i_R = i$, and

$$v_R = \frac{1}{2}i_R = \frac{1}{2}i = \frac{1}{2}\left(2\frac{dv}{dt}\right) = \frac{dv}{dt}$$

Finally, by KVL we have that $v_o = -v_R$. Hence,

$$v_o = -\frac{dv}{dt}$$

In other words, the output voltge $v_o(t)$ is the derivative of the input voltage $v(t)$ (multiplied by the constant -1). Because of this, we call such a circuit a **differentiator**.

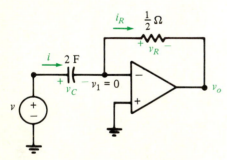

Fig. 3.15 Differentiator circuit.

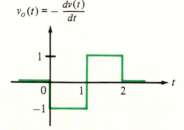

Fig. 3.16 Output voltage of the given differentiator.

For the case that $v(t)$ is the function given in Fig. 3.13, the output voltage $v_o(t)$ is as given in Fig. 3.16.

Since no physical dielectric is perfect, an actual capacitor will allow a certain amount of current (perhaps extremely small), called **leakage current**, through it.

For this reason, we can model a physical capacitor as an ideal capacitor in parallel with a resistance. This leakage resistance is inversely proportional to the capacitance. Depending on the construction of capacitors, values can range from a few picofarads (10^{-12} F) up to 10,000 μF and more. The product of leakage resistance and capacitance typically lies between 10 and 10^6 Ω-F. The **working voltage** of a capacitor—that is, the maximum voltage that can be applied to the capacitor without damaging it or breaking down the dielectric—can be anything from a few volts to hundreds, or even thousands, of volts.

The capacitor, like the operational amplifier and the resistor, has the valuable property of miniaturization—it can be made part of integrated circuits. On the other hand, because semiconductors do not possess the appropriate magnetic properties, and because of size limitations, inductors are not, in general, readily adaptable to IC form. This, however, does not diminish the importance of the inductor—your radio, television set, and various kinds of telecommunications equipment employ many of them.

3.3 INTEGRAL RELATIONSHIPS

Given any fixed value t_0, the integral of the real function $f(t)$ is

$$\int_{-\infty}^{t_0} f(t) \, dt$$

which is a real number that represents the net area under the curve $f(t)$ for the segment of the t-axis that extends from $-\infty$ to t_0. Since t_0 can be any real number between $-\infty$ and ∞, let us instead use for the upper limit on the integral the variable t. In this way, the integral will not result in a specific number, but rather will result in a function of the variable t. Let us call this function $g(t)$; that is, we define

$$g(t) = \int_{-\infty}^{t} f(t) \, dt$$

Specifically then, the area under the curve $f(t)$ between $-\infty$ and t_0 is

$$g(t_0) = \int_{-\infty}^{t_0} f(t) \, dt$$

Therefore, for all $t > t_0$, we have that

$$g(t) = \int_{-\infty}^{t} f(t) \, dt = \int_{-\infty}^{t_0} f(t) \, dt + \int_{t_0}^{t} f(t) \, dt$$

or

$$g(t) = g(t_0) + \int_{t_0}^{t} f(t) \, dt \qquad (3.3)$$

Now let us return to the inductor. With reference to Fig. 3.1, we know that

$$v = L \frac{di}{dt}$$

This is the differential relationship between voltage and current for an inductor. From this, we may now obtain an integral relationship between voltage and current for an inductor. To do this, first integrate both sides of this equation with respect to time, choosing t_0 and t as the lower and upper limits respectively. Doing this we get

$$\int_{t_o}^{t} v(t)\,dt = \int_{t_o}^{t} L\frac{di(t)}{dt}\,dt$$

By the chain rule of calculus, we can change the variable of integration for the term on the right. Since $t = t_0$ implies that $i(t) = i(t_0)$, we have

$$\int_{t_0}^{t} v(t)\,dt = \int_{i(t_0)}^{i(t)} L\,di(t) = Li(t)\,\Big|_{i(t_0)}^{i(t)} = L[i(t) - i(t_0)]$$

Hence,

$$i(t) - i(t_0) = \frac{1}{L}\int_{t_0}^{t} v(t)\,dt$$

or

$$i(t) = i(t_0) + \frac{1}{L}\int_{t_0}^{t} v(t)\,dt \qquad (3.4)$$

But by Equation (3.3), for $t > t_0$, we can write Equation (3.4) as

$$i(t) = \frac{1}{L}\int_{-\infty}^{t} v(t)\,dt \qquad (3.5)$$

Since Equations (3.4) and (3.5) are equivalent, either one is referred to as the integral relationship between voltage and current for an inductor.

Example 3.8

For the circuit shown in Fig. 3.17, let us find the inductor current i given that the voltage v is as described in Fig. 3.18.

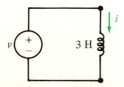

Fig. 3.17 Circuit with an inductor.

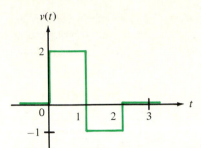

Fig. 3.18 Waveform of applied voltage.

For $t \leq 0$, we have that

$$i(t) = \frac{1}{L} \int_{-\infty}^{t} v(t) \, dt = \frac{1}{3} \int_{-\infty}^{t} 0 \, dt = 0 \text{ A}$$

For $0 < t \leq 1$ s,

$$i(t) = \frac{1}{L} \int_{-\infty}^{t} v(t) \, dt = i(t_0) + \frac{1}{L} \int_{t_0}^{t} v(t) \, dt = i(0) + \frac{1}{L} \int_{0}^{t} v(t) \, dt$$

and since $i(t) = 0$ for $t \leq 0$,

$$i(t) = 0 + \frac{1}{3} \int_{0}^{t} 2 \, dt = \frac{1}{3} (2t) \bigg|_{0}^{t} = \frac{2}{3} t \text{ A}$$

For $1 < t \leq 2$ s,

$$i(t) = \frac{1}{L} \int_{-\infty}^{t} v(t) \, dt = i(t_0) + \frac{1}{L} \int_{t_0}^{t} v(t) \, dt = i(1) + \frac{1}{L} \int_{1}^{t} v(t) \, dt$$

and since $i(t) = \frac{2}{3} t$ for $0 < t \leq 1$ s,

$$i(t) = \frac{2}{3} (1) + \frac{1}{3} \int_{1}^{t} (-1) \, dt = \frac{2}{3} - \frac{1}{3} t \bigg|_{1}^{t} = \frac{2}{3} - \frac{1}{3} (t - 1)$$

$$= -\frac{1}{3} t + 1 \text{ A}$$

For $t > 2$ s,

$$i(t) = i(t_0) + \frac{1}{L} \int_{t_0}^{t} v(t) \, dt = i(2) + \frac{1}{L} \int_{2}^{t} v(t) \, dt$$

and since $i(t) = -\frac{1}{3} t + 1$ for $1 < t \leq 2$ s,

$$i(t) = \left[-\frac{1}{3} (2) + 1 \right] + \frac{1}{3} \int_{2}^{t} 0 \, dt = \frac{1}{3} \text{ A}$$

Thus, in summary, we have that the current is as given in Fig. 3.19.

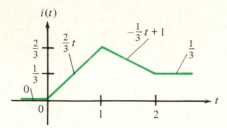

Fig. 3.19 Resulting inductor current.

For Example 3.8, we see that for time $t < 0$, the voltage applied to the inductor is zero, and as we would expect, the resulting current through the inductor is zero. For time $0 \leq t < 2$ s, the voltage is nonzero, and as our analysis tells us, so is the current. But look—for time $t > 2$ s, the voltage applied is again zero, but the current is not. It remains constant forever! Why?

For time $t > 2$ s, the voltage source has a value of zero volts, and thus behaves as a short circuit. Unlike a resistor, an inductor can have zero volts across it and, at the same time, have a nonzero current through it (remember that an inductor acts as a short circuit to dc). So, there is no paradox.

The energy stored in the inductor is shown in Fig. 3.20. We see that for $t < 0$, since the current is zero, the energy stored is zero. When the input voltage is positive, for $0 \leq t < 1$ s, the energy stored in the inductor increases with time. When the input voltage goes negative, for $1 \leq t < 2$ s, the energy stored decreases until time $t = 2$ s, where it is $\frac{1}{6}$ J. For $t \geq 2$ s, however, when the voltage applied is zero, we have the equivalent situation shown in Fig. 3.21. Since the inductor and the voltage source are ideal, no energy is dissipated, and the energy stays constant for $t \geq 2$ s. This causes the current to stay constant for $t \geq 2$ s.

$$w_L(t) = \frac{1}{2} L i^2(t) = \frac{3}{2} i^2(t)$$

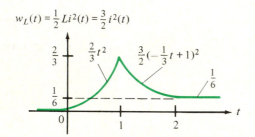

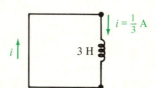

Fig. 3.20 Energy stored in the inductor.   **Fig. 3.21 Equivalent circuit for $t \geq 2$ s.**

Since all physical inductors and voltage sources have some associated resistance, in a corresponding actual circuit the current $i(t)$ and the energy stored $w_L(t)$ would eventually become zero. The more resistance, the sooner zero values would be reached.

To return to the capacitor with the voltage and current indicated as shown in Fig. 3.9, we know that

$$i = C\frac{dv}{dt}$$

and this is the differential relationship between current and voltage. In a manner identical to that discussed for the inductor, we can obtain the equivalent integral relationships between voltage and current for a capacitor that are given by

$$v(t) = v(t_0) + \frac{1}{C}\int_{t_0}^{t} i(t)\, dt \tag{3.6}$$

and

$$v(t) = \frac{1}{C}\int_{-\infty}^{t} i(t)\, dt \tag{3.7}$$

Example 3.9

For the circuit shown in Fig. 3.22, let us determine v given that i is described by Fig. 3.23. In this case, let us consider three intervals of time.

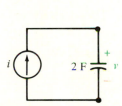

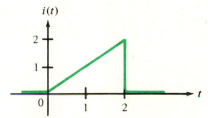

Fig. 3.22 Circuit with a capacitor.   **Fig. 3.23 Waveform of applied current.**

For $t \le 0$,

$$v(t) = \frac{1}{C}\int_{-\infty}^{t} i(t)\, dt = \frac{1}{2}\int_{-\infty}^{t} 0\, dt = 0 \text{ V}$$

For $0 < t \le 2$ s,

$$v(t) = \frac{1}{C}\int_{-\infty}^{t} i(t)\, dt = v(0) + \frac{1}{C}\int_{0}^{t} i(t)\, dt$$

$$= 0 + \frac{1}{2}\int_{0}^{t} t\, dt = \frac{1}{2}\left(\frac{t^2}{2}\right)\Bigg|_{0}^{t} = \frac{t^2}{4} \text{ V}$$

For $t > 2$ s,

$$v(t) = \frac{1}{C} \int_{-\infty}^{t} i(t)\, dt = v(2) + \frac{1}{2} \int_{2}^{t} i(t)\, dt$$

and since $v(t) = t^2/4$ for $0 < t \le 2$ s,

$$v(t) = \frac{2^2}{4} + \frac{1}{2} \int_{2}^{t} 0\, dt = 1\ \text{V}$$

In summary, then, the voltage $v(t)$ is as shown in Fig. 3.24

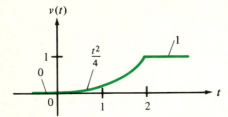

Fig. 3.24 Resulting capacitor voltage.

In Example 3.9, for time $t \le 0$, the current applied to the capacitor is zero, and the resulting voltage across it is, therefore, zero. For $0 < t < 2$ s when the current is nonzero, so is the voltage. For $t \ge 2$ s, the current is again zero, but the voltage remains nonzero—in particular, it stays a constant 1 V.

The energy stored in the capacitor is

$$w_C(t) = \frac{1}{2} C v^2(t) = v^2(t) = \begin{cases} 0 & \text{for} \quad -\infty < t < 0 \\ t^4/16 & \text{for} \quad 0 \le t < 2\ \text{s} \\ 1 & \text{for} \quad 2 \le t < \infty \end{cases}$$

Thus, the energy stored in the capacitor increases from 0 to 1 J in the time interval $0 \le t < 2$ s. But for $t \ge 2$ s, since the current source has a value of zero, it behaves as an open circuit. Therefore, there is no way for the capacitor to discharge; that is, there is no element that can absorb the energy stored in the capacitor. Consequently, the energy stored and the voltage across the capacitor remain constant for $t \ge 2$ s.

Since a physical capacitor has some associated leakage resistance (it may be very large), and a practical current source has an internal resistance, in a corresponding actual circuit, the energy stored and the voltage across the capacitor would eventually become zero—the smaller the effective resistance, the sooner zero is reached.

Example 3.10

For the op-amp circuit shown in Fig. 3.25, since $v_1 = 0$, then $i = 2v_s$. Since the input terminals of the op amp draw no current,

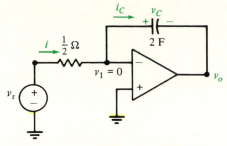

Fig. 3.25 Integrator circuit.

$$i_C = i = 2\,v_s$$

Thus,

$$v_C = \frac{1}{C} \int_{-\infty}^{t} i_C(t)\, dt = \frac{1}{2} \int_{-\infty}^{t} 2v_s(t)\, dt = \int_{-\infty}^{t} v_s(t)\, dt$$

By KVL, $v_o = -v_C$. Hence,

$$v_o(t) = -\int_{-\infty}^{t} v_s(t)\, dt$$

and the output voltage is the integral of the input voltage (multiplied by the constant -1). For this reason such a circuit is known as an **integrator.** This type of circuit is extremely useful—it is the backbone of the analog computer.

Again, consider the inductor shown in Fig. 3.1. Since

$$i(t) = \frac{1}{L} \int_{-\infty}^{t} v(t)\, dt$$

then the current through the inductor at time $t = a$ is

$$i(a) = \frac{1}{L} \int_{-\infty}^{a} v(t)\, dt$$

For any positive, real number ϵ, the current through the inductor at time $t = a + \epsilon$ is

$$i(a + \epsilon) = \frac{1}{L} \int_{-\infty}^{a+\epsilon} v(t)\, dt = \frac{1}{L} \int_{-\infty}^{a} v(t)\, dt + \frac{1}{L} \int_{a}^{a+\epsilon} v(t)\, dt$$

$$= i(a) + \frac{1}{L} \int_{a}^{a+\epsilon} v(t)\, dt$$

If ϵ gets arbitrarily small, then

$$\frac{1}{L} \int_a^{a+\epsilon} v(t)\, dt$$

gets arbitrarily small and $i(a + \epsilon)$ gets arbitrarily close to $i(a)$. We conclude, therefore, that *the current through an inductor cannot change instantaneously*.

Since for a capacitor, depicted in Fig. 3.9, it is true that

$$v(t) = \frac{1}{C} \int_{-\infty}^t i(t)\, dt$$

proceeding in a manner as above, we can deduce that *the voltage across a capacitor cannot change instantaneously*.

3.4 IMPORTANT CIRCUIT CONCEPTS

Just as we analyzed circuits containing independent and dependent sources and resistors (i.e., resistive circuits) via nodal analysis or mesh analysis, so can we write a set of node or mesh equations for circuits that contain inductors and capacitors in addition to resistors and sources. The procedures are as was described for the case of resistive circuits; the difference is that for inductors and capacitors, the appropriate relationships between voltage and current are used in place of Ohm's law.

Example 3.11

Let us write the node equations for the circuit shown in Fig. 3.26. At node v_1, we have that

$$v_1 = 6 \cos 3t$$

At node v_2, by KCL,

$$i_1 - i_2 + i_3 = 0$$

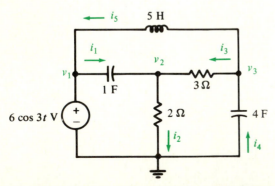

Fig. 3.26 Circuit with resistors, capacitors, and an inductor.

from which

$$\frac{d}{dt}(v_1 - v_2) - \frac{1}{2}v_2 + \frac{1}{3}(v_3 - v_2) = 0 \tag{3.8}$$

At node v_3, by KCL,

$$i_3 - i_4 + i_5 = 0$$

from which

$$\frac{1}{3}(v_3 - v_2) + 4\frac{dv_3}{dt} + \frac{1}{5}\int_{-\infty}^{t} (v_3 - v_1) \, dt = 0 \tag{3.9}$$

Equation (3.8), whose variables are voltages, is a **differential equation** since it contains variables and their derivatives. Equation (3.9) is an **integrodifferential equation** since it contains an integral as well as derivatives.

The circuit in Fig. 3.27 is identical to the one in Fig. 3.26. Let us write the mesh equations for this circuit. For mesh i_a, by KVL,

$$v_a + v_b = 6 \cos 3t$$

from which

$$\int_{-\infty}^{t} (i_a - i_b) \, dt + 2(i_a - i_c) = 6 \cos 3t$$

For mesh i_b, by KVL,

$$-v_a + v_c + v_e = 0$$

from which

$$-\int_{-\infty}^{t} (i_a - i_b) \, dt + 3(i_b - i_c) + 5\frac{di_b}{dt} = 0$$

For mesh i_c, by KVL,

$$-v_b - v_c + v_d = 0$$

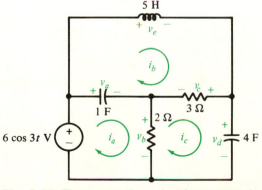

Fig. 3.27 The preceding circuit repeated.

from which

$$-2(i_a - i_c) - 3(i_b - i_c) + \frac{1}{4} \int_{-\infty}^{t} i_c \, dt = 0$$

This last equation is an **integral equation** since it contains variables and their integrals.

■

Writing the node or mesh equations for a circuit, as was done in the preceding example, is not difficult. Finding the solutions of such equations, however, is another matter, and, in general, is no simple task. We shall soon see how to analyze certain simple circuits by solving simple differential equations in one variable. For more complicated circuits, we will simplify the analysis by utilizing additional concepts to be discussed later.

For the capacitor and the inductor and their associated descriptions shown in Fig. 3.28, all of these relationships between voltage and current are linear relationships. For instance, suppose that for either element, the voltage v is determined from the current i. Then when $i = i_1 + i_2$, the resulting voltage $v = v_1 + v_2$, where v_1 is the voltage due solely to current i_1, and v_2 is the voltage due solely to current i_2. Also, the voltage due to the current Ki (that is, the current i scaled by the constant factor K) is Kv (the voltage due to current i scaled by the same constant K). Consequently, the principle of superposition is also valid for circuits with capacitors and inductors, if the above current–voltage relationships are employed.

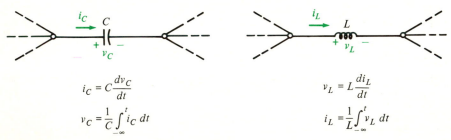

$$i_C = C \frac{dv_C}{dt} \qquad\qquad v_L = L \frac{di_L}{dt}$$

$$v_C = \frac{1}{C} \int_{-\infty}^{t} i_C \, dt \qquad\qquad i_L = \frac{1}{L} \int_{-\infty}^{t} v_L \, dt$$

Fig. 3.28 A capacitor and an inductor.

Duality

Let us formalize our discussion of duality, which was mentioned briefly earlier (see Examples 1.10 and 1.11).

Consider the two circuits shown in Fig. 3.29. Using mesh analysis for circuit A and nodal analysis for circuit B, we obtain the following results:

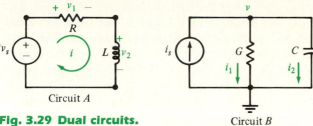

Fig. 3.29 Dual circuits.

Circuit A

Circuit B

$$\text{Circuit } A \qquad\qquad \text{Circuit } B$$

$$v_s = v_1 + v_2 \qquad i_s = i_1 + i_2$$

$$v_s = Ri + L\frac{di}{dt} \qquad i_s = Gv + C\frac{dv}{dt}$$

In other words, both circuits are described by the same equations:

$$x_s = x_1 + x_2$$

$$x_s = a_1 y + a_2\frac{dy}{dt}$$

Note that a variable that is a current in one circuit is a voltage in the other, and vice versa. For these two circuits, this result is not a coincidence, but rather is due to the concept known as **duality.**

In actuality, circuits A and B are **dual circuits** when the values $R = G$ and $L = C$ and the functions $v_s = i_s$, since the following roles are interchanged:

series $\leftrightarrow$ parallel
voltage $\leftrightarrow$ current
resistance $\leftrightarrow$ conductance
capacitance $\leftrightarrow$ inductance

In summary, the relationships between a circuit and its dual are given in the following table.

Circuit	Dual Circuit
Series connection	Parallel connection
Parallel connection	Series connection
Voltage of x volts	Current of x amperes
Current of x amperes	Voltage of x volts
Resistance of x ohms	Conductance of x mhos
Conductance of x mhos	Resistance of x ohms
Capacitance of x farads	Inductance of x henries
Inductance of x henries	Capacitance of x farads

The usefulness of duality lies in the fact that once a circuit is analyzed, its dual is, in essence, analyzed also—two for the price of one! It may even be advantageous to construct the dual of a given circuit rather than work with the given circuit itself. Note that if circuit *B* is the dual of circuit *A*, then taking the dual of circuit *B* in essence results in circuit *A*. Not every circuit, however, has a dual. A circuit has a dual if, and only if, it is a planar network.

Given a series-parallel network, its dual can be found by inspection using the above table. However, a circuit may not be series-parallel. We now present the rules for obtaining the dual of a planar circuit, regardless of whether or not it is a series-parallel network.

Rule 1. Inside of each mesh, including the infinite region surrounding the circuit, place a node.

Rule 2. Suppose two of these nodes, say nodes *a* and *b*, are in adjacent meshes. Then there is at least one element in the boundary common to these two meshes. Place the dual of each common element between nodes *a* and *b*.

Example 3.12

To find the dual of the circuit shown in Fig. 3.26 by the rules given above, we proceed as shown in Fig. 3.30. The dual is the circuit drawn in green, and this dual circuit is redrawn in Fig. 3.31.

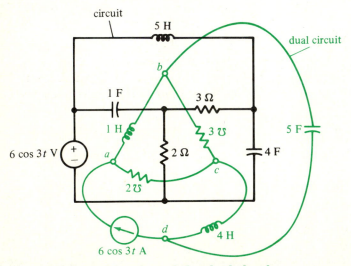

Fig. 3.30 Determination of the dual circuit.

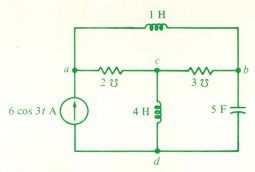

Fig. 3.31 Dual circuit redrawn.

3.5 FIRST-ORDER CIRCUITS

Consider the simple *RC* circuit shown in Fig. 3.32. In this circuit is an ideal switch *S*, which is **closed** (a short circuit) for time $t < 0$. Since for $t < 0$ this circuit is a dc circuit, the capacitor behaves as an open circuit. Thus, by voltage division,

$$v(t) = \frac{RV}{R + R_g} \qquad \text{for} \quad t < 0$$

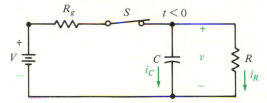

Fig. 3.32 *RC* circuit with a switch.

Now suppose that the switch *S* **opens** (becomes an open circuit) instantaneously at time $t = 0$ and remains open for $t > 0$ as shown in Fig. 3.33. Since the voltage across the capacitor cannot change instantaneously, then the voltage remains the same at $t = 0$, that is,

$$v(0) = \frac{RV}{R + R_g}$$

But what is $v(t)$ for $t > 0$? To answer this question, refer to Fig. 3.33. By KCL,

$$i_C + i_R = 0$$

from which

$$C\frac{dv}{dt} + \frac{v}{R} = 0 \qquad\qquad\qquad \textbf{(3.10)}$$

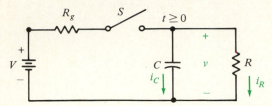

Fig. 3.33 The circuit after the switch is opened.

In Equation (3.10) there appears the variable $v(t)$ we wish to determine, as well as its first derivative $dv(t)/dt$. For this reason, we say that this equation is a **first-order differential equation.** Since the circuit in Fig. 3.33 is described by a first-order differential equation, we call it a **first-order circuit.** Since every nonzero term in Equation (3.10) is of the first degree in the dependent variable v and its derivative, we call it an **homogeneous, linear** differential equation. Note that the coefficients of v and its derivative are constants. To solve Equation (3.10), that is, to find the function $v(t)$ that satisfies the equation and the initial (or boundary) condition of $v(0)$, we proceed as follows.

The Natural Response

From the differential equation (3.10), we have that

$$\frac{dv}{dt} = -\frac{1}{RC}v$$

Dividing both sides by v (called **separating variables**), we get

$$\frac{1}{v}\frac{dv}{dt} = -\frac{1}{RC}$$

Integrating both sides of this equation with respect to time, we have that

$$\int \frac{1}{v}\frac{dv}{dt}\,dt = \int -\frac{1}{RC}dt$$

By the chain rule of calculus, we can write

$$\int \frac{1}{v}\frac{dv}{dt}\,dt = \int \frac{dv}{v} = \int -\frac{1}{RC}dt$$

Integrating, we get

$$\ln v(t) = -\frac{t}{RC} + K$$

where K is a constant of integration. From this equation, taking powers of e, the result is

$$v(t) = e^{-t/RC + K} = e^{-t/RC}e^{K}$$

Setting $t = 0$, we obtain

$$v(0) = e^0 e^K = e^K$$

Substitution of this value of e^K into the preceding equation yields

$$v(t) = v(0)e^{-t/RC} \qquad \text{for} \quad t \geq 0 \tag{3.11}$$

Thus, this is the solution to the differential equation (3.10) subject to the constraint of the initial condition; that is, the voltage across the capacitor at time $t = 0$ is $v(0) = RV/(R + R_g)$. A sketch of the function $v(t)$ versus t for $t \geq 0$ is shown in Fig. 3.34.

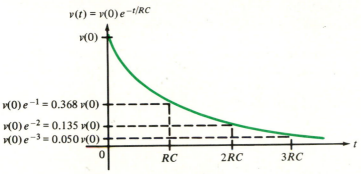

Fig. 3.34 Voltage across the capacitor.

Of course, a variable is a variable is a variable—so even if we change the variable's name from $v(t)$ to $x(t)$, by going through the same routine, we get the same results. In other words, the solution to the homogeneous, first-order, linear differential equation (with constant coefficients)

$$\frac{dx(t)}{dt} + ax(t) = 0 \tag{3.12}$$

subject to the initial (or boundary) condition $x(0)$ is

$$x(t) = x(0)e^{-at} \qquad \text{for} \quad t \geq 0 \tag{3.13}$$

Note that the resistor current can now be determined for $t \geq 0$ since

$$i_R(t) = \frac{v(t)}{R} = \frac{v(0)}{R} e^{-t/RC}$$

Furthermore, the capacitor current for $t \geq 0$ is

$$i_C(t) = -i_R(t) = -\frac{v(0)}{R} e^{-t/RC}$$

or alternatively,

$$i_C(t) = C\frac{dv(t)}{dt} = C\frac{d}{dt}[v(0)e^{-t/RC}] = C\left(-\frac{1}{RC}\right)v(0)e^{-t/RC}$$

$$= -\frac{v(0)}{R}e^{-t/RC}$$

From these results, we see that initially (at $t = 0$) the capacitor is charged to $v(0)$ volts, and for $t > 0$, the capacitor discharges through the resistor exponentially. From the formula

$$w_C(t) = \tfrac{1}{2}Cv^2(t)$$

we see that the energy stored in the capacitor at $t = 0$ is

$$w_C(0) = \tfrac{1}{2}Cv^2(0) \quad \text{joules}$$

Since the voltage goes to zero as time goes to infinity, the energy stored in the capacitor also goes to zero as time goes to infinity. Where does this energy go?

The power absorbed by the resistor is

$$p_R(t) = Ri_R^2(t) = R\left(\frac{v(0)}{R}e^{-t/RC}\right)^2 = \frac{v^2(0)}{R}e^{-2t/RC}$$

Therefore, the total energy absorbed by the resistor is

$$w_R = \int_0^\infty p_R(t)\,dt = \int_0^\infty \frac{v^2(0)}{R}e^{-2t/RC}\,dt = -\frac{RC}{2}\frac{v^2(0)}{R}e^{-2t/RC}\Big|_0^\infty$$

$$= -\frac{C}{2}v^2(0)[0 - 1] = \frac{1}{2}Cv^2(0) = w_C(0)$$

Thus, we see that the energy initially stored in the capacitor is eventually dissipated as heat by the resistor.

We now see that the voltage and current in the circuit shown in Fig. 3.33 have the form

$$f(t) = Ke^{-t/\tau}$$

where K and τ are constants. The constant τ is called the **time constant** and, since the power of e should be a dimensionless number, the units of τ, therefore, are seconds. For the voltage and current expressions above, the time constant is $\tau = RC$. Thus, we see that the product of resistance and capacitance has seconds as units, that is, ohms $\times$ farads $=$ seconds. This fact can be verified from $R = v/i$ and $C = q/v$, which have the respective units volts per ampere $=$ volts per coulomb per second and coulombs per volt. Hence, the product RC has the second as its unit.

From Fig. 3.34, we see that as t goes to infinity, $v(t)$ goes to zero. Although $v(t)$ never identically equals zero for any finite value of t, when $t = 3\tau$, the value of $v(t)$ is down to only about 5% of the value of $v(t)$ at $t = 0$. Thus, after just a few time constants, the value of the function $v(t)$ is practically zero.

For the circuit given in Fig. 3.33, for $t \geq 0$, the independent source has no effect on the circuit—only the initial condition $v(0)$ has an effect on the circuit for $t \geq 0$. For this reason, we say that the voltage and current that we determined for $t \geq 0$ are **natural responses.**

Now consider the circuit shown in Fig. 3.35. In this circuit, the switch S is closed for $t < 0$. Since an inductor is a short circuit to dc, the voltage $v = 0$, and hence the current through R is zero. Thus,

$$i(t) = \frac{V}{R_g} \quad \text{for} \quad t < 0$$

If the switch S opens at time $t = 0$ and remains open for $t > 0$ as shown in Fig. 3.36, since the current through an inductor cannot change instantaneously, then $i(0) = V/R_g$. For $t \geq 0$, by KVL,

$$L\frac{di}{dt} + Ri = 0$$

or

$$\frac{di}{dt} + \frac{R}{L}i = 0 \tag{3.14}$$

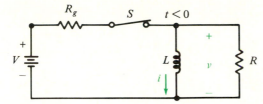

Fig. 3.35 *RL* **circuit with a switch.**

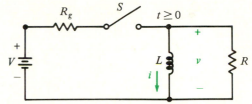

Fig. 3.36 The circuit after the switch is opened.

Since this equation has the form of Equation (3.12), then the solution of Equation (3.14) has the form of Equation (3.13). In other words, the solution of differential Equation (3.14) is

$$i(t) = i(0)e^{-Rt/L} \quad \text{for} \quad t \geq 0$$

A sketch of $i(t)$ versus t for $t \geq 0$ is shown in Fig. 3.37. Note that in this case the time constant is $\tau = L/R$. Thus, dividing henries by ohms yields seconds.

For the simple *RL* circuit given in Fig. 3.36, the voltage $v(t)$ is

$$v(t) = L\frac{di(t)}{dt} = L\frac{d}{dt}[i(0)e^{-Rt/L}]$$

$$= -Ri(0)e^{-Rt/L} \quad \text{for} \quad t \geq 0$$

Alternatively, we have that

$$v(t) = -Ri(t) = -Ri(0)e^{-Rt/L} \quad \text{for} \quad t \geq 0$$

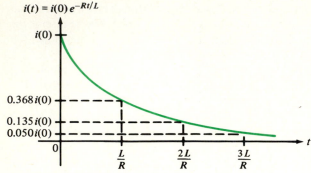

$i(t) = i(0)\,e^{-Rt/L}$

Fig. 3.37 Current through the inductor.

The energy stored in the inductor initially (at $t = 0$) is determined from

$$w_L(t) = \tfrac{1}{2}Li^2(t)$$

to be

$$w_L(0) = \tfrac{1}{2}Li^2(0) \quad \text{joules}$$

and as time goes to infinity, since the current in the inductor goes to zero exponentially, the energy stored in the inductor also goes to zero.

The power absorbed by the resistor is

$$p_R(t) = Ri^2(t) = R[i(0)e^{-Rt/L}]^2 = Ri^2(0)e^{-2Rt/L}$$

and, therefore, the total energy absorbed by the resistor is

$$w_R = \int_0^\infty p_R(t)\,dt = \int_0^\infty Ri^2(0)e^{-2Rt/L}\,dt = -\frac{L}{2R}Ri^2(0)e^{-2Rt/L}\Big|_0^\infty$$

$$= -\frac{L}{2}i^2(0)[0 - 1] = \frac{1}{2}Li^2(0) = w_L(0)$$

Thus, we see that the energy initially stored in the inductor is eventually dissipated as heat by the resistor.

Example 3.13

Consider the circuit shown in Fig. 3.38(a), where the applied voltage $v_s(t)$ is depicted in Fig. 3.38(b). Let us determine the voltage $v(t)$ across the capacitor for all time.

Applying KCL at node v_1, we obtain

$$\frac{v_1 - v}{2} = i + 2i = 3i = 3\left(\frac{v_s - v_1}{3}\right)$$

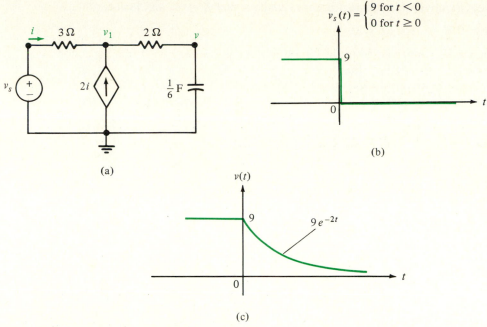

$$v_s(t) = \begin{cases} 9 \text{ for } t < 0 \\ 0 \text{ for } t \geq 0 \end{cases}$$

(b)

(a)

$9 e^{-2t}$

(c)

Fig. 3.38 Example of a natural response.

Simplifying this expression, we get

$$3v_1 = v + 2v_s$$

Applying KCL at node v, we get

$$\frac{1}{6}\frac{dv}{dt} + \frac{v - v_1}{2} = 0$$

from which

$$\frac{dv}{dt} + 3v - 3v_1 = 0$$

Substituting the fact that $3v_1 = v + 2v_s$ into this equation yields

$$\frac{dv}{dt} + 2v = 2v_s$$

For $t < 0$, $v_s(t) = 9$ V and the circuit is a dc circuit. Thus, the voltage across the capacitor is a constant, that is, the voltage has the form $v(t) = K$. Substituting this into the above differential equation results in

$$\frac{dK}{dt} + 2K = 2(9) = 18$$

Since the derivative of a constant is zero, we have that $K = 9$, and hence,

$$v(t) = 9 \text{ V} \qquad \text{for} \quad t < 0$$

This fact can be confirmed from the circuit by noting that for $t < 0$, the capacitor behaves as an open circuit. Thus, $i + 2i = 3i = 0$. This implies that $i = 0$, and, therefore, $v = v_1 = v_s = 9$ V.

Since the voltage across a capacitor cannot change instantaneously, we have that $v(0) = 9$ V. For $t \geq 0$, $v_s(t) = 0$, and the above differential equation becomes

$$\frac{dv}{dt} + 2v = 0$$

The solution of this equation is

$$v(t) = v(0)e^{-2t} = 9e^{-2t} \qquad \text{for} \quad t \geq 0$$

A sketch of this natural response is shown in Fig. 3.38(c).

The Complete Response

Now that we have discussed the responses of first-order circuits that have no applied inputs and nonzero initial (or boundary) conditions, let us consider the case of a response due to a nonzero input.

For the simple RC circuit shown in Fig. 3.39, the switch S moves (instantaneously) from a to b at time $t = 0$. We see that the voltage $v_s(t) = 0$ for $t < 0$, and $v_s(t) = V$ for $t \geq 0$. Thus, let us define a function $u(t)$, called the **unit step function,** by

$$u(t) = \begin{cases} 0 & \text{for} \quad t < 0 \\ 1 & \text{for} \quad t \geq 0 \end{cases}$$

Hence, we can write $v_s(t)$ as $v_s(t) = Vu(t)$. This function is depicted in Fig. 3.40.

Fig. 3.39 *RC* **circuit with an applied step of voltage.**   **Fig. 3.40 Step function.**

From Fig. 3.39 we see that the voltage $Vu(t)$ is applied to a series RC combination. Thus, we can call the resulting voltage $v(t)$ and current $i(t)$ **step responses.** Before we determine a step response, say $v(t)$, mathematically, let us try to predict what will happen. The capacitor is originally uncharged and the input voltage $v_s(t)$ is zero for $t < 0$. Since there is no voltage to produce a current that will charge the capacitor, it should be obvious that $v(t) = 0$ for $t < 0$. At time $t = 0$, when the input voltage goes from 0 to V volts, since the voltage across the capacitor cannot change instantaneously (it acts as a short circuit to an instantaneous change of voltage), $v(0) = 0$. By KVL, $v_R(0) = V - 0 = V$ volts, and $i(0) = v_R(0)/R = V/R$. So, the voltage across the resistor changes instantaneously, as does the current through the resistor, which is the current through the capacitor. The nonzero current through the capacitor begins to charge it to some nonzero value. Since the input voltage remains constant for $t > 0$, by KVL, an increase in the capacitor voltage means a decrease in the resistor voltage. This, in turn, means a reduction in the resistor current—which is the capacitor current—so that the capacitor charges at a slower rate than before. The charging process continues until the capacitor is completely charged. After a long, long time, the input voltage seems more and more like a constant (dc), so as time goes to infinity, the capacitor behaves as an open circuit. Therefore, the final voltage across the capacitor is V volts. For the case of a natural response, we have seen that a capacitor discharges exponentially. Thus, we can take an educated guess that it will charge up exponentially as well.

To summarize, for $t < 0$ and $t = 0$, the voltage across the capacitor is zero. For $t > 0$, the voltage across the capacitor increases at a slower and slower rate until it is completely charged to V volts. By KVL, the voltage across the resistor is zero for $t < 0$, it jumps to V volts instantaneously at $t = 0$, and decreases exponentially to zero as time increases. Furthermore, the current through the resistor—which equals the current through the capacitor—is proportional to the voltage across the resistor.

Let us now confirm our intuition by a mathematical analysis. Using nodal analysis, by KCL,

$$\frac{Vu(t) - v_C}{R} = C\frac{dv_C}{dt}$$

from which

$$\frac{dv_C}{dt} + \frac{1}{RC}v_C = \frac{V}{RC}u(t) \tag{3.15}$$

Since $u(t) = 0$ for $t < 0$, and $u(t) = 1$ for $t \geq 0$, let us consider differential Equation (3.15) for the two time intervals $t < 0$ and $t \geq 0$.

First, for $t < 0$, the right-hand side of the differential equation becomes zero. Since we want to find the response to the input under the circumstance that there is no initial voltage across the capacitor, and since the input is zero before time $t = 0$, the solution is $v_C(t) = 0$ for $t < 0$.

Next, for $t \geq 0$, Equation (3.15) becomes

$$\frac{dv_C}{dt} + \frac{1}{RC}v_C = \frac{V}{RC} \tag{3.16}$$

This equation has the more general form

$$\frac{dx(t)}{dt} + ax(t) = f(t) \tag{3.17}$$

where $x(t) = v_C(t)$, $a = 1/RC$, and $f(t) = V/RC$. To find the solution of this equation, multiply both sides by e^{at}. Therefore, we have

$$e^{at}\frac{dx(t)}{dt} + e^{at}ax(t) = e^{at}f(t)$$

Then

$$\frac{d}{dt}[e^{at}x(t)] = e^{at}\frac{dx(t)}{dt} + e^{at}ax(t) = e^{at}f(t)$$

Integrating both sides of this equation with respect to t, we get

$$\int \frac{d[e^{at}x(t)]}{dt}\,dt = \int e^{at}f(t)\,dt = \int d[e^{at}x(t)]$$

or

$$e^{at}x(t) = \int e^{at}f(t)\,dt + A$$

where A is a constant of integration. Multiplying both sides of this equation by e^{-at}, we obtain the solution to Equation (3.17), which is

$$x(t) = e^{-at}\int e^{at}f(t)\,dt + Ae^{-at}$$

where the constant A is determined from the initial condition. Thus, we see that, in general, the solution of differential Equation (3.17), which is called the **complete response,** consists of the sum of two parts. The first part is

$$e^{-at}\int e^{at}f(t)\,dt$$

which basically is determined by the function $f(t)$. Since $f(t)$ results from the input of a circuit, we call this part of the solution the **forced response** (also known as the **steady-state response**), and we refer to $f(t)$ as the **forcing function.** The other part of the solution is

$$Ae^{-at}$$

which we recognize has the form of the **natural response.** (It is also called the **transient response.**)

Now let us be specific and consider the case that the forcing function is a constant, say b. Then differential Equation (3.17) becomes

$$\frac{dx(t)}{dt} + ax(t) = b$$

The forced response $x_f(t)$ is

$$x_f(t) = e^{-at} \int e^{at}b \, dt = e^{-at}\left(\frac{1}{a}e^{at}\right)b = \frac{b}{a}$$

which is a constant, while the natural response is $x_n(t) = Ae^{-at}$. Thus, the complete response is

$$x(t) = x_f(t) + x_n(t) = \frac{b}{a} + Ae^{-at}$$

where the constant A is determined from the initial condition.

We now have the solution to Equation (3.16):

$$v_C(t) = \frac{V/RC}{1/RC} + Ae^{-t/RC} = V + Ae^{-t/RC} \tag{3.18}$$

But, as $v_C(t) = 0$ for $t < 0$, since the voltage across a capacitor cannot change instantaneously, $v_C(0) = 0$. Setting $t = 0$ in Equation (3.18), we get

$$v_C(0) = V + Ae^{-0} = 0 \qquad \Rightarrow \qquad A = -V$$

Thus, the voltage across the capacitor is

$$v_C(t) = V - Ve^{-t/RC} = V(1 - e^{-t/RC}) \qquad \text{for} \quad t \geq 0$$

Combining this result with the fact that $v_C(t) = 0$ for $t < 0$, we can express $v_C(t)$ for all t as the product

$$v_C(t) = V(1 - e^{-t/RC})u(t) \tag{3.19}$$

Having determined the capacitor voltage, we can now find the capacitor current (which is equal to the resistor current in this case). For $t < 0$,

$$i(t) = C\frac{dv_C(t)}{dt} = 0$$

For $t \geq 0$,

$$i(t) = C\frac{d}{dt}[V(1 - e^{-t/RC})] = CV(-e^{-t/RC})\left(-\frac{1}{RC}\right) = \frac{V}{R}e^{-t/RC}$$

Thus, we can write the single expression for all time:

$$i(t) = \frac{V}{R} e^{-t/RC} u(t) \tag{3.20}$$

Sketches of $v_C(t)$ and $i(t)$ are shown in Fig. 3.41.

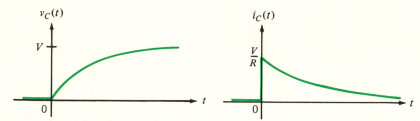

Fig. 3.41 Waveforms of the resulting capacitor voltage and current.

Again consider the circuit in Fig. 3.39, but suppose that the switch S returns to position a at time $t = 1$ s. Under this circumstance, the voltage $v_s(t)$ is the voltage pulse shown in Fig. 3.42. To determine the voltage $v_C(t)$ across the capacitor, for $t < 1$ s, we proceed as above. In other words, $v_C(t) = 0$ for $t < 0$, and $v_C(t) = V(1 - e^{-t/RC})$ for $0 \le t < 1$ s. Since the voltage across a capacitor cannot change instantaneously, $v_C(1) = V(1 - e^{-1/RC})$. But for $t \ge 1$ s, we have to determine a natural response. In other words, for $t \ge 1$ s, as discussed in the section on natural responses, we can get Equation (3.10). A similar approach to the solution of this equation as taken previously results in

$$v_C(t) = v_C(1)e^{-(t-1)/RC} = V(1 - e^{-1/RC})e^{-(t-1)/RC} \qquad \text{for} \quad t \ge 1 \text{ s}$$

A sketch of the voltage $v_C(t)$ is shown in Fig. 3.43.

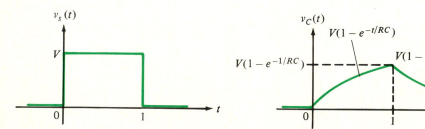

Fig. 3.42 Voltage pulse.   **Fig. 3.43 Voltage response to input pulse.**

Although this last problem employed a complete response (for $t < 1$ s) and a natural response (for $t \ge 1$ s), we can use an alternative approach. Recall that previously it was stated that inductors and capacitors are also linear elements, so that we can use the properties of linearity on circuits with such elements. In particular, we can express the pulse $v_s(t)$ as the sum of the two voltage steps $v_a(t) = Vu(t)$ and $v_b(t) = -Vu(t-1)$ as shown in Fig. 3.44. By linearity, the response to $v_s(t) = v_a(t) + v_b(t)$ is the sum of the response to $v_a(t)$ and the re-

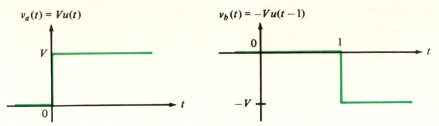

Fig. 3.44 Components of the given voltage pulse.

sponse to $v_b(t)$. But we have already seen that the response to $v_a(t)$ is

$$v_{Ca}(t) = V(1 - e^{-t/RC})u(t)$$

Also, by linearity, if the input is scaled by a constant, the output is also scaled by the same constant. In particular, the response to $-Vu(t)$ is $-V(1 - e^{-t/RC})u(t)$.

We are now in a position to use another property, known as **time invariance,** which is attributable to the fact that the differential equations describing our circuits have constant coefficients. (Such circuits are called **time-invariant** circuits.) This property says that delaying an input by a certain amount of time delays the response by the same amount of time. In particular, the response to $v_b(t) = -Vu(t - 1)$, and this is the input $-Vu(t)$ delayed by 1 s, is

$$v_{Cb}(t) = -V(1 - e^{-(t-1)/RC})u(t - 1)$$

Thus, by linearity, we have that the response to $v_s(t) = v_a(t) + v_b(t)$ is

$$v_C(t) = v_{Ca}(t) + v_{Cb}(t) = V(1 - e^{-t/RC})u(t) - V(1 - e^{-(t-1)/RC})u(t - 1)$$

To compare this answer with our previous results, note that for $t < 0$, $u(t) = u(t - 1) = 0$, so $v_C(t) = 0$. Also, for $0 \le t < 1$ s, $u(t) = 1$ and $u(t - 1) = 0$, so $v_C(t) = V(1 - e^{-t/RC})$. But for $t \ge 1$ s, $u(t) = u(t - 1) = 1$. We then have that

$$v_C(t) = V(1 - e^{-t/RC}) - V(1 - e^{-(t-1)/RC})$$
$$= V - Ve^{-t/RC} - V + Ve^{-(t-1)/RC}$$
$$= V(-e^{-t/RC}e^{1/RC}e^{-1/RC} + e^{-(t-1)/RC})$$
$$= V(1 - e^{-1/RC})e^{-(t-1)/RC} \qquad \text{for} \quad t \ge 1 \text{ s}$$

which agrees with our previous analysis (see Fig. 3.43).

From Equation (3.20), the current due to $v_a(t)$ is

$$i_a(t) = \frac{V}{R}e^{-t/RC}u(t)$$

By the linearity and time-invariance properties, the current due to $v_b(t)$ is

$$i_b(t) = -\frac{V}{R}e^{-(t-1)/RC}u(t - 1)$$

Thus, by linearity, the current due to $v_s(t) = v_a(t) + v_b(t)$ is

$$i(t) = i_a(t) + i_b(t) = \frac{V}{R}e^{-t/RC}u(t) - \frac{V}{R}e^{-(t-1)/RC}u(t-1)$$

For $t < 0$, clearly $i(t) = 0$, while for $0 \le t < 1$ s, $i(t) = (V/R)e^{-t/RC}$. For $t \ge 1$ s,

$$i(t) = \frac{V}{R}e^{-t/RC} - \frac{V}{R}e^{-(t-1)/RC} = \frac{V}{R}(e^{-t/RC}e^{1/RC}e^{-1/RC} - e^{-(t-1)/RC})$$

$$= \frac{V}{R}(e^{-1/RC} - 1)e^{-(t-1)/RC}$$

A sketch of $i(t)$ is shown in Fig. 3.45.

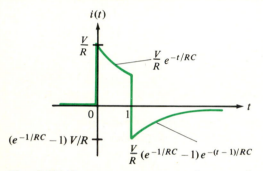

Fig. 3.45 Current response to input pulse.

Example 3.14

For the circuit shown in Fig. 3.38(a), we have already seen (Example 3.13) that

$$\frac{dv}{dt} + 2v = 2v_s$$

Now suppose that the applied voltage $v_s(t)$ is as described by Fig. 3.46(a). For $t < 0$, when $v_s(t) = 9$ V, we have already seen in Example 3.13 that $v(t) = 9$ V. Furthermore, $v(0) = 9$ V as well. However, for $t \ge 0$, we have that $v_s(t) = 5$ V, so the circuit is described by the differential equation

$$\frac{dv}{dt} + 2v = 2v_s = 10 \quad \text{for} \quad t \ge 0$$

The solution to this equation is

$$v(t) = \tfrac{10}{2} + Ae^{-2t} = 5 + Ae^{-2t}$$

Using the fact that $v(0) = 9$ V, substititution of $t = 0$ into this expression yields

$$v(0) = 5 + Ae^{-0} = 9 \quad \Rightarrow \quad A = 4$$

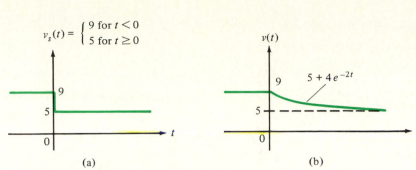

$$v_s(t) = \begin{cases} 9 \text{ for } t < 0 \\ 5 \text{ for } t \geq 0 \end{cases}$$

(a)

(b)

Fig. 3.46 Example of a complete response.

Therefore,

$$v(t) = 5 + 4e^{-2t} \qquad \text{for} \quad t \geq 0$$

A plot of the response $v(t)$ is shown in Fig. 3.46(b).

An alternative approach to the analysis performed in Example 3.14 can be taken by using the fact that the applied voltage can be expressed as

$$v_s(t) = 9 - 4\,u(t) = v_1(t) + v_2(t)$$

where $v_1(t) = 9$ and $v_2(t) = -4\,u(t)$. As has already been discussed, the voltage $v_a(t)$ across the capacitor due to an applied constant voltage $v_1(t) = 9$ V is $v_a(t) = 9$ V.

On the other hand, to determine the voltage $v_b(t)$ across the capacitor due to an applied step of voltage $v_2(t) = -4u(t)$, we use the differential equation

$$\frac{dv_b}{dt} + 2v_b = 2v_2(t) = -8u(t)$$

For $t < 0$ [when $u(t) = 0$], the solution is $v_b(t) = 0$. Since $v_b(t)$ is the voltage across a capacitor, $v_b(0) = 0$.

For $t \geq 0$, the differential equation becomes

$$\frac{dv_b}{dt} + 2v_b = -8$$

and the solution of this has the form

$$v_b(t) = \frac{-8}{2} + Ae^{-2t} = -4 + Ae^{-2t}$$

Substituting $t = 0$ into this expression yields

$$v_b(0) = -4 + Ae^{-0} = 0 \quad \Rightarrow \quad A = 4$$

Thus,

$$v_b(t) = -4 + 4e^{-2t} \quad \text{for} \quad t \geq 0$$

Combining this with the fact that $v_b(t) = 0$ for $t < 0$, we can express the response to $v_2(t) = -4u(t)$ as

$$v_b(t) = (-4 + 4e^{-2t})u(t) = -4(1 - e^{-2t})u(t)$$

Using the property of linearity, we deduce that the response to $v_s(t) = 9 - 4u(t)$ $= v_1(t) + v_2(t)$ is

$$v(t) = v_a(t) + v_b(t) = 9 - 4(1 - e^{-2t})u(t)$$

Clearly, for $t < 0$ [when $u(t) = 0$], $v(t) = 9$; while for $t \geq 0$ [when $u(t) = 1$],

$$v(t) = 9 - 4(1 - e^{-2t}) = 9 - 4 + 4e^{-2t}$$
$$= 5 + 4e^{-2t}$$

This confirms the validity of the alternative approach.

3.6 SECOND-ORDER CIRCUITS

Having dealt with circuits containing either a single inductor or capacitor, we will now consider the case of two energy-storage elements in a circuit. Most such networks, called **second-order circuits,** are described by second-order, linear differential equations. We shall see that the complexity of analysis increases significantly when compared with that encountered with first-order circuits. It is because of this fact that we do not extend such an approach to higher-order circuits, but instead take a different point of view in the following chapter.

The Natural Response

Suppose that for the series RLC circuit shown in Fig. 3.47, the initial current through the inductor is $i(0)$ and the initial voltage across the capacitor is $v(0)$. By KVL,

$$v + Ri + L\frac{di}{dt} = 0$$

Since $i = C\,dv/dt$, we have that

$$v + RC\frac{dv}{dt} + LC\frac{d^2v}{dt^2} = 0$$

from which

$$\frac{d^2v}{dt^2} + \frac{R}{L}\frac{dv}{dt} + \frac{1}{LC}v = 0 \tag{3.21}$$

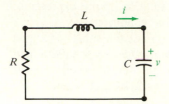

Fig. 3.47 Series *RLC* circuit.

This equation has the following general form:

$$\frac{d^2y(t)}{dt^2} + 2\alpha\frac{dy(t)}{dt} + \omega_n^2 y(t) = 0 \tag{3.22}$$

We have seen that the solution to a first-order homogeneous differential equation is $y(t) = Ae^{st}$, where A and s are constants. However, is this a solution to Equation (3.22)? To find out, substitute this expression into the equation. The result is

$$\frac{d^2}{dt^2}(Ae^{st}) + 2\alpha\frac{d}{dt}(Ae^{st}) + \omega_n^2(Ae^{st}) = 0 \tag{3.23}$$

from which

$$s^2 Ae^{st} + 2\alpha s Ae^{st} + \omega_n^2 Ae^{st} = 0$$

Dividing both sides by Ae^{st} yields

$$s^2 + 2\alpha s + \omega_n^2 = 0 \tag{3.24}$$

If this equality holds, then so does the one given in Equation (3.23). By the quadratic formula, this equality holds when

$$s = \frac{-2\alpha \pm \sqrt{4\alpha^2 - 4\omega_n^2}}{2} = -\alpha \pm \sqrt{\alpha^2 - \omega_n^2}$$

In other words, the two values of s that satisfy Equation (3.24) are

$$s_1 = -\alpha - \sqrt{\alpha^2 - \omega_n^2} \quad \text{and} \quad s_2 = -\alpha + \sqrt{\alpha^2 - \omega_n^2}$$

Thus, $A_1 e^{s_1 t}$ and $A_2 e^{s_2 t}$ satisfy Equation (3.22). As a consequence of this fact, it follows that their sum

$$y(t) = A_1 e^{s_1 t} + A_2 e^{s_2 t}$$

satisfies Equation (3.22). Since this is the most general expression (it contains both

$A_1 e^{s_1 t}$ and $A_2 e^{s_2 t}$) that satisfies Equation (3.22), it is the solution. In this expression, A_1 and A_2 are arbitrary constants determined from initial conditions.

Note that, depending on the relative values of α and ω_n, the values of s_1 and s_2 can be either real or complex numbers. If $\alpha > \omega_n$, then $\alpha^2 - \omega_n^2 > 0$, so s_1 and s_2 are real numbers. This condition is referred to as the **overdamped** case. If $\alpha < \omega_n$, then $\alpha^2 - \omega_n^2 < 0$ (equivalently, $\omega_n^2 - \alpha^2 > 0$), so that s_1 and s_2 are complex numbers. In particular,

$$s_1 = -\alpha - \sqrt{a^2 - \omega_n^2} = -\alpha - \sqrt{-(\omega_n^2 - \alpha^2)} = -\alpha - j\omega_d$$

$$s_2 = -\alpha + \sqrt{a^2 - \omega_n^2} = -\alpha + \sqrt{-(\omega_n^2 - \alpha^2)} = -\alpha + j\omega_d$$

where $j = \sqrt{-1}$ and $\omega_d = \sqrt{\omega_n^2 - \alpha^2}$. This condition is referred to as the **underdamped** case. Finally, the special case that $\alpha = \omega_n$, called the **critically damped** case, results in $s_1 = s_2 = -\alpha$.

Returning to the series *RLC* circuit shown in Fig. 3.47, we have for the second-order differential Equation (3.21)

$$2\alpha = \frac{R}{L} \quad \Rightarrow \quad \alpha = \frac{R}{2L} \quad \text{and} \quad \omega_n^2 = \frac{1}{LC} \quad \Rightarrow \quad \omega_n = \frac{1}{\sqrt{LC}}$$

Suppose that $R = 5\ \Omega$, $L = \frac{1}{2}$ H, and $C = \frac{1}{8}$ F. Also, suppose that $i(0) = 1$ A and $v(0) = 2$ V. Then

$$\alpha = \frac{R}{2L} = 5 > \omega_n = \frac{1}{\sqrt{LC}} = 4$$

so the circuit is overdamped. Since

$$s_1 = -\alpha - \sqrt{a^2 - \omega_n^2} = -8 \quad \text{and} \quad s_2 = -\alpha + \sqrt{a^2 - \omega_n^2} = -2$$

the expression for the voltage $v(t)$ has the form

$$v(t) = A_1 e^{-8t} + A_2 e^{-2t}$$

Setting $t = 0$ and using the initial condition $v(0) = 2$, we get the equation

$$2 = A_1 + A_2 \tag{3.25}$$

Taking the derivative of $v(t)$ yields

$$\frac{dv(t)}{dt} = -8A_1 e^{-8t} - 2A_2 e^{-2t} \tag{3.26}$$

The second initial condition given is $i(0) = 1$, not

$$\left.\frac{dv(t)}{dt}\right|_{t=0} = \frac{dv(0)}{dt}$$

However, referring to the circuit given in Fig. 3.47, we have that

$$i(t) = C\frac{dv(t)}{dt} \quad \Rightarrow \quad i(0) = \frac{1}{8}\frac{dv(0)}{dt} = 1$$

from which

$$\frac{dv(0)}{dt} = 8$$

Setting $t = 0$, Equation (3.26) becomes

$$\frac{dv(0)}{dt} = -8A_1 - 2A_2 = 8 \qquad \text{(3.27)}$$

The solution of the simultaneous equations [(3.25) and (3.27)] is

$$A_1 = -2 \qquad \text{and} \qquad A_2 = 4$$

Thus, the solution of Equation (3.21) is

$$v(t) = -2e^{-8t} + 4e^{-2t} \qquad \text{for} \quad t \geq 0$$

We then have that

$$i(t) = C\frac{dv(t)}{dt} = \frac{1}{8}\frac{d}{dt}(-2e^{-8t} + 4e^{-2t})$$

$$= 2e^{-8t} - e^{-2t} \qquad \text{for} \quad t \geq 0$$

From our expressions for $i(t)$ and $v(t)$, setting $t = 0$, we get

$$v(0) = -2e^{-0} + 4e^{-0} = -2 + 4 = 2 \text{ V}$$

$$i(0) = 2e^{-0} - e^{-0} = 2 - 1 = 1 \text{ A}$$

as required. Plots of $v(t)$ and $i(t)$ for $t \geq 0$ are shown in Fig. 3.48.

   The initial current in the inductor, and hence through each element, is a clockwise current. The initial voltage across the capacitor is a positive value with the polarity indicated. The capacitor, therefore, tends to discharge by producing a counterclockwise current. The net effect is that, due to the inductor current, the capacitor will first charge to a greater voltage before it begins to discharge. The current through the inductor goes to zero when $t \approx 0.116$ s, at which time the capacitor is charged to its maximum value of approximately 2.38 V. After this time, the current becomes counterclockwise (negative in value) and the energy stored is eventually dissipated in the resistor as heat. Note that if the direction of the initial current had the opposite direction (or if the polarity of the initial voltage was opposite), the capacitor would not have charged to a greater value before it discharged.

   In the second of the three cases (underdamping) to be considered, we will see quite different waveforms. Of course, we could change the values of R, L, and C for the series RLC circuit just discussed to get the underdamped case ($\omega_n > \alpha$)— the initial conditions $v(0)$ and $i(0)$ have no effect on the damping of a circuit. However, let us discuss the underdamped case with a different circuit.

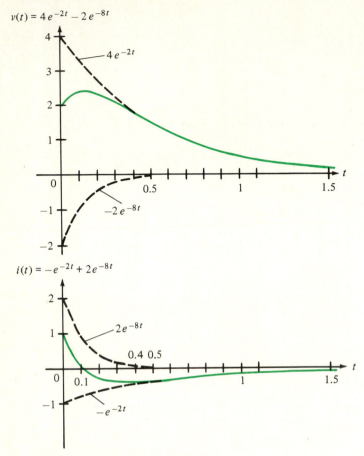

$v(t) = 4e^{-2t} - 2e^{-8t}$

$i(t) = -e^{-2t} + 2e^{-8t}$

Fig. 3.48 Overdamped responses.

The Parallel *RLC* Circuit

For the parallel *RLC* circuit shown in Fig. 3.49, the initial conditions of the circuit are $i(0)$ and $v(0)$. By KCL,

$$\frac{v}{R} + i + C\frac{dv}{dt} = 0$$

Substituting $v = L\, di/dt$ into this equation, we get

$$\frac{d^2i}{dt^2} + \frac{1}{RC}\frac{di}{dt} + \frac{1}{LC}i = 0$$

which has the form of Equation (3.22), where

$$\alpha = \frac{1}{2RC} \quad \text{and} \quad \omega_n = \frac{1}{\sqrt{LC}}$$

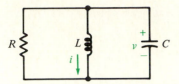

Fig. 3.49 Parallel *RLC* circuit.

(Note that these results could have been obtained from the series *RLC* circuit in Fig. 3.47 by duality.)

Suppose that for this circuit $R = \frac{1}{12}\ \Omega$, $L = \frac{1}{50}$ H, and $C = 2$ F, and the initial conditions are $i(0) = 1$ A and $v(0) = -0.14$ V. Then $\alpha = 3$ and $\omega_n = 5$, so $\omega_n > \alpha$ and the circuit is underdamped. Since

$$\omega_d = \sqrt{\omega_n^2 - \alpha^2} = \sqrt{25 - 9} = 4$$

then

$$s_1 = -\alpha - j\omega_d = -3 - j4 \qquad \text{and} \qquad s_2 = -\alpha + j\omega_d = -3 + j4$$

and the expression for the current $i(t)$ has the form

$$i(t) = A_1 e^{(-3 - j4)t} + A_2 e^{(-3 + j4)t}$$

$$= A_1 e^{-3t} e^{-j4t} + A_2 e^{-3t} e^{j4t}$$

$$= e^{-3t}(A_1 e^{-j4t} + A_2 e^{j4t})$$

Using **Euler's formula** (we will derive this very important result later),

$$\boxed{e^{j\theta} = \cos\theta + j\sin\theta}$$

we get

$$i(t) = e^{-3t}(A_1 \cos 4t - jA_1 \sin 4t + A_2 \cos 4t + jA_2 \sin 4t)$$

$$= e^{-3t}[(A_1 + A_2)\cos 4t + j(-A_1 + A_2)\sin 4t]$$

$$= e^{-3t}(B_1 \cos 4t + B_2 \sin 4t)$$

Letting $t = 0$, we get

$$i(0) = e^{-0}(B_1 \cos 0 + B_2 \sin 0) \quad \Rightarrow \quad 1 = B_1$$

Taking the derivative of $i(t)$ results in

$$\frac{di(t)}{dt} = e^{-3t}(4B_2 \cos 4t - 4B_1 \sin 4t) - 3e^{-3t}(B_1 \cos 4t + B_2 \sin 4t)$$

and letting $t = 0$

$$\frac{di(0)}{dt} = 4B_2 - 3B_1 = 4B_2 - 3 \quad \Rightarrow \quad B_2 = \frac{1}{4}\left(\frac{di(0)}{dt} + 3\right)$$

For the circuit, though, we have that

$$L\frac{di(t)}{dt} = v(t) \quad \Rightarrow \quad \frac{di(t)}{dt} = \frac{1}{L}v(t)$$

so

$$\frac{di(0)}{dt} = 50v(0) = 50(-0.14) = -7$$

Thus,

$$B_2 = \tfrac{1}{4}(-7 + 3) = -1$$

and the expression for the current is

$$i(t) = e^{-3t}(\cos 4t - \sin 4t) \qquad \text{for} \quad t \geq 0$$

Before attempting to sketch this function, let us derive an alternative form for this type of expression.

In the present situation—the underdamped case—the solution to Equation (3.22) has the form

$$y(t) = A_1 e^{s_1 t} + A_2 e^{s_2 t} = A_1 e^{-\alpha t} e^{-j\omega_d t} + A_2 e^{-\alpha t} e^{j\omega_d t}$$

$$= e^{-\alpha t}(A_1 e^{-j\omega_d t} + A_2 e^{j\omega_d t})$$

where $\omega_d = \sqrt{\omega_n^2 - \alpha^2}$. Using Euler's formula and combining constants, we have just seen an example of the fact that for the underdamped case, we can rewrite $y(t)$ in the form

$$y(t) = e^{-\alpha t}(B_1 \cos \omega_d t + B_2 \sin \omega_d t) \tag{3.28}$$

Even so, it is possible further to combine the two sinusoids that have **frequency** ω_d into a single sinusoid of frequency ω_d. One possibility is $B \cos(\omega_d t - \phi)$, where we call B the **amplitude** of the sinusoid and ϕ is the **phase angle**.* Since the angle of a sinusoid is measured in radians and the unit of time is the second, the unit of ω_d is radians per second (rad/s).

Now we want to determine under what conditions the following equality holds:

$$B_1 \cos \omega_d t + B_2 \sin \omega_d t = B \cos(\omega_d t - \phi) \tag{3.29}$$

Using the trigonometric identity for the cosine of the difference of two angles, we get

$$B_1 \cos \omega_d t + B_2 \sin \omega_d t = B(\cos \omega_d t \cos \phi + \sin \omega_d t \sin \phi)$$

$$= B \cos \phi \cos \omega_d t + B \sin \phi \sin \omega_d t$$

*Among other possibilities are $D \sin(\omega_d t + \theta)$ and $E \cos(\omega_d t + \psi)$.

and equality holds when

$$B_1 = B \cos \phi \qquad \text{and} \qquad B_2 = B \sin \phi$$

Thus,

$$\frac{B_2}{B_1} = \frac{B \sin \phi}{B \cos \phi} = \tan \phi \quad \Rightarrow \quad \phi = \tan^{-1}\left(\frac{B_2}{B_1}\right)$$

The angle ϕ can be depicted by the right triangle shown in Fig. 3.50. We then have that

$$B_1 = B \cos \phi \quad \Rightarrow \quad B = \frac{B_1}{\cos \phi} = \frac{B_1}{B_1/\sqrt{B_1^2 + B_2^2}} = \sqrt{B_1^2 + B_2^2}$$

In summary, the conditions for which Equation (3.29) holds are

$$B = \sqrt{B_1^2 + B_2^2} \qquad \text{and} \qquad \phi = \tan^{-1}\left(\frac{B_2}{B_1}\right)$$

and the expression given by Equation (3.28) can be rewritten as

$$y(t) = Be^{-\alpha t} \cos(\omega_d t - \phi) \tag{3.30}$$

where $\omega_d = \sqrt{\omega_n^2 - \alpha^2}$. A typical sketch of $y(t)$ for $t \geq 0$ is shown in Fig. 3.51. Such a curve is called a **damped sinusoid**.

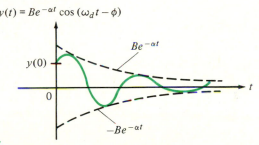

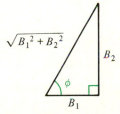

Fig. 3.50 Relationship between B_1, B_2, and ϕ.

Fig. 3.51 Damped sinusoid.

As a consequence of this discussion, we call α the **damping factor,** since its value determines how fast the sinusoid's amplitude diminishes. We call ω_d the **damped natural frequency,** or **damped frequency** for short. For the case that the damping factor $\alpha = 0$, we have that $\omega_d = \sqrt{\omega_n^2 - \alpha^2} = \omega_n$, and we refer to ω_n as the **undamped natural frequency,** or **undamped frequency*** for short.

---

*Although some texts refer to ω_n as the "resonance frequency," except for very simple circuits such as the series and parallel *RLC* circuits, the resonance frequency (to be discussed in Chapter 4) of the circuit is generally different from the undamped frequency.

In the present example, we have that

$$i(t) = e^{-3t}(\cos 4t - \sin 4t) = e^{-3t}(B_1 \cos 4t + B_2 \sin 4t)$$

so

$$B = \sqrt{B_1^2 + B_2^2} = \sqrt{2} \quad \text{and} \quad \phi = \tan^{-1} \frac{B_2}{B_1} = \tan^{-1}\left(\frac{-1}{1}\right) = -\frac{\pi}{4} \quad \text{rad}$$

Thus, we can also write the current as

$$i(t) = \sqrt{2}e^{-3t} \cos(4t + \pi/4) \quad \text{for} \quad t \geq 0$$

The voltage $v(t)$ is

$$v(t) = \frac{1}{50}\frac{d}{dt}[e^{-3t}(\cos 4t - \sin 4t)]$$

$$= \frac{1}{50}[e^{-3t}(-4 \sin 4t - 4 \cos 4t) - 3e^{-3t}(\cos 4t - \sin 4t)]$$

$$= \frac{1}{50}e^{-3t}[-7 \cos 4t - \sin 4t]$$

In this case,

$$B = \sqrt{(-7)^2 + (-1)^2} = 5\sqrt{2}$$

and

$$\phi = \tan^{-1}\left(\frac{-1}{-7}\right) = 3.28 = -3.0 \quad \text{rad}$$

so

$$v(t) = \frac{\sqrt{2}}{10}e^{-3t} \cos(4t + 3.0) \quad \text{for} \quad t \geq 0$$

What's that? Your calculator says that $\tan^{-1}(-1/-7)$ is equal to 0.142 radian (8 degrees)? If it does, be careful. Your calculator cannot distinguish between $\tan^{-1}(-1/-7)$ and $\tan^{-1}(1/7)$—these are different quantities! And it is the former we require in this case. By inspection of Fig. 3.52 we can see that $\tan^{-1}(1/7) = 0.142$ rad and $\tan^{-1}(-1/-7) = 3.28 = -3.0$ rad.

Plots of $i(t)$ and $v(t)$ for $t \geq 0$ are shown in Fig. 3.53(a) and 3.53(b) respectively.

What happens in the underdamped case is that the energy that is stored in the inductor and capacitor eventually gets dissipated by the resistor. However, the energy does not simply go from the L and C directly to the R. Instead, it is transferred back and forth between the two energy storage elements, with the resistance taking its toll during the process. The result of this is a response that is oscillatory (changes sign more than once). In particular, an underdamped response is the product of a real exponential and a sinusoid, that is, a damped sinusoid.

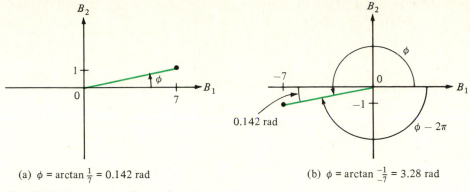

(a) $\phi = \arctan \frac{1}{7} = 0.142$ rad

(b) $\phi = \arctan \frac{-1}{-7} = 3.28$ rad

Fig. 3.52 Determination of arctangent.

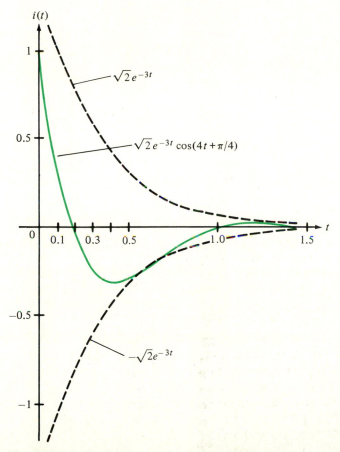

Fig. 3.53(a) Underdamped current response.

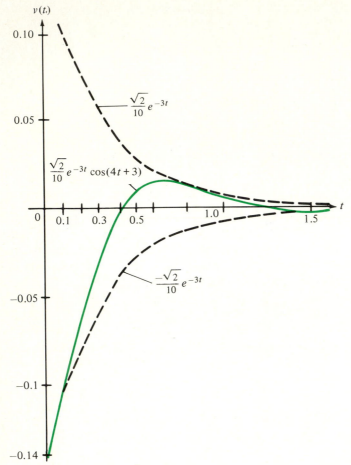

Fig. 3.53(b) Underdamped voltage response.

In the example above, although oscillatory, the responses (current and voltage) become negligible before too many oscillations occurred. This is due to the fact that ω_n is not much larger than α (i.e., the circuit is slightly underdamped). By reducing the damping (making α smaller), the circuit becomes more underdamped, and more oscillations occur before the responses get close to zero. A sketch of an underdamped response where α is relatively small is shown in Fig. 3.54. For the parallel *RLC* circuit when $R = \infty$ or for the series *RLC* circuit when $R = 0$, we have the underdamped case with no damping ($\alpha = 0$). A consequence of this is a response that will be perfectly sinusoidal, and thus will never decrease in amplitude. Such a situation cannot be physically constructed with ordinary capacitors and inductors. There are, however, electronic ways for producing such a result. A device that accomplishes this is known as an **oscillator.**

Having considered the case of overdamping ($\alpha > \omega_n$) and underdamping ($\omega_n > \alpha$), it is now time to investigate the critical damping condition ($\alpha = \omega_n$). We will do this while we are analyzing a new circuit.

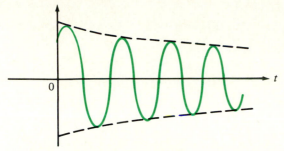

Fig. 3.54 Damped sinusoid having a relatively small damping factor.

A Series-Parallel *RLC* Circuit

We have seen for the series *RLC* circuit and the parallel *RLC* circuit that the natural frequency in both cases is $\omega_n = 1/\sqrt{LC}$. As a consequence, it is frequently assumed that this is a general result for any circuit containing one inductor and one capacitor. The circuit in Fig. 3.55 will show that such an assumption is unwarranted. Applying KVL to the mesh on the left, we get

$$R_1 i + L\frac{di}{dt} + v = 0 \tag{3.31}$$

and applying KCL at the node common to the inductor and the capacitor, we obtain

$$i = C\frac{dv}{dt} + \frac{v}{R_2} \tag{3.32}$$

Substituting Equation (3.32) into Equation (3.31) yields

$$R_1\left(C\frac{dv}{dt} + \frac{v}{R_2}\right) + L\frac{d}{dt}\left(C\frac{dv}{dt} + \frac{v}{R_2}\right) + v = 0$$

from which

$$LC\frac{d^2v}{dt^2} + \left(R_1 C + \frac{L}{R_2}\right)\frac{dv}{dt} + \left(\frac{R_1}{R_2} + 1\right)v = 0$$

or

$$\frac{d^2v}{dt^2} + \left(\frac{R_1}{L} + \frac{1}{R_2 C}\right)\frac{dv}{dt} + \frac{R_1 + R_2}{R_2 LC}v = 0 \tag{3.33}$$

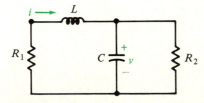

Fig. 3.55 Series-parallel *RLC* circuit.

Thus, for this circuit,

$$\alpha = \frac{R_1}{2L} + \frac{1}{2R_2C} \qquad \text{and} \qquad \omega_n = \sqrt{\frac{R_1 + R_2}{R_2LC}}$$

Suppose that $R_1 = 3 \ \Omega$, $R_2 = 1 \ \Omega$, $L = 1$ H, and $C = 1$ F. Then

$$\alpha = \tfrac{3}{2} + \tfrac{1}{2} = 2 \qquad \text{and} \qquad \omega_n = \sqrt{\tfrac{4}{1}} = 2$$

and since $\alpha = \omega_n$, the circuit is critically damped. However,

$$y(t) = A_1 e^{s_1 t} + A_2 e^{s_2 t} = A_1 e^{-\alpha t} + A_2 e^{-\alpha t} = A e^{-\alpha t}$$

is not the solution to Equation (3.22) for this case. This is so because the two initial conditions $i(0)$ and $v(0)$ cannot be satisfied with the one arbitrary constant A. To determine what the solution is, let us return to Equation (3.22). Using the critically-damped condition $\alpha = \omega_n$, Equation (3.22) becomes

$$\frac{d^2 y(t)}{dt^2} + 2\alpha \frac{dy(t)}{dt} + \alpha^2 y(t) = 0$$

or

$$\frac{d}{dt}\left[\frac{dy(t)}{dt} + \alpha y(t)\right] + \alpha \left[\frac{dy(t)}{dt} + \alpha y(t)\right] = 0 \qquad (3.34)$$

If we define the function $f(t)$ by

$$f(t) = \frac{dy(t)}{dt} + \alpha y(t)$$

then Equation (3.34) becomes

$$\frac{df(t)}{dt} + \alpha f(t) = 0$$

which is a first-order differential equation whose solution is

$$f(t) = A_1 e^{-\alpha t}$$

Thus,

$$\frac{dy(t)}{dt} + \alpha y(t) = A_1 e^{-\alpha t}$$

Multiplying both sides by $e^{\alpha t}$ yields

$$e^{\alpha t}\frac{dy(t)}{dt} + e^{\alpha t}\alpha y(t) = A_1$$

or

$$\frac{d}{dt}[e^{\alpha t} y(t)] = A_1$$

Integrating both sides with respect to t results in

$$\int \frac{d[e^{\alpha t}y(t)]}{dt} dt = \int A_1 dt$$

or

$$e^{\alpha t}y(t) = A_1 t + A_2$$

where A_2 is a constant of integration. Multiplying both sides by $e^{-\alpha t}$ gives

$$y(t) = A_1 te^{-\alpha t} + A_2 e^{-\alpha t} = (A_1 t + A_2)e^{-\alpha t}$$

and this is the solution of Equation (3.22) for the case of critical damping.

Returning now to the series-parallel RLC circuit given in Fig. 3.55, the expression for $v(t)$, that is, the solution of Equation (3.33) for the case of critical damping, is

$$v(t) = A_1 te^{-2t} + A_2 e^{-2t}$$

Given the initial conditions $i(0) = 1$ A and $v(0) = 2$ V, then setting $t = 0$ yields

$$v(0) = 0 + A_2 = 2 \quad \Rightarrow \quad A_2 = 2$$

Next,

$$\frac{dv(t)}{dt} = -2A_1 te^{-2t} + A_1 e^{-2t} - 2A_2 e^{-2t}$$

so

$$\frac{dv(0)}{dt} = A_1 - 2A_2 = A_1 - 4$$

From Equation (3.32),

$$\frac{dv(t)}{dt} = -\frac{v(t)}{R_2 C} + \frac{i(t)}{C} = -v(t) + i(t)$$

Setting $t = 0$

$$\frac{dv(0)}{dt} = -v(0) + i(0) = -2 + 1 = -1$$

so

$$A_1 - 4 = -1 \quad \Rightarrow \quad A_1 = 3$$

Therefore, the expression for $v(t)$ is

$$v(t) = 3te^{-2t} + 2e^{-2t} = (3t + 2)e^{-2t} \quad \text{for} \quad t \geq 0$$

In addition, from Equation (3.32)

$$i(t) = 1\frac{d}{dt}[(3t + 2)e^{-2t}] + (3t + 2)e^{-2t}$$

$$= -2(3t + 2)e^{-2t} + 3e^{-2t} + (3t + 2)e^{-2t}$$

$$= -(3t + 2)e^{-2t} + 3e^{-2t} = (-3t + 1)e^{-2t} \qquad \text{for} \quad t \geq 0$$

Sketches of $v(t)$ and $i(t)$ for $t \geq 0$ are shown in Fig. 3.56.

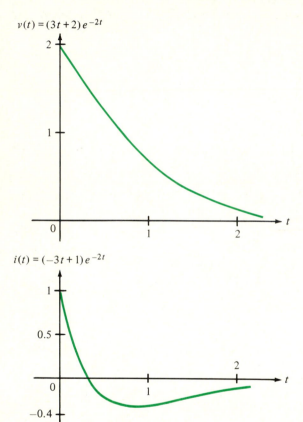

Fig. 3.56 Critically damped responses.

Circuits with Nonzero Inputs

Having considered the natural response for some second-order circuits, let us now look at second-order circuits with nonzero inputs. We will limit our discussion to constant forcing functions. Other types of forcing functions will be dealt with in the next two chapters when additional concepts are introduced.

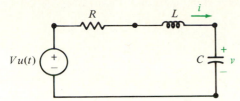

Fig. 3.57 Step of voltage applied to a series *RLC* circuit.

Consider the series *RLC* circuit shown in Fig. 3.57 whose input is a voltage step function. By KVL,

$$Vu(t) = Ri + L\frac{di}{dt} + v$$

Substituting $i = C\dfrac{dv}{dt}$ into this equation, we can obtain

$$\frac{d^2v}{dt^2} + \frac{R}{L}\frac{dv}{dt} + \frac{1}{LC}v = \frac{V}{LC}u(t) \tag{3.35}$$

For $t < 0$, the right-hand side of this equation becomes zero, and for zero initial conditions the solution is $v(t) = 0$. However, for $t \geq 0$, Equation (3.35) becomes

$$\frac{d^2v}{dt^2} + \frac{R}{L}\frac{dv}{dt} + \frac{1}{LC}v = \frac{V}{LC} \tag{3.36}$$

As in the case of first-order differential equations, the solution $v(t)$ consists of two parts—the forced response $v_f(t)$ and the natural response $v_n(t)$, that is, the solution is of the form

$$v(t) = v_f(t) + v_n(t)$$

where $v_n(t)$ is the solution to

$$\frac{d^2v}{dt^2} + 2\alpha\frac{dv}{dt} + \omega_n^2 v = 0$$

where $\alpha = R/2L$ and $\omega_n = 1/\sqrt{LC}$.

In this case, the forcing function is a constant. Thus, the forced response is a constant, say K. If the natural response $v_n(t)$ is substituted into the left side of Equation (3.36), the left side of the equation becomes zero. Thus, the forced response $v_f(t) = K$ must satisfy Equation (3.36), that is,

$$\frac{d^2K}{dt^2} + \frac{R}{L}\frac{dK}{dt} + \frac{1}{LC}K = \frac{V}{LC} \quad \Rightarrow \quad \frac{1}{LC}K = \frac{V}{LC} \quad \Rightarrow \quad K = V$$

Hence, the solution to Equation (3.36) is as follows:
For the overdamped case:

$$v(t) = V + A_1 e^{s_1 t} + A_2 e^{s_2 t}$$

For the underdamped case:

$$v(t) = V + A_1 e^{-\alpha t} \cos \omega_d t + A_2 e^{-\alpha t} \sin \omega_d t$$

For the critically-damped case:

$$v(t) = V + A_1 t e^{-\alpha t} + A_2 e^{-\alpha t}$$

where the constants A_1 and A_2 are determined from the circuit topology and the initial conditions $i(0) = v(0) = 0$.

For the circuit in Fig. 3.57, since no energy initially is stored in the inductor and capacitor, when the input voltage is zero, the voltage $v(t)$ and the current $i(t)$ will be zero. At $t = 0$, when the input voltage becomes V volts, since the voltage across a capacitor and the current through an inductor cannot change instantaneously, $v(0) = 0$ and $i(0) = 0$. After a long time, the input acts as a constant, the inductor behaves as a short circuit, and the capacitor behaves as an open circuit. Thus, eventually, the voltage across the capacitor will be V volts and the current through the inductor will be zero amperes. The shape of the voltage and current waveforms between initial and final values will depend upon whether the circuit is overdamped, underdamped, or critically damped.

Example 3.15

For the series RLC circuit given in Fig. 3.57, the underdamped case results when $R = 12\ \Omega$, $L = 2$ H, and $C = \frac{1}{50}$ F, since $\alpha = 3$ and $\omega_n = 5$. Also, $\omega_d = 4$. If $V = \frac{2}{5}$ V, then the solution of Equation (3.36) is

$$v(t) = \tfrac{2}{5} + A_1 e^{-3t} \cos 4t + A_2 e^{-3t} \sin 4t$$

Setting $t = 0$ yields

$$v(0) = \tfrac{2}{5} + A_1 = 0 \quad \Rightarrow \quad A_1 = -\tfrac{2}{5}$$

Taking the derivative of $v(t)$, we get

$$\frac{dv(t)}{dt} = -3A_1 e^{-3t} \cos 4t - 4A_1 e^{-3t} \sin 4t - 3A_2 e^{-3t} \sin 4t + 4A_2 e^{-3t} \cos 4t$$

so

$$\frac{dv(0)}{dt} = -3A_1 + 4A_2 = \frac{6}{5} + 4A_2$$

Since $i(t) = C\, dv(t)/dt$, then

$$\frac{dv(0)}{dt} = \frac{1}{C} i(0) = 0 = \frac{6}{5} + 4A_2 \quad \Rightarrow \quad A_2 = -\frac{3}{10}$$

Therefore, for $t \geq 0$,

$$v(t) = \tfrac{2}{5} - \tfrac{2}{5} e^{-3t} \cos 4t - \tfrac{3}{10} e^{-3t} \sin 4t$$

$$= \tfrac{2}{5} - \tfrac{1}{10}e^{-3t}(4 \cos 4t + 3 \sin 4t)$$

$$= \tfrac{2}{5} - \tfrac{1}{2}e^{-3t} \cos(4t - 0.64)$$

Combining this with the fact that $v(t) = 0$ for $t < 0$ results in the voltage step response,

$$v(t) = [\tfrac{2}{5} - \tfrac{1}{2}e^{-3t} \cos(4t - 0.64)]u(t)$$

and, therefore, the current for $t \geq 0$ is

$$i(t) = C\frac{dv(t)}{dt} = \frac{1}{50}\frac{d}{dt}\left(\frac{2}{5} - \frac{2}{5}e^{-3t} \cos 4t - \frac{3}{10}e^{-3t} \sin 4t\right)$$

$$= \frac{1}{50}\left(\frac{6}{5}e^{-3t} \cos 4t + \frac{8}{5}e^{-3t} \sin 4t + \frac{9}{10}e^{-3t} \sin 4t - \frac{12}{10}e^{-3t} \cos 4t\right)$$

$$= \frac{1}{50}\left(\frac{25}{10}e^{-3t} \sin 4t\right) \qquad \text{for} \quad t \geq 0$$

Thus, the current step response is

$$i(t) = (\tfrac{1}{20}e^{-3t} \sin 4t)u(t)$$

Sketches of $v(t)$ and $i(t)$ are shown in Fig. 3.58.

For the parallel RLC circuit shown in Fig. 3.59, by KCL,

$$\frac{v}{R} + i + C\frac{dv}{dt} = i_s$$

Substituting $v = L\, di/dt$ into this equation, we get

$$\frac{d^2i}{dt^2} + \frac{1}{RC}\frac{di}{dt} + \frac{1}{LC}i = \frac{1}{LC}i_s \tag{3.37}$$

Again, the solution of this equation has the form

$$i(t) = i_f(t) + i_n(t)$$

where $i_f(t)$ is the forced response and $i_n(t)$ is the natural response. In determining the natural response—whether the circuit is overdamped, underdamped, or critically damped—we use the fact that for a parallel RLC circuit

$$\alpha = \frac{1}{2RC} \qquad \text{and} \qquad \omega_n = \frac{1}{\sqrt{LC}}$$

Suppose that $R = 6\ \Omega$, $L = 7$ H, $C = \tfrac{1}{42}$ F, $i_s(t) = 6\, u(t)$, $v(0) = 0$, and $i(0) = -4$ A. Then Equation (3.37) becomes

$$\frac{d^2i}{dt^2} + 7\frac{di}{dt} + 6i = 36\, u(t)$$

and

$$\alpha = \tfrac{7}{2} > \omega_n = \sqrt{6}$$

$v(t) = [0.4 - 0.5 \, e^{-3t} \cos(4t - 0.64)] \, u(t)$

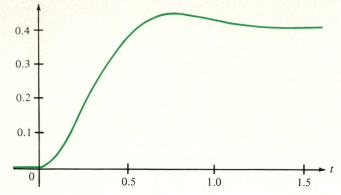

$i(t) = [0.05 \, e^{-3t} \sin 4t] \, u(t)$

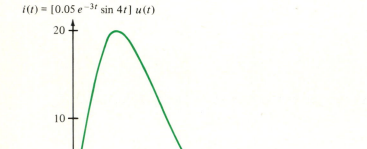

Fig. 3.58 Step responses of an underdamped series *RLC* circuit.

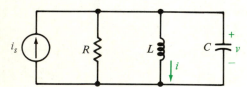

Fig. 3.59 Parallel *RLC* circuit with an applied current.

so the circuit is overdamped. Since

$$s = -\alpha \pm \sqrt{\alpha^2 - \omega_n^2} = -\tfrac{7}{2} \pm \sqrt{\tfrac{49}{4} - 6} = -\tfrac{7}{2} \pm \tfrac{5}{2}$$

then

$$s_1 = -6 \quad \text{and} \quad s_2 = -1$$

and, therefore, the natural response has the form

$$i_n(t) = A_1 e^{-6t} + A_2 e^{-t}$$

For $t \geq 0$, the forcing function is a constant, so the forced response is a constant K that can be obtained by substituting K into the differential equation. Also, from the circuit, for the dc case we have that $i_f(t) = 6$. Hence, the complete response has the form

$$i(t) = 6 + A_1 e^{-6t} + A_2 e^{-t}$$

Setting $t = 0$, we get

$$i(0) = 6 + A_1 + A_2 = -4 \quad \Rightarrow \quad A_1 + A_2 = -10 \qquad \text{(3.38)}$$

But

$$\frac{di(t)}{dt} = -6A_1 e^{-6t} - A_2 e^{-t} \quad \Rightarrow \quad \frac{di(0)}{dt} = -6A_1 - A_2$$

From the circuit, $v(t) = L\, di(t)/dt$. Thus, setting $t = 0$, we get

$$\frac{di(0)}{dt} = \frac{1}{7} v(0) = 0 = -6A_1 - A_2 \qquad \text{(3.39)}$$

From Equations (3.38) and (3.39), we get $A_1 = 2$ and $A_2 = -12$. Therefore, the complete response is

$$i(t) = 6 + 2e^{-6t} - 12e^{-t} \qquad \text{for} \quad t \geq 0$$

SUMMARY

1. The voltage across an inductor is directly proportional to the derivative of the current through it.

2. An inductor behaves as a short circuit to dc.

3. The current through a capacitor is directly proportional to the derivative of the voltage across it.

4. A capacitor behaves as an open circuit to dc.

5. The current through an inductor is directly proportional to the integral of the voltage across it.

6. The voltage across a capacitor is directly proportional to the integral of the current through it.

7. The energy stored in an inductor is $Li^2/2$ joules, while the energy stored in a capacitor is $Cv^2/2$ joules.

8. The current through an inductor cannot change instantaneously. The voltage across a capacitor cannot change instantaneously.

9. A planar circuit and its dual are, in essence, described by the same equations.

10. Writing node and mesh equations for circuits containing inductors and capacitors is done as was done for resistive circuits. Except for simple circuits, the solutions of such equations will be avoided.

11. A circuit containing either a single inductor or a single capacitor is described by a first-order differential equation.

12. If no independent source is present, the differential equation describing a circuit is homogeneous and the solution is the natural response.

13. The complete response is the sum of two terms—one is the forced response and the other has the form of the natural response.

14. The natural response of a second-order circuit is either overdamped, underdamped, or critically damped.

15. An overdamped response is the sum of two (real) exponentials.

16. An underdamped response is a damped sinusoid.

17. A critically-damped response is the sum of an exponential and another exponential (with the same time constant) multiplied by time.

18. As with the case of first-order circuits, the complete response of a second-order circuit is the sum of two terms—one is the forced response and the other has the form of the natural response.

Problems for Section 3.1

3.1 Given the circuit shown in Fig. P3.1(a), suppose that the current $i(t)$ is given by the function in Fig. P3.1(b). Sketch $v(t)$, $w_L(t)$, $p_R(t)$, $v_R(t)$, and $v_s(t)$.

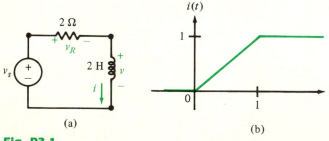

(a)   (b)

Fig. P3.1

3.2 Repeat Problem 3.1 for the current given by the function in Fig. P3.2.

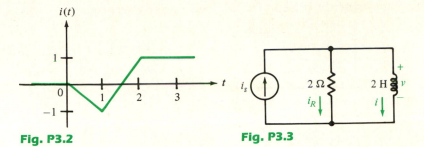

Fig. P3.2   **Fig. P3.3**

3.3 Given the circuit shown in Fig. P3.3, suppose that the current $i(t)$ is given by the function in Fig. P3.1(b). Sketch $v(t)$, $w_L(t)$, $p_R(t)$, $i_R(t)$, and $i_s(t)$.

3.4 Repeat Problem 3.3 for the current given in Fig. P3.2.

3.5 Given the circuit and current shown in Figs. P3.5(a) and P3.5(b), respectively, sketch $v(t)$.

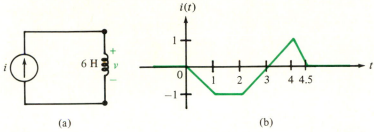

(a)   (b)

Fig. P3.5

3.6 For the circuit and current shown in Figs. P3.6(a) and P3.6(b), respectively, sketch $v_R(t)$, $v_L(t)$, and $v(t)$.

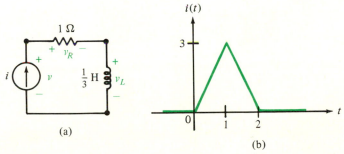

(a)   (b)

Fig. P3.6

Problems for Section 3.2

3.7 Given the circuit and voltage shown in Figs. P3.7(a) and P3.7(b), respectively, sketch $i(t)$, $w_C(t)$, $p_R(t)$, $v_R(t)$, and $v_s(t)$.

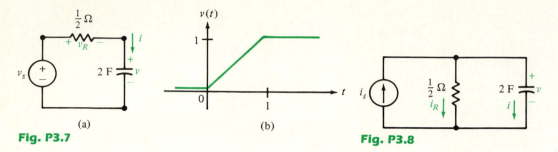

(a)

Fig. P3.7

(b)

Fig. P3.8

3.8 Given the circuit shown in Fig. P3.8, suppose that the voltage $v(t)$ is described by the function given in Fig. P3.7(b). Find $i(t)$, $w_C(t)$, $p_R(t)$, $i_R(t)$, and $i_s(t)$, and sketch these functions.

3.9 Given the op-amp circuit shown in Fig. P3.9, suppose that $v(t)$ is described by the function given in Fig. P3.7(b). Sketch $i(t)$, $i_R(t)$, $v_R(t)$, $v_s(t)$, and $v_o(t)$.

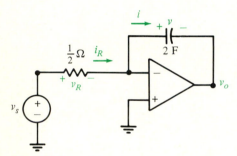

Fig. P3.9

3.10 Repeat Problem 3.9 for the case that there is an additional $\frac{1}{2}$-Ω resistor in parallel with the capacitor.

3.11 Repeat Problem 3.9 for the op-amp circuit shown in Fig. P3.11.

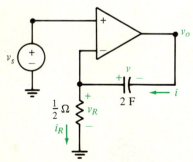

Fig. P3.11

3.12 For the op-amp circuit given in Fig. P3.11, place an additional $\frac{1}{2}$-Ω resistor in parallel with the capacitor and repeat Problem 3.9.

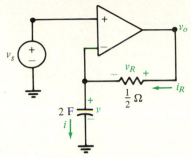

Fig. P3.13

3.13 Repeat Problem 3.9 for the op-amp circuit shown in Fig. P3.13.

3.14 For the op-amp circuit given in Fig. P3.13, place an additional $\frac{1}{2}$-Ω resistor in parallel with the capacitor and repeat Problem 3.9.

Problems for Section 3.3

3.15 The voltage $v(t)$ across a 2-F capacitor at time $t = 1$ s is $\frac{1}{4}$ V. Given that the current through the capacitor is $i(t) = t$ for $1 \le t \le 2$ s and $i(t) = 0$ for $2 < t < \infty$, find $v(t)$ for $t \ge 1$ s.

3.16 The current $i(t)$ through a 3-H inductor at time $t = 1$ s is $\frac{2}{3}$ A. Given that the voltage across the inductor is $v(t) = -1$ for $1 \le t \le 2$ s and $v(t) = 0$ for $2 < t < \infty$, find $i(t)$ for $t \ge 1$ s.

3.17 For the circuit and voltage $v(t)$ described in Figs. P3.17(a) and P3.17(b), respectively, sketch $i(t)$.

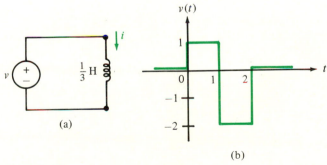

(a)

(b)

Fig. P3.17

3.18 Show the following:
(a) Inductors in series can be combined as in Fig. P3.18(a).
(b) Inductors in parallel can be combined as in Fig. P3.18(b).
(c) Capacitors in parallel can be combined as in Fig. P3.18(c).
(d) Capacitors in series can be combined as in Fig. P3.18(d).

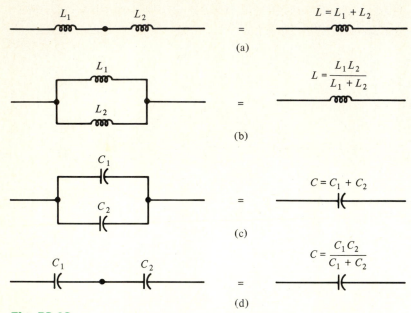

Fig. P3.18

3.19 For the circuit and voltage $v(t)$ shown in Figs. P3.19(a) and P3.19(b), respectively, sketch $i(t)$, $v_R(t)$, and $v_s(t)$.

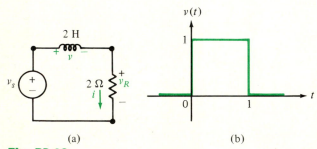

(a)   (b)

Fig. P3.19

3.20 For the circuit shown in Fig. P3.20, suppose that $v(t)$ is the voltage given in Fig. P3.19(b). Sketch $i_R(t)$, $i_L(t)$, and $i(t)$.

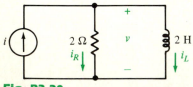

Fig. P3.20

3.21 For the op-amp circuit given in Fig. P3.21, suppose that $v(t)$ is the voltage given in Fig. P3.19(b). Sketch $i_R(t)$, $i_C(t)$, $v_C(t)$, and $v_o(t)$.

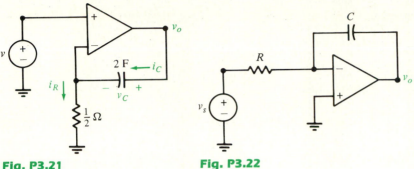

Fig. P3.21   **Fig. P3.22**

3.22 For the integrator circuit given in Fig. P3.22, show that

$$v_o(t) = -\frac{1}{RC} \int_{-\infty}^{t} v_s(t)\, dt$$

Problems for Section 3.4

3.23 For the circuit shown in Fig. P3.23, write a set of node equations in the variables v_1, v_2, and v_3.

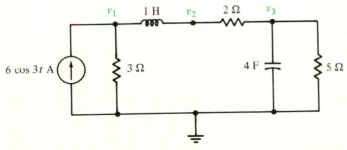

Fig. P3.23

3.24 Repeat Problem 3.23 for the circuit shown in Fig. P3.24.

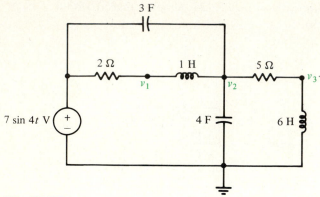

Fig. P3.24

3.25 Assuming clockwise mesh currents, write the mesh equations for the circuit given in Fig. 3.31.

3.26 Find the dual of the circuit given in Fig. P3.23.

3.27 Find the dual of the circuit given in Fig. P3.24.

3.28 Find the dual of the circuit given in Fig. P3.28.

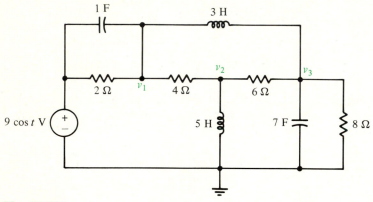

Fig. P3.28

Problems for Section 3.5

3.29 For the circuit shown in Fig. P3.29, the switch opens at time $t = 0$. Write a differential equation in $v(t)$ for $t \geq 0$. Find $v(t)$ and $i(t)$ for all time and sketch these functions.

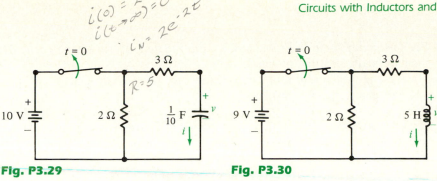

Fig. P3.29   Fig. P3.30

3.30 For the circuit shown in Fig. P3.30, the switch opens at time $t = 0$. Write a differential equation in $i(t)$ for $t \geq 0$. Find $i(t)$ and $v(t)$ for all time and sketch these functions.

3.31 Repeat Problem 3.29 for the circuit given in Fig. P3.31.

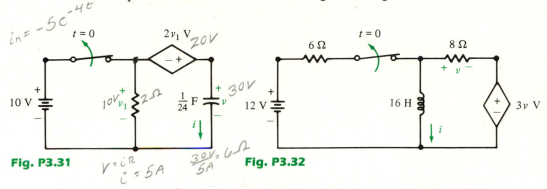

Fig. P3.31   Fig. P3.32

3.32 Repeat Problem 3.30 for the circuit shown in Fig. P3.32.

3.33 For the circuit given in Fig. P3.33, find $v(t)$ and $i(t)$ for all time and sketch these functions.

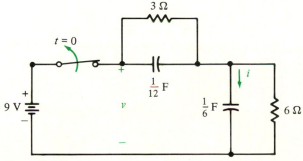

Fig. P3.33

3.34 For the circuit given in Problem 3.33, change the value of the $\frac{1}{12}$-F capacitor to $\frac{1}{3}$ F and repeat the problem.

3.35 For the circuit given in Problem 3.33, replace the $\frac{1}{6}$-F capacitor by a 12-H inductor and repeat the problem.

3.36 For the circuit given in Problem 3.33, replace the $\frac{1}{6}$-F capacitor by a $1\frac{1}{2}$-H inductor and repeat the problem.

3.37 The step responses $v_C(t)$ and $i(t)$ for the series RC circuit shown in Fig. P3.37(a) are given by Equations (3.19) and (3.20) respectively. Use duality to determine the step responses $i_L(t)$ and $v(t)$ for the parallel GL circuit given in Fig. P3.37(b).

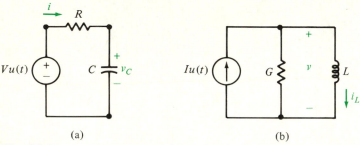

(a)　　　　　　　　　　　　　(b)

Fig. P3.37

3.38 Use Thévenin's theorem to find the step response $v(t)$ for the circuit shown in Fig. P3.38.

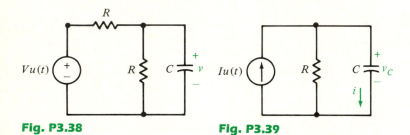

Fig. P3.38　　　　　**Fig. P3.39**

3.39 Apply Norton's theorem to the circuit given in Fig. P3.37(a) to obtain the step responses $v_C(t)$ and $i(t)$ for the parallel RC circuit shown in Fig. P3.39.

3.40 For the series RL circuit shown in Fig. P3.40, suppose that $v_s(t) = Vu(t)$. Write a differential equation in the variable $i(t)$, and find the step responses $i(t)$ and $v(t)$. Sketch these functions.

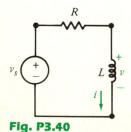

Fig. P3.40

3.41 For the circuit given in Fig. P3.40, suppose that $v_s(t) = Vu(t) - Vu(t-1)$ as shown in Fig. 3.42. Find the responses $v(t)$ and $i(t)$. Sketch these functions.

3.42 Find the step response $v(t)$ for the circuit shown in Fig. P3.42.

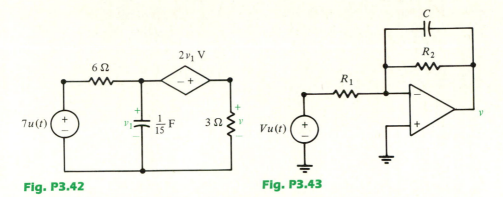

Fig. P3.42   **Fig. P3.43**

3.43 For the op-amp circuit given in Fig. P3.43, find the step response $v(t)$.

3.44 For the circuit shown in Fig. P3.44, find the step response $v(t)$.

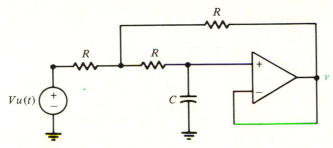

Fig. P3.44

Problems for Section 3.6

3.45 For the circuit shown in Fig. P3.45, find $v(t)$ and $i(t)$ for all time.

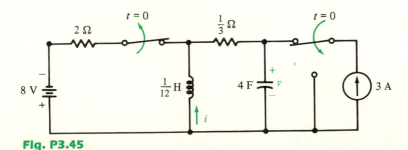

Fig. P3.45

3.46 For the circuit shown in Fig. P3.46, find $v(t)$ for all time.

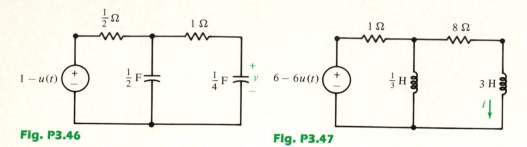

Fig. P3.46

Fig. P3.47

3.47 For the circuit shown in Fig. P3.47, find $i(t)$ for all time.
3.48 For the circuit shown in Fig. P3.48, find $v(t)$ for all time.

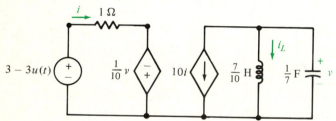

Fig. P3.48

3.49 For the circuit shown in Fig. P3.49, find $v(t)$ and $i(t)$ for all t.

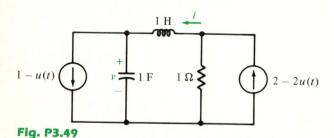

Fig. P3.49

3.50 For the circuit shown in Fig. P3.50, find $v(t)$ and $i(t)$ for all t.

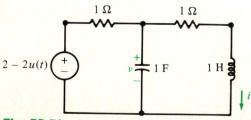

Fig. P3.50

3.51 For the circuit shown in Fig. P3.51, find i(t) and v(t) for all *t*.

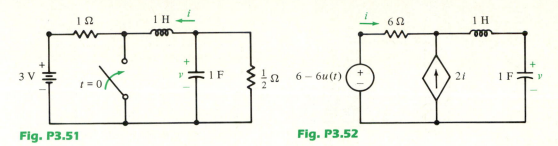

Fig. P3.51 **Fig. P3.52**

3.52 For the circuit shown in Fig. P3.52, find $v(t)$ for all *t*.

3.53 For the circuit shown in Fig. 3.57, suppose that $V = 1$ V, $R = 16\ \Omega$, $L = 4$ H, and $C = \frac{1}{16}$ F. Find the step response $v(t)$.

3.54 For the circuit shown in Fig. P3.46, change the designation of the independent voltage source to $u(t)$ and find the step response $v(t)$.

3.55 For the circuit shown in Fig. P3.50, change the designation of the independent voltage source to $u(t)$ and find the step response $i(t)$.

3.56 For the circuit shown in Fig. P3.52, change the designation of the independent voltage source to $u(t)$ and find the step response $v(t)$.

3.57 For the circuit shown in Fig. P3.48, change the designation of the independent voltage source to $u(t)$ and find the step response $i_L(t)$.

Chapter 4

Sinusoidal Analysis

INTRODUCTION

Step functions are useful in determining the responses of circuits when they are first turned on or when sudden or irregular changes occur in the input; this is called **transient analysis.** However, to see how a circuit responds to a regular or repetitive input—i.e., for **steady-state analysis**—the function that is, by far, the most useful is the sinusoid.

The sinusoid is an extremely important and ubiquitous function. The shape of ordinary household voltage is sinusoidal. Consumer radio transmissions are either AM, in which the **a**mplitude of a sinusoid is changed or **m**odulated according to some information signal, or FM, in which the **f**requency of a sinusoid is **m**odulated. [Consumer television uses AM for the picture (video) and FM for the sound (audio).] For these reasons and more, sinusoidal analysis is a fundamental topic in the study of electric circuits.

In this chapter, we shall see that although we can analyze sinusoidal circuits with the techniques discussed previously, by generalizing the sinusoid, we can perform analysis in a more simplified manner that avoids the direct solution of differential equations.

In our previous analyses, our major concern was with determining voltages and currents. In many applications (e.g., electric utilities), the energy or power supplied and that absorbed are extremely important quantities. Knowing the instantaneous power is useful, since many electric and electronic devices have maximum instantaneous or "peak" power ratings, which, for satisfactory operation, should not be exceeded. By averaging instantaneous power, we obtain average power, which is the average rate at which energy is supplied or absorbed. Power usages vary from a fraction of a watt for small electronic circuits to millions of watts for large electric utilities.

4.1 TIME-DOMAIN ANALYSIS

Of all the functions encountered in electrical engineering, perhaps the most important is the sinusoid.

From trigonometry, recall the plot of $\cos x$ versus the angle x (whose units are radians) shown in Fig. 4.1(a). By changing the angle from x to ωt, we get the plot shown in Fig. 4.1(b), where t is time in seconds, so that ω must be in radians per second (rad/s).* We say that ω is the **radian** or **angular frequency** of $\cos \omega t$. To

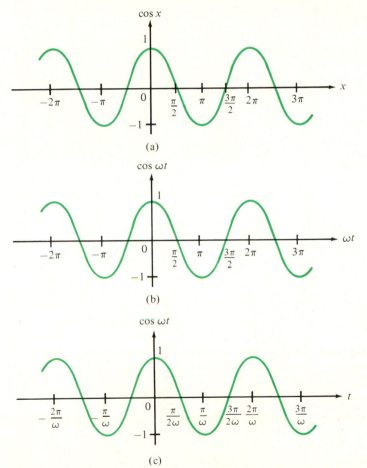

(a)

(b)

(c)

Fig. 4.1 Cosine function versus angle (a and b) and versus time (c).

---

*It has been suggested that the term "radians per second" be replaced by "metz"—abbreviated Mz—in honor of the American electrical engineer Charles P. Steinmetz (1865–1923).

plot cos ωt versus time t, divide the numbers on the horizontal axis (abscissa) by ω. The result is shown in Fig. 4.1(c). From this plot, it is easy to see that this function goes through a complete cycle in $2\pi/\omega$ seconds. We call the time to complete one cycle the **period** of the sinusoid and denote it by T, that is, the number of seconds per cycle is

$$T = \frac{2\pi}{\omega} \quad \text{seconds}$$

The inverse of this quantity is the number of cycles that occur in one second. That is, the number of cycles per second is

$$\frac{1}{T} = \frac{\omega}{2\pi} \quad \frac{1}{\text{seconds}}$$

The number of cycles per second is designated by f and is called the **cyclical** or **ordinary frequency** of the sinusoid. The term "cycles per second" has been replaced by "hertz" (Hz).* Thus, the relationships between cyclical and radian frequencies are

$$f = \frac{\omega}{2\pi} \quad \text{Hz}$$

and

$$\omega = 2\pi f \quad \text{rad/s}$$

To obtain a plot of the general sinusoid $\cos(\omega t + \phi)$ versus t, reconsider the plot of cos ωt versus ωt. Replacing ωt by $\omega t + \phi$, where ϕ is a positive quantity, corresponds to translating (or shifting) the sinusoid to the left by the amount ϕ; that is, we obtain the plot shown in Fig. 4.2. Then in dividing by ω, we get the plot shown in Fig. 4.3. Of course, if ϕ is a negative quantity, the shift is to the right. We call ϕ the **phase angle** or simply **angle** of the sinusoid. Note that when $\phi = -\pi/2$ radians (−90 degrees),

$$\cos(\omega t - \pi/2) = \sin \omega t$$

Similarly,

$$\sin(\omega t + \pi/2) = \cos \omega t$$

Consider a circuit with a sinusoidal input. In general, the response consists of the natural response and the forced response. In particular, let us concentrate on the

---

*Named for the German physicist Heinrich Hertz (1857–1894).

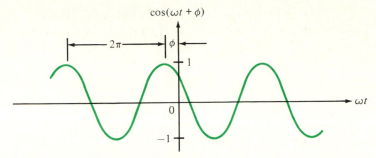

Fig. 4.2 Sinusoid versus angle.

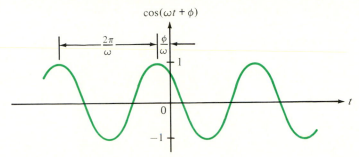

Fig. 4.3 Sinusoid versus time.

forced or steady-state response. We know that the integral or derivative of a sinusoid that has a given frequency is again a sinusoid of that frequency. For a resistor, the voltage and the current differ by a constant, while for an inductor and a capacitor, the relationships between voltage and current are given in terms of integrals and derivatives. Thus, for each of these elements, a sinusoidal input produces a sinusoidal output. Since adding sinusoids of frequency ω results in a single sinusoid of frequency ω [for example, see Equation (3.29)], we can deduce that the forced response of an *RLC* circuit whose input is a sinusoid of frequency ω is also a sinusoid of frequency ω.

Example 4.1

Consider the sinusoidal circuit shown in Fig. 4.4. Since

$$v_s = 1i + v_o \quad \text{and} \quad i = \frac{1}{2}\frac{dv_o}{dt}$$

then

$$\frac{dv_o}{dt} + 2v_o = 12 \sin 2t \qquad \qquad (4.1)$$

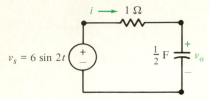

Fig. 4.4 Simple sinusoidal circuit.

Since the forced response is a sinusoid of frequency $\omega = 2$ rad/s, it has the general form $A \cos(2t + \phi)$. However, to avoid using the trigonometric identity for the cosine of the sum of two angles, let us use the alternative form [see Equation (3.29)]:

$$v_o(t) = A_1 \cos 2t + A_2 \sin 2t$$

Substituting this into Equation (4.1) results in

$$(-2A_1 + 2A_2) \sin 2t + (2A_1 + 2A_2) \cos 2t = 12 \sin 2t$$

Equating coefficients yields the simultaneous equations

$$-2A_1 + 2A_2 = 12 \qquad \text{and} \qquad 2A_1 + 2A_2 = 0$$

the solution of which is $A_1 = -3$ and $A_2 = 3$. Hence, the forced response is

$$v_o(t) = -3 \cos 2t + 3 \sin 2t$$

By Equation (3.29),

$$v_o(t) = 3\sqrt{2} \cos(2t - 3\pi/4)$$

Also,

$$v_o(t) = 3\sqrt{2} \cos(2t - \pi/2 - \pi/4) = 3\sqrt{2} \sin(2t - \pi/4)$$

A plot of the functions $v_s(t)$ and $v_o(t)$ versus ωt is shown in Fig. 4.5. Since $v_s(t)$ reaches its peak before $v_o(t)$, we can say that $v_s(t)$ **leads** $v_o(t)$ by $\pi/4$ radians (45°) or that $v_o(t)$ **lags** $v_s(t)$ by $\pi/4$ radians. Since the output of the circuit lags the input, the given circuit is an example of what is called a **lag network**.

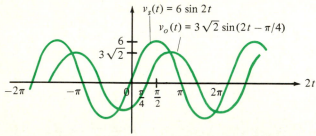

Fig. 4.5 Output voltage $v_o(t)$ lags input voltage $v_s(t)$ by 45°.

The current in this circuit is

$$i(t) = C \frac{dv_o(t)}{dt} = \frac{1}{2} \frac{dv_o(t)}{dt} = \frac{1}{2} (6 \sin 2t + 6 \cos 2t)$$

$$= 3 \cos 2t + 3 \sin 2t = 3\sqrt{2} \cos(2t - \pi/4)$$

$$= 3\sqrt{2} \cos(2t - \pi/2 + \pi/4) = 3\sqrt{2} \sin(2t + \pi/4)$$

Thus, the current $i(t)$ leads the voltage $v_s(t)$ by 45 degrees, and $i(t)$ leads the voltage $v_o(t)$ by 90°.

■

As indicated by the preceding example, sinusoidal analysis is straightforward. However, the amount of arithmetic required can be extremely tedious. A lot of drudgery involved in analyzing sinusoidal circuits can be eliminated if we utilize the concept of a complex sinusoid and the property of linearity. Such an approach not only will be numerically advantageous, but will give rise to concepts that will prove invaluable in the future. To proceed, though, it will be necessary to study complex numbers.

4.2 COMPLEX NUMBERS

As mentioned prevously, we designate the constant $\sqrt{-1}$ by j, that is

$$j = \sqrt{-1}$$

Taking powers of j, we get

$$j^2 = (\sqrt{-1})^2 = -1$$

$$j^3 = j^2 j = -j$$

$$j^4 = j^2 j^2 = (-1)(-1) = 1$$

$$j^5 = j^4 j = j$$

etc.

A **complex number** **A** is a number of the form

$$\mathbf{A} = a + jb$$

where a and b are real numbers. We call a the **real part** of **A**, denoted

$$\mathrm{Re}(\mathbf{A}) = a$$

and we call b the **imaginary part** of **A**, denoted

$$\mathrm{Im}(\mathbf{A}) = b$$

If $a = 0$, we say that **A** is **purely imaginary**, and if $b = 0$, we say that **A** is **purely real.**

Two complex numbers $A_1 = a_1 + jb_1$ and $A_2 = a_2 + jb_2$ are **equal** if both

$$a_1 = a_2$$

and

$$b_1 = b_2$$

The **sum** of two complex numbers is

$$A_1 + A_2 = (a_1 + jb_1) + (a_2 + jb_2) = (a_1 + a_2) + j(b_1 + b_2)$$

and the **difference** is

$$A_1 - A_2 = (a_1 + jb_1) - (a_2 + jb_2) = (a_1 - a_2) + j(b_1 - b_2)$$

The **complex conjugate** of $A = a + jb$ is denoted by A^* and is defined as

$$A^* = a - jb$$

The **product** of two complex numbers is

$$A_1 A_2 = (a_1 + jb_1)(a_2 + jb_2) = a_1 a_2 + ja_1 b_2 + ja_2 b_1 + j^2 b_1 b_2$$
$$= (a_1 a_2 - b_1 b_2) + j(a_1 b_2 + a_2 b_1)$$

and the **quotient** is

$$\frac{A_1}{A_2} = \frac{a_1 + jb_1}{a_2 + jb_2} = \frac{(a_1 + jb_1)(a_2 - jb_2)}{(a_2 + jb_2)(a_2 - jb_2)}$$
$$= \frac{a_1 a_2 + b_1 b_2}{a_2^2 + b_2^2} + j\frac{b_1 a_2 - a_1 b_2}{a_2^2 + b_2^2}$$

Although addition and subtraction of complex numbers are simple operations, multiplication and division are a little more complicated. This is a consequence of the form of a complex number, that is, a complex number A can be written as $A = a + jb$. We call this the **rectangular form** of the complex number A. Fortunately, there is an alternative form for a complex number that lends itself very nicely to multiplication and division, although not to addition and subtraction. To develop this form, we first derive Euler's[†] formula, which was mentioned earlier.

Let θ be a real variable. Define the function $f(\theta)$ by

$$f(\theta) = \cos \theta + j \sin \theta$$

Taking the derivative with respect to θ yields

$$\frac{df(\theta)}{d\theta} = -\sin \theta + j \cos \theta = j^2 \sin \theta + j \cos \theta$$
$$= j (\cos \theta + j \sin \theta) = jf(\theta)$$

[†]Named for the Swiss mathematician Leonhard Euler (1707–1783).

from which

$$j = \frac{1}{f(\theta)} \frac{df(\theta)}{d\theta}$$

Integrating both sides with respect to θ results in

$$\int j \, d\theta = \int \frac{1}{f(\theta)} \frac{df(\theta)}{d\theta} \, d\theta = \int \frac{df(\theta)}{f(\theta)}$$

Integrating yields

$$j\theta + K = \ln f(\theta) = \ln(\cos \theta + j \sin \theta)$$

where K is a constant of integration. Setting $\theta = 0$ gives

$$K = \ln 1 = 0$$

so

$$j\theta = \ln(\cos \theta + j \sin \theta)$$

Taking powers of e, we get

$$\boxed{e^{j\theta} = \cos \theta + j \sin \theta}$$

(4.2)

and this result is known as **Euler's formula.** Replacing θ by $-\theta$, we obtain another version:

$$\boxed{e^{-j\theta} = \cos \theta - j \sin \theta}$$

(4.3)

Thus, we see that $e^{j\theta}$ is a complex number such that

$$\text{Re}(e^{j\theta}) = \cos \theta \qquad \text{and} \qquad \text{Im}(e^{j\theta}) = \sin \theta$$

By adding and subtracting Equations (4.2) and (4.3), we can get, respectively,

$$\boxed{\cos \theta = \frac{1}{2}(e^{j\theta} + e^{-j\theta}) \qquad \text{and} \qquad \sin \theta = \frac{1}{2j}(e^{j\theta} - e^{-j\theta})}$$

(4.4)

If M is a constant, then

$$Me^{j\theta} = M(\cos \theta + j \sin \theta) = M \cos \theta + jM \sin \theta$$

is a complex number where

$$\text{Re}(Me^{j\theta}) = M \cos \theta \qquad \text{and} \qquad \text{Im}(Me^{j\theta}) = M \sin \theta$$

Given a (nonzero) complex number $\mathbf{A} = a + jb$, is there a complex number of the form $Me^{j\theta}$ (where $M > 0$) that equals $\mathbf{A}$? If there is, then

$$a + jb = M \cos \theta + jM \sin \theta$$

or

$$a = M \cos \theta \quad \text{and} \quad b = M \sin \theta$$

so

$$\cos \theta = a/M \quad \text{and} \quad \sin \theta = b/M$$

from which

$$\tan \theta = \frac{\sin \theta}{\cos \theta} = \frac{b/M}{a/M} = \frac{b}{a} \quad \Rightarrow \quad \theta = \tan^{-1}\left(\frac{b}{a}\right)$$

Pictorially, we have the right triangle shown in Fig. 4.6. Thus,

$$a = M \cos \theta \quad \Rightarrow \quad M = \frac{a}{\cos \theta} = \frac{a}{a/\sqrt{a^2 + b^2}} = \sqrt{a^2 + b^2}$$

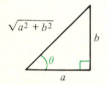

Fig. 4.6 Right triangle describing angle θ.

In summary,

$$\boxed{a + jb = Me^{j\theta}}$$

when

$$\boxed{\begin{array}{c} M = \sqrt{a^2 + b^2} \\ \theta = \tan^{-1}(b/a) \end{array}} \quad \text{or} \quad \boxed{\begin{array}{c} a = M \cos \theta \\ b = M \sin \theta \end{array}}$$

We say that $Me^{j\theta}$ is the **exponential** or **polar form** of a (nonzero) complex number, where the positive number M is called the **magnitude** and θ is called the **angle** (or **argument**) of the complex number. We denote the magnitude of $\mathbf{A}$ by $|\mathbf{A}|$ and the angle of $\mathbf{A}$ by $\text{ang}(\mathbf{A})$.

Note that by using the polar forms of the complex numbers $\mathbf{A}_1 = M_1 e^{j\theta_1}$ and $\mathbf{A}_2 = M_2 e^{j\theta_2}$, the product is

$$\mathbf{A}_1 \mathbf{A}_2 = (M_1 e^{j\theta_1})(M_2 e^{j\theta_2}) = M_1 M_2 e^{j(\theta_1 + \theta_2)}$$

Thus,

$$|\mathbf{A}_1\mathbf{A}_2| = M_1M_2 = |\mathbf{A}_1||\mathbf{A}_2|$$

and

$$\text{ang}(\mathbf{A}_1\mathbf{A}_2) = \theta_1 + \theta_2 = \text{ang}(\mathbf{A}_1) + \text{ang}(\mathbf{A}_2)$$

The quotient is

$$\frac{\mathbf{A}_1}{\mathbf{A}_2} = \frac{M_1e^{j\theta_1}}{M_2e^{j\theta_2}} = \frac{M_1}{M_2}e^{j(\theta_1 - \theta_2)}$$

Thus,

$$\left|\frac{\mathbf{A}_1}{\mathbf{A}_2}\right| = \frac{M_1}{M_2} = \frac{|\mathbf{A}_1|}{|\mathbf{A}_2|}$$

and

$$\text{ang}\left(\frac{\mathbf{A}_1}{\mathbf{A}_2}\right) = \theta_1 - \theta_2 = \text{ang}(\mathbf{A}_1) - \text{ang}(\mathbf{A}_2)$$

Note that

$$(Me^{j\theta})^* = (M \cos \theta + jM \sin \theta)^* = M \cos \theta - jM \sin \theta = Me^{-j\theta}$$

Furthermore, if $\mathbf{A}_1 = \mathbf{A}_2$, then $M_1 = M_2$ and $\theta_1 = \theta_2$.

Example 4.2

If

$$\mathbf{A}_1 = -3 + j4 = M_1e^{j\theta_1}$$

then

$$|\mathbf{A}_1| = M_1 = \sqrt{(-3)^2 + (4)^2} = 5$$

and

$$\theta_1 = \tan^{-1}(4/-3) = 2.21 \text{ rad} = 126.9°$$

while if

$$\mathbf{A}_2 = -5 - j12 = M_2e^{j\theta_2}$$

then

$$|\mathbf{A}_2| = M_2 = \sqrt{(-5)^2 + (-12)^2} = 13$$

and

$$\theta_2 = \tan^{-1}(-12/-5) = -1.97 \text{ rad} = -112.6°$$

Thus,

$$\mathbf{A_1 A_2} = 5 \, e^{j126.9°} \, 13 e^{-j112.6°}$$

$$= 5(13)e^{j(126.9° - 112.6°)} = 65e^{j14.3°}$$

Also,

$$\mathbf{A_1 A_2} = 65 \cos(14.3°) + j65 \sin(14.3°) = 63 + j16$$

4.3 FREQUENCY-DOMAIN ANALYSIS

Given an *RLC* circuit whose input is the sinusoid $A \cos(\omega t + \theta)$, the forced response is a sinusoid of frequency ω, say $B \cos(\omega t + \phi)$. If the input is shifted by 90°, that is, if the input is $A \sin(\omega t + \theta)$, then by the time-invariance property, the corresponding response is $B \sin(\omega t + \phi)$. Furthermore, if the input is scaled by a constant K, then by the property of linearity, so will be the response; that is, the response to $KA \sin(\omega t + \theta)$ is $KB \sin(\omega t + \phi)$. Again, by linearity, the response to

$$A \cos(\omega t + \theta) + KA \sin(\omega t + \theta)$$

is

$$B \cos(\omega t + \phi) + KB \sin(\omega t + \phi)$$

Consider now the special case that the constant $K = \sqrt{-1} = j$. Thus, the response to

$$A \cos(\omega t + \theta) + jA \sin(\omega t + \theta) = Ae^{j(\omega t + \theta)}$$

is

$$B \cos(\omega t + \phi) + jB \sin(\omega t + \phi) = Be^{j(\omega t + \phi)}$$

We, therefore, see that whether we wish to find the response to $A \cos(\omega t + \theta)$ or the response to the **complex sinusoid** $Ae^{j(\omega t + \theta)}$, we must determine the same two constants B and ϕ.

Although it may be somewhat easier computationally to find the response to a (real) sinusoid by considering the corresponding complex sinusoidal case instead, the determination of the appropriate differential equation is still required. By introducing the appropriate concepts, however, we will be able to eliminate this requirement, and the result will be the simplification of sinusoidal analysis.

Just as the sinusoid $A \cos(\omega t + \theta)$ has frequency ω, amplitude A, and phase angle θ, we can say that the complex sinusoid

$$Ae^{j(\omega t + \theta)} = Ae^{j\theta} e^{j\omega t}$$

also has frequency ω, amplitude A, and phase angle θ. Let us denote the complex number

$$\mathbf{A} = Ae^{j\theta}$$

by

$$\mathbf{A} = A\underline{/\theta}$$

which is known as the **phasor** representation of the (real or complex) sinusoid.

Why do we do this? We begin to answer this question by considering a resistor. If the current is a complex sinusoid, say,

$$i = Ie^{j(\omega t + \theta)}$$

then so is the voltage, say,

$$v = Ve^{j(\omega t + \phi)}$$

Thus, by Ohm's law,

$$v = Ri$$

$$Ve^{j(\omega t + \phi)} = RIe^{j(\omega t + \phi)}$$

$$Ve^{j\omega t}e^{j\phi} = RIe^{j\omega t}e^{j\theta}$$

and dividing by $e^{j\omega t}$ yields

$$Ve^{j\phi} = RIe^{j\theta}$$

From this equation, we deduce that $V = RI$ and $\phi = \theta$. Using the phasor forms of the voltage and current, we can write

$$V\underline{/\phi} = RI\underline{/\theta}$$

Thus, we have the phasor equation

$$\boxed{\mathbf{V} = R\mathbf{I}}$$

where $\mathbf{V} = V\underline{/\phi}$ and $\mathbf{I} = I\underline{/\theta}$. Since $\phi = \theta$, the voltage and the current are in phase for a resistor.

For an inductor, if

$$i = Ie^{j(\omega t + \theta)}$$

and

$$v = Ve^{j(\omega t + \phi)}$$

then

$$v = L\frac{di}{dt}$$

$$Ve^{j(\omega t + \phi)} = L\frac{d}{dt}\left(Ie^{j(\omega t + \theta)}\right)$$

$$Ve^{j\omega t}\,e^{j\phi} = j\omega LIe^{j\omega t}e^{j\theta}$$

so

$$Ve^{j\phi} = j\omega LIe^{j\theta}$$

and using the phasor form of voltage and current

$$V\underline{/\phi} = j\omega LI\underline{/\theta}$$

Therefore,

$$\boxed{\mathbf{V} = j\omega L\mathbf{I}}$$

where $\mathbf{V} = V\underline{/\phi}$ and $\mathbf{I} = I\underline{/\theta}$. Note that since

$$\text{ang}(\mathbf{V}) = \text{ang}(j\omega L\mathbf{I}) = \text{ang}(j\omega L) + \text{ang}(\mathbf{I})$$

then

$$\phi = 90° + \theta$$

Thus, we see that for an inductor, the voltage leads the current by 90°, or equivalently, the current lags the voltage by 90°.

For a capacitor, if

$$i = Ie^{j(\omega t + \theta)}$$

and

$$v = Ve^{j(\omega t + \phi)}$$

then

$$i = C\frac{dv}{dt}$$

$$Ie^{j(\omega t + \theta)} = C\frac{d}{dt}(Ve^{j(\omega t + \phi)})$$

$$Ie^{j\omega t}e^{j\theta} = j\omega CVe^{j\omega t}e^{j\phi}$$

so

$$Ie^{j\theta} = j\omega CVe^{j\phi}$$

and

$$I\underline{/\theta} = j\omega CV\underline{/\phi} \quad \Rightarrow \quad \mathbf{I} = j\omega C\mathbf{V}$$

from which

$$\boxed{\mathbf{V} = \frac{1}{j\omega C}\mathbf{I}}$$

Using reasoning as was used for the inductor, we deduce that for a capacitor, the current leads the voltage (the voltage lags the current) by 90°. Specifically,

$$\theta = 90° + \phi$$

or

$$\phi = \theta - 90°$$

(If you would like an easy way to remember whether current leads or lags voltage for a capacitor or an inductor, just think of the word "capacitor" as the key. Let the first three letters stand for "**c**urrent **a**nd **p**otential," thus indicating that current comes before (leads) voltage for a capacitor. This, of course, implies that current lags voltage for an inductor.)

Since a phasor is a complex number, we can represent it as a point in the complex-number plane, and therefore, specify it by an arrow directed from the origin to the point. For example, for a resistor R having a voltage phasor $\mathbf{V} = V\underline{/\phi}$ and a current phasor $\mathbf{I} = I\underline{/\phi}$, a diagram characterizing the relationship between these two phasors is as shown in Fig. 4.7(a). We refer to this as a **phasor diagram.** Note that because $\phi = \theta$ for a resistor, the voltage and current phasors are colinear.

The phasor diagram for an L-henry inductor is shown in Fig. 4.7(b). Given the voltage phasor $\mathbf{V} = V\underline{/\phi}$ and the current phasor $\mathbf{I} = I\underline{/\theta}$, the phasor diagram

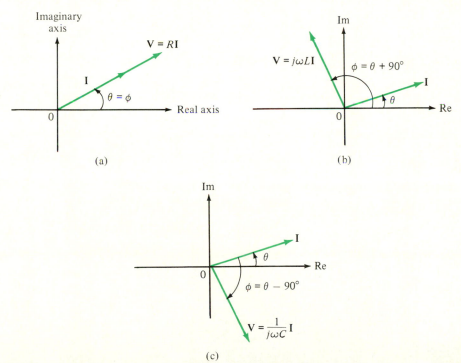

Fig. 4.7 Phasor diagrams for the (a) resistor, (b) inductor, and (c) capacitor.

displays the fact that the voltage leads the current by 90°—this is indicated by the voltage phasor having an angle that is 90° greater than the current phasor.

Figure 4.7(c) shows the phasor diagram for a C-farad capacitor whose voltage phasor is $\mathbf{V} = V\underline{/\phi}$ and whose current phasor is $\mathbf{I} = I\underline{/\theta}$. In this case, the current angle is 90° greater than the voltage angle (i.e., $\theta = 90° + \phi$), so the current leads the voltage by 90°.

Impedance and Admittance

We have seen that for the resistor, inductor, and capacitor, we get a phasor equation of the form

$$\mathbf{V} = \mathbf{ZI}$$

and we call the quantity $\mathbf{Z}$ the **impedance** of the element. In other words, the impedance of an R-ohm resistor, L-henry inductor, and C-farad capacitor are, respectively,

$$\mathbf{Z}_R = R, \qquad \mathbf{Z}_L = j\omega L, \qquad \mathbf{Z}_C = \frac{1}{j\omega C}$$

The equation

$$\mathbf{V} = \mathbf{Z}\,\mathbf{I}$$

is the phasor version of Ohm's law. From this equation, we have that impedance is the ratio of voltage phasor to current phasor, that is,

$$\mathbf{Z} = \mathbf{V}/\mathbf{I}$$

Furthermore,

$$\mathbf{I} = \mathbf{V}/\mathbf{Z}$$

In this expression, the reciprocal of impedance appears. As this can occur frequently, it is useful to give it a name. We call the ratio of current phasor to voltage phasor **admittance** and denote it by the symbol $\mathbf{Y}$, that is,

$$\mathbf{Y} = \mathbf{I}/\mathbf{V}$$

from which

$$\mathbf{Y}_R = \frac{1}{R}, \qquad \mathbf{Y}_L = 1/j\omega L, \qquad \mathbf{Y}_C = j\omega C$$

Thus,

$$\mathbf{I} = \mathbf{YV} \qquad \text{and} \qquad \mathbf{V} = \mathbf{I}/\mathbf{Y}$$

As for the resistive case, the units of impedance are ohms and those of admittance are mhos (or siemens).

Example 4.3

Given a sinusoid whose frequency is $\omega = 100$ rad/s, for a 25-Ω resistor

$$\mathbf{Z}_R = 25\ \Omega, \qquad \mathbf{Y}_R = \tfrac{1}{25} = 0.04\ \mho$$

for a $\frac{1}{20}$-H inductor

$$\mathbf{Z}_L = j5\ \Omega, \quad \mathbf{Y}_L = \frac{1}{j5} = -j(0.2)\ \mho$$

and for a $\frac{1}{50}$-F capacitor

$$\mathbf{Z}_C = \frac{1}{j2} = -j(0.5)\ \Omega, \quad \mathbf{Y}_C = j2\ \mho$$

Suppose that for a sinusoidal *RLC* circuit, writing KVL around a loop results in the equation

$$v_1(t) + v_2(t) + v_3(t) + \cdots + v_n(t) = 0$$

For the case of a real sinusoid, this equation takes the form

$$V_1 \cos(\omega t + \theta_1) + V_2 \cos(\omega t + \theta_2) + \cdots + V_n \cos(\omega t + \theta_n) = 0$$

while for the case of a complex sinusoid,

$$V_1 e^{j(\omega t + \theta_1)} + V_2 e^{j(\omega t + \theta_2)} + \cdots + V_n e^{j(\omega t + \theta_n)} = 0$$

Dividing both sides of the preceding equation by $e^{j\omega t}$ yields

$$V_1 e^{j\theta_1} + V_2 e^{j\theta_2} + \cdots + V_n e^{j\theta_n} = 0$$

or using phasor notation

$$V_1 \underline{/\theta_1} + V_2 \underline{/\theta_2} + \cdots + V_n \underline{/\theta_n} = 0$$

so

$$\mathbf{V}_1 + \mathbf{V}_2 + \cdots + \mathbf{V}_n = 0$$

where $\mathbf{V}_i = V_i \underline{/\theta_i}$. Thus, we see that in addition to voltage functions of time, KVL holds for voltage phasors. We similarly can draw the conclusion that KCL holds for current phasors as well as for time functions.

For inductors and capacitors, as well as resistors, the phasor relationships between voltage and current are

$$\mathbf{V} = \mathbf{Z}\,\mathbf{I}, \qquad \mathbf{I} = \mathbf{V}/\mathbf{Z}, \qquad \mathbf{Z} = \mathbf{V}/\mathbf{I}$$

That is, these are the phasor versions of Ohm's law. By using KVL and KCL, we can analyze a sinusoidal circuit without resorting to differential equations—we simply look at the circuit from the phasor point of view and then proceed as was done for resistive circuits using nodal or mesh analysis. For resistive circuits, we have to manipulate real numbers; for sinusoidal circuits, complex-number arithmetic is required.

If we analyze a circuit by using time functions and the various differential and integral relationships between voltage and current—and, hence, are required to solve differential equations—we say that we are working in the **time domain.** However, if we have a sinusoidal circuit, we can use the phasor representations of the time functions and the impedances (or admittances) of the various (nonsource) elements or combinations of elements. This is said to be the **frequency-domain** approach.

Making the transformation from the time to the frequency domain is a simple matter. Just express a function of time by its phasor representation and resistors, inductors, and capacitors (or combinations of them) by their impedances or admittances. Proceed as for resistive circuits to find the phasor representations of the desired responses. The transformation back to the time domain then is done by inspection.

Example 4.4

Reconsider the circuit given in Fig. 4.4. This circuit [note that $6\cos(2t - 90°) = 6\sin 2t$] is shown in the frequency domain in Fig. 4.8. By KVL,

$$\mathbf{V}_s = 1\mathbf{I} - j\mathbf{I}$$

$$6\underline{/-90°} = (1 - j)\mathbf{I} = \sqrt{2}\underline{/-45°}\,\mathbf{I}$$

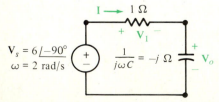

Fig. 4.8 Frequency-domain representation of circuit in Fig. 4.4.

from which

$$I = \frac{6\underline{/-90°}}{\sqrt{2}\underline{/-45°}} = 3\sqrt{2}\underline{/-45°}$$

Also,

$$V_o = -jI = (1\underline{/-90°})(3\sqrt{2}\underline{/-45°}) = 3\sqrt{2}\underline{/-135°}$$

The phasor diagram for this circuit is shown in Fig. 4.9. From this diagram, we note that the capacitor voltage lags applied voltage by 45°. Therefore, if the capacitor voltage is considered to be the output voltage, then the circuit is a lag network. Note, however, that the current leads the applied voltage by 45°. Since $V_1 = 1I = I$, then the voltage across the resistor leads the applied voltage by 45°. Consequently, if the voltage across the resistor is considered to be the output, then the given circuit is a **lead network.**

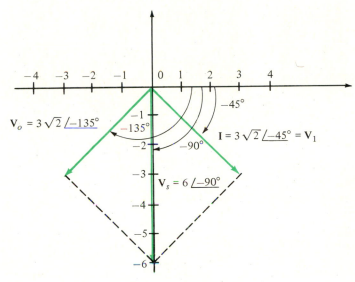

Fig. 4.9 Phasor diagram for given circuit.

For this circuit, by KVL,

$$V_s = V_1 + V_o$$

The phasor diagram shows that, indeed, V_s is the (complex-number) sum of V_o and V_1.

Having found the voltage phasor V_o and the current phasor I, we can write the corresponding time functions, respectively,

$$v_o(t) = 3\sqrt{2} \cos(2t - 135°) \quad \text{and} \quad i(t) = 3\sqrt{2} \cos(2t - 45°)$$

Note that we can also obtain $\mathbf{V}_o$ by voltage division as follows:

$$\mathbf{V}_o = \frac{\mathbf{Z}_C}{\mathbf{Z}_C + \mathbf{Z}_R} \mathbf{V}_s = \frac{-j}{-j + 1} (6\underline{/-90°}) = \frac{1\underline{/-90°}}{\sqrt{2}\underline{/-45°}} (6\underline{/-90°}) = 3\sqrt{2}\underline{/-135°}$$

Some Frequency-Domain Principles

The voltage-divider formula used in Example 4.4 can be derived as was done for resistive circuits by considering the situation depicted in Fig. 4.10. By KVL,

$$\mathbf{V} = \mathbf{Z}_1\mathbf{I} + \mathbf{Z}_2\mathbf{I} = (\mathbf{Z}_1 + \mathbf{Z}_2)\mathbf{I}$$

from which

$$\mathbf{I} = \frac{\mathbf{V}}{\mathbf{Z}_1 + \mathbf{Z}_2} \tag{4.5}$$

and since $\mathbf{V}_2 = \mathbf{Z}_2\mathbf{I}$, then

$$\mathbf{V}_2 = \frac{\mathbf{Z}_2}{\mathbf{Z}_1 + \mathbf{Z}_2} \mathbf{V}$$

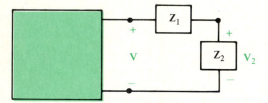

Fig. 4.10 Voltage division.

Note also that the current through the series connection of $\mathbf{Z}_1$ and $\mathbf{Z}_2$ is given by Equation (4.5). That is, the series combination behaves as a single impedance $\mathbf{Z}$, where $\mathbf{Z} = \mathbf{Z}_1 + \mathbf{Z}_2$. This is shown in Fig. 4.11.

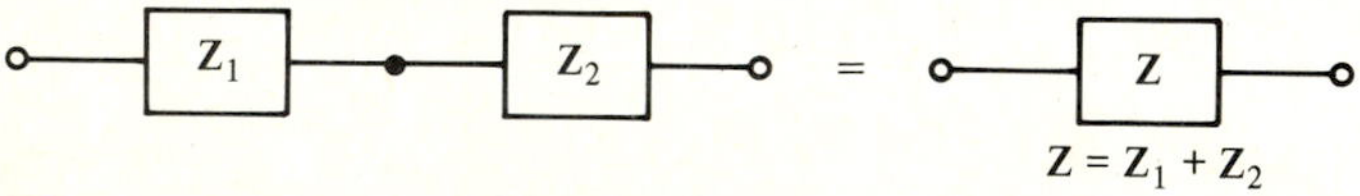

$$Z = Z_1 + Z_2$$

Fig. 4.11 Combining impedances in series.

Consider now the case of current division depicted in Fig. 4.12. Since

$$\mathbf{I} = \frac{\mathbf{V}}{\mathbf{Z}_1} + \frac{\mathbf{V}}{\mathbf{Z}_2} = \left(\frac{1}{\mathbf{Z}_1} + \frac{1}{\mathbf{Z}_2}\right)\mathbf{V}$$

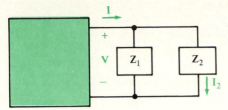

Fig. 4.12 Current division.

then

$$V = \frac{I}{(1/\mathbf{Z}_1) + (1/\mathbf{Z}_2)} = \frac{\mathbf{Z}_1\mathbf{Z}_2 I}{\mathbf{Z}_1 + \mathbf{Z}_2} \tag{4.6}$$

But $\mathbf{I}_2 = \mathbf{V}/\mathbf{Z}_2$, so

$$\mathbf{I}_2 = \frac{\mathbf{Z}_1}{\mathbf{Z}_1 + \mathbf{Z}_2} \mathbf{I}$$

is the current-division formula. From Equation (4.6), we have that the impedance of the parallel connection of $\mathbf{Z}_1$ and $\mathbf{Z}_2$ is

$$\mathbf{Z} = \frac{\mathbf{V}}{\mathbf{I}} = \frac{1}{(1/\mathbf{Z}_1) + (1/\mathbf{Z}_2)} = \frac{\mathbf{Z}_1\mathbf{Z}_2}{\mathbf{Z}_1 + \mathbf{Z}_2}$$

Thus, two impedances in parallel behave as a single impedance $\mathbf{Z}$, where $1/\mathbf{Z} = 1/\mathbf{Z}_1 + 1/\mathbf{Z}_2$. This is shown pictorially in Fig. 4.13. We now conclude that impedances in series and parallel combine as resistances do. Using dual arguments, we may deduce that admittances in parallel and series combine as conductances do. This is demonstrated in Fig. 4.14.

In addition to voltage and current division, other resistive network concepts apply equally to sinusoidal circuits, if such networks are described in the frequency domain, that is, in terms of phasors and impedances (or admittances). One very important example is Thévenin's theorem, which says, in essence, that (with a few exceptions, of course) a circuit is equivalent to an appropriate voltage source in series with an appropriate impedance. This situation is depicted in Fig. 4.15.

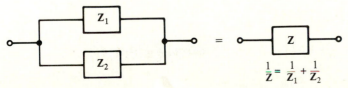

Fig. 4.13 Combining impedances in parallel.

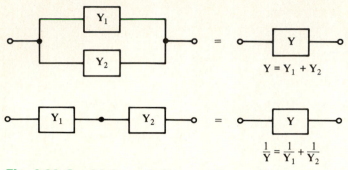

Fig. 4.14 Combining admittances in parallel and series.

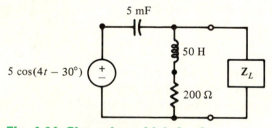

Fig. 4.15 Thévenin's theorem.

Example 4.5

For the circuit shown in Fig. 4.16, we can determine the effect of the circuit on the load impedance $\mathbf{Z}_L$ by employing Thévenin's theorem.

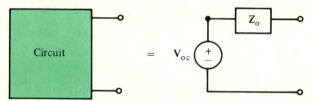

Fig. 4.16 Given sinusoidal circuit.

We first determine $\mathbf{V}_{oc}$ from the circuit in Fig. 4.17. By nodal analysis, at the node labeled $\mathbf{V}_{oc}$, by KCL,

$$\frac{(5\underline{/-30°}) - \mathbf{V}_{oc}}{-j50} = \frac{\mathbf{V}_{oc} - \mathbf{V}}{j200}$$

from which

$$3\mathbf{V}_{oc} + \mathbf{V} = 20\underline{/-30°} \qquad (4.7)$$

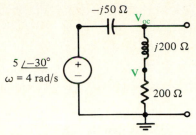

Fig. 4.17 Determination of open-circuit voltage.

At the node labeled $\mathbf{V}$, we get

$$\frac{\mathbf{V}_{oc} - \mathbf{V}}{j200} = \frac{\mathbf{V}}{200}$$

from which

$$\mathbf{V}_{oc} = \mathbf{V} + j\mathbf{V} = (1 + j)\mathbf{V} \qquad \Rightarrow \qquad \mathbf{V} = \frac{1}{1 + j}\,\mathbf{V}_{oc}$$

Substituting this result into Equation (4.7) yields

$$3\mathbf{V}_{oc} + \frac{1}{1 + j}\,\mathbf{V}_{oc} = 20\underline{/-30°} = \frac{4 + j3}{1 + j}\,\mathbf{V}_{oc}$$

from which

$$\mathbf{V}_{oc} = \frac{1 + j}{4 + j3}\,(20\underline{/-30°}) = \frac{\sqrt{2}\underline{/45°}}{5\underline{/36.9°}}\,(20\underline{/-30°}) = 4\sqrt{2}\underline{/-21.9°}$$

Thus,

$$v_{oc}(t) = 4\sqrt{2}\,\cos(4t - 21.9°)$$

To find Thévenin-equivalent (output) impedance $\mathbf{Z}_o$, set the independent source to zero. The resulting circuit is shown in Fig. 4.18. We can find $\mathbf{Z}_o$ by combining impedances in series and parallel. In particular (the symbol $\parallel$ means

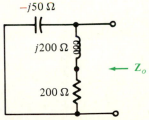

Fig. 4.18 Determination of Thévenin-equivalent (output) impedance.

"in parallel with"),

$$\mathbf{Z}_o = (\mathbf{Z}_L + \mathbf{Z}_R) \| \mathbf{Z}_C$$

$$= \frac{(j200 + 200)(-j50)}{j200 + 200 - j50} = \frac{200\,(1 - j)}{(4 + j3)} = 8 - j56 = 40\sqrt{2}\underline{/-81.9°}\ \Omega$$

In the frequency domain, we have, therefore, the Thévenin-equivalent circuit shown in Fig. 4.19. In representing this circuit in the time domain, we can use the *RLC* combination from which we determined $\mathbf{Z}_o$, thus yielding the circuit in Fig. 4.20(a). Alternatively, we can use the fact that $\mathbf{Z}_o = 8 - j56\ \Omega$ has the form of a resistor in series with a capacitor, that is,

$$\mathbf{Z}_o = R + \frac{1}{j\omega C} = R - \frac{j}{\omega C}$$

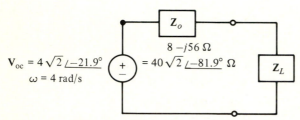

$$\mathbf{V}_{oc} = 4\sqrt{2}\underline{/-21.9°}$$
$$\omega = 4\ \text{rad/s}$$

Fig. 4.19 Thévenin-equivalent circuit in the frequency domain.

Thus,

$$R = 8\ \Omega \qquad \text{and} \qquad 1/\omega C = 56 \qquad \Rightarrow \qquad C = 1/224 = 4.46\ \text{mF}$$

and an alternative equivalent circuit is as shown in Fig. 4.20(b).

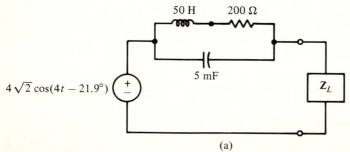

(a)

Fig. 4.20 Thévenin-equivalent circuits in the time domain.

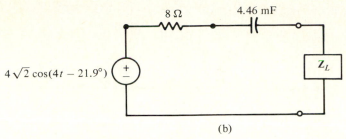

(b)

Fig. 4.20 *(cont.)*

For the impedance $\mathbf{Z} = a + jb$, we call the real part a the **resistance** and the imaginary part b the **reactance**. For the case that b is negative (as in Example 4.5), we say that the reactance is **capacitive** and denote it by X_C. If b is positive, we say that the reactance is **inductive** and denote it by X_L. In other words, an impedance has the form $\mathbf{Z} = R + jX$.

Example 4.6

Now let us demonstrate how we can employ mesh analysis and Cramer's rule. First consider the sinusoidal circuit shown in Fig. 4.21(a), which contains a dependent source. In making a transformation to the frequency domain and arbitrarily assigning clockwise mesh currents, we obtain the circuit shown in Fig. 4.21(b). By KVL, for mesh $\mathbf{I}_1$, we have that

$$6\mathbf{I}_1 - j3(\mathbf{I}_1 - \mathbf{I}_2) = 9$$

and the first mesh equation becomes

$$(2 - j)\mathbf{I}_1 + j\mathbf{I}_2 = 3$$

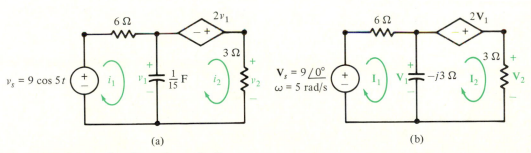

(a)   (b)

Fig. 4.21 Example of mesh analysis.

For mesh $\mathbf{I}_2$, by KVL,

$$-\mathbf{V}_1 - 2\mathbf{V}_1 + 3\mathbf{I}_2 = 0$$

from which

$$j3(\mathbf{I}_1 - \mathbf{I}_2) + \mathbf{I}_2 = 0$$

so the second mesh equation becomes

$$j3\mathbf{I}_1 + (1 - j3)\mathbf{I}_2 = 0$$

Calculating determinants yields

$$\mathbf{D} = \begin{vmatrix} 2 - j & j \\ j3 & 1 - j3 \end{vmatrix} = (2 - j)(1 - j3) - (j3)(j) = 2 - j7$$

$$\mathbf{D}_1 = \begin{vmatrix} 3 & j \\ 0 & 1 - j3 \end{vmatrix} = 3(1 - j3), \qquad \mathbf{D}_2 = \begin{vmatrix} 2 - j & 3 \\ j3 & 0 \end{vmatrix} = -j9$$

By Cramer's rule,

$$\mathbf{I}_1 = \frac{\mathbf{D}_1}{\mathbf{D}} = \frac{3(1 - j3)}{2 - j7} = \frac{3(23 + j)}{53} = 1.30\underline{/2.5°}$$

$$\mathbf{I}_2 = \frac{\mathbf{D}_2}{\mathbf{D}} = \frac{-j9}{2 - j7} = \frac{-j9(2 + j7)}{(2 - j7)(2 + j7)} = \frac{9(7 - j2)}{53} = 1.24\underline{/-16°}$$

Therefore,

$$i_1(t) = 1.30 \cos(5t + 2.5°)$$

and

$$i_2(t) = 1.24 \cos(5t - 16°)$$

4.4 AVERAGE POWER

We previously defined power to be the product of voltage and current. For the case that voltage and current are constants (the dc case), the instantaneous power is equal to the average value of the power. This result is a consequence of the fact that the instantaneous and average values of a constant are equal. A sinusoid, however, has an average value of zero. This does not mean that power due to a sinusoid is zero. We shall see what effect sinusoids have in terms of power.

Although our prime interest will be in the sinusoid, we shall also be able to discuss average power for the case of repetitive or periodic nonsinusoidal waveforms by using the notion of effective values. Yet it is the sinusoid that receives the greatest attention, since it is in this form that electric power is produced and distributed.

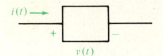

Fig. 4.22 Arbitrary element.

Recall that for the arbitrary element shown in Fig. 4.22 the instantaneous power absorbed by the element is $p(t) = v(t)i(t)$. For the case that the voltage is a sinusoid, say $v(t) = V \cos(\omega t + \phi_1)$, the current will also be sinusoidal, say $i(t) = I \cos(\omega t + \phi_2)$. Then the instantaneous power absorbed by the element is

$$p(t) = [V \cos(\omega t + \phi_1)][I \cos(\omega t + \phi_2)]$$

Using the trigonometric identity

$$(\cos \alpha)(\cos \beta) = \tfrac{1}{2} \cos(\alpha - \beta) + \tfrac{1}{2} \cos(\alpha + \beta)$$

we get

$$p(t) = VI\left[\frac{1}{2} \cos(\phi_1 - \phi_2) + \frac{1}{2} \cos(2\omega t + \phi_1 + \phi_2)\right]$$

$$= \frac{VI}{2} \cos(\phi_1 - \phi_2) + \frac{VI}{2} \cos(2\omega t + \phi_1 + \phi_2)$$

If this is the instantaneous power, what is the average power P? The second term on the right in the last equation is a sinusoid of radian frequency 2ω. We know that the average value of a sinusoid is zero—this can be verified with calculus. Thus, the **average power** absorbed is

$$P = \tfrac{1}{2} VI \cos (\phi_1 - \phi_2)$$

or

$$\boxed{P = \tfrac{1}{2} VI \cos \theta}$$

where $\theta = \phi_1 - \phi_2$ is the difference in phase angles between voltage and current.

For the special case of a resistor, since the voltage and current are in phase, then the average power absorbed by the resistor is

$$P_R = \tfrac{1}{2} VI \cos 0 = \tfrac{1}{2} VI$$

From the fact that $V = RI$, we get

$$\boxed{P_R = \tfrac{1}{2} VI = \tfrac{1}{2} RI^2 = \tfrac{1}{2} V^2/R}$$

For an impedance that is purely imaginary (reactive), that is, $\mathbf{Z} = jX$, where X is a positive or a negative number, as depicted in Fig. 4.23, let us find the average

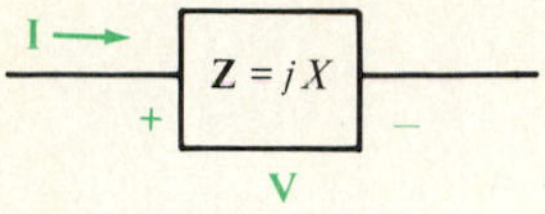

Fig. 4.23 Arbitrary reactance.

power absorbed by such an element. Since

$$\mathbf{Z} = \frac{\mathbf{V}}{\mathbf{I}} = \frac{V\underline{/\phi_1}}{I\underline{/\phi_2}} = \frac{V}{I}\underline{/(\phi_1 - \phi_2)}$$

then for positive X, $jX = X\underline{/90°}$, while for negative X, $jX = -j(-X) = -X\underline{/-90°}$. Thus, we have

$$\mathbf{Z} = |X|\ \underline{/\pm90°} = \frac{V}{I}\ \underline{/(\phi_1 - \phi_2)}$$

and the average power P_X absorbed by this impedance is

$$P_X = \tfrac{1}{2}\ VI\ \cos(\phi_1 - \phi_2) = \tfrac{1}{2}\ VI\ \cos(\pm90°) = 0$$

In other words, purely reactive impedances, such as (ideal) inductors and capacitors or combinations of the two, absorb zero average power, and are referred to as **lossless elements**.

Example 4.7

For the *RC* circuit shown in Fig. 4.24(a), the corresponding frequency-domain representation is shown in Fig. 4.24(b). Since

$$\mathbf{I} = \frac{\mathbf{V}_s}{\mathbf{Z}} = \frac{6\underline{/-90°}}{1 - j} = \frac{6\underline{/-90°}}{\sqrt{2}\underline{/-45°}} = 3\sqrt{2}\underline{/-45°}$$

then the average power absorbed by the series *RC* impedance (supplied by the voltage source) is

$$P = \tfrac{1}{2}\ VI\ \cos\theta = \tfrac{1}{2}(6)(3\sqrt{2})\ \cos(-90° + 45°) = 9\sqrt{2}\ \cos(-45°) = 9\ \text{W}$$

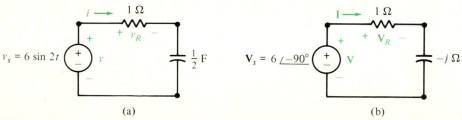

(a)   (b)

Fig. 4.24 Sinusoidal circuit in (a) the time domain and (b) the frequency domain.

By voltage division, the voltage across the resistor is

$$\mathbf{V}_R = \frac{\mathbf{Z}_R}{\mathbf{Z}_R + \mathbf{Z}_C}\, \mathbf{V}_s = \frac{1}{1 - j}\, (6\underline{/-90°}) = \frac{6\underline{/-90°}}{\sqrt{2}\underline{/-45°}} = \frac{6}{\sqrt{2}}\,\underline{/-45°} = V_R\underline{/\phi_R}$$

and the power absorbed by the resistor is

$$P_R = \frac{1}{2}\frac{V_R^2}{R} = \frac{1}{2}\left(\frac{18}{1}\right) = 9 \text{ W}$$

Of course, this conclusion could have been obtained using the fact that the impedance of a capacitor is purely imaginary, so the power absorbed by the capacitor is
$P_C = 0$, and all of the power supplied by the source, therefore, must be absorbed
by the resistor.

Given a circuit with a load impedance $\mathbf{Z}_L = R_L + jX_L$, as depicted in Fig.
4.25, what impedance $\mathbf{Z}_L$ will absorb the maximum amount of power? In a manner
similar to that for resistive circuits (but mathematically slightly more involved), it
can be shown that the load that absorbs the maximum power is

$$\mathbf{Z}_L = R_L + jX_L = R_o - jX_o = \mathbf{Z}_o^*$$

where $\mathbf{Z}_o = R_o + jX_o$ is the Thévenin-equivalent (output) impedance of the circuit.

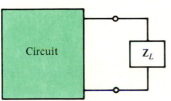

Fig. 4.25 Sinusoidal circuit and its associated load.

For the case that $\mathbf{Z}_L$ is restricted to be purely real (resistive), it can be shown
that the value of $\mathbf{Z}_L = R_L$ that absorbs the maximum amount of power (among all
resistive loads) is

$$R_L = \sqrt{R_o^2 + X_o^2} = |\mathbf{Z}_o|$$

We have seen that for a resistor R with a sinusoidal current through it, say,
$i(t) = I \cos(\omega t + \phi)$, the average power absorbed by the resistor is $I^2 R/2$. What
dc current I_e would yield the same power absorption by R? Since the power

absorbed by R due to a direct current I_e is $I_e^2 R$, we get the equality

$$I_e^2 R = \tfrac{1}{2} I^2 R$$

from which

$$I_e = \frac{I}{\sqrt{2}} \approx 0.707 \, I$$

Thus, the constant current $I_e = I/\sqrt{2}$ and the sinusoidal current $i(t) = I \cos(\omega t + \phi)$ result in the same average power absorption by R. For this reason, we say that $I/\sqrt{2}$ is the **effective value** of the sinusoid whose amplitude is I.

Since the voltage across R is

$$v(t) = Ri(t) = RI \cos(\omega t + \phi) = V \cos(\omega t + \phi)$$

then the effective value of $v(t)$ is $V_e = V/\sqrt{2}$. Futhermore, the average power absorbed by R is

$$P_R = \frac{VI}{2} = \frac{V}{\sqrt{2}} \frac{I}{\sqrt{2}} = V_e I_e$$

Alternatively, since $V = RI$, then

$$P_R = \frac{RI^2}{2} = R \frac{I}{\sqrt{2}} \frac{I}{\sqrt{2}} = RI_e^2$$

and

$$P_R = \frac{V^2}{2R} = \frac{1}{R} \frac{V}{\sqrt{2}} \frac{V}{\sqrt{2}} = \frac{V_e^2}{R}$$

In summary, given a resistor R having current $i(t) = I \cos(\omega t + \phi)$ through it and voltage $v(t) = V \cos(\omega t + \phi)$ across it, then the average power absorbed by the resistor is

$$P_R = V_e I_e = I_e^2 R = \frac{V_e^2}{R}$$

where $V_e = V/\sqrt{2}$ and $I_e = I/\sqrt{2}$ are the effective values of the voltage and the current, respectively.

For an arbitrary element

$$P = \tfrac{1}{2} VI \cos\theta = V_e I_e \cos\theta$$

(4.8)

where θ is the difference in phase between the voltage and the current.

Example 4.8

Ordinary household voltage is designated typically as 115 V ac, 60 Hz. The term "ac," which stands for **alternating current,** indicates a sinusoid, and 60 Hz is the frequency of the sinusoid. However, the 115 V does not refer to the amplitude of the sinusoid, but rather to its effective value. Thus, the amplitude of the sinusoid is $115\sqrt{2} \approx 163$ V.

◼

Again, for a resistor R, consider the case that the current (or the voltage) is nonsinusoidal. Since the instantaneous power absorbed by the resistor is $p(t) = v(t)i(t)$, then by Ohm's law, we get

$$p(t) = Ri^2(t)$$

Summing the instantaneous power from time t_1 to t_2 and dividing by the width of the interval $t_2 - t_1$ yields the average power absorbed in the interval, that is,

$$P = \frac{1}{t_2 - t_1} \int_{t_1}^{t_2} p(t)\ dt = \frac{1}{t_2 - t_1} \int_{t_1}^{t_2} Ri^2(t)\ dt$$

Suppose that the current (and hence the voltage) is "repetitive" or "periodic," and the period is T. Then the average power absorbed in one period is

$$P = \frac{1}{T} \int_{t_o}^{t_o + T} p(t)\ dt = \frac{1}{T} \int_{t_o}^{t_o + T} Ri^2(t)\ dt$$

where t_o is an arbitrary value. Again, let I_e be the dc current that results in the same amount of average power absorption. Thus,

$$RI_e^2 = \frac{1}{T} \int_{t_o}^{t_o + T} Ri^2(t)\ dt$$

from which the expression for I_e, the effective value of $i(t)$, is

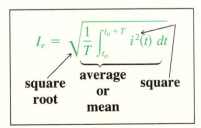

and for this reason the effective value is also known as the **root-mean-square** or **rms** value. Similarly, the effective or rms value of $v(t)$ is

$$V_e = \sqrt{\frac{1}{T} \int_{t_o}^{t_o+T} \overline{v^2(t)}\, dt}$$

It is a routine exercise in integral calculus to confirm that the rms value of the sinusoid $A \cos(\omega t + \phi)$, whose period is $T = 2\pi/\omega$, is $A/\sqrt{2}$.

Example 4.9

The function $f(t)$ shown in Fig. 4.26 is periodic, where the period is $T = 4$ s. Although the average or mean value of this function is zero (why?), the rms value of $f(t)$ is

$$\sqrt{\frac{1}{T} \int_{-2}^{2} f^2(t)\, dt} = \sqrt{\frac{1}{4}\left[\int_{-2}^{-1} 1^2\, dt + \int_{-1}^{1} (-t)^2\, dt + \int_{1}^{2} (-1)^2\, dt \right]}$$

$$= \sqrt{\frac{1}{4}\left[t \Big|_{-2}^{-1} + \frac{t^3}{3}\Big|_{-1}^{1} + t\Big|_{1}^{2} \right]} = \sqrt{\frac{1}{4}\left(1 + \frac{2}{3} + 1\right)} = \sqrt{\frac{2}{3}}$$

In other words, if the current through a resistor R is $i(t) = f(t)$, then the average power absorbed by R is

$$P = RI_e^2 = R\left(\sqrt{\frac{2}{3}}\right)^2 = \frac{2R}{3} \text{ W}$$

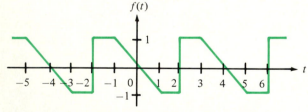

Fig. 4.26 Nonsinusoidal, periodic waveform.

4.5 IMPORTANT POWER CONCEPTS

Consider the arbitrary load, as depicted in Fig. 4.27, where $v(t) = V \cos(\omega t + \phi_1)$ and $i(t) = I \cos(\omega t + \phi_2)$. Then the average power absorbed by the load is, repeating Equation (4.8),

$$P = \tfrac{1}{2}VI \cos \theta = V_e I_e \cos \theta \qquad (4.8)$$

where $\theta = \phi_1 - \phi_2$. If the load is purely resistive, then the average power it

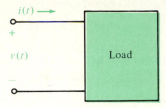

Fig. 4.27 Arbitrary load.

absorbs is $P_R = V_e I_e$. However, if the load is not purely resistive, even though it might "appear" at first glance that the average power absorbed is $V_e I_e$, the average power absorbed is given by Equation (4.8). We call $V_e I_e$ the **apparent power** and use the unit **volt-ampere** (VA) to distinguish this quantity from the actual power (the average power) that is absorbed by the load. The quantity $\cos \theta$, called the **power factor** and abbreviated pf, is the ratio of average power to apparent power, that is,

$$\text{pf} = \frac{\text{average power}}{\text{apparent power}} = \cos \theta = \frac{P}{V_e I_e}$$

The angle θ is referred to as the **power-factor angle.**

The impedance of the load is

$$\mathbf{Z} = \frac{\mathbf{V}}{\mathbf{I}} = \frac{V\underline{/\phi_1}}{I\underline{/\phi_2}} = \frac{V}{I}\underline{/(\phi_1 - \phi_2)} = \frac{V}{I}\underline{/\theta}$$

so that the angle of the load is the pf angle. The general form of a load impedance is $\mathbf{Z} = R + jX$. If $X > 0$, we say that the load is **inductive**; if $X < 0$, we say it is **capacitive.** Assuming a positive resistance R, then for an inductive load

$$\theta = \text{ang } \mathbf{Z} = \tan^{-1}\frac{X}{R} > 0$$

that is, the pf angle is positive; while for a capacitive load

$$\theta = \text{ang } \mathbf{Z} = \tan^{-1}\frac{X}{R} < 0$$

that is, the pf angle is negative. The former case ($\phi_1 > \phi_2$) implies that the current lags the voltage, whereas the latter case ($\phi_1 < \phi_2$) implies that the current leads the voltage. In the vernacular of the power industry, we refer to the pf as being **lagging** or **leading** respectively. In other words, if the current lags the voltage for a load (the inductive case), the result is a lagging pf; if the current leads the voltage (the capacitive case), the result is a leading pf.

You may ask, "Why talk about lagging and leading pf at all? Why not just talk about the impedance of the load?" One reason is that, in some applications (e.g., power systems), a load may not consist simply of ordinary inductors, capacitors, and resistors. In the following example, the load is an electric motor.

Example 4.10

A 1000-W electric motor is connected to a 200-V rms ac, 60-Hz source, and the result is a lagging pf of 0.8. Since the pf is lagging, the angle of the current is less than the voltage angle by $\cos^{-1}(0.8) = 36.9°$. Arbitrarily selecting the voltage angle to be zero, we can depict the given situation as shown in Fig. 4.28. From Equation (4.8), we have

$$I_e = \frac{P}{V_e \cos \theta} = \frac{1000}{200 \,(0.8)} = \frac{25}{4} = 6.25$$

so that

$$\mathbf{I}_m = 6.25\underline{/-36.9°} = 5 - j3.75$$

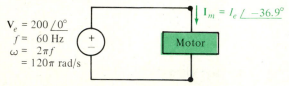

$\mathbf{V}_e = 200\underline{/0°}$
$f = 60$ Hz
$\omega = 2\pi f$
$\quad = 120\pi$ rad/s

Fig. 4.28 Example of a lagging pf.

Of course, since the motor absorbs 1000 W of average power, then the source supplies this power. But suppose that a 28 μF capacitor is placed in parallel with the motor. Since the current through the motor does not change, we have the situation shown in Fig. 4.29. Since

$$\mathbf{I}_C = \frac{\mathbf{V}_e}{\mathbf{Z}_C} = j\omega C\mathbf{V} = (1\underline{/90°})(120\pi)(28 \times 10^{-6})(200\underline{/0°}) = 2.11\underline{/90°}$$

$$= j2.11$$

then

$$\mathbf{I} = \mathbf{I}_m + \mathbf{I}_C = 5 - j1.64 = 5.26\underline{/-18.1°}$$

and the pf for the new load (the motor in parallel with the capacitor) is $\cos(18.1°) = 0.95$, and since the pf angle is positive, the pf is lagging.

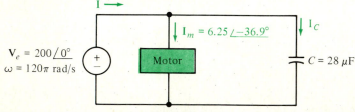

Fig. 4.29 Power-factor correction.

We now see that placing a capacitor in parallel with the motor increases the lagging pf from 0.8 to 0.95. (This is an example of a **power-factor correction.**) Since capacitors absorb zero average power, and since the motor absorbs 1000 W in either case, the voltage source supplies 1000 W in both cases. Note, however, that originally 6.25 A of rms current is drawn from the voltage source, while after the capacitor is connected in parallel, the rms current is 5.26 A—a decrease of about 16%.

■

There are situations in which current changes such as the one in Example 4.10 have significant implications. One important case is when the voltage source is an electric utility company generator and the load is some demand by a consumer. Between the two is the power transmission line, which is not ideal and has some associated nonzero resistance.

Example 4.11

Suppose that a large consumer of electricity requires 10 kW of power by using 230 V rms at a pf angle of 60° lagging, that is, a pf of 0.5 lagging. We conclude that the current drawn by the load has an effective value of

$$I_e = \frac{P}{V_e \cos \theta} = \frac{10{,}000}{230(0.5)} \approx 87 \text{ A}$$

If the transmission-line resistance is $R = 0.1 \ \Omega$, then, as indicated in Fig. 4.30,

$$\mathbf{V}_e = 0.1 \ \mathbf{I}_L + \mathbf{V}_L$$

$$\approx (8.7 \underline{/-60°}) + (230 \underline{/0°}) = 234.5 \underline{/-1.8°}$$

Thus, the voltage that must be generated by the power company must have an effective value of $V_e \approx 234.5$ V. Also, the line loss is $I_e^2 R = 756$ W. Although the utility gets paid for 10 kW of power, it has to produce 10,756 W.

If the consumer can correct the pf from 0.5 to 0.9 lagging, the load current will become $I_e \approx 48$ A rms, so the line loss will be approximately 233 W—which

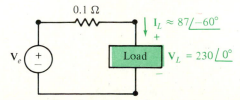

Fig. 4.30 Inclusion of transmission-line resistance.

is about 30% of the loss at the lower pf. It is for this reason that an electric utility takes pf into consideration when dealing with large consumers.

An average household uses relatively small amounts of power and typically has a reasonably high pf. Thus, pf is not a consideration in home electricity usage.

Complex Power

Previously, we found it computationally convenient to introduce the abstract concept of a complex sinusoid—a real sinusoid is just a special case. So, too, we now discuss and apply the notion of "complex power."

Again consider the arbitrary load depicted in Fig. 4.31, where $\mathbf{V} = V\underline{/\phi_1}$, $\mathbf{I} = I\underline{/\phi_2}$, $V_e = V/\sqrt{2}$, $I_e = I/\sqrt{2}$. Since $P = V_e I_e \cos\theta$, and $\cos\theta = \mathrm{Re}[e^{j\theta}]$, then

$$P = V_e I_e (\mathrm{Re}[e^{j\theta}]) = V_e I_e (\mathrm{Re}[e^{j(\phi_1-\phi_2)}]) = V_e I_e (\mathrm{Re}[e^{j\phi_1}e^{-j\phi_2}])$$

$$= \mathrm{Re}[V_e I_e e^{j\phi_1}e^{-j\phi_2}] = \mathrm{Re}[(V_e e^{j\phi_1})(I_e e^{-j\phi_2})]$$

$$= \mathrm{Re}[(V_e e^{j\phi_1})(I_e e^{j\phi_2})^*]$$

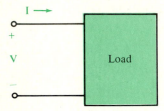

Fig. 4.31 Arbitrary load.

That is, the average power absorbed by the load is equal to the real part of some complex quantity. Defining $\mathbf{V}_{\mathrm{rms}}$ and $\mathbf{I}_{\mathrm{rms}}$, respectively, by

$$\mathbf{V}_{\mathrm{rms}} = V_e e^{j\phi_1} \quad \text{and} \quad \mathbf{I}_{\mathrm{rms}} = I_e e^{j\phi_2}$$

then

$$P = \mathrm{Re}[\mathbf{V}_{\mathrm{rms}}\mathbf{I}_{\mathrm{rms}}^*]$$

That is, the average power absorbed is the real part of the quantity

$$\boxed{\mathbf{P} = \mathbf{V}_{\mathrm{rms}}\mathbf{I}_{\mathrm{rms}}^*}$$

which is called the **complex power** absorbed by the load, that is,

$$\mathbf{P} = (V_e e^{j\phi_1})(I_e e^{-j\phi_2}) = V_e I_e e^{j(\phi_1-\phi_2)} = V_e I_e e^{j\theta}$$

Thus, we see that the magnitude of the complex power is the apparent power $|\mathbf{P}| = V_e I_e$, and the angle of the complex power is the pf angle. By Euler's formula, we can write the complex power in rectangular form as follows:

$$\mathbf{P} = V_e I_e (\cos\theta + j\sin\theta) = V_e I_e \cos\theta + j V_e I_e \sin\theta = P + jQ$$

where $P = V_e I_e \cos\theta$ is the average power absorbed by the load—also called the **real power**—and $Q = V_e I_e \sin\theta$ is called the **reactive power**. As mentioned before, the unit of P is the watt (W). The unit of Q is the **var**—short for **volt-ampere reactive (VAR)**. The unit of complex power $\mathbf{P}$, like its magnitude $|\mathbf{P}|$ (apparent power), is the volt-ampere (VA). Thus, we have

$$V_e I_e = |\mathbf{P}| = \sqrt{P^2 + Q^2}$$

and the angle of $\mathbf{P}$ is the pf angle θ, so

$$\theta = \tan^{-1}(Q/P)$$

From the triangle—called a **power triangle**—shown in Fig. 4.32, we deduce that the pf is

$$\cos\theta = \frac{P}{\sqrt{P^2 + Q^2}} = \frac{P}{|\mathbf{P}|}$$

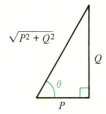

Fig. 4.32 Power triangle.

Example 4.12

For the circuit shown in Fig. 4.33, the source is given in terms of the ordinary (non-rms) phasor representation. The impedance loading the voltage source is

$$\mathbf{Z}_1 = \frac{\mathbf{V}}{\mathbf{I}} = \frac{R[(1/j\omega C) + j\omega L]}{R + (1/j\omega C) + j\omega L} = \frac{R(1 - \omega^2 LC)}{1 - \omega^2 LC + j\omega RC}$$

$$= \frac{40(1-4)}{(1-4)+j4} = \frac{-120}{-3+j4} = \frac{120}{3-j4} = \frac{120}{5\underline{/-53°}} = 24\underline{/53°}$$

Thus,

$$\mathbf{I} = \frac{\mathbf{V}}{\mathbf{Z}_1} = \frac{240\sqrt{2}\underline{/0°}}{24\underline{/53°}} = 10\sqrt{2}\underline{/-53°}$$

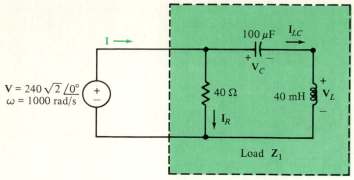

Fig. 4.33 Voltage source and its associated load.

so that the complex power supplied by the source is

$$\mathbf{P} = \mathbf{V}_{\text{rms}} \mathbf{I}_{\text{rms}}^* = (240\underline{/0°})(10\underline{/53°}) = 2400\underline{/53°} = 1440 + j1920 \text{ VA}$$

Hence, the apparent power is $|\mathbf{P}| = 2400$ VA, the pf angle is $\text{ang}(\mathbf{P}) = 53°$, and thus the pf is $\cos 53° = 0.6$ lagging. The real power is $P = 1440$ W and the reactive power is $Q = 1920$ VAR.

Since a pf that is too small may cost money, let us try to increase the pf. As $\text{pf} = \cos \theta = P/|\mathbf{P}|$, one way to increase the pf is to increase the real power P. But real power is costly. Instead, let us place another load in parallel with the given one. This will not affect the voltage and current of the original load, but it will have an effect on the current through the voltage source. The resulting situation is depicted in Fig. 4.34, where $\mathbf{Z}_1 = 24\underline{/53°}$, $\mathbf{I}_1 = 10\sqrt{2}\underline{/-53°}$, and $\mathbf{P}_1 = 2400\underline{/53°} = 1440 + j1920$. Suppose we wish to raise the lagging pf to 0.9. Since $\cos^{-1}(0.9) = 25.84°$, let us call the current drain on the source $\mathbf{I} = \sqrt{2} I\underline{/-25.84°}$. The resulting complex power supplied by the source is

$$\mathbf{P} = \mathbf{V}_{\text{rms}} \mathbf{I}_{\text{rms}}^* = (240\underline{/0°})(I\underline{/25.84°}) = 240I\underline{/25.84°} \text{ VA}$$

As we wish the real power to remain the same as before, we have

$$P = |\mathbf{P}| \cos \theta = (240I)(0.9) = 1440 \text{ W}$$

from which

$$I = \frac{1440}{(240)(0.9)} = 6.67 \quad \Rightarrow \quad \mathbf{I} = 6.67\sqrt{2}\underline{/-25.84°} \text{ A}$$

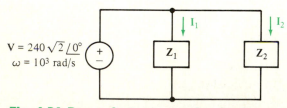

Fig. 4.34 Power-factor correction for $\mathbf{Z}_1$.

By KCL,

$$\mathbf{I}_2 = \mathbf{I} - \mathbf{I}_1 = (6.67\sqrt{2}\underline{/-25.84°}) - (10\sqrt{2}\underline{/-53°}) = 5.1\sqrt{2}\underline{/90°} \text{ A}$$

Thus,

$$\mathbf{Z}_2 = \frac{\mathbf{V}}{\mathbf{I}_2} = \frac{240\underline{/0°}}{5.1\underline{/90°}} = 47\underline{/-90°} = -j47 \ \Omega$$

This is purely reactive, which might have been expected because of the constraint that there be no additional real power supplied by the source. This impedance can be realized with a capacitor as follows:

$$\frac{1}{j\omega C} = \frac{-j}{\omega C} = -j47 \quad \Rightarrow \quad C = \frac{1}{47\omega} = \frac{1}{47(10^3)} = 21.3 \ \mu\text{F}$$

Return now to the original circuit given in Fig. 4.33. The current through the resistor is

$$\mathbf{I}_R = \mathbf{V}/40 = 6\sqrt{2}\underline{/0°} = 6\sqrt{2} \text{ A}$$

and the complex power absorbed by the resistor is

$$\mathbf{P}_R = (240\underline{/0°})(6\underline{/0°}) = 1440 \text{ W}$$

which, of course, is purely real. The current through the series LC combination is

$$\mathbf{I}_{LC} = \frac{1}{(1/j\omega C) + j\omega L} = \frac{240\sqrt{2}\underline{/0°}}{(10/j) + j40} = -j8\sqrt{2} = 8\sqrt{2}\underline{/-90°} \text{ A}$$

The voltage across the capacitor is

$$\mathbf{V}_C = \frac{1}{j\omega C}\mathbf{I}_{LC} = \frac{-j8\sqrt{2}}{j10^{-1}} = -80\sqrt{2} = 80\sqrt{2}\underline{/180°} \text{ V}$$

and the voltage across the inductor is

$$\mathbf{V}_L = j\omega L \mathbf{I}_{LC} = j40(-j8\sqrt{2}) = 320\sqrt{2} \text{ V}$$

The complex power[†] absorbed by the capacitor is

$$\mathbf{P}_C = \tfrac{1}{2}\mathbf{V}_C \mathbf{I}_{LC}^* = (80\underline{/180°})(8\underline{/90°}) = 640\underline{/-90°} = -j640 \text{ VA}$$

and the complex power absorbed by the inductor is

$$\mathbf{P}_L = \tfrac{1}{2}\mathbf{V}_L \mathbf{I}_{LC}^* = (320\underline{/0°})(8\underline{/90°}) = 2560\underline{/90°} = j2560 \text{ VA}$$

The sum of the complex powers absorbed by the resistor, capacitor, and inductor is

$$\mathbf{P}_R + \mathbf{P}_C + \mathbf{P}_L = 1440 - j640 + j2560 = 1440 + j1920 \text{ VA}$$

[†] In terms of ordinary non-rms phasors $\mathbf{V}$ and $\mathbf{I}$, the formula for complex power is

$$\mathbf{P} = \tfrac{1}{2}\mathbf{VI}^*$$

and this is exactly the complex power supplied by the source. This result is a consequence of the fact that, as with the case of real power, complex power is conserved.

4.6 POLYPHASE CIRCUITS

Single-Phase, Three-Wire Circuits

Previously, we mentioned that ordinary household voltage is sinusoidal, having an approximate rms value of 115 V and a frequency of 60 Hz. Let us discuss the subject further.

Coming from a power line into a home's service panel or fuse box are three wires. One wire is bare; it has no insulation. This is the **neutral** or **ground wire**, so called since it must be connected to a water pipe or a rod (or both) that goes into the ground. (The earth is the reference potential of zero volts.) Typically, another wire is red and a third is black. The voltage between the red and neutral wires (plus sign at the red wire) is $v_{rn}(t) = 115\sqrt{2}\cos(120\pi t + \phi)$—that is, 115 V (rms) ac, 60 Hz. The voltage $v_{bn}(t)$ between the black and the neutral wires (plus at the black wire) is $v_{bn}(t) = -v_{rn}(t)$. This means that the voltage between the neutral and black wires (with the plus at the neutral wire) is $v_{nb}(t) = -v_{bn}(t) = v_{rn}(t)$, and, therefore, these two voltages have the same phase angle. As these voltages remain fairly constant for a wide variety of practical loads (appliances), we can reasonably approximate the **single-phase, three-wire source** just described as shown in Fig. 4.35, where $v_{rn}(t) = v_{nb}(t) = v(t) = 115\sqrt{2}\cos(120\pi t + \phi)$. By KVL, we have that $v_{rb}(t) = 230\sqrt{2}\cos(120\pi t + \phi)$. Thus, in addition to two 115-V ac, 60-Hz sources, there is a supply of 230 V (rms) ac, 60 Hz. This higher voltage is useful for higher power appliances, since for a given higher power requirement, a higher voltage uses less current than a lower voltage. The result is a lower line loss.

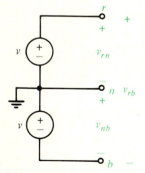

Fig. 4.35 Single-phase, three-wire source.

The single-phase, three-wire circuit shown in Fig. 4.36 (which is represented in the frequency domain) consists of a single-phase, three-wire source and three

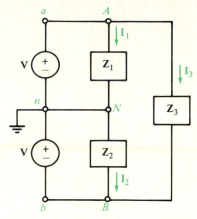

Fig. 4.36 Single-phase, three-wire circuit.

load impedances. The short circuits between nodes a and A and between nodes b and B are the **lines** of the circuit. The short circuit between nodes n and N is the **neutral wire** of the circuit. It may be that the lines have a nonzero impedance $\mathbf{Z}_g$ and the neutral wire has a nonzero impedance $\mathbf{Z}_n$.

Example 4.13

For the single-phase, three-wire circuit given in Fig. 4.36, suppose that $\mathbf{Z}_1 = 60 \ \Omega$, $\mathbf{Z}_2 = 80 \ \Omega$, $\mathbf{Z}_3 = 10 + j15 \ \Omega$, and $\mathbf{V} = 115\underline{/0°}$. Then

$$\mathbf{I}_1 = \frac{\mathbf{V}}{\mathbf{Z}} = \frac{115\underline{/0°}}{60} = 1.917\underline{/0°}$$

$$\mathbf{I}_2 = \frac{\mathbf{V}}{\mathbf{Z}} = \frac{115\underline{/0°}}{80} = 1.438\underline{/0°}$$

$$\mathbf{I}_3 = \frac{2\mathbf{V}}{\mathbf{Z}} = \frac{230\underline{/0°}}{10 + j15} = \frac{230\underline{/0°}}{18.03\underline{/56.31°}} = 12.76\underline{/-56.31°}$$

The power delivered to the loads is $P_1 + P_2 + P_3$, where

$$P_1 = |\mathbf{I}_1|^2 60 = 220.4 \text{ W}$$

$$P_2 = |\mathbf{I}_2|^2 80 = 165.3 \text{ W}$$

$$P_3 = |\mathbf{I}_3|^3 10 = 1627.7 \text{ W}$$

Thus,

$$P_1 + P_2 + P_3 = 2013.4 \text{ W}$$

Since

$$\mathbf{I}_1 + \mathbf{I}_3 = 1.917\underline{/0°} + 12.76\underline{/-56.31°} = 13.915\underline{/-49.73°} \text{ A}$$

$$\mathbf{I}_2 + \mathbf{I}_3 = 1.438\underline{/0°} + 12.76\underline{/-56.31°} = 13.61\underline{/-51.27°} \text{ A}$$

as a check, the power supplied by the upper source is

$$|\mathbf{V}||\mathbf{I}_1 + \mathbf{I}_3| \cos \theta_1 = (115)(13.915) \cos(49.73°) = 1034.4 \text{ W}$$

and the power supplied by the lower source is

$$|\mathbf{V}||\mathbf{I}_2 + \mathbf{I}_3| = (115)(13.61) \cos(51.27°) = 979.2 \text{ W}$$

for a total of 2013.6 W. The apparent discrepancy between this value and $P_1 + P_2 + P_3$ given above is the result of round-off error.

■

Three-Phase Circuits (Systems)

A single-phase, three-wire circuit contains a source that produces two sinusoidal voltages that have the same amplitude and the same phase angle. A circuit that contains a source that produces (sinusoidal) voltages with different phases is called a **polyphase system.** The importance of this concept lies in the fact that most of the generation and distribution of electric power in the United States is accomplished with polyphase systems. The most common polyphase system is the balanced three-phase system, which has the property that it supplies constant instantaneous power. This results in less vibration of the rotating machinery used to generate electric power.

A **Y (wye)-connected three-phase source** in the frequency domain is shown in Fig. 4.37. Terminals a, b, and c are called the **line terminals** and n is called the **neutral terminal.** The source is said to be **balanced** if the voltages $\mathbf{V}_{an}$, $\mathbf{V}_{bn}$, and $\mathbf{V}_{cn}$, called **phase voltages,** have the same amplitude and sum to zero, that is, if

$$|\mathbf{V}_{an}| = |\mathbf{V}_{bn}| = |\mathbf{V}_{cn}| \qquad \text{and} \qquad \mathbf{V}_{an} + \mathbf{V}_{bn} + \mathbf{V}_{cn} = 0$$

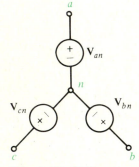

Fig. 4.37 A Y-connected three-phase source.

Suppose that the amplitude of the sinusoids is V. If we arbitrarily select the angle of $\mathbf{V}_{an}$ to be zero, that is, if $\mathbf{V}_{an} = V\underline{/0°}$, then the two situations that result in a balanced source are as follows:

Case I

$\mathbf{V}_{an} = V\underline{/0°}$

$\mathbf{V}_{bn} = V\underline{/-120°}$

$\mathbf{V}_{cn} = V\underline{/-240°} = V\underline{/120°}$

Case II

$\mathbf{V}_{an} = V\underline{/0°}$

$\mathbf{V}_{bn} = V\underline{/120°} = V\underline{/-240°}$

$\mathbf{V}_{cn} = V\underline{/240°} = V\underline{/-120°}$

For Case I, $v_{an}(t)$ leads $v_{bn}(t)$ by 120°, and $v_{bn}(t)$ leads $v_{cn}(t)$ by 120°. It is, therefore, called a **positive** or **abc phase sequence.** Similarly, Case II is called a **negative** or **acb phase sequence.** Clearly, a negative phase sequence can be converted to a positive phase sequence simply by relabeling the terminals. Thus, we need only consider positive phase sequences.

It is a simple matter to determine the voltages between the line terminals—called **line voltages.** By KVL,

$$\mathbf{V}_{ab} = \mathbf{V}_{an} - \mathbf{V}_{bn} = V\underline{/0°} - V\underline{/-120°} = \sqrt{3}\ V\underline{/30°}$$

Similarly,

$$\mathbf{V}_{bc} = \sqrt{3}\ V\underline{/-90°} \quad \text{and} \quad \mathbf{V}_{ca} = \sqrt{3}\ V\underline{/-210°} = \sqrt{3}\ V\underline{/150°}$$

Let us now connect our balanced source to a **balanced Y-connected three-phase load** as shown in Fig. 4.38. The short circuits between nodes a and A, between nodes b and B, and between nodes c and C are the **lines** of the system. If we denote the current through $\mathbf{Z}$ from terminal x to y by $\mathbf{I}_{xy}$, then the line currents are

$$\mathbf{I}_{AN} = \frac{\mathbf{V}_{an}}{\mathbf{Z}} = \frac{V\underline{/0°}}{\mathbf{Z}} = \frac{V}{\mathbf{Z}}$$

$$\mathbf{I}_{BN} = \frac{\mathbf{V}_{bn}}{\mathbf{Z}} = \frac{V\underline{/-120°}}{\mathbf{Z}} = \frac{V}{\mathbf{Z}}(1\underline{/-120°}) = \mathbf{I}_{AN}(1\underline{/-120°})$$

$$\mathbf{I}_{CN} = \frac{\mathbf{V}_{cn}}{\mathbf{Z}} = \frac{V\underline{/-240°}}{\mathbf{Z}} = \frac{V}{\mathbf{Z}}(1\underline{/-240°}) = \mathbf{I}_{AN}(1\underline{/120°})$$

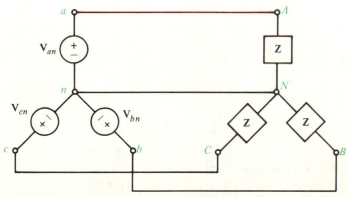

Fig. 4.38 Balanced Y-Y-connected three-phase system.

By KCL,

$$\mathbf{I}_{Nn} = \mathbf{I}_{AN} + \mathbf{I}_{BN} + \mathbf{I}_{CN} = \mathbf{I}_{AN} + \mathbf{I}_{AN}(1\underline{/-120°}) + \mathbf{I}_{AN}(1\underline{/120°}) = 0$$

so we see that there is no current in the "neutral wire" (the short circuit between terminals n and N). We deduce that the neutral wire can be removed without affecting the voltages and currents.

If the lines all have the same impedance, the effective load is still balanced and so the neutral current is zero, and the neutral wire can again be removed. If the current in the neutral wire is zero when the neutral wire's impedance is both zero and infinity, we may deduce that the current is also zero for any value in between. For the case of a balanced source, balanced load, and balanced (equal) line imped-ances (called a **balanced system**), whether or not there is a neutral wire connecting terminals n and N, it is convenient to think of these two points as being connected by a short circuit. In this way, we can analyze the system on a "per-phase" basis, that is, as three separate single-phase problems. This can greatly simplify the analysis of the system.

Example 4.14

For the balanced three-phase system given in Fig. 4.38, suppose that $\mathbf{V}_{an} = 120\underline{/0°}$ is an rms phasor. Assuming a positive phase sequence, then $\mathbf{V}_{bn} = 120\underline{/-120°}$ and $\mathbf{V}_{cn} = 120\underline{/120°}$. Thus, the rms line voltage phasors are

$$\mathbf{V}_{ab} = 120\sqrt{3}\underline{/30°}, \qquad \mathbf{V}_{bc} = 120\sqrt{3}\underline{/-90°}, \qquad \mathbf{V}_{ca} = 120\sqrt{3}\underline{/150°}$$

and the line voltage is $120\sqrt{3} = 208$ V rms.

If $\mathbf{Z} = 30 + j40 = 50\underline{/53°}$ Ω, then the rms line current $\mathbf{I}_{aA} = \mathbf{I}_{AN}$ is

$$\mathbf{I}_{aA} = \frac{\mathbf{V}_{an}}{\mathbf{Z}} = \frac{120\underline{/0°}}{50\underline{/53°}} = 2.4\underline{/-53°} \text{ A}$$

and the power absorbed by the single load $\mathbf{Z}$ between terminals A and N is

$$P_{AN} = |\mathbf{V}_{an}|\,|\mathbf{I}_{AN}| \cos 53° = 120(2.4)(0.6) = 172.8 \text{ W}$$

so that the total power absorbed by the three-phase load is $3(172.8) = 518.4$ W.

The powers mentioned above are average powers. How does this compare with the instantaneous power? For the single load between A and N, the instanta-neous power absorbed is

$$p_{AN}(t) = v_{an}(t)\, i_{AN}(t)$$
$$= [120\sqrt{2} \cos \omega t][2.4\sqrt{2} \cos(\omega t - 53°)]$$
$$= 576 \cos \omega t \cos(\omega t - 53°)$$

which is a time-varying quantity. However, the total instantaneous power absorbed by the three-phase load is

$$p(t) = p_{AN}(t) + p_{BN}(t) + p_{CN}(t)$$

where

$$p_{BN}(t) = v_{bn}(t) \, i_{BN}(t)$$
$$= [120\sqrt{2} \cos(\omega t - 120°)][2.4\sqrt{2} \cos(\omega t - 53° - 120°)]$$

and

$$p_{CN}(t) = v_{cn}(t) \, i_{CN}(t)$$
$$= [120\sqrt{2} \cos(\omega t - 240°)][2.4\sqrt{2} \cos(\omega t - 53° - 240°)]$$

Using the trigonometric identity

$$\cos \alpha \cos \beta = \tfrac{1}{2} \cos(\alpha + \beta) + \tfrac{1}{2} \cos(\alpha - \beta)$$

we get

$$p(t) = 288 [\cos(2\omega t - 53°) + \cos 53°]$$
$$+ 288 [\cos(2\omega t - 53° - 240°) + \cos 53°]$$
$$+ 288 [\cos(2\omega t - 53° - 120°) + \cos 53°]$$

It is a routine matter to show that since the above three sinusoids are spaced 120° apart, they sum to zero. Thus,

$$p(t) = 288 [3 \cos 53°] = 518.4 \text{ W}$$

so we see that the instantaneous power absorbed by the load (supplied by the source) is constant! This result holds for any balanced Y-Y-connected three-phase system, which is one reason for its importance.

<hr>

Example 4.15

Suppose now that the balanced three-phase system given in Fig. 4.38 has a line voltage of 250 V rms and the total power absorbed by the load is 600 W at a leading pf angle of 30°. Thus, the per-phase voltage—for example, V_{an}—has the rms magnitude $|V_{an}| = |V_{ab}|/\sqrt{3} = 250/\sqrt{3}$. Since the per-phase power absorbed is 200 W, from the fact that $P_{AN} = |V_{an}||I_{AN}| \cos \theta$, we have that the rms line current has the magnitude

$$|I_{AN}| = \frac{P_{AN}}{|V_{an}| \cos \theta} = \frac{200}{(250/\sqrt{3})(\sqrt{3}/2)} = 1.6 \text{ A}$$

The per-phase impedance has magnitude

$$Z = \frac{|V_{an}|}{|I_{AN}|} = \frac{250/\sqrt{3}}{1.6} = 90 \ \Omega$$

and since the pf is leading, the current leads the voltage so

$$\mathbf{Z} = 90\underline{/-30°} = 78 - j45 \ \Omega$$

Delta Connections

More common than a balanced Y-connected three-phase load is a Δ (**delta**)-**connected** load. A Y-connected source with a balanced Δ-connected load is shown in Fig. 4.39. We see that the individual loads are connected directly across the lines, and, consequently, it is relatively easier to add or remove one of the components of a Δ-connected load than with a Y-connected load. As a matter of fact, it may not even be possible to do so for a Y-connected load since the neutral terminal may not be accessible.

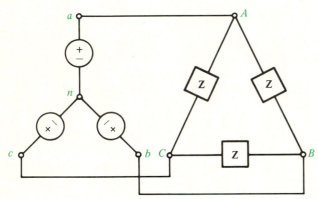

Fig. 4.39 Three-phase system with a balanced Δ-connected load.

Suppose that the voltage between terminals x and y (plus that at x) is denoted by $\mathbf{V}_{xy}$. For a balanced source

$$\mathbf{V}_{an} = V\underline{/0°}, \qquad \mathbf{V}_{bn} = V\underline{/-120°}, \qquad \mathbf{V}_{cn} = V\underline{/-240°}$$

and as was shown previously, the line voltages are

$$\mathbf{V}_{ab} = \sqrt{3}\,V\underline{/30°}, \qquad \mathbf{V}_{bc} = \sqrt{3}\,V\underline{/-90°}, \qquad \mathbf{V}_{ca} = \sqrt{3}\,V\underline{/-210°}$$

which are the phase voltages for a Δ-connected load. The phase currents are

$$\mathbf{I}_{AB} = \frac{\mathbf{V}_{ab}}{\mathbf{Z}}, \qquad \mathbf{I}_{BC} = \frac{\mathbf{V}_{bc}}{\mathbf{Z}}, \qquad \mathbf{I}_{CA} = \frac{\mathbf{V}_{ca}}{\mathbf{Z}}$$

Thus, by KCL, the line current $\mathbf{I}_{aA}$ is

$$\mathbf{I}_{aA} = \mathbf{I}_{AB} - \mathbf{I}_{CA} = \frac{\mathbf{V}_{ab}}{\mathbf{Z}} - \frac{\mathbf{V}_{ca}}{\mathbf{Z}} = \frac{1}{\mathbf{Z}}(\mathbf{V}_{ab} - \mathbf{V}_{ca})$$

$$\mathbf{I}_{aA} = \frac{1}{\mathbf{Z}}(\sqrt{3}\,V\underline{/30°} - \sqrt{3}\,V\underline{/-210°}) = \frac{3V\underline{/0°}}{\mathbf{Z}}$$

Similarly,

$$\mathbf{I}_{bB} = \frac{3V\underline{/-120°}}{\mathbf{Z}}, \qquad \mathbf{I}_{cC} = \frac{3V\underline{/-240°}}{\mathbf{Z}}$$

Since the magnitude of a phase current—for example $\mathbf{I}_{AB}$—is $|\mathbf{I}_{AB}| = |\mathbf{V}_{ab}|/|\mathbf{Z}| = \sqrt{3}\,V/|\mathbf{Z}|$, and the magnitude of a line current—for example, $\mathbf{I}_{aA}$—is $|\mathbf{I}_{aA}| = 3V/|\mathbf{Z}|$, then

$$|\mathbf{I}_{aA}| = \sqrt{3}\,|\mathbf{I}_{AB}|$$

That is, a line current is $\sqrt{3}$ times as great as a phase current. Furthermore, since the phase currents are 120° apart, they are a balanced set, as are the line currents.

Example 4.16

Suppose that, for the balanced three-phase system given in Fig. 4.39, the line voltage is 250 V rms and the total power absorbed is 600 W at a leading pf angle of 30°. Since the per-phase power absorbed is 200 W, from the fact that $P_{AB} = |\mathbf{V}_{ab}||\mathbf{I}_{AB}|\cos\theta$, we have that the rms phase current—for example, $\mathbf{I}_{AB}$—has magnitude

$$|\mathbf{I}_{AB}| = \frac{P_{AB}}{V_{ab}\cos\theta} = \frac{200}{250(\sqrt{3}/2)} = 0.92\ \text{A}$$

while the rms line current—for example, $\mathbf{I}_{aA}$—has magnitude

$$|\mathbf{I}_{aA}| = \sqrt{3}\,|\mathbf{I}_{AB}| = 1.6\ \text{A}$$

The per-phase impedance has magnitude

$$|\mathbf{Z}| = \frac{|\mathbf{V}_{ab}|}{|\mathbf{I}_{AB}|} = \frac{250}{0.92} = 271\ \Omega$$

and since the pf is leading, the current leads to voltage, so

$$\mathbf{Z} = 271\underline{/-30°} = 234 - j135\ \Omega$$

∎

Suppose now that a balanced three-phase load is Y-connected, the magnitude of the line voltage is V_L volts rms, and the line current is I_L amperes rms. If the power P_1 absorbed per phase is due to an rms voltage V and an rms current I with pf angle θ, then

$$P_1 = VI\cos\theta = \frac{V_L}{\sqrt{3}}I_L\cos\theta$$

and, therefore, the total power absorbed is

$$P = 3P_1 = \sqrt{3}\, V_L I_L \cos\theta$$

For the case of a balanced Δ-connected load,

$$P_1 = VI \cos\theta = V_L\left(\frac{I_L}{\sqrt{3}}\right)\cos\theta$$

and again

$$P = 3P_1 = \sqrt{3}\, V_L I_L \cos\theta$$

Thus, we see that regardless of whether a balanced load is Y-connected or Δ-connected, in terms of the line voltage, line current, and load impedance phase angle, we can use the same formula for the total power absorbed by the load:

$$\boxed{P = \sqrt{3}\, V_L I_L \cos\theta}$$

In our discussions, we have assumed that the source is balanced and Y-connected. Although the same conclusions can be obtained for a balanced Δ-connected source, for practical reasons, such sources are uncommon. If a Δ-connected source is not balanced exactly, a large current can circulate around the loop formed by the elements comprising the delta. The result is the reduction of the source's current-delivering capacity and an increase in the system's losses.

One might feel that a Y connection is equivalent to a Δ connection, and vice versa. But under what conditions is equivalence obtained? To answer this question, consider the Δ and Y connections shown in Fig. 4.40. For the Δ connection, the mesh equations are

$$\mathbf{V}_1 = \mathbf{Z}_{AC}\mathbf{I}_1 - \mathbf{Z}_{AC}\mathbf{I}_3 \tag{4.9}$$

$$\mathbf{V}_2 = \mathbf{Z}_{BC}\mathbf{I}_2 - \mathbf{Z}_{BC}\mathbf{I}_3 \tag{4.10}$$

$$0 = -\mathbf{Z}_{AC}\mathbf{I}_1 - \mathbf{Z}_{BC}\mathbf{I}_2 + (\mathbf{Z}_{AB} + \mathbf{Z}_{BC} + \mathbf{Z}_{AC})\mathbf{I}_3 \tag{4.11}$$

From Equation (4.11),

$$\mathbf{I}_3 = \frac{\mathbf{Z}_{AC}\mathbf{I}_1}{\mathbf{Z}_{AB} + \mathbf{Z}_{BC} + \mathbf{Z}_{AC}} + \frac{\mathbf{Z}_{BC}\mathbf{I}_2}{\mathbf{Z}_{AB} + \mathbf{Z}_{BC} + \mathbf{Z}_{AC}}$$

and on substituting this expression for $\mathbf{I}_3$ into Equations (4.9) and (4.10), we get, respectively,

$$\mathbf{V}_1 = \frac{\mathbf{Z}_{AB}\mathbf{Z}_{AC} + \mathbf{Z}_{AC}\mathbf{Z}_{BC}}{\mathbf{Z}_{AB} + \mathbf{Z}_{BC} + \mathbf{Z}_{AC}}\,\mathbf{I}_1 - \frac{\mathbf{Z}_{AC}\mathbf{Z}_{BC}}{\mathbf{Z}_{AB} + \mathbf{Z}_{BC} + \mathbf{Z}_{AC}}\,\mathbf{I}_2$$

and

$$\mathbf{V}_2 = \frac{-\mathbf{Z}_{AC}\mathbf{Z}_{BC}}{\mathbf{Z}_{AB} + \mathbf{Z}_{BC} + \mathbf{Z}_{AC}}\,\mathbf{I}_1 - \frac{\mathbf{Z}_{AB}\mathbf{Z}_{BC} + \mathbf{Z}_{AC}\mathbf{Z}_{BC}}{\mathbf{Z}_{AB} + \mathbf{Z}_{BC} + \mathbf{Z}_{AC}}\,\mathbf{I}_2$$

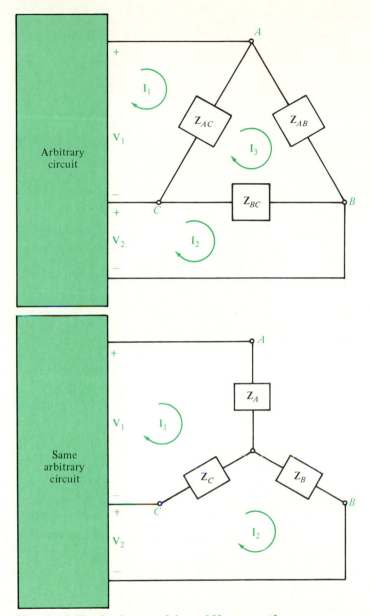

Fig. 4.40 Equivalence of Δ and Y connections.

The mesh equations for the Y connection are

$$\mathbf{V}_1 = (\mathbf{Z}_A + \mathbf{Z}_C)\,\mathbf{I}_1 - \mathbf{Z}_C\mathbf{I}_2$$

and

$$\mathbf{V}_2 = -\mathbf{Z}_C\mathbf{I}_1 + (\mathbf{Z}_B + \mathbf{Z}_C)\,\mathbf{I}_2$$

Since the two connections are equivalent when the values of V_1 and V_2 and I_1 and I_2 are the same for both cases, by equating coefficients we get

$$\mathbf{Z}_A = \frac{\mathbf{Z}_{AB}\mathbf{Z}_{AC}}{\mathbf{Z}_{AB} + \mathbf{Z}_{BC} + \mathbf{Z}_{AC}}, \qquad \mathbf{Z}_B = \frac{\mathbf{Z}_{AB}\mathbf{Z}_{BC}}{\mathbf{Z}_{AB} + \mathbf{Z}_{BC} + \mathbf{Z}_{AC}},$$

$$\mathbf{Z}_C = \frac{\mathbf{Z}_{AC}\mathbf{Z}_{BC}}{\mathbf{Z}_{AB} + \mathbf{Z}_{BC} + \mathbf{Z}_{AC}}$$

and these are the conditions for finding the equivalent Y connection given a Δ connection. By using dual arguments, we can derive the conditions for finding the equivalent Δ connection given a Y connection. (In this case, we can use admittances instead of impedances.) They are

$$\mathbf{Y}_{AB} = \frac{\mathbf{Y}_A\mathbf{Y}_B}{\mathbf{Y}_A + \mathbf{Y}_B + \mathbf{Y}_C}, \qquad \mathbf{Y}_{BC} = \frac{\mathbf{Y}_B\mathbf{Y}_C}{\mathbf{Y}_A + \mathbf{Y}_B + \mathbf{Y}_C}, \qquad \mathbf{Y}_{AC} = \frac{\mathbf{Y}_A\mathbf{Y}_C}{\mathbf{Y}_A + \mathbf{Y}_B + \mathbf{Y}_C}$$

Alternatively, it is a simple matter to express these conditions in terms of impedances as follows:

$$\mathbf{Z}_{AB} = \frac{\mathbf{Z}_A\mathbf{Z}_B + \mathbf{Z}_B\mathbf{Z}_C + \mathbf{Z}_A\mathbf{Z}_C}{\mathbf{Z}_C}, \qquad \mathbf{Z}_{BC} = \frac{\mathbf{Z}_A\mathbf{Z}_B + \mathbf{Z}_B\mathbf{Z}_C + \mathbf{Z}_A\mathbf{Z}_C}{\mathbf{Z}_A},$$

$$\mathbf{Z}_{AC} = \frac{\mathbf{Z}_A\mathbf{Z}_B + \mathbf{Z}_B\mathbf{Z}_C + \mathbf{Z}_A\mathbf{Z}_C}{\mathbf{Z}_B}$$

Going from a Y connection to its equivalent Δ connection, or vice versa, is called a **Y-Δ (wye–delta) transformation** or **Δ-Y (delta–wye) transformation** respectively. Since the equivalence is with respect to the three terminals A, B, and C, the equivalence holds for any pair of them also.

Example 4.17

We can find the impedance (resistance) Z of the bridge circuit shown in Fig. 4.41 by using nodal or mesh analysis. Instead, let us use a Δ-Y transformation on the Δ formed by the $\frac{1}{5}$-Ω, $\frac{1}{2}$-Ω, and 1-Ω resistors. In doing so, we get

$$\mathbf{Z}_A = \frac{(1)(\frac{1}{2})}{1 + \frac{1}{5} + \frac{1}{2}} = \tfrac{5}{17}\,\Omega, \qquad \mathbf{Z}_B = \frac{(1)(\frac{1}{5})}{1 + \frac{1}{5} + \frac{1}{2}} = \tfrac{2}{17}\,\Omega,$$

$$\mathbf{Z}_C = \frac{(\frac{1}{2})(\frac{1}{5})}{1 + \frac{1}{5} + \frac{1}{2}} = \tfrac{1}{17}\,\Omega$$

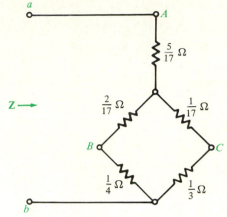

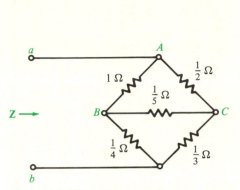

Fig. 4.41 Circuit containing a Δ connection.

Fig. 4.42 Circuit resulting from a Δ-Y transformation.

As far as terminals a and b are concerned, the bridge circuit given in Fig. 4.41 is equivalent, therefore, to the configuration shown in Fig. 4.42. The $\frac{1}{17}$-Ω and $\frac{1}{3}$-Ω resistors in series have a resistance of $\frac{1}{17} + \frac{1}{3} = \frac{20}{51}$ Ω, while the $\frac{2}{17}$-Ω and $\frac{1}{4}$-Ω resistors in series have a resistance of $\frac{2}{17} + \frac{1}{4} = \frac{25}{68}$ Ω. The resulting parallel combination has a resistance of

$$\frac{\left(\frac{20}{51}\right)\left(\frac{25}{68}\right)}{\frac{20}{51} + \frac{25}{68}} = \frac{100}{527} \ \Omega$$

and this in series with $\frac{5}{17}$ Ω has a resistance of

$$\mathbf{Z} = \tfrac{100}{527} + \tfrac{5}{17} = \tfrac{255}{527} = 0.48 \approx 0.5 \ \Omega$$

Power Measurements

The **wattmeter** is a device that measures the average power absorbed by a two-terminal load. Such a device is depicted in Fig. 4.43. The wattmeter, which is used typically at frequencies between a few hertz and a few hundred hertz, has two coils. One, the current coil, has a very low impedance (ideally zero), and the other, the voltage coil, has a very high impedance (ideally infinite). The current coil is connected in series with the load and the voltage coil is connected in parallel with the load. When the current and the voltage shown for the load are both positive valued, the wattmeter has an upscale deflection and reads the average power absorbed by the load. The same result occurs when both the current and voltage are negative valued.

Having seen how to measure the power absorbed by a single load, how can we use the wattmeter to measure the power absorbed by a three-phase load? For a balanced load, if we can connect a wattmeter to one phase of the load as shown in

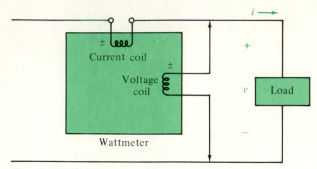

Fig. 4.43 Wattmeter connected to a load.

Fig. 4.43, then the total power absorbed is three times the reading. For an unbalanced load, if three separate wattmeters can be connected, then the total power absorbed is the sum of the the three readings. However, a Y-connected load may not have its neutral terminal accessible. In such a case, although the current coil can be connected properly, the voltage coil cannot. On the other hand, for a Δ-connected load such as a three-phase rotating machine, only the three terminals of the load are accessible. This means that the voltage coil can be connected appropriately, but the current coil cannot. But don't give up, there is a way to measure the power absorbed by a three-phase load, whether balanced or unbalanced, with the use of wattmeters. To see how, consider the three wattmeters

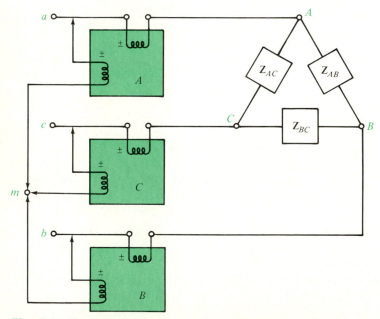

Fig. 4.44 Power measurement for a three-phase load.

and Δ-connected load shown in Fig. 4.44. In this situation, the current coils of the wattmeters are connected in series with the lines aA, bB, and cC. The voltage coils are connected between a line and some arbitrary point m. The average power read on wattmeter A is

$$P_A = \frac{1}{T} \int_0^T v_{am} i_{aA} \, dt$$

and similar expressions are obtained for wattmeters B and C. Summing these average powers, we get

$$P = P_A + P_B + P_C = \frac{1}{T} \int_0^T v_{am} i_{aA} \, dt + \frac{1}{T} \int_0^T v_{bm} i_{bB} \, dt + \frac{1}{T} \int_0^T v_{cm} i_{cC} \, dt$$

$$= \frac{1}{T} \int_0^T (v_{am} i_{aA} + v_{bm} i_{bB} + v_{cm} i_{cC}) \, dt$$

But, by KCL,

$$i_{aA} = i_{AB} + i_{AC}, \qquad i_{bB} = -i_{AB} + i_{BC}, \qquad i_{cC} = -i_{AC} - i_{BC}$$

Upon substituting these expressions for the line currents into the preceding integral, we obtain

$$P = \frac{1}{T} \int_0^T [(v_{am} - v_{bm}) i_{AB} + (v_{bm} - v_{cm}) i_{BC} + (v_{am} - v_{cm}) i_{AC}] \, dt$$

and by KVL,

$$P = \frac{1}{T} \int_0^T (v_{ab} i_{AB} + v_{bc} i_{BC} + v_{ac} i_{AC}) \, dt$$

$$= \frac{1}{T} \int_0^T v_{ab} i_{AB} \, dt + \frac{1}{T} \int_0^T v_{bc} i_{BC} \, dt + \frac{1}{T} \int_0^T v_{ac} i_{AC} \, dt$$

which is the total power absorbed by the three-phase load.

Since point m was chosen arbitrarily, we can place it anywhere without affecting the end result—the sum of the three wattmeter readings is the power absorbed by the three-phase load. The judicious choice of point m to be on line cC means that wattmeter C will have zero volts across its voltage coil and, hence, will read 0 W. For such a case, wattmeter C is superfluous, and the sum of the readings of wattmeters A and B is equal to the total power absorbed by the three-phase load. The resulting configuration, shown in Fig. 4.45, is an example of what is known as the **two-wattmeter method** for measuring power. It can also be applied when a three-phase load is Y-connected, and as for the case of a Δ connection, the load can be unbalanced as well as balanced. Furthermore, it is immaterial whether or not the source is balanced—the variables utilized are line voltages and line currents.

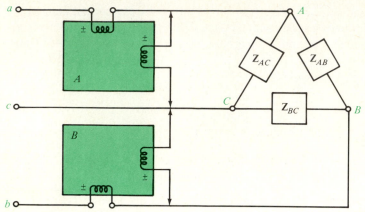

Fig. 4.45 Two-wattmeter method for measuring power.

SUMMARY

1. If the input of a linear, time-invariant circuit is a sinusoid, the response is a sinusoid of the same frequency.

2. A complex number $\mathbf{c}$ can be written in rectangular form as $\mathbf{c} = a + jb$, where a is the real part of $\mathbf{c}$ and b is the imaginary part of $\mathbf{c}$. It can also be written in polar form as $\mathbf{c} = Me^{j\theta}$, where M (also denoted $|\mathbf{c}|$) is a positive number called the magnitude of $\mathbf{c}$ and θ is the angle of $\mathbf{c}$.

3. Finding the amplitude and phase angle of a sinusoidal steady-state response can be accomplished with either real or complex sinusoids.

4. Using the concepts of phasors and impedance (or admittance), sinusoidal circuits can be analyzed in the frequency domain in a manner analogous to that for resistive circuits by using the phasor versions of KCL, KVL, and nodal or mesh analysis.

5. Important circuit concepts such as the principle of superposition and Thévenin's theorem are also applicable in the frequency domain.

6. The instantaneous power absorbed by an element is equal to the product of the voltage across it and the current through it.

7. The average power absorbed by a resistance R having a sinusoidal current of amplitude I and voltage of amplitude V is

$$P_R = VI/2 = RI^2/2 = V^2/2R$$

8. The average power absorbed by a capacitance or an inductance is zero.

9. A circuit whose Thévenin-equivalent (output) impedance is $\mathbf{Z}_o$ transfers maximum power to a load $\mathbf{Z}_L$ when $\mathbf{Z}_L$ is equal to the complex conjugate of $\mathbf{Z}_o$.

10. For the case in which $\mathbf{Z}_L$ is restricted to be purely resistive, maximum power is transferred when $\mathbf{Z}_L$ equals the magnitude of $\mathbf{Z}_o$.

11. The effective or rms value of a sinusoid of amplitude A is $A/\sqrt{2}$.

12. The average power absorbed by a resistance R having a current whose effective value is I_e and a voltage whose effective value is V_e is

$$P_R = V_e I_e = I_e^2 R = V_e^2/R$$

13. The power factor (pf) is the ratio of average power to apparent power.

14. If current lags voltage, the pf is lagging. If current leads voltage, the pf is leading.

15. Average or real power can be generalized with the notion of complex power.

16. The ordinary household uses a single-phase, three-wire electrical system.

17. The most common polyphase electrical system is the balanced three-phase system.

18. Three-phase sources are generally Y-connected, and three-phase loads are generally Δ-connected.

19. The device commonly used to measure power is the wattmeter.

20. Three-phase load power measurements can be taken with the two-wattmeter method.

Problems for Section 4.2

polar also

4.1 Find the exponential form of the following complex numbers given in rectangular form.

(a) $4 + j7$	**(e)** 4
(b) $3 - j5$	**(f)** -5
(c) $-2 + j3$	**(g)** $j7$
(d) $-1 - j6$	**(h)** $-j2$

4.2 Find the rectangular form of the following complex numbers given in exponential form.

(a) $3e^{j70°}$	**(e)** $6e^{j90°}$
(b) $2e^{j120°}$	**(f)** $e^{-j90°}$
(c) $5e^{-j60°}$	**(g)** $2e^{j180°}$
(d) $4e^{-j150°}$	**(h)** $2e^{-j180°}$

4.3 Find the rectangular form of the sum $\mathbf{A}_1 + \mathbf{A}_2$ given that

(a) $\mathbf{A}_1 = 3e^{j30°}$, $\mathbf{A}_2 = 4e^{j60°}$

(b) $\mathbf{A}_1 = 3e^{j30°}$, $\mathbf{A}_2 = 4e^{-j30°}$

(c) $\mathbf{A}_1 = 5e^{-j60°}$, $\mathbf{A}_2 = 2e^{j120°}$

(d) $\mathbf{A}_1 = 4e^{j45°}$, $\mathbf{A}_2 = 2e^{-j90°}$

Problems for Section 4.3

4.4 For the sinusoidal circuit shown in Fig. P4.4, find $v(t)$ by using voltage division. Draw a phasor diagram. Is this circuit a lag network or a lead network?

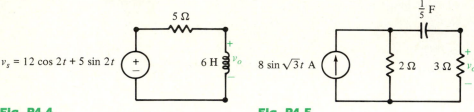

Fig. P4.4   **Fig. P4.5**

4.5 For the circuit given in Fig. P4.5, find $v(t)$ by using current division.

4.6 For the circuit shown in Fig. 4.21(a), find $v_2(t)$. Draw a phasor diagram for the circuit.

4.7 A voltage of $v(t) = 17 \cos 3t$ V is applied to a series RLC circuit. If $R = \frac{5}{3}\ \Omega$, $L = 5$ H, and $C = \frac{1}{25}$ F, what is the voltage $v_C(t)$ across the capacitor?

4.8 For the RLC connection given in Fig. P4.8, find the impedance Z when (a) $\omega = 1$, (b) $\omega = 4$, and (c) $\omega = 8$ rad/s.

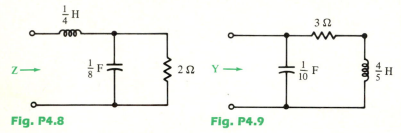

Fig. P4.8   **Fig. P4.9**

4.9 For the RLC connection shown in Fig. P4.9, find the admittance Y when (a) $\omega = 1$, (b) $\omega = 5$, and (c) $\omega = 10$ rad/s.

4.10 For the circuit given in Fig. P4.10, find $v_o(t)$ by first replacing that portion of the circuit in the shaded box by its Thévenin equivalent.

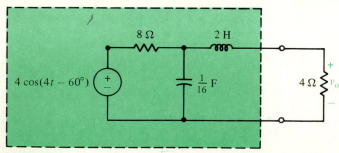

Fig. P4.10

4.11 Find the Thévenin equivalent (to the left of terminals a and b) of the circuit shown in Fig. P4.11.

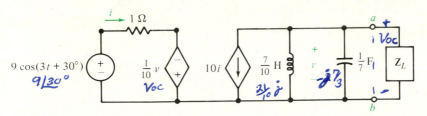

Fig. P4.11

4.12 For the op-amp circuit shown in Fig. P4.12, find $v_o(t)$.

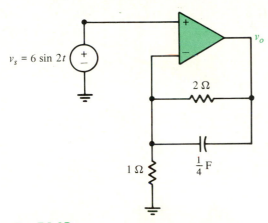

Fig. P4.12

4.13 For the op-amp circuit given in Fig. P4.13, find $v_o(t)$.

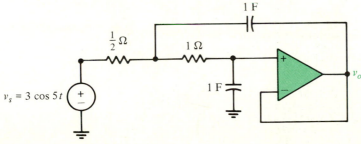

Fig. P4.13

4.14 For the op-amp circuit shown in Fig. P4.14, find $v_o(t)$.

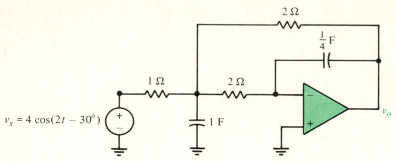

$v_s = 4\cos(2t - 30°)$

Fig. P4.14

4.15 For the circuit shown in Fig. P4.15, find the currents I_1 and I_2.

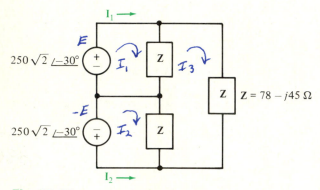

$250\sqrt{2}\,\underline{/-30°}$

$Z = 78 - j45\ \Omega$

Fig. P4.15

4.16 Use mesh analysis to find I_1 and I_2 for the circuit given in Fig. P4.16.

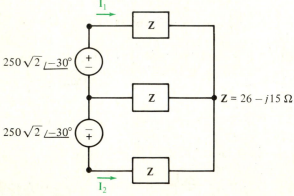

$250\sqrt{2}\,\underline{/-30°}$

$Z = 26 - j15\ \Omega$

Fig. P4.16

Problems for Section 4.4

4.17 For the *RLC* circuit shown in Fig. P4.17, find the average power absorbed by the 4-Ω load resistor for the case that (a) $C = \frac{1}{8}$ F; (b) $C = \frac{1}{18}$ F; (c) $C = \frac{1}{32}$ F.

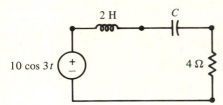

2 H   C

10 cos 3t   4 Ω

Fig. P4.17

4.18 Given the circuit shown in Fig. P4.10:
 (a) Find the average power absorbed by each resistor.
 (b) Replace the 4-Ω load resistor by an impedance that absorbs the maximum average power, and determine this maximum power.
 (c) Replace the 4-Ω load resistor with a pure resistance that absorbs the maximum power for resistive loads, and determine this power.

4.19 For the circuit shown in Fig. P4.11:
 (a) Find the load impedance $\mathbf{Z}_L$ that absorbs the maximum average power, and determine this power.
 (b) Find the load resistance that absorbs the maximum power for resistive loads, and determine this power.

4.20 For the op-amp circuit given in Fig. P4.12, find the average power absorbed by each resistor.

4.21 For the op-amp circuit given in Fig. P4.13, find the average power absorbed by each resistor.

4.22 For the op-amp circuit given in Fig. P4.14, find the average power absorbed by each resistor.

4.23 For the circuit given in Fig. P4.15:
 (a) Find the average power absorbed by each impedance.
 (b) Find the average power supplied by each source.
 (c) Find the instantaneous power supplied by each source.
 (d) Find the total instantaneous power supplied by both sources.

4.24 For the circuit given in Fig. P4.16:
 (a) Find the average power absorbed by each impedance.
 (b) Find the average power supplied by each source.
 (c) Find the instantaneous power supplied by each source.
 (d) Find the total instantaneous power supplied by both sources.

4.25 For the op-amp circuit shown in Fig. P4.25:
 (a) Find the average power absorbed by each resistor for the case that $v_s(t) = \cos(\omega t + \phi)$.

(b) Find the average power supplied by the independent source.
(c) Answer the question: Why doesn't the power supplied in (b) equal the power absorbed in (a)?

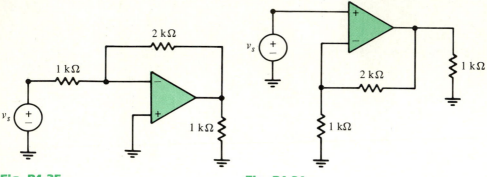

Fig. P4.25   **Fig. P4.26**

4.26 Repeat Problem 4.25 for the op-amp circuit shown in Fig. P4.26.
4.27 Find the average power absorbed by the op amp (a) in Fig. P4.25; (b) in Fig. P4.26.
4.28 For each function given in Fig. P4.28, find the rms value.

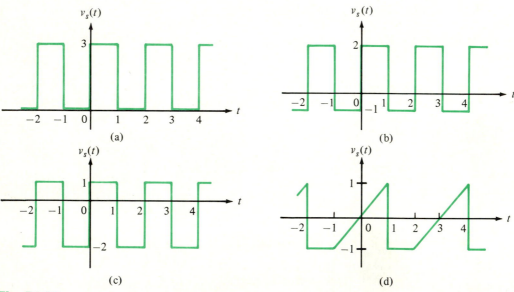

Fig. P4.28

4.29 Repeat Problem 4.28 for the exponential functions shown in Fig. P4.29.

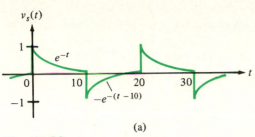

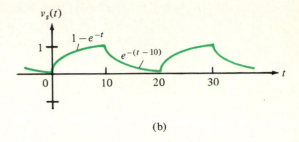

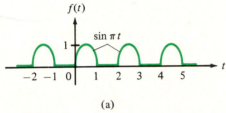

(a)

(b)

Fig. P4.29

4.30 Find the rms values of the "rectified" sine waves shown in Fig. P4.30. [*Hint:* $\sin^2 x = \frac{1}{2}(1 - \cos 2x)$.]

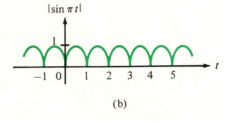

(a)

(b)

Fig. P4.30

Problems for Section 4.5

4.31 The load shown in Fig. P4.31 operates at 60 Hz.
  (a) What are the pf and the pf angle of this load?
  (b) Is the pf leading or lagging?
  (c) To what value should the capacitor be changed to get a lagging pf of 0.8?

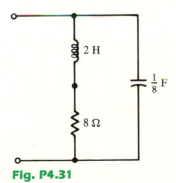

Fig. P4.31

4.32 An 115-V rms, 60-Hz electric hair dryer absorbs 500 W at a lagging pf of 0.95. What rms-valued current is drawn by this dryer?

4.33 Two loads, which are connected in parallel, operate at 230 V rms. One load absorbs 500 W at a pf of 0.8 lagging, and the other absorbs 1000 W at a pf of 0.9 lagging.

(a) What is the current drawn by the combined load?

(b) Find the pf of the combined load.

(c) Is this pf leading or lagging?

4.34 The combined load given in Problem 4.33 is connected to a source with an internal resistance of 1 Ω. What is the rms value of the source?

4.35 For the combined load given in Problem 4.33, a third load, which operates at 1500 W at a pf of 0.9 leading, is connected in parallel.

(a) What is the current drawn by the resulting composite load?

(b) Find the pf of the composite load.

(c) Is this pf leading or lagging?

4.36 The parallel connection of two 115-V rms, 60-Hz loads operates at 2000 W with a lagging pf of 0.95. If one load absorbs 1200 W at a pf of 0.8 lagging:

(a) What are the power absorbed and the pf angle of the other load?

(b) What reactive element should be placed in parallel with the load to result in an overall unity pf?

4.37 For the circuit shown in Fig. P4.37:

(a) Find the complex power supplied by the source.

(b) Find the apparent power supplied by the source.

(c) Find the power factor and determine whether it is leading or lagging.

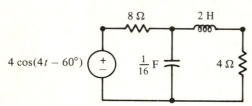

Fig. P4.37

4.38 For the circuit given in Fig. P4.38, the mesh currents shown are $\mathbf{I}_1 = 2\sqrt{2}\angle{-105°}$ and $\mathbf{I}_2 = \sqrt{2}\angle{-105°}$.

(a) What is the complex power absorbed by the capacitor?

(b) What is the real power absorbed by the capacitor?

(c) What is the complex power absorbed by the resistor?

(d) What is the real power absorbed by the resistor?

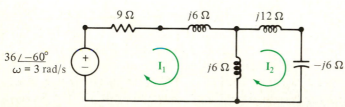

Fig. P4.38

4.39 For the circuit given in Fig. P4.38:
(a) Find the complex power absorbed by each inductor, the capacitor, and the resistor.
(b) Find the complex power supplied by the source.
(c) Is the complex power absorbed equal to the complex power supplied (is complex power conserved)?

4.40 For the circuit given in Fig. P4.38:
(a) Find the apparent power supplied by the source.
(b) Find the apparent power absorbed by each of the remaining elements.
(c) Is the apparent power absorbed equal to the apparent power supplied (is apparent power conserved)?

4.41 A load, which operates at 440 V rms, draws 5 A rms at a leading pf of 0.95. Determine:
(a) the complex power absorbed by the load,
(b) the apparent power absorbed by the load,
(c) the impedance of the load.

Problems for Section 4.6

4.42 For the single-phase, three-wire circuit shown in Fig. P4.42, find the power supplied by each source if $Z_1 = 60\ \Omega$, $Z_2 = 80\ \Omega$, $Z_3 = 40\ \Omega$ and (a) $R_g = R_n = 0\ \Omega$; (b) $R_g = 1\ \Omega$, $R_n = 2\ \Omega$.

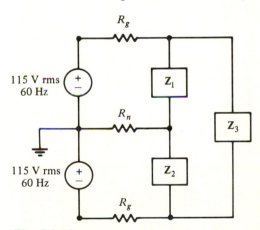

Fig. P4.42

4.43 For the single-phase, three-wire circuit given in Fig. P4.42, suppose $R_g = R_n = 0\ \Omega$. Assuming Z_1 absorbs 500 W at a lagging pf of 0.8, Z_2 absorbs 1000 W at a lagging pf of 0.9, and Z_3 absorbs 1500 W at a leading pf of 0.95, find the currents through the sources.

4.44 A balanced Y-Y three-phase system has 130-V rms phase voltages and per-phase impedance of $Z = 12 + j12\ \Omega$. Given that the line impedance is zero, find the line currents and the total power absorbed by the load.

4.45 Repeat Problem 4.44 for the case that the line impedances are 1 Ω.

4.46 A balanced Y-Y three-phase system has 210-V rms, 60-Hz line voltages. If the total load absorbs 3 kW of power at a lagging pf of 0.85:
(a) Find the per-phase impedance.
(b) What value capacitors should be placed in parallel with the per-phase impedances to result in a 0.95 lagging pf?

4.47 A balanced, three-phase Y-connected source, whose phase voltages are 115 V rms, has an unbalanced Y-connected load. Given that $\mathbf{Z}_{AN} = 3 + j4$ Ω, $\mathbf{Z}_{BN} = 10$ Ω, and $\mathbf{Z}_{CN} = 5 + j12$ Ω, for the case that the lines and the neutral wire have zero impedances, find the line currents and the total power absorbed by the load.

4.48 Repeat Problem 4.47 for the case that there is no neutral wire.

4.49 A balanced Y-Δ three-phase system has $\mathbf{V}_{an} = 130\underline{/0°}$ V rms and $Z = 4\sqrt{2}\underline{/45°}$ Ω. Given that the line impedance is zero, find the line currents and the total power absorbed by the load.

4.50 Repeat Problem 4.49 for the case that the line impedances are 1 Ω.

4.51 A balanced, three-phase Y-connected source with 230-V rms line voltages has an unbalanced Δ-connected load. The load impedances are $\mathbf{Z}_{AB} = 8$ Ω, $\mathbf{Z}_{BC} = 4 + j3$ Ω, and $\mathbf{Z}_{AC} = 12 - j5$ Ω. For the case that the lines have zero impedances, find the line currents and the total power absorbed by the load.

4.52 Repeat Problem 4.46 for a balanced Y-Δ three-phase system.

4.53 A balanced, three-phase Y-connected source with 120-V rms line voltages is loaded with a balanced Y-connection having $3 + j4$ Ω per phase and a balanced Δ connection having $5 - j12$ Ω per phase. Find the total power absorbed by the load and the pf.

4.54 A balanced, three-phase Δ-connected load has a per-phase impedance of $\mathbf{Z} = 36 + j36$ Ω. For the two-wattmeter connection given in Fig. 4.46, find the two readings for the case of line voltages of $130\sqrt{3}$ V rms produced by a balanced Y-connected source.

4.55 For the circuit given in Fig. P4.55, find the wattmeter readings P_A and P_B.

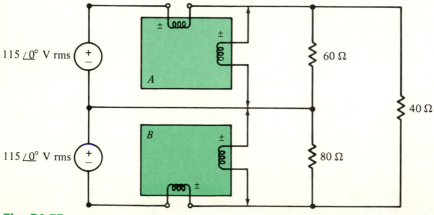

115 $\underline{/0°}$ V rms

60 Ω

A

40 Ω

115 $\underline{/0°}$ V rms

B

80 Ω

Fig. P4.55

4.56 Find the wattmeter readings P_A and P_B for the circuit shown in Fig. P4.56.

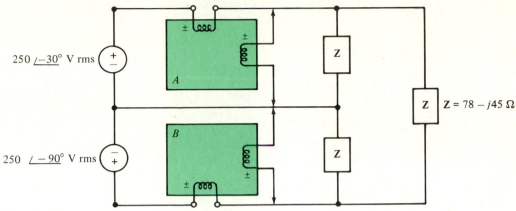

250 $\angle-30°$ V rms

250 $\angle-90°$ V rms

$Z = 78 - j45\ \Omega$

Fig. P4.56

4.57 Find the wattmeter readings P_A and P_B for the circuit given in Fig. P4.57.

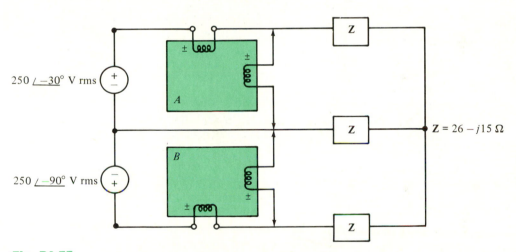

250 $\angle-30°$ V rms

250 $\angle-90°$ V rms

$Z = 26 - j15\ \Omega$

Fig. P4.57

4.58 The Y-Δ three-phase system given in Problem 4.51 has two wattmeters connected as shown in Fig. 4.46. Find the readings of these wattmeters.

Chapter 5

Circuit and System Concepts

INTRODUCTION

As the impedances of capacitors and inductors are frequency dependent, so is the impedance of an arbitrary RLC circuit. This dependence, which can be depicted graphically by plots of magnitude and phase angle versus frequency, constitutes the frequency response of the impedance. Impedance, which is the ratio of voltage to current (phasors), is only one network function whose frequency response can be displayed. It is the frequency response of a network that describes the network's behavior to all sinusoidal forcing functions, and nonsinusoidal forcing functions that are comprised of sinusoids.

An important frequency is that for which a network function or parameter reaches a maximum value. In certain simple networks, this occurs when an impedance or admittance is purely real—a condition known as resonance. In such a situation, we can quantitatively describe the frequency selectivity of the network with a figure of merit known as the quality factor. We can also talk about the notion of bandwidth.

Just as the concept of complex sinusoids allowed us to treat sinusoidal circuits in a simple manner, the concept of complex frequency does a similar job for damped sinusoidal circuits. This generalization not only allows for the easy determination of forced responses to sinusoidal and damped sinusoidal forcing functions, but indicates the form of the natural response as well.

A **system** is an interconnection of components or devices that perform some objective. The components of a system can be electrical, mechanical, thermal, chemical, etc. Electrical systems are commonly used for communications, computation, control, and instrumentation. An electric circuit is an example of

232

an electrical system, in particular, a linear circuit is an example of a **linear system.** Systems described by linear differential equations are linear systems.

The description of a linear circuit or system resulting from a damped-sinusoidal excitation may first appear to be of limited usefulness. However, with the use of a very important mathematical tool—the Laplace transform—such circuit and system descriptions can be used to obtain complete responses to many types of common excitations. By using Laplace transforms, functions of time are transformed to functions of frequency. Consequently, frequency-domain analysis can be performed, thereby eliminating the necessity to solve differential equations.

5.1 FREQUENCY RESPONSE

For the parallel connection of a resistor R and capacitor C shown in Fig. 5.1, the impedance $\mathbf{Z}$ of the combination is given by

$$\mathbf{Z} = \frac{\mathbf{Z}_R \mathbf{Z}_C}{\mathbf{Z}_R + \mathbf{Z}_C} = \frac{R(1/j\omega C)}{R + (1/j\omega C)} = \frac{R}{1 + j\omega RC}$$

$$= \frac{R}{\sqrt{1 + (\omega RC)^2}} \underline{/-\tan^{-1}(\omega RC)} = |\mathbf{Z}|\underline{/\theta}$$

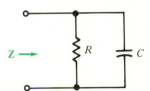

Fig. 5.1 Parallel *RC* impedance.

Since $\mathbf{Z}$ is a complex function—the notation $\mathbf{Z}(j\omega)$ is often used—and, therefore, can be expressed in terms of a magnitude and an angle as $\mathbf{Z} = |\mathbf{Z}| \underline{/\theta}$, both the magnitude $|\mathbf{Z}|$ and the angle θ are, in general, functions of ω. We, therefore, may plot $|\mathbf{Z}|$ versus ω and θ versus ω. In obtaining the former, note that when $\omega = 0$, the magnitude is $|\mathbf{Z}| = R$; when $\omega = 1/RC$, then $|\mathbf{Z}| = R/\sqrt{2} \approx 0.707\,R$. Also, when $\omega \to \infty$, then $|\mathbf{Z}| \to 0$. Thus, we have the plot shown in Fig. 5.2, which we call the **amplitude response** of $\mathbf{Z}$. The plot of θ versus ω is known as the **phase response** of $\mathbf{Z}$, and for this example, it is shown in Fig. 5.3. In this plot, we see that when $\omega = 0$, then $\theta = -\tan^{-1} \omega RC = 0$, and when $\omega = 1/RC$, then $\theta = -\tan^{-1} \omega RC = -45°$. Furthermore, when $\omega \to \infty$, then $\theta \to -90°$. The amplitude and phase responses comprise the **frequency response** of the impedance $\mathbf{Z}$.

Although a frequency response consists of both an amplitude response and a phase response, quite often the amplitude response alone is referred to as the frequency response.

$$|\mathbf{Z}| = \frac{R}{\sqrt{1 + (\omega RC)^2}}$$

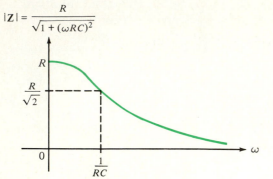

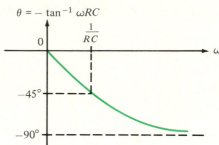

Fig. 5.2 Amplitude response.　　　**Fig. 5.3 Phase response.**

For the circuit shown in Fig. 5.4, $\mathbf{V} = \mathbf{ZI}$ where $\mathbf{Z}$ is the impedance of the parallel RC combination. From complex–number arithmetic,

$$|\mathbf{V}| = |\mathbf{ZI}| = |\mathbf{Z}|\,|\mathbf{I}|$$

Given the current $\mathbf{I}$, then $|\mathbf{V}|$ is maximum when $|\mathbf{Z}|$ is maximum. This occurs when $\omega = 0$ where $|\mathbf{Z}| = R$, and the average power absorbed by the resistor is

$$P_0 = \frac{1}{2}\frac{|\mathbf{V}|^2}{R} = \frac{1}{2}\frac{R^2|\mathbf{I}|^2}{R} = \frac{1}{2}R|\mathbf{I}|^2$$

However, when $\omega = 1/RC = \omega_c$, then $|\mathbf{Z}| = R/\sqrt{2}$, so the power absorbed is

$$P_1 = \frac{1}{2}\frac{(R/\sqrt{2})^2|\mathbf{I}|^2}{R} = \frac{1}{2}\left(\frac{R|\mathbf{I}|^2}{2}\right) = \frac{1}{2}P_0$$

That is, P_1 is half of the power P_0. For this reason, we say that $\omega_c = 1/RC$ is the **half-power frequency** or **cutoff frequency.** In general, given an amplitude response whose maximum value is M, the half-power frequencies are those frequencies for which the magnitude is $M/\sqrt{2}$.

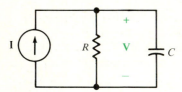

Fig. 5.4 Current applied to parallel RC impedance.

Example 5.1

For the circuit shown in Fig. 5.5, the input voltage (phasor) is $\mathbf{V}_1$ and the output voltage is $\mathbf{V}_2$. By voltage division,

$$\mathbf{V}_2 = \frac{R}{R + (1/j\omega C)} \mathbf{V}_1 \quad \Rightarrow \quad \frac{\mathbf{V}_2}{\mathbf{V}_1} = \frac{j\omega RC}{1 + j\omega RC}$$

Denoting the ratio of output voltage to input voltage by $\mathbf{H}(j\omega)$, that is, $\mathbf{H}(j\omega) = \mathbf{V}_2/\mathbf{V}_1$, then

$$|\mathbf{H}(j\omega)| = \left|\frac{\mathbf{V}_2}{\mathbf{V}_1}\right| = \frac{|\mathbf{V}_2|}{|\mathbf{V}_1|} = \frac{\omega RC}{\sqrt{1 + (\omega RC)^2}}$$

For $\omega = 0$, we have that $|\mathbf{H}(j\omega)| = 0$. To determine the value of $|\mathbf{H}(j\omega)|$ as $\omega \to \infty$, note that

$$|\mathbf{H}(j\omega)| = \sqrt{\frac{(\omega RC)^2}{1 + (\omega RC)^2}} = \frac{1}{\sqrt{1 + 1/(\omega RC)^2}}$$

Thus, $|\mathbf{H}(j\omega)| \to 1$ as $\omega \to \infty$. Also, note that for $\omega = 1/RC$, $|\mathbf{H}(j\omega)| = 1/\sqrt{2}$. The amplitude response of the function $\mathbf{H}(j\omega)$—called a **system function** or a **transfer function**—is shown in Fig. 5.6. Clearly, $\omega_c = 1/RC$ is the half-power frequency.

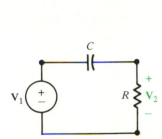

Fig. 5.5 Simple high-pass filter.

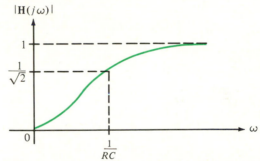

Fig. 5.6 Amplitude response of a simple high-pass filter.

It is possible to obtain a rough sketch of the amplitude response for the circuit in Fig. 5.5 without writing an analytical expression for the transfer function $\mathbf{H}(j\omega)$. For $\omega = 0$—the dc case—the capacitor is an open circuit, so $\mathbf{V}_2$, and consequently $|\mathbf{H}(j\omega)| = 0$. For small values of ω, the impedance of the capacitor is still large relative to R, so by voltage division $\mathbf{V}_2$ is relatively small, as is $|\mathbf{H}(j\omega)|$. However, as frequency (ω) increases, the impedance of the capacitor decreases, so that the values of $\mathbf{V}_2$ and $|\mathbf{H}(j\omega)|$ increase. In the limit, as frequency becomes infinite, the impedance of the capacitor becomes zero, so $\mathbf{V}_2 \to \mathbf{V}_1 \Rightarrow |\mathbf{H}(j\omega)| \to 1$.

From the amplitude response of $\mathbf{H}(j\omega)$, we see that if the input voltage is a sinusoid, then the amplitude of the output sinusoid will be relatively small for low frequencies (i.e., frequencies below ω_c). Conversely, for high frequencies (i.e., frequencies above ω_c), the amplitude of the output sinusoid approximately equals the amplitude of the input sinusoid. In other words, the high frequencies are "passed" by the *RC* circuit, while the low frequencies are "stopped." For this reason, we say that the circuit is an example of a simple **high-pass filter.** The amplitude response shown in Fig. 5.7(a) is indicative of a simple **low-pass filter,** and the one depicted in Fig. 5.7(b) describes a simple **bandpass filter.**

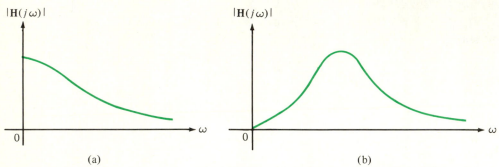

(a)   (b)

Fig. 5.7 Amplitude responses of (a) simple low-pass filter, and (b) simple bandpass filter.

The magnitude $|\mathbf{V}_2/\mathbf{V}_1|$ of a voltage transfer function, which, in general, is a function of frequency, is also called the **voltage gain** of the system. Although we can, as discussed above, directly plot gain versus frequency to obtain the amplitude response, in practice it is convenient to use logarithmic scales for gain and frequency, rather than linear scales.

The power gain for a system is $\log_{10} P_2/P_1$, and the units are **bels** (B), named in honor of Alexander Graham Bell. Multiplying this quantity by 10 gives the gain in **decibels** (dB). In other words,

$$\text{gain} = 10 \log_{10} \frac{P_2}{P_1} \quad \text{dB}$$

For a given resistance, since power is proportional to the square of the voltage, we have that

$$\text{gain} = 10 \log_{10} \frac{V_2^2}{V_1^2} = 20 \log_{10} \frac{V_2}{V_1} \quad \text{dB}$$

A plot of gain in decibels versus frequency on a logarithmic scale, and a plot of the angle (on a linear scale) versus frequency, again on a logarithmic scale, constitute a **Bode plot** or **diagram.***

---

*Named for the American electrical engineer Hendrick Bode (1905–   ).

Example 5.2

Consider the simple low-pass filter shown in Fig. 5.8. Since

$$\frac{\mathbf{V}_2}{\mathbf{V}_1} = \frac{1/j\omega C}{(1/j\omega C) + R} = \frac{1}{1 + j\omega RC} = \frac{1}{\sqrt{1 + (\omega RC)^2}} \underline{/-\tan^{-1}\omega RC}$$

the gain in decibels is given by

$$20 \log_{10}\left|\frac{\mathbf{V}_2}{\mathbf{V}_1}\right| = 20 \log_{10}\frac{1}{\sqrt{1 + (\omega RC)^2}} = -20 \log_{10}\sqrt{1 + (\omega RC)^2}$$

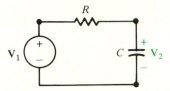

Fig. 5.8 Simple low-pass filter.

Note that for very low frequencies ($\omega \ll 1/RC$), the gain is approximately constant and approximately equal to $-20 \log_{10}\sqrt{1} = 0$ dB. For very high frequencies ($\omega \gg 1/RC$), the gain is approximately given by $-20 \log_{10}\omega RC$ dB. A plot of this function versus frequency on a logarithmic scale is a straight line with a slope of -20 dB per decade—a decade is an increase in frequency by a factor of 10. For the half-power frequency $\omega_c = 1/RC$, the gain is $-20 \log_{10}\sqrt{2} = -3.01 \approx -3$ dB. A sketch of gain versus frequency is shown in Fig. 5.9. The two straight lines, 0 and $-20 \log_{10}\omega RC$, are called **asymptotes,** and the fre-

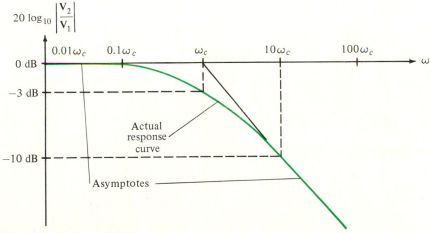

Fig. 5.9 Amplitude response.

quency at which they intersect (the half-power frequency ω_c) is called the **corner frequency** or **break frequency.**

A sketch of the angle versus frequency is shown in Fig. 5.10. A major motivation for using Bode plots is the fact that the response of a product of transfer functions is equal to the sum of the individual transfer-function responses.

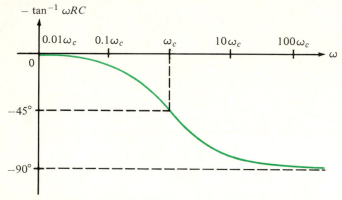

Fig. 5.10 Phase response.

5.2 RESONANCE

For the parallel *RLC* circuit shown in Fig. 5.11, the admittance "seen" by the current source is $\mathbf{Y} = \mathbf{I}/\mathbf{V}$, where

$$\mathbf{Y} = \mathbf{Y}_R + \mathbf{Y}_L + \mathbf{Y}_C = \frac{1}{R} + \frac{1}{j\omega L} + j\omega C = \frac{1}{R} + j\left(\omega C - \frac{1}{\omega L}\right)$$

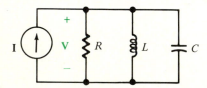

Fig. 5.11 Parallel *RLC* circuit.

From this expression, we can see that there is some frequency ω_r for which the imaginary part of $\mathbf{Y}$ will be zero, that is,

$$\omega_r C - \frac{1}{\omega_r L} = 0 \quad \Rightarrow \quad \omega_r C = \frac{1}{\omega_r L} \quad \Rightarrow \quad \omega_r = \frac{1}{\sqrt{LC}}$$

We say that a circuit with at least one capacitor and one inductor is in **resonance** or is **resonant** when the imaginary part of its admittance (or impedance) is equal to zero. The circuit given in Fig. 5.11 is in resonance when $\omega_r = 1/\sqrt{LC}$, and this is the **resonance frequency.** Since

$$|\mathbf{V}| = \left|\frac{\mathbf{I}}{\mathbf{Y}}\right| = \frac{|\mathbf{I}|}{|\mathbf{Y}|}$$

from the fact that $|\mathbf{Y}|$ is minimum at resonance, we conclude that $|\mathbf{V}|$ is maximum at the resonance frequency $\omega_r = 1/\sqrt{LC}$ and is given by

$$|\mathbf{V}| = \frac{|\mathbf{I}|}{1/R} = R|\mathbf{I}|$$

For $\omega = 0$, the impedance of the inductor is zero; thus, so is the voltage $\mathbf{V}$. As $\omega \to \infty$, the impedance of the capacitor $1/j\omega C \to 0$, and again $\mathbf{V} \to 0$. A sketch of $|\mathbf{V}|$ versus ω is shown in Fig. 5.12 where ω_1 and ω_2 are the "lower" and "upper" half-power frequencies respectively.

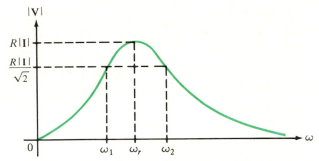

Fig. 5.12 Voltage magnitude versus frequency.

Since, at resonance, the parallel RLC combination acts simply as the resistance R, the parallel LC combination—known as a **tank circuit**—behaves as an open circuit.

We define the **bandwidth** BW of the resonant circuit in terms of the half-power frequencies by

$$BW = \omega_2 - \omega_1$$

Note that the smaller the bandwidth BW is, the "sharper" or "narrower" is the amplitude response.

Another conventional quantity that describes sharpness is the **quality factor,** designated Q, which is defined at the resonance frequency to be

$$Q = 2\pi\left(\frac{\text{maximum energy stored}}{\text{total energy lost in a period}}\right)$$

Since energy is stored in capacitors and inductors, the maximum energy stored is designated $[w_C(t) + w_L(t)]_{max}$. Since energy is dissipated by resistors, the energy lost in a period T is $P_R T$, where P_R is the resistive power, that is,

$$Q = \frac{2\pi[w_C(t) + w_L(t)]_{max}}{P_R T}$$

Now let us determine the quality factor for the parallel RLC circuit given in Fig. 5.11. Assume that the current source is given by $i(t) = I\cos\omega_r t$, where $\omega_r = 1/\sqrt{LC}$. At resonance, $\mathbf{Y} = 1/R$, so

$$v(t) = Ri(t) = RI\cos\omega_r t$$

Then the energy stored in the capacitor is

$$w_C(t) = \tfrac{1}{2}Cv^2(t) = \tfrac{1}{2}C(RI\cos\omega_r t)^2 = \tfrac{1}{2}CR^2 I^2\cos^2\omega_r t$$

Furthermore, since the inductor current $\mathbf{I}_L$ (going down) is $\mathbf{I}_L = \mathbf{V}/j\omega L$, then at resonance, since $v(t) = RI\cos\omega_r t$,

$$\mathbf{I}_L = \frac{RI\underline{/0°}}{\omega_r L\underline{/90°}} = \frac{RI}{\omega_r L}\underline{/-90°}$$

so

$$i_L(t) = \frac{RI}{\omega_r L}\cos(\omega_r t - 90°) = \frac{RI}{\omega_r L}\sin\omega_r t$$

at resonance. The fact that the LC tank portion of the circuit behaves as an open circuit at resonance does not mean that there are no currents in the inductor and capacitor. As a matter of fact, the current, given above, circulates around the loop formed by the inductor and capacitor.

At resonance, the energy stored in the inductor is

$$w_L(t) = \frac{1}{2}Li_L^2(t) = \frac{1}{2}\frac{R^2 I^2}{\omega_r^2 L}\sin^2\omega_r t = \frac{1}{2}CR^2 I^2\sin^2\omega_r t$$

since $\omega_r^2 = 1/LC$. Therefore, the total energy stored is

$$w_C(t) + w_L(t) = \tfrac{1}{2}CR^2 I^2(\cos^2\omega_r t + \sin^2\omega_r t) = \tfrac{1}{2}CR^2 I^2$$

which is a constant. Since the power absorbed by a resistor is $P_R = I^2 R/2$, then the energy lost in a period is

$$P_R T = \frac{1}{2}I^2 RT = \frac{1}{2}I^2 R\left(\frac{2\pi}{\omega_r}\right) = \frac{\pi I^2 R}{\omega_r}$$

Thus, the quality factor of the given parallel RLC circuit is

$$Q = \frac{2\pi(\tfrac{1}{2}CR^2 I^2)}{\pi I^2 R/\omega_r} = \omega_r RC$$

Since $\omega_r C = 1/\omega_r L$, alternatively, we can write

$$Q = \omega_r RC = R/\omega_r L = R\sqrt{C/L}$$

Again considering the expression for the admittance of the circuit in Fig. 5.11, we have that

$$\mathbf{Y} = \frac{1}{R} + j\left(\omega C - \frac{1}{\omega L}\right) = \frac{1}{R} + j\left(\frac{\omega C \omega_r R}{\omega_r R} - \frac{\omega_r R}{\omega L \omega_r R}\right)$$

$$= \frac{1}{R} + j\frac{1}{R}\left(\frac{\omega}{\omega_r}Q - \frac{\omega_r}{\omega}Q\right) = \frac{1}{R}\left[1 + jQ\left(\frac{\omega}{\omega_r} - \frac{\omega_r}{\omega}\right)\right]$$

At the half-power frequencies ω_1 and ω_2,

$$|\mathbf{V}| = \frac{R|\mathbf{I}|}{\sqrt{2}} = \frac{|\mathbf{I}|}{|\mathbf{Y}|} \quad \Rightarrow \quad |\mathbf{Y}| = \frac{\sqrt{2}}{R}$$

This condition occurs when

$$Q\left(\frac{\omega}{\omega_r} - \frac{\omega_r}{\omega}\right) = \pm 1 \tag{5.1}$$

For the case that the right-hand side of Equation (5.1) is $+1$, then

$$Q\left(\frac{\omega^2 - \omega_r^2}{\omega_r \omega}\right) = 1$$

from which

$$Q\omega^2 - \omega_r \omega - Q\omega_r^2 = 0$$

By the quadratic formula, we get a positive value and a negative value of ω that satisfy this equation. Since the half-power frequencies are positive quantities, we select the positive value. Thus, one of the half-power frequencies is

$$\frac{\omega_r}{2Q} + \omega_r\sqrt{\left(\frac{1}{2Q}\right)^2 + 1}$$

For the case that the right-hand side of Equation (5.1) is -1, as above we can show that the resulting half-power frequency is

$$-\frac{\omega_r}{2Q} + \omega_r\sqrt{\left(\frac{1}{2Q}\right)^2 + 1}$$

Thus, we see that the latter half-power frequency is ω_1 while the former is ω_2; consequently, the bandwidth for the parallel RLC circuit is

$$BW = \omega_2 - \omega_1 = \omega_r/Q \tag{5.2}$$

We also have that

$$BW = \omega_r/\omega_r RC = 1/RC$$

Thus, we see that when the Q of the circuit is relatively small, the bandwidth BW is relatively large; whereas if the Q is relatively large, then the BW is relatively small. The latter case indicates a sharper, or more frequency-selective, amplitude response.

From Equation (5.2), we have that $Q = \omega_r/BW$ for the parallel *RLC* circuit. This ratio of resonance frequency to bandwidth is sometimes defined to be the **selectivity** of a circuit. For a parallel *RLC* circuit, we see that the circuit's quality factor and selectivity are equal. In general, though, the selectivity and the quality factor of a circuit will not be the same.

Series Resonance

Let us now consider the case of a series *RLC* circuit as shown in Fig. 5.13. The impedance $\mathbf{Z} = \mathbf{V}/\mathbf{I}$ is given by

$$\mathbf{Z} = R + j\omega L + \frac{1}{j\omega C} = R + j\left(\omega L - \frac{1}{\omega C}\right)$$

so the resonance frequency ω_r is that frequency for which $\omega_r L - 1/\omega_r C = 0$.

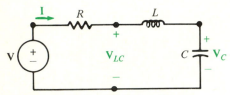

Fig. 5.13 Series *RLC* circuit.

Thus,

$$\omega_r = 1/\sqrt{LC}$$

as was the case for the parallel *RLC* circuit.

By duality, we have that the quality factor of the series *RLC* circuit is

$$Q = \omega_r L/R$$

and since $\omega_r L = 1/\omega_r C$, then also

$$Q = 1/\omega_r RC = \sqrt{L/C}/R$$

In a process similar to that for the parallel *RLC* circuit, it can be shown that the bandwidth for the series *RLC* circuit is again

$$BW = \omega_r/Q = R/L$$

so that the selectivity and the quality factor are equal.

Example 5.3

For the series *RLC* circuit shown in Fig. 5.13, where $R = \frac{5}{3}\ \Omega$, $L = 5$ H, and $C = \frac{1}{25}$ F, it can be shown (see Problem 4.7) that when $v(t) = 17\cos 3t$, then

$v_C(t) = 5\sqrt{17} \cos(3t - 166°)$. For this circuit, the resonance frequency is $\omega_r = 1/\sqrt{LC} = \sqrt{5}$ rad/s. Thus, the quality factor is

$$Q = \frac{\omega_r L}{R} = \frac{5\sqrt{5}}{5/3} = 3\sqrt{5}$$

and the bandwidth is

$$BW = \frac{R}{L} = \frac{5/3}{5} = \frac{1}{3} \text{ rad/s}$$

A derivation of the formulas for the half-power frequencies for the series RLC circuit can be accomplished as was done for the parallel RLC circuit. What result are the same expressions. Thus, the lower half-power frequency is

$$\omega_1 = -\frac{\omega_r}{2Q} + \omega_r \sqrt{\left(\frac{1}{2Q}\right)^2 + 1} = 2.082 \text{ rad/s}$$

and the upper half-power frequency is

$$\omega_2 = \frac{\omega_r}{2Q} + \omega_r \sqrt{\left(\frac{1}{2Q}\right)^2 + 1} = 2.415 \text{ rad/s}$$

And as a check on the above result, the bandwidth is

$$BW = \omega_2 - \omega_1 = 0.333 \text{ rad/s}$$

Suppose now that the frequency of the voltage source equals the circuit's resonance frequency, say $v(t) = 17 \cos \sqrt{5}t$. Since $|\mathbf{I}| = |\mathbf{V}|/|\mathbf{Z}|$, at resonance

$$|\mathbf{I}| = \frac{|\mathbf{V}|}{5/3} = \frac{3}{5}(17) = \frac{51}{5} \text{ A}$$

Then

$$|\mathbf{V}_C| = |\mathbf{Z}_C||\mathbf{I}| = \frac{1}{\omega C}|\mathbf{I}| = \frac{25}{\sqrt{5}}\left(\frac{51}{5}\right) = 51\sqrt{5} \text{ V}$$

Note that although the amplitude of the input sinusoid is $|\mathbf{V}| = 17$ V, the amplitude of the output voltage is $|\mathbf{V}_C| = 51\sqrt{5}$ V. That is, the output voltage magnitude is $Q = 3\sqrt{5}$ times larger than the input. Futhermore, since

$$\mathbf{V}_{LC} = (\mathbf{Z}_L + \mathbf{Z}_C)\mathbf{I} = \left(j\omega L + \frac{1}{j\omega C}\right)I = \left(j5\sqrt{5} - \frac{j}{\sqrt{5}/25}\right)\mathbf{I} = 0$$

then the series combination of the inductor and capacitor acts as a short circuit, although the voltage across each element is nonzero.

Just as we defined the quality factor Q for a resonant circuit, we can define it for a reactance in series or parallel with a resistance. To demonstrate this fact, consider the series case for an inductor L as shown in Fig. 5.14. Let us assume that $\mathbf{I} = I\underline{/0°}$. Then the energy stored is given by

$$w(t) = \tfrac{1}{2}Li^2(t) = \tfrac{1}{2}LI^2 \cos^2 \omega t$$

which has the maximum value

$$w_m = \tfrac{1}{2}LI^2$$

and since $X_s = \omega L \quad \Rightarrow \quad L = X_s/\omega$, then

$$w_m = \frac{X_s}{2\omega}I^2$$

The energy lost per period is

$$P_sT = \frac{1}{2}I^2R_s\left(\frac{2\pi}{\omega}\right) = \frac{\pi I^2 R_s}{\omega}$$

Therefore, the quality factor is

$$Q = \frac{2\pi(X_s I^2/2\omega)}{\pi I^2 R_s/\omega} = \frac{X_s}{R_s}$$

Fig. 5.14 Series inductive reactance. 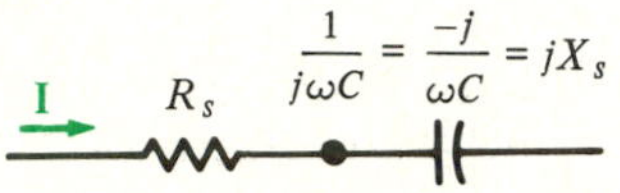   **Fig. 5.15 Series capacitive reactance.**

For the case of a resistor in series with a capacitor C as shown in Fig. 5.15, proceeding as above, we may obtain the fact that $Q = |X_s|/R_s$. Hence, in general, for a resistance in series with a reactance X_s, the quality factor is

$$Q = \frac{|X_s|}{R_s}$$

For the case of a resistance in parallel with a reactance X_p as shown in Fig. 5.16, it can be shown that the quality factor is

$$Q = \frac{R_p}{|X_p|}$$

Consider the resonant circuit shown in Fig. 5.17, where the admittance is

$$\mathbf{Y} = j\omega C + \frac{1}{R_s + j\omega L}$$

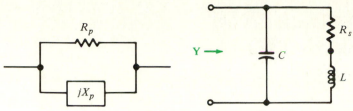

Fig. 5.16 Parallel reactance. Fig. 5.17 High-Q resonant circuit.

Suppose that the series reactance, consisting of the inductor L and resistor R_s, has a large (at least 20) quality factor (called a "high-Q coil"). Then

$$\frac{X_L}{R_s} \gg 1 \quad \Rightarrow \quad \frac{\omega L}{R_s} \gg 1 \quad \Rightarrow \quad \omega L \gg R_s$$

Then the admittance is approximated by

$$\mathbf{Y} \approx j\omega C + \frac{1}{j\omega L}$$

Thus, the resonance frequency is approximated by $\omega_r \approx 1\sqrt{LC}$. The impedance is

$$\mathbf{Z} = \frac{(R_s + j\omega L)(1/j\omega C)}{R_s + j\omega L + 1/j\omega C} \approx \frac{(j\omega L)(1/j\omega C)}{R_s + j\omega L + 1/j\omega C}$$

for a high-Q series reactance. Thus, the admittance is approximated by

$$\mathbf{Y} \approx \frac{C}{L}\left(R_s + j\omega L + \frac{1}{j\omega C}\right) = \frac{R_s C}{L} + j\omega C + \frac{1}{j\omega L}$$

which is the expression for a parallel RLC circuit, where $R = L/R_s C$. Hence, for a high-Q coil, the two admittances shown in Fig. 5.18 are approximately equal.

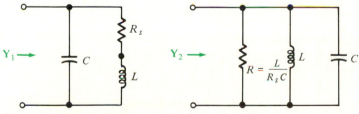

Fig. 5.18 Approximate equivalent circuits.

Example 5.4

The practical tank circuit shown in Fig. 5.19 is to be in resonance at 1 MHz, that is, $\omega_r = 2\pi \times 10^6$ rad/s. Assuming a high-Q coil, $\omega_r \approx 1/\sqrt{LC}$, from which

$$L \approx \frac{1}{\omega_r^2 C} = \frac{1}{(4\pi^2 \times 10^{12})(500 \times 10^{-12})} = 50.7 \ \mu\text{H}$$

The quality factor of the series reactance at the resonance frequency is

$$\frac{X_s}{10} = \frac{\omega_r L}{10} = \frac{318}{10} = 31.8$$

which is much greater than one, as was assumed. Since

$$\frac{L}{R_s C} = 10,132 \approx 10,000$$

then the tank circuit given in Fig. 5.19 can be approximated by the parallel *RLC* circuit shown in Fig. 5.20. The quality factor of this parallel *RLC* circuit is

$$Q = \omega_r RC = (2\pi \times 10^6)(10 \times 10^3)(500 \times 10^{-12}) = 31.4$$

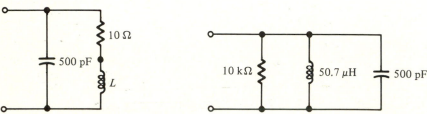

Fig. 5.19 Practical tank circuit.   **Fig. 5.20 Equivalent of practical tank circuit.**

5.3 COMPLEX FREQUENCY

In the preceding chapter, our major preoccupation was with the sinusoid. By generalizing real sinusoids to complex sinusoids, we developed the notion of phasors with which we were able to find forced responses of sinusoidal circuits without having to deal with differential equations. As a consequence, we were able to perform analyses in the frequency domain much more easily than in the time domain.

We are now in a position to go one step further. By extending the above-mentioned concepts, we can employ frequency-domain analysis for damped sinusoidal circuits, that is, circuits having forcing functions of the form $Ae^{\sigma t} \cos(\omega t + \theta)$.

For the damped sinusoidal function $Ae^{\sigma t} \cos(\omega t + \theta)$, we know that the frequency ω has radians per second as its unit. Since the power of the natural logarithm base e is to be dimensionless, the units of the (usually nonpositive) damping factor σ have been designated **nepers per second,** where the **neper*** (Np) is a dimensionless unit. For this reason, σ is referred to as the **neper frequency.**

---

*Named for the Scottish mathematician John Napier (1550–1617).

Now suppose that the input to an *RLC* circuit has the form $Ae^{\sigma t}\cos(\omega t + \theta)$. Using an argument similar to that for the real sinusoidal case, the forced response will have the form $Be^{\sigma t}\cos(\omega t + \phi)$. Thus, by the property of linearity, the response to

$$Ae^{\sigma t}\cos(\omega t + \theta) + jAe^{\sigma t}\sin(\omega t + \theta) = Ae^{\sigma t}e^{j(\omega t + \theta)}$$
$$= Ae^{j\theta}e^{(\sigma + j\omega)t}$$
$$= Ae^{j\theta}e^{st}$$
$$= \mathbf{V}_{in}e^{st}$$

where $\mathbf{V}_{in} = Ae^{j\theta}$ and $s = \sigma + j\omega$, which is called **complex frequency,** has the form

$$Be^{\sigma t}\cos(\omega t + \phi) + jBe^{\sigma t}\sin(\omega t + \phi) = Be^{\sigma t}e^{j(\omega t + \phi)}$$
$$= Be^{j\phi}e^{(\sigma + j\omega)t}$$
$$= Be^{j\phi}e^{st}$$
$$= \mathbf{V}_o e^{st}$$

where $\mathbf{V}_o = Be^{j\phi}$. Hence, knowing the response to $Ae^{j\theta}e^{st}$ is tantamount to knowing the response to $Ae^{\sigma t}\cos(\omega t + \theta)$. That is, we can determine B and ϕ by finding the response either to $Ae^{j\theta}e^{st}$ or to $Ae^{\sigma t}\cos(\omega t + \theta)$.

Just as we associated the real sinusoid $A\cos(\omega t + \theta)$ or the complex sinusoid $Ae^{j(\omega t + \theta)}$ with the phasor $\mathbf{A} = Ae^{j\theta}$, so we can associate the damped sinusoid $Ae^{\sigma t}\cos(\omega t + \theta)$ or the damped complex sinusoid

$$Ae^{\sigma t}e^{j(\omega t + \theta)} = Ae^{j\theta}e^{st}$$

with the phasor $\mathbf{A} = Ae^{j\theta}$ also. However, for the latter case, $s = \sigma + j\omega$ must be implicit, whereas only ω need be for the former.

For a resistor R having voltage v across it and current i through it, if the current is a damped complex sinusoid, say

$$i = Ie^{\sigma t}e^{j(\omega t + \theta)} = Ie^{j\theta}e^{st}$$

then the voltage is also a damped complex sinusoid, say

$$v = Ve^{\sigma t}e^{j(\omega t + \phi)} = Ve^{j\phi}e^{st}$$

Since $v = Ri$, then

$$Ve^{j\phi}e^{st} = RIe^{j\theta}e^{st}$$

and dividing by e^{st} yields

$$Ve^{j\phi} = RIe^{j\theta}$$

from which we make the identities $V = RI$ and $\phi = \theta$. Using phasor notation,

$$V\underline{/\phi} = RI\underline{/\theta} \quad \Rightarrow \quad \mathbf{V} = R\mathbf{I}$$

where $\mathbf{V} = V\underline{/\phi}$ and $\mathbf{I} = I\underline{/\theta}$. Defining the **impedance** $\mathbf{Z}_R$ of the resistance as the ratio of voltage phasor to current phasor, we have

$$\mathbf{Z}_R = \frac{\mathbf{V}}{\mathbf{I}} = R$$

For an inductor L, if $i = Ie^{j\theta}e^{st}$ and $v = Ve^{j\phi}e^{st}$, since $v = L\,di/dt$, then

$$Ve^{j\phi}e^{st} = L\frac{d}{dt}(Ie^{j\theta}e^{st}) = LsIe^{j\theta}e^{st}$$

Dividing by e^{st} gives

$$Ve^{j\phi} = LsIe^{j\theta}$$

or

$$V\underline{/\phi} = LsI\underline{/\theta} \qquad \Rightarrow \qquad \mathbf{V} = Ls\mathbf{I}$$

Thus, the impedance $\mathbf{Z}_L$ of an inductor is

$$\mathbf{Z}_L = \frac{\mathbf{V}}{\mathbf{I}} = Ls$$

For a capacitor C, since $i = C\,dv/dt$, then

$$Ie^{j\theta}e^{st} = C\frac{d}{dt}(Ve^{j\phi}e^{st}) = CsVe^{j\phi}e^{st}$$

from which

$$Ie^{j\theta} = CsVe^{j\phi}$$

so

$$I\underline{/\theta} = CsV\underline{/\phi} \qquad \Rightarrow \qquad \mathbf{I} = Cs\mathbf{V}$$

The impedance of a capacitor is, therefore,

$$\mathbf{Z}_C = \frac{\mathbf{V}}{\mathbf{I}} = \frac{1}{Cs}$$

From the definition of impedance, we see that we have the equations

$$\mathbf{V} = \mathbf{ZI} \qquad \text{and} \qquad \mathbf{I} = \mathbf{V}/\mathbf{Z}$$

This latter equation suggests the usefulness of the reciprocal of impedance. We, therefore, define the ratio of current phasor to voltage phasor as **admittance**, denoted **Y**. Thus, we have the element admittances

$$\mathbf{Y}_R = \frac{1}{R}, \qquad \mathbf{Y}_L = \frac{1}{Ls}, \qquad \mathbf{Y}_C = Cs$$

and the alternative forms of Ohm's law:

$$\mathbf{Y} = \mathbf{I}/\mathbf{V}, \qquad \mathbf{I} = \mathbf{YV}, \qquad \mathbf{V} = \mathbf{I}/\mathbf{Y}$$

Just as we analyzed sinusoidal circuits with the use of (sinusoidal) phasors and impedance, so can we analyze damped sinusoidal circuits. For the former case, impedance is a function of ω—denoted $\mathbf{Z}(j\omega)$, and for the latter, it is a function of σ and ω—denoted $\mathbf{Z}(s)$.

Example 5.5

For the circuit given in Fig. 5.21, using phasor and impedance notation, we get the circuit shown in Fig. 5.22. By KCL,

$$\frac{\mathbf{V}_C - \mathbf{V}}{2s} + \frac{\mathbf{V}_C}{4} + \frac{\mathbf{V}_C}{4/s} = 0$$

or

$$2(\mathbf{V}_C - \mathbf{V}) + s\mathbf{V}_C + s^2\mathbf{V}_C = 0$$

$$(s^2 + s + 2)\mathbf{V}_C = 2\mathbf{V}$$

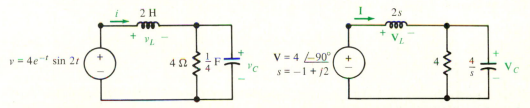

Fig. 5.21 Damped-sinusoidal circuit.   **Fig. 5.22 Circuit in the frequency domain.**

Thus,

$$\mathbf{V}_C = \frac{2\mathbf{V}}{s^2 + s + 2}$$

Alternatively, we can find $\mathbf{V}_C$ as follows: The parallel *RC* combination has the impedance

$$\mathbf{Z}_{RC} = \frac{4(4/s)}{4 + 4/s} = \frac{4}{s + 1}$$

By voltage division, we have that

$$\mathbf{V}_C = \frac{\mathbf{Z}_{RC}}{\mathbf{Z}_{RC} + \mathbf{Z}_L} \mathbf{V} = \frac{4/(s + 1)}{4/(s + 1) + 2s} \mathbf{V} = \frac{2\mathbf{V}}{s^2 + s + 2}$$

From the fact that $\mathbf{V} = 4\underline{/-90°}$ and $s = -1 + j2$, we get

$$\mathbf{V}_C = \frac{2(4\underline{/-90°})}{(-1 + j2)^2 + (-1 + j2) + 2} = 2\sqrt{2}\underline{/45°}$$

Thus, the forced response is

$$v_C(t) = 2\sqrt{2}\, e^{-t} \cos(2t + 45°)$$

By removing the capacitor and applying Thévenin's theorem, we obtain the circuits shown in Fig. 5.23(a) and (b). By voltage division

$$\mathbf{V}_{oc} = \frac{4\mathbf{V}}{4 + 2s} = \frac{2\mathbf{V}}{s + 2}$$

and combining the resistor and inductor in parallel,

$$\mathbf{Z}_o = \frac{2s(4)}{2s + 4} = \frac{4s}{s + 2}$$

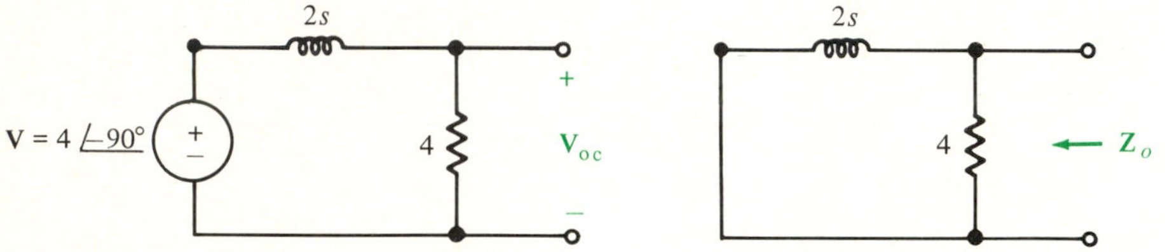

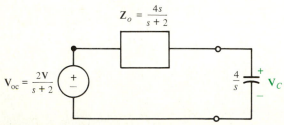

(a) Open-circuit voltage   (b) Output impedance

Fig. 5.23 Determination of Thévenin-equivalent circuit.

Hence, we get the Thévenin-equivalent circuit shown in Fig. 5.24. Again, by voltage division

$$\mathbf{V}_C = \frac{(4/s)\mathbf{V}_{oc}}{4/s + 4s/(s + 2)} = \frac{(s + 2)\mathbf{V}_{oc}}{(s + 2) + s^2} = \frac{2\mathbf{V}}{s^2 + s + 2}$$

as was obtained previously.

$$\mathbf{Z}_o = \frac{4s}{s + 2}$$

$$\mathbf{V}_{oc} = \frac{2\mathbf{V}}{s + 2}$$

Fig. 5.24 Thévenin-equivalent circuit.

Poles and Zeros

If we define the voltage transfer function $\mathbf{H}_C(s)$ to be the ratio of $\mathbf{V}_C$ to $\mathbf{V}$, we then have that for the circuit given in Fig. 5.22:

$$\mathbf{H}_C(s) = \frac{\mathbf{V}_C}{\mathbf{V}} = \frac{2}{s^2 + s + 2}$$

Factoring the denominator, we can rewrite $\mathbf{H}_C(s)$ as

$$\mathbf{H}_C(s) = \frac{2}{(s + 1/2 - j\sqrt{7}/2)(s + 1/2 + j\sqrt{7}/2)}$$

Thus, we see for the special case that

$$s = -\frac{1}{2} + j\frac{\sqrt{7}}{2} \qquad \text{or} \qquad s = -\frac{1}{2} - j\frac{\sqrt{7}}{2}$$

the transfer function becomes infinite—which implies that the forced response is infinite. We say that the function $\mathbf{H}_C(s)$ has a **pole** at each of these two complex frequencies.

To find the voltage transfer function $\mathbf{H}_L(s) = \mathbf{V}_L/\mathbf{V}$ for the circuit in Fig. 5.22, we can again use voltage division as follows. From

$$\mathbf{V}_L = \frac{2s\mathbf{V}}{2s + 4/(s + 1)} = \frac{2s(s + 1)\mathbf{V}}{2s(s + 1) + 4}$$

we get

$$\mathbf{H}_L(s) = \frac{\mathbf{V}_L}{\mathbf{V}} = \frac{s(s + 1)}{s^2 + s + 2}$$

Therefore, $\mathbf{H}_L(s)$ has the same poles as $\mathbf{H}_C(s)$. However, there are two frequencies, $s = 0$ and $s = -1$, for which $\mathbf{H}_L(s) = 0$. For this reason, we say that the network function $\mathbf{H}_L(s)$ has **zeros** at $s = 0$ and at $s = -1$.

Since $s = \sigma + j\omega$ is a complex quantity, we can indicate any value of s as a point in a complex-number plane. If we label the horizontal axis σ and the vertical axis $j\omega$, the result is known as the **s-plane.** If we denote a pole by the symbol "×" and a zero by the symbol "○," then depicting the poles and zeros of a function in the s-plane is referred to as a **pole-zero plot.** The pole-zero plot for $\mathbf{H}_L(s)$ is shown in Fig. 5.25. The pole-zero plot for $\mathbf{H}_C(s)$ can be obtained from this by removing the two zeros on the σ (real) axis, that is, the pole-zero plot for $\mathbf{H}_C(s)$ does not have any (finite) zeros.

Analysis of the circuit in Fig. 5.21 results in the differential equation

$$\frac{d^2 v_C(t)}{dt^2} + \frac{dv_C(t)}{dt} + 2v_C(t) = 2v(t)$$

The solution to this equation when $v(t) = 0$ is the natural response. As was done previously, we get the natural response by assuming a solution of the form Ae^{st}.

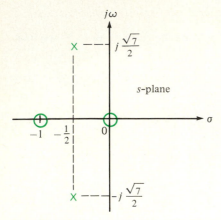

Fig. 5.25 Pole-zero plot.

Doing so, we obtain two values for s that satisfy the equation

$$s^2 + s + 2 = 0$$

Using the quadratic formula, we get

$$s = \frac{-1 \pm \sqrt{1-8}}{2} = -\frac{1}{2} \pm j\frac{\sqrt{7}}{2}$$

In other words, the natural response $v_n(t)$ has the form

$$v_n(t) = A_1 e^{s_1 t} + A_2 e^{s_2 t}$$

where

$$s_1 = -\frac{1}{2} + j\frac{\sqrt{7}}{2} \qquad \text{and} \qquad s_2 = -\frac{1}{2} - j\frac{\sqrt{7}}{2}$$

But note, these are precisely the poles of the voltage transfer function $\mathbf{H}_C(s) = \mathbf{V}_C/\mathbf{V}$. Thus, and this is true in general, the poles of the transfer function specify the natural response—provided that there was no cancellation of a common pole and zero. In this example, the complex values for s_1 and s_2, that is, the complex poles, indicate that this second-order circuit is underdamped. Thus, the natural response has the more common alternative forms

$$v_n(t) = B_1 e^{-\alpha t} \cos \omega_d t + B_2 e^{-\alpha t} \sin \omega_d t$$

$$= B e^{-\alpha t} \cos(\omega_d t - \phi)$$

and from the above differential equation, $\alpha = \frac{1}{2}$, $\omega_n = \sqrt{2}$, and

$$\omega_d = \sqrt{\omega_n^2 - \alpha^2} = \sqrt{2 - 1/4} = \sqrt{7}/2$$

Hence, for an input of

$$v(t) = 4e^{-t} \sin 2t \, u(t)$$

the complete response has the form (see Example 5.5 for the forced response)

$$v_C(t) = Be^{-t/2} \cos\left(\frac{\sqrt{7}}{2}t - \phi\right) + 2\sqrt{2}e^{-t}\cos(2t + \pi/4)$$

The constants B and ϕ (or B_1 and B_2) are determined from the initial conditions $i(0)$ and $v_C(0)$.

If in the above discussion, the poles of $\mathbf{H}_C(s)$ were real, say $s_1 = -a_1$ and $s_2 = -a_2$, instead of complex, then the form of the natural response would be that of the overdamped case, that is,

$$v_n(t) = A_1 e^{-a_1 t} + A_2 e^{-a_2 t}$$

Finally, if the two poles of $\mathbf{H}_C(s)$ were real and equal (called a **double pole**), say $s_1 = s_2 = -\alpha$, then the form of the natural response would be that of the critically damped case, that is,

$$v_n(t) = A_1 e^{-\alpha t} + A_2 t e^{-\alpha t}$$

Again let us return to the circuit given in Fig. 5.22 and consider yet another network function for the same input variable. This time, suppose that the output variable is the inductor current $\mathbf{I}$. Since the impedance seen by the voltage source is

$$\mathbf{Z} = \mathbf{Z}_L + \mathbf{Z}_{RC} = 2s + \frac{4}{s+1} = \frac{2(s^2 + s + 2)}{s+1}$$

and since $\mathbf{Z} = \mathbf{V}/\mathbf{I}$, then the transfer function of interest $\mathbf{I}/\mathbf{V}$ is just the admittance $\mathbf{Y}(s)$ seen by the source, that is,

$$\mathbf{Y}(s) = \frac{\mathbf{I}}{\mathbf{V}} = \frac{\frac{1}{2}(s+1)}{s^2 + s + 2}$$

and again the poles of $\mathbf{Y}(s)$ are the same as the poles of $\mathbf{H}_C(s)$ and $\mathbf{H}_L(s)$. The pole-zero plot for $\mathbf{Y}(s)$ can be obtained from the pole-zero plot for $\mathbf{H}_L(s)$ by removing the zero at the origin of the s-plane.

It is not just coincidence that the poles for the network functions $\mathbf{V}_C/\mathbf{V}$, $\mathbf{V}_L/\mathbf{V}$, and $\mathbf{I}/\mathbf{V}$ determined above are the same. Provided that one portion of a circuit is not physically separated from the rest, each transfer function (defined as the ratio of an output to a given input) will have the same poles regardless of which voltage or current is chosen as the output variable. This should not be surprising though, for the poles are those values of s, called **natural frequencies,** that determine the natural response. And since the form of the natural response is the same throughout a (nonseparated) circuit, the poles of any transfer function (with respect to a given input variable) will be the same.

5.4 INTRODUCTION TO SYSTEMS

As was demonstrated in Section 5.3, a linear circuit (or any system described by a linear differential equation with constant coefficients, for that matter) having an input, say $x(t) = \mathbf{X}e^{st}$, will have a forced response of the form $y(t) = \mathbf{Y}e^{st}$. Substituting these terms into the general form of the differential equation describing the circuit (or system)

$$a_n \frac{d^n y(t)}{dt^n} + a_{n-1} \frac{d^{n-1} y(t)}{dt^{n-1}} + \cdots + a_1 \frac{dy(t)}{dt} + a_0 y(t)$$

$$= b_m \frac{d^m x(t)}{dt^m} + b_{m-1} \frac{d^{m-1} x(t)}{dt^{m-1}} + \cdots + b_1 \frac{dx(t)}{dt} + b_0 x(t) \tag{5.3}$$

results in

$$(a_n s^n + a_{n-1} s^{n-1} + \cdots + a_1 s + a_0) \mathbf{Y} e^{st}$$

$$= (b_m s^m + b_{m-1} s^{m-1} + \cdots + b_1 s + b_0) \mathbf{X} e^{st}$$

From this expression, we find that the general form of the transfer function $\mathbf{H}(s)$ is

$$\mathbf{H}(s) = \frac{\mathbf{Y}}{\mathbf{X}} = \frac{b_m s^m + b_{m-1} s^{m-1} + \cdots + b_1 s + b_0}{a_n s^n + a_{n-1} s^{n-1} + \cdots + a_1 s + a_0} \tag{5.4}$$

That is, $\mathbf{Y}/\mathbf{X}$ is a ratio of polynomials in s with real coefficients.

A result from mathematics theory is that a polynomial with real coefficients can be factored into a product of quadratics of the form $as^2 + bs + c$, where a, b, and c are real numbers. From the quadratic formula, such a term can further be factored into the product $(s - s_1)(s - s_2)$, where

$$s_1 = \frac{-b + \sqrt{b^2 - 4ac}}{2a} \quad \text{and} \quad s_2 = \frac{-b - \sqrt{b^2 - 4ac}}{2a}$$

For the case that $b^2 \geq 4ac$, the numbers s_1 and s_2 are real. However, if $b^2 < 4ac$, then s_1 and s_2 are complex numbers and are conjugates. As a consequence of this, we can express the transfer function, or any other network function for that matter, in the form

$$\mathbf{H}(s) = \frac{K(s - z_1)(s - z_2) \cdots (s - z_m)}{(s - p_1)(s - p_2) \cdots (s - p_n)} \tag{5.5}$$

where $z_1, z_2, \ldots, z_m$ are the zeros and $p_1, p_2, \ldots, p_n$ are the poles. The zeros and poles can be either real or complex, but complex zeros or poles occur in conjugate pairs.

For the case of distinct poles, the natural response $y_n(t)$ has the form

$$y_n(t) = A_1 e^{p_1 t} + A_2 e^{p_2 t} + \cdots + A_n e^{p_n t}$$

If pole p_i is at the origin of the s-plane, that is, if $p_i = 0$, then the term in the natural response corresponding to it is $A_i e^0 = A_i$, a constant. If p_i is on the negative real

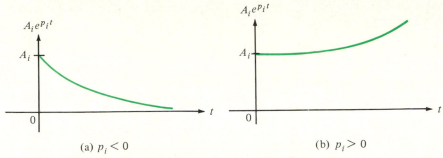

(a) $p_i < 0$ (b) $p_i > 0$

Fig. 5.26 Natural-response forms for a real pole.

axis or the positive real axis of the s-plane, then $A_i e^{p_i t}$ is a decaying exponential or increasing exponential, respectively, as shown in Fig. 5.26. If p_i is a complex number, then there is a pole $p_j = p_i^*$. As we saw in the discussion on second-order circuits, when $p_i = \sigma + j\omega$, the corresponding terms in the natural response can be combined in the form $A e^{\sigma t} \cos(\omega t + \theta)$. If the conjugate pair of poles is in the left half of the s-plane (i.e., $\sigma < 0$), this term is a damped sinusoid. If the pair is on the $j\omega$-axis ($\sigma = 0$), the term is a (pure) sinusoid. If the pair is in the right half of the s-plane, the term is an increasing sinusoid. These three cases are depicted in Fig. 5.27.

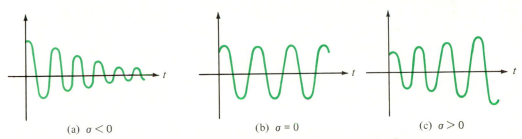

(a) $\sigma < 0$ (b) $\sigma = 0$ (c) $\sigma > 0$

Fig. 5.27 Natural-response forms for complex-conjugate poles.

As may be surmised from this discussion, a circuit that has a "stable" operation does not have a network function with poles in the right half of the s-plane. In some applications, poles on the imaginary axis are undesirable, and only poles in the left-half s-plane are allowable.

Block Diagrams

Consider a linear system (i.e., a system described by a linear differential equation) whose input is **X** and whose output is **Y** as depicted by the "block" in Fig. 5.28. The transfer function of the system is

$$\mathbf{H}(s) = \frac{\mathbf{Y}}{\mathbf{X}}$$

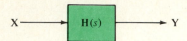

Fig. 5.28 Linear system.

Suppose that the transfer function of the system is $\mathbf{H}(s) = 1/s$. Then from Equation (5.4), we have that $a_1 = b_0 = 1$ and $a_0 = a_2 = a_3 = \cdots = a_n = b_1 = b_2 = \cdots = b_m = 0$. Thus, the differential equation describing the system is

$$\frac{dy(t)}{dt} = x(t)$$

Therefore,

$$\int \frac{dy(t)}{dt} \, dt = \int x(t) \, dt = y(t)$$

In other words, the output of the system is the integral of the input. For this reason, we call this simple system an **integrator** and can represent a device that performs integration as in Fig. 5.29.

Fig. 5.29 Integrator.

Another simple system is shown in Fig. 5.30. In this case, the output is just the sum of the two inputs $\mathbf{X}_1$ and $\mathbf{X}_2$. For this reason, this device is called an **adder** or **summer.** Although the adder in Fig. 5.30 has two inputs, an adder can have three or more inputs as well.

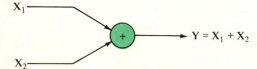

Fig. 5.30 Two-input adder.

For the simple system shown in Fig. 5.31, the output equals the input scaled by the constant A. Thus, we call this device a **scaler** or **amplifier.**

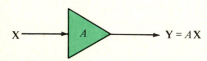

Fig. 5.31 Scaler.

Using the three types of devices described (integrators, adders, and scalers), we now can "simulate" or "build" a linear system described by a transfer function having the form of Equation (5.4), provided that $n \geq m$.

Example 5.6

Let us construct a system whose transfer function is

$$\mathbf{H}(s) = \frac{\mathbf{Y}}{\mathbf{X}} = \frac{s + 2}{3s^2 + 4s + 5} \tag{5.6}$$

From this expression, we have that

$$(3s^2 + 4s + 5)\mathbf{Y} = (s + 2)\mathbf{X}$$

$$3s^2\mathbf{Y} = s\mathbf{X} + 2\mathbf{X} - 4s\mathbf{Y} - 5\mathbf{Y}$$

$$\mathbf{Y} = \frac{1}{3s}\mathbf{X} + \frac{2}{3s^2}\mathbf{X} - \frac{4}{3s}\mathbf{Y} - \frac{5}{3s^2}\mathbf{Y}$$

This last equation can be realized as shown in Fig. 5.32. We deduce, therefore, that the system given in Fig. 5.32 has the transfer function specified by Equation (5.6).

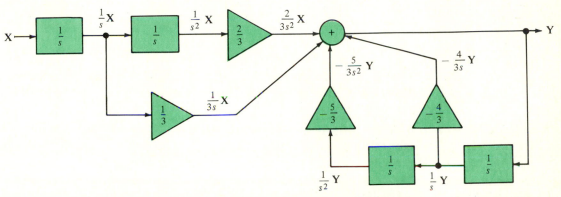

Fig. 5.32 Realization (simulation) of given transfer function.

We did not indicate what type of system (electrical, mechanical, etc.) was described by Equation (5.6). However, regardless of what type of system it is, we can simulate it with the system shown in Fig. 5.32. Thus, if the original system is cumbersome or not easily accessible, we can test or study its responses to various inputs by applying these inputs to the simulated system in Fig. 5.32. This is precisely what the analog computer is all about. An **analog computer** is an electronic system that contains integrators, adders, and scalers, and on which we can simulate linear systems.

In our discussion of operational amplifiers, we saw how we could use op amps and other circuit elements to construct integrators, adders, and scalers. Yet, as in Fig. 5.32, for example, we do not need to indicate specifically how these devices are fabricated, and instead we merely represent them by appropriately labeled "blocks" (or circles or triangles). For this reason, we can refer to Fig. 5.32 as a **block diagram** of the system described by Equation (5.6).

In general, there may be more than one way to represent a system in terms of various building blocks, and blocks other than integrators, adders, and scalers can be utilized. For instance, it should be relatively easy by now to show that the simple low-pass filter given in Fig. 5.33(a) has the voltage transfer function

$$\mathbf{H}(s) = \frac{\mathbf{V}_2}{\mathbf{V}_1} = \frac{1}{RCs + 1}$$

and we can represent this simple system by the single block shown in Fig. 5.33(b).

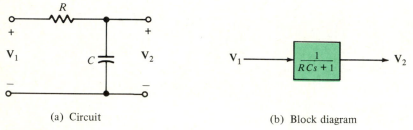

(a) Circuit   (b) Block diagram

Fig. 5.33 Simple low-pass filter.

Example 5.7

Consider the system with input **X** and output **Y** indicated by the block diagram given in Fig. 5.34. Since the output of the block labeled **H** is **HY**, then the output

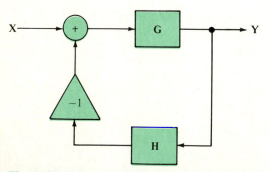

Fig. 5.34 System with negative feedback.

of the adder is $X - HY$. Hence, an expression for the overall system output Y is

$$Y = G(X - HY)$$

from which the output is given by

$$Y = \frac{G}{1 + GH} X$$

and the system transfer function is

$$\frac{Y}{X} = \frac{G}{1 + GH}$$

Feedback

Note that, in the preceding example, a portion (HY) of the output Y is "fed back" and subtracted from the input X and applied to G. (The output of G is the overall system output Y.) This, then, is an example of **feedback,** and since a portion of the output is subtracted from the input, the feedback is **negative.** If a portion of the output were added to the input, we would have **positive** feedback.

There are two major applications of feedback in electrical systems. One is to change the characteristics or performance of a system to obtain a desirable effect, and the other is to control a system (such a system is called an **automatic control system,** or **control system** for short).

Example 5.8

To illustrate how negative feedback can be used to improve system performance, consider an amplifier whose transfer function (gain) $G = 1100$. Suppose that, because of a temperature change, aging of components, faulty repair, or a combination of these and other factors, the gain drops to $G' = 900$. This is a change of 18 percent of the original value.

Now, however, suppose that the amplifier $G = 1100$ is part of a feedback system as depicted in Fig. 5.34. Further, suppose that the feedback element H is a simple resistive voltage-divider network with voltage transfer function $H = \frac{1}{110}$. Then the gain of the overall system (amplifier) is

$$G_f = \frac{G}{1 + GH} = \frac{1100}{1 + 1100(1/110)} = 100$$

If the gain of G drops to $G' = 900$, the overall system gain is

$$G'_f = \frac{G'}{1 + G'H} = \frac{900}{1 + 900(1/110)} = 98$$

and this is a change of only 2 percent. Thus, we see that although the amplifier **G** = 1100 without feedback has more gain than the feedback amplifier $\mathbf{G}_f$ = 100, it is the latter whose gain is much less sensitive to possible parameter variations. The fact that the feedback amplifier has less gain than the amplifier without feedback is not usually considered a problem as further stages of amplification (with feedback) can be added if needed.

In the preceding example, we saw how negative feedback reduced the sensitivity of amplifier gain by reducing its dependence on various device parameters. Negative feedback can offer other advantages as well. It can also reduce the sensitivity of amplifier gain to frequency-dependent elements such as inductors and capacitors. In addition, negative feedback can reduce the effect of disturbances resulting from distortion, external interference, or internal noise. Network parameters such as circuit time constants and input and output impedance also can be modified with the use of negative feedback.

Not only are there advantages to negative feedback, but positive feedback also has useful applications, as is illustrated in the following example.

Example 5.9

Consider the positive-feedback system shown in Fig. 5.35, which contains a simple high-pass filter $\mathbf{G}(s) = 2s/(s + 1)$ and whose feedback element is a simple low-pass filter $\mathbf{H}(s) = 1/(s + 1)$. As was done for the feedback system in Fig. 5.34, we can easily obtain

$$\mathbf{G}_f(s) = \frac{\mathbf{Y}}{\mathbf{X}} = \frac{\mathbf{G}(s)}{1 - \mathbf{G}(s)\mathbf{H}(s)}$$

from which

$$\mathbf{G}_f(s) = \frac{2s/(s + 1)}{1 - [2s/(s + 1)][1/(s + 1)]} = \frac{2s(s + 1)}{s^2 + 1} = \frac{2s(s + 1)}{(s - j)(s + j)}$$

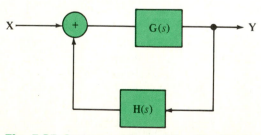

Fig. 5.35 System with positive feedback.

Hence, the transfer function for this feedback system has poles on the imaginary axis of the s-plane (at $s = j$ and $s = -j$). From our discussion at the beginning of this section (see Fig. 5.27), we see that this system will produce a sinusoid of frequency $\omega = 1$ rad/s. Such a device is called an **oscillator,** and there are a number of electronic circuits for constructing one (see Section 10.2).

So far we have seen two specific examples demonstrating how feedback can be used to obtain desirable performance from a system. We shall now describe how feedback can be employed for automatic control. The feedback system (albeit nonlinear) that will be discussed is the phase-locked loop—a device that, in recent years, has seen increasing use in FM demodulation.

A **phase-locked loop** (PLL) is an electronic feedback control system that can be described by the block diagram shown in Fig. 5.36. The PLL contains a **voltage-controlled oscillator** (VCO), which is an oscillator whose frequency is controlled by a dc voltage coming from the low-pass filter. When an input is applied to the PLL, its frequency and phase are compared with the frequency and phase of the VCO by the **phase detector.** The output of the phase detector is an "error voltage" that is proportional to the difference between the frequencies and phases of the input and VCO. The output of the phase detector goes to a low-pass filter to remove higher-frequency components. The output of the low-pass filter, which is the output of the system, changes the frequency of the VCO such that the difference between the input and the VCO is reduced. This process continues until the input and the VCO have the same frequency. At this point, we say that the PLL is **locked** or is in **phase lock.** There is a difference between the phases of the input and VCO that produces an error voltage that keeps the PLL locked.

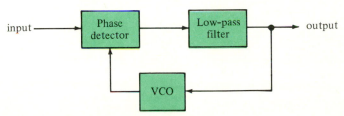

Fig. 5.36 Phase-locked loop (PLL).

If the input to the PLL shown in Fig. 5.36 is an FM signal, then the output is the demodulated signal. In addition to FM demodulation, the PLL has numerous other practical applications. For some of these, the output of the VCO is the output of the system. Under this circumstance, the connection between the VCO and the phase detector is the feedback element.

Although the concept of the PLL was originated by British scientists in 1932, its utilization for most applications was economically unfeasible until its appearance as an integrated-circuit (IC) package in the 1970s.

5.5 THE LAPLACE TRANSFORM

In Section 5.4 we saw that the transfer function $\mathbf{H}(s)$ for a linear circuit or system is a ratio of polynomials in the complex-frequency variable s [see Equations (5.4) and (5.5)]. Such a circuit or system characterization was obtained by considering forced responses to damped sinusoids. Although this may seem like a very restrictive characterization, we will now see how we can generalize it so that complete responses to arbitrary inputs can be obtained without the need for writing and solving differential equations.

In order to accomplish this, the transformation of circuits from the time domain to the frequency domain will not be done by using phasors (for the sinusoidal case or the damped-sinusoidal case), but instead will be done with the use of a more sophisticated mathematical transformation—the Laplace transform.

Definition and Properties

Given a function of time $f(t)$, we define its **Laplace transform***—designated either $\mathscr{L}[f(t)]$ or $\mathbf{F}(s)$—to be

$$\mathscr{L}[f(t)] = \mathbf{F}(s) = \int_0^\infty f(t)e^{-st}\,dt \qquad (5.7)$$

where $s = \sigma + j\omega$. Because the lower limit of the integral in the definition of the Laplace transform is zero, the Laplace transform treats a function $f(t)$ as if $f(t) = 0$ for $t < 0$. Consequently, we will consider only such functions.

Example 5.10

Recall the unit step function $u(t)$ defined by

$$u(t) = \begin{cases} 0 & \text{for} & t < 0 \\ 1 & \text{for} & t \geq 0 \end{cases}$$

Then

$$\mathscr{L}[u(t)] = \int_0^\infty u(t)e^{-st}\,dt = \int_0^\infty e^{-st}\,dt = \left. -\frac{1}{s}e^{-st}\right|_0^\infty = -\frac{1}{s}(0 - 1)$$

$$\mathscr{L}[u(t)] = \frac{1}{s}$$

For the decaying exponential $e^{-at}\,u(t)$, where $a > 0$, we have that

---

*Named for the French mathematician Marquis Pierre Simon de Laplace (1749–1827).

$$\mathscr{L}[e^{-at}u(t)] = \int_0^\infty e^{-at}u(t)e^{-st}\,dt = \int_0^\infty e^{-(s+a)t}\,dt$$

$$= -\frac{1}{s+a}e^{-(s+a)t}\Big|_0^\infty = -\frac{1}{s+a}(0-1)$$

$$= \frac{1}{s+a}$$

Although the Laplace transform of a function $f(t)$ can be obtained by using the defining integral [Equation (5.7)], sometimes it is more convenient to use some of the properties of this transform. We will now derive some of the more useful properties.

If $f(t) = f_1(t) + f_2(t)$, then

$$\mathscr{L}[f(t)] = \int_0^\infty f(t)e^{-st}\,dt = \int_0^\infty [f_1(t) + f_2(t)]e^{-st}\,dt$$

$$= \int_0^\infty [f_1(t)e^{-st} + f_2(t)e^{-st}]\,dt$$

$$= \int_0^\infty f_1(t)e^{-st}\,dt + \int_0^\infty f_2(t)e^{-st}\,dt$$

$$= \mathscr{L}[f_1(t)] + \mathscr{L}[f_2(t)] \tag{5.8}$$

In other words, the Laplace transform of a sum of functions is equal to the sum of the transforms of the individual functions.

If K is a constant, then

$$\mathscr{L}[Kf(t)] = \int_0^\infty Kf(t)e^{-st}\,dt = K\int_0^\infty f(t)e^{-st}\,dt$$

$$= K\mathscr{L}[f(t)] \tag{5.9}$$

In other words, if a function is scaled by a constant, then the Laplace transform of the function is scaled by the same constant.

The properties of the Laplace transform given by Equations (5.8) and (5.9) collectively are referred to as the **linearity property** of the Laplace transform. We also say that the Laplace transform is a **linear transformation.**

Example 5.11

Let us find the Laplace transform of

$$f(t) = 3(1 - e^{-2t})u(t)$$

Since we can express this function in the form

$$f(t) = (3 - 3e^{-2t})u(t) = 3u(t) - 3e^{-2t}u(t) = f_1(t) + f_2(t)$$

where

$$f_1(t) = 3u(t) \qquad \text{and} \qquad f_2(t) = -3e^{-2t}u(t)$$

then by the linearity property of the Laplace transform

$$\mathcal{L}[f(t)] = \mathcal{L}[3\,u(t)] + \mathcal{L}[-3e^{-2t}u(t)]$$

$$= 3\mathcal{L}[u(t)] - 3\mathcal{L}[e^{-2t}u(t)]$$

$$= 3\left(\frac{1}{s}\right) - 3\left(\frac{1}{s+2}\right) = \frac{3}{s} - \frac{3}{s+2}$$

$$= \frac{3(s+2) - 3s}{s(s+2)} = \frac{6}{s(s+2)}$$

■

Now recall Euler's formula

$$e^{j\theta} = \cos\theta + j\sin\theta \tag{5.10}$$

Replacing θ by $-\theta$, we get

$$e^{-j\theta} = \cos(-\theta) + j\sin(-\theta)$$

Since $\cos(-\theta) = \cos\theta$ and $\sin(-\theta) = -\sin\theta$, then

$$e^{-j\theta} = \cos\theta - j\sin\theta \tag{5.11}$$

By adding Equations (5.10) and (5.11), we get

$$\cos\theta = \frac{1}{2}(e^{j\theta} + e^{-j\theta}) \tag{5.12}$$

and subtracting Equation (5.11) from Equation (5.10), we obtain

$$\sin\theta = \frac{1}{2j}(e^{j\theta} - e^{-j\theta}) \tag{5.13}$$

These results can be used to find the Laplace transforms of sinusoids.

Example 5.12

Given $f(t) = \cos\omega_0 t\, u(t)$. Then

$$\mathcal{L}[\cos\omega_0 t\, u(t)] = \mathcal{L}[\tfrac{1}{2}(e^{j\omega_0 t} + e^{-j\omega_0 t})u(t)]$$

$$= \tfrac{1}{2}\mathcal{L}[e^{j\omega_0 t}u(t) + e^{-j\omega_0 t}u(t)]$$

$$= \tfrac{1}{2}\mathcal{L}[e^{j\omega_0 t}u(t)] + \tfrac{1}{2}\mathcal{L}[e^{-j\omega_0 t}u(t)]$$

Just as $\mathcal{L}[e^{-at}u(t)] = 1/(s + a)$, so too

$$\mathcal{L}[e^{-s_0 t}u(t)] = \frac{1}{s + s_0}$$

where s_0 is a complex number. Therefore,

$$\mathcal{L}[\cos \omega_0 t \, u(t)] = \frac{1}{2}\left(\frac{1}{s - j\omega_0}\right) + \frac{1}{2}\left(\frac{1}{s + j\omega_0}\right)$$

$$= \frac{1}{2}\frac{s + j\omega_0 + s - j\omega_0}{(s - j\omega_0)(s + j\omega_0)}$$

$$= \frac{s}{s^2 + \omega_0^2}$$

Another property of the Laplace transform involves the derivative of a function. Specifically,

$$\mathcal{L}\left[\frac{df(t)}{dt}\right] = \int_0^\infty \frac{df(t)}{dt} e^{-st} \, dt = \int_0^\infty e^{-st} \frac{df(t)}{dt} \, dt$$

We may employ the formula for integration by parts

$$\int_a^b u \, dv = uv \Big|_a^b - \int_a^b v \, du$$

by selecting

$$u = e^{-st} \quad \text{and} \quad dv = \frac{df(t)}{dt} \, dt = df(t)$$

Then

$$du = -se^{-st} \, dt \quad \text{and} \quad v = f(t)$$

Thus,

$$\mathcal{L}\left[\frac{f(t)}{dt}\right] = e^{-at} f(t) \Big|_0^\infty - \int_0^\infty f(t)[-se^{-st}] \, dt$$

$$= 0 - f(0) + s \int_0^\infty f(t)e^{-st} \, dt$$

$$= -f(0) + s\mathcal{L}[f(t)]$$

This result is known as the **differentiation property** of the Laplace transform.

Example 5.13

Since

$$\frac{d[\sin \omega_0 t\, u(t)]}{dt} = \omega_0 \cos \omega_0 t\, u(t)$$

then

$$\mathcal{L}\left(\frac{d[\sin \omega_0 t\, u(t)]}{dt}\right) = \omega_0 \mathcal{L}[\cos \omega_0 t\, u(t)]$$

$$- \sin 0\, u(0) + s\mathcal{L}[\sin \omega_0 t\, u(t)] = \omega_0 \frac{s}{s^2 + \omega_0^2}$$

and therefore,

$$\mathcal{L}[\sin \omega_0 t\, u(t)] = \frac{\omega_0}{s^2 + \omega_0^2}$$

∎

In order to find $\mathcal{L}[d^2 f(t)/dt^2]$, we can apply the differentiation property twice to obtain

$$\mathcal{L}\left[\frac{d^2 f(t)}{dt^2}\right] = -\frac{df(0)}{dt} - sf(0) + s^2\mathcal{L}[f(t)]$$

where $df(0)/dt$ is the derivative of $f(t)$ evaluated at $t = 0$.

Formulas for the Laplace transforms of higher-order derivatives can be obtained by repeated applications of the differentiation property.

Another important property of the Laplace transform is obtained as follows: Suppose that $\mathbf{F}(s) = \mathcal{L}[f(t)]$. Then

$$\mathcal{L}[e^{-at}f(t)] = \int_0^\infty e^{-at}f(t)e^{-st}\, dt = \int_0^\infty f(t)e^{-(s+a)t}\, dt$$

$$= \mathbf{F}(s + a)$$

In other words, $\mathcal{L}[e^{-at}f(t)]$ can be obtained from $\mathcal{L}[f(t)]$—simply replace each s in $\mathcal{L}[f(t)]$ by $s + a$. This result is referred to as the **complex-translation property** of the Laplace transform.

Example 5.14

Since $\mathcal{L}[u(t)] = \mathbf{F}(s) = 1/s$, then by the complex-translation property

$$\mathcal{L}[e^{-at}u(t)] = \mathbf{F}(s + a) = \frac{1}{s + a}$$

Furthermore, since

$$\mathcal{L}[\cos \omega_0 t \, u(t)] = \frac{s}{s^2 + \omega_0^2}$$

then

$$\mathcal{L}[e^{-at} \cos \omega_0 t \, u(t)] = \frac{s + a}{(s + a)^2 + \omega_0^2}$$

Also, since

$$\mathcal{L}[\sin \omega_0 t \, u(t)] = \frac{\omega_0}{s^2 + \omega_0^2}$$

then

$$\mathcal{L}[e^{-at} \sin \omega_0 t \, u(t)] = \frac{\omega_0}{(s + a)^2 + \omega_0^2}$$

∎

Given that $\mathcal{L}[f(t)] = \mathbf{F}(s)$, then

$$\frac{d\mathbf{F}(s)}{ds} = \frac{d}{ds} \int_0^\infty f(t) e^{-st} \, dt = \int_0^\infty f(t) \frac{d[e^{-st}]}{ds} \, dt$$

$$= \int_0^\infty f(t)[-t e^{-st}] \, dt = -\int_0^\infty t f(t) e^{-st} \, dt = -\mathcal{L}[tf(t)]$$

Thus,

$$\mathcal{L}[tf(t)] = -\frac{d\mathbf{F}(s)}{ds} = -\frac{d}{ds}\{\mathcal{L}[f(t)]\}$$

and this is known as the **complex-differentiation property** of the Laplace transform.

Example 5.15

Since $\mathcal{L}[e^{-at} u(t)] = 1/(s + a)$, then

$$\mathcal{L}[te^{-at} u(t)] = -\frac{d}{ds}\left(\frac{1}{s + a}\right) = \frac{1}{(s + a)^2}$$

Setting $a = 0$, we get

$$\mathcal{L}[tu(t)] = 1/s^2$$

∎

A summary of some of the properties of the Laplace transform, as well as the transforms of common functions, is given by Table 5.1.

Table 5.1 Table of Laplace Transforms

f(t)	Property	F(s)
$f(t)$	Definition	$\int_0^\infty f(t)e^{-st}\,dt$
$f_1(t) + f_2(t)$	Linearity	$\mathbf{F}_1(s) + \mathbf{F}_2(s)$
$Kf(t)$	Linearity	$K\mathbf{F}(s)$
$\dfrac{df(t)}{dt}$	Differentiation	$s\,\mathbf{F}(s) - f(0)$
$\dfrac{d^2f(t)}{dt^2}$	Differentiation	$s^2\mathbf{F}(s) - sf(0) - \dfrac{df(0)}{dt}$
$\displaystyle\int_0^t f(t)\,dt$	Integration	$\dfrac{1}{s}\mathbf{F}(s)$
$tf(t)$	Complex differentiation	$-\dfrac{d\mathbf{F}(s)}{ds}$
$e^{-at}f(t)$	Complex translation	$\mathbf{F}(s + a)$
$f(t - a)\,u(t - a)$	Real translation	$e^{-as}\mathbf{F}(s)$
$u(t)$		$\dfrac{1}{s}$
$e^{-at}u(t)$		$\dfrac{1}{s + a}$
$\cos \omega_0 t\, u(t)$		$\dfrac{s}{s^2 + \omega_0^2}$
$\sin \omega_0 t\, u(t)$		$\dfrac{\omega_0}{s^2 + \omega_0^2}$
$e^{-at}\cos \omega_0 t\, u(t)$		$\dfrac{s + a}{(s + a)^2 + \omega_0^2}$
$e^{-at}\sin \omega_0 t\, u(t)$		$\dfrac{\omega_0}{(s + a)^2 + \omega_0^2}$
$t\, u(t)$		$\dfrac{1}{s^2}$
$te^{-at}u(t)$		$\dfrac{1}{(s + a)^2}$
$t^n e^{-at}u(t)$		$\dfrac{n!}{(s + a)^{n+1}}$

Inverse Laplace Transforms

Given a function $F(s)$, in order to determine the function $f(t)$ such that $\mathscr{L}[f(t)] = F(s)$, we must take the **inverse Laplace transform** of $F(s)$—which is denoted as $\mathscr{L}^{-1}[F(s)] = f(t)$. Although this can be done with the use of a mathematical formula, such an approach requires the use of advanced mathematics. Therefore, we will take inverse Laplace transforms simply by inspecting the table of Laplace transforms (Table 5.1) to see what function $f(t)$ has the Laplace transform $F(s)$. If $F(s)$ is not in the table, we will decompose it into functions which are in the table or are readily obtainable by using the properties of Laplace transforms.

Just as the Laplace transform is a linear transformation, so too the inverse Laplace transform is a linear transformation. Specifically, if $F(s) = F_1(s) + F_2(s)$, then

$$\mathscr{L}^{-1}[F(s)] = \mathscr{L}^{-1}[F_1(s) + F_2(s)] = \mathscr{L}^{-1}[F_1(s)] + \mathscr{L}^{-1}[F_2(s)]$$

Furthermore,

$$\mathscr{L}^{-1}[KF(s)] = K\,\mathscr{L}^{-1}[F(s)]$$

Example 5.16

Let us determine $f_1(t)$ given that its Laplace transform is

$$\mathscr{L}[f_1(t)] = F_1(s) = \frac{14s + 23}{s^2 + 4s + 5}$$

By completing the square for the denominator, this function can be written in the form

$$F_1(s) = \frac{14s + 23}{(s + 2)^2 + 1^2} = \frac{14(s + 2) - 5}{(s + 2)^2 + 1^2}$$

$$= \frac{14(s + 2)}{(s + 2)^2 + 1^2} - \frac{5(1)}{(s + 2)^2 + 1^2}$$

Since the inverse Laplace transform of a sum is equal to the sum of the individual inverse Laplace transforms, by using Table 5.1, we get

$$f_1(t) = 14e^{-2t} \cos t\, u(t) - 5e^{-2t} \sin t\, u(t)$$

Next let us find $f_2(t)$ given that

$$\mathscr{L}[f_2(t)] = F_2(s) = \frac{14s + 23}{s^2 + 5s + 4}$$

In this case the denominator cannot be put into the form $(s + a)^2 + \omega_0^2$, where ω_0 is a real number. However, it is true that

$$F_2(s) = \frac{14s + 23}{s^2 + 5s + 4} = \frac{3}{s + 1} + \frac{11}{s + 4} \tag{5.14}$$

Therefore, from Table 5.1, we obtain

$$f_2(t) = 3e^{-t}u(t) + 11e^{-4t}u(t)$$

In Example 5.16, Equation (5.14) indicates that $\mathbf{F}_2(s)$ can be expressed as the sum of two functions, each of which is in the form of a function in Table 5.1. There is a systematic method for decomposing a function into a sum of simpler functions—such a decomposition is called a **partial-fraction expansion**, and we now describe a procedure for obtaining it.

Suppose we are given a function $\mathbf{F}(s) = \mathbf{N}(s)/\mathbf{D}(s)$, where $\mathbf{N}(s)$ and $\mathbf{D}(s)$ are polynomials in s with real coefficients. If the roots of $\mathbf{D}(s)$ are $s_1, s_2, s_3, \ldots s_n$, we can write $\mathbf{F}(s)$ in the form

$$\mathbf{F}(s) = \frac{\mathbf{N}(s)}{\mathbf{D}(s)} = \frac{\mathbf{N}(s)}{(s - s_1)(s - s_2)(s - s_3) \cdots (s - s_n)} \tag{5.15}$$

If the degree of $\mathbf{D}(s)$ is greater than the degree of $\mathbf{N}(s)$, and if the roots of $\mathbf{D}(s)$—i.e., the poles of $\mathbf{F}(s)$—are distinct, then we may write

$$\mathbf{F}(s) = \frac{K_1}{s - s_1} + \frac{K_2}{s - s_2} + \frac{K_3}{s - s_3} + \cdots + \frac{K_n}{s - s_n} \tag{5.16}$$

To find K_1, simply multiply both sides of Equation (5.16) by $s - s_1$, which yields

$$(s - s_1)\mathbf{F}(s) = K_1 + \frac{K_2(s - s_1)}{s - s_2} + \frac{K_3(s - s_1)}{s - s_3} + \cdots + \frac{K_n(s - s_1)}{s - s_n}$$

If we set $s = s_1$, then this equation becomes

$$(s - s_1)\mathbf{F}(s)\Big|_{s = s_1} = K_1$$

Similarly,

$$(s - s_2)\mathbf{F}(s)\Big|_{s = s_2} = K_2$$

etc.

Example 5.17

Let us take the partial-fraction expansion of

$$\mathbf{F}(s) = \frac{2s^2 + 11s + 19}{(s + 1)(s + 2)(s + 3)} = \frac{K_1}{s + 1} + \frac{K_2}{s + 2} + \frac{K_3}{s + 3}$$

Multiplying this expression by $s + 1$ and then setting $s = -1$, we get

$$K_1 = \frac{2s^2 + 11s + 19}{(s + 2)(s + 3)}\bigg|_{s = -1} = \frac{2(-1)^2 + 11(-1) + 19}{(-1 + 2)(-1 + 3)} = 5$$

Multiplying by $s + 2$ and then setting $s = -2$, we obtain

$$K_2 = \frac{2s^2 + 11s + 19}{(s + 1)(s + 3)}\bigg|_{s = -2} = \frac{2(-2)^2 + 11(-2) + 19}{(-2 + 1)(-2 + 3)} = -5$$

Finally,

$$K_3 = \frac{2s^2 + 11 + 19}{(s + 1)(s + 2)}\bigg|_{s = -3} = \frac{2(-3)^2 + 11(-3) + 19}{(-3 + 1)(-3 + 2)} = 2$$

Hence,

$$\mathbf{F}(s) = \frac{5}{s + 1} - \frac{5}{s + 2} + \frac{2}{s + 3}$$

and from Table 5.1, the inverse Laplace transform of $\mathbf{F}(s)$ is

$$f(t) = 5e^{-t}u(t) - 5e^{-2t}u(t) + 2e^{-3t}u(t)$$

In Example 5.17, the function $\mathbf{F}(s)$ has poles that are both distinct and real. In general, however, the poles of $\mathbf{F}(s)$ can be complex—occurring in conjugate pairs—and even nondistinct. Although there are formal procedures for taking partial-fraction expansions for these cases as well, we will study only circuits or systems whose complexity is such that we may deal with these cases by utilizing the technique for distinct, real poles.

Example 5.18

Given the following function $\mathbf{F}_1(s)$ with a double pole:

$$\mathbf{F}_1(s) = \frac{s^2 + 5s + 6}{s(s + 1)^2}$$

Let us determine the term K_0 associated with the pole at the origin of the s-plane by writing

$$\mathbf{F}_1(s) = \frac{K_0}{s} + \mathbf{G}_1(s)$$

Then

$$K_0 = s\mathbf{F}(s)\bigg|_{s=0} = \frac{s^2 + 5s + 6}{(s + 1)^2}\bigg|_{s = 0} = 6$$

Thus, we can write

$$\mathbf{G}_1(s) = \mathbf{F}_1(s) - \frac{K_0}{s} = \frac{s^2 + 5s + 6}{s(s+1)^2} - \frac{6}{s} = \frac{-5s - 7}{(s+1)^2}$$

$$= \frac{-5(s+1) - 2}{(s+1)^2} = -\frac{5}{s+1} - \frac{2}{(s+1)^2}$$

Hence,

$$\mathbf{F}_1(s) = \frac{6}{s} - \frac{5}{s+1} - \frac{2}{(s+1)^2}$$

and from Table 5.1, we have that

$$f_1(t) = 6u(t) - 5e^{-t}u(t) - 2te^{-t}u(t)$$

Next consider the following function $\mathbf{F}_2(s)$ with complex poles at $s = -1 - j2$ and $s = -1 + j2$.

$$\mathbf{F}_2(s) = \frac{5s^2 + 30s - 15}{s(s^2 + 2s + 5)} = \frac{5s^2 + 30s - 15}{s(s+1+j2)(s+1-j2)}$$

However, we need not use the rightmost expression since we can write

$$\mathbf{F}_2(s) = \frac{K_0}{s} + \mathbf{G}_2(s)$$

where

$$K_0 = s\mathbf{F}_2(s)\bigg|_{s=0} = \frac{5s^2 + 30s - 15}{s^2 + 2s + 5}\bigg|_{s=0} = -3$$

Thus,

$$\mathbf{G}_2(s) = \mathbf{F}_2(s) - \frac{K_0}{s} = \frac{5s^2 + 30s - 15}{s(s^2 + 2s + 5)} + \frac{3}{s} = \frac{8s + 36}{s^2 + 2s + 5}$$

$$= \frac{8s + 36}{(s+1)^2 + 2^2} = \frac{8(s+1) + 14(2)}{(s+1)^2 + 2^2}$$

Therefore,

$$\mathbf{F}_2(s) = -\frac{3}{s} + \frac{8(s+1)}{(s+1)^2 + 2^2} + \frac{14(2)}{(s+1)^2 + 2^2}$$

and from Table 5.1,

$$f_2(t) = -3u(t) + 8e^{-t}\cos 2t\, u(t) + 14e^{-t}\sin 2t\, u(t)$$

■

Application to Circuits and Systems

One very important application of the Laplace transform is to the solution of differential equations.

Example 5.19

Suppose that we wish to solve the linear, second-order differential equation

$$\frac{d^2x(t)}{dt^2} + 3\frac{dx(t)}{dt} + 2x(t) = 4e^{-3t}u(t)$$

subject to the initial conditions $x(0) = 2$ and $dx(0)/dt = -1$. Taking the Laplace transform of this differential equation, if we let $\mathbf{X}(s) = \mathcal{L}[x(t)]$, then

$$\left[s^2\mathbf{X}(s) - sx(0) - \frac{dx(0)}{dt} \right] + 3[s\mathbf{X}(s) - x(0)] + 2\mathbf{X}(s) = \frac{4}{s+3}$$

Then

$$s^2\mathbf{X}(s) - 2s + 1 + 3s\mathbf{X}(s) - 6 + 2\mathbf{X}(s) = \frac{4}{s+3}$$

or

$$(s^2 + 3s + 2)\mathbf{X}(s) = \frac{4}{s+3} + 2s + 5 = \frac{2s^2 + 11s + 19}{s+3}$$

Thus,

$$\mathbf{X}(s) = \frac{2s^2 + 11s + 19}{(s+1)(s+2)(s+3)}$$

By Example 5.17,

$$x(t) = (5e^{-t} - 5e^{-2t} + 2e^{-3t})u(t)$$

Of course, we can write the differential equation or equations which characterize a circuit and then use Laplace transforms to solve such equations. However, we can avoid writing differential equations if we employ Laplace-transform (frequency-domain) concepts directly. Let's see how this is done.

For a resistor having a value of R ohms, we know that

$$v(t) = Ri(t)$$

Taking the Laplace transform of both sides of this equation results in

$$\mathbf{V}(s) = R\mathbf{I}(s)$$

where $\mathbf{V}(s) = \mathcal{L}[v(t)]$ and $\mathbf{I}(s) = \mathcal{L}[i(t)]$. Defining the impedance $\mathbf{Z}_R(s)$ of the resistor to be the ratio of voltage transform to current transform, we get

$$\mathbf{Z}_R(s) = \frac{\mathbf{V}(s)}{\mathbf{I}(s)} = R$$

The circuit symbols of a resistor in the time domain and in the frequency (transform) domain are shown in Fig. 5.37(a) and (b) respectively.

(a) Time domain   (b) Frequency domain
Fig. 5.37 Resistor circuit symbols.

For an inductor having a value of L henries,

$$v(t) = L\frac{di(t)}{dt}$$

Taking the Laplace transform, we get

$$\mathbf{V}(s) = L[s\mathbf{I}(s) - i(0)]$$

$$\mathbf{V}(s) = Ls\mathbf{I}(s) - Li(0) \tag{5.17}$$

or

$$\mathbf{I}(s) = \frac{1}{Ls}\mathbf{V}(s) + \frac{i(0)}{s} \tag{5.18}$$

For the case of zero initial conditions, $i(0) = 0$ and thus,

$$\mathbf{V}(s) = Ls\mathbf{I}(s)$$

We then define the impedance $\mathbf{Z}_L(s)$ of the inductor to be the ratio of voltage transform to current transform when the initial current is zero. Thus,

$$\mathbf{Z}_L(s) = \frac{\mathbf{V}(s)}{\mathbf{I}(s)} = Ls$$

The time-domain circuit symbol for an inductor is shown in Fig. 5.38(a), while Fig. 5.38(b) shows the circuit symbol in the frequency domain when the initial current is zero. For the case that the initial current is not necessarily zero, the parallel connection shown in Fig. 5.38(c) models the frequency-domain description given by Equation 5.18. Note that if $i(0) = 0$, the model in Fig. 5.38(c) is equivalent to Fig. 5.38(b).

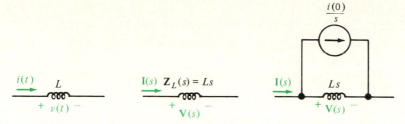

(a) Time domain   (b) Frequency domain—
                      zero initial current

(c) Frequency domain—
     nonzero initial current

Fig. 5.38 Inductor circuit symbols.

For a capacitor having a value of C farads,

$$i(t) = C \frac{dv(t)}{dt}$$

so

$$\mathbf{I}(s) = C[s\mathbf{V}(s) - v(0)]$$
$$\mathbf{I}(s) = Cs\mathbf{V}(s) - Cv(0) \tag{5.19}$$

or

$$\mathbf{V}(s) = \frac{1}{Cs}\mathbf{I}(s) + \frac{v(0)}{s} \tag{5.20}$$

The impedance $\mathbf{Z}_C(s)$ of the capacitor (for zero initial voltage) is

$$\mathbf{Z}_C(s) = \frac{\mathbf{V}(s)}{\mathbf{I}(s)} = \frac{1}{Cs}$$

The circuit symbols for a capacitor in the time domain and the frequency domain, with zero and nonzero initial conditions, are shown in Fig. 5.39.

Since we have the same expressions for the impedances of resistors, inductors, and capacitors as we had for the damped-sinusoidal case, we can use the same circuit analysis techniques—the difference being that we use the Laplace transforms of time functions rather than their phasor representations.

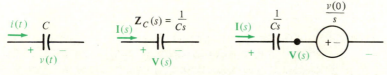

(a) Time domain   (b) Frequency domain—
                      zero initial voltage

(c) Frequency domain—
     nonzero initial voltage

Fig. 5.39 Capacitor circuit symbols.

Example 5.20

Given that the series *RLC* circuit shown in Fig. 5.40(a) has zero initial conditions, let us find the step response $v(t)$. Fig. 5.40(b) shows the circuit in the frequency domain. Of course this circuit can be analyzed by using either mesh analysis or nodal analysis. However, even more simply, by voltage division we can write

$$\mathbf{V}(s) = \frac{50/s}{50/s + 2s + 12} \mathbf{V}_g(s) = \frac{50}{2s^2 + 12s + 50}\left(\frac{1}{s}\right)$$

or

$$\mathbf{V}(s) = \frac{25}{s(s^2 + 6s + 25)} = \frac{K_0}{s} + \mathbf{F}(s)$$

where

$$K_0 = s\mathbf{V}(s)\big|_{s=0} = \frac{25}{s^2 + 6s + 25}\bigg|_{s=0} = 1$$

Thus,

$$\mathbf{F}(s) = \mathbf{V}(s) - \frac{K_0}{s} = \frac{25}{s(s^2 + 6s + 25)} - \frac{1}{s} = \frac{-s - 6}{s^2 + 6s + 25}$$

Hence,

$$\mathbf{V}(s) = \frac{1}{s} - \frac{s + 6}{s^2 + 6s + 25} = \frac{1}{s} - \frac{s + 6}{(s + 3)^2 + 4^2}$$

$$= \frac{1}{s} - \frac{s + 3}{(s + 3)^2 + 4^2} - \frac{\frac{3}{4}(4)}{(s + 3)^2 + 4^2}$$

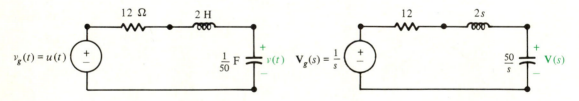

(a) Time domain   (b) Frequency domain

Fig. 5.40 A series *RLC* circuit.

Therefore, from Table 5.1,

$$v(t) = u(t) - e^{-3t} \cos 4t \, u(t) - \tfrac{3}{4}e^{-3t} \sin 4t \, u(t)$$

$$= [1 - e^{-3t}(\cos 4t + \tfrac{3}{4} \sin 4t)] \, u(t)$$

and this is the complete response (i.e., forced and natural responses).

For a linear circuit with zero initial conditions, if the input is scaled by a constant, then the response is scaled by the same constant. Thus, the response to $v_g(t) = \frac{2}{5}u(t)$ is

$$\frac{2}{5}v(t) = \frac{2}{5}u(t) - \frac{2}{5}e^{-3t}\cos 4t\, u(t) - \frac{3}{10}e^{-3t}\sin 4t\, u(t)$$

(see Example 3.15).

∎

Having analyzed a circuit with zero initial conditions, let us now consider the case of a circuit with nonzero initial conditions.

Example 5.21

Suppose that we wish to find $v(t)$ for the circuit shown in Fig. 5.41(a) subject to the initial condition $v(0) = 9$ V. In the frequency domain, the circuit is as shown in Fig. 5.41(b). Note that the voltage (transform) $V(s)$ that is to be determined is the voltage across the series combination of a $\frac{1}{6}$-F capacitor and a voltage source having a value of $v(0)/s = 9/s$.

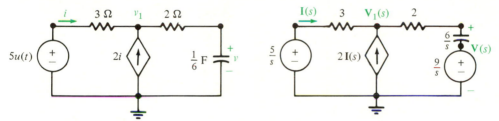

Fig. 5.41 Circuit with nonzero initial condition.

By KCL at the node labeled $V_1(s)$,

$$I(s) + 2I(s) = \frac{V_1(s) - V(s)}{2}$$

or

$$3I(s) = 3\left[\frac{5/s - V_1(s)}{3}\right] = \frac{V_1(s) - V(s)}{2}$$

Simplifying this expression, we get

$$3V_1(s) - V(s) = \frac{10}{s} \tag{5.21}$$

Again, by KCL,

$$\frac{\mathbf{V}_1(s) - \mathbf{V}(s)}{2} = \frac{\mathbf{V}(s) - 9/s}{6/s}$$

from which

$$3\mathbf{V}_1(s) - (s + 3)\mathbf{V}(s) = -9 \qquad (5.22)$$

Subtracting Equation (5.22) from Equation (5.21) yields

$$-\mathbf{V}(s) + (s + 3)\mathbf{V}(s) = \frac{10}{s} + 9$$

or

$$(s + 2)\mathbf{V}(s) = \frac{9s + 10}{s}$$

Therefore,

$$\mathbf{V}(s) = \frac{9s + 10}{s(s + 2)} = \frac{5}{s} + \frac{4}{s + 2}$$

Hence,

$$v(t) = 5u(t) + 4e^{-2t}u(t) = (5 + 4e^{-2t})u(t)$$

(see Example 3.14).

■

In Chapter 4 we saw how to find forced sinusoidal responses by using phasors. Let us now see an example of how to determine the complete response to a sinusoidal source (which is zero for $t < 0$) by using Laplace-transform techniques.

Example 5.22

The circuit in Fig. 5.42(a) has a sinusoidal voltage excitation which begins at time $t = 0$. Fig. 5.42(b) shows the frequency-domain representation of the circuit. By KVL,

$$\mathbf{V}_g = 1\mathbf{I}(s) + \frac{2}{s}\mathbf{I}(s) = \frac{s + 2}{s}\mathbf{I}(s)$$

Thus,

$$\mathbf{I}(s) = \frac{s}{s + 2}\mathbf{V}_g = \left(\frac{s}{s + 2}\right)\left(\frac{12}{s^2 + 2^2}\right) = \frac{12s}{(s + 2)(s^2 + 4)}$$

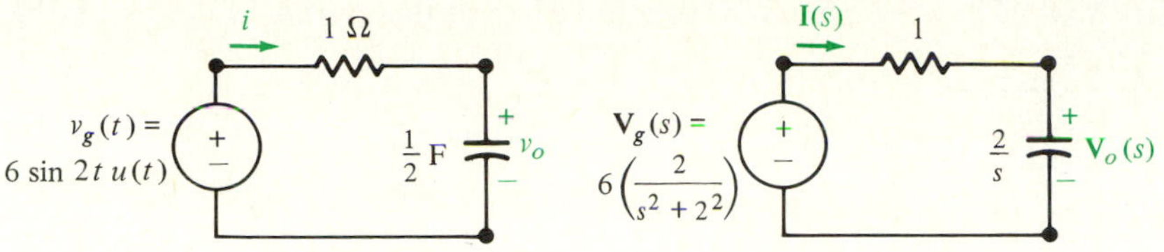

(a) Time domain   (b) Frequency domain

Fig. 5.42 A simple sinusoidal circuit.

However,

$$\frac{12s}{(s + 2)(s^2 + 4)} = \frac{K}{s + 2} + \mathbf{F}(s)$$

where

$$K = \frac{12s}{s^2 + 4}\bigg|_{s = -2} = -3$$

Therefore,

$$\mathbf{F}(s) = \frac{12s}{(s + 2)(s^2 + 4)} - \frac{K}{s + 2} = \frac{12s}{(s + 2)(s^2 + 4)} + \frac{3}{s + 2} = \frac{3s + 6}{s^2 + 4}$$

Hence,

$$\mathbf{I}(s) = -\frac{3}{s + 2} + \frac{3s}{s^2 + 2^2} + \frac{3(2)}{s^2 + 2^2}$$

and the resulting current is

$$i(t) = -3e^{-2t}u(t) + 3 \cos 2t \, u(t) + 3 \sin 2t \, u(t)$$

This response is a complete response—the natural response is $i_n(t) = -3e^{-2t}u(t)$ and the forced response is $i_f(t) = 3 \cos 2t \, u(t) + 3 \sin 2t \, u(t) = 3\sqrt{2} \cos (2t - 45°) \, u(t)$.

If only the forced (steady-state) response is of interest,* for sinusoidal circuits Laplace transform techniques can be avoided by taking the simpler phasor-analysis approach (see Example 4.4).

Suppose that the input to a linear system has a Laplace transform of $\mathbf{X}(s)$, and suppose that the Laplace transform of the output, given that all the initial conditions

---

*The natural (transient) response is negligible after just a few time constants—the time constant is one-half second for this example.

are zero, is $\mathbf{Y}(s)$. Then the **transfer function $\mathbf{H}(s)$** of the system is defined to be

$$\mathbf{H}(s) = \frac{\mathbf{Y}(s)}{\mathbf{X}(s)}$$

If the transfer function of a linear system is known, then when the input is specified, the output transform can be determined from the equation

$$\mathbf{Y}(s) = \mathbf{H}(s)\mathbf{X}(s)$$

Taking the inverse Laplace transform of $\mathbf{Y}(s)$ yields the corresponding output $y(t)$ in the time domain.

Example 5.23

Given a simple low-pass filter with a voltage transfer function of

$$\mathbf{H}(s) = \frac{\mathbf{V}_2(s)}{\mathbf{V}_1(s)} = \frac{3}{s + 3}$$

let us find the output voltage $v_2(t)$ given that the input voltage is $v_1(t) = 2e^{-3t}u(t)$.
  Since

$$\mathbf{V}_1(s) = \mathcal{L}[2e^{-3t}u(t)] = \frac{2}{s + 3}$$

then

$$\mathbf{V}_2(s) = \mathbf{H}(s)\mathbf{V}_1(s) = \left(\frac{3}{s + 3}\right)\left(\frac{2}{s + 3}\right) = \frac{6}{(s + 3)^2}$$

and

$$v_2(t) = 6te^{-3t}u(t)$$

In this case, the forced and natural responses combine into the single term $6te^{-3t}u(t)$. This is a consequence of exciting the system at its pole; i.e., the pole of $\mathbf{V}_1(s)$ is the same as the pole of $\mathbf{H}(s)$.

■

SUMMARY

1. The frequency response of a circuit consists of the amplitude response and the phase response.

2. The frequencies at which the amplitude response drops to $1/\sqrt{2}$ of its maximum value are the half-power frequencies.

3. The frequencies at which an impedance (or admittance) is purely real are the resonance frequencies of the impedance (or admittance).

4. The bandwidth and the quality factor are measures of the sharpness of an amplitude response.

5. An impedance can be expressed as a function of the complex frequency s, as can other parameters like the voltage transfer function.

6. The poles and zeros of a ratio of polynomials in s can be depicted in the s-plane with a pole-zero plot.

7. If there are no cancellations of common poles and zeros, the poles of a network function indicate the form of the natural response.

8. A linear system can be simulated with integrators, adders, and scalers (i.e., with an analog computer).

9. Systems are often represented by block diagrams.

10. Feedback can improve system performance and can be used for purposes of control.

11. The Laplace transform is a linear transformation that can be used to solve linear differential equations or analyze linear circuits.

12. The inverse Laplace transform can be found by using a table of transforms and various transform properties, as well as partial-fraction expansions.

13. The impedance of an R-ohm resistor is R, of an L-henry inductor is Ls, and of a C-farad capacitor is $1/Cs$.

14. An inductor (or a capacitor) with a nonzero initial condition can be modeled by an independent source and an inductor (or capacitor) with a zero initial condition.

15. Circuit analysis using Laplace transforms results in complete (both forced and natural) responses.

Problems for Section 5.1

5.1 For the circuit given in Fig. P5.1, sketch the amplitude and phase responses of $\mathbf{V}_1/\mathbf{V}_s$. What type of filter is this?

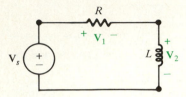

Fig. P5.1

5.2 Repeat Problem 5.1 for V_2/V_s.

5.3 Sketch the amplitude response of V_2/V_1 for each of the op-amp circuits in Fig. P5.3. Indicate the half-power frequency.

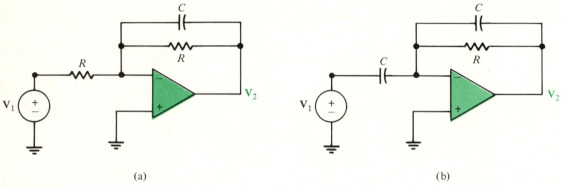

(a) (b)

Fig. P5.3

5.4 Repeat Problem 5.3 for each of the circuits shown in Fig. P5.4.

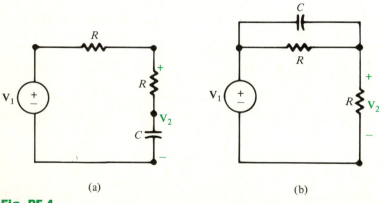

(a) (b)

Fig. P5.4

5.5 Repeat Problem 5.3 for each of the op-amp circuits in Fig. P5.5.

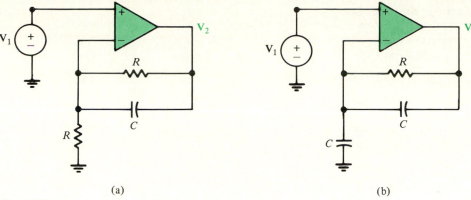

Fig. P5.5

5.6 For the op-amp circuit shown in Fig. P5.6, sketch the amplitude response of V_2/V_1, indicating the half-power frequencies. What type of filter is this circuit?

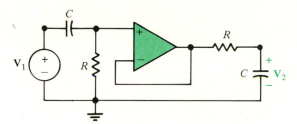

Fig. P5.6

5.7 For the circuit given in Fig. 5.5, suppose $R = 5.1$ kΩ and $C = 0.01$ μF. On semilog paper (the vertical axis has a linear scale and the horizontal axis has a logarithmic scale), draw the Bode plot of V_2/V_1.

5.8 Repeat Problem 5.7 for the circuit given in Fig. 5.8.

5.9 For the series RLC circuit shown in Fig. P5.9, draw the Bode plot of V_2/V_1 for the case that $R = 100$ Ω.

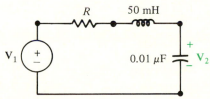

Fig. P5.9

5.10 Repeat Problem 5.9 for the case that $R = 2$ kΩ.

Problems for Section 5.2

5.11 Find a formula for the resonance frequency of the admittance shown in Fig. P5.11.

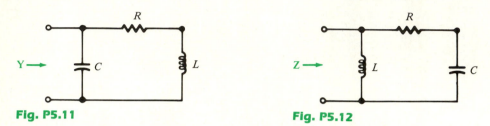

Fig. P5.11

Fig. P5.12

5.12 Find a formula for the resonance frequency of the impedance shown in Fig. P5.12.

5.13 Repeat Problem 5.12 for the impedance given in Fig. P5.13.

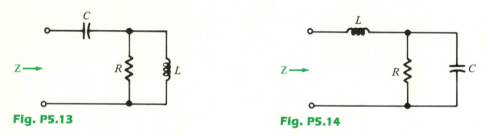

Fig. P5.13

Fig. P5.14

5.14 Repeat Problem 5.12 for the impedance given in Fig. P5.14.

5.15 The resonance frequency of the circuit shown in Fig. P5.15 is $\omega_r = 3$ rad/s. Find the quality factor Q of the circuit.

Fig. P5.15

Fig. P5.16

5.16 The resonance frequency of the circuit shown in Fig. P5.16 is $\omega_r = 2$ rad/s. Find the quality factor Q of the circuit.

5.17 A 10-Ω resistor and a 2-H inductor are connected in series, and $\omega = 50$ rad/s.

(a) What is the Q of this series connection?

(b) What parallel *RL* connection has the same admittance as the series connection?

(c) What is the Q of this parallel connection?

5.18 A 10-Ω resistor and a 2-H inductor are connected in parallel, and $\omega = 50$ rad/s.

(a) What is the Q of this parallel connection?

(b) What series *RL* connection has the same impedance as the parallel connection?

(c) What is the Q of this series connection?

5.19 Given the practical tank circuit shown in Fig. P5.11 and that $R = 25\ \Omega$, $L = 0.62$ mH, and $C = 20$ pF, approximate this admittance by a parallel *RLC* circuit. What is the quality factor of the parallel circuit?

5.20 An AM radio has a parallel *RLC* circuit for tuning in stations. Suppose that $R = 1.24$ MΩ, $L = 0.62$ mH, and $C = 20$ pF.

(a) What is the resonance frequency in hertz?

(b) What is the quality factor of the circuit?

(c) What is the bandwidth in hertz?

(d) If the current into the parallel connection due to a station at 1430 kHz is 5 μA rms, what is the resulting voltage at that frequency?

(e) Repeat part (d) for a station at 1450 kHz.

Problems for Section 5.3

5.21 For the series *RLC* circuit shown in Fig. P5.21, suppose that $R = \frac{5}{3}\ \Omega$, $L = 5$ H, and $C = \frac{1}{25}$ F. Draw a pole-zero plot for (a) $\mathbf{I}/\mathbf{V}$; (b) $\mathbf{V}_R/\mathbf{V}$; (c) $\mathbf{V}_L/\mathbf{V}$; (d) $\mathbf{V}_C/\mathbf{V}$.

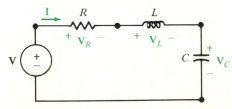

Fig. P5.21

5.22 Repeat Problem 5.21 for the case that $R = 2\ \Omega$, $L = \frac{1}{2}$ H, and $C = 2$ F.

5.23 Repeat Problem 5.21 for the case that $R = 2\ \Omega$, $L = 2$ H, and $C = 2$ F.

5.24 Repeat Problem 5.21 for the *RLC* circuit given in Fig. P5.24.

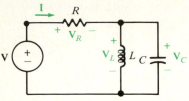

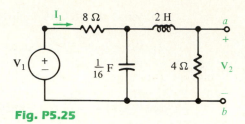

Fig. P5.24   **Fig. P5.25**

5.25 Given the circuit shown in Fig. P5.25:

(a) Find the impedance V_1/I_1.

(b) Find the voltage transfer function V_2/V_1.

(c) Find the Thévenin-equivalent circuit with respect to terminals a and b.

(d) Draw the pole-zero plots for parts (a) and (b).

5.26 Repeat Problem 5.25 for the circuit given in Fig. P5.26.

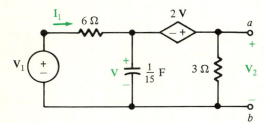

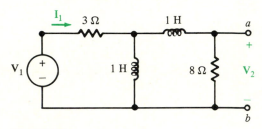

Fig. P5.26   **Fig. P5.27**

5.27 Repeat Problem 5.25 for the circuit shown in Fig. P5.27.

5.28 Given the op-amp circuit shown in Fig. P5.28:

(a) Find the voltage transfer function V_2/V_1.

(b) Find the impedance V_1/I_1.

(c) Draw the pole-zero plots for parts (a) and (b).

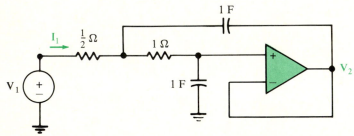

Fig. P5.28

5.29 Repeat Problem 5.28 for the op-amp circuit given in Fig. P5.29.

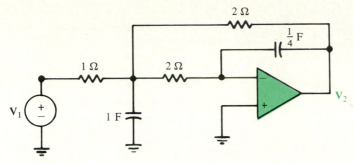

Fig. P5.29

5.30 For the op-amp circuit given in Fig. P5.29, change the capacitor to $\frac{1}{16}$ F, and repeat the problem.

Problems for Section 5.4

5.31 Use integrators, adders, and scalers to simulate the transfer function

$$H(s) = \frac{15s}{s^2 + 15s + 5}$$

5.32 Repeat Problem 5.31 for the transfer function

$$H(s) = \frac{s^2 + 5}{s^2 + 15s + 5}$$

5.33 Find the transfer function $H(s) = Y/X$ of the system shown in Fig. P5.33.

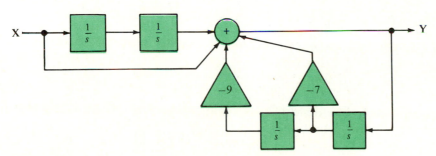

Fig. P5.33

5.34 Find the transfer function $H(s) = Y/X$ of the system shown in Fig. P5.34.

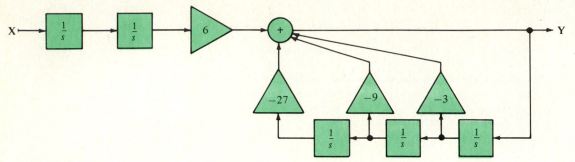

Fig. P5.34

5.35 For the feedback system shown in Fig. 5.35, find the transfer function $\mathbf{Y}/\mathbf{X}$ given that $\mathbf{G}(s) = (s + 1)/(s + 2)$ and $\mathbf{H}(s) = 1/(s + 3)$.

5.36 For the feedback system shown in Fig. 5.35, suppose that $\mathbf{G}(s) = (s + 1)/(s + 2)$. Determine $\mathbf{H}(s)$ such that the resulting transfer function is $\mathbf{Y}/\mathbf{X} = (s + 1)(s + 5)/(s + 3)^2$.

5.37 For the feedback system given in Fig. 5.35, suppose that $\mathbf{H}(s) = (s + 1)/(s + 2)$. Determine $\mathbf{G}(s)$ such that the resulting transfer function is $\mathbf{Y}/\mathbf{X} = (s + 2)^2/(s + 1)(s + 4)$.

5.38 For the feedback system given in Fig. 5.35, suppose that $\mathbf{G}(s) = 4s(s + 1)/(s + 2)^2$ and $\mathbf{H}(s) = 1/(s + 1)$. At what frequency will the system oscillate?

Problems for Section 5.5

5.39 Find the Laplace transform of each of the following functions:

(a) $(2e^{-8t} - e^{-2t})u(t)$

(b) $(6 + 2e^{-6t} - 12e^{-t})u(t)$

(c) $(2 + 3t)e^{-2t}u(t)$

(d) $(\cos 4t - \sin 4t)e^{-3t}u(t)$

5.40 Find the Laplace transform of each of the following functions:

(a) $\sin(\omega_0 t - \phi)u(t)$

(b) $\cos(\omega_0 t - \phi)u(t)$

(c) $e^{-at}\sin(\omega_0 t - \phi)\,u(t)$

(d) $e^{-at}\cos(\omega_0 t - \phi)\,u(t)$

5.41 Find the inverse Laplace transform of each of the following functions:

(a) $\dfrac{600}{s(s + 10)(s + 30)}$

(b) $\dfrac{60(s + 4)}{s(s + 2)(s + 12)}$

5.42 Find the inverse Laplace transform of each of the following functions:

(a) $\dfrac{12s}{(s + 3)(s^2 + 9)}$

(b) $\dfrac{4(s^2 + 1)}{s(s^2 + 4)}$

5.43 Find the inverse Laplace transform of each of the following functions:

(a) $\dfrac{(s + 2)(s + 3)}{s(s + 1)^2}$

(b) $\dfrac{6(s + 1)^3}{s^4}$

5.44 Find the solution to the differential equation

$$\frac{d^2x(t)}{dt^2} + 7\frac{dx(t)}{dt} + 6x(t) = 36u(t)$$

subject to the initial conditions $dx(0)/dt = 0$ and $x(0) = -4$.

5.45 Find the solution to the differential equation

$$\frac{d^2x(t)}{dt^2} + 3\frac{dx(t)}{dt} + 2x(t) = 20 \cos 2t \, u(t)$$

subject to the zero initial conditions $dx(0)/dt = x(0) = 0$.

5.46 For the series *RLC* circuit shown in Fig. P5.46, find $v(t)$ and $i(t)$ given that $v(0) = 2$ V and $i(0) = 1$ A.

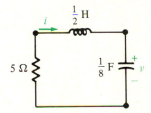

Fig. P5.46

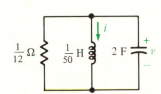

Fig. P5.47

5.47 For the parallel *RLC* circuit shown in Fig. P5.47, find $v(t)$ and $i(t)$ given that $v(0) = -0.14$ V and $i(0) = 1$ A.

5.48 For the series-parallel circuit shown in Fig. P5.48, find $v(t)$ and $i(t)$ given that $v(0) = 2$ V and $i(0) = 1$ A.

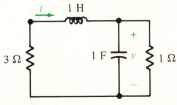

Fig. P5.48

5.49 Given the circuit with zero initial conditions shown in Fig. P5.49, find $v(t)$ for the case that $v_g(t) = u(t)$.

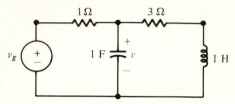

Fig. P5.49

5.50 Repeat Problem 5.49 for the case that $v_g(t) = \frac{4}{3}(1 - e^{-3t})\, u(t)$.

5.51 For the series RLC circuit given in Fig. 5.40, let $R = 2\ \Omega$, $L = 1$ H, and $C = 1$ F. Let $i(t)$ be the clockwise current. Find $v(t)$ when $i(0) = \frac{1}{2}$ A and $v(0) = 0$.

5.52 Given the parallel RLC circuit shown in Fig. P5.52, find $v(t)$ and $i(t)$ for the case that $v(0) = 0$ and $i(0) = -4$ A.

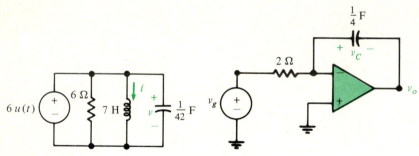

Fig. P5.52   **Fig. P5.53**

5.53 For the op-amp circuit shown in Fig. P5.53, suppose that $v_C(0) = 4$ V. Find $v_o(t)$ given that

(a) $v_g(t) = e^{-3t}u(t)$

(b) $v_g(t) = \cos 2t\, u(t)$.

5.54 Repeat Problem 5.53 for the circuit shown in Fig. P5.54.

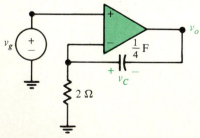

Fig. P5.54

5.55 For the circuit shown in Fig. P5.55, find $v(t)$ given that the initial conditions are zero.

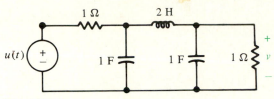

Fig. P5.55

5.56 Given that the transfer function of a linear system is $\mathbf{H}(s) = 1/(s + 2)$, find the output $y(t)$ when the input $x(t)$ is
  (a) $u(t)$
  (b) $e^{-t}u(t)$
  (c) $(1 - e^{-t})u(t)$
  (d) $e^{-2t}u(t)$

5.57 Repeat Problem 5.56 for the case that $\mathbf{H}(s) = s/(s + 2)$.

5.58 Given that the transfer function of a linear system is $\mathbf{H}(s) = (s - 1)/(s + 10)$, find the input $x(t)$ when the output $y(t)$ is
  (a) $(-1 + 2e^{-t})u(t)$
  (b) $(-2e^{-t} + 3e^{-2t})u(t)$
  (c) $(1 - 11t)e^{-10t}u(t)$
  (d) $(1 - 2t)e^{-t}u(t)$

5.59 Given that the input to a linear system is $x(t) = e^{-t}u(t)$, find the transfer function $\mathbf{H}(s)$ when the output $y(t)$ is
  (a) $e^{-2t}u(t)$
  (b) $\sin t\, u(t)$
  (c) $e^{-t}\sin t\, u(t)$
  (d) $te^{-t}u(t)$
  (e) $(e^{-t} - e^{-2t})u(t)$

5.60 Repeat Problem 5.59 for the case that $x(t) = \cos t\, u(t)$.

PART II
Electronics

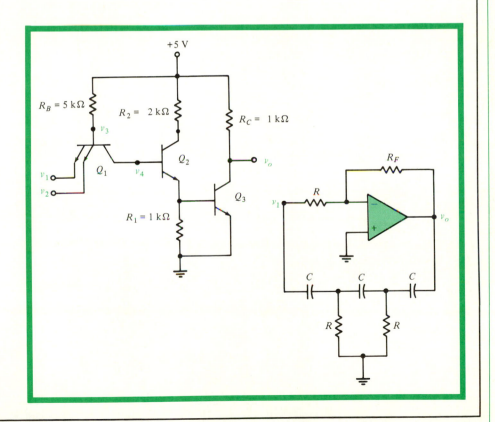

Chapter 6

Diodes

INTRODUCTION

We begin our study of electronics with a discussion of semiconductor diodes. This topic naturally evolves into the subject of the bipolar junction transistor (BJT), which will be the focus of Chapter 7. Although a BJT is inherently a current-controlled device, a field-effect transistor (FET), to be studied in Chapter 8, is inherently a voltage-controlled device. Throughout the chapters on diodes, BJTs, and FETs, numerous examples and applications are presented. Chapters 9 and 10 consider applications not covered in the chapters preceding them.

Unlike elements encountered previously in this book, the diode is a nonlinear element. In terms of analysis, nonlinear elements tend to make circuits more complicated. However, with the judicious use of approximations and assumptions (or educated guesses), we will be able to deal with nonlinear circuits with not too much more effort than is required for linear circuits. This is not a great price to pay considering that innumerable important devices employ nonlinear elements, and these devices would be impossible to fabricate using only linear elements. Even though we will be dealing with nonlinear circuits, we will utilize some of the linear analysis techniques studied in earlier chapters.

Unlike with electric circuits, an in-depth study of electronics requires more emphasis on physical principles. It is for this reason that we begin this chapter with a discussion of the electric properties of some types of materials.

6.1 SEMICONDUCTORS

The flow of charge (current) in a metal results from the movement of electrons. An electron is a negatively charged particle having charge magnitude $q = 1.60 \times 10^{-19}$ C. In a metal, the atoms consist of outer, or valence, electrons that are free to move, and positively charged ions, each of which consists of the nucleus and the tightly bound inner electrons of the atom. Figure 6.1 is a two-dimensional depiction of the ions and the valence, or free electrons, of a metal. Because of thermal energy, the free electrons are continually in motion. The direction of the motion of an electron changes after each inelastic collision with a heavy (essentially stationary) ion. We call the average distance between collisions the **mean free path.** If no voltage is applied to a metal, then the random motion of the electrons results in zero average current.

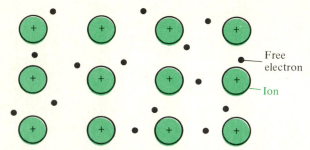

Free
electron

Ion

Fig. 6.1 Positive ions and free electrons of a metal.

The application of a voltage to a metal will result in an **electric field** $\mathscr{E}$ measured in volts per meter (V/m). Consider the case that $\mathscr{E}$ is a constant. Then, on the average, electrons will go from a given potential to a higher potential at a speed, denoted u, called the **drift speed.** This quantity is directly proportional to the electric field, that is,

$$u = \mu \mathscr{E} \tag{6.1}$$

where the proportionality constant μ is known as the **mobility** of the electrons. Since the units of u and $\mathscr{E}$ are m/s and V/m, respectively, the unit of μ is m^2/V-s.

Figure 6.2 shows a length-L section of a metallic conductor that has a cross-sectional area of A square meters. Let N be the number of free electrons in this section of the conductor, and suppose that these N electrons move to the left with a drift speed of u. If it takes T seconds for an electron to travel L meters, then from the drift speed $u = L/T$, we deduce that $T = L/u$. But a flow to the left of N electrons in T seconds means a current i going to the right, where

$$i = \frac{Nq}{T} = \frac{Nq}{L/u} = \frac{Nqu}{L}$$

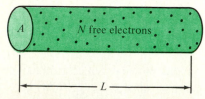

A N free electrons

L

Fig. 6.2 Section of a metallic conductor.

We define **current density** J by

$$J = \frac{i}{A} \tag{6.2}$$

Thus, we have that

$$J = \frac{i}{A} = \frac{Nqu}{LA} \tag{6.3}$$

Since the volume of the conductor section shown in Fig. 6.2 is LA, and since the number of free electrons in this volume is N, then the density of free electrons, called the **free-electron concentration,** is $n = N/LA$. Since N is a number, then the units of n are $1/m^3$ or m^{-3}. Hence, Equation (6.3) can be rewritten

$$J = nqu \tag{6.4}$$

The product nq, known as **charge density,** has as units C/m^3.

From Equations (6.1) and (6.4), we have that

$$J = nq\mu\mathcal{E} = \sigma\mathcal{E} \tag{6.5}$$

where $\sigma = nq\mu$ is known as the **conductivity** of the metal. The units of σ are $(\Omega\text{-m})^{-1}$ or $\mho/m$. The reciprocal of conductivity is called **resistivity,** denoted by ρ, and has units of Ω-m. In other words, $\rho = 1/\sigma$.

Example 6.1

For the conductor depicted in Fig. 6.2, from Equation (6.2)

$$i = JA = \sigma\mathcal{E}A = \frac{\sigma\mathcal{E}AL}{L} = \frac{\sigma A}{L}\mathcal{E}L$$

As the unit of $\mathcal{E}L$ is $(V/m)(m) = V$, then $v = \mathcal{E}L$ is the voltage across the length-L conductor. Then, by Ohm's law,

$$i = \frac{\sigma A}{L}v = \frac{1}{R}v$$

where the resistance of the conductor is $R = L/\sigma A = \rho L/A$.

■

Even though the SI unit for temperature is the **kelvin*** (K), because manufacturers typically use **degrees Celsius**[†] (°C), we shall use both temperature mea-

---

*Formerly "degrees Kelvin" (°K), named for the British physicist William Thompson, first Baron Kelvin (1824–1907).

[†]Named for the Swedish astronomer Anders Celsius (1701–1744), it is also known as "degrees centigrade."

surements. If T_K is the temperature in K and T_C is the temperature in °C, then the relationship between the two is given simply by

$$T_K = T_C + 273$$

Example 6.2

A length of copper wire with a cross-sectional area of 1 square millimeter (mm² or 10^{-6} m²) carries a current of 0.4 A. Given that the conductivity of copper at 20°C (293 K) is 5.78×10^7 ℧/m, and the free-electron concentration is 8.43×10^{28} m⁻³. Since $J = nqu$, we have that the drift speed is

$$u = \frac{J}{nq} = \frac{i}{nqA}$$

$$= \frac{0.4}{(8.43 \times 10^{28})(1.6 \times 10^{-19})(10^{-6})} = 2.97 \times 10^{-5} \text{ m/s}$$

From the fact that $\sigma = nq\mu$, the mobility of the free electrons is

$$\mu = \frac{\sigma}{nq} = \frac{5.78 \times 10^7}{(8.43 \times 10^{28})(1.6 \times 10^{-19})} = 4.29 \times 10^{-3} \text{ m}^2/\text{V-s}$$

Since $u = \mu\mathscr{E}$, the electric field strength is

$$\mathscr{E} = \frac{u}{\mu} = \frac{2.97 \times 10^{-5}}{4.29 \times 10^{-3}} = 6.92 \times 10^{-3} \text{ V/m}$$

From the formula obtained in the preceding example, we can calculate the resistance of a 1-meter length of wire. We have

$$R = \frac{L}{\sigma A} = \frac{1}{(5.78 \times 10^7)(10^{-6})} = 0.0173 \ \Omega$$

From the equation $\sigma = nq\mu$, we see that conductivity σ is proportional to the concentration of free electrons n. A good conductor, such as copper, has a value of n on the order of 10^{28} to 10^{29}. Conversely, an **insulator** has a value of n on the order of 10^7. Materials such as silicon (Si) and germanium (Ge) have conductivities between these values and are known as **semiconductors.** At 300 K, for silicon $n = 1.5 \times 10^{16}$ m⁻³ and for germanium $n = 2.5 \times 10^{19}$ m⁻³.

A semiconductor such as silicon or germanium has a crystal lattice structure as illustrated two-dimensionally in Fig. 6.3. An atom has four valence electrons—each of which is shared with an adjacent atom. This is called a **covalent bond.** In Fig. 6.3, the nucleus and inner electrons of an atom are designated by a circle

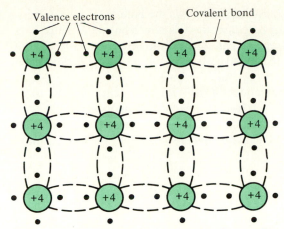

Fig. 6.3 Crystal lattice structure of a semiconductor.

labeled "+4" since this is a positively charged ion, and a valence electron is designated by a solid dot. Since the valence electrons are shared with adjacent atoms, these electrons are tightly bound to the atoms—and the crystal has a low conductivity.

At low temperatures, this type of material acts as an insulator. However, at room temperatures, thermal energy can break the covalent bond by dislodging a valence electron. The resulting vacated space, designated by a hollow dot in Fig. 6.4, is called a **hole.** The combination of the hole and the resulting **free electron** is known as a **hole-electron pair.**

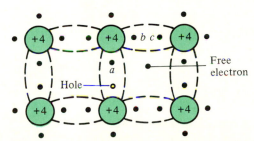

Fig. 6.4 Hole-electron pair in a semiconductor.

Relative to an electron, 1 joule (J) is a very large amount of energy. For this reason, we define the **electron volt** (eV) by

$$1 \text{ eV} = 1.60 \times 10^{-19} \text{ J}$$

The energy required to form a hole-electron pair is 1.1 eV for silicon and 0.72 eV for germanium at room temperature.

Once a hole has been created, another valence electron can fill in this hole and, thus, produce a different hole. This process can continue, giving the effect of a hole in motion. For example, if the electron at point a in Fig. 6.4 fills in the hole, there will be a hole at point a. If the electron at b then fills the hole at a, the result is a hole at b. If the electron at c fills the hole at b, there is now a hole at c. Thus, even though it is electrons that are moving, we can think of the original hole as moving to a, then to b, and then to c. Consequently, we can think of this phenomenon in terms of a flow of (free) positive charge, called a **hole current.**

In a pure (called **intrinsic**) semiconductor the number of holes is equal to the number of free electrons. Thus, the free-electron concentration n and the **hole concentration,** designated p, must be equal, that is $n = p$. We call this number the **intrinsic concentration** of the semiconductor and designate it by n_i. That is, for an intrinsic semiconductor,

$$n_i = n = p$$

Recombination occurs when a free electron fills in a hole. As a result of thermal energy, new hole-electron pairs are continually being generated, and old ones recombine at the same rate.

In a metal, current is conducted only by negative charges (electrons). For this reason, we say that metal is **unipolar.** However, a semiconductor such as silicon or germanium is said to be **bipolar** since current can be conducted by both negative charges (electrons) and positive charges (holes).

Repeating Equation (6.5), we have for a metal that the current density J is given by

$$J = n\mu q \mathscr{E} = \sigma \mathscr{E} \qquad (6.5)$$

where n is the free-electron concentration, μ is the free-electron mobility, $q = 1.60 \times 10^{-19}$ C is the charge, σ is the conductivity of the metal, and $\mathscr{E}$ is the electric field.

For a semiconductor, let μ_n denote the mobility of the free electrons and μ_p the mobility of the holes. Under the influence of an electric field, electrons and holes will move in opposite directions. The consequence of this is electron and hole currents in the same direction. Generalizing Equation (6.5), the resulting current density J is given by

$$J = (n\mu_n + p\mu_p)q \mathscr{E} = \sigma \mathscr{E} \qquad (6.6)$$

where $\sigma = (n\mu_n + p\mu_p)q$ is the conductivity of the semiconductor, n is the free-electron concentration, p is the hole concentration, and $q = 1.60 \times 10^{-19}$ C is the charge. (Remember that for an intrinsic semiconductor, $p = n = n_i$.)

Example 6.3

At 300 K, the intrinsic concentration of silicon is 1.5×10^{16} m^{-3}. Furthermore, the free-electron mobility is 0.13 m^2/V-s and the hole mobility is 0.05 m^2/V-s. Thus, the (intrinsic) conductivity of Si is

$$\sigma = (n\mu_n + p\mu_p)q = (n_i\mu_n + n_i\mu_p)q = n_i(\mu_n + \mu_p)q$$

$$= (1.5 \times 10^{16})(0.13 + 0.05)(1.6 \times 10^{-19}) = 4.32 \times 10^{-4} \; \mho/m$$

and the resistivity of Si is

$$\rho = \frac{1}{\sigma} = 2.315 \times 10^3 = 2315 \ \Omega\text{-m}$$

In Example 6.3, we determined the conductivity of silicon (at 300 K) to be 4.32×10^{-4} ℧/m. A metal such as copper has a much higher conductivity—about 5.8×10^7 ℧/m. For a metal, as temperature increases, random thermal motion of the free electrons increases, the time between collisions decreases, and electron mobility decreases. As a consequence, the conductivity of a metal decreases as temperature increases. However, for a semiconductor, as temperature increases, more hole-electron pairs are generated so that the free-electron and hole concentrations increase. The result is an increase in the conductivity of the semiconductor.

6.2 DOPED SEMICONDUCTORS

Not only does the conductivity of silicon or germanium increase as temperature increases, but the conductivity of semiconductors can be dramatically increased by adding small amounts of appropriate impurities. Such a process is referred to as **doping,** and we can say that the resulting semiconductor is **doped.**

To see the consequences of doping, first consider the case that a small amount of an element, called an **impurity,** with five valence electrons (e.g., arsenic or phosphorus) is added to a semiconductor such as silicon or germanium. As shown in Fig. 6.5, what results is a doped semiconductor with a covalent bonding. However, because each impurity atom has five valence electrons, only four of which are used for the covalent bond, there is an excess free electron for each impurity atom. Since such an impurity "donates" free electrons (negative charges), we call the

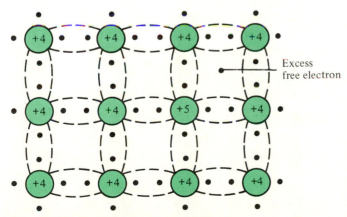

Fig. 6.5 An *n*-type semiconductor.

impurity a **donor** and refer to the impure semiconductor as an ***n*-type semiconductor**—the "*n*" signifying donor or **negative**.

For the case in which silicon or germanium is doped with an impurity having three valence electrons (e.g., boron or gallium), what results is a doped semiconductor, as illustrated in Fig. 6.6, which again has a covalent bonding. In this case, however, because it has only three valence electrons, each impurity atom produces an excess hole. Since a hole will "accept" a free electron, we call the impurity an **acceptor** and refer to the doped semiconductor as a ***p*-type semiconductor**—the "*p*" signifying acceptor or **positive**.

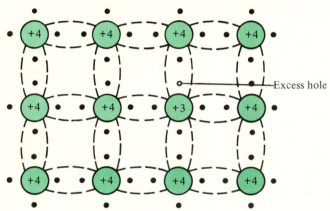

Fig. 6.6 A *p*-type semiconductor.

In a doped semiconductor, the current will be carried predominantly either by (free) electrons or by holes. In an *n*-type semiconductor, electrons are the **majority carriers** and holes are the **minority carriers.** The converse is true for a *p*-type semiconductor. Both holes and free electrons are said to be **mobile carriers.**

To see how doping affects the conductivity of a semiconductor, we now present the **mass-action law.** This law states that regardless of the amount of doping for a semiconductor, under thermal equilibrium

$$np = n_i^2$$

(6.7)

where n is the free-electron concentration, p is the hole concentration, and n_i is the intrinsic concentration of the semiconductor.

As was seen in Fig. 6.5, when a semiconductor is doped with donor atoms, we get the situation shown in Fig. 6.7. We can consider the boxed region around the donor atom shown to be an immobile positive ion. If the concentration of donor atoms is N_D, then at ordinary temperatures, the concentration of immobile positive ions will be N_D also. Similarly, if a semiconductor is doped with acceptor atoms having a concentration of N_A, then the resulting concentration of immobile negative

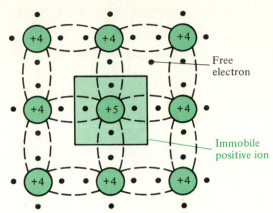

Free electron

Immobile positive ion

Fig. 6.7 Immobile positive ion.

ions will be N_A as well. In general, for a semiconductor with n-type and p-type impurities, since the semiconductor must be electrically neutral,

$$N_D + p = N_A + n \tag{6.8}$$

where p is the hole concentration and n is the (free-) electron concentration.

For an n-type semiconductor, $N_A = 0$, and Equation (6.8) becomes $N_D + p = n$. Also, for an n-type semiconductor, there are many more free electrons than holes ($n \gg p$). Thus, we deduce that $N_D = n - p \approx n$, and by the mass-action law (Equation 6.7),

$$p = \frac{n_i^2}{n} \approx \frac{n_i^2}{N_D}$$

In summary, for an n-type semiconductor

$$\boxed{n \approx N_D \quad \text{and} \quad p \approx \frac{n_i^2}{N_D}} \tag{6.9}$$

Similarly, for a p-type semiconductor

$$\boxed{p \approx N_A \quad \text{and} \quad n \approx \frac{n_i^2}{N_A}} \tag{6.10}$$

We are now in a position to see how doping a semiconductor affects its conductivity.

Example 6.4

At 300 K, the intrinsic concentration of Si is 1.5×10^{16} m^{-3}. Furthermore, the free-electron mobility is 0.13 m^2/V-s and the hole mobility is 0.05 m^2/V-s. In the preceding example, we calculated the conductivity as 4.32×10^{-4} ℧/m.

Suppose that silicon is doped with two parts per 10^8 of a donor impurity. Given that there are 5×10^{28} Si atoms/m^3, then the concentration of donor atoms is

$$N_D = (5 \times 10^{28})\left(\frac{2}{10^8}\right) = 10^{21} \text{ m}^{-3}$$

From Equations (6.9),

$$n \approx N_D = 10^{21} \text{ m}^{-3}$$

and

$$p \approx \frac{n_i^2}{N_D} = \frac{(1.5 \times 10^{16})^2}{10^{21}} = 2.25 \times 10^{11} \text{ m}^{-3}$$

Thus, the conductivity of the doped silicon is

$$\sigma = (n\mu_n + p\mu_p)q$$
$$\approx [(10^{21})(0.13) + (2.25 \times 10^{11})(0.05)](1.6 \times 10^{-19}) = 20.8 \text{ ℧/m}$$

Hence, by doping the semiconductor with two parts per 100 million, the conductivity increases by a factor of

$$\frac{20.8}{4.32 \times 10^{-4}} \approx 48,000$$

The *p-n* Junction

We have seen that a current in a semiconductor can be obtained as a result of the influence of an electric field [Equation (6.6)]. Yet there is another way in which electric charge can be transported in a semiconductor. If there is a nonuniform concentration of electric charge, then the charge will move from a region of a given concentration to one of lower concentration. This phenomenon is known as **diffusion.** (It is analogous to the diffusion of gases.) The movement of charge is not due to the repulsion of like charges, but rather is a result of a statistical phenomenon.

In general, the concentration of holes p in a semiconductor is not constant but rather is a function of distance x. The hole-concentration rate of change dp/dx is known as the **concentration gradient** for holes.

The hole current density J_p due to diffusion is proportional to the concentration gradient. In particular,

$$J_p = -qD_p \frac{dp}{dx} \tag{6.11}$$

where D_p, called the **diffusion constant** for holes, has square meters per second (m^2/s) as units. Similarly, for the case of (free) electrons, the electron current density J_n is given by

$$J_n = qD_n \frac{dn}{dx} \tag{6.12}$$

where dn/dx is the concentration gradient for electrons, and the diffusion constant for electrons D_n also has units m^2/s.

Diffusion and mobility are not independent of each other, and in fact are related by the **Einstein equation,** *

$$\frac{D_n}{\mu_n} = \frac{D_p}{\mu_p} = \frac{kT}{q} \tag{6.13}$$

where $k = 1.381 \times 10^{-23}$ J/K is the **Boltzmann constant,** [†] T is the temperature (in kelvins), and $q = 1.60 \times 10^{-19}$ C. Since the units of kT/q are J/C $=$ V, we define the quantity V_T by

$$V_T = \frac{kT}{q} = \frac{T}{11,586} \approx \frac{T}{11,600} \tag{6.14}$$

and refer to V_T as the "volt equivalent of temperature."

Given a semiconductor with hole and electron concentrations p and n, respectively, if an electric field $\mathscr{E}$ is also present, then the total current density for holes is [see Equations (6.5), (6.6), and (6.11)]

$$J_p = p\mu_p q \mathscr{E} - qD_p \frac{dp}{dx} \tag{6.15}$$

while for electrons the total current density is [see Equations (6.5), (6.6), and (6.12)]

$$J_n = n\mu_n q \mathscr{E} + qD_n \frac{dn}{dx} \tag{6.16}$$

Consider a doped semiconductor with a nonuniform hole concentration p. Such a semiconductor is said to be **graded.** Suppose that there are no external con-

---

*Named for the American (German-born) physicist Albert Einstein (1879–1955).
[†]Named for the Austrian physicist Ludwig Boltzmann (1844–1906).

nections to the semiconductor. Since p is nonuniform, there is a hole diffusion current. But nothing is connected to the semiconductor—so in order to cancel the diffusion current, there must be a current due to hole drift that is equal and opposite to the diffusion current. However, the existence of a drift current means there is an electric field; otherwise the drift is random and averages to zero.

From Equation (6.15), since the net current is zero, we have

$$0 = p\mu_p q\mathscr{E} - qD_p \frac{dp}{dx}$$

from which

$$\mathscr{E} = \frac{D_p}{p\mu_p} \frac{dp}{dx}$$

From Equations (6.13) and (6.14), this equation becomes

$$\mathscr{E} = \frac{V_T}{p} \frac{dp}{dx}$$

However, there is a relationship between potential v and electric field $\mathscr{E}$— $\mathscr{E} = -dv/dx$. Using this equation, we have

$$-\frac{dv}{dx} = \frac{V_T}{p} \frac{dp}{dx}$$

Integrating both sides of this expression with respect to distance x between points x_1 and x_2, we obtain

$$\int_{x_1}^{x_2} -\frac{dv}{dx} dx = \int_{x_1}^{x_2} \frac{V_T}{p} \frac{dp}{dx} dx$$

Applying the chain rule of calculus, this integral equation becomes

$$\int_{v_1}^{v_2} - dv = \int_{p_1}^{p_2} V_T \frac{dp}{p}$$

where v_1 and v_2 are the potentials at x_1 and x_2, respectively; and p_1 and p_2 are the hole concentrations at x_1 and x_2, respectively. Evaluating these integrals, we find that

$$v_2 - v_1 = V_T(\ln p_1 - \ln p_2)$$

or

$$v_{21} = V_T \ln \frac{p_1}{p_2} \tag{6.17}$$

where we define $v_{21} = v_2 - v_1$. Note, therefore, that the difference in potential between points x_1 and x_2 depends on the hole concentrations at the two points and not on the distance between the points.

From Equation (6.17), we have

$$\ln \frac{p_1}{p_2} = \frac{v_{21}}{V_T}$$

Taking powers of e, we get

$$\frac{p_1}{p_2} = e^{v_{21}/V_T}$$

from which

$$p_1 = p_2 e^{v_{21}/V_T} \tag{6.18}$$

Similarly, for the case of electrons

$$n_1 = n_2 e^{-v_{21}/V_T} \tag{6.19}$$

Taking the product of Equations (6.18) and (6.19) results in

$$n_1 p_1 = (n_2 e^{-v_{21}/V_T})(p_2 e^{v_{21}/V_T}) = n_2 p_2$$

Thus, we see that (under thermal equilibrium) the product of n and p is always the same and is independent of position and the amount of doping. Since $n = p = n_i$ for an intrinsic semiconductor, the product of n and p must be $np = n_i^2$—which is the mass-action law [Equation (6.7)].

Now consider the case that one side of a semiconductor is p-type material and the other side is n-type material as depicted in Fig. 6.8. We shall deal with the situation that there are no external connections (i.e., the open-circuit case). The boundary between the two sides of the semiconductor is called a **p-n junction,** and we say that the semiconductor is **step graded.**

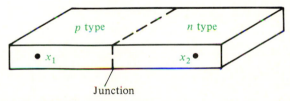

Fig. 6.8 A *p-n* junction.

Suppose that for the p-type material the hole concentration p_p is uniform. From Equations (6.10), then

$$p_p \approx N_A \quad \text{and} \quad n_p \approx \frac{n_i^2}{N_A}$$

In addition, suppose that for the n-type material, the free-electron concentration n_n is uniform. From Equations (6.9), then

$$n_n \approx N_D \quad \text{and} \quad p_n \approx \frac{n_i^2}{N_D}$$

In general, there is a difference in concentrations between the two sides with the concentration at the junction zero. From Equation (6.17), the potential difference between the two sides of the semiconductor is

$$v_{21} = V_T \ln \frac{p_1}{p_2} = V_T \ln \frac{p_p}{p_n} = V_T \ln \frac{N_A}{n_i^2/N_D}$$

and defining the **contact potential (difference)** v_0 to be $v_0 = v_{21}$, therefore, we have

$$v_0 = V_T \ln \frac{N_A N_D}{n_i^2} \qquad\qquad (6.20)$$

Use of a similar argument for free-electron concentrations will result in the same expression for v_0. The order of magnitude for v_0 typically is a few tenths of a volt (see Problems 6.13–6.18).

The contact potential of a *p-n* junction cannot cause an external current; if external connections are made, what results is the creation of two new contact potentials that negate the original one.

Example 6.5

The *p* side of germanium *p-n* junction has a hole concentration of 2×10^{21} m^{-3}, while the *n* side has a free-electron concentration of 5×10^{22} m^{-3}. Therefore, from Equation (6.20), the contact potential at 300 K of this *p-n* junction is

$$v_0 = V_T \ln \frac{N_A N_D}{n_i^2} = \frac{300}{11,586} \ln \frac{(2 \times 10^{21})(5 \times 10^{22})}{(2.5 \times 10^{19})^2} = 0.31 \text{ V}$$

6.3 THE JUNCTION DIODE

Let us now look at the *p-n* junction in further detail. Figure 6.9 shows a two-dimensional depiction of a *p-n* junction, where the *p*-type material is on the left-hand side of the semiconductor and the *n*-type material is on the right-hand side. Since there is a concentration gradient across the junction, diffusion results— holes diffuse to the right and free electrons diffuse to the left. The net effect is a diffusion current directed to the right. When they meet, in the vicinity of the junction, holes and free electrons combine and are neutralized. This results in uncovered ions in the neighborhood of the junction. Since this region is depleted of mobile carriers (holes and free electrons), it is called the **depletion region.** It is also referred to as the **space-charge region** and the **transition region.** (The width of this region typically is of the order of 5×10^{-7} m.)

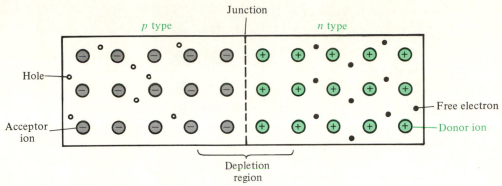

Fig. 6.9 A *p-n* junction.

The ions in the depletion region establish an electric field directed from right to left. This, in turn, produces a drift to the left of holes, and a drift to the right of free electrons—that is, a conduction (drift) current directed to the left. If there are no external connections to the junction (i.e., the open-circuit case), the diffusion current directed to the right and the conduction current directed to the left must sum to a net current of zero.

If we connect **metal** (or **ohmic**) **contacts** to a *p-n* junction as illustrated in Fig. 6.10(a), we have a circuit element known as a **junction diode.** The circuit symbol for this element is given in Fig. 6.10(b). The *p* side of the junction diode is called the **anode** and the *n* side is called the **cathode.**

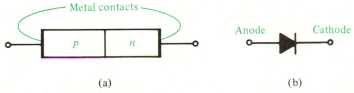

Fig. 6.10 Junction diode.

Let us connect a battery (dc voltage source) to a junction diode as shown in Fig. 6.11. Here the positive terminal of the battery is connected to the cathode (*n* side) and the negative terminal is connected to the anode (*p* side). Under this condition, known as **reverse bias,** holes have a tendency to diffuse to the right and drift to the left. Conversely, electrons tend to diffuse to the left and drift to the right. The result is a wide depletion region, and a potential across the junction of $v_0 + V$, where v_0 is the contact potential [Equation (6.20)] and V is the battery potential.

In the *n*-type material, holes are minority carriers—there are relatively few holes, and they are continually being thermally generated and recombining with free electrons. Some of these holes drift to the left into the depletion region, where they are pushed across by the electric field. A similar situation exists in the *p*-type material where free electrons are the minority carriers. Here the electron drift is to the right. The net effect is a small current I_o, called the **reverse saturation current,**

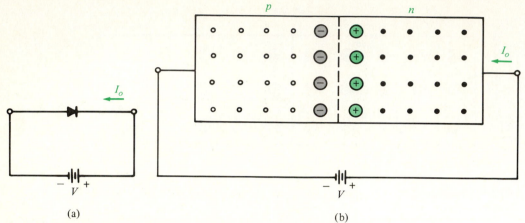

Fig. 6.11 Reverse-biased junction diode.

through the diode directed to the left. For the most part, this current is not dependent on the value of V, but rather depends on temperature. A higher temperature means that more hole-electron pairs are generated, and more minority carriers produce a greater value for I_o. [Typically, at 300 K, I_o is in the microampere (μA), or 10^{-6} A, range for germanium, and in the nanoampere (nA), or 10^{-9} A, range for silicon.] However, if V is made sufficiently large, the junction will "break down" (see Section 6.6) and a large current can result.

Now suppose that the polarity of the battery shown in Fig. 6.11 is reversed. In other words, suppose that the positive terminal of the battery is connected to the anode (p side) and the negative terminal is connected to the cathode (n side). Under this condition, known as **forward bias,** holes tend to diffuse to the right. Furthermore, free electrons tend to diffuse to the left. If V is large enough (a few tenths of a volt), diffusion overcomes the opposing tendency to drift, and holes (majority carriers) from the p-type material cross the junction and are injected into the n-type material, where they recombine with free electrons. In addition, free electrons (majority carriers) from the n-type material are injected into the p-type material, where they recombine with holes. The net result of this is a current directed to the right that is sustained by the continual thermal generation of majority carriers.

Although the applied voltage V narrows the depletion region, the depletion region never disappears. A narrower depletion region means more current, but the current is limited by the "bulk" resistance of the semiconductor crystal material and the resistance of the metal contacts.

Shown in Fig. 6.12 is a (junction) diode that has a voltage v across it and a current i through it. The polarity of v and the direction of i are conventional. The relationship between i and v is given by

$$i = I_o(e^{v/\eta V_T} - 1)$$

(6.21)

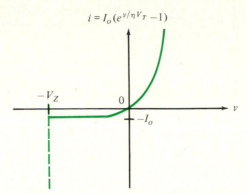

$$i = I_o(e^{v/\eta V_T} - 1)$$

Fig. 6.12 Diode.   **Fig. 6.13 The *i-v* characteristic curve of a diode.**

where I_o is the reverse saturation current, $V_T = T/11{,}586$ is the volt equivalent of temperature, and the constant η (eta) is given by

$$\eta = \begin{cases} 1 & \text{for Ge} \\ 2 & \text{for Si} \end{cases}$$

A typical plot of current i versus voltage v, called the ***i-v* characteristic curve,** for a diode is shown in Fig. 6.13.

If the diode is forward biased and $v > 0.2$ V, then at room temperature a good approximation (for both silicon and germanium) of Equation (6.21) is

$$i \approx I_o e^{v/\eta V_T} \tag{6.22}$$

If the diode is reverse biased and $-V_z < v < -0.2$ V, then a good approximation is

$$i \approx -I_o \tag{6.23}$$

If the voltage across the diode is made negative enough, that is, for a large enough reverse bias, the diode will break down and allow negative current (more negative than $-I_o$) through it. From the *i-v* characteristic given in Fig. 6.13, we see that when $v = -V_z$, the diode breaks down, and the dashed line indicates that a negative (reverse) current results. Note that when $i < -I_o$, the voltage across the diode is essentially constant ($v = -V_z$). We discuss the phenomenon of diode breakdown further in Section 6.6.

In Equation (6.21), v is the independent variable and i is the dependent variable. That is, given a value for v, it is a simple matter to determine the corresponding value for i. From Equation (6.21), we have

$$i + I_o = I_o e^{v/\eta V_T}$$

so

$$e^{v/\eta V_T} = \frac{i + I_o}{I_o} = \frac{i}{I_o} + 1$$

and

$$\frac{v}{\eta V_T} = \ln\left(\frac{i}{I_o} + 1\right)$$

Thus,

$$v = \eta V_T \ln\left(\frac{i}{I_o} + 1\right)$$

(6.24)

In this expression, i is the independent variable and v is the dependent variable. However, since the natural logarithm exists only for positive numbers, Equation (6.24) is valid only for the case that $i > -I_o$.

Example 6.6

At 300 K, a 1N4153 silicon diode has a breakdown voltage of $V_Z = 75$ V and a reverse saturation current of $I_o = 10$ nA. If the voltage across the diode is $v = 0.6$ V, then by Equation (6.21), the current through it is

$$i = I_o(e^{v/\eta V_T} - 1) = (10 \times 10^{-9})(e^{(0.6)(11,586)/(2)(300)} - 1) = 1.08 \text{ mA}$$

while if $v = 0.7$ V, then

$$i = (10 \times 10^{-9})(e^{(0.7)(11,586)/(2)(300)} - 1) = 7.42 \text{ mA}$$

Conversely, if the current through the diode is $i = 5$ mA, then by Equation (6.24), the voltage across it is

$$v = \eta V_T \ln\left(\frac{i}{I_o} + 1\right) = 2\left(\frac{300}{11,586}\right)\ln\left(\frac{5 \times 10^{-3}}{10 \times 10^{-9}} + 1\right) = 0.68 \text{ V}$$

Diode Behavior

Now let us inspect the i-v characteristic (see Fig. 6.13) of a diode more closely. For the case that the diode is forward biased ($i > 0$, $v > 0$), Fig. 6.14 shows i-v characteristics for typical germanium and silicon diodes. For germanium, η is half the value that it is for silicon, while I_o is of the order of 1000 times larger for germanium than for silicon. As can be seen in Fig. 6.14, these facts mean that, for a given voltage, the germanium diode will have a much larger current than the

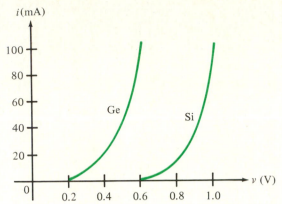

Fig. 6.14 Typical germanium (Ge) and silicon (Si) i-v characteristic curves.

silicon diode. Alternatively, for a given current, the germanium diode will have a smaller voltage than the silicon diode.

From Fig. 6.14, note that the current through the germanium diode is very small* for voltage values of less than 0.2 V. In other words, there is no significant current through the germanium diode until the voltage across it exceeds 0.2 V. For this reason, we say that 0.2 V is the **cut-in** or **threshold voltage** V_γ of the germanium diode. Furthermore, inspection of Fig. 6.14 reveals that the approximate value for the cut-in voltage of the silicon diode is $V_\gamma = 0.6$ V.

It has been determined experimentally that for a temperature increase of 1°C (an increase of 1 K), in both germanium and silicon diodes the reverse saturation current increases by about 7 percent. This means that for an increase in temperature of 10°C, the reverse saturation current increases by a factor of $(1.07)^{10} \approx 2$. In other words, I_o is a function of temperature T, and I_o approximately doubles for every rise of 10°C. To express this fact in mathematical form, if $I_o(T_a)$ is the reverse saturation current at temperature T_a, then the reverse saturation current at temperature T_b is given approximately by the equation

$$I_o(T_b) = 2^{(T_b - T_a)/10} I_o(T_a) \qquad (6.25)$$

Note that in Equation (6.25), we can use either °C or K (but not both together) as the unit for temperature.

From the fact that the reverse saturation current of a diode is a function of temperature, it is apparent that, for a given voltage, the diode current is also temperature dependent. Specifically, if temperature increases, then I_o increases. Even though the exponent in Equation (6.21) decreases as T increases, I_o increases

---

*By "very small" we mean a value of less than 1 percent of the maximum current rating of the diode.

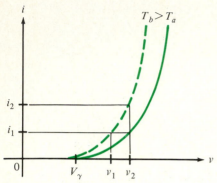

Fig. 6.15 Temperature dependence of diode *i-v* characteristic.

at a higher rate. The result is that for a given voltage v, the diode current i increases as T increases. This phenomenon is illustrated in Fig. 6.15, where the solid curve is the i-v characteristic of a diode (either germanium or silicon) at temperature T_a, and the dashed curve is the i-v characteristic of the same diode at temperature $T_b > T_a$. From Fig. 6.15, we see that if the diode voltage is $v = v_2$, then the diode current is $i = i_1$ for temperature T_a, and $i = i_2$ for $T_b > T_a$. To maintain the current $i = i_1$ when the temperature increases from T_a to T_b, the diode voltage should be lowered from $v = v_2$ to $v = v_1$. In other words, by decreasing the voltage appropriately, we can compensate for the increase in temperature. Specifically, at room temperature (for either germanium or silicon), a constant current can be maintained if the voltage is decreased by approximately 2.5 mV for each 1°C increase in temperature. In mathematical terms,

$$\frac{dv}{dT} \approx -2.5 \text{ mV/°C} \tag{6.26}$$

Thus, for Fig. 6.15, we have when $i = i_1$ that $v_2 - v_1 \approx -2.5(T_a - T_b)$ mV, or

$$v_2 - v_1 \approx 2.5(T_b - T_a) \times 10^{-3} \text{ V} \tag{6.27}$$

Example 6.7

For the 1N4153 silicon diode given in Example 6.6, the reverse saturation current at 300 K is $I_o(300) = 10$ nA. From Equation (6.25), at 316 K the reverse saturation current is

$$I_o(316) = 2^{(316 - 300)/10} I_o(300) = 2^{1.6}(10 \times 10^{-9}) = 30.3 \text{ nA}$$

In Example 6.6, we determined that at 300 K when $v = 0.7$ V, then $i = 7.42$ mA. To have the same current at 316 K, by Equation (6.27), the voltage must be reduced by

$$v_2 - v_1 \approx 2.5(316 - 300) \times 10^{-3} = 0.04 \text{ V}$$

that is, at 316 K for $i = 7.42$ mA, the voltage should be $v \approx 0.7 - 0.04 = 0.66$ V.

Now let us look at the simple diode circuit shown in Fig. 6.16. We can attempt to analyze this circuit by writing some equations, and begin with the equation relating the current and voltage for the diode:

$$i = I_o(e^{v/\eta V_T} - 1) \tag{6.21}$$

where it is assumed that I_o, η, and T are given. By KVL, we can also write

$$v_1 = v + Ri \tag{6.28}$$

where it is assumed that v_1 and R are given. Thus, we have two equations, (6.21) and (6.28), and two unknowns. However, Equation (6.21) is nonlinear, so that the techniques used for linear, algebraic simultaneous equations cannot be employed. For example, substituting Equation (6.21) into Equation (6.28) results in a transcendental equation that has no analytical solution.

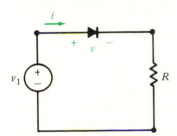

Fig. 6.16 Simple diode circuit.

Fortunately, though, there are other analysis techniques that can be utilized. Suppose that v_1 has a positive value. Since current goes through a resistor from a given potential to a lower potential, we should recognize for Fig. 6.16 that i, and hence v, are positive; that is, the diode is forward biased. Next, by using equation (6.21) and a calculator, we can determine enough points to sketch an i-v (characteristic) curve for the diode. But from Equation (6.28) we have that

$$i = -\frac{1}{R}v + \frac{v_1}{R} \tag{6.29}$$

This is the equation of a straight line of slope $-1/R$ and vertical-axis intercept v_1/R. This straight line, called the **load line**, can be plotted using the same axes as for the diode curve as is shown in Fig. 6.17. Since any point on the diode curve satisfies Equation (6.21), and any point on the load line satisfies Equation (6.29), the intersection of both plots satisfies both equations simultaneously. The point of

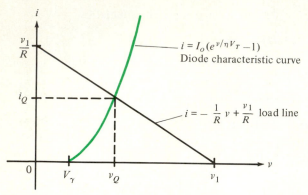

Fig. 6.17 Graphical analysis of a simple diode circuit.

intersection is designated Q and is called the **quiescent operating point** (or **operating point,** or Q **point** for short). Thus, the solution to the equations, which is the current i_Q and the voltage v_Q, has been obtained graphically instead of analytically.

Instead of a graphical approach, with the aid of a hand-held calculator, a simple numerical method can be employed. The technique is an iterative process that begins by assuming a partial numerical solution, which is used to calculate the remainder of the solution. This, in turn, can be used to refine the original assumption. The process is repeated until a desired accuracy is obtained.

Example 6.8

For the diode circuit given in Fig. 6.16, $v_1 = 3$ V, $R = 1$ kΩ, and the silicon diode has a reverse saturation current of 10 nA at 300 K. Let us assume that $v = 0.6$ V. Then, by Ohm's law,

$$i = \frac{-v + v_1}{R} = \frac{-0.6 + 3}{10^3} = 2.4 \text{ mA}$$

Now let us use this value of current to refine the value of the voltage. By Equation (6.24)

$$v = \eta V_T \ln\left(\frac{i}{I_o} + 1\right) = 2\left(\frac{300}{11{,}586}\right)\ln\left(\frac{2.4 \times 10^{-3}}{10 \times 10^{-9}} + 1\right) = 0.6416 \text{ V}$$

and using this value of the voltage, we can refine the value of the current from

$$i = \frac{-v + v_1}{10^3} = \frac{-0.6416 + 3}{10^3} = 2.358 \text{ mA}$$

Performing one more iteration,

$$v = 2\left(\frac{300}{11{,}586}\right)\ln\left(\frac{2.358 \times 10^{-3}}{10 \times 10^{-9}} + 1\right) = 0.6406 \text{ V}$$

and

$$i = \frac{-0.6406 + 3}{10^3} = 2.359 \text{ mA}$$

Since this value of the current changed only in the fourth significant digit, from a practical point of view, we need not proceed any further.

■

6.4 THE IDEAL DIODE

Figure 6.13 shows a typical i-v characteristic curve for a diode. The relationship between current and voltage, given by Equation (6.21), is clearly nonlinear and is analytically cumbersome. There are many situations in which a simplified description for a diode can be used for analysis purposes without sacrificing reasonable accuracy. Because of this, we define an **ideal diode** to be a two-terminal device whose i-v characteristic is that given by Fig. 6.18. The circuit symbol for the ideal diode shown in Fig. 6.19(a) differs slightly (it is hollow, not solid) from the nonideal diode (see Fig. 6.12). For the case that the current through an ideal diode is in the forward direction (i.e., for forward bias), the voltage across the diode is zero—that is, the diode behaves as a short circuit. For the case that the voltage across an ideal diode is nonpositive (i.e., for reverse bias), the current through the diode is zero—that is, the diode behaves as an open circuit. These situations are depicted in Figs. 6.19(b) and (c) respectively.

Fig. 6.18 Ideal diode _I-V_ characteristic.

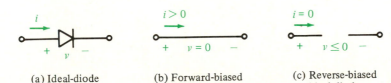

(a) Ideal-diode symbol (b) Forward-biased ideal diode (c) Reverse-biased ideal diode

Fig. 6.19 Ideal-diode behavior.

Since an ideal diode, as well as a nonideal diode, is a nonlinear element, we cannot immediately apply linear circuit analysis techniques to circuits containing diodes. Instead, what we shall do is, under certain conditions, assume diodes to be forward biased or reverse biased. For the former case, we say that a diode is **ON,** while for the latter case, we say that a diode is **OFF.** After assuming that diodes are ON or OFF, we replace them by short circuits or open circuits respectively. Next, we can apply known analysis techniques to the resulting linear circuits. If the current through an ON diode is calculated to be nonpositive, the assumption about its state was incorrect, and it must be assumed to be OFF. After changing assumptions, the analysis must be repeated from the beginning. Similarly, if the voltage across an OFF diode is calculated to be positive, its assumption of being OFF was incorrect, and it should be assumed to be ON. Again, the analysis must begin anew. If all diodes assumed to be ON are found to have positive currents, and if all diodes assumed to be OFF are found to have nonpositive voltages, then the assumptions about them were right, and the circuit currents and voltages calculated are correct. Diode circuit analysis often is shortened by using intuition or educated guessing as to whether a diode should be assumed to be OFF or ON.

Example 6.9

Given the simple ideal-diode circuit shown in Fig. 6.20, let us first consider the case of a positive input voltage, that is, $v_s > 0$. Since current goes through a resistor from a given potential to a lower potential, let us assume that D is ON. Replacing D by a short circuit, we get the circuit shown in Fig. 6.21. By Ohm's law, we have that $i = v_s/R$. Since $v_s > 0$, then $i > 0$, and the assumption that D is ON is correct.

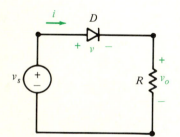

Fig. 6.20 Simple ideal-diode circuit.

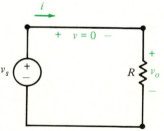

Fig. 6.21 Simple diode circuit with D assumed ON.

Next consider the case that $v_s \leq 0$. For this situation, let us assume that D is OFF. Replacing D by an open circuit yields the circuit shown in Fig. 6.22. By KVL, $v = v_s - Ri = v_s$ since $i = 0$. Since $v_s \leq 0$, then $v \leq 0$ and the assumption that D is OFF is correct.

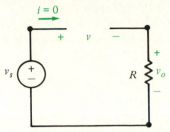

Fig. 6.22 Simple diode circuit with *D* assumed OFF.

From the discussion above, when $v_s > 0$, the output voltage is $v_o = v_s$. When $v_s \leq 0$, $v_o = Ri = 0$. Therefore, the output voltage v_o is given by

$$v_o = \begin{cases} 0 & \text{for} \quad v_s \leq 0 \\ v_s & \text{for} \quad v_s > 0 \end{cases}$$

Consider the case that the input voltage v_s is a sinusoid as shown in Fig. 6.23(a). Then the resulting output voltage v_o is the **half-wave rectified** sinusoid shown in Fig. 6.23(b). We call the simple diode circuit shown in Fig. 6.20 a **half-wave rectifier.**

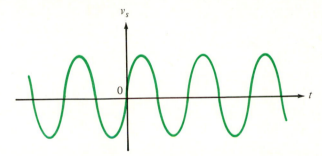

(a) Input to half-wave rectifier

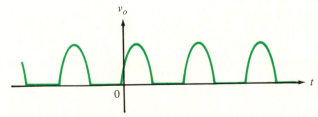

(b) Output of half-wave rectifier

Fig. 6.23 Input and output for half-wave rectifier circuit.

Having seen one application of diodes, let us now look at another.

Example 6.10

In addition to an ideal diode, the circuit shown in Fig. 6.24 contains two independent voltage sources. If the input voltage is $v_s > 0$, then v_s tends to forward bias D. However, the 3-V source tends to reverse bias D. Let us find out which voltage source dominates.

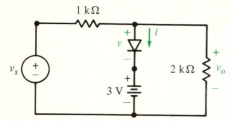

Fig. 6.24 Diode clipper circuit.

We begin by assuming that D is ON. From the resulting circuit shown in Fig. 6.25, by Ohm's law,

$$i_1 = \frac{v_s - 3}{10^3} \quad \text{and} \quad i_2 = \frac{3}{2 \times 10^3} = 1.5 \text{ mA}$$

and by KCL,

$$i = i_1 - i_2 = \frac{v_s}{10^3} - 3 \times 10^{-3} - 1.5 \times 10^{-3} = \frac{v_s}{10^3} - 4.5 \times 10^{-3}$$

Thus, we see that the diode is ON ($i > 0$) provided that $v_s > 4.5$ V. We may, therefore, now deduce for $v_s \le 4.5$ V that D will be OFF. When D is ON, by KVL, we have that the output voltage $v_o = 3$ V.

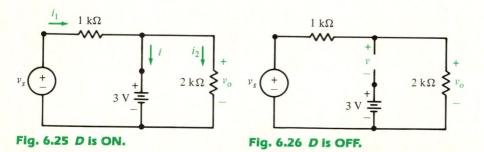

Fig. 6.25 _D_ is ON.   **Fig. 6.26 _D_ is OFF.**

To determine the output voltage when D is OFF, we need to analyze the circuit shown in Fig. 6.26. By voltage division, we have that the output voltage is

$$v_o = \frac{2 \times 10^3}{2 \times 10^3 + 1 \times 10^3} v_s = \frac{2}{3} v_s$$

As a matter of redundancy, by KVL,

$$v = v_o - 3 = \tfrac{2}{3} v_s - 3$$

and D is OFF ($v \le 0$) provided that $(2v_s/3) - 3 \le 0 \Rightarrow v_s \le 4.5$ V, as was determined above.

In summary, for the diode circuit given in Fig. 6.24, the output voltage v_o is given by

$$v_o = \begin{cases} \dfrac{2}{3} v_s & \text{for} \quad v_s \le 4.5 \text{ V} \\[2mm] 3 & \text{for} \quad v_s > 4.5 \text{ V} \end{cases}$$

Suppose now that the input voltage v_s is the sinusoid with an amplitude of 6 V as shown in Fig. 6.27(a). Then the output voltage v_o is the "clipped" sinusoid shown in Fig. 6.27(b). Because of this, we say that the circuit shown in Fig. 6.24 is an example of a diode **clipper circuit.**

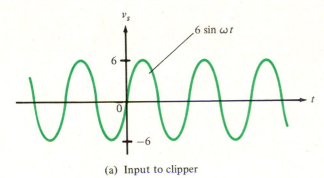

(a) Input to clipper

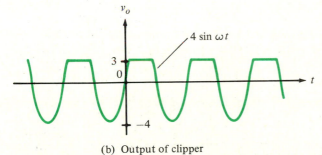

(b) Output of clipper

Fig. 6.27 Input and output of diode clipper circuit.

Application to Digital Logic Circuits

A major use of electronics is for digital logic circuits that are employed in computers, control systems, data processing systems, and digital communication systems. Typically, a digital circuit has an input and an output that are equal to one of only two possible voltages—a low voltage and a high voltage. Because there are only two possible states, one can be designated by a 0 and the other by a 1. If we use the 0 to represent the low voltage and the 1 to represent the high voltage, the result is known as **positive logic.** Conversely, if 0 represents the high voltage and 1 represents the low voltage, then we have **negative logic.** The importance and applications of digital logic to digital systems will be studied in detail in the portion of this book on digital systems (Chapters 11, 12, and 13). Now, however, we will see how ideal diodes can be used to fabricate some types of digital logic circuits.

We begin by considering the two-input diode circuit shown Fig. 6.28. Let us choose the case of positive logic, where the low voltage (logical 0) is 0 volts and the high voltage (logical 1) is V volts.

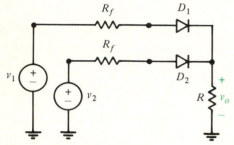

Fig. 6.28 Diode OR gate.

If $v_1 = v_2 = 0$, then there is no voltage to produce any current. Since no current goes through R, we have that $v_o = 0$.

If $v_1 = 0$ and $v_2 = V$, let us assume that D_1 is OFF and D_2 is ON. This means that we must analyze the circuit shown in Fig. 6.29. By KVL,

$$V = R_f i_2 + R i_2 = (R_f + R) i_2$$

from which

$$i_2 = \frac{V}{R_f + R}$$

Since $i_2 > 0$, then D_2 is indeed ON. Furthermore, by voltage division,

$$v_o = \frac{R}{R + R_f} V \qquad\qquad (6.30)$$

By KVL, $v_a = -v_o < 0$, and this confirms that D_1 is OFF. For the case that $R \gg R_f$, from Equation (6.30), we have that $v_o \approx V$.

Similarly, if $v_1 = V$ and $v_2 = 0$, we deduce that D_2 is OFF, D_1 is ON, and for $R \gg R_f$, $v_o \approx V$.

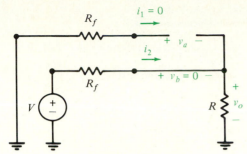

Fig. 6.29 D_1 is **OFF,** D_2 is **ON.**

Finally, if $v_1 = v_2 = V$, let us assume that both D_1 and D_2 are ON. Thus, we need to analyze the circuit shown in Fig. 6.30. By KVL,

$$V = R_f i_1 + R(i_1 + i_2) = (R_f + R)i_1 + Ri_2 \tag{6.31}$$

and

$$V = R_f i_2 + R(i_1 + i_2) = Ri_1 + (R_f + R)i_2 \tag{6.32}$$

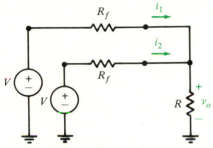

Fig. 6.30 D_1 and D_2 are **ON.**

Solving these two equations for i_1, by Cramer's rule,

$$i_1 = \frac{(R_f + R)V - RV}{(R_f + R)^2 - R^2} = \frac{R_f V}{R_f^2 + 2R_f R} = \frac{V}{R_f + 2R} \tag{6.33}$$

By symmetry (or by Cramer's rule, again), $i_2 = i_1$. Since $i_1 = i_2 > 0$, then indeed D_1 and D_2 are ON. Furthermore,

$$v_o = R(i_1 + i_2) = R\left(\frac{2V}{R_f + 2R}\right) = \frac{R}{R + R_f/2}V$$

and for $R \gg R_f$, we have that $v_o \approx V$.

In summary, for the case that $R \gg R_f$, the circuit given in Fig. 6.28 yields the table of voltages and the corresponding logic table shown, respectively, in Table 6.1(a) and 6.1(b).

Table 6.1(a) Voltage Table

v_1	v_2	v_o
0	0	0
0	V	V
V	0	V
V	V	V

Table 6.1(b) Logic Table

v_1	v_2	v_o
0	0	0
0	1	1
1	0	1
1	1	1

The operation described by logic table 6.1(b) is called the **OR operation** since the output is 1 if either one input OR the other is 1. For this reason, the circuit shown in Fig. 6.28 is known as an **OR gate,** and it is an example of a **diode logic** (DL) circuit.

A more general version of a (positive-logic) DL OR gate is shown in Fig. 6.31. In this case, the low voltage (logical 0) is designated by $V(0)$. Similarly, the high voltage (logical 1) can be designated by $V(1)$. For the case that $R \gg R_f$, analysis of the logic circuit in Fig. 6.31 yields the logic table given by Table 6.1(b). (See Problem 6.47.)

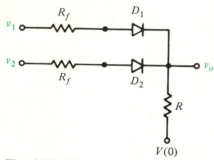

Fig. 6.31 DL two-input OR gate.

Although Figs. 6.28 and 6.31 show two-input DL OR gates, OR gates with more than two inputs are easily obtained—for each new input, simply place another series connection of a resistor R_f and a diode between that new input and the output.

A second important DL gate is shown in Fig. 6.32. Again we deal with positive logic, where the low voltage is 0 volts and the high voltage is V volts.

If $v_1 = v_2 = 0$, because of the battery, let us assume that D_1 and D_2 are ON. Therefore, from the resulting circuit in Fig. 6.33, by KVL,

$$V = R(i_1 + i_2) + R_f i_1 = (R + R_f)i_1 + Ri_2$$

and

$$V = R(i_1 + i_2) + R_f i_2 = Ri_1 + (R + R_f)i_2$$

Note that these are the Equations (6.31) and (6.32) obtained previously. Thus, the

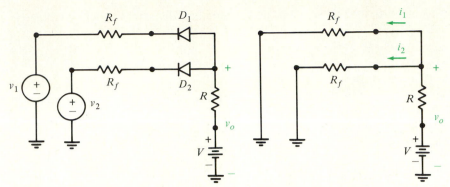

Fig. 6.32 Diode AND gate.   **Fig. 6.33 D_1 and D_2 are ON.**

solution is given by Equation (6.33), that is,

$$i_1 = i_2 = \frac{V}{R_f + 2R}$$

and since $i_1 = i_2 > 0$, then D_1 and D_2 are indeed ON. Furthermore, by voltage division (after both resistors labeled R_f are combined in parallel), we have

$$v_o = \frac{R_f/2}{R_f/2 + R}V = \frac{R_f}{R_f + 2R}V$$

and for $R \gg R_f$, then $v_o \approx 0$.

For the case that $v_1 = 0$ and $v_2 = V$, we assume that D_1 is ON and D_2 is OFF. From the resulting circuit in Fig. 6.34, by KVL, we have

$$V = Ri_1 + R_f i_1 = (R + R_f)i_1 \qquad \Rightarrow \qquad i_1 = \frac{V}{R + R_f}$$

Since $i_1 > 0$, then indeed D_1 is ON. Furthermore, by KVL,

$$v_b = -Ri_1 + V - V - R_f i_2 = -Ri_1 = \frac{-RV}{R + R_f}$$

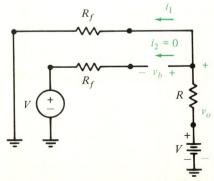

Fig. 6.34 D_1 is ON, D_2 is OFF.

and since $v_b < 0$, then D_2 is indeed OFF. By voltage division, the output voltage is

$$v_o = \frac{R_f}{R_f + R} V$$

so for $R \gg R_f$, $v_o \approx 0$.

Similarly, if $v_1 = V$ and $v_2 = 0$, then D_1 is OFF and D_2 is ON, and when $R \gg R_f$, the output voltage is $v_o \approx 0$.

Finally, if $v_1 = v_2 = V$, let us assume that D_1 and D_2 are OFF. From Fig. 6.35, since $i_1 = i_2 = 0$, by KVL,

$$v_a = -(i_1 + i_2)R + V - V - R_f i_1 = 0$$

and similarly, $v_b = 0$. Thus, D_1 and D_2 are indeed OFF, and the output voltage is

$$v_o = -R(i_1 + i_2) + V = V$$

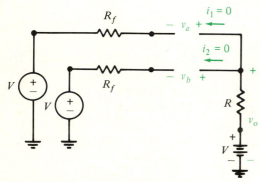

Fig. 6.35 D_1 and D_2 are OFF.

In summary, for the case that $R \gg R_f$, the circuit given in Fig. 6.32 yields the table of voltages and the corresponding logic table shown, respectively, in Table 6.2(a) and Table 6.2(b).

Table 6.2(a) Voltage Table

v_1	v_2	v_o
0	0	0
0	V	0
V	0	0
V	V	V

Table 6.2(b) Logic Table

v_1	v_2	v_o
0	0	0
0	1	0
1	0	0
1	1	1

The operation described by logic table 6.2(b) is called the **AND operation** since the output is 1 only if one input AND the other are both 1. For this reason, the circuit shown in Fig. 6.32 is known as a DL **AND gate**.

A more general version of a (positive-logic) DL AND gate is shown in Fig. 6.36, where $V(1)$ designates the high voltage. For the case that $R \gg R_f$, this circuit yields logic table 6.2(b). (See Problem 6.48.) As with the case of the OR gate, the number of inputs to an AND gate can be increased.

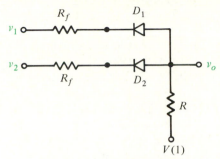

Fig. 6.36 DL two-input AND gate.

Suppose that negative logic is used for the DL circuit given in Fig. 6.32, that is, the low voltage of 0 volts is the logical 1 and the high voltage of V volts is the logical 0. Thus, according to the voltage table 6.2(a), the corresponding logic table is as given by Table 6.3. However, this is the same as logic table 6.1(b)—the rows are presented in reverse order—which describes the OR operation. Thus, we see that a negative-logic AND gate is a positive-logic OR gate. Similarly, it is true that a negative-logic OR gate is a positive-logic AND gate. These statements hold not only for the specific cases shown in Figs. 6.28 and 6.32, but for the more general versions shown in Figs. 6.31 and 6.36 as well. Specifically, if the circuit given in Fig. 6.31 is to be a negative-logic AND gate, the node labeled $V(0)$ must be changed to $V(1)$. Similarly, if the circuit given in Fig. 6.36 is to be a negative-logic OR gate, the node labeled $V(1)$ must be changed to $V(0)$.

Table 6.3

v_1	v_2	v_o
1	1	1
1	0	1
0	1	1
0	0	0

6.5 NONIDEAL-DIODE MODELS

Although the ideal model of a diode can be used with relatively good accuracy in the analysis and design of different types of electronic circuits, there may be occasions when a more accurate model of a diode is needed. Therefore, we now discuss how to model nonideal diodes, and we will see some examples of analysis using these models.

The **static** or **dc resistance** R of a (nonideal) diode that is operating at current i_Q and voltage v_Q is defined to be

$$R = \frac{v_Q}{i_Q}$$

For example, a diode operating at a current of $i_Q = 1.2$ mA and a voltage of $v_Q = 0.6$ V has a dc resistance of $R = 0.6/(1.2 \times 10^{-3}) = 500\ \Omega$. If the same diode operates instead at $i_Q = 8.28$ mA and $v_Q = 0.7$ V, then the dc resistance is $R = 0.7/(8.28 \times 10^{-3}) = 84.5\ \Omega$.

The dc resistance of a diode is useful when a diode operates constantly at one particular current and voltage. However, since the current and voltage of a diode typically vary, of greater importance is the **dynamic** or **ac** (or **incremental**) **resistance** r of a diode, which is defined at a given operating point by

$$r = \frac{dv}{di} \tag{6.34}$$

Similarly, we can define the **ac conductance** g by

$$g = \frac{1}{r} = \frac{di}{dv} \tag{6.35}$$

For a diode, we know that

$$i = I_o(e^{v/\eta V_T} - 1) = I_o e^{v/\eta V_T} - I_o \tag{6.36}$$

Thus,

$$g = \frac{di}{dv} = \frac{1}{\eta V_T} I_o e^{v/\eta V_T}$$

However, from Equation (6.36),

$$I_o e^{v/\eta V_T} = i + I_o$$

Therefore, the ac conductance of a diode is

$$g = \frac{di}{dv} = \frac{i + I_o}{\eta V_T} \tag{6.37}$$

so that the ac resistance of a diode is given by

$$\boxed{r = \frac{dv}{di} = \frac{\eta V_T}{i + I_o}} \tag{6.38}$$

For the case that $i \gg I_o$, a good approximation for the ac resistance of a diode at room temperature (295 K) is

$$r = \frac{dv}{di} \approx \frac{\eta V_T}{i} = \begin{cases} \dfrac{0.025}{i} & \text{for Ge} \\[2mm] \dfrac{0.05}{i} & \text{for Si} \end{cases}$$

Example 6.11

A silicon diode operating at 1.2 mA has an ac resistance at 300 K of

$$r \approx \frac{\eta V_T}{i} = \frac{2(300/11,586)}{1.2 \times 10^{-3}} = 43 \ \Omega$$

If the same diode operates at 8.28 mA, then at the same temperature, the ac resistance is

$$r \approx \frac{2(300/11,586)}{8.28 \times 10^{-3}} = 6.25 \ \Omega$$

■

This last example has demonstrated that the ac resistance of a diode, which can vary widely, depends on the current through the diode.

The dashed curve shown in Fig. 6.37 is a typical i-v characteristic of a diode. Since the current is very small when the diode is reverse biased (if the actual scale were used, the dashed curve in the third quadrant would be even closer to the negative-voltage axis), as for the ideal case, we will assume that the diode behaves as an open circuit (this situation corresponds to the heavy solid line drawn on the negative-voltage axis). Now, however, for the case that the diode is forward biased, let us approximate the dashed curve in the first quadrant by a straight line of slope $1/R_f$ as shown. [Since the slope of the straight line has units amperes per volt (A/V), the slope is designated by the reciprocal of resistance.] The value of this

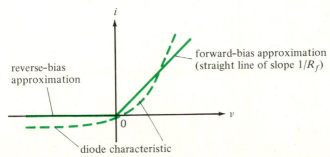

Fig. 6.37 Approximation of diode i-v characteristic.

forward resistance R_f is a typical ac resistance of the forward-biased diode. (For the preceding example, a value of R_f in the neighborhood of 20 Ω is reasonable.)

For the ideal case, a reverse-biased (OFF) diode behaves as an open circuit when $v \leq 0$, and a forward biased (ON) diode behaves as a short circuit when $i > 0$. According to the above discussion, a better approximation is to treat a forward-biased diode ($i > 0$) as a resistance R_f. We still treat the reverse-biased diode ($v \leq 0$) as an open circuit. Consequently, we can model the diode shown in Fig. 6.38(a) simply by the series connection of an ideal diode and a resistance R_f as shown in Fig. 6.38(b); and this model is more accurate than just an ideal diode. (For an even more accurate model, see Problem 6.39.) Of course, because they are in series, we can interchange the positions of the ideal diode and R_f in Fig. 6.38(b) without changing the effect of the model.

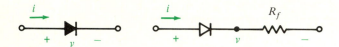

(a) Diode   (b) Diode model using ideal diode and resistance

Fig. 6.38 A diode and its simple model.

As an example, consider the OR gate shown in Fig. 6.39(a). With the diode model depicted in Fig. 6.38, the OR gate can be analyzed using the ideal-diode circuit shown in Fig. 6.39(b). However, in that circuit, we can combine R_s and R_f into a single resistance. The resulting circuit, therefore, can be treated identically to the OR gate given in Fig. 6.28, and this further demonstrates the usefulness of the concept of the ideal diode.

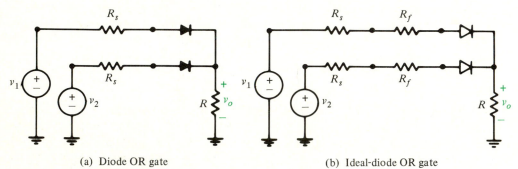

(a) Diode OR gate   (b) Ideal-diode OR gate

Fig. 6.39 Use of simple diode model.

More Accurate Models

We now can go one step further and get an even better (but still relatively simple) approximation of a diode. To do this, we enlarge the i-v characteristic curve of a forward-biased diode so that the cut-in voltage V_y becomes apparent. Such a situation is shown in Fig. 6.40, where the i-v characteristic is the dashed curve. Let

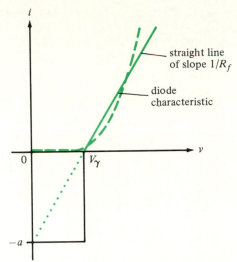

Fig. 6.40 Further approximation of diode *i-v* characteristic.

us approximate this dashed curve by a straight line of slope $1/R_f$ as shown. This line intersects the horizontal axis at $v = V_\gamma$ and is extended, in dotted form, to the vertical axis, where it intersects at some point $i = -a$. (This is done for computational purposes.) The equation of the solid line has the form $i = mv + b$, where m is the slope (i.e., $m = 1/R_f$) and b is the vertical-axis intercept (i.e., $b = -a$). Thus we have

$$i = \frac{1}{R_f}v - a \qquad\qquad (6.39)$$

However, from Fig. 6.40, we see that the slope is $1/R_f = a/V_\gamma$. Thus, $a = V_\gamma/R_f$, and substituting this fact into Equation (6.39) yields

$$i = \frac{1}{R_f}v - \frac{V_\gamma}{R_f} \quad\Rightarrow\quad v = V_\gamma + R_f i \qquad\qquad (6.40)$$

It is easy to see that these equivalent relationships between current i and voltage v are realized by the simple series connection shown in Fig. 6.41. In other words, a forward-biased diode behaves approximately as a series connection of a battery V_γ and a resistance R_f, where V_γ is the cut-in voltage of the diode and R_f is a typical ac resistance of the diode. If the diode is reversed biased, it behaves approximately as an open circuit. Thus, a diode model more accurate than the one shown in Fig. 6.38(b) is the one shown in Fig. 6.42(b)—it consists of an ideal diode, a battery

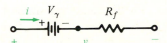

Fig. 6.41 Realization of $v = V_\gamma + R_f i$.

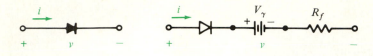

(a) Diode   (b) Diode model using battery

Fig. 6.42 A diode and an accurate model.

V_γ, and a resistance R_f. For this situation, however, the diode is OFF if $v \le V_\gamma$. As before, though, the diode is ON if $i > 0$. (For an even more accurate model, see Problem 6.40.)

Example 6.12

For the diode circuit shown in Fig. 6.43, D_1 is silicon with a cut-in voltage of 0.6 V and forward resistance of 10 Ω, while D_2 is germanium with a cut-in voltage of 0.2 V and a forward resistance of 20 Ω. Suppose that $R = 180 \Omega$. Let us assume that D_1 and D_2 are both ON. Then, using the diode model given in Fig. 6.42, when the diodes are ON, the given circuit is equivalent to the one shown in Fig. 6.44. By KVL, we have that

$$6 = 180(i_1 + i_2) + 10i_1 + 0.6 \quad \Rightarrow \quad 190i_1 + 180i_2 = 5.4$$

and

$$6 = 180(i_1 + i_2) + 20i_2 + 0.2 \quad \Rightarrow \quad 180i_1 + 200i_2 = 5.8$$

Solving these equations, we get $i_1 = 6.43$ mA and $i_2 = 23.2$ mA. Since $i_1 > 0$ and $i_2 > 0$, D_1 and D_2 are indeed ON. Thus,

$$v = 10i_1 + 0.6 = 10(6.43 \times 10^{-3}) + 0.6 = 0.664 \text{ V}$$

Suppose now that R is changed to 560 Ω. Again, let us assume that D_1 and D_2 are both ON. Therefore, by KVL,

$$6 = 560(i_1 + i_2) + 10i_1 + 0.6 \quad \Rightarrow \quad 570i_1 + 560i_2 = 5.4$$

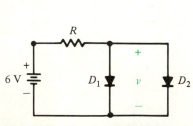

Fig. 6.43 Circuit with silicon and germanium diodes.

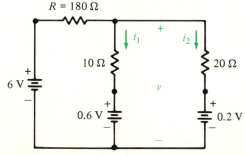

Fig. 6.44 Given circuit with $R = 180 \Omega$ and D_1, D_2 ON.

and

$$6 = 560(i_1 + i_2) + 20i_2 + 0.2 \quad \Rightarrow \quad 560i_1 + 580i_2 = 5.8$$

Solving these equations, we get $i_1 = -6.82$ mA and $i_2 = 16.6$ mA. Since $i_1 < 0$, our assumption that D_1 is ON is incorrect. Now we assume that D_1 is OFF and D_2 is ON. Replacing D_1 by its reverse-bias equivalent circuit, we get the circuit shown in Fig. 6.45. By KVL,

$$6 = 560i_2 + 20i_2 + 0.2 \quad \Rightarrow \quad i_2 = 10 \text{ mA}$$

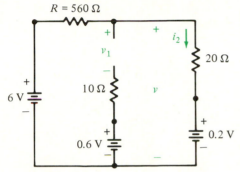

Fig. 6.45 Given circuit with $R = 560\ \Omega$ and D_1 OFF, D_2 ON.

Since $i_2 > 0$, D_2 is indeed ON. Furthermore,

$$v = 20i_2 + 0.2 = 20(10 \times 10^{-3}) + 0.2 = 0.4 \text{ V}$$

and because $v = 0.4$ V < 0.6 V $= V_y$, D_1 is indeed OFF.

■

On numerous occasions, especially in the analysis of digital circuits, an approximate diode model based on that given in Fig. 6.42 is used. It is assumed that the forward resistance is zero, and to compensate for this assumption, a slightly larger value than V_y is used for the battery voltage—typically 0.3 V for germanium and 0.7 V for silicon. Using such a model for a silicon diode, for example, if the diode is ON ($i > 0$), then the diode voltage is $v = 0.7$ V. Conversely, if the diode is OFF ($v < 0.7$ V), then the diode behaves as an open circuit.

Example 6.13

For the circuit shown in Fig. 6.46(a), the diodes are silicon. If a diode is conducting a current ($i > 0$), we will suppose that the corresponding voltage across it is $v = 0.7$ V.

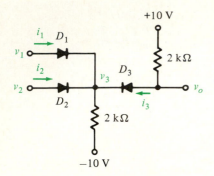

v_1	v_2	v_o
2	2	2
2	8	8
8	2	8
8	8	8

(b) Voltage table for circuit

(a) Digital circuit with three diodes

Fig. 6.46 DL OR gate.

First consider the case that the input voltages are $v_1 = v_2 = 2$ V. Let us assume that D_1 and D_2 are both ON. Under this circumstance, by KVL, $v_3 = -0.7 + v_1 = -0.7 + 2 = 1.3$ V < 10 V. Thus, let us also assume that D_3 is ON. Therefore, the output voltage is $v_o = 0.7 + v_3 = 0.7 + 1.3 = 2$ V. To validate our assumptions, note that

$$i_3 = \frac{10 - v_o}{2 \times 10^3} = \frac{10 - 2}{2 \times 10^3} = 4 \text{ mA}$$

and

$$i_1 + i_2 + i_3 = \frac{v_3 + 10}{2 \times 10^3} = \frac{1.3 + 10}{2 \times 10^3} = 5.65 \times 10^{-3} \text{ A}$$

so

$$i_1 + i_2 = 5.65 \times 10^{-3} - i_3 = 1.65 \text{ mA}$$

By symmetry, half of this current is i_1 and half is i_2. Hence, since $i_1 > 0$, $i_2 > 0$, and $i_3 > 0$, all the diodes are confirmed to be ON.

Next consider the case that $v_1 = 2$ V and $v_2 = 8$ V. Now let us assume that D_2 is ON. Thus, $v_3 = -0.7 + v_2 = 7.3$ V. Since the voltage across D_1 is $v_1 - v_3 = 2 - 7.3 = -5.3$ V < 0.7 V, let us assume that D_1 is OFF. Also, since 10 V $> v_3 = 7.3$ V, let us assume that D_3 is ON. Thus, $v_o = 0.7 + v_3 = 8$ V. To confirm our assumptions, we have that

$$i_3 = \frac{10 - v_o}{2 \times 10^3} = \frac{10 - 8}{2 \times 10^3} = 1 \text{ mA}$$

and if D_1 is OFF,

$$i_2 + i_3 = \frac{v_3 + 10}{2 \times 10^3} = \frac{7.3 + 10}{2 \times 10^3} = 8.65 \times 10^{-3} \text{ A}$$

so

$$i_2 = 8.65 \times 10^{-3} - i_3 = 7.65 \text{ mA}$$

Since $i_2 > 0$ and $i_3 > 0$, then D_2 and D_3 are indeed ON. As stated above, the voltage across D_1 is $v_1 - v_3 = -5.3$ V < 0.7 V, so D_1 is indeed OFF.

By symmetry, we deduce for $v_1 = 8$ V and $v_2 = 2$ V that D_1 and D_3 are ON, while D_2 is OFF, and again $v_o = 8$ V.

Finally, for the case that $v_1 = v_2 = 8$ V, let us assume that D_1 and D_2 are both ON. Then $v_3 = -0.7 + 8 = 7.3$ V < 10 V. Therefore, let us assume that D_3 is also ON. Thus, $v_o = 0.7 + v_3 = 8$ V. To confirm that all the diodes are ON, note that

$$i_3 = \frac{10 - v_o}{2 \times 10^3} = \frac{10 - 8}{2 \times 10^3} = 1 \text{ mA}$$

Also,

$$i_1 + i_2 + i_3 = \frac{v_3 + 10}{2 \times 10^3} = \frac{7.3 + 10}{2 \times 10^3} = 8.65 \times 10^{-3} \text{ A}$$

so

$$i_1 + i_2 = 8.65 \times 10^{-3} - i_3 = 7.65 \text{ mA}$$

Again by symmetry, half of this current is i_1 and half is i_2. Hence, since $i_1 > 0$, $i_2 > 0$, and $i_3 > 0$, all three diodes are indeed ON.

The corresponding table of voltages given in Fig. 6.46(b) indicates that the given circuit is an OR gate, where the low voltage (2 V) is the logical 0 and the high voltage (8 V) is the logical 1.

6.6 ZENER DIODES

As was mentioned in Section 6.3, if a diode is sufficiently reverse biased, that is, if the voltage is made negative enough, the diode will allow the conduction of a wide range of currents in the reverse direction. Furthermore, the diode voltage remains approximately constant at some value, say $-V_Z$, over this current range. This property, known as **breakdown,** is illustrated in the reverse-bias i-v diode characteristic shown in Fig. 6.47. The value V_Z is called the **(reverse) breakdown voltage** of the diode.

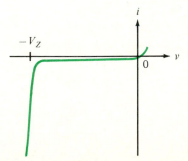

Fig. 6.47 Reverse-bias portion of diode *i-v* characteristic curve.

The phenomenon of diode breakdown is attributable to two types of mechanisms:

1. In a semiconductor diode, holes and free electrons are continually being thermally generated. For the case of reverse bias, a small reverse saturation current is due to minority carriers—and this current is, in essence, independent of applied voltage (before the breakdown voltage is reached). But increasing the reverse-bias voltage causes the depletion region to widen and makes the junction potential greater. Minority carriers crossing the junction, therefore, acquire greater energy. If this energy is large enough, the collision of a carrier with an ion will disrupt a covalent bond, thereby generating a new hole-electron pair. The new carriers produced in this manner can have sufficient energy, as a result of further collisions, to generate additional carriers. This cumulative process, which allows large reverse currents, is known as the **avalanche effect** or **avalanche breakdown,** and V_Z is called the **avalanche (breakdown) voltage.**

2. When the strength of an electric field across the junction of a semiconductor becomes sufficiently large (about 2×10^7 V/m), electrons are pulled from their covalent bonds, thus creating hole-electron pairs. As with the case of the avalanche effect, the increase in carriers can result in large reverse currents. However, unlike the avalanche effect, the carrier increase is not the result of collisions, but of electric field strength. This is known as the **Zener* effect** or **Zener breakdown.** Consequently, V_Z is also referred to as the **Zener breakdown voltage.**

For heavily doped diodes, Zener breakdown occurs at about 6 V or less, while for lightly doped diodes the breakdown voltage, due predominantly to the avalanche effect, is higher. Diodes that are designed to be used in the breakdown state are called **avalanche, breakdown,** or **Zener diodes**—the last is the most common terminology. This term is used even in cases where the avalanche effect predominates. Silicon Zener diodes have breakdown voltages in the range of a few volts to a few hundred volts. Furthermore, they are able to dissipate up to 50 W of power. The circuit symbol for a Zener diode is shown in Fig. 6.48.

Fig. 6.48 Zener diode circuit symbol.

Example 6.14

The circuit shown in Fig. 6.49 contains two silicon Zener diodes D_1 and D_2 with reverse saturation currents of 10 nA and 20 nA, respectively, at 300 K, and both diodes have breakdown voltages of 10 V.

---

*Named for the American physicist Clarence Zener.

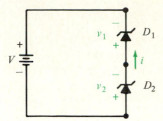

Fig. 6.49 Circuit with two Zener diodes.

First suppose that $V = 6$ V. Although this voltage reverse biases the diodes, it is not large enough to cause either diode to operate in the breakdown state. Since the diodes are in series, the current through them is the same. But $I_{o1} = 10$ nA and $I_{o2} = 20$ nA. Thus, assume that $i = -10$ nA; $i = -20$ nA implies that D_1 is in the breakdown state—which is a contradiction. Since $i = -I_{o1}$, the voltage v_1 is somewhere in the range of $-10 < v_1 < -0.2$ [see Equation (6.23)]. However, Equation (6.24), for D_2,

$$v_2 = \eta V_T \ln\left(\frac{i}{I_{o2}} + 1\right) = (2)\left(\frac{300}{11{,}586}\right)\ln\left(\frac{-10 \times 10^{-9}}{20 \times 10^{-9}} + 1\right) = -0.036 \text{ V}$$

Thus, by KVL,

$$v_1 = -v_2 - V = 0.036 - 6 = -5.964 \text{ V}$$

Now suppose that $V = 12$ V. If we again assume that $i = -10$ nA, as above, we get that $v_2 = -0.036$ V, from which $v_1 = -v_2 - V = 0.036 - 12 = -11.964$ V. However, this is not possible since the breakdown voltage of D_1 is 10 V (i.e., breakdown occurs for $v_1 = -10$ V). Thus, we deduce that D_1 is in the breakdown state ($v_1 = -10$ V), and, therefore, $v_2 = -V - v_1 = -12 + 10 = -2$ V. Furthermore, since $-10 < v_2 < -0.2$, then $i = -I_{o2} = -20$ nA.

Zener diodes are commonly used in "voltage regulators"—circuits that keep load voltages essentially constant despite variations in supply voltage and/or load resistance. In such applications, to ensure operation in the breakdown state, the minimum (reverse) current is chosen to be 10 percent of the maximum (reverse) current.

Example 6.15

A Zener diode has a breakdown voltage of 6 V and a maximum current rating of 50 mA. This diode is used in the voltage-regulator circuit shown in Fig. 6.50. Here v_s is the supply voltage and R_L is the load resistance. Since the maximum (reverse) current is 50 mA, the minimum (reverse) current is chosen to be $0.1(50 \times 10^{-3}) = 5$ mA.

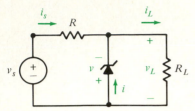

Fig. 6.50 Zener diode voltage-regulator circuit.

Let us assume that the diode is in the breakdown state, that is, $v = -6$ V and -50 mA $\leq i \leq -5$ mA. If $v_s = 24$ V, then the supply current i_s is

$$i_s = \frac{24 - 6}{R} = \frac{18}{R}$$

Since the maximum reverse diode current (which is obtained when $R_L = \infty$) is 50 mA, we must have that

$$\frac{18}{R} \leq 50 \times 10^{-3} \qquad \Rightarrow \qquad R \geq 360 \ \Omega$$

We arbitrarily choose $R = 500 \ \Omega$. Thus, the supply current is $i_s = 18/R = 18/500 = 36$ mA. Since the load current is $i_L = v_L/R_L = 6/R_L$ and $i \leq -5$ mA, then

$$i = i_L - i_s = \frac{6}{R_L} - 36 \times 10^{-3} \leq -5 \times 10^{-3} \qquad \Rightarrow \qquad R_L \geq 194 \ \Omega$$

In summary, if $v_s = 24$ V and $R = 500 \ \Omega$, then the load voltage is regulated at $v_L = 6$ V when $R_L \geq 194 \ \Omega$.

Now suppose that again $R = 500 \ \Omega$. For the case that $R_L = 1$ kΩ, the load current is $i_L = v_L/R_L = 6/(1 \times 10^3) = 6$ mA. Since

$$i = i_L - i_s = 6 \times 10^{-3} - \frac{v_s - 6}{500} \leq -5 \times 10^{-3}$$

then $v_s \geq 11.5$ V. Furthermore, since $i \geq -50$ mA, then $v_s \leq 34$ V. In summary, if $R = 500 \ \Omega$ and $R_L = 1$ kΩ, then the load voltage is regulated at $v_L = 6$ V when 11.5 V $\leq v_s \leq 34$ V.

∎

Zener Diode Characteristics

In the preceding two examples, we assumed that the voltage across a Zener diode operating in the breakdown state was a constant value and did not depend upon the diode current. Such a situation can be modeled by a battery V_Z. However, a more accurate model of a Zener diode in the breakdown state is a battery V_Z in series with a resistance R_Z. These ideal and nonideal models of a Zener diode in the breakdown

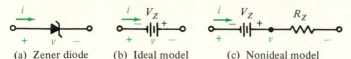

(a) Zener diode   (b) Ideal model   (c) Nonideal model

Fig. 6.51 Models of a Zener diode in the breakdown state.

state are shown in Fig. 6.51. The resistance R_Z normally ranges from several ohms to several hundred ohms. Generally, Zener diodes with breakdown voltages between 6 and 10 V have relatively small values of R_Z, whereas those with breakdown voltages below 6 V and above 10 V have relatively large values of R_Z.

When a Zener diode is not operating in the breakdown state, it behaves as an ordinary diode, and we can model it as was done previously for the diode. In particular, an ideal Zener diode (which is symbolized in Fig. 6.52(a)—note that the symbol is hollow, not solid), when in the breakdown state, behaves as the battery shown in Fig. 6.51(b), and when not in the breakdown state, behaves as the ideal diode depicted in Fig. 6.19. The i-v characteristic corresponding to an ideal Zener diode, therefore, is as shown in Fig. 6.52(b). (See Problem 6.41.)

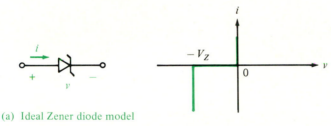

(a) Ideal Zener diode model

(b) Ideal Zener diode $i-v$ characteristic

Fig. 6.52 Symbol and *i-v* characteristic for the ideal Zener diode.

Example 6.16

The ideal Zener diode circuit given in Fig. 6.53 has the input voltage $v_s = 24 \sin \omega t$ shown in Fig. 6.54(a). Suppose that the Zener diodes, connected back to back, have a breakdown voltage of 12 V.

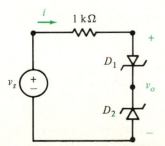

Fig. 6.53 Circuit with back-to-back ideal Zener diodes.

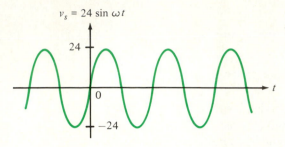

(a) Sinusoidal input voltage

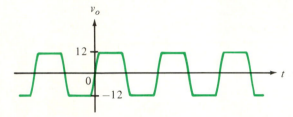

(b) Nonsinusoidal output voltage

Fig. 6.54 Input and output voltages for given Zener diode circuit.

If $0 \le v_s < 12$ V, since D_2 cannot be in the breakdown state, then D_2 behaves as an open circuit and $i = 0$. Therefore, by KVL, the output voltage is $v_o = -1000i + v_s = v_s$. On the other hand, if $v_s \ge 12$ V, then D_2 will be in the breakdown state and D_1 will behave as a short circuit. Thus, the output voltage is $v_o = 12$ V. By symmetry, we deduce that if $-12 < v_s \le 0$, then $v_o = v_s$; and if $v_s \le -12$ V, then $v_o = -12$ V. Hence, when the input is the sinusoidal voltage given in Fig. 6.54(a), the output voltage is the clipped sinusoid shown in Fig. 6.54(b).

As with other diode characteristics, the breakdown voltage V_Z of a diode is also temperature dependent. The variation in V_Z is specified by its percentage change per degree Celsius. Such a **temperature coefficient** of a Zener diode typically lies in the range of ± 0.1 percent/°C. If diode breakdown is due predominantly to the avalanche effect ($V_Z > 6$ V), the temperature coefficient is positive. Conversely, if breakdown involves the Zener effect ($V_Z < 6$ V), the temperature coefficient is negative.

Example 6.17

A Zener diode with a breakdown voltage of 12 V at 25°C has a temperature coefficient of +0.075 percent/°C. Therefore, the increase in the breakdown voltage per degree Celsius is

$$\left(\frac{0.075}{100}\right)(12) = 0.009 \text{ V/°C}$$

Thus, if the temperature of the diode rises to 45°C, then the breakdown voltage increases by

$$(0.009)(45 - 25) = 0.18 \text{ V}$$

Hence, at 45°C, the Zener diode has a breakdown voltage of

$$V_Z = 12 + 0.18 = 12.18 \text{ V}$$

6.7 EFFECTS OF CAPACITANCE

A common diode application is to rectify a sinusoid as depicted in Fig. 6.23. A reason for this rectification can be illustrated by placing a capacitor C in parallel with the resistor R in the half-wave rectifier circuit shown in Fig. 6.20. The resulting circuit is shown in Fig. 6.55. Suppose that for $t < 0$, the input voltage $v_s = 0$, and for $t \geq 0$, $v_s = V \sin \omega t$, as shown in Fig. 6.56(a). Clearly, for $t \leq 0$, since the input voltage $v_s = 0$, then $v_o = 0$. For $0 < t \leq \pi/2\omega$, $v_s > 0$ and the (ideal) diode is forward biased. Thus, in this interval of time $v_o = v_s$. This means that the capacitor is charged up to V volts when $t = \pi/2\omega$. Now, after $t = \pi/2\omega$,

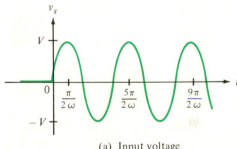

(a) Input voltage

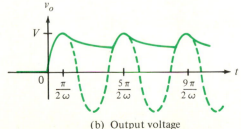

(b) Output voltage

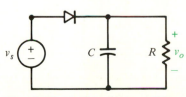

Fig. 6.55 Half-wave rectifier with added capacitor.

Fig. 6.56 Input and output of peak detector.

the input voltage v_s decreases and becomes less than V. However, since the voltage across the capacitor is $v_o = V$, the diode is reverse biased. The voltage across capacitor C, though, will discharge (exponentially, of course) through resistor R. If the time constant is much larger than the period of the sinusoid, that is, if $RC \gg T = 2\pi/\omega$, then while the diode is reverse biased, the voltage v_o will be approximately equal to V. The diode will remain reverse biased until, somewhere near $t = 5\pi/2\omega$, the voltage v_s increases enough to forward bias the diode. Then, as v_s increases to its maximum value, the capacitor will again charge up to V volts at time $t = 5\pi/2\omega$. This quickly charging and slowly discharging process is continually repeated, and the result is the output voltage v_o shown in Fig. 6.56(b). Since the time constant RC is not infinite, the voltage v_o fluctuates with time. This voltage variation, known as **ripple,** can be made smaller by making the time constant larger.

As we have now seen, the circuit given in Fig. 6.55 converts a sinusoidal voltage to a nearly constant voltage whose value is approximately equal to the peak value of the input voltage. Because of this, we refer to such a circuit as a **peak detector.** Although this circuit can be used as part of a dc power supply, the same type of circuit can be employed for demodulating or detecting AM signals (e.g., in ordinary AM radio receivers or in the video sections of television sets). For this reason, a peak detector is also known as an **envelope detector.**

Now consider the diode circuit shown in Fig. 6.57, where the applied (input) voltage v_s is the sinusoid $V \sin \omega t$ shown in Fig. 6.58(a). The first time that the

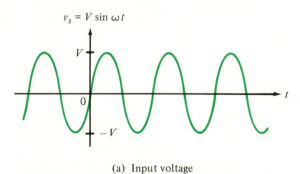

$v_s = V \sin \omega t$

(a) Input voltage

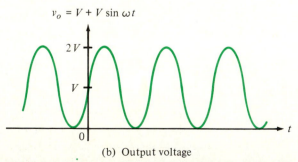

$v_o = V + V \sin \omega t$

(b) Output voltage

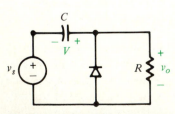

Fig. 6.57 Clamping circuit.

Fig. 6.58 Input and output of clamping circuit.

input voltage goes negative, the diode will be forward biased and the capacitor will charge to V volts with the polarity indicated. Since the voltage across the capacitor cannot discharge through the diode, it is possible only for it to discharge through the resistor R. If $RC \gg T = 2\pi/\omega$, then the voltage across the capacitor will discharge only slightly before it is again fully charged to V volts with the polarity shown. Hence, for a large time constant RC, the voltage across the capacitor will remain, in essence, constant at a value of V volts. By KVL, the resulting output voltage is $v_o = V + v_s$. From a plot of this voltage as shown in Fig. 6.58, we see that the output has its minimum value "clamped" to 0 volts. For this reason, the circuit given in Fig. 6.57 is called a **clamping circuit.** Note that the clamping circuit took a purely ac voltage (the input v_s) and added a constant (V) to it. In some types of applications (e.g., television sets), such a circuit thus is called a **dc restorer.**

By combining the peak detector of Fig. 6.55 with the clamping circuit given in Fig. 6.57, we obtain the circuit shown in Fig. 6.59. Suppose that the input to this circuit is $v_s = V \sin \omega t$ as in Fig. 6.58(a). For the case that $RC \gg T = 2\pi/\omega$, the clamping portion of the circuit will produce a voltage whose waveform is as shown in Fig. 6.58(b). As a consequence, the peak-detecting portion of the circuit will cause the output voltage to be $v_o \approx 2V$. Since this is the voltage from peak to peak for the input, the circuit is called a **peak-to-peak detector.** Furthermore, given the input v_s, since the output voltage is twice as large as that which would be obtained by using a peak detector, the circuit is sometimes called a **voltage doubler.**

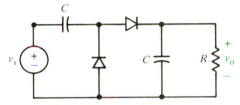

Fig. 6.59 Peak-to-peak detector.

Diode Capacitance

Having considered some external effects of capacitance, we now turn our attention to inherent diode capacitances. By definition, current i is related to charge q by $i = dq/dt$. Since the relationship between current i and voltage v for a capacitor C is $i = Cdv/dt$, we have that $Cdv/dt = dq/dt$. Using the chain rule of calculus, we deduce that $C = dq/dv$.

We know that the depletion region of a p-n junction has charge and a potential difference across it. From elementary physics, we know that a pair of parallel conducting plates with area A that are separated by a distance d has a capacitance $C = \epsilon A/d$, where ϵ is the **permittivity** (measured in F/m) of the material between the plates. So, too, a diode has a capacitance C_T, called the **transition capacitance** (or **depletion-region capacitance,** or **space-charge capacitance**) given by

$$C_T = \frac{\epsilon A}{W}$$

(6.41)

where ϵ is the permittivity of the semiconductor material, A is the area of the junction, and W is the width of the depletion region.

Since the depletion-region width changes with applied voltage (the larger the reverse bias, the wider is the depletion region), the value of the transition capacitance depends on the diode voltage. For the case of reverse bias, values of C_T generally range between 5 and 500 pF. Diodes that are constructed specifically for reverse-bias use as voltage-variable capacitors are called **varactors** or **varicaps.** These devices are used, for example, for electronically tuning television sets.

For the case of forward bias, the depletion region gets narrower, and, therefore, C_T increases. However, in the forward-bias condition, the transition capacitance is negligible compared with another type of diode capacitance. Therefore, let us consider a forward-biased diode in some detail.

When a diode is forward biased, holes (majority carriers) go from the p side to the n side (and vice versa for free electrons). When they enter the n side, holes (minority carriers) encounter many free electrons with which they can recombine. The average time that a hole exists before it recombines with a free electron is the **mean lifetime** and is denoted by τ_p (τ_n for free electrons). Since the flow rate of charge (mobile carriers) is current, then the flow of charge q having average lifetime τ yields a diode current of

$$i = \frac{q}{\tau} = I_o(e^{v/\eta V_T} - 1) \approx I_o e^{v/\eta V_T}$$

(6.42)

where the approximation is quite good for $v \geq 0.2$ V. Thus, we have that

$$q \approx \tau I_o e^{v/\eta V_T}$$

and the resulting capacitance C_D is

$$C_D = \frac{dq}{dv} \approx \frac{\tau I_o}{\eta V_T} e^{v/\eta V_T}$$

Substituting the approximation given in Equation (6.42) into the preceding approximation, we have that

$$C_D \approx \frac{\tau i}{\eta V_T}$$

(6.43)

Since it is known for the case of forward bias that the current through a diode is due almost entirely to diffusion (i.e., drift current is negligible), the capacitance C_D

is called the **diffusion capacitance** (also known as the **storage capacitance**) of the diode.

Example 6.18

A forward-biased silicon diode has a current of 30 mA and a mobile carrier mean lifetime of 25 μs. Then the diffusion capacitance of the diode at 300 K is

$$C_D \approx \frac{\tau i}{\eta V_T} = \frac{(25 \times 10^{-6})(30 \times 10^{-3})}{2(300)/11{,}586} = 14.5 \ \mu\text{F}$$

Although this typical value of capacitance is relatively large (much larger than the transition capacitance), recall that the forward resistance R_f of a diode is typically small. Consequently, depending upon the circuit application of the diode, the time constant $R_f C_D$ may be small enough that the diffusion capacitance can be ignored.

Diode Switching Times

Because of its inherent capacitance, a (nonideal) diode cannot be driven from a reverse-biased condition to the forward-biased state (or vice versa) instantaneously—a certain nonzero time is required. For instance, consider the simple diode circuit shown in Fig. 6.60(a), where the applied voltage v_s is given in Fig. 6.60(b). Initially, the applied voltage is $v_s = -V_R$ and the diode is reverse biased. Thus, the current is $i = -I_o$. At time $t = t_1$, the applied voltage instantaneously changes to $v_s = V_F$; this will forward bias the diode. If $R_L \gg R_f$ and $V_F \gg V_\gamma$, then the current increases from $i = -I_o$ to $i \approx V_F/R_L$ as depicted in Fig. 6.60(c). The time required for the current to go from about 10 percent of V_F/R_L to 90 percent of V_F/R_L is called the **forward recovery time** t_{fr} of the diode. This time is dependent upon the relatively small transition capacitance and is usually negligible compared with the time required for the diode to switch from forward to reverse bias. We now discuss this case.

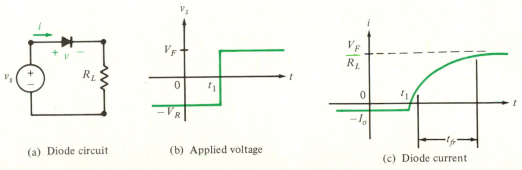

(a) Diode circuit   (b) Applied voltage   (c) Diode current

Fig. 6.60 Diode switching from reverse to forward bias.

Suppose that an applied voltage results in a forward biased diode. Although the situation we discuss is for holes, a similar description applies to free electrons. Holes (majority carriers) in the p-type material cross the junction into the n-type material, where they are minority carriers. The original hole concentration in the n side is attributable to thermal energy. The holes injected into the n-type material, therefore, increase the hole concentration. (The deeper a hole goes into the n side, the more chance there is of recombination. Thus, the hole concentration is greater closer to the junction.) When the applied voltage changes polarity so as to reverse bias the diode, the excess (injected) holes in the n side return to the p side. Hence, until these excess holes (minority carriers) are removed from the n-type material, there is a significant current in the reverse direction. After the elimination of the excess minority carriers, the reverse current decays to its final value of I_o.

To see the consequences of this discussion, suppose that the diode circuit given in Fig. 6.60(a) has the applied voltage shown in Fig. 6.61(a). Initially, the applied voltage is $v_s = V_F$ and the diode is forward biased. If $R_L \gg R_f$ and $V_F \gg V_\gamma$, then the current is $i \approx V_F/R_L$. At time $t = t_1$, the applied voltage instantaneously changes to $v_s = -V_R$, and this will reverse bias the diode. However, due to the excess minority carriers, the current becomes $i = -V_R/R_L$, and this current remains until time $t = t_2$, when the excess minority carriers are removed. After this has occurred, the current will increase until time $t = t_3$ when $i = -I_o$. This behavior is summarized in Fig. 6.61(b). The time difference $t_2 - t_1$ is called the **storage time** t_s of the diode (it is the time during which the excess holes are "stored" in the n side and the excess free electrons are "stored" in the p side). The time difference $t_3 - t_2$, called the **transition time** t_t, is dependent upon the relatively large diffusion capacitance. The sum $t_s + t_t$ is called the **reverse recovery time** t_{rr} of the diode. Diode reverse recovery times typically range from 1 ns to 1 μs—generally the larger the current-carrying capability of a diode, the larger is the value of t_{rr}.

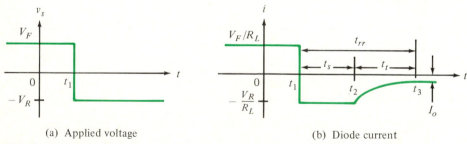

(a) Applied voltage   (b) Diode current

Fig. 6.61 Diode switching from forward to reverse bias.

SUMMARY

1. Semiconductors such as silicon and germanium have covalent bonds. Thermal energy can disrupt covalent bonds, thus creating hole-electron pairs.

2. A metal is unipolar—current is conducted by only one type of charge (electrons). A semiconductor such as germanium or silicon is bipolar—current is conducted by two types of charges (electrons and holes).

3. For a metal, conductivity decreases as temperature increases. For a semiconductor, conductivity increases as temperature increases.

4. Doping an intrinsic semiconductor with donor atoms produces an *n*-type semiconductor. Doping with acceptor atoms produces a *p*-type semiconductor.

5. In an *n*-type semiconductor, electrons are the majority carriers and holes are the minority carriers. Conversely, in a *p*-type semiconductor, the majority carriers are holes and the minority carriers are electrons.

6. Doping can significantly increase the conductivity of a semiconductor.

7. Current in a semiconductor is due to diffusion as well as drift.

8. The contact potential of a *p-n* junction is approximately 0.2 V for germanium and 0.6 V for silicon.

9. A *p-n* junction with appropriate metal contacts forms a junction diode.

10. A forward-biased (junction) diode can easily pass current, whereas a reverse-biased diode allows only a very small current.

11. Diode characteristics are temperature sensitive; for example, the reverse saturation current approximately doubles for each 10-degree increase in temperature.

12. Some diode circuits can be analyzed either graphically or numerically.

13. A forward-biased ideal diode behaves as a short circuit; a reverse-biased ideal diode acts as an open circuit.

14. A nonideal diode can be accurately modeled with an ideal diode, a battery, and a resistor.

15. A diode that is excessively reverse biased will break down. Diode breakdown is due to the avalanche effect and the Zener effect. Zener diodes are fabricated to be used in the breakdown state.

16. The capacitance of a *p-n* junction is predominantly the transition capacitance when a diode is reverse biased, and essentially the diffusion capacitance when a diode is forward biased. Diode switching times are dependent upon these capacitances.

17. Varactor diodes are voltage-variable capacitors.

18. Diodes can be used to design such circuits as rectifiers, clippers, voltage regulators, peak or envelope detectors, clampers, voltage doublers, electronic tuners, and digital logic circuits such as AND gates and OR gates.

Problems for Section 6.1

6.1 Ordinary household copper wire has a cross-sectional area of 3.32×10^{-6} m^2. Given that the free-electron concentration of copper is 8.43×10^{28} m^{-3}, find the current corresponding to a typical drift speed of 4×10^{-4} m/s.

6.2 A copper wire 1 millimeter (mm) in diameter has a resistance of 2.2×10^{-2} Ω/m and a current of 2 A. Given that the free-electron concentration of copper is 8.43×10^{28} m^{-3}, determine (a) the current density, (b) the drift velocity, (c) the mobility, and (d) the conductivity.

6.3 An aluminum wire of diameter 2 mm has a conductivity of 2.9×10^7 ℧/m and a current of 50 mA. Find the voltage across a 1-meter length of this wire.

6.4 A bar of pure silicon has a length of 1 cm (10^{-2} m) and cross-sectional area of 1 mm^2. At 300 K, the intrinsic concentration is 1.5×10^{16} m^{-3}. The free-electron and hole mobilities are 0.13 m^2/V-s and 0.05 m^2/V-s respectively. Find (a) the conductivity and (b) the resistance of the bar.

6.5 For the silicon bar given in Problem 6.4, at 310 K, the intrinsic concentration is 3.15×10^{16} m^{-3}. Assume the same mobilities and repeat the problem.

6.6 A bar of pure germanium has a length of 1 cm and a cross-sectional area of 1 mm^2. At 300 K, the intrinsic concentration is 2.5×10^{19} m^{-3}. The free-electron and hole mobilities are 0.38 m^2/V-s and 0.18 m^2/V-s respectively. Find (a) the conductivity and (b) the resistance of the bar.

Problems for Section 6.2

6.7 The bar of silicon given in Problem 6.4 is doped with two parts per 10^8 of an acceptor impurity. Determine (a) the conductivity and (b) the resistance of the resulting p-type semiconductor bar.

6.8 Determine the concentration of acceptor impurity required to dope silicon at 300 K such that the resulting conductivity is 20.8 ℧/m. How many parts per 10^8 is this doping?

6.9 The germanium bar given in Problem 6.6 is doped with one part per 10^8 of an acceptor impurity. Given that there are 4.4×10^{28} Ge atoms/m^3, determine (a) the conductivity and (b) the resistance of the resulting p-type semiconductor bar.

6.10 Repeat Problem 6.9 for the case of a donor impurity.

6.11 With reference to Problem 6.9, determine the parts per 10^9 of doping that results in a bar resistance of 1 kΩ.

6.12 Find the diffusion constants for holes and electrons at 300 K for (a) silicon and (b) germanium. (*Hint*: See Problem 6.6 for Ge mobilities.)

6.13 A piece of silicon has 5×10^{28} atoms/m^3. If one side is doped with one part per 10^8 of an acceptor impurity, and the other side is doped with two parts per 10^7 of a donor impurity, what is the contact potential across the p-n

junction at 300 K, given that the intrinsic concentration of silicon is 1.5×10^{16} m^{-3}?

6.14 Repeat Problem 6.13 for a piece of germanium (4.4×10^{28} atoms/m^3) at 300 K, given that the intrinsic concentration of germanium is 2.5×10^{19} m^{-3}.

6.15 The p side of a silicon p-n junction has a conductivity of 50 ℧/m, while the n side has a conductivity of 100 ℧/m. Determine the contact potential across the junction at 300 K, given that the electron and hole mobilities are 0.13 m^2/V-s and 0.05 m^2/V-s respectively. (*Hint*: Assume that the majority-carrier concentration is much greater than the minority-carrier concentration.)

6.16 Repeat Problem 6.15 for a germanium p-n junction with electron and hole mobilities of 0.38 m^2/V-s and 0.18 m^2/V-s respectively.

6.17 With reference to Problem 6.13, how many parts per 10^7 of donor impurity will result in a contact potential of 0.7 V?

6.18 With reference to Problem 6.14, how many parts per 10^7 of donor impurity will result in a contact potential of 0.3 V?

Problems for Section 6.3

6.19 A silicon diode has a reverse saturation current of 5 nA at 300 K.
  (a) Find the current for the case that the forward-bias voltage is 0.7 V.
  (b) Find the forward-bias voltage that results in a current of 15 mA.
  (c) Suppose that this diode is used in the circuit shown in Fig. 6.16. If $v_1 = 6$ V, what value of R results in a current of 15 mA?

6.20 A germanium diode has a reverse saturation current of 10 μA at 300 K.
  (a) Find the current for the case that the forward-bias voltage is 0.3 V.
  (b) Find the forward-bias voltage that results in a current of 0.5 A.
  (c) Suppose that this diode is used in the circuit shown in Fig. 6.16. If $v_1 = 12$ V, what value of R results in a current of 0.5 A?

6.21 A silicon diode is forward biased at 20 mA and 0.8 V at 300 K. Find the reverse saturation current of the diode at (a) 300 K, (b) 310 K, (c) 316 K, and (d) 284 K.

6.22 For the diode described in Problem 6.21, find the forward-bias voltage for which the current is 20 mA when the temperature is (a) 310 K, (b) 316 K, and (c) 284 K.

6.23 A germanium diode is forward biased at 0.5 A and 0.3 V at 300 K. Find the reverse saturation current of the diode at (a) 300 K, (b) 310 K, (c) 324 K, and (d) 276 K.

6.24 For the diode described in Problem 6.23, find the forward-bias voltage for which the current is 0.5 A when the temperature is (a) 310 K, (b) 324 K, and (c) 276 K.

6.25 A silicon diode has a reverse saturation current of 5 nA at 300 K. Determine the temperature for the case that the forward-bias voltage and current are 0.8 V and 35 mA, respectively.

6.26 A germanium diode has a reverse saturation current of 10 μA at 300 K. Determine the temperature for the case that the forward-bias voltage and current are 0.3 V and 1.2 A respectively.

6.27 For the diode circuit shown in Fig. P6.27, although the two silicon diodes D_1 and D_2 have identical characteristics, the reverse saturation current is unknown. The 6-V battery forward biases D_1 and reverse biases D_2. Find v_1 and v_2 at 300 K.

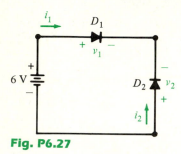

Fig. P6.27

6.28 For the diode circuit given in Fig. P6.27, D_1 is silicon with a reverse saturation current of 5 nA and D_2 is germanium with a reverse saturation current of 10 μA. Find v_1 and v_2 at 300 K.

6.29 For the diode circuit shown in Fig. P6.29, D_1 and D_2 are silicon diodes with reverse saturation currents of 5 nA and 10 nA, respectively, at 300 K. Given that both diodes are reverse biased, find v_1 and v_2.

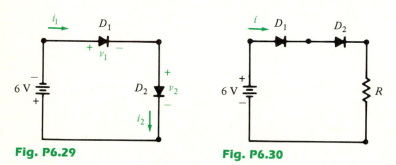

Fig. P6.29　　　　**Fig. P6.30**

6.30 For the diode circuit shown in Fig. P6.30, D_1 and D_2 are silicon diodes having reverse saturation currents of 5 nA and 10 nA, respectively, at 300 K. Given that both diodes are forward biased, find the value of R for which the current is 15 mA.

6.31 A germanium diode has a reverse saturation current of 20 μA. at 300 K.
(a) Plot an *i-v* characteristic curve on a piece of graph paper.

(b) This diode is used in the circuit given in Fig. 6.16, where $v_1 = 1$ V and $R = 0.5 \ \Omega$. Use the i-v curve obtained in part (a) to determine graphically the current and voltage of the diode.

6.32 Use the numerical method described in Example 6.8 to find the voltage and current for the diode referred to in Problem 6.31. (Begin with a voltage of 0.2 V.)

Problems for Section 6.4

6.33 Given that the input voltage v_s is sinusoidal, sketch the resulting output voltage v_o for the diode circuit shown in (a) Fig. P6.33(a) and (b) Fig. P6.33(b).

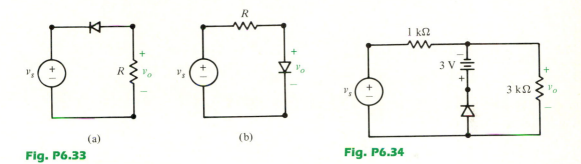

(a)   (b)

Fig. P6.33   **Fig. P6.34**

6.34 The input voltage v_s to the clipper circuit shown in Fig. P6.34 is a sinusoid with amplitude 12 V. Sketch the output voltage v_o.

6.35 Repeat Problem 6.34 for the clipper circuit shown in Fig. P6.35.

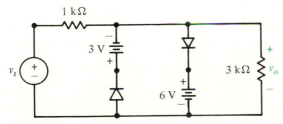

Fig. P6.35

6.36 For the circuit given in Problem 6.34, reverse the polarity of the 3-V source and repeat the problem.

6.37 The input voltage v_s to the ideal-diode circuit shown in Fig. P6.37 is a sinusoid with amplitude 6 V. Sketch the output voltage v_o.

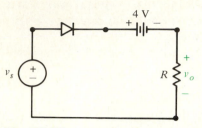

Fig. P6.37

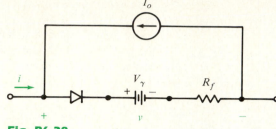

Fig. P6.38

6.38 Sketch the i-v characteristic of the ideal-diode circuit shown in Fig. P6.38, given that $I_o > 0$, $R_f = 0$, and $V_\gamma = 0$.

6.39 Repeat Problem 6.38 for the case that (a) $I_o > 0$, $R_f > 0$, $V_\gamma = 0$; (b) $I_o > 0$, $R_f = 0$, $V_\gamma > 0$.

6.40 Repeat Problem 6.38 for the case that $I_o > 0$, $R_f > 0$, and $V_\gamma > 0$.

6.41 Sketch the i-v characteristic of the ideal-diode circuit shown in Fig. P6.41 given that $V_Z > 0$ and (a) $R_Z = 0$; (b) $R_Z > 0$.

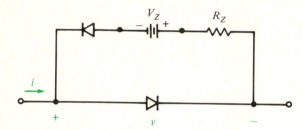

Fig. P6.41

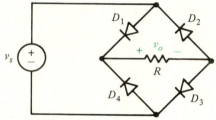

Fig. P6.42

6.42 The circuit shown in Fig. P6.42 is known as a **full-wave** (or **bridge**) **rectifier circuit.** Sketch the output voltage v_o when the input voltage v_s is a sinusoid. Determine which diodes are ON and which are OFF when (a) $v_s > 0$ and (b) $v_s < 0$.

6.43 The circuit shown in Fig. P6.43 is also an example of a full-wave rectifier. Sketch the output voltage v_o when $v_s = V \sin \omega t$.

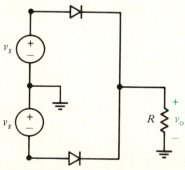

Fig. P6.43

6.44 When a diode is used in a rectifier circuit, the maximum reverse-bias voltage across that diode is known as the **peak inverse voltage** (PIV) of the diode. (Of course, the PIV must be less than the breakdown voltage of the diode.) If $v_s = V \cos \omega t$, find the PIV of the rectifier circuit given in (a) Fig. 6.20, (b) Fig. P6.42, and (c) Fig. P6.43.

6.45 For the ideal-diode circuit given in Fig. 6.28, let $R_f = 1 \text{ k}\Omega$ and $R = 9 \text{ k}\Omega$. Find the output voltage v_o when the input voltages are (a) $v_1 = v_2 = 0$; (b) $v_1 = 0$, $v_2 = 10 \text{ V}$; (c) $v_1 = v_2 = 10 \text{ V}$.

6.46 Repeat Problem 6.45 for the ideal-diode circuit in Fig. 6.32 given that $V = 10 \text{ V}$.

6.47 For the OR gate given in Fig. 6.31, let the low voltage be $V(0)$ and the high voltage be $V(1)$. Show that when $R \gg R_f$, the logic table that results is given by Table 6.1(b).

6.48 For the AND gate given in Fig. 6.36, let the low voltage be $V(0)$ and the high voltage be $V(1)$. Show that when $R \gg R_f$, the logic table that results is as given by Table 6.2(b).

6.49 A general positive-logic OR gate is shown in Fig. P6.49. For this DL gate, the low voltage is 0 volts and the high voltage is V volts. Find the output voltage v_o when $v_1 = v_2 = 0$.

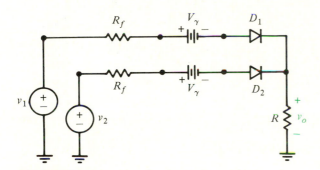

Fig. P6.49

6.50 For the OR gate given in Fig. P6.49, find the output voltage v_o when $v_1 = 0$ and $v_2 = V$ (and, hence, when $v_1 = V$ and $v_2 = 0$) given that $R \gg R_f$ and $V \gg V_\gamma$.

6.51 For the OR gate given in Fig. P6.49, find the approximate output voltage v_o when $v_1 = v_2 = V$, given that $R \gg R_f$ and $V \gg V_\gamma$.

6.52 A general positive-logic AND gate is shown in Fig. P6.52. For this DL gate, the low voltage is 0 volts and the high voltage is V volts. Find the approximate output voltage v_o when $v_1 = v_2 = 0$ given that $R \gg R_f$ and $V \gg V_\gamma$.

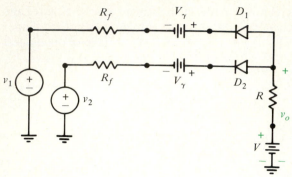

Fig. P6.52

6.53 For the AND gate given in Fig. P6.52, find the approximate output voltage v_o when $v_1 = 0$ and $v_2 = V$ (and, hence, when $v_1 = V$ and $v_2 = 0$) given that $R \gg R_f$ and $V \gg V_\gamma$.

6.54 For the AND gate given in Fig. P6.52, find the output voltage v_o when $v_1 = v_2 = V$.

Problems for Section 6.5

6.55 A silicon diode with a reverse saturation current of 5 nA at 300 K has a forward-bias voltage of 0.7 V.
(a) Find the dc resistance of the diode.
(b) Find the ac resistance of the diode.
(c) Repeat part (a) for 310 K.
(d) Repeat part (b) for 310 K.

6.56 A germanium diode with a reverse saturation current of 10 μA at 300 K has a forward-bias voltage of 0.3 V.
(a) Find the dc resistance of the diode.
(b) Find the ac resistance of the diode.
(c) Repeat part (a) for 310 K.
(d) Repeat part (b) for 310 K.

6.57 For the diode circuit shown in Fig. P6.57, D_1 is germanium with a cut-in voltage of 0.2 V and D_2 is silicon with a cut-in voltage of 0.6 V. Both diodes have a forward resistance of 10 Ω. Find the range of values of v_s for which

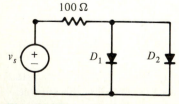

Fig. P6.57

(a) D_1 and D_2 are both OFF, (b) D_1 and D_2 are both ON. (c) D_1 is ON and D_2 is OFF.

6.58 For the diode circuit shown in Fig. P6.58, D_1 is germanium with a cut-in voltage of 0.2 V and a forward resistance of 10 Ω, and D_2 is silicon with a cut-in voltage of 0.6 V and a forward resistance of 30 Ω.

(a) Find v_o when $v_s = 3$ V.

(b) Find the value of v_s that just turns ON D_2, that is, the value of v_s for which $v_o = 0.6$ V.

(c) Find v_o when $v_s = 6$ V.

(d) Find v_o when $v_s = -3$ V.

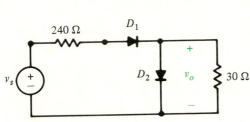

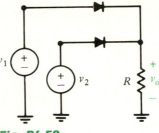

Fig. P6.58

Fig. P6.59

6.59 For the diode circuit shown in Fig. P6.59, D_1 and D_2 each has a cut-in voltage of V_γ and forward resistance of R_f. Verify that this circuit is an OR gate when the low voltage is 0, the high voltage is $V \gg V_\gamma$, and $R \gg R_f$.

6.60 Repeat Problem 6.59 for the AND gate shown in Fig. P6.60.

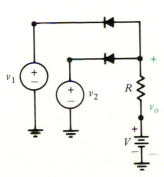

Fig. P6.60

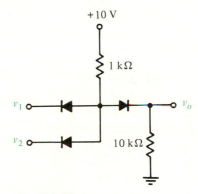

Fig. P6.61

6.61 The diodes in the circuit shown in Fig. P6.61 are silicon. Assume that when a diode conducts, its voltage is 0.7 V and $R_f = 0$.

(a) Find v_o when $v_1 = v_2 = 2$ V.

(b) Find v_o when $v_1 = 2$ V and $v_2 = 6$ V.

(c) Find v_o when $v_1 = v_2 = 6$ V.

(d) What type of logic circuit is this?

6.62 For the circuit given in Fig. 6.46, if the 2-kΩ resistor connected to the output is replaced by a resistor whose value is too small, the circuit no longer will function as an OR gate. Find the minimum resistance allowable for proper OR-gate operation.

6.63 For the circuit given in Fig. 6.46, if the 2-kΩ resistor connected to the cathodes of the diodes is replaced by a resistor whose value is too large, the circuit no longer will function as an OR gate. Find the maximum resistance allowable for proper OR-gate operation.

6.64 For the circuit given in Fig. P6.61, if the 10-kΩ resistor is replaced by a resistor whose value is too small, the circuit no longer will function as an AND gate. Find the minimum resistance allowable for proper AND-gate operation.

Problems for Section 6.6

6.65 The identical silicon Zener diodes in the circuit shown in Fig. P6.65 have reverse saturation currents of 10 nA at 300 K. Find i, v_1, and v_2 when the breakdown voltage is (a) 9 V and (b) 5 V.

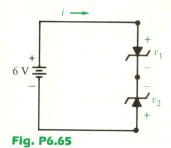

Fig. P6.65

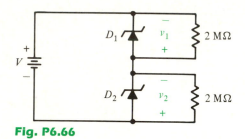

Fig. P6.66

6.66 For the Zener diode circuit shown in Fig. P6.66, at 300 K, D_1 and D_2 have reverse saturation currents of 1 μA and 2 μA, respectively. If both diodes have a breakdown voltage of 10 V, find v_1 and v_2 when (a) $V = 6$ V and (b) $V = 12$ V.

6.67 The Zener diode in the voltage-regulator circuit given in Fig. 6.50 has a breakdown voltage of 9 V and is to operate with a reverse current between 10 and 100 mA. Given that $R = 200$ Ω;
   (a) Find the range of the load resistance R_L that results in a 9-V load voltage when $v_s = 24$ V.
   (b) Find the range of the supply voltage v_s that results in a 9-V load voltage when $R_L = 600$ Ω.

6.68 The Zener diode in the voltage-regulator circuit given in Fig. 6.50 has a breakdown voltage of 12 V. Suppose that the diode (reverse) current is to be 10 mA.
   (a) If $v_s = 24$ V and $R_L = 2$ kΩ, find R.

(b) If $R = 2$ kΩ and $R_L = 2$ kΩ, find v_s.

(c) If $v_s = 24$ V and $R = 300$ Ω, find R_L.

6.69 Repeat Problem 6.68 using the Zener diode model given in Fig. 6.51(c), where $R_Z = 200$ Ω.

6.70 The circuit shown in Fig. P6.70 contains a 6-V Zener diode, that is, a diode with a breakdown voltage of 6 V. Suppose that the supply voltage v_s is a half-wave rectified sine wave [see Fig. 6.23(b)] with an amplitude of 12 V.

(a) Use the ideal Zener diode model given in Fig. 6.51(b) to obtain a sketch of v_L.

(b) Use the model given in Fig. 6.51(c) with $R_Z = 100$ Ω to obtain a sketch of v_L.

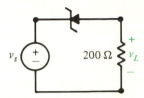

Fig. P6.70

6.71 For the circuit shown in Fig. 6.53, replace the ideal Zener diode D_1 by an (ordinary) ideal diode (anode on top). Given that the breakdown voltage for D_2 is 12 V, sketch v_o for the case that $v_s = 24 \sin \omega t$.

Problems for Section 6.7

6.72 For the circuit given in Fig. 6.55, reverse the diode. Sketch the output voltage v_o of the resulting circuit when the intput is the voltage v_s given by Fig. 6.56(a) and $RC \gg T = 2\pi/\omega$.

6.73 For the diode circuit given in Fig. P6.42, place a capacitor C in parallel with the resistor R. Sketch the resulting output voltage v_o when the supply voltage v_s is as given by Fig. 6.56(a) and $RC \gg T = 2\pi/\omega$.

6.74 Repeat Problem 6.73 for the circuit given in Fig. P6.43.

6.75 For the clamping circuit given in Fig. 6.57, sketch the output voltage v_o when

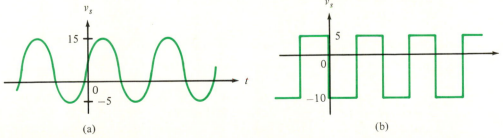

(a)   (b)

Fig. P6.75

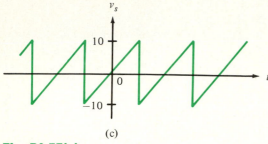

(c)

Fig. P6.75(c)

the input is the voltage v_s (and $RC \gg T$) shown in (a) Fig. P6.75(a), (b) Fig. P6.75(b), and (c) Fig. P6.75(c).

6.76 Reverse the diode in the clamping circuit given in Fig. 6.57 and repeat Problem 6.75.

6.77 Consider the clamping circuit shown in Fig. P6.77. Given that the input voltage is $v_s = V \cos \omega t$ and $RC \gg T$, sketch the output voltage v_o when (a) $V_1 = V/2$ and (b) $V_1 = -V/2$.

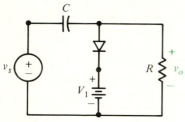

Fig. P6.77

6.78 For the clamping circuit given in Problem 6.77, reverse the diode and repeat the problem.

6.79 Find the output voltage v_o of the peak-to-peak detector given in Fig. 6.59, when $RC \gg T$ and the input v_s is the voltage shown in (a) Fig. P6.75(a), (b) Fig. P6.75(b), and (c) Fig. P6.75(c).

6.80 A silicon diode with permittivity 1.04×10^{-10} F/m is reverse biased such that the depletion region has a width of 4×10^{-6} m.
  (a) If the cross section of the diode is a square measuring 1 mm on a side, what is the value of the transition capacitance?
  (b) What cross-sectional area would result in a transition capacitance of 15 pF?

6.81 A silicon diode with a reverse saturation current of 16 nA at 300 K has a forward-bias voltage of 0.7 V and a mean carrier lifetime of 25 μs.
  (a) Find the diffusion capacitance.
  (b) Find the voltage that will result in a diffusion capacitance of 10 μF.

Chapter 7

Bipolar Junction Transistors (BJTs)

INTRODUCTION

In the preceding chapter, we studied the diode (a *p-n* junction) and discussed
many of its practical applications. By sandwiching a narrow section of *n*-type
semiconductor between two sections of *p*-type semiconductor (or vice versa),
we obtain a bipolar junction transistor (BJT), or transistor for short. Such a device does not act simply as two *p-n* junctions, but instead displays a more elaborate behavior that has far-reaching consequences. In this chapter, we study
the various characteristics and behavior of such a device. We also see how transistors can be utilized to design digital logic circuits in both discrete and
integrated-circuit (IC) form. The importance and applications of digital logic circuits are covered in Chapters 11, 12, and 13.

7.1 THE *p-n-p* TRANSISTOR

By sandwiching a narrow section of n-type material between two sections of p-type
material as illustrated in Fig. 7.1(a), we obtain a p-n-p **bipolar junction transistor**
(BJT). The three sections of the BJT are known as the **emitter** (E), the **base** (B),
and the **collector** (C). The circuit symbol for a p-n-p BJT is shown in Fig. 7.1(b).
If the p-type semiconductor is replaced by an n-type semiconductor, and vice
versa, what results is an n-p-n BJT as depicted in Fig. 7.2(a). The circuit symbol
for an n-p-n BJT is given in Fig. 7.2(b). Note that for both cases (p-n-p and n-p-n),
by convention, the emitter current i_E, the base current i_B, and the collector current
i_C are directed into the device. We denote the voltage between terminals a and b
(with the plus at terminal a) by v_{ab}. For example, the voltage between the emitter

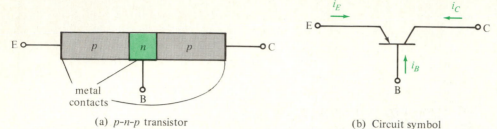

(a) *p-n-p* transistor　　　　　　　　　(b) Circuit symbol

Fig. 7.1 A *p-n-p* BJT and its circuit symbol.

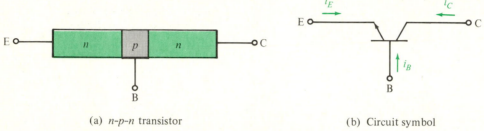

(a) *n-p-n* transistor　　　　　　　　　(b) Circuit symbol

Fig. 7.2 An *n-p-n* BJT and its circuit symbol.

and the base (with the plus at the emitter) is v_{EB}, and also, $v_{EB} = -v_{BE}$. The only difference between the circuit symbols for the *p-n-p* and *n-p-n* transistors is the direction of the arrowhead on the emitter. As is the case for any *p-n* junction, the arrowhead points from the *p*-type to the *n*-type material.

Some of the important operating characteristics of a *p-n-p* transistor are seen when the emitter–base junction (or emitter junction for short) is forward biased and the collector–base junction (or collector junction for short) is reverse biased. This situation is depicted in Fig. 7.3. Here, the supply voltage V_{EE} forward biases the emitter junction and produces an emitter current i_E, while V_{CC} reverse biases the

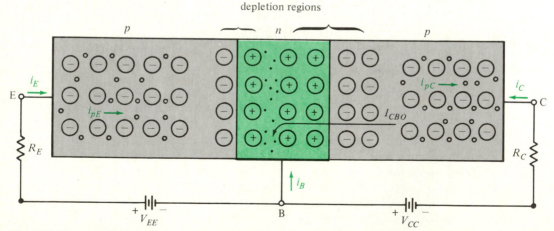

Fig. 7.3 A *p-n-p* BJT with emitter junction forward biased and collector junction reverse biased.

collector junction. As mentioned in the preceding chapter, the current through a p-n junction is due almost entirely to diffusion—drift current being negligible. Thus, the emitter current i_E has two components—a hole diffusion current i_{pE} going from emitter to base, and an electron diffusion current i_{nE} going from base to emitter. That is, $i_E = i_{pE} + i_{nE}$. However, because the emitter in a BJT is much more heavily doped than the base, the emitter current is due almost entirely to the diffusion of holes, that is, $i_E \approx i_{pE}$. Holes are "emitted" from the emitter and enter the n-type base. Although some recombine with electrons, the vast majority of these holes are accelerated across the collector–base junction by its electric field. The holes are then "collected" by the collector. Let us denote the hole current from base to collector by i_{pC}. Further, let us denote that fraction of the emitted holes that makes it to the collector by α. In other words, the collector hole current i_{pC} is a fraction α of the emitter hole current $i_{pE} = i_E$ (for simplicity, we have changed the approximation to an equality), that is,

$$i_{pC} = \alpha i_E$$

Temporarily, consider the case that the emitter is open-circuited ($R_E = \infty$). Thus, we must have that $i_E = 0$. Even though the collector junction is reverse biased, there is a nonzero collector current. Just as a reverse-biased p-n junction has a reverse saturation current going from the n-type to the p-type semiconductor, there is a reverse-bias current going from base to collector. This current is attributable mainly to thermally generated minority carriers (holes going from base to collector and free electrons going from collector to base). Some additional carriers may be produced in the collector–base depletion region as a result of collisions, as is the situation for the avalanche effect. Also, there is a small current due to charge that moves along the surface of the device—this is known as **(surface) leakage current**, and it is proportional to the voltage across the junction. The total reverse current, denoted I_{CBO} (the current from **c**ollector to **b**ase with the emitter **o**pen), is called the **reverse collector saturation current**, and, by convention, is directed from collector to base. For a p-n-p transistor, therefore, I_{CBO} is a negative number. Typically, at 300 K, I_{CBO} is of the order of nanoamperes for silicon and of microamperes for germanium. And, as was the case for the reverse saturation current of a junction diode, I_{CBO} approximately doubles for every 10-degree rise in temperature for both silicon and germanium [see Equation (6.25)].

With the emitter junction again forward biased (and the collector junction still reverse biased), we have that the collector current i_C is given by

$$i_C = I_{CBO} - i_{pC} = I_{CBO} - \alpha i_E \tag{7.1}$$

Thus,

$$\alpha i_E = I_{CBO} - i_C \tag{7.2}$$

from which

$$\boxed{\alpha = \frac{I_{CBO} - i_C}{i_E}} \tag{7.3}$$

This quantity α (the fraction of emitted holes that gets collected) is called the **large-signal current gain** of the transistor, and typically ranges between 0.95 and 0.998. Furthermore, the value of α depends upon such variables as the collector–base voltage v_{CB} and the temperature T.

Regions of Operation

For the situation just described, where the emitter junction is forward biased and the collector junction is reverse biased, we say that the transistor is biased in, or operates in, the **active region** (also called the **linear region**). That is, for active-region operation, Equations (7.1)–(7.3) are valid. In particular, even though it can change with v_{CB}, the quantity α is close to unity. Further, because I_{CBO} is normally small, since $i_C = -\alpha i_E + I_{CBO}$, we see that the collector current i_C depends on the emitter current i_E and is, in essence, independent of v_{CB}. In addition, a reasonable approximation relating the collector current to the emitter current is

$$i_C \approx -\alpha i_E \tag{7.4}$$

Let us redraw Fig. 7.3 using the circuit symbol of a *p-n-p* transistor as shown in Fig. 7.4. Similar to a forward-biased *p-n* junction, a forward-biased emitter–base junction has an i_E-v_{EB} characteristic as shown in Fig. 7.5. Also, as for the case of a junction diode, a small change in voltage can produce a large change in current. Let us now see the consequences of this fact.

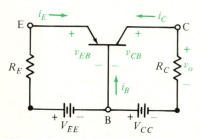

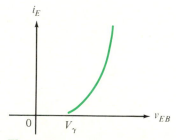

Fig. 7.4 A *p-n-p* transistor biased in the active region.

Fig. 7.5 Emitter–base junction i_E-v_{EB} characteristic.

Since a change in emitter current Δi_E will produce a change in collector current Δi_C, let us define the **small-signal current gain** α_{ac} of the transistor by

$$\alpha_{ac} = \Delta i_C / \Delta i_E \tag{7.5}$$

with v_{CB} held constant. From the circuit in Fig. 7.4, $v_o = -R_C i_C$. Thus,

$$\Delta v_o = -R_C \Delta i_C \tag{7.6}$$

However, from Equation (7.5), $\Delta i_C = \alpha_{ac} \Delta i_E$, and substituting this into Equation

(7.6) yields

$$\Delta v_o = -\alpha_{ac} R_C \Delta i_E$$

This quantity can be many times the value of Δv_{EB} (as we see quite soon).

Just as we defined the ac resistance of a diode [see Equation (6.34)], so too we define the ac (incremental) resistance r_e of the emitter–base junction by

$$r_e = \frac{\Delta v_{EB}}{\Delta i_E} \tag{7.7}$$

As for the case of the junction diode, a good approximation of r_e for silicon at room temperature is $r_e \approx 0.05/i_E$. From Equation (7.7) we have

$$\Delta v_{EB} = r_e \Delta i_E$$

Taking the ratio of the output voltage change Δv_o to the change in the voltage applied to the transistor Δv_{EB}, we obtain an ac (incremental) voltage gain A of

$$A = \frac{\Delta v_o}{\Delta v_{EB}} = \frac{-\alpha_{ac} R_C \Delta i_E}{r_e \Delta i_E} = \frac{-\alpha_{ac} R_C}{r_e} \tag{7.8}$$

Using the typical numerical values of $i_E = 2$ mA, $R_C = 4$ kΩ, and $\alpha_{ac} = -0.975$, we have

$$r_e \approx \frac{0.05}{i_E} = \frac{0.05}{2 \times 10^{-3}} = 25 \ \Omega$$

and

$$A = \frac{-\alpha_{ac} R_C}{r_e} = \frac{(0.975)(4 \times 10^3)}{25} = 156$$

This is an example of how a transistor (operating in the active region) can be utilized as an ac voltage amplifier. (A transistor also can be used to amplify current and for power gain.) For the circuit given in Fig. 7.4, we can consider the emitter the input and the collector the output. Since the base is common to both the input and the output portions of the circuit, we refer to this as a **common-base** (CB) configuration.

We have just seen a demonstration of how a current (i_E) through a low input resistance (r_e) is transferred to a current (i_C) through a high output resistance (R_C). It was because such a device behaved as a "*trans*fer res*istor*" that it became known as a **transistor.**

Transistor Characteristics

When a transistor is in the active region, $i_C = -\alpha i_E + I_{CBO}$ [Equation (7.1)]. The term I_{CBO} is the current through the reverse-biased collector–base (p-n) junction. Since, by definition, I_{CBO} is directed from collector to base, then $-I_{CBO}$ is the current going from base to collector. (This current corresponds to the reverse saturation current I_o going from cathode to anode through a reverse-biased junction diode.) Thus, as for the case of a diode, we can write a general expression relating the current from collector to base that is due to the voltage v_{CB} between collector and base. That current is

$$-I_{CBO}(e^{v_{CB}/\eta V_T} - 1)$$

Thus, the general form of Equation (7.1) is

$$i_C = -\alpha i_E - I_{CBO}(e^{v_{CB}/\eta V_T} - 1)$$

or

$$i_C = -\alpha i_E + I_{CBO}(1 - e^{v_{CB}/\eta V_T}) \tag{7.9}$$

Clearly, when v_{CB} is negative (the collector junction is reverse biased) and $|v_{CB}| \gg \eta V_T$, then Equation (7.9) becomes Equation (7.1).

For the CB circuit given in Fig. 7.4, when the transistor is in the active region, the emitter junction is forward biased and the collector junction is reverse biased. When the emitter junction is forward biased, v_{EB} and i_E are positive quantities. Most of i_E goes to the collector (the fraction is α) and the remainder goes down the base. This implies that i_C and i_B are negative quantities. Also, when the collector junction is reverse biased, then v_{CB} is less than the cut-in voltage V_γ.

From Equation (7.9), we see that the collector current i_C is dependent upon the emitter current i_E and the collector–base voltage v_{CB}. A plot of i_C versus v_{CB} for different values of i_E for a typical silicon p-n-p transistor is shown in Fig. 7.6. This family of curves is called the (CB) **collector static characteristics** of the transistor. For the case that the emitter has no current, that is, $i_E = 0$, when $v_{CB} < 0$, we have $i_C = I_{CBO}$ (a negative number whose magnitude is exaggerated on the vertical axis). When $i_E = 0$ and $v_{CB} > 0$, then $i_C > 0$, and since it lies below the horizontal axis, this situation is not depicted by the collector characteristics.

For the case that $v_{CB} < V_\gamma = 0.6$ V (for silicon), the collector junction is

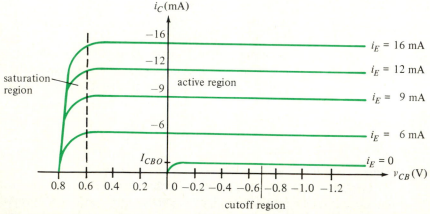

Fig. 7.6 Collector static characteristics (CB case) for a p-n-p silicon BJT.

reverse biased. In addition, if $i_E > 0$, then the emitter junction is forward biased. Thus, the region to the right of the dashed line $v_{CB} = 0.6$ V and above the curve $i_E = 0$ corresponds to the active region.

For the case that $v_{CB} > V_\gamma = 0.6$ V, the collector junction is forward biased. For the region $0.6 < v_{CB} < 0.8$ V, both the collector and emitter junctions are forward biased, and this is known as the **saturation region.** Finally, the region on and below the curve $i_E = 0$ corresponds to the case that the emitter and collector junctions are both reverse biased—and this is called the **cutoff region.** The very important notions of transistor saturation and cutoff will be discussed in detail later.

Another transistor characteristic was mentioned earlier (see Fig. 7.5) and is a plot of i_E versus v_{EB} (or vice versa) for different values of v_{CB}. A typical example of these **emitter static characteristics** for a silicon *p-n-p* transistor is shown in Fig. 7.7. Note that when the emitter junction is forward biased, a nominal value for v_{EB} is 0.6 V (for silicon). The curve corresponding to an open collector is typical of a silicon junction diode. But why should this characteristic change with v_{CB}? Recall that, as the reverse bias of a junction increases, its depletion region increases. Thus, as the reverse bias of the collector junction increases, the depletion region widens and, in effect, narrows the base. Hence, the hole concentration gradient within the base increases, and this, in turn, means a larger emitter current [see Equation (6.11)], given the same value of v_{EB}.

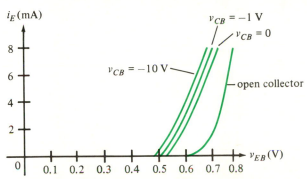

Fig. 7.7 Emitter static characteristics for a *p-n-p* silicon BJT.

The fact that the base width effectively decreases as the collector junction reverse-bias voltage increases means that there is less of a chance for recombination in the base. Thus, if $v_{CB} < 0$, as $|v_{CB}|$ increases, a greater fraction of the holes entering the base will make it to the collector. Hence, for a reverse-biased collector junction, α increases as $|v_{CB}|$ increases. This, in turn, means a small increase (typically less than 1 percent) in $|i_C|$ with $|v_{CB}|$. Note that for the collector characteristics given in Fig. 7.6, therefore, in the active region, the curves relating i_C to v_{CB} are, in essence, horizontal.

For a CB configuration, the emitter static characteristics of a transistor (e.g., Fig. 7.7) are sometimes called the **input characteristics,** and the collector static characteristics (e.g., Fig. 7.6) are referred to as the **output characteristics.**

Example 7.1

Consider the silicon *p-n-p* transistor CB circuit shown in Fig. 7.8. For the input portion of the circuit, by KVL, we have

$$v_{EB} - 6 + (1.5 \times 10^3)i_E = 0$$

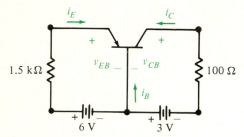

Fig. 7.8 Silicon *p-n-p* transistor CB circuit.

Let us assume that the transistor is in the active region, and consequently use the typical value $v_{EB} = 0.6$ V. Doing so, we obtain

$$i_E = \frac{6 - v_{EB}}{1.5 \times 10^3} = \frac{5.4}{1.5 \times 10^3} = 3.6 \text{ mA}$$

Since $i_E > 0$, this confirms that the emitter junction is forward biased.

For the output portion of the circuit, by KVL, we have

$$v_{CB} = -100i_C - 3$$

One way to proceed is to plot the load line corresponding to this equation on the collector characteristics (if available) of the transistor. However, using the facts that $i_C \approx -\alpha i_E$ [Equation (7.4)] and $\alpha \approx 1$, let us employ the approximation $i_C \approx -i_E = -3.6$ mA. Therefore, we have

$$v_{CB} = -100i_C - 3 \approx -100(-3.6 \times 10^{-3}) - 3 = -2.64 \text{ V}$$

Since $v_{CB} < V_\gamma = 0.6$ V, this confirms that the collector junction is reverse biased. Thus, the transistor is indeed biased in the active region.

■

So far we have discussed the *p-n-p* transistor. For the case of an *n-p-n* transistor, the values of all of the currents and voltages are opposite in sign to those of a *p-n-p* transistor. For example, an *n-p-n* transistor operating in the active region will have negative values for i_E and v_{EB}, while i_B and i_C will be positive quantities. Note, too, that if the *p-n-p* BJT in Fig. 7.4 were replaced by an *n-p-n* transistor, for active-region operation, the polarities of the supply voltages V_{CC} and V_{EE} would have to be reversed.

7.2 THE *n-p-n* TRANSISTOR

Having discussed the common-base (CB) configuration for a *p-n-p* transistor, let us consider the *n-p-n* BJT in the **common-emitter** (CE) configuration shown in Fig. 7.9. As was mentioned previously, the transistor is biased in the active region when the emitter junction is forward biased and the collector junction is reverse biased. Clearly, the former is accomplished with the supply voltage V_{BB}. But when is the latter condition satisfied?

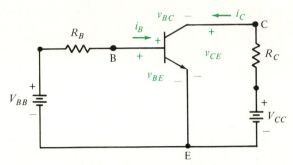

Fig. 7.9 An *n-p-n* transistor CE circuit.

For an *n-p-n* transistor, the collector junction is reverse biased when $v_{BC} < V_\gamma$ (typically 0.5 V for silicon). By KVL, $v_{BC} = v_{BE} - v_{CE}$. Thus, the collector junction is reverse biased when

$$v_{BE} - v_{CE} < V_\gamma$$

or when

$$v_{CE} > v_{BE} - V_\gamma$$

For the CB configuration, the collector (output) characteristics are plots of collector (output) current i_C versus collector–base (output) voltage v_{CB} for different values of emitter (input) current i_E. For the CE configuration, the collector (output) characteristics are plots of collector (output) current i_C versus collector–emitter (output) voltage v_{CE} for different values of base (input) current i_B. The collector (static) characteristics for a typical silicon *n-p-n* BJT in the CE configuration are shown in Fig. 7.10. As discussed above, the collector junction of an *n-p-n* transistor is reverse biased when $v_{CE} > v_{BE} - V_\gamma$. Typically, the collector junction of a silicon BJT is reverse biased when $v_{CE} > 0.2$ V.* Since the emitter junction is forward biased when $i_B > 0$, the portion of the characteristics above the curve $i_B = 0$ and to the right of the dashed line $v_{CE} = 0.2$ V is the active region. The area to the left of the dashed line $v_{CE} = 0.2$ V (this corresponds to the case that the collector junction is forward biased) and above the curve $i_B = 0$ (this corresponds

---

*Some texts use 0.3 V.

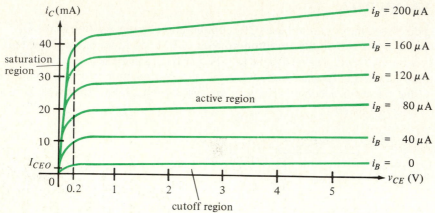

Fig. 7.10 Collector characteristics for a silicon *n-p-n* CE transistor.

to the case that the emitter junction is forward biased) is the saturation region. Finally, for the case that $i_B \leq 0$, the emitter junction is reverse biased, so the region to the right of the dashed line $v_{CE} = 0.2$ V (which corresponds to a reverse-biased collector junction) that is bounded above by the curve $i_B = 0$ is the cutoff region. For the case that $i_B = 0$, the resulting collector current is I_{CEO}, since by definition I_{CEO} is the current that goes from collector to emitter with the base open ($i_B = 0$). As in the case of I_{CBO}, the value of I_{CEO} on the vertical axis is exaggerated.

For a BJT operating in the active region, by Equation (7.1), we have

$$i_C = I_{CBO} - \alpha i_E$$

However, by KCL, $i_E = -i_B - i_C$. Therefore,

$$i_C = I_{CBO} - \alpha(-i_B - i_C) = I_{CBO} + \alpha i_B + \alpha i_C$$

from which

$$i_C = \frac{\alpha}{1 - \alpha} i_B + \frac{1}{1 - \alpha} I_{CBO} \tag{7.10}$$

We now define the **large-signal CE current gain*** β by

$$\boxed{\beta = \frac{\alpha}{1 - \alpha}} \tag{7.11}$$

From this definition, we can write

$$\frac{1}{1 - \alpha} = \frac{1 - \alpha + \alpha}{1 - \alpha} = \frac{1 - \alpha}{1 - \alpha} + \frac{\alpha}{1 - \alpha} = 1 + \beta \tag{7.12}$$

---

*α is the large-signal CB current gain.

Substituting this fact into Equation (7.10) results in

$$i_C = \beta i_B + (1 + \beta)I_{CBO} \qquad \qquad (7.13)$$

To get an idea of some typical values of β, consider the case that $\alpha = 0.98$. Then by Equation (7.11)

$$\beta = \frac{\alpha}{1 - \alpha} = \frac{0.98}{1 - 0.98} = 49$$

As another example, if $\alpha = 0.99$, then

$$\beta = \frac{0.99}{1 - 0.99} = 99$$

As has been indicated here, typically $\beta \gg 1$, so a reasonable approximation is $1 + \beta \approx \beta$. Furthermore, for a transistor biased in the active region, generally $i_B \gg I_{CBO}$. For these reasons, Equation (7.13) yields

$$i_C \approx \beta i_B \qquad \text{or} \qquad \beta \approx \frac{i_C}{i_B}$$

For example, for a transistor that has the collector characteristics given in Fig. 7.10, when $i_B = 120 \ \mu A$ and $v_{CE} = 4$ V, then $i_c = 30$ mA. Under this condition,

$$\beta \approx \frac{i_C}{i_B} = \frac{30 \times 10^{-3}}{120 \times 10^{-6}} = 250$$

Again from Equation (7.1), we know that

$$i_C = I_{CBO} - \alpha i_E \approx -\alpha i_E$$

Since α changes only slightly as v_{CB} changes (remember, the effective base width changes), then for a given value of i_E, there will be only a small change in the value of i_C as v_{CB} changes. Because of this, CB collector characteristic curves (Fig. 7.6) are nearly horizontal. However, since

$$i_C \approx \beta i_B \qquad \text{and} \qquad \beta = \frac{\alpha}{1 - \alpha}$$

then a slight change in α due to a change in v_{CB}, which, in turn, means a change in $v_{CE} = v_{CB} + v_{BE}$, can cause a large change in β. In the numerical example above, an approximately 1 percent change in α from 0.98 to 0.99 corresponds to more than a 100 percent change in β from 49 to 99. It is because of this that CE collector characteristic curves (Fig. 7.10) are by no means horizontal.

For the CE configuration, the base (input) characteristics are plots of base (input) current i_B versus base-emitter (input) voltage v_{BE} for different values of collector-emitter (output) voltage v_{CE}. The base (static) characteristics of a typical silicon n-p-n BJT in the CE configuration are shown in Fig. 7.11. When there is

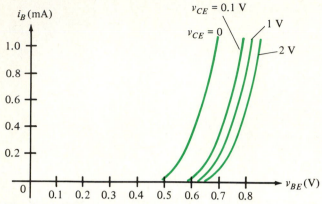

Fig. 7.11 Base characteristics for a silicon *n-p-n* CE transistor.

a short circuit between the collector and emitter ($v_{CE} = 0$), the resulting i_B-v_{BE} curve is typical of a forward-biased diode. By holding v_{BE} constant, if v_{CE} increases, then $v_{CB} = v_{CE} - v_{BE}$ increases. This, in turn, decreases the effective base width. The result is less hole–electron recombination in the base, and less base current. In other words, as v_{CE} is increased, an i_B-v_{BE} curve moves to the right.

Since typical values for i_B are in the range of 0.1–0.2 mA (100–200 μA), quite often for a silicon *n-p-n* transistor a value of $v_{BE} = 0.7$ V is used for a forward-biased emitter junction. (For the case of germanium, $v_{BE} = 0.2$ V.)

Simple Transistor Circuits

Let us now analyze some simple transistor circuits.

Example 7.2

The BJT circuit shown in Fig. 7.12 has a silicon *n-p-n* transistor with $\beta = 100$ and $I_{CBO} = 25$ nA. By KVL and Ohm's law, we have that

$$i_B = \frac{10 - v_{BE}}{330 \times 10^3}$$

Let us assume that the transistor is biased in the active region. Using the approximation $v_{BE} = 0.7$ V, we obtain

$$i_B = \frac{10 - 0.7}{330 \times 10^3} = 2.82 \times 10^{-5} = 28.2 \ \mu A$$

Since $i_B = 28.2 \ \mu A \gg I_{CBO} = 25$ nA, then

$$i_C \approx \beta i_B = 100(2.82 \times 10^{-5}) = 2.82 \text{ mA}$$

The fact that $i_B > 0$ indicates that the emitter junction is indeed forward biased.

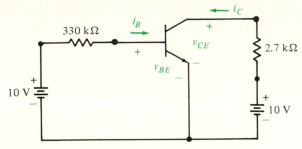

Fig. 7.12 A CE transistor circuit.

To check that the collector junction is reverse biased, by KVL,

$$v_{CE} = -2.7 \times 10^3 i_C + 10 \qquad\qquad (7.14)$$
$$= (-2.7 \times 10^3)(2.82 \times 10^{-3}) + 10 = 2.39 \text{ V}$$

Since $v_{CE} > 0.2$ V, the collector junction is indeed reverse biased. A simple confirmation is obtained by KVL from

$$v_{BC} = v_{BE} - v_{CE} = 0.7 - 2.39 = -1.69 \text{ V} < V_\gamma$$

Hence, the transistor is indeed operating in the active region.

If the collector characteristics of the transistor are available, a graphical approach can be taken. Plot the load line corresponding to Equation (7.14) on the collector characteristics. This line intersects the curve $i_B = 28.2\ \mu$A (and such a curve probably will have to be interpolated) at the operating point whose ordinate is the value for i_C and whose abscissa is the value for v_{CE}.

Consider the case that the 2.7-kΩ collector resistor shown in Fig. 7.12 is changed to a 4.7-kΩ resistor. Proceeding as above, we again get that $i_C \approx \beta i_B = 2.82$ mA. However, for this situation

$$v_{CE} = -4.7 \times 10^3 i_C + 10 = (-4.7 \times 10^3)(2.82 \times 10^{-3}) + 10 = -3.25 \text{ V}$$

and

$$v_{BC} = v_{BE} - v_{CE} = 0.7 + 3.25 = 3.95 \text{ V}$$

Using either the fact that $v_{CE} < 0.2$ V or $v_{BC} > V_\gamma$, we see that the collector junction is not reversed biased, and, therefore, the transistor is not in the active region.

Example 7.3

Suppose that for the circuit given in Fig. 7.12, a 1-kΩ resistor is placed in series with the emitter. The resulting circuit is shown in Fig. 7.13. By KVL, we have

$$10 = 330 \times 10^3 i_B + v_{BE} + 1000(i_B + i_C)$$

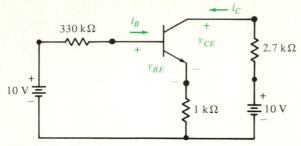

Fig. 7.13 Circuit resulting from modification of preceding circuit.

If we assume that the transistor is operating in the active region, then using $v_{BE} = 0.7$ V and $i_C \approx \beta i_B = 100 i_B$, we have that

$$10 = 330 \times 10^3 i_B + 0.7 + 1000(i_B + 100 i_B)$$

Solving for i_B results in $i_B = 21.58$ μA. Since $i_B \gg I_{CBO} = 25$ nA, then the use of $i_C \approx \beta i_B$ is justified. Thus,

$$i_C \approx \beta i_B = 100(21.58 \times 10^{-6}) = 2.16 \text{ mA}$$

Futhermore,

$$v_{CE} = -2.7 \times 10^3 i_C + 10 - 1000(i_B + i_C)$$
$$= (-2.7 \times 10^3)(2.16 \times 10^{-3}) + 10 - 1000(21.58 \times 10^{-6} + 2.16 \times 10^{-3})$$
$$= 1.99 \text{ V}$$

Since $i_B > 0$, the emitter junction is indeed forward biased, and since $v_{CE} > 0.2$ V, the collector junction is indeed reverse biased. This confirms that the transistor is in the active region.

Example 7.4

The circuit shown in Fig. 7.14 has two *n-p-n* transistors Q_1 and Q_2. Suppose that the respective current gains are $\beta_1 = 50$ and $\beta_2 = 100$. Let us analyze this circuit neglecting reverse collector saturation currents. First note that the original circuit given in Fig. 7.14 is equivalent to the circuit shown in Fig. 7.15. For the circuit shown in Fig. 7.15, we can apply Thévenin's theorem to the portion to the left of transistor Q_1. The resulting circuit is shown in Fig. 7.16.

Assuming that Q_1 and Q_2 are in the active region, we will use

$$v_{BE_1} = v_{BE_2} = 0.7 \text{ V}, \qquad i_{C_1} \approx \beta_1 i_{B_1} = 50 i_{B_1}, \qquad i_{C_2} \approx \beta_2 i_{B_2} = 100 i_{B_2}$$

By KCL, we have that

$$i_{B_2} = i_{B_1} + i_{C_1} = i_{B_1} + 50 i_{B_1} = 51 i_{B_1}$$

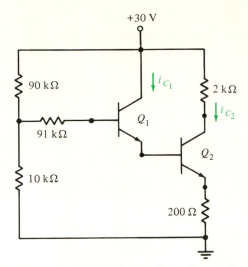

Fig. 7.14 Original circuit to be analyzed.

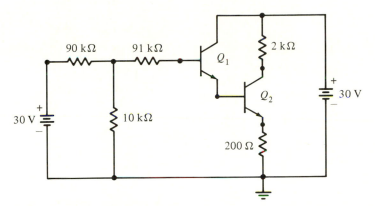

Fig. 7.15 Equivalent circuit.

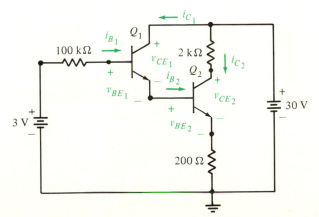

Fig. 7.16 Equivalent circuit obtained after applying Thévenin's theorem.

Furthermore, by KVL,

$$3 = 100 \times 10^3 i_{B_1} + v_{BE_1} + v_{BE_2} + 200(i_{B_2} + i_{C_2})$$

$$= 100 \times 10^3 i_{B_1} + 0.7 + 0.7 + 200(51 i_{B_1} + 5100 i_{B_1})$$

from which

$$i_{B_1} = \frac{3 - 1.4}{100 \times 10^3 + 200(5151)} = 1.42 \ \mu A$$

Therefore,

$$i_{C_1} = 50 i_{B_1} = 71 \ \mu A$$

$$i_{B_2} = 51 i_{B_1} = 72.4 \ \mu A$$

$$i_{C_2} = 100 i_{B_2} = 7.24 \ mA$$

By KVL,

$$v_{CE_1} = 30 - 200(i_{B_2} + i_{C_2}) - v_{BE_2}$$

$$= 30 - 200(72.4 \times 10^{-6} + 7.24 \times 10^{-3}) - 0.7 = 27.8 \ V$$

and

$$v_{CE_2} = -2 \times 10^3 i_{C_2} + 30 - 200(i_{B_2} + i_{C_2})$$

$$= -2 \times 10^3 (7.24 \times 10^{-3}) + 30 - 200(72.4 \times 10^{-6} + 7.24 \times 10^{-3})$$

$$= 14.1 \ V$$

Since $i_{B_1} > 0$, $i_{B_2} > 0$, $v_{CE_1} > 0.2$ V, and $v_{CE_2} > 0.2$ V, then both Q_1 and Q_2 are biased in the active region.

7.3 CUTOFF AND SATURATION

Previously, we saw that a transistor is in the active region when the emitter junction is forward biased and the collector junction is reverse biased. We mentioned that a transistor whose collector junction is reverse biased, and whose emitter junction is also reverse biased, is in the cutoff region—we also say that the transistor is **cut off.** A transistor whose emitter junction is forward biased and whose collector junction is also forward biased is in the saturation region—we also say that the transistor is **saturated.** For linear applications such as amplification, a transistor operates in the active region. Conversely, transistors typically operate mainly in the cutoff and saturation regions for such digital-circuit applications as logic gates.

Consider the simple n-p-n silicon transistor CE circuit shown in Fig. 7.17. For this circuit, we have that $i_C = (V_{CC} - v_{CE})/R_C = -v_{CE}/R_C + V_{CC}/R_C$. A plot of this equation on the collector characteristics of the transistor gives us the load line shown in Fig. 7.18. With reference to the input characteristics shown in Fig. 7.11,

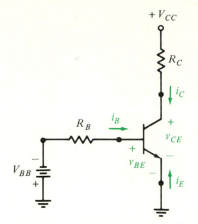

Fig. 7.17 Transistor biased in the cutoff region.

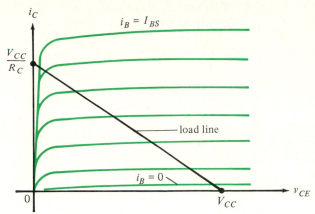

Fig. 7.18 Output characteristics and load line.

for there to be a significant base current, we should have $v_{BE} \geq 0.5$ V. For the case that $v_{BE} < 0.5$ V, we can consider the base current to be $i_B = 0$ and the emitter junction to be reverse biased. As a consequence of this, from the collector (output) characteristics given in Fig. 7.18, we see that, in essence, $i_C = 0$ and $v_{CE} = V_{CC}$. Furthermore, by KCL, $i_E = -i_B - i_C = 0$. Because $v_{BE} < 0.5$ V, then

$$v_{BC} = v_{BE} - v_{CE} < 0.5 - V_{CC}$$

and so the collector junction is reverse biased as well. In summary, if the supply voltage V_{BB} is such that $v_{BE} < 0.5$ V, then the silicon transistor in Fig. 7.17 will be cut off ($i_B = i_C = i_E = 0$ and $v_{CE} = V_{CC}$).

As we have just stated, when a transistor is cut off, ideally all its currents are zero. In particular, I_{CBO} is treated as being negligible. However, as was mentioned previously, I_{CBO} is temperature dependent, approximately doubling for each $10°C$ increase. If we take the effect of I_{CBO} into consideration, then at cutoff, by KVL, we have that

$$v_{BE} = -R_B i_B - V_{BB} = R_B I_{CBO} - V_{BB} \tag{7.15}$$

Although the transistor will remain at cutoff for $v_{BE} < 0.5$ V, to ensure a margin of safety, it may be desirable to use the condition that $v_{BE} \leq 0$. Therefore, from Equation (7.15), we want

$$v_{BE} = R_B I_{CBO} - V_{BB} \leq 0$$

from which

$$V_{BB} \geq R_B I_{CBO} \tag{7.16}$$

For example, if $R_B = 200$ kΩ and $I_{CBO} = 50$ μA due to an elevated temperature, then for the transistor in the circuit shown in Fig. 7.17 to remain safely in cutoff, from inequality (7.16), we should have

$$V_{BB} \geq R_B I_{CBO} = (200 \times 10^3)(50 \times 10^{-6}) = 10 \text{ V}$$

Perhaps an overzealous individual may want to make V_{BB} very large to guarantee cutoff. However, there is a limit as to how much the emitter junction can be reverse biased. The maximum reverse-bias voltage, called the **emitter–base breakdown voltage,** is designated BV_{EBO} since it is measured from the emitter to the base with the collector open ($i_C = 0$). Values of BV_{EBO} typically range from volts to tens of volts. Exceeding BV_{EBO} can result in excessive current (and damage to the transistor) and undesirable operating characteristics.

Another type of breakdown occurs when v_{CE} is made too large. Specifically, the **collector–emitter breakdown voltage,** designated BV_{CEO}, is the maximum voltage that can be applied from the collector to the emitter with the base open ($i_B = 0$). If this value of voltage is exceeded, then excessive collector current can result, as indicated by the collector characteristics shown in Fig. 7.19. Inspection of the curve for $i_B = 0$ reveals that $BV_{CEO} \approx 40$ V. For the case of a transistor in the CB configuration, the output breakdown voltage is BV_{CBO}—the maximum voltage from collector to base with the emitter open ($i_E = 0$). Generally, BV_{CBO} is greater than 50 V, and usually BV_{CEO} is about half of BV_{CBO}.

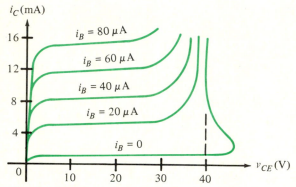

Fig. 7.19 Collector (output) characteristics, including the effects of breakdown.

The excessive current resulting from transistor breakdown can be due to either the avalanche effect or to another phenomenon known as **reach-through** (or **punch-through**). This mechanism is a result of a widened collector–base junction depletion region. If this junction is sufficiently reverse biased, then the depletion region can completely spread across the narrow base and "reach through" to the emitter junction. The consequence of this can be a large current. The breakdown voltage limitation of a particular transistor is determined by the phenomenon, either the avalanche effect or reach-through, which occurs at the lower voltage.

dc Current Gain

In Section 7.2, we saw how the large-signal (CE) current gain β was reasonably approximated by $\beta \approx i_C/i_B$. Therefore, let us now designate this current ratio by its own notation. Specifically, we define the **dc current gain** h_{FE} (also called the

dc beta β_{dc}) by

$$h_{FE} = \frac{i_C}{i_B}$$

(7.17)

and, as a consequence of this definition, we have that $h_{FE} \approx \beta$.

Not only does h_{FE} change with temperature (it increases as temperature increases), but for a given temperature, h_{FE} depends on the collector current i_C. A typical variation in h_{FE} is displayed in Fig. 7.20. Despite this characteristic, for the sake of simplicity, in our calculations we shall treat h_{FE} as a constant.

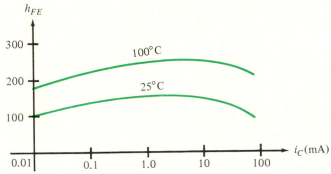

Fig. 7.20 Dependence of h_{FE} on i_C and on temperature. (Note the logarithmic horizontal axis.)

Again consider the circuit shown in Fig. 7.17 and the associated output characteristics and load line given in Fig. 7.18. If the polarity of V_{BB} is reversed, then this supply voltage will forward bias (when $V_{BB} > 0.5$ V) the emitter junction and make $i_B > 0$. This means that the operation of the transistor goes (along the load line) from cutoff to the active region. By further increasing V_{BB}, and hence i_B, the transistor eventually will enter the saturation region. In particular, when $i_B = I_{BS}$, we see that there is sufficient base current to put the transistor into saturation. For this case, the collector current is $i_C \approx V_{CC}/R_C$ and the corresponding collector–emitter voltage is designated $v_{CE(sat)}$—typically 0.2 V for silicon. Increasing i_B even more just drives the transistor further into saturation without appreciably affecting i_C and v_{CE}, that is, we still have that $i_C \approx V_{CC}/R_C$ and $v_{CE} = v_{CE(sat)} \approx 0.2$ V.

When there is sufficient base current to saturate a transistor, typically the corresponding base–emitter voltage, designated $v_{BE(sat)}$, is $v_{BE(sat)} = 0.8$ V for silicon. (For germanium, we use $v_{BE(sat)} = 0.3$ V and $v_{CE(sat)} = 0.1$ V.) We see that for a silicon transistor in saturation $i_B > 0$ implies that the emitter junction is forward biased, and $v_{BC} = v_{BE(sat)} - v_{CE(sat)} = 0.8 - 0.2 = 0.6$ V means that the collector junction is forward biased.

When a transistor is in saturation, we shall say that it is **ON**; when it is in cutoff, we shall say that it is **OFF**. In summary, then, when a transistor is ON, it conducts current ($\approx V_{CC}/R_C$) from collector to emitter, and the voltage between collector and emitter (≈ 0.2 V) is close to 0 V. In other words, it behaves approximately as a short circuit between collector and emitter. Conversely, when a transistor is OFF, it will not conduct any significant current between collector and emitter, and, therefore, it acts essentially as an open circuit. Hence, we see that a transistor operating in saturation and cutoff is an electronic switch that is turned ON and OFF, respectively, by base current.

Example 7.5

Given the CE circuit in Fig. 7.21, the silicon transistor has $h_{FE} = 100$. Suppose we assume that the transistor is in the active region. For this case, $v_{BE} = 0.7$ V, so we have that

$$i_B = \frac{10 - 0.7}{100 \times 10^3} = 93 \ \mu A \qquad \Rightarrow \qquad i_C = h_{FE} i_B = 100(93 \times 10^{-6}) = 9.3 \text{ mA}$$

Since

$$v_{CE} = -2 \times 10^3 i_C + 10 = -8.6 < 0.2 \text{ V}$$

our assumption that the transistor was biased in the active region was incorrect. Therefore, let us assume that the transistor is operating in the saturation region, that is, $v_{BE} = v_{BE(sat)} = 0.8$ V and $v_{CE} = v_{CE(sat)} = 0.2$ V. By Ohm's law, the saturation currents are

$$i_B = \frac{10 - 0.8}{100 \times 10^3} = 92 \ \mu A \qquad \text{and} \qquad i_C = \frac{10 - 0.2}{2 \times 10^3} = 4.9 \text{ mA}$$

On the border between the active and saturation regions, we have that

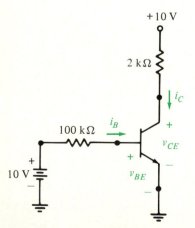

Fig. 7.21 A CE transistor circuit.

$h_{FE} = i_C/i_B$ [Equation (7.17)]. Thus, if i_C is the saturation current, then the minimum current required to border on the saturation region is $i_B = i_C/h_{FE}$. In other words, for the transistor to go into saturation, if i_C is the saturation current, then we must have that

$$i_B \geq \frac{i_C}{h_{FE}} \tag{7.18}$$

For this example,

$$i_B = 92 \ \mu A > \frac{i_C}{h_{FE}} = \frac{4.9 \times 10^{-3}}{100} = 49 \ \mu A$$

so the transistor is indeed in saturation.

Suppose now that h_{FE} for the transistor was not given. What is the range of values for h_{FE} that will result in saturation-region operation? Again assuming that the transistor is in saturation, we get $i_B = 92 \ \mu A$ and $i_C = 4.9$ mA. From inequality (7.18) for saturation-region operation, we must have that

$$h_{FE} \geq \frac{i_C}{i_B} = \frac{4.9 \times 10^{-3}}{92 \times 10^{-6}} = 53.3 \approx 53$$

Example 7.6

Given the circuit shown in Fig. 7.22, where silicon transistors Q_1 and Q_2 have dc betas of h_{FE_1} and h_{FE_2}, respectively, let us verify that Q_1 is in saturation (ON) and Q_2 is in cutoff (OFF), and determine the minimum value of h_{FE_1} that will turn ON Q_1.

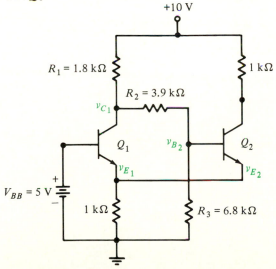

Fig. 7.22 Transistor circuit with Q_1 ON and Q_2 OFF.

If Q_2 is assumed to be OFF, then all the currents into Q_2 are zero ($i_{B_2} = i_{C_2} = i_{E_2} = 0$). Therefore, we can remove Q_2 and analyze the resulting circuit shown in Fig. 7.23. By applying Thévenin's theorem between the collector of Q_1 and the reference node, we obtain the circuit shown in Fig. 7.24. If Q_1 is assumed to be ON, then $v_{BE_1} = 0.8$ V and $v_{CE_1} = 0.2$ V. By KVL, we have

$$5 = 0.8 + 1000(i_{B_1} + i_{C_1})$$

from which

$$1000i_{B_1} + 1000i_{C_1} = 4.2$$

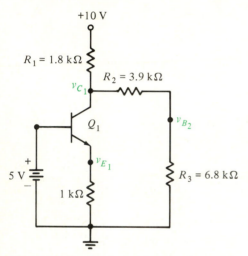

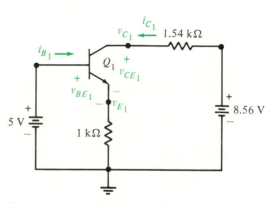

Fig. 7.23 Equivalent circuit after OFF transistor is removed.

Fig. 7.24 Circuit obtained by applying Thévenin's theorem.

Also,

$$8.56 = 1540i_{C_1} + 0.2 + 1000(i_{B_1} + i_{C_1})$$

from which

$$1000i_{B_1} + 2540i_{C_1} = 8.36$$

Solving these equations, we get $i_{B_1} = 1.5$ mA and $i_{C_1} = 2.7$ mA. Thus, for Q_1 to be in saturation, we require that

$$h_{FE_1} \geq \frac{i_{C_1}}{i_{B_1}} = \frac{2.7 \times 10^{-3}}{1.5 \times 10^{-3}} = 1.8$$

When this condition is met, we have that

$$v_{E_1} = 1000(i_{B_1} + i_{C_1}) = 1000(1.5 \times 10^{-3} + 2.7 \times 10^{-3}) = 4.2 \text{ V}$$

and

$$v_{C_1} = v_{CE_1} + v_{E_1} = 0.2 + 4.2 = 4.4 \text{ V}$$

With reference to the circuit shown in Fig. 7.23, by voltage division,

$$v_{B_2} = \frac{R_3}{R_2 + R_3} v_{C_1} = \frac{6800}{3900 + 6800}(4.4) = 2.80 \text{ V}$$

With reference to the circuit shown in Fig. 7.22, $v_{E_1} = v_{E_2}$, so

$$v_{BE_2} = v_{B_2} - v_{E_2} = 2.8 - 4.2 = -1.4 < 0.5 \text{ V}$$

and Q_2 is indeed OFF.

Transistor Switching Times

Since a transistor is a physical device, it cannot change state (go from cutoff to saturation, or vice versa) instantaneously. To discuss what happens, therefore, let us again consider the simple CE transistor circuit shown in Fig. 7.25, where $V_{CC} \gg 0.2$ V and the applied voltage v_a is as given in Fig. 7.26. For time $t < 0$, since $v_a = -V_1 < 0$, the transistor is cut off and $i_C = 0$. At $t = 0$, the applied voltage instantaneously changes to $v_a = V_2$. Suppose that V_2 is sufficiently large to saturate the transistor. When the transistor is in saturation, $i_C \approx V_{CC}/R_C = I_S$. However, the transistor will not go from cutoff to saturation instantaneously. Instead, the current will increase from $i_C = 0$ to $i_C = I_S$ as depicted in Fig. 7.27. The time it takes for the collector current to increase from $i_C = 0$ to $i_C = 0.1I_S$ is called the **delay time** t_d. The time it takes for this current to rise from $i_C = 0.1I_S$ to $i_C = 0.9I_S$ is known as the **rise time** t_r. We then define the **turn-on time** t_{ON} to be $t_{ON} = t_d + t_r$.

At time $t = T$, the applied voltage instantaneously changes from $v_a = V_2$ to $v_a = -V_1$. As a consequence of this, the transistor will go from saturation to cutoff—but not instantaneously. The time it takes for the collector current to drop from $i_C = I_S$ to $i_C = 0.9I_S$ is called the **storage time** t_s, while the time it takes this

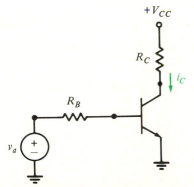

Fig. 7.25 Simple CE transistor circuit.

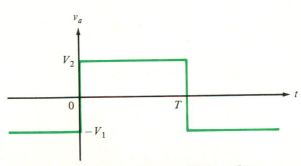

Fig. 7.26 Voltage applied to CE transistor circuit.

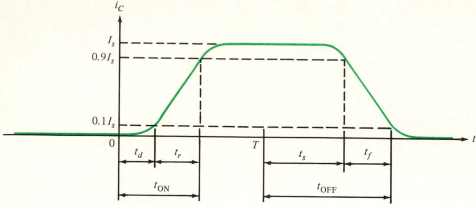

Fig. 7.27 Transistor switching times.

current to fall from $i_C = 0.9I_s$ to $i_C = 0.1I_s$ is known as the **fall time** t_f. We then define the **turn-off time** t_{OFF} to be $t_{OFF} = t_s + t_f$.

When the transistor is cut off, both the collector and emitter junctions are reverse biased. For the emitter junction to be forward biased, first a certain amount of time is required for the emitter-junction transition capacitance to be charged up. Time is also required for minority carriers to diffuse across the base and then enter the collector. This produces the delay time t_d (of the order of nanoseconds). Before it can go into saturation, the transistor goes (via the load line—see Fig. 7.18) into the active region. The time required for the transistor to go through the active region from cutoff to saturation accounts for the rise time t_r (of the order of nanoseconds to tens of nanoseconds).

When a transistor is in saturation, both junctions are forward biased and conducting. As a consequence of this, an excess of minority carriers is stored in the base. For the collector junction to be reverse biased, this large amount of minority-carrier charge must first be removed. The time required for the removal of the excess charge determines the storage time t_s (of the order of hundreds of nanoseconds). Before it can be cut off, the transistor goes into the active region. The time required for the transistor to go through the active region from saturation to cutoff accounts for the fall time t_f (of the order of tens of nanoseconds).

7.4 APPLICATIONS TO DIGITAL LOGIC CIRCUITS

A simple but very important digital logic circuit is the **inverter** or **NOT gate.** The output of such a device is the logical inverse of the input. In other words, if the input is a logical 0, then the output is a logical 1; if the input is a logical 1, then the output is a logical 0. An example of a simple NOT gate is shown in Fig. 7.28, where the low voltage is 0.2 V and the high voltage is 6 V.

For the case that the input voltage is $v_1 = 0.2$ V, since $v_1 < 0.5$ V, the input tends to turn OFF the transistor. In addition, the negative supply voltage $(-6\ \text{V})$

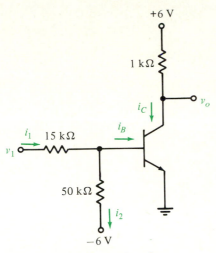

Fig. 7.28 Inverter circuit.

also tends to turn OFF the transistor. Therefore, let us assume that the transistor is OFF for $v_1 = 0.2$ V. Thus, $i_B = 0$ and

$$i_1 = i_2 = \frac{0.2 - (-6)}{15 \times 10^3 + 50 \times 10^3} = \frac{6.2}{65 \times 10^3} = 95.4 \ \mu A$$

Hence,

$$v_{BE} = 50 \times 10^3 (95.4 \times 10^{-6}) - 6 = -1.23 < 0.5 \text{ V}$$

and the transistor is indeed cut off. Since $i_C = 0$, the resulting output voltage is

$$v_o = -1000 i_C + 6 = 6 \text{ V}$$

For the case that the input voltage is $v_1 = 6$ V, the input tends to turn ON the transistor. Although the negative supply voltage tends to turn OFF the transistor, it is connected to a greater resistance than the input voltage. Therefore, let us assume that the transistor is ON for $v_1 = 6$ V. Thus, $v_{BE} = 0.8$ V and

$$i_B = i_1 - i_2 = \frac{6 - 0.8}{15 \times 10^3} - \frac{0.8 - (-6)}{50 \times 10^3} = 0.21 \text{ mA}$$

Also,

$$i_C = \frac{6 - 0.2}{1000} = 5.8 \text{ mA}$$

Hence, for the transistor to be in saturation, the transistor must have

$$h_{FE} \geq \frac{i_C}{i_B} = \frac{5.8 \times 10^{-3}}{0.21 \times 10^{-3}} = 27.6$$

When this inequality is satisfied, the transistor is indeed in saturation, and the resulting output voltage is

$$v_o = v_{CE(sat)} = 0.2 \text{ V}$$

We have now confirmed that the given circuit is a NOT gate. The logic table for a NOT gate with input v_1 and output v_o is given in Table 7.1. For such a situation, we say that the output v_o is the **complement** of the input v_1, and we use a bar over a variable to denote the complement. Thus, we would write $v_o = \bar{v}_1$.

Table 7.1 NOT-Gate Logic Table

v_1	v_o
0	1
1	0

Diode-Transistor Logic (DTL)

Recall the logic table for a two-input AND gate (Table 6.2), which is repeated as Table 7.2(a). If we invert the output of an AND gate, the resulting combination of an AND gate followed by a NOT gate, described by Table 7.2(b), is called a **NAND gate**—NAND being a contraction of NOT-AND. This very important digital logic circuit can be constructed by connecting a transistor NOT gate (see Fig. 7.28) to a diode AND gate (see Fig. 6.32). The resulting (positive-logic) NAND gate shown in Fig. 7.29, which utilizes diodes and a transistor, is an example of **diode-transistor logic** (DTL).

In the analysis of the given NAND gate, we will assume that a reverse-biased diode is an open circuit, and a forward-biased diode has a voltage of 0.7 V across it. Furthermore, the low voltage (logical 0) is 0.2 V and the high voltage (logical 1) is 6 V.

For the case that $v_1 = v_2 = 0.2$ V, the positive supply voltage (6 V) tends to forward bias the diodes. Although the negative supply voltage (-6 V) tends to reverse bias the diodes, the resistance in series with the negative supply voltage is greater than that for the positive supply voltage. Therefore, let us assume that D_1 and D_2 are ON. Thus, since the voltage across a diode is 0.7 V, then

$$v_3 = 0.7 + 0.2 = 0.9 \text{ V}$$

Table 7.2(a) AND-Gate Table

v_1	v_2	v_o
0	0	0
0	1	0
1	0	0
1	1	1

Table 7.2(b) NAND-Gate Table

v_1	v_2	v_o
0	0	1
0	1	1
1	0	1
1	1	0

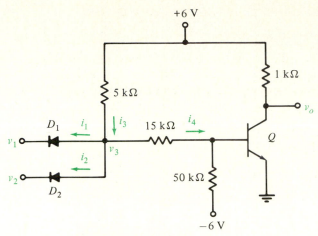

Fig. 7.29 A DTL NAND gate.

Although this voltage tends to turn ON the transistor Q, the negative supply voltage tends to turn OFF Q. Therefore, let us assume that Q is OFF. Thus, $i_B = 0$ and

$$i_4 = \frac{0.9 - (-6)}{15 \times 10^3 + 50 \times 10^3} = 0.106 \text{ mA}$$

Hence,

$$v_{BE} = 50 \times 10^3(0.106 \times 10^{-3}) - 6 = -0.7 \text{ V}$$

and Q is indeed cut off. Also, the total diode current is

$$i_1 + i_2 = i_3 - i_4 = \frac{6 - 0.9}{5 \times 10^{-3}} - 0.106 \times 10^{-3} = 0.914 \text{ mA}$$

By symmetry, for identical diodes, $i_1 = i_2 = 0.457$ mA, and since $i_1 > 0$, $i_2 > 0$, then D_1 and D_2 are indeed ON. Finally, since Q is OFF, then $i_C = 0$ and $v_o = 6$ V.

For the case that $v_1 = 0.2$ V and $v_2 = 6$ V, as above, let us assume that D_1 is ON. Therefore, again $v_3 = 0.7 + 0.2 = 0.9$ V. This means that the voltage across D_2 is $v_3 - v_2 = 0.9 - 6 = -5.1 < 0.7$ V, so when D_1 is ON, then D_2 is OFF. As above, we again find that Q is OFF, and $v_o = 6$ V. Now, however, $i_1 = 0.914$ mA > 0 so D_1 is indeed ON, and D_2 is indeed OFF.

By symmetry, for the case that $v_1 = 6$ V and $v_2 = 0.2$ V, we deduce that D_1 is OFF, D_2 is ON, Q is OFF, and $v_o = 6$ V.

Finally, consider the case that $v_1 = v_2 = 6$ V. Since the supply voltages are not higher in potential than either input, let us assume that D_1 and D_2 are OFF ($i_1 = i_2 = 0$). Since Q tends to be turned ON by 6 V through 5 kΩ + 15 kΩ = 20 kΩ, and tends to be turned OFF by −6 V through 50 kΩ, let us assume that Q is ON. Since $i_1 = i_2 = 0$ and $v_{BE} = 0.8$ V, then

$$i_3 = i_4 = \frac{6 - 0.8}{5 \times 10^3 + 15 \times 10^3} = 0.26 \text{ mA}$$

and

$$i_B = 0.26 \times 10^{-3} + \frac{-6 - 0.8}{50 \times 10^3} = 0.124 \text{ mA}$$

When Q is ON, the collector current is

$$i_C = \frac{6 - 0.2}{1000} = 5.8 \text{ mA}$$

Hence, Q is indeed in saturation when

$$h_{FE} \geq \frac{i_C}{i_B} = \frac{5.8 \times 10^{-3}}{0.124 \times 10^{-3}} = 46.8$$

Furthermore,

$$v_3 = -5 \times 10^3 i_3 + 6 = -5 \times 10^3 (0.26 \times 10^{-3}) + 6 = 4.7 \text{ V}$$

so that the voltage across the diodes is

$$v_3 - 6 = 4.7 - 6 = -1.3 \text{ V}$$

Thus, D_1 and D_2 are indeed OFF, and when $h_{FE} \geq 46.8$, Q is in saturation and $v_o = 0.2$ V.

This then confirms that the DTL circuit given in Fig. 7.29 is a NAND gate.

DTL gates commonly appear in **integrated-circuit** (IC) form (to be discussed later). Under this circumstance, because of economics, resistance values are limited to 30 kΩ and capacitance values to 100 pF. On the other hand, diodes and transistors can be fabricated inexpensively. Because of this, large resistive values such as those in the NAND gate given in Fig. 7.29 are to be avoided for IC designs.

Example 7.7

Let us verify that the DTL circuit shown in Fig. 7.30 is a NAND gate when h_{FE} is sufficiently large. Note, too, that only a single 5-V supply voltage is used. In this case, therefore, although the low voltage is still 0.2 V, the high voltage is 5 V.

For the case that $v_1 = v_2 = 0.2$ V, the 5-V supply voltage tends to forward bias D_1 and D_2. Therefore, let us assume that D_1 and D_2 are ON. Thus, $v_3 = 0.7 + 0.2 = 0.9$ V. However, this voltage is not large enough to turn ON D_3 and D_4, and so we shall assume that D_3 and D_4 are OFF. Under this assumption,

$$i_1 + i_2 = i_3 = \frac{5 - 0.9}{5 \times 10^3} = 0.82 \text{ mA}$$

and by symmetry, for identical diodes, $i_1 = i_2 = 0.41$ mA > 0. Hence, D_1 and D_2 are indeed ON. Since $i_4 = 0$, then $i_B = 0$. Consequently, $v_{BE} = 0$ and the voltage across D_3 and D_4 is

$$v_3 - v_{BE} = 0.9 < 0.7 + 0.7 = 1.4 \text{ V}$$

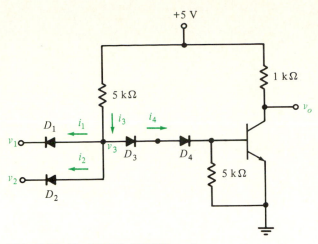

Fig. 7.30 An IC DTL NAND gate.

and so D_3 and D_4 indeed are OFF. Furthermore, $v_{BE} = 0 < 0.5$ V means that Q is OFF, and, hence, $v_o = 5$ V.

For the case that $v_1 = 0.2$ V and $v_2 = 5$ V, we again assume that D_1 is ON. Again, $v_3 = 0.7 + 0.2 = 0.9$ V. As above, this means that we assume D_3 and D_4 are OFF. Furthermore, the voltage across D_2 is $v_3 - v_2 = 0.9 - 5 = -4.1 < 0.7$ V, so we assume D_2 is OFF. This means that $i_2 = i_4 = 0$ and

$$i_1 = i_3 = \frac{5 - 0.9}{5 \times 10^3} = 0.82 \text{ mA} > 0$$

Thus, D_1 is indeed ON, and D_2, D_3, D_4 are indeed OFF. Again, since $i_4 = 0$, we deduce that Q is OFF and $v_o = 5$ V.

For the case that $v_1 = 5$ V and $v_2 = 0.2$ V, just as in the preceding case, we find that D_2 is ON; D_1, D_3, D_4, and Q are OFF; and $v_o = 5$ V.

Finally, for the case that $v_1 = v_2 = 5$ V, since there is no potential difference between either input and the supply voltage, let us assume that D_1 and D_2 are OFF ($i_1 = i_2 = 0$). This means that the supply voltage has a tendency to turn ON D_3, D_4, and Q. Therefore, let us make these assumptions. By KVL, we have that

$$v_3 = 0.7 + 0.7 + 0.8 = 2.2 \text{ V}$$

and since $v_3 - 5 = 2.2 - 5 = -2.8 < 0.7$ V, then D_1 and D_2 are indeed OFF. Furthermore,

$$i_3 = \frac{5 - 2.2}{5 \times 10^3} = 0.56 \text{ mA}$$

Since $i_4 = i_3 > 0$, then D_3 and D_4 are indeed ON. In addition, if Q is ON, then

$$i_B = i_4 - \frac{v_{BE}}{5 \times 10^3} = 0.56 \times 10^{-3} - \frac{0.8}{5 \times 10^3} = 0.4 \text{ mA} \qquad \textbf{(7.19)}$$

and

$$i_C = \frac{5 - 0.2}{1000} = 4.8 \text{ mA}$$

Thus, for Q indeed to be ON, we require that

$$h_{FE} \geq \frac{i_C}{i_B} = \frac{4.8 \times 10^{-3}}{0.4 \times 10^{-3}} = 12$$

and when Q is in saturation, $v_o = 0.2$ V. This confirms that the DTL circuit given in Fig. 7.30 is a NAND gate.

Let us now determine the power dissipated by the given NAND gate. To do this, let us find the power supplied by the 5-V source. For the case that the output is 0.2 V (logical 0), D_1 and D_2 are OFF and Q is ON. Using previous results, the current through the 5-V power supply is

$$i_s = i_3 + i_C = 0.56 \times 10^{-3} + 4.8 \times 10^{-3} = 5.36 \text{ mA}$$

and the power supplied, therefore, is

$$P_0 = 5(5.36 \times 10^{-3}) = 26.8 \text{ mW}$$

For the case that the output is 5 V (logical 1), D_1 and/or D_2 is ON and Q is OFF. Using previous results,

$$i_s = i_3 + i_C = 0.82 \times 10^{-3} + 0 = 0.82 \text{ mA}$$

and the power supplied, therefore, is

$$P_1 = 5(0.82 \times 10^{-3}) = 4.1 \text{ mW}$$

Assuming that the NAND gate spends the same amount of time in the 0 state that it does in the 1 state, the average power supplied is

$$P = \frac{P_0 + P_1}{2} = \frac{26.8 \times 10^{-3} + 4.1 \times 10^{-3}}{2} = 15.45 \text{ mW}$$

If we ignore the small amount of power (of the order of 0.1 mW) absorbed by the stage(s) preceding the NAND gate that turns ON D_1 and/or D_2, then the power supplied by the 5-V source is equal to the power dissipated by the gate. Thus, the average power dissipated by the NAND gate given in Fig. 7.30 is 15.45 mW.

Noise Margin and Fan-out

Noise voltages in circuits can be caused by such things as power supply voltage spikes and transients resulting from switching. The minimum noise voltage at the input of a digital logic gate that causes the output to deviate from its proper value is known as the **noise margin.** Specifically, if the output of a gate should be in the

0 state, the noise margin is denoted $NM(0)$. In other words, if the input changes by $NM(0)$ volts, the output no longer will be in the 0 state (this does not necessarily mean that the output will then be in the 1 state—only that the output is not in the 0 state as it should be). Conversely, if the output of a gate should be in the 1 state, the noise margin is denoted $NM(1)$.

For the NAND gate discussed in Example 7.7, the output is in the 0 state (Q is ON and $v_o = 0.2$ V) when both $v_1 = 5$ V and $v_2 = 5$ V. Under this circumstance, D_3 and D_4 are ON. The voltage v_3 required for D_3, D_4, and Q to be ON is $v_3 = 0.7 + 0.7 + 0.8 = 2.2$ V. If v_3 drops below 2.2 V, there will be insufficient voltage to turn ON Q. This can occur when either input voltage falls below 1.5 V, for then the corresponding diode would conduct and v_3 would drop below 2.2 V. In other words, an improper output voltage results if an input decreases by more than $5 - 1.5 = 3.5$ V. Thus, we have that $NM(0) = -3.5$ V (the negative sign indicates a voltage decrease). A noise voltage with a magnitude of 3.5 V is unlikely to occur for the given circuit.

On the other hand, the output of the NAND gate is in the 1 state (Q is OFF and $v_o = 5$ V) when either $v_1 = 0.2$ V or $v_2 = 0.2$ V. Under either condition, the corresponding diode is ON and $v_3 = 0.7 + 0.2 = 0.9$ V. Furthermore, D_3 and D_4 are OFF. For both of these diodes to conduct and for Q not to be OFF ($v_{BE} \geq 0.5$ V), v_3 has to increase to $v_3 = 0.7 + 0.7 + 0.5 = 1.9$ V. This can occur when the lowest input voltage rises to 1.2 V, for then v_3 would increase to $0.7 + 1.2 = 1.9$ V. In other words, an improper output voltage results if the 0.2-V input(s) increases by $NM(1) = 1.2 - 0.2 = 1$ V. Using one diode in place of the series connection of D_3 and D_4 would reduce the noise margin by 0.7 V to $NM(1) = 0.3$ V, while adding a third diode in series would increase the noise margin by 0.7 V to $NM(1) = 1.7$ V.

In our discussions so far, we have considered the situation in which a digital logic circuit was unloaded—that is, nothing was connected to the output of the circuit. Let us now investigate the condition that a gate is loaded with N similar gates. Specifically, consider the case of the NAND gate given in Fig. 7.30. Shown in Fig. 7.31 is such a gate, and only the input portion of one of the N gates which loads it is explicitly depicted.

If $v_1 = 0.2$ V or $v_2 = 0.2$ V, then Q will be OFF and v_o will tend to be 5 V. This, in turn, means that D_1' is OFF. Similarly, the other diodes also connected to v_o will be OFF. Thus, $v_o = 5$ V.

Now, however, consider the case that $v_1 = v_2 = 5$ V. Under this circumstance, Q is ON and $v_o = 0.2$ V. This, in turn, means that D_1' is ON and $v_3' = 0.7 + 0.2 = 0.9$ V. As a result,

$$i_3' = \frac{5 - 0.9}{5 \times 10^3} = 0.82 \text{ mA}$$

We know that for Q to be ON, $h_{FE} \geq i_C/i_B$. The base current is the same value as was obtained previously—$i_B = 0.4$ mA [Equation (7.19)]. However, the collector current is the sum of i_R and the current coming from the load. This load current is maximum when, for each of the N output stages, only one of the diodes conducts;

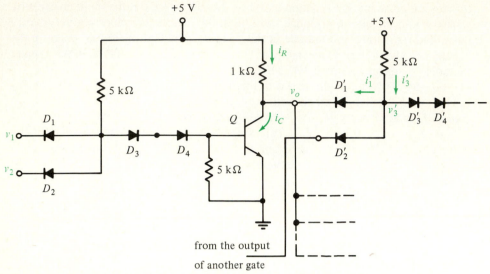

Fig. 7.31 Two-input NAND gate loaded by *N* similar gates.

for example, for the first stage shown, the maximum contribution to the collector current is $i_1' = i_3' = 0.82$ mA. This current is called the **standard load.** Since there are N stages, the maximum collector current is

$$i_C = i_R + Ni_1' = \frac{5 - 0.2}{1000} + N(0.82 \times 10^{-3})$$

$$= 4.8 \times 10^{-3} + 0.82 \times 10^{-3} N$$

To guarantee that Q is ON, we must have that $h_{FE} \geq i_C / i_B$ or $h_{FE} i_B \geq i_C$. Specifically, if we are given that $h_{FE} = 40$, then for proper operation

$$h_{FE} i_B \geq i_C$$

$$40(0.4 \times 10^{-3}) \geq 4.8 \times 10^{-3} + 0.82 \times 10^{-3} N$$

from which

$$\frac{16 - 4.8}{0.82} \geq N \quad \Rightarrow \quad N \leq 13.7$$

The maximum number of similar output gates that can load a given gate, and still have that gate function properly, is called the **fan-out** of the gate. Since fan-out must be an integer, for the circuit under consideration, the fan-out is $N = 13$. The number of inputs to a gate is known as **fan-in.** In this case, the fan-in is 2.

One way significantly to increase the fan-out of the NAND gate under discussion is to replace diode D_3 by a transistor Q', as shown in Fig. 7.32. Also shown is a portion of the first gate of the fan-out.

For the case that at least one input is 0.2 V, the corresponding diode is ON and

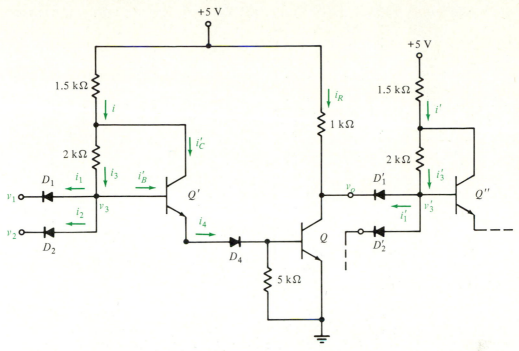

Fig. 7.32 A NAND gate with significantly increased fan-out.

$v_3 = 0.7 + 0.2 = 0.9$ V. Since this potential is not high enough to turn ON Q', D_4, and Q, then Q is OFF and $v_o = 5$ V. (The diodes in the load that are connected to v_o also are OFF.)

For the case that $v_1 = v_2 = 5$ V, diodes D_1 and D_2 are OFF ($i_1 = i_2 = 0$). Since the 5-V supply tends to turn ON Q', let us assume that Q and D_4 are ON. We will not assume that Q' is in saturation because a positive current i_3 results in

$$v'_{BC} = -2000i_3 < 0.5 \text{ V}$$

and this means that the collector junction of Q' is reverse biased. Thus, let us assume that Q' is in the active region. Under these assumptions,

$$v_3 = 0.7 + 0.7 + 0.8 = 2.2 \text{ V}$$

Since $i_1 = i_2 = 0$, then $i'_B = i_3$. Furthermore,

$$i = i'_B + i'_C = i'_B + h'_{FE}i'_B = (1 + h'_{FE})i'_B$$

Specifically, for the case that $h'_{FE} = 40$, by KVL,

$$5 = 1.5 \times 10^3 i + 2 \times 10^3 i'_B + v_3 = 1.5 \times 10^3 (41i'_B) + 2 \times 10^3 i'_B + 2.2$$

from which

$$i'_B = \frac{5 - 2.2}{61.5 \times 10^3 + 2 \times 10^3} = 44 \text{ }\mu\text{A}$$

Since $i'_B > 0$ and $v'_{BC} = -2000i'_B < 0.5$ V, then Q' is indeed in the active region. Also,

$$i'_C = h'_{FE}i'_B = 40(44 \times 10^{-6}) = 1.76 \text{ mA}$$

Since the current $i_4 = i'_B + i'_C > 0$, then D_4 is indeed ON. We also have that the base current for Q is

$$i_B = i'_B + i'_C - \frac{v_{BE}}{5 \times 10^3} = 44 \times 10^{-6} + 1.76 \times 10^{-3} - \frac{0.8}{5 \times 10^3}$$

$$= 1.64 \text{ mA}$$

If Q is ON, then $v_o = 0.2$ V and D'_1 is ON. This, in turn, means that $v'_3 = 0.7 + 0.2 = 0.9$ V and Q'' is OFF. Thus, the standard-load current is

$$i'_1 = \frac{5 - 0.9}{1.5 \times 10^3 + 2 \times 10^3} = 1.17 \text{ mA}$$

and the collector current for a fan-out of N is $i_C = i_R + Ni'_1$. Given that $h_{FE} = 40$, for Q indeed to be in saturation, we must have that

$$h_{FE}i_B \geq i_C = i_R + Ni'_1$$

$$40(1.64 \times 10^{-3}) \geq \frac{5 - 0.2}{1000} + N(1.17 \times 10^{-3})$$

from which

$$\frac{65.6 - 4.8}{1.17} \geq N \quad \Rightarrow \quad N \leq 51.97$$

Hence, the fan-out for the NAND gate given in Fig. 7.32 is $N = 51$, and a dramatic increase in fan-out has been achieved.

7.5 INTEGRATED-CIRCUIT (IC) LOGIC

Up to this point, we have been treating diodes, transistors, and resistors as distinct, separate elements, that is, as **discrete components.** However, a circuit having such elements can be fabricated on a single, small* silicon chip, and what results is called an **integrated circuit** (IC). The same processes used to fabricate discrete diodes and transistors are used to make ICs.

The number of gates on a digital IC determines the category in which such an IC is placed. Thirty or fewer gates on a chip is known as **small-scale integration** (SSI). Between 30 and 300 gates is called **medium-scale integration** (MSI). **Large-scale integration** (LSI) refers to between 300 and 3000 gates, while **very-large-scale integration** (VLSI) denotes more than 3000 gates.

---

*A typical cross section is 50 by 50 mils, where 1 mil = 0.001 inch = 0.0254 mm.

A simplified version of an IC-chip *n-p-n* transistor is shown in Fig. 7.33(a) and its circuit symbol in Fig. 7.33(b). Here we see that the device is fabricated on a *p*-type **substrate** (or foundation) in a vertical manner. By embedding more than one *n*-type region into the *p*-type base, we obtain a multiple-emitter *n-p-n* transistor as depicted in Fig. 7.34(a) for the case of two emitters. The corresponding circuit symbol for such a device is shown in Fig. 7.34(b).

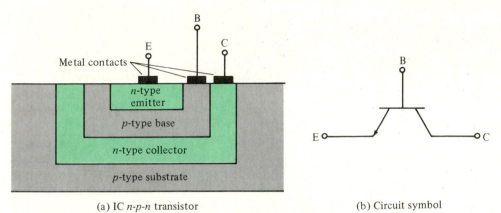

(a) IC *n-p-n* transistor   (b) Circuit symbol

Fig. 7.33 An IC *n-p-n* transistor and its circuit symbol.

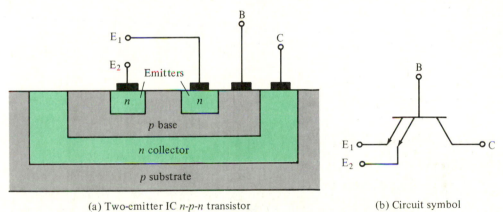

(a) Two-emitter IC *n-p-n* transistor   (b) Circuit symbol

Fig. 7.34 A two-emitter IC *n-p-n* transistor and its circuit symbol.

Transistor-Transistor Logic (TTL)

By using a multiple-emitter transistor to replace the input diodes, we can construct an IC NAND gate that uses only transistors (and resistors). Such a circuit is shown in Fig. 7.35 and is an example of **transistor-transistor logic,** abbreviated TTL (or T^2L). This type of logic is faster (smaller switching times) than DTL. Now let us see how the NAND gate given in Fig. 7.35 operates.

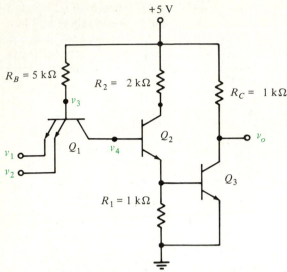

Fig. 7.35 A TTL NAND gate.

If either input is 0.2 V, then the corresponding emitter junction will be forward biased by the 5-V supply voltage. If we assume active region operation for Q_1, then $v_3 = 0.7 + 0.2 = 0.9$ V, while if we assume operation in the saturation region, then $v_3 = 0.8 + 0.2 = 1$ V. In either case, v_3 is not large enough to forward bias the base–collector junction of Q_1 and the base–emitter junctions of both Q_2 and Q_3 (a minimum of 1.5 V is required). Thus, all three junctions are reverse biased, and Q_2 and Q_3 are OFF. Even though Q_2 is OFF, there exists a small current I_{CBO_2} and

$$I_{CBO_2} = -i_{B_2} = i_{C_1}$$

Depending on whether Q_1 is in the active or saturation region,

$$i_{B_1} = \frac{5 - v_3}{R_B} = \frac{5 - 0.9}{5000} = 0.82 \text{ mA}$$

or

$$i_{B_1} = \frac{5 - v_3}{R_B} = \frac{5 - 1}{5000} = 0.8 \text{ mA}$$

respectively. However, if $h_{FE_1} \geq i_{C_1}/i_{B_1}$, then Q_1 will be in saturation. Since $i_{C_1} = I_{CBO_2}$ is very small compared with either value of i_{B_1}, then indeed Q_1 is ON as only a very small value of h_{FE_1} is required. Thus, for the case that, say, $v_1 = 0.2$ V, we have

$$v_4 = v_{CE_1} + v_1 = 0.2 + 0.2 = 0.4 \text{ V}$$

which confirms that Q_2 and Q_3 are OFF. Hence, $v_o = 5$ V.

Now consider the case that $v_1 = v_2 = 5$ V. Since the supply voltage and the input voltages are equal, the emitter junctions of Q_1 are reverse biased. However, the 5-V supply voltage is large enough to forward bias the collector junction of Q_1. But this is a transistor operation that we have not yet mentioned—emitter junction reverse biased and collector junction forward biased. Since this is like active-region operation with the emitter and collector reversing roles, it is known as **inverted-mode (active-region) operation.** Because a transistor is fabricated for (normal) active-region operation, when used in the inverted mode it has a small dc beta $h_{FE(im)}$—typically less than unity.

When Q_1 operates in the inverted mode, if the base current is i_{B_1}, then the total emitter current (corresponding to the inverted collector) is $h_{FE(im)}i_{B_1}$. Thus, the collector current is

$$i_{C_1} = -i_{B_1} - h_{FE(im)}i_{B_1} = -(1 + h_{FE(im)})i_{B_1}$$

and

$$i_{B_2} = -i_{C_1} = (1 + h_{FE(im)})i_{B_1} \qquad (7.20)$$

Since this current (due to the 5-V supply) tends to turn ON Q_2 and Q_3, let us assume that they are in saturation. Then

$$v_3 = v_{BC_1} + v_{BE_2} + v_{BE_3} = 0.7 + 0.8 + 0.8 = 2.3 \text{ V}$$

and

$$i_{B_1} = \frac{5 - 2.3}{5000} = 0.54 \text{ mA}$$

Even neglecting the inverted-mode beta ($h_{FE(im)} = 0$), we have from Equation (7.20) at a minimum that

$$i_{B_2} = i_{B_1} = 0.54 \text{ mA}$$

Since

$$i_{C_2} = \frac{5 - (0.2 + 0.8)}{2000} = 2 \text{ mA}$$

then for Q_2 to be ON, we require that

$$h_{FE_2} \geq \frac{i_{C_2}}{i_{B_2}} = \frac{2 \times 10^{-3}}{0.54 \times 10^{-3}} = 3.7$$

Furthermore, since

$$i_{B_3} = i_{B_2} + i_{C_2} - \frac{0.8}{1000} = (0.54 + 2 - 0.8) \times 10^{-3} = 1.74 \text{ mA}$$

and

$$i_{C_3} = \frac{5 - 0.2}{1000} = 4.8 \text{ mA}$$

then for Q_3 to be ON, we require that

$$h_{FE_3} \geq \frac{i_{C_3}}{i_{B_3}} = \frac{4.8 \times 10^{-3}}{1.74 \times 10^{-3}} = 2.76$$

Hence, when these conditions for h_{FE_2} and h_{FE_3} are satisfied, Q_2 and Q_3 will be ON, and $v_o = 0.2$ V.

This concludes the demonstration that the TTL circuit given in Fig. 7.35 is a NAND gate.

Increasing Switching Speed

Recall that when a transistor goes from saturation to cutoff, there is a significant storage of time t_s that is attributable to the storage of minority carriers in the base. For the TTL NAND gate given in Fig. 7.35, when $v_1 = v_2 = 5$ V, Q_1 is in the inverted-mode active region while Q_2 and Q_3 are ON. Under these conditions,

$$v_4 = v_{BE_2} + v_{BE_3} = 0.8 + 0.8 = 1.6 \text{ V}$$

When an input becomes 0.2 V, then the corresponding emitter junction of Q_1 is forward biased, and $v_3 = 0.7 + 0.2 = 0.9$ V. This means that the voltage across the collector junction of Q_1 (because Q_2 is still ON during the storage time) is

$$v_3 - v_4 = 0.9 - 1.6 = -0.7 < V_\gamma = 0.5 \text{ V}$$

so the collector junction is reverse biased. Thus, Q_1 is in the active region and its collector current is

$$i_{C_1} = h_{FE_1} i_{B_1} = h_{FE_1} \left(\frac{5 - 0.9}{5000} \right) = 8.2 \times 10^{-4} h_{FE_1} = -i_{B_2}$$

For the case that $h_{FE_1} = 50$, then $i_{B_2} = -41$ mA. This relatively large current quickly removes stored charge from Q_2 and Q_3 and, subsequently, Q_1 will turn ON while Q_2 and Q_3 will turn OFF. Now we can see why, for saturating logic, TTL has the highest speed.

For the TTL NAND gate given in Fig. 7.35, the output transistor Q_3 has a collector resistance of $R_C = 1$ kΩ. When Q_3 goes from ON to OFF, v_o goes from 0.2 to 5 V. However, the time required for this transition is affected in the following manner.

There is an effective capacitance C_L loading Q_3, which is, in essence, due to the junctions which are loading the gate and stray wiring capacitance. Although the capacitance C_L charges to 5 V with a time constant of $R_C C_L$, the resulting charge time may be longer than desired. This time constant can be reduced by decreasing R_C, but, as a consequence, there is an increase in collector current, and, hence, power dissipation—an undesirable effect. By modifying the given NAND gate as shown in Fig. 7.36, the charge time can be reduced.

For this circuit, which utilizes a **totem-pole** connection of transistors Q_3 and Q_4, when both input voltages are high, as discussed above, Q_1 is in the inverted

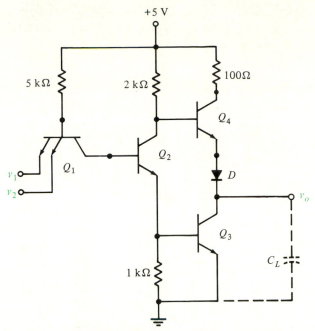

Fig. 7.36 A TTL NAND gate with totem-pole output stage.

mode while Q_2 and Q_3 are ON. Suppose that the capacitance C_L has already discharged to $v_o = 0.2$ V. Then the collector voltage of Q_2 is

$$v_{C_2} = v_{CE_2} + v_{BE_3} = 0.2 + 0.8 = 1 \text{ V} = v_{B_4}$$

and this is the voltage at the base of Q_4. For a moment, suppose that the diode D is replaced by a short circuit. Under this circumstance,

$$v_{E_4} = v_{CE_3} = 0.2 \text{ V}$$

and

$$v_{BE_4} = v_{B_4} - v_{E_4} = 1 - 0.2 = 0.8 \text{ V}$$

which implies that Q_4 is ON. The collector current that results is

$$i_{C_4} = \frac{5 - v_{C_4}}{100} = \frac{5 - (v_{CE_4} + v_{CE_3})}{100} = \frac{5 - (0.2 + 0.2)}{100} = 46 \text{ mA}$$

Since this current is excessive, and therefore wasteful, insert diode D as shown. The voltage across the emitter junction of Q_4 and D is

$$v_{B_4} - v_{C_3} = 1 - 0.2 = 0.8 \text{ V}$$

Because this voltage is less than the sum of their cut-in voltages, Q_4 and D are OFF.

When either input changes to 0.2 V, as discussed previously, Q_1 goes into saturation while Q_2 and Q_3 go OFF. However, because of the capacitance C_L, the output voltage does not instantaneously rise to the high voltage. While the output voltage is still $v_o = 0.2$ V, since Q_2 is OFF, then $v_{B_4} = v_{C_2} = 5$ V. This will tend to turn ON Q_4 and D. Assuming this, we have that

$$v_{B_4} = v_{BE_4} + v_D + v_o = 0.8 + 0.7 + 0.2 = 1.7 \text{ V}$$

Since Q_2 is OFF,

$$i_{B_4} = \frac{5 - v_{C_2}}{2000} = \frac{5 - v_{B_4}}{2000} = \frac{5 - 1.7}{2000} = 1.65 \text{ mA}$$

and since Q_3 is OFF,

$$i_{C_4} = \frac{5 - v_{C_4}}{100} = \frac{5 - (v_{CE_4} + v_D + v_o)}{100} = \frac{5 - (0.2 + 0.7 + 0.2)}{100}$$

$$= 39 \text{ mA}$$

Thus, for Q_4 indeed to be in saturation, we require that

$$h_{FE_4} \geq \frac{i_{C_4}}{i_{B_4}} = \frac{39 \times 10^{-3}}{1.65 \times 10^{-3}} = 23.6$$

and the fact that $i_{B_4} + i_{C_4} = 1.65 \times 10^{-3} + 39 \times 10^{-3} = 40.65$ mA > 0 confirms that D is ON. Since Q_3 is OFF ($i_{C_3} = 0$), the current $i_{B_4} + i_{C_4}$ rapidly charges the capacitance C_L. The time constant is RC_L, where R is the sum of the 100-Ω collector resistance of Q_4, the saturation resistance of Q_4 (i.e., the resistance between the collector and emitter of Q_4 when it is ON—typically tens of ohms or less), and the forward resistance of the diode. As v_o increases,

$$i_{C_4} = \frac{5 - (v_{CE_4} + v_D + v_o)}{100}$$

decreases. When v_o becomes sufficiently large, then $v_{B_4} - v_o = 5 - v_o$ will not be large enough to forward bias the emitter junction of Q_4 and D. This occurs when

$$5 - v_o = v_{BE_4(\text{cut-in})} + v_{D(\text{cut-in})} = 0.5 + 0.6 = 1.1 \text{ V}$$

from which

$$v_o = 5 - 1.1 = 3.9 \text{ V}$$

and this is the high voltage for the NAND gate.

Now suppose that either input is the low voltage 0.2 V and the output voltage is $v_o = 3.9$ V (Q_3 is OFF). If the inputs become $v_1 = v_2 = 3.9$ V, since the 5-V supply voltage is large enough to forward bias the collector junction of Q_1 and turn ON Q_2 and Q_3, then the voltage at the base of Q_1 is

$$v_{B_1} = v_{BC_1} + v_{BE_2} + v_{BE_3} = 0.7 + 0.8 + 0.8 = 2.3 \text{ V}$$

Since $v_{B_1} - 3.9 = 2.3 - 3.9 = -1.6 < 0.5$ V, the emitter junctions are reverse biased and Q_1 is in the inverted mode. As discussed above, under these conditions, D and Q_4 will turn OFF. In the process of turning ON, Q_3 has a large collector current, which results in C_L discharging quickly.

The transistor Q_4 in the TTL NAND gate given in Fig. 7.36 is known as an **active pull-up.** The 1-kΩ resistor in series with the collector of Q_3 in the NAND gate given in Fig. 7.35 is called a **passive pull-up.** The term "pull-up" is used because the capacitance C_L is charged up, or its voltage is pulled up, by current going through these elements.

For the NAND gate given in Fig. 7.36, the diode D is in series with the emitter of Q_4. If this diode instead is placed in series with the base of Q_4, the operation of the TTL gate will remain unchanged. But why are such gates referred to as TTL if they contain diodes? They are because such diodes can be realized by a base–emitter junction of some other transistor, for example.

Schottky TTL

When a metal is connected to a semiconductor, the resulting contact can be either **ohmic (nonrectifying)** or **rectifying.** In previous discussions, the implicit assumption was that a metal–semiconductor junction was ohmic. However, the formation of such a contact requires an appropriate fabrication process. As an example, for attaching aluminum to n-type silicon, the region of the silicon adjacent to the aluminum is made highly conductive by doping it heavily—denoted as n^+. This results in an ohmic contact.

Aluminum in contact with n-type silicon that is not heavily doped acts as a p-type impurity. As a consequence of this fact, such a metal–semiconductor contact is rectifying and behaves in a manner similar to that of an ordinary p-n junction. This type of junction, however, is known as a **Schottky*** (or **Schottky-barrier**) **diode.** The circuit symbol for this device is shown in Fig. 7.37. The i-v characteristic for a Schottky diode is like that of an ordinary p-n junction diode. A major distinction, however, is that the cut-in voltage for a Schottky diode is about 0.3 V as compared with 0.6 V for a silicon junction diode. Furthermore, a typical forward-bias voltage for the former is about 0.4 V, while it is about 0.7 V for the latter.

Fig. 7.37 Schottky diode.

For an aluminum n-type silicon Schottky diode, a current in the forward direction is essentially due to the flow of electrons from the n-type silicon to the metal. In essence, therefore, a Schottky diode is a majority-carrier device and there is no storage of minority carriers as takes place with an ordinary junction diode. Consequently, the storage time of a Schottky diode is negligible; hence, Schottky diodes have higher switching speeds than do junction diodes.

---

*Named for the Swiss physicist Walter Schottky.

Consider a BJT where a Schottky diode is connected between the base and the collector as shown in Fig. 7.38. Suppose that an attempt is made to saturate the transistor. As the base current increases, the base–emitter voltage increases and the collector–emitter voltage decreases. This, in turn, will forward bias the Schottky diode and the base–collector voltage will be "clamped" at about $v_{BC} = 0.4$ V. Assuming that the BJT is in saturation, we get

$$v_{CE} = -v_{BC} + v_{BE} = -0.4 + 0.8 = 0.4 \text{ V}$$

But $v_{CE} > 0.2$ V is a contradiction to our assumption, and, therefore, the transistor is not in saturation. Hence, the Schottky diode clamp prevents the BJT from going into saturation.

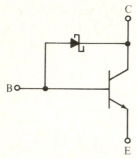

Fig. 7.38 A BJT with a Schottky diode clamp.

A BJT, along with a Schottky diode clamp, can be fabricated as shown in Fig. 7.39(a). Such a device is known as a **Schottky** (or **Schottky-clamped**) **transistor,** and its circuit symbol is given in Fig. 7.39(b).

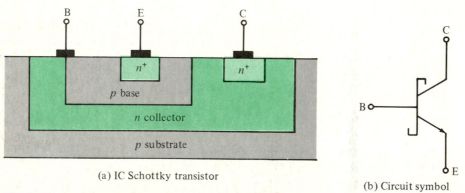

(a) IC Schottky transistor

(b) Circuit symbol

Fig. 7.39 An IC Schottky transistor and its circuit symbol.

Since a Schottky transistor cannot go into saturation, its storage time is eliminated, and, therefore, the turn-off time is reduced. A consequence of this is that Schottky TTL has a higher switching speed than does ordinary TTL. A Schottky

TTL version of the TTL NAND gate given in Fig. 7.36 is shown in Fig. 7.40. As mentioned previously, the diode D in Fig. 7.36 can be placed in series with the base of Q_4. For the NAND gate in Fig. 7.40, the diode is replaced by the Schottky transistor Q_5. Furthermore, since $v_{CE_4} = v_{BE_4} + v_{CE_6}$, transistor Q_4 in Fig. 7.40 cannot saturate, so an ordinary BJT can be used.

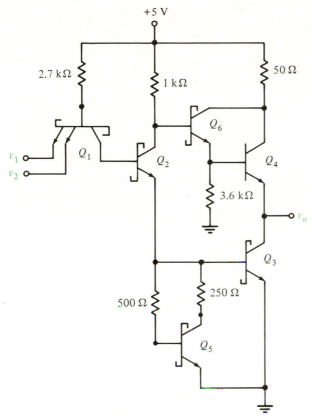

Fig. 7.40 Schottky TTL NAND gate.

Although the speed of the Schottky TTL NAND gate shown in Fig. 7.40 is about three times as fast as the TTL NAND gate given in Fig. 7.36, the former dissipates about twice as much power as does the latter. Where power dissipation has priority over speed, **low-power Schottky TTL** can be employed. For such a NAND gate, shown in Fig. 7.41, the resistance values are significantly larger. Although the speed of such a gate is about the same as for standard TTL, the power dissipation for the gate in Fig. 7.41 is approximately 20 percent of that for the gate in Fig. 7.36.

Despite the fact that Schottky TTL is fast logic, another type of nonsaturating logic (ECL) is even faster. This topic will be discussed in Section 7.6.

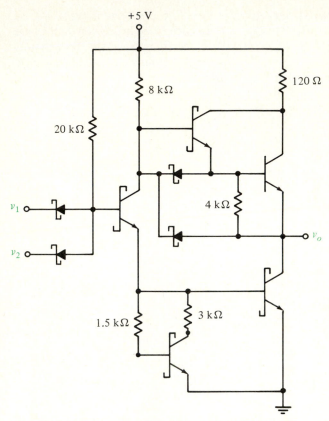

Fig. 7.41 Low-power Schottky TTL NAND gate.

RTL and DCTL

Recall the logic table for a two-input OR gate (Table 6.1), which is repeated as Table 7.3(a). If we invert the output of an OR gate, the resulting combination of an OR gate followed by a NOT gate, described by Table 7.3(b), is called a **NOR gate**—NOR being a contraction of NOT-OR. Such a digital logic circuit can be realized by the **resistor-transistor logic** (RTL) gate shown in Fig. 7.42.

Table 7.3(a) OR-Gate Table

v_1	v_2	v_o
0	0	0
0	1	1
1	0	1
1	1	1

Table 7.3(b) NOR-Gate Table

v_1	v_2	v_o
0	0	1
0	1	0
1	0	0
1	1	0

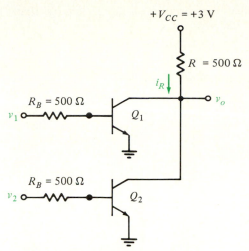

Fig. 7.42 An RTL NOR gate.

For the case that $v_1 = v_2 = 0.2$ V, both Q_1 and Q_2 must be OFF. Thus, $i_{C_1} = i_{C_2} = 0$ and $i_R = i_{C_1} + i_{C_2} = 0$, so

$$v_o = -Ri_R + 3 = 3 \text{ V}$$

If $v_1 = 0.2$ V and $v_2 = 3$ V, then Q_1 is OFF. Assume that Q_2 is ON. Then

$$i_{B_2} = \frac{v_2 - v_{BE_2}}{R_B} = \frac{3 - 0.8}{500} = 4.4 \text{ mA}$$

and

$$i_{C_2} = i_R = \frac{3 - 0.2}{500} = 5.6 \text{ mA}$$

For Q_2 to be in saturation, therefore, we must have that

$$h_{FE_2} \geq \frac{i_{C_2}}{i_{B_2}} = \frac{5.6 \times 10^{-3}}{4.4 \times 10^{-3}} = 1.27$$

Similarly, when $v_1 = 3$ V and $v_2 = 0.2$ V, for Q_1 to be ON and Q_2 to be OFF, we require that $h_{FE_1} \geq 1.27$.

Finally, for the case that $v_1 = v_2 = 3$ V, assume that Q_1 and Q_2 are ON. Under this circumstance, as above, $i_{B_1} = i_{B_2} = 4.4$ mA. Furthermore, if Q_1 and Q_2 are identical, then, as above, $i_R = 5.6$ mA, and $i_{C_1} = i_{C_2} = i_R/2 = 2.8$ mA. Thus, Q_1 and Q_2 are in saturation when

$$h_{FE_1} = h_{FE_2} \geq \frac{i_{C_1}}{i_{B_1}} = \frac{2.8 \times 10^{-3}}{4.4 \times 10^{-3}} = 0.64$$

Even if Q_1 and Q_2 are not identical, the maximum collector current is $i_R = 5.6$ mA.

Thus, to guarantee that Q_1 and Q_2 are ON, we should have for Q_1 and Q_2 that

$$h_{FE} \geq \frac{5.6 \times 10^{-3}}{4.4 \times 10^{-3}} = 1.27$$

Hence, we see that the RTL circuit given in Fig. 7.42 satisfies the logic table for a (positive-logic) NOR gate.

Now consider the situation in which the RTL NOR gate given in Fig. 7.42 is loaded by ten identical gates. When either v_1 or v_2 is high, $v_o = 0.2$ V and the ten load transistors Q_1' through Q_{10}' will be OFF. However, when $v_1 = v_2 = 0.2$ V, both Q_1 and Q_2 will be OFF and v_o will be high. Consequently, Q_1' through Q_{10}' are to be ON. Thus, by KCL,

$$i_R = i_{C_1} + i_{C_2} + i_{B_1}' + i_{B_2}' + \cdots + i_{B_{10}}'$$

$$\frac{3 - v_o}{500} = 0 + 0 + \frac{v_o - 0.8}{500} + \frac{v_o - 0.8}{500} + \cdots + \frac{v_o - 0.8}{500}$$

$$= 10\left(\frac{v_o - 0.8}{500}\right)$$

from which

$$3 - v_o = 10v_o - 8 \quad \Rightarrow \quad v_o = 1 \text{ V}$$

Thus, the high voltage is 1 V. Furthermore, for each transistor in saturation

$$i_B = \frac{v_o - v_{BE}}{R_B} = \frac{1 - 0.8}{500} = 0.4 \text{ mA}$$

and

$$i_C = \frac{3 - 0.2}{500} = 5.6 \text{ mA}$$

Hence, we require for each transistor that

$$h_{FE} \geq \frac{i_C}{i_B} = \frac{5.6 \times 10^{-3}}{0.4 \times 10^{-3}} = 14$$

Since TTL is superior to RTL in switching speed and fan-out, the former is the overwhelming choice in practice and the latter sees little application. However, RTL can be modified to make it increase fan-out. This is done by making the base resistance $R_B = 0$. What results is known as **direct-coupled transistor logic** (DCTL). An example of this type of logic is illustrated by the DCTL NOR gate and its fan-out shown in Fig. 7.43.

For the case that $v_1 = v_2 = 0.2$ V (the low voltage), Q_1 and Q_2 are OFF. Under this circumstance, the output voltage v_o will be the high voltage. This high voltage, then, will turn ON the fan-out transistors Q_1' through Q_N'. But, because of the direct coupling, we have that the high voltage is $v_o = 0.8$ V. Thus, we see that the high voltage is independent of V_{CC} (when $V_{CC} > 0.8$ V).

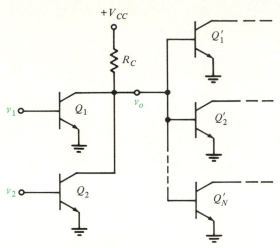

Fig. 7.43 A DCTL NOR gate and its fan-out.

For the case that an input is high, say $v_1 = 0.8$ V, Q_1 is ON and $v_o = 0.2$ V. Under this condition, the fan-out transistors will be OFF. We have now confirmed NOR-gate operation.

The advantages of DCTL are that low-voltage supplies (as low as $V_{CC} = 1.5$ V) and transistors with low breakdown voltages can be used, and the power dissipation is low. Unfortunately, DCTL has numerous disadvantages. To begin with, since $R_B = 0$, there are large base currents, and hence, large amounts of stored charge in the bases of the transistors—this means a reduction in switching speed. Since the difference between the high and low voltages of DCTL is only 0.6 V, the noise margins are low, and such logic is susceptible to noise. Another problem can arise when the fan-out transistors do not have identical characteristics. If one fan-out transistor has a lower value of base–emitter saturation voltage $v_{BE(\text{sat})}$ than another, the former will draw more base current than the latter, and there may be insufficient current for the latter to go into saturation. In such a situation, we say that the former transistor **hogs** the current. Needless to say, current hogging can result in improper gate operation. Although there are other disadvantages, these are usually sufficient to preclude the use of DCTL that employs BJTs. However, as we see in the next chapter, this type of logic sees important applications with field-effect transistors (FETs). Even before then, we will see how BJT DCTL preceded the emergence of the very important integrated-injection logic (I^2L).

7.6 ADDITIONAL IMPORTANT IC LOGIC

Consider the symmetrical BJT circuit shown in Fig. 7.44, where the voltage applied to the base of Q_2 is some reference voltage V_R. If v_1 has a value in the interval $-V_{EE} + 0.7 < v_1 < V_{CC} + 0.5$, then v_1 will tend to forward bias the emitter junction and reverse bias the collector junction of Q_1, that is, put Q_1 in the active region.

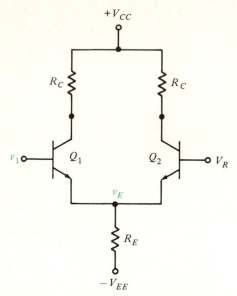

Fig. 7.44 A symmetrical BJT circuit.

Suppose that v_1 is such that Q_1 is in the active region and v_1 exceeds V_R by 0.2 V, that is, $v_1 > V_R + 0.2$. From this inequality,

$$V_R - v_1 < -0.2$$

Since

$$v_{BE_2} = V_R - v_E = V_R - (-0.7 + v_1) = 0.7 + V_R - v_1$$

from the above inequality,

$$v_{BE_2} = 0.7 + V_R - v_1 < 0.7 - 0.2 = 0.5 \text{ V}$$

and this means that Q_2 is cut off. Conversely, if V_R has a value in the range $-V_{EE} + 0.7 < V_R < V_{CC} + 0.5$, then V_R will tend to put Q_2 in the active region. In addition, if V_R exceeds v_1 by 0.2 V, then, using reasoning as above, when Q_2 is in the active region, Q_1 will be cut off.

Emitter-Coupled Logic (ECL)

We now describe another type of nonsaturating logic, called **emitter-coupled logic** (ECL) or **current-mode logic** (CML), which is based on the circuit shown in Fig. 7.44. With ECL, as described above, when one transistor is cut off, the other transistor is not in saturation but is in the active region. Because of this kind of operation, as for the case of Schottky TTL, the problem of stored charge is eliminated. The consequence of this is that ECL is the fastest type of logic.

Let us use the basic circuit given in Fig. 7.44 to try to construct a NOR gate. We will do this by adding another transistor, as shown in Fig. 7.45. For the case

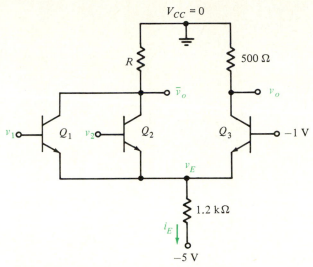

Fig. 7.45 Possible ECL NOR gate.

that either input voltage is high, the corresponding transistor should be in the active region and Q_3 will be OFF. Then, $v_o = -500i_{C_3} + V_{CC} = 0 + 0 = 0$ should be the high voltage. Conversely, if v_1 and v_2 are low, then Q_3 should be in the active region while Q_1 and Q_2 will be OFF. Then the output $v_o = -500i_{C_3}$ is the low voltage. Such a behavior would result in OR-gate operation. Note that the voltage at the collectors of Q_1 and Q_2 is the complement of v_o—that is why it is denoted $\bar{v}_o$. Therefore, if v_o is the output of an OR gate, then $\bar{v}_o$ would be the output of a NOR gate. Let us now determine whether or not the circuit given in Fig. 7.45 is indeed a NOR/OR gate.

If both input voltages are low, assume that Q_1 and Q_2 are OFF and Q_3 is in the active region. Then

$$v_E = -v_{BE_3} - 1 = -1.7 \text{ V}$$

and

$$i_E = \frac{v_E - (-5)}{1200} = \frac{-1.7 + 5}{1200} = 2.75 \text{ mA}$$

Since

$$i_E = i_{C_3} + i_{B_3} = h_{FE_3}i_{B_3} + i_{B_3} \approx h_{FE_3}i_{B_3} = i_{C_3}$$

then

$$v_o = -500i_{C_3} \approx -500(2.75 \times 10^{-3}) = -1.375 \text{ V}$$

and this is the low voltage. Since

$$v_{BE_1} = v_{BE_2} = v_1 - v_E = -1.375 - (-1.7) = 0.325 < 0.5 \text{ V}$$

then Q_1 and Q_2 are indeed cut off. Furthermore,

$$v_{BC_3} = -1 - (-1.375) = 0.375 < 0.5 \text{ V}$$

so that the collector junction of Q_3 is reverse biased, and Q_3 indeed is in the active region.

Now consider the case that at least one input voltage, say v_1, is high. Assume that Q_1 is in the active region and Q_3 is cut off. Under these circumstances, $v_o = -500i_{C_3} + 0 = 0$ is the high voltage. Also,

$$v_E = -v_{BE_1} + v_1 = -0.7 + 0 = -0.7 \text{ V}$$

Since

$$v_{BE_3} = -1 - v_E = -1 + 0.7 = -0.3 < 0.5 \text{ V}$$

then Q_3 is indeed cut off. Furthermore,

$$i_E = \frac{v_E - (-5)}{1200} = \frac{-0.7 + 5}{1200} = 3.58 \text{ mA}$$

Since $i_{C_1} + i_{C_2} \gg i_{B_1} + i_{B_2}$, then $i_{C_1} + i_{C_2} \approx i_E$. Because we wish to have $\overline{v}_o = -1.375 \text{ V}$ (the low voltage), then

$$\overline{v}_o = -R(i_{C_1} + i_{C_2}) \approx -Ri_E$$

from which

$$R \approx -\frac{\overline{v}_o}{i_E} = \frac{1.375}{3.58 \times 10^{-3}} = 384 \text{ } \Omega$$

But note that

$$v_{BC_1} = v_1 - \overline{v}_o = 0 - (-1.375) = 1.375 > 0.5 \text{ V}$$

This means that the collector junction of Q_1 is not reverse biased, and, therefore, contrary to our assumption, Q_1 is not in the active region. Because of this kind of problem, we will change the value of the collector resistance for Q_3 and modify the given circuit by adding two transistors as shown in Fig. 7.46.

For this circuit, if both input voltages are low, again assume that Q_1 and Q_2 are OFF and Q_3 is in the active region. As was the case for the preceding circuit, $v_E = -1.7 \text{ V}$ and $i_E = 2.75 \text{ mA}$. Also, $i_{C_3} \approx i_E$. Ignoring i_{B_5}, we have

$$v_{C_3} = -300i_{C_3} \approx -300i_E = -300(2.75 \times 10^{-3}) = -0.825 \text{ V}$$

This voltage is large enough (relative to -5 V) to forward bias the emitter junction of Q_5. Since

$$v_{BC_5} = v_{C_3} = -0.825 < 0.5 \text{ V}$$

then the collector junction of Q_5 is reverse biased. Thus, Q_5 is in the active region, and, therefore,

$$v_o = -v_{BE_5} + v_{C_3} = -0.7 - 0.825 = -1.525 \text{ V}$$

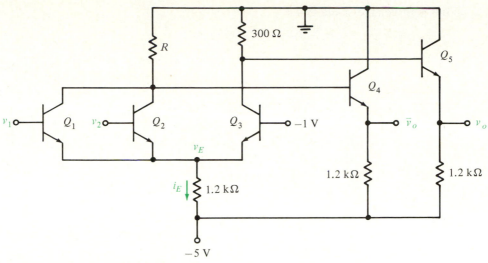

Fig. 7.46 An ECL NOR/OR gate.

is the low voltage. For active-region operation, typically $i_{C_5} \gg i_{B_5}$, and, therefore,

$$i_{C_5} \approx i_{B_5} + i_{C_5} = \frac{v_o - (-5)}{1200} = \frac{-1.525 + 5}{1200} = 2.9 \text{ mA}$$

Since $i_{C_5} = 2.9$ mA $\gg i_{B_5}$, then also $i_E = 2.75$ mA $\gg i_{B_5}$, and ignoring i_{B_5} in the calculation of v_{C_3} above is a reasonable thing to do. Since

$$v_{BE_1} = v_{BE_2} = v_1 - v_E = -1.525 - (-1.7) = 0.175 < 0.5 \text{ V}$$

then Q_1 and Q_2 are indeed OFF. Also, since

$$v_{BC_3} = -1 - v_{C_3} = -1 - (-0.825) = -0.175 < 0.5 \text{ V}$$

then the collector junction of Q_3 is reverse biased, and Q_3 indeed is in the active region.

Now consider the case that at least one input voltage, say v_1, is high. Assume that Q_1 is in the active region and Q_3 is cut off. Ignoring i_{B_5}, we have that

$$v_{C_3} = -300 i_{C_3} = 0$$

This voltage is large enough (relative to -5 V) to forward bias the emitter junction of Q_5. Since

$$v_{BC_5} = v_{C_3} = 0 < 0.5 \text{ V}$$

then the collector junction of Q_5 is reverse biased. Thus, Q_5 is in the active region, and, therefore,

$$v_o = -v_{BE_5} + v_{C_3} = -0.7 + 0 = -0.7 \text{ V}$$

is the high voltage. Note that

$$i_{C_5} \approx i_{B_5} + i_{C_5} = \frac{v_o - (-5)}{1200} = \frac{-0.7 + 5}{1200} = 3.58 \text{ mA} \gg i_{B_5}$$

For a typical value such as $h_{FE_5} = 50$,

$$i_{B_5} = \frac{i_{C_5}}{h_{FE_5}} = \frac{3.58 \times 10^{-3}}{50} = 71.6 \text{ }\mu\text{A}$$

So even if we did not ignore i_{B_5} in calculating v_{C_3}, we would obtain $v_{C_3} = -300(i_{C_3} + i_{B_5}) = -300(71.6 \times 10^{-6}) = -0.02 \approx 0$. Thus, ignoring i_{B_5} is justified. We also have that

$$v_E = -v_{BE_1} + v_1 = -0.7 - 0.7 = -1.4 \text{ V}$$

Since

$$v_{BE_3} = -1 - v_E = -1 + 1.4 = 0.4 < 0.5 \text{ V}$$

then Q_3 is indeed cut off. Furthermore,

$$i_E = \frac{v_E - (-5)}{1200} = \frac{-1.4 + 5}{1200} = 3 \text{ mA}$$

Previously, we determined that when Q_3 was in the active region, we had $v_{C_3} = -0.825$ V. Therefore, now that Q_1 is in the active region, we wish to have $v_{C_1} = -0.825$ V. Thus, ignoring base currents,

$$v_{C_1} = -Ri_E$$

from which

$$R = -\frac{v_{C_1}}{i_E} = \frac{0.825}{3 \times 10^{-3}} = 275 \text{ }\Omega$$

Furthermore, since

$$v_{BC_1} = v_1 - v_{C_1} = -0.7 - (-0.825) = 0.125 < 0.5 \text{ V}$$

the collector junction of Q_1 is reverse biased, and Q_1 is indeed in the active region. Note too that Q_4, like Q_5, is always in the active region.

This concludes our demonstration that, for the ECL gate given in Fig. 7.46, v_o is the output of an OR gate, while $\bar{v}_o$ is the output of a NOR gate.

Integrated-Injection Logic (I^2L)

As mentioned previously, a major disadvantage of DCTL is the problem of current hogging. Fortunately, this difficulty was eliminated with the development of another type of logic known as **integrated-injection logic** (I^2L or IIL). To see how I^2L can evolve from DCTL, let us first consider the DCTL NOR gate and its two-transistor load shown in Fig. 7.47 (compare with Fig. 7.43). This circuit is to be fabricated on a single IC chip.

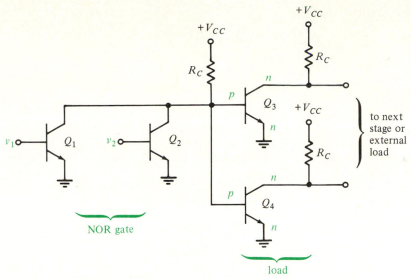

Fig. 7.47 A DCTL NOR gate and its load.

For the given circuit, all four BJTs are *n-p-n* transistors. But, since the bases of Q_3 and Q_4 are connected together, they can be merged by using the same *p*-type region for both transistors. Similarly, both emitters of Q_3 and Q_4 can be merged into a single *n*-type region. In essence, what results is a two-collector transistor. (Such a device can be fabricated in a manner similar to the two-emitter transistor illustrated in Fig. 7.34.) Of course, *m*-collector (*m* greater than two) transistors can also be fabricated. Because of the space-saving construction of such devices, a typical value for h_{FE} (total collector current divided by the base current) is 5, whereas 50 is typical for a regular BJT.

We have now seen how we can reduce the space requirements of the circuit given in Fig. 7.47 by merging regions for the load transistors. Because IC resistors require much more space than do IC transistors, let us now concentrate on the collector resistor R_C. Transistor Q_1 or Q_2 will saturate when its respective input is high, and, in either case, Q_3 and Q_4 will be cut off. However, when v_1 and v_2 are low, Q_1 and Q_2 will be OFF. Consequently, Q_3 and Q_4 will be ON. The current required to turn ON Q_3 and Q_4 goes through R_C. Thus, we can eliminate R_C if we replace it with some other supplier of current. For this reason, let us replace R_C by the *p-n-p* transistor shown in Fig. 7.48. Such a transistor is called a **current injector.** Only the *p-n-p* transistor shown in Fig. 7.48 is to replace the resistor R_C. The resistor R and the supply voltage V_{CC} are external to the chip. In other words, even though there originally may be many resistors R_C, each one of them is replaced by a *p-n-p* transistor, and all such transistors use the same external resistor R and supply voltage V_{CC}. This supply voltage will forward bias the emitter junction of the *p-n-p* transistor, producing an emitter current of

$$i_E = \frac{V_{CC} - v_{EB}}{R}$$

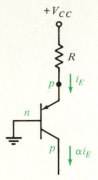

Fig. 7.48 Current injector.

Therefore, when connected to a load, the resulting current coming out of the collector is approximately αi_E.

Let us now redraw the circuit given in Fig. 7.47, replacing Q_3 and Q_4 with a multiple (two)-collector transistor and the resistors labeled R_C with p-n-p current injectors. The resulting circuit is shown in Fig. 7.49. But note that the collector of the leftmost p-n-p transistor and the base of the multiple-collector transistor are both p-type regions, and, hence, can be merged. In addition, the base of the p-n-p transistor and the emitter of the multiple-collector transistor can be merged into the same n-type region. Consequently, even more space can be saved. It is because of the merging of regions that I^2L is sometimes referred to as **merged-transistor logic** (MTL). Because of its small space requirements, I^2L has application in LSI and VLSI circuits such as calculators, digital wristwatches, and microprocessors.

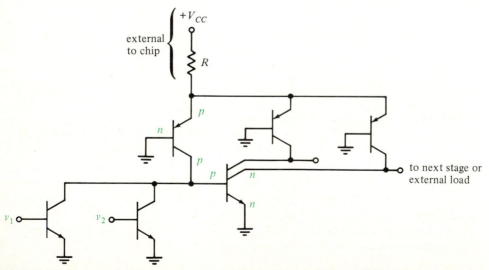

Fig. 7.49 Circuit obtained after substitution of multiple-collector transistor and current injectors.

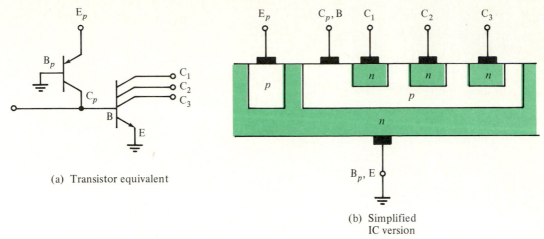

(a) Transistor equivalent

(b) Simplified IC version

Fig. 7.50 Basic I²L unit.

The basic I²L unit (building block) consists of a multiple-collector *n-p-n* transistor to whose base is connected a *p-n-p* current-injector transistor as shown in Fig. 7.50(a) for the case of three collectors. A simplified IC version of this building block is shown in Fig. 7.50(b). For the sake of simplicity, however, as shown in Fig. 7.51, the circuit symbol of this basic I²L gate uses an independent current source *I* to represent the current injector.

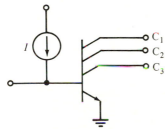

Fig. 7.51 Circuit symbol for the basic I²L gate.

To form an I²L inverter, we require only single-collector *n-p-n* transistors. Specifically, Fig. 7.52 shows an I²L inverter and its load. When the input voltage is $v_1 = 0.2$ V, then Q_1 is OFF—and the leftmost injected current I goes through the source of v_1 (a transistor in saturation). Since Q_1 is OFF, the injected current I for Q_2 turns ON that transistor. Thus, $v_o = v_{BE_2} = 0.8$ V. Conversely, when the source of v_1 is a transistor in cutoff, then assume that the injected current $I = i_{B_1}$ turns ON Q_1 and the input voltage is $v_1 = v_{BE_1} = 0.8$ V. When Q_1 is in saturation, $v_o = 0.2$ V. This implies that Q_2 is OFF, so $i_{C_1} = I$. Therefore, for Q_1 to be in saturation, we require that

$$h_{FE_1} \geq \frac{i_{C_1}}{i_{B_1}} = \frac{I}{I} = 1$$

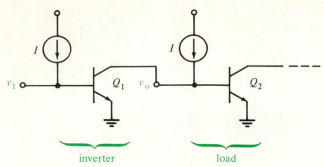

Fig. 7.52 An I²L inverter (NOT gate).

When the output voltage of an inverter changes from low voltage to high, the input capacitance of the following stage must charge up correspondingly. Conversely, this capacitance must discharge when the inverter changes from high voltage to low. This charging and discharging, which determine the switching times (logic speed), depend on the injection current I. Increasing I results in faster logic, but at a price—more power is required.

Example 7.8

Consider the I²L circuit shown in Fig. 7.53. For the case that $v_1 = v_2 = 0.2$ V, then Q_1 and Q_2 are OFF. Since $i_{C_1} = i_{C_2} = 0$, then $i_{B_3} = I$ and Q_3 is ON. Thus, $v_o = v_{BE_3} = 0.8$ V.

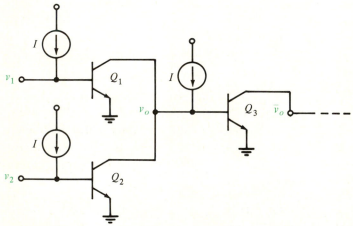

Fig. 7.53 An I²L NOR gate.

When $v_1 = 0.2$ V and $v_2 = 0.8$ V, then Q_1 is OFF and Q_2 is ON. Thus, $v_o = 0.2$ V (and Q_3 is OFF). Due to the symmetry of the circuit, when

$v_1 = 0.8$ V and $v_2 = 0.2$ V, then Q_1 is ON and Q_2 is OFF. Again, $v_o = 0.2$ V (and Q_3 is OFF).

Finally, when $v_1 = v_2 = 0.8$ V, then Q_1 and Q_2 are ON, and $v_o = 0.2$ V (and Q_3 is OFF).

From these results, we now see that the I^2L circuit given in Fig. 7.53 is a NOR gate when v_o is taken as the output. Since the following stage is an inverter, if $\bar{v}_o$ is taken as the output, an OR gate is obtained.

■

SUMMARY

1. Sandwiching a narrow section of n-type semiconductor between two sections of p-type semiconductor yields a p-n-p bipolar junction transistor (BJT). If the n- and p-type materials are interchanged, an n-p-n BJT results.

2. The terminals of a BJT are the base (B), collector (C), and emitter (E).

3. The common-base (CB) large-signal current gain α of a BJT typically is in the range from 0.95 to 0.998. The common-emitter (CE) large-signal current gain β of a BJT often lies between 50 and 200.

4. A BJT is in the active region when its emitter junction is forward biased and its collector junction is reverse biased. In this case, $i_C \approx -\alpha i_E$ and $i_C \approx \beta i_B$.

5. A BJT is cut off (OFF) when its emitter and collector junctions are reverse biased. In this case, $i_B = i_C = i_E \approx 0$.

6. A BJT is in saturation (ON) when its emitter and collector junctions are forward biased.

7. A BJT is in the inverted mode when its emitter junction is reverse biased and its collector junction is forward biased.

8. The CB collector characteristics of a BJT are plots of collector current versus collector–base voltage for different values of emitter current. The CE collector characteristics of a BJT are plots of collector current versus collector–emitter voltage for different values of base current.

9. For a silicon BJT in saturation, $v_{CE} \approx 0.2$ V. A silicon BJT is cut off when $v_{BE} < 0.5$ V and $v_{BC} < 0.5$ V.

10. The reverse collector saturation current I_{CBO} of a BJT approximately doubles for every 10°C rise in temperature.

11. The dc current-gain h_{FE} of a BJT is the ratio of collector current to base current ($h_{FE} \approx \beta$). For a BJT to be in saturation, a necessary condition is that $h_{FE} i_B \geq i_C$.

12. The sum of the delay time t_d and the rise time t_r of a BJT is the turn-on time t_{ON}. The sum of the storage time t_s and the fall time t_f is the turn-off time t_{OFF}. These times determine the switching speed of the BJT.

13. Transistors and diodes can be used to design such digital logic circuits as NOT, NAND, NOR, AND, and OR gates.

14. Noise margin is the input noise voltage that causes an improper output voltage. Fan-out is the maximum number of similar gates that can load a given gate. Fan-in is the number of inputs to gate.

15. Diode-transistor logic (DTL) gates employ both diodes and transistors, while transistor-transistor logic (TTL) utilizes transistors in place of diodes.

16. Although resistor-transistor logic (RTL) and BJT direct-coupled transistor logic (DCTL) are little used, their study is useful for understanding other types of logic (e.g., I^2L).

17. Schottky TTL and emitter-coupled logic (ECL) do not require transistors to saturate. ECL is the fastest type of logic.

18. Integrated-injection logic (I^2L) saves considerable space by employing current-injecting transistors in place of resistors. Because of this, I^2L is suitable for large-scale integration (LSI) and very-large-scale integration (VLSI).

Problems for Section 7.1

7.1 For the circuit shown in Fig. 7.4, $R_C = R_E = 1$ kΩ and $V_{CC} = V_{EE} = 5$ V. Assume that the transistor is in the active region and that $\alpha \approx 1$. Find (a) i_E and (b) v_{CB}.

7.2 For the circuit given in Fig. 7.4, $V_{CC} = V_{EE} = 5$ V. Assume that the transistor is in the active region and that $\alpha \approx 1$.

(a) Find R_E such that $i_E = 5$ mA.

(b) Given that $i_E = 5$ mA, find R_C such that $v_{CB} = -1$ V.

(c) Given that $i_E = 5$ mA, find the maximum value of R_C such that the transistor is in the active region.

7.3 For the circuit given in Fig. 7.4, $R_C = R_E = 1$ kΩ. Assume that the transistor is in the active region and that $\alpha \approx 1$.

(a) Find V_{EE} such that $i_E = 4$ mA.

(b) Given that $i_E = 4$ mA, find V_{CC} such that $v_{CB} = -2$ V.

(c) Given that $i_E = 4$ mA, find the minimum value of V_{CC} such that the transistor is in the active region.

7.4 For the circuit given in Fig. 7.4, assume that the transistor is in the active region and the small-signal current gain is $\alpha_{ac} \approx -1$. Find the ac voltage gain $A = \Delta v_o / \Delta v_{EB}$ when (a) $R_C = R_E = 1$ kΩ, $V_{CC} = V_{EE} = 5$ V; (b) $R_C = 800$ Ω, $R_E = 880$ Ω, $V_{CC} = V_{EE} = 5$ V; and (c) $R_C = R_E = 1$ kΩ, $V_{CC} = 6$ V, $V_{EE} = 4.6$ V.

7.5 For the circuit given in Fig. 7.4, $R_C = 1.8$ kΩ, $R_E = 180$ Ω, and $V_{CC} = 12$ V. Assume that the transistor is in the active region and that $\alpha \approx 1$.
(a) Find v_o when $V_{EE} = 1.41$ V.
(b) Find v_o when $V_{EE} = 1.5$ V.
(c) Use the results of (a) and (b) to determine the ac voltage gain $\Delta v_o / \Delta V_{EE}$.

7.6 The transistor in the circuit shown in Fig. 7.4 has the collector characteristics given in Fig. 7.6. Assuming $R_C = 75$ Ω and $V_{CC} = 1.2$ V, graphically determine v_{CB} when i_E is (a) 6 mA, (b) 9 mA, (c) 12 mA, and (d) 16 mA.

7.7 The transistor in the circuit shown in Fig. 7.4 has the emitter characteristics given in Fig. 7.7. Given that $R_C = 250$ Ω, $R_E = 500$ Ω, $V_{CC} = 8$ V, and $i_E = 8$ mA, (a) graphically determine v_{EB} and (b) find V_{EE}.

Problems for Section 7.2

7.8 For the circuit given in Fig. 7.13, replace the 2.7-kΩ resistor by a 4.7-kΩ resistor and determine whether or not the transistor is in the active region.

7.9 The circuit shown in Fig. 7.9 has $V_{BB} = V_{CC} = 10$ V. Assuming the transistor has $\beta = 100$ and negligible I_{CBO}, (a) find the maximum value of R_C for active-region operation when $R_B = 330$ kΩ, and (b) find the minimum value of R_B for active-region operation when $R_C = 2.7$ kΩ.

7.10 For the circuit given in Fig. 7.14, what is the maximum value to which the 2-kΩ resistor can be changed such that Q_2 will still be in the active region?

7.11 The silicon transistor in the circuit shown in Fig. P7.11 has $\beta = 40$. Verify that this transistor is biased in the active region by finding (a) i_B, (b) i_C, and (c) v_{CE}.

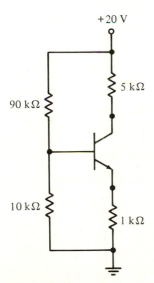

Fig. P7.11

7.12 Repeat Problem 7.11 for the circuit shown in Fig. P7.12.

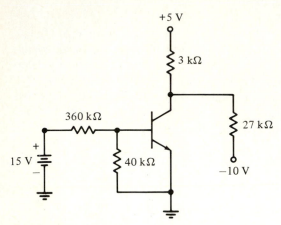

Fig. P7.12

7.13 Repeat Problem 7.11 for the circuit shown in (a) Fig. P7.13(a) and (b) Fig. P7.13(b).

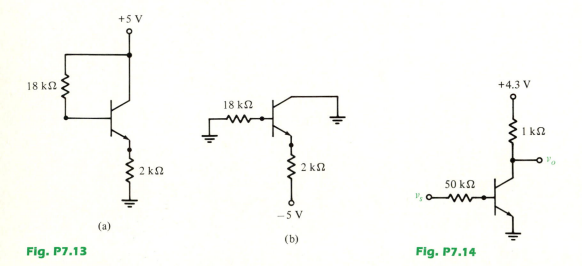

Fig. P7.13

Fig. P7.14

7.14 For the circuit shown in Fig. P7.14, the silicon transistor has $\beta = 50$. Given that the transistor operates in the active region, find v_o when (a) $v_s = 1$ V and (b) $v_s = 4$ V.

7.15 The silicon transistor in the circuit shown in Fig. P7.15 is in the active region. If $i_B = 50$ μA and $i_C = 4.25$ mA, find (a) β, (b) V_{BB}, and (c) v_{CE}.

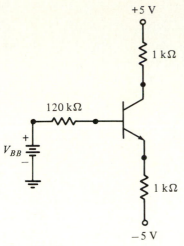

Fig. P7.15

7.16 For the circuit shown in Fig. P7.15, the silicon transistor has $\beta = 51$. If $V_{BB} = 0$, verify that the transistor operates in the active region by finding (a) i_B, (b) i_C, and (c) v_{CE}.

7.17 The silicon transistor in the circuit shown in Fig. P7.17 has $\beta = 100$. Determine the value of the collector resistor R_C such that the transistor is in the active region and $v_{CE} = 10$ V.

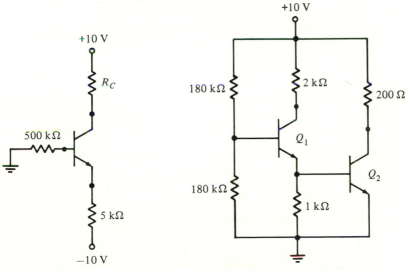

Fig. P7.17   **Fig. P7.18**

7.18 The silicon transistors Q_1 and Q_2 in the circuit shown in Fig. P7.18 have $\beta_1 = \beta_2 = 40$. Verify that both transistors are in the active region by finding (a) i_{B_1}, (b) i_{B_2}, (c) v_{CE_1}, and (d) v_{CE_2}.

Problems for Section 7.3

7.19 For the circuit given in Fig. 7.21, find (a) the minimum value of h_{FE} required to saturate the transistor when the 2-kΩ collector resistor is changed to 5.6 kΩ, and (b) the minimum value of collector resistance for which the transistor will saturate when $h_{FE} = 100$.

7.20 The silicon transistor in the circuit shown in Fig. P7.20 has $h_{FE} = 100$. Determine whether or not the transistor is in saturation when (a) $R_E = 0$ Ω and (b) $R_E = 1.5$ kΩ.

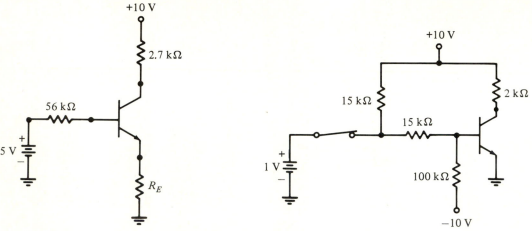

Fig. P7.20   **Fig. P7.21**

7.21 For the circuit shown in Fig. P7.21, (a) verify that the silicon transistor is OFF when the switch is closed, and (b) find the minimum value of h_{FE} for which the transistor is ON when the switch is open.

7.22 The silicon transistors Q_1 and Q_2 in the circuit shown in Fig. P7.22 have dc betas of h_{FE_1} and h_{FE_2} respectively.

(a) If $V_{BB} = 5$ V, what is the minimum value of h_{FE_1} for which Q_1 will saturate?

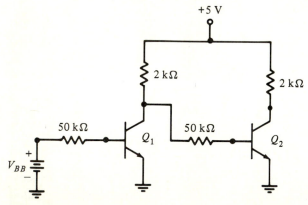

Fig. P7.22

(b) If $V_{BB} = 0$, what is the minimum value of h_{FE_2} that will saturate Q_2?

7.23 For the circuit given in Fig. 7.22, let $V_{BB} = 6$ V, $R_1 = 1$ kΩ, $R_2 = 3.3$ kΩ, and $R_3 = 8.2$ kΩ. Verify that Q_1 is ON and Q_2 is OFF by (a) finding the minimum value of h_{FE_1} that will saturate Q_1, and (b) determining v_{BE_2}.

7.24 The silicon transistors Q_1 and Q_2 in the circuit shown in Fig. P7.24 are identical. Determine the minimum value of h_{FE} required to saturate Q_1 when (a) $v_1 = 5$ V, $v_2 = 0.2$ V; and (b) $v_1 = v_2 = 5$ V.

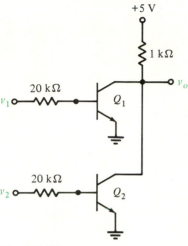

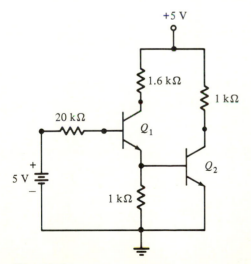

Fig. P7.24

7.25 For the silicon transistors Q_1 and Q_2 in the circuit given in Fig. P7.24, $h_{FE_1} = h_{FE_2} = 30$. Find v_o when (a) $v_1 = v_2 = 0.2$ V; (b) $v_1 = 0.2$ V, $v_2 = 5$ V; and (c) $v_1 = v_2 = 5$ V.

7.26 Find the minimum values of h_{FE_1} and h_{FE_2} for the silicon transistors Q_1 and

Fig. P7.26

Q_2, respectively, in the circuit shown in Fig. P7.26 such that both transistors will be in saturation.

Problems for Section 7.4

7.27 For the inverter circuit given in Fig. 7.28, the low voltage is 0.2 V and the high voltage is 6 V. Determine the minimum value of h_{FE} required for proper operation if the collector resistor is changed to (a) 500 Ω and (b) 2 kΩ.

7.28 For the inverter circuit given in Fig. 7.28, the low voltage is 0.2 V and the high voltage is 6 V. Determine the minimum value of collector resistance required for proper operation when h_{FE} is (a) 50 and (b) 100.

7.29 The DTL circuit shown in Fig. P7.29 is an example of a (positive-logic) **NOR gate** (NOR is a contraction of NOT-OR). If the low voltage is 0.2 V and the high voltage is 5 V, find v_o when $h_{FE} = 50$ and (a) $v_1 = v_2 = 0.2$ V; (b) $v_1 = 0.2$ V, $v_2 = 5$ V; and (c) $v_1 = v_2 = 5$ V.

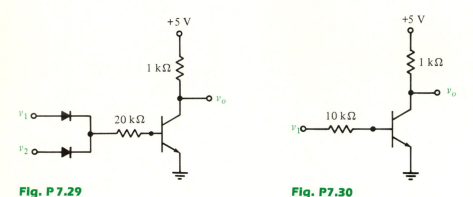

Fig. P 7.29   **Fig. P7.30**

7.30 The circuit shown in Fig. P7.30 is an inverter (NOT gate). The low voltage for this gate is 0.2 V. Suppose that the gate is loaded by ten identical inverters.

(a) Find the high voltage; that is, with the load connected, find v_o when $v_1 = 0.2$ V.

(b) Find the minimum value of h_{FE} required for proper operation.

(c) Find the fan-out when $h_{FE} = 50$.

7.31 For the NAND gate given in Fig. 7.32, find (a) the power dissipated by the gate and (b) the noise margins $NM(0)$ and $NM(1)$.

7.32 By modifying the NAND gate given in Fig. 7.32 as shown in Fig. P7.32, higher noise margins are obtained. [This is desirable for operation in high-noise environments and is an example of a **high-threshold logic (HTL)** gate.] The low voltage for the gate is 0.2 V and the high voltage is 15 V. The breakdown voltage of the Zener diode is 6 V.

(a) If $h_{FE} = 50$ for Q_1, what is the minimum value of h_{FE} for Q_2?

(b) Find the power dissipated by this gate.

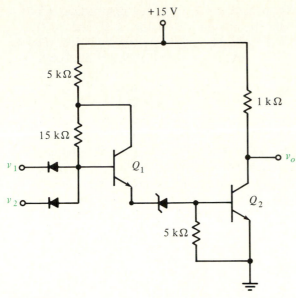

Fig. P7.32

7.33 For the NAND gate given in Fig. P7.32, find (a) the noise margins $NM(0)$ and $NM(1)$ and (b) the fan-out when $h_{FE} = 50$ for Q_1 and Q_2.

7.34 Suppose that the diode D_4 of the NAND gate given in Fig. 7.32 is replaced by the base–emitter junction of transistor Q_2 as shown in Fig. P7.34. The low

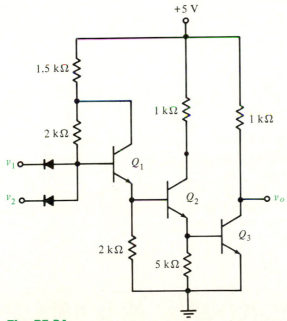

Fig. P7.34

voltage for the resulting NAND gate is 0.2 V and the high voltage is 5 V. If $h_{FE} = 50$ for Q_1, find the minimum value of h_{FE} for (a) Q_2 and (b) Q_3.

7.35 For the NAND gate given in Fig. P7.34, find (a) the power dissipated by the gate, (b) the noise margins, and (c) the fan-out when $h_{FE} = 50$ for Q_1, Q_2, and Q_3.

7.36 The NOR gate given in Fig. P7.29 is loaded by 20 identical gates. The low voltage for this gate is 0.2 V.

(a) Find the high voltage; that is, with the load connected, find v_o when $v_1 = v_2 = 0.2$ V.

(b) Find the minimum value of h_{FE} required for proper operation.

(c) Find the fan-out when $h_{FE} = 50$.

7.37 In addition to the DTL gate given in Fig. P7.29, a NOR gate can be constructed by connecting the collectors of two inverters together as shown in Fig. P7.37. This is an example of **wired logic** (or **collector logic**), so called because the collectors of Q_1 and Q_2 are wired together. (Note that this NOR gate requires no diodes.) Given that $h_{FE} = 50$ for Q_1 and Q_2, find v_o when (a) $v_1 = v_2 = 0.2$ V; (b) $v_1 = 0.2$ V, $v_2 = 5$ V; and (c) $v_1 = v_2 = 5$ V.

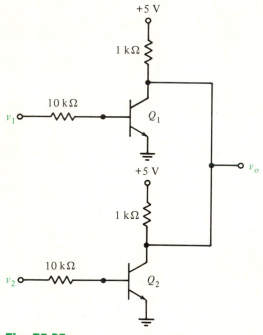

Fig. P7.37

Problems for Section 7.5

7.38 Find the power dissipated by the TTL NAND gate given in Fig. 7.35.

7.39 For the TTL NAND gate shown in Fig. 7.35, let $R_B = 2.7$ kΩ, $R_C = 500$ Ω, $R_1 = 100$ Ω, and $R_2 = 500$ Ω.

(a) Given that $h_{FE(im)} = 0.6$ for Q_1, find the minimum values of h_{FE} for Q_2 and Q_3.

(b) Find the fan-out when $h_{FE} = 50$ for Q_3.

(c) Find the power dissipated by the gate.

7.40 For the TTL NAND gate given in Fig. 7.36, suppose that $h_{FE(im)} = 0.5$.

(a) Find the minimum values of h_{FE} for Q_2 and Q_3.

(b) Determine the fan-out when $h_{FE} = 40$ for Q_3.

7.41 Find the noise margins $NM(0)$ and $NM(1)$ for the TTL NAND gate given in (a) Fig. 7.35 and (b) Fig. 7.36.

7.42 The TTL circuit shown in Fig. P7.42 has transistors with $h_{FE} = 50$.

(a) Find v_o when $v_1 = v_2 = 0.2$ V.

(b) Find v_o when $v_1 = 5$ V.

(c) What type of TTL gate is this circuit?

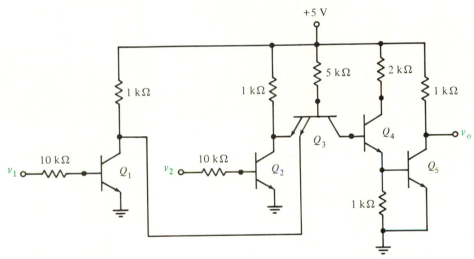

Fig. P7.42

7.43 Determine the fan-out of the RTL NOR gate in Fig. 7.42 given that the transistors have $h_{FE} = 50$.

7.44 For the RTL NOR gate given in Fig. 7.42, let $R_B = 20$ kΩ, $R = 5$ kΩ, and $V_{CC} = 5$ V. Suppose that this gate is loaded by ten identical gates.

(a) Determine the high voltage.

(b) Find the minimum value of h_{FE} for the transistors.

(c) Find the noise margins $NM(0)$ and $NM(1)$.

7.45 The wired-logic (see Problem 7.37) circuit shown in Fig. P7.45 has transistors with $h_{FE} = 50$.

(a) Find v_o when $v_1 = v_2 = 5$ V.

(b) Find v_o when $v_1 = 0.2$ V.

(c) What type of RTL gate is this circuit?

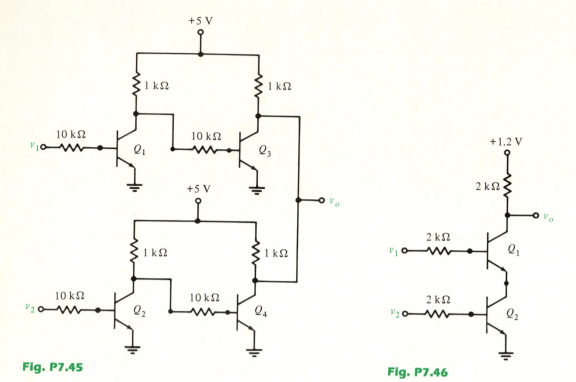

Fig. P7.45

Fig. P7.46

7.46 The RTL circuit shown in Fig. P7.46 is a NAND gate. The low voltage is 0.4 V and the high voltage is 1.2 V.

(a) For each combination of input voltages, determine whether each transistor is OFF, ON, or neither (e.g., if $i_B > 0$ but $i_C = 0$, Q behaves as a diode so it is neither OFF nor ON).

(b) Find the minimum values of h_{FE} for Q_1 and Q_2.

Problems for Section 7.6

7.47 Suppose that the 500-Ω resistance in the ECL circuit given in Fig. 7.45 is changed to 100 Ω. Why will this new circuit not result in NOR/OR-gate operation?

7.48 Given the ECL NOR/OR gate shown in Fig. P7.48:

(a) When v_1 and v_2 are low, assume that Q_1 and Q_2 are cut off and Q_3 is in the active region. Neglect base currents and determine the low voltage.

(b) When v_1 is high, assume that Q_1 is in the active region and Q_3 is cut off. Neglect base currents and determine the high voltage.

(c) Determine R such that v_o and $\bar{v}_o$ take on the same pair of values.

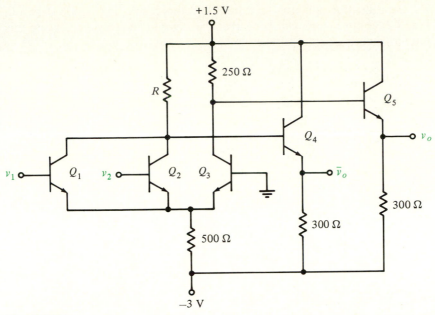

Fig. P7.48

7.49 Determine the noise margins $NM(0)$ and $NM(1)$ for the ECL NOR gate given in (a) Fig. 7.46 and (b) Fig. P7.48.

7.50 Find the power dissipated by the ECL gate given in Fig. 7.46.

7.51 Given the I^2L circuit shown in Fig. P7.51, determine the type of gate this is when the output is taken to be (a) v_o and (b) $\bar{v}_o$.

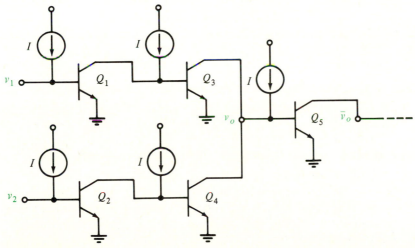

Fig. P7.51

7.52 For the sake of simplicity, the current injectors for the I²L circuit shown in Fig. P7.52 are not shown. Construct the logic table corresponding to this circuit when the output is taken to be (a) v_o and (b) $\bar{v}_o$.

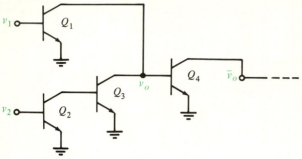

Fig. P7.52

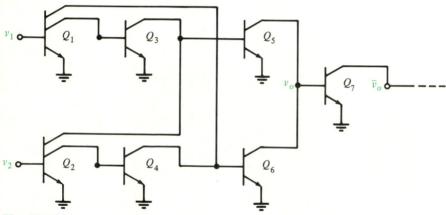

Fig. P7.53

7.53 Repeat Problem 7.52 for the I²L circuit shown in Fig. P7.53.
7.54 Repeat Problem 7.52 for the I²L circuit shown in Fig. P7.54.

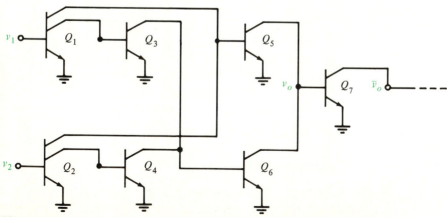

Fig. P7.54

Chapter 8

Field-Effect
Transistors (FETs)

INTRODUCTION

Unlike the bipolar junction transistor (BJT) studied in Chapter 7, another type of transistor—the field-effect transistor (FET)—is a unipolar device; that is, current results from the flow of majority carriers only. There are two types of FETs—the junction field-effect transistor (JFET) and the metal-oxide-semiconductor field-effect transistor (MOSFET).

Whereas the BJT is a current-controlled device, the FET is a voltage-controlled device. A voltage controls an electric field (hence, the terminology "field effect"), which, in turn, controls a current. An FET inherently possesses a high input resistance; therefore, in digital circuit applications, FET logic generally has a higher fan-out than BJT logic. Although BJTs operate at higher speeds than FETs, the FET is inherently less noisy than the BJT.

FETs, especially MOSFETs, are simpler to fabricate and require less space than do BJTs. Also, MOSFET logic does not involve the need for resistors and diodes. As a consequence of this, MOSFETs find widespread use in large-scale integration (LSI) and very-large-scale integration (VLSI) applications, such as hand-held calculators, digital wristwatches, microprocessors, and memories. For more general-purpose logic design, complementary MOSFETs (CMOS) offer greater speed and lower power consumption.

8.1 THE JUNCTION FIELD-EFFECT TRANSISTOR (JFET)

A side view of a simplified version of a **junction field-effect transistor** (JFET) is shown in Fig. 8.1(a). In particular, this device is an **n-channel** JFET. The corresponding circuit symbol is given in Fig. 8.1(b). [Interchanging the n- and p-type

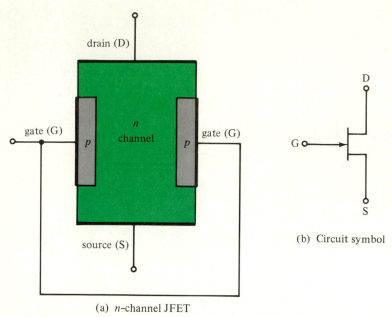

(a) *n*-channel JFET

(b) Circuit symbol

Fig. 8.1 An *n*-channel JFET and its circuit symbol.

semiconductors results in a *p*-channel JFET, and its circuit symbol is obtained from the one in Fig. 8.1(b) simply by reversing the direction of the arrow.] In essence, an *n*-channel JFET is made by taking a bar of *n*-type semiconductor (silicon) and forming heavily doped *p*-type regions, called the **gate** (G), on both sides. This results in two *p-n* junctions. Metal contacts are made to the gate as well as to the ends of the *n*-type semiconductor. One end of this *n*-type semiconductor, known as the **channel,** is called the **drain** (D) and the other end is called the **source** (S) of the JFET. Because of its physical symmetry, either end of the channel can be the drain and the other end the source. However, the end in which majority carriers (electrons for an *n*-type semiconductor) enter the channel is designated the source, and the end from which they leave the channel is the drain. Unlike the BJT, the JFET is a unipolar device—current is due to majority carriers only. As we shall see, current going from drain to source (i.e., electrons going from source to drain) through the channel can be controlled by a voltage applied between gate and source.

To understand the behavior of a JFET, consider the circuit shown in Fig. 8.2, where all of the JFET's currents and voltages are indicated. For the case that $V_{DD} = V_{GG} = 0$, all external currents and voltages, of course, will be zero. However, there are two *p-n* junctions with their associated depletion regions as depicted in Fig. 8.3. For the case that $V_{DD} = v_{DS}$ is a small (positive) number, a small (positive) drain current i_D results. Since $v_{DS} = V_{DD}$ is small, the drain voltage v_D (with respect to the reference) is approximately equal to the source voltage v_S, that is, $v_D \approx v_S = 0$. However, if V_{GG} is increased, since $v_{GS} = -V_{GG}$, then the *p-n* junctions become more reverse biased. This means that the depletion regions get wider, and because $v_D \approx v_S = 0$, they essentially are symmetrical. This widening

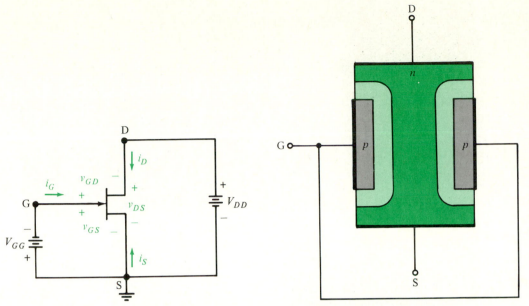

Fig. 8.2 An *n*-channel JFET circuit.

Fig. 8.3 The depletion regions in an *n*-channel JFET for small values of v_{DS}.

of the depletion regions, in effect, narrows the channel. By sufficiently increasing V_{GG}, the depletion regions will widen enough to eliminate the channel. The value of $v_{GS} = -V_{GG}$ for which such a situation occurs is called the **pinchoff voltage** V_p of the JFET. Note that for an *n*-channel JFET, V_p is a negative number. The magnitude of V_p typically lies in the range from 2 to 6 V.

Let us now again consider the case that $v_{GS} = 0$ ($V_{GG} = 0$ V). As mentioned above, if $V_{DD} = v_{DS} = 0$, then $i_D = 0$ and $v_D = v_S = 0$. Thus, the voltage across each *p-n* junction is zero, which means that they are reverse biased. (We say that the gate is reverse biased.) If $V_{DD} > 0$, but is still small, then the drain current i_D is small and $v_D \approx v_S = 0$. Thus, the gate remains reverse biased, so $i_G = 0$. As long as $V_{DD} = v_{DS} = v_D - v_S$ remains small (i.e., $v_D \approx v_S$), the depletion regions will be essentially symmetrical and i_D will be proportional to v_{DS}—that is, the channel will behave as a linear resistor. However, if $V_{DD} = v_{DS} = v_D - v_S$ gets larger, then i_D gets larger and v_D becomes greater than v_S. Since, when $v_{GS} = 0$,

$$v_{GD} = v_{GS} - v_{DS} = 0 - v_{DS} = -v_{DS}$$

the gate-drain voltage is negative while the gate-source voltage is zero. Thus, the portion of the *p-n* junction between the gate and drain is more reverse biased than the portion between the gate and source. As a consequence of this, the channel is narrower at the drain end than at the source end. This situation is illustrated in Fig. 8.4 for the condition that $0 < v_{DS} < |V_p|$. As v_{DS} increases, i_D increases, the channel narrows, and the resistance of the channel increases. Because of the manner in which the channel configuration changes, the nature of the channel resistance is nonlinear.

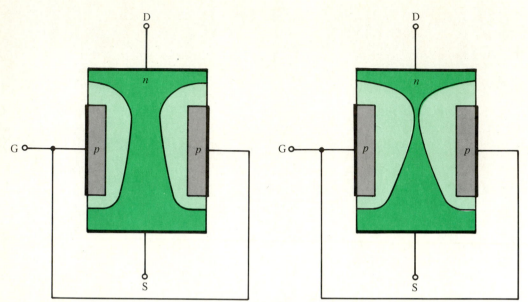

Fig. 8.4 The depletion regions in an n-channel JFET for $0 < v_{DS} < |V_p|$.

Fig. 8.5 The depletion regions in an n-channel JFET for $v_{DS} = -V_p$.

If v_{DS} is made large enough, the reverse-bias gate-drain voltage will be sufficient to pinch off the channel. By definition, this occurs when $v_{GD} = V_p$, that is, when

$$v_{GD} = v_{GS} - v_{DS} = 0 - v_{DS} = V_p \tag{8.1}$$

or $v_{DS} = -V_p$. For this condition, as illustrated in Fig. 8.5, even though it is technically "pinched off," the channel is not completely eliminated, but instead is very narrow. This allows a nonzero drain current i_D. If $v_{DS} > -V_p$, then the channel is still pinched off (very narrow), but it is narrow over a greater distance. As v_{DS} increases, the drain current i_D tends to increase. However, the increasing resistance due to the reduction in the width of the channel tends to decrease the drain current. The net effect is that for $v_{DS} \geq -V_p$, the drain current i_D is approximately constant. Because of this, the drain current is said to **saturate.** (The term "saturation" when used for a JFET has a different meaning than when it is used for a BJT.)

JFET Characteristics

The discussion above can be summarized graphically by the plot of i_D versus v_{DS} (for the case that $v_{GS} = 0$) shown in Fig. 8.6. From this plot, we see that for small values of v_{DS}, the (channel of the) device behaves as a linear resistor, and as v_{DS} increases, the channel resistance becomes nonlinear. When $v_{DS} \geq -V_p$ and the channel is pinched off, the drain current remains essentially constant as v_{DS} varies.

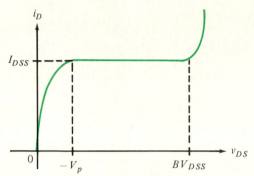

Fig. 8.6 Typical plot of i_D versus v_{DS} when $v_{GS} = 0$ for an n-channel JFET.

The value of this constant (saturation) current is designated I_{DSS} to indicate the current from drain to source with the gate short-circuited to the source ($v_{GS} = 0$). For a JFET, the current I_{DSS} can range from tenths of a milliampere to hundreds of milliamperes.

Throughout the above discussion, due to reverse-biased p-n junctions, we had that $i_G = 0$. However, practically speaking, we know that a p-n junction will break down if too large a reverse-bias voltage is applied. As indicated in Fig. 8.6, if v_{DS} increases too much, an excessive reverse bias will occur and junction breakdown can result in a large drain current (going from drain to gate). The value of v_{DS} for which this occurs is designated BV_{DSS} to indicate the breakdown voltage from drain to source with the gate short-circuited ($v_{GS} = 0$). This voltage typically ranges from 20 to 50 V.

Let us return to the JFET circuit given in Fig. 8.2. We have already seen [Equation (8.1)] that when $v_{GS} = 0$, the channel is pinched off when $v_{GD} = -v_{DS} = V_p$ or $v_{DS} = -V_p$. However, when $V_{GG} > 0$, that is, $v_{GS} < 0$, then the channel is pinched off when

$$v_{GD} = v_{GS} - v_{DS} = V_p$$

or when

$$v_{DS} = v_{GS} - V_p$$

Thus, by decreasing v_{GS}, the value of v_{DS} required to pinch off the channel decreases. In particular, Fig. 8.7 shows a set of curves of i_D versus v_{DS} for different values of v_{GS} for an n-channel JFET. We refer to these curves as the **drain (static) characteristics** or **output characteristics** of the JFET. For this device, $I_{DSS} = 5$ mA and the pinchoff voltage is $V_p = -4$ V. Thus, when $v_{GS} = 0$ V, the channel pinches off when $v_{DS} = -V_p = 4$ V. However, if $v_{GS} = -1$ V, then the channel pinches off when $v_{DS} = v_{GS} - V_p = -1 + 4 = 3$ V. Similarly, when $v_{GS} = -2$ V, pinchoff occurs when $v_{DS} = -2 + 4 = 2$ V, and so on. The dashed curve shown in Fig. 8.7 corresponds to the equation $v_{DS} = v_{GS} - V_p$. To the right of this curve ($v_{DS} > v_{GS} - V_p$) the channel of the JFET is pinched off, and to the left of this curve

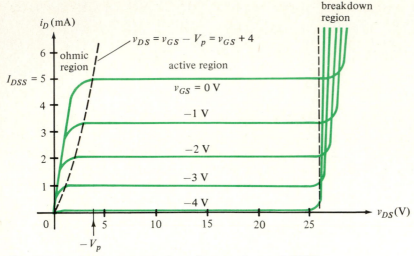

Fig. 8.7 Drain characteristics of an *n*-channel JFET.

($v_{DS} < v_{GS} - V_p$) it is not. The region to the left of this curve is known as the **ohmic region** (or **nonsaturation region***).

Not only does a value of $v_{GS} < 0$ reduce the value of v_{DS} required to pinch off the channel, but, as indicated in Fig. 8.7, it also reduces the value of v_{DS} required for junction breakdown and establishes a **breakdown region** as shown. The region between the ohmic and breakdown regions is called the **active region** (or **pinchoff region** or **saturation region**). Note that when the gate is sufficiently reverse biased, the channel will be eliminated—this, of course, occurs when $v_{GS} = V_p$. Under this circumstance, increasing v_{DS} to any value in the active region will not be sufficient to produce a drain current. Thus, we have that $i_D = 0$ and we say that the JFET is in **cutoff.** (In an actual device, due to leakage, the drain current of a cutoff JFET is small, but nonzero. However, we will use the approximation $i_D = 0$.)

In the discussions above, we were concerned only with the situation in which the gate is reverse biased. Ideally, then, $i_G = 0$. However, practically speaking, there is a small gate current even when the JFET is cut off. Although such a current is of the order of a few nanoamperes or less at 25°C, it approximately doubles for every 10°C rise in temperature. For the sake of simplicity, we will assume the ideal case that $i_G = 0$.

When a JFET operates in the ohmic region, it can be shown that

$$i_D = I_{DSS}\left[2\left(1 - \frac{v_{GS}}{V_p}\right)\frac{v_{DS}}{-V_p} - \left(\frac{v_{DS}}{V_p}\right)^2\right] \tag{8.2}$$

---

*This terminology is unfortunate since the corresponding region for a BJT is called the saturation region. To avoid confusion, we will use the term "ohmic region."

For small values of v_{DS}, we have $-V_p \gg v_{DS}$. This, in turn, means that $(v_{DS}/V_p)^2$ is small. Thus,

$$i_D \approx I_{DSS}\left[2\left(1 - \frac{v_{GS}}{V_p}\right)\left(\frac{v_{DS}}{-V_p}\right)\right] = \frac{2I_{DSS}}{-V_p}\left(1 - \frac{v_{GS}}{V_p}\right)v_{DS}$$

from which

$$\frac{i_D}{v_{DS}} \approx \frac{2I_{DSS}}{-V_p}\left(1 - \frac{v_{GS}}{V_p}\right) = \frac{2I_{DSS}(v_{GS} - V_p)}{V_p^2}$$

Therefore, for small values of v_{DS}, when the channel behaves as a linear resistor, r_{DS}, the approximate value of this resistance is

$$r_{DS} = \frac{v_{DS}}{i_D} \approx \frac{V_p^2}{2I_{DSS}(v_{GS} - V_p)} \qquad (8.3)$$

Typically, this resistance ranges from several ohms to several hundred ohms.

It can also be shown, when a JFET is in the active region, that

$$i_D = I_{DSS}\left(1 - \frac{v_{GS}}{V_p}\right)^2 \qquad (8.4)$$

The corresponding plot of i_D versus v_{GS} shown in Fig. 8.8 is known as the **transfer characteristic** of the JFET. For an n-channel JFET, in the active region $v_{GS} \leq 0$. However, when $v_{GS} \leq V_p$, the JFET is cut off and $i_D = 0$. For this reason, V_p is also known as the **gate-source cutoff voltage,** designated $V_{GS(OFF)}$. Ideally, the transfer characteristic of a JFET is independent of the value of v_{DS}, but practically speaking, this is not necessarily the case.

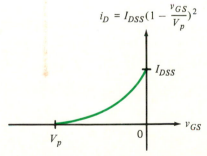

Fig. 8.8 An *n*-channel JFET transfer characteristic.

As shown in the JFET drain characteristics given in Fig. 8.7, the boundary between the active and ohmic regions is described by

$$v_{DS} = v_{GS} - V_p$$

Substituting $v_{GS} = v_{DS} + V_p$ into either Equation (8.3) or (8.4)—they both must satisfy the boundary condition—we find that

$$i_D = \frac{I_{DSS}}{V_p^2} v_{DS}^2$$

which is the equation of a parabola.

Simple JFET Circuits

Let us now analyze some simple JFET circuits.

Example 8.1

For the simple circuit shown in Fig. 8.9, the JFET has $I_{DSS} = 8$ mA and $V_p = -4$ V. Since the gate is short-circuited to the source, $v_{GS} = 0$ V. By Ohm's law, we have that

$$i_D = \frac{18 - v_{DS}}{500} = -\frac{1}{500} v_{DS} + 36 \times 10^{-3}$$

The load line corresponding to this equation can be plotted on the drain characteristics of the JFET, if available. The intersection of the load line and the curve $v_{GS} = 0$ will graphically yield the values of i_D and v_{DS}. Furthermore, the region of operation of the JFET can be determined graphically. However, let us suppose that the drain characteristics are not available. As a consequence, let us assume that

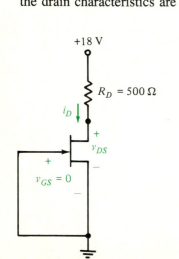

Fig. 8.9 Simple *n*-channel JFET circuit.

the JFET is in the active region. Then from Equation (8.4) we have

$$i_D = 8 \times 10^{-3}\left(1 - \frac{0}{-4}\right)^2 = 8 \text{ mA} = I_{DSS}$$

By KVL,

$$v_{DS} = -500(8 \times 10^{-3}) + 18 = 14 \text{ V}$$

Since $v_{DS} = 14 > v_{GS} - V_p = 0 - (-4) = 4$, the JFET is indeed in the active region.

Now suppose that the resistance R_D is changed to 2 kΩ. Again, if we assume that the JFET is in the active region, we get $i_D = 8$ mA and

$$v_{DS} = -2000(8 \times 10^{-3}) + 18 = 2 \text{ V}$$

But since $v_{DS} < v_{GS} - V_p$, our assumption that the JFET is in the active region is incorrect. Therefore, let us assume that the JFET is in the ohmic region. From Equation (8.2) we have that

$$i_D = 8 \times 10^{-3}\left[2\left(1 - \frac{0}{-4}\right)\frac{v_{DS}}{4} - \left(\frac{v_{DS}}{-4}\right)^2\right] = 4 \times 10^{-3}v_{DS} - 0.5 \times 10^{-3}v_{DS}^2$$

And from the circuit, we have that

$$i_D = \frac{18 - v_{DS}}{2000} = 9 \times 10^{-3} - 0.5 \times 10^{-3}v_{DS}$$

Combining these two equations, we get

$$4v_{DS} - 0.5v_{DS}^2 = 9 - 0.5v_{DS}$$

from which

$$v_{DS}^2 - 9v_{DS} + 18 = 0$$

Solving this quadratic equation, we get the two solutions $v_{DS_1} = 3$ V and $v_{DS_2} = 6$ V. However, for the JFET to be in the ohmic region, we require that $v_{DS} < v_{GS} - V_p = 0 - (-4) = 4$ V. Thus, we select as the solution $v_{DS} = 3$ V, and this confirms that the JFET is in the ohmic region. Futhermore, we have that the drain current is

$$i_D = \frac{18 - 3}{2000} = 7.5 \text{ mA}$$

Example 8.2

Suppose that for the circuit given in Fig. 8.9, we place a 1-kΩ resistor in series with the source of the JFET as shown in Fig. 8.10. Again $I_{DSS} = 8$ mA and

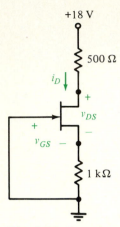

Fig. 8.10 Preceding circuit with added source resistance.

$V_p = -4$ V. Assuming that the gate is reverse biased ($v_{GS} \leq 0$), then $i_G = 0$. Thus,

$$i_D = -\frac{v_{GS}}{1000} \tag{8.5}$$

If the transfer characteristic of the JFET is available, the line described by this equation can be plotted using the same axes. The intersection of this line and the transfer-characteristic curve gives the graphical solution of the values for i_D and v_{GS}. But suppose that the transfer characteristic is not given.

Assuming that the JFET is in the active region, from Equation (8.4) we have

$$i_D = 8 \times 10^{-3}\left(1 - \frac{v_{GS}}{-4}\right)^2 = 8 \times 10^{-3}\left(1 + \frac{v_{GS}}{2} + \frac{v_{GS}^2}{16}\right)$$

Combining this with Equation (8.5), we get

$$-v_{GS} = 8\left(1 + \frac{v_{GS}}{2} + \frac{v_{GS}^2}{16}\right) = 8 + 4v_{GS} + \frac{1}{2}v_{GS}^2$$

from which

$$v_{GS}^2 + 10v_{GS} + 16 = 0$$

The two solutions to this quadratic equation are $v_{GS_1} = -2$ V and $v_{GS_2} = -8$ V. If the JFET is to be in the active region, then $V_p = -4 < v_{GS} \leq 0$. Thus, the solution we seek is $v_{GS} = -2$ V, and the gate is reverse biased as assumed. Substituting this value into Equation (8.5) yields $i_D = 2$ mA. Furthermore, by KVL,

$$v_{DS} = -500i_D + 18 - 1000i_D = -1500(2 \times 10^{-3}) + 18 = 15 \text{ V}$$

and since $v_{DS} = 15 > v_{GS} - V_p = -2 + 4 = 2$, then the JFET is indeed in the active region.

An alternative approach to the analysis of the circuit can be taken. By KVL, or from Equation (8.5), we have

$$v_{GS} = -1000i_D$$

Substituting this fact into Equation (8.4) yields

$$i_D = 8 \times 10^{-3}\left(1 - \frac{-1000i_D}{-4}\right)^2 = 8 \times 10^{-3}(1 - 500i_D + 62{,}500i_D^2)$$

from which

$$i_D^2 - 10 \times 10^{-3}i_D + 16 \times 10^{-6} = 0$$

The solutions to this equation are $i_{D_1} = 2 \times 10^{-3}$ and $i_{D_2} = 8 \times 10^{-3}$. For the JFET to be in the active region, $0 < i_D \le I_{DSS} = 8$ mA. However, if we select $i_D = 8$ mA, then

$$v_{GS} = -1000i_D = -1000(8 \times 10^{-3}) = -8 \text{ V}$$

Since $v_{GS} = -8 < V_p = -4$, the JFET is not in the active region—a contradiction. Therefore, we choose $i_D = 2$ mA. This value of drain current results in

$$v_{GS} = -1000i_D = -1000(2 \times 10^{-3}) = -2 \text{ V}$$

and

$$v_{DS} = -500i_D + 18 - 1000i_D = 15 \text{ V}$$

Since $V_p = -4 < v_{GS} = -2 < 0$ and $v_{DS} = 15 > v_{GS} - V_p = 2$, the JFET is indeed in the active region.

Although this section has discussed only n-channel JFETs, p-channel JFETs are treated in a similar manner. Figure 8.11 shows the circuit symbol for a p-channel JFET. Note that the currents and voltages are designated just as they are for an n-channel JFET. This means that the currents and voltages for a p-channel JFET are opposite in sign to those of the corresponding n-channel JFET. For example,

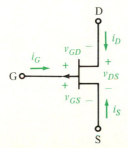

Fig. 8.11 Circuit symbol for a p-channel JFET.

for an n-channel JFET, the drain characteristics are plots of positive-valued drain current (i_D) versus positive-valued drain-source voltage (v_{DS}) for different non-positive values of gate-source voltage (v_{GS}). For a p-channel JFET, though, the drain characteristics have negative values for i_D and v_{DS}, while v_{GS} takes on non-negative values. Figure 8.12 shows typical drain characteristics for a p-channel JFET.

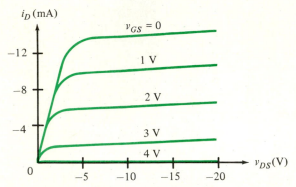

Fig. 8.12 Drain characteristics for a p-channel JFET.

8.2 METAL-OXIDE-SEMICONDUCTOR FIELD-EFFECT TRANSISTORS (MOSFETs)

To function properly, the channel of a JFET is narrowed or depleted of majority charge carriers. Because of this fact, we say that a JFET operates in a **depletion mode,** and we refer to it as a **depletion-type** FET. The JFET, however, is not the only type of depletion FET.

The Depletion MOSFET

To see how another kind of depletion FET can be fabricated, we begin with a body or foundation of p-type silicon, called a **substrate,** and form two heavily-doped n-type regions, denoted n^+. An n-type channel, labeled n, is then formed between the two n^+ regions, as shown in Fig. 8.13. In addition to the substrate (SB), metal contacts are made to the two n^+ regions—these are the drain (D) and the source (S). Next, the channel is covered with a thin insulating oxide layer of silicon dioxide (SiO_2). The gate (G) is obtained by depositing a metal (aluminum) on the oxide layer as is illustrated. Such a device is known as a **metal-oxide-semiconductor field-effect transistor** (MOSFET). In particular, the transistor shown in Fig. 8.13 is an **n-channel** MOSFET (also denoted NMOS). If the n- and p-type materials are interchanged, a **p-channel** MOSFET (also denoted PMOS) results. Because there is a layer of insulation between the gate and the channel, a MOSFET is also known as an **insulated-gate field-effect transistor** (IGFET).

For normal JFET operation, the gate is reverse biased, and the resistance between the gate and source is typically many megohms. However, for a MOSFET,

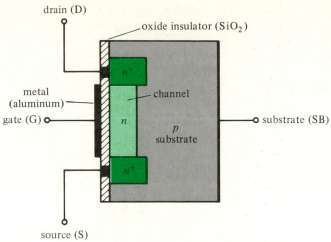

Fig. 8.13 An *n*-channel depletion MOSFET.

because of the insulation between the gate and the substrate, the resistance between the gate and the source is extremely high—typically ranging from 10^{10} to 10^{15} Ω. Therefore, as we did for a JFET, we will take as a given that $i_G = 0$ for a MOSFET.

Typically, the substrate and the source of a MOSFET are connected together. If v_{GS} is made negative, then positive charges are induced in the *n*-type channel, thereby effectively narrowing it, as is the case for a JFET. In addition, for varying values of v_{DS}, the behavior of the device is similar to that of a JFET. Since the MOSFET depicted in Fig. 8.13 can operate in the depletion mode, it is called a **depletion MOSFET.**

When operating in the depletion mode, a depletion MOSFET has characteristics like those of a JFET. However, because of the insulation between the gate and the semiconductor, positive values of v_{GS} are allowable. When v_{GS} is made positive, additional negative charges are induced in the *n*-type channel, thereby effectively widening it. When the channel is widened by or enhanced with majority charge carriers, we say that it operates in an **enhancement mode.** Operation of a depletion MOSFET in the enhancement mode is similar to that in the depletion mode. Consequently, even though in the depletion mode it behaves similarly to a JFET, a depletion MOSFET's characteristics are extended versions of those of a JFET (because positive values of v_{GS} are included). This is demonstrated by the typical *n*-channel depletion MOSFET drain and transfer characteristics shown in Fig. 8.14. Equations (8.2) and (8.4), which describe operation in the ohmic and active regions, respectively, again are valid for depletion MOSFETs.

More than one circuit symbol is used for an *n*-channel depletion MOSFET; two of the more common are shown in Fig. 8.15. The typical situation in which the substrate is connected to the source is depicted in Fig. 8.15(a), and the simplified circuit symbol is given by Fig. 8.15(b). We shall give preference to the latter symbol. By reversing the direction of the arrow in either circuit symbol, we obtain the circuit symbol for a *p*-channel depletion MOSFET.

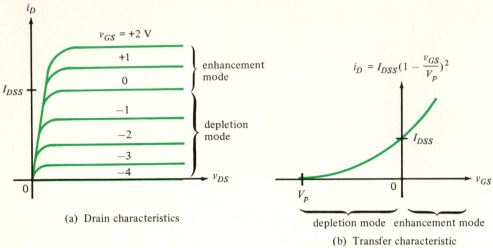

(a) Drain characteristics

(b) Transfer characteristic

Fig. 8.14 Characteristics of a typical *n*-channel depletion MOSFET.

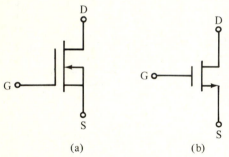

(a)   (b)

Fig. 8.15 Circuit symbols for an *n*-channel depletion MOSFET.

Example 8.3

Consider the simple MOSFET circuit shown in Fig. 8.16, where the depletion NMOS transistor has $I_{DSS} = 8$ mA and $V_p = -2$ V. By inspection, $v_{GS} = 1$ V. If the MOSFET is in the active region, then from Equation (8.4)

$$i_D = I_{DSS}\left(1 - \frac{v_{GS}}{V_p}\right)^2 = 8 \times 10^{-3}\left(1 - \frac{1}{-2}\right)^2 = 18 \text{ mA}$$

Thus,

$$v_{DS} = -500(18 \times 10^{-3}) + 16 = 7 \text{ V}$$

and since $v_{DS} = 7 > v_{GS} - V_p = 1 + 2 = 3$, the MOSFET is indeed in the active region.

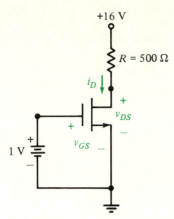

Fig. 8.16 Simple MOSFET circuit.

Suppose that the 500-Ω resistance is changed to 750 Ω. If we again assume that the MOSFET is in the active region, then, as above, $i_D = 18$ mA and

$$v_{DS} = -750(18 \times 10^{-3}) + 16 = 2.5 \text{ V}$$

However, since $v_{DS} = 2.5 < v_{GS} - V_p = 3$, then, contrary to our assumption, the MOSFET is not in the active region. Let us assume that it is in the ohmic region. From Equation (8.2), we have

$$i_D = I_{DSS}\left[2\left(1 - \frac{v_{GS}}{V_p}\right)\frac{v_{DS}}{-V_p} - \left(\frac{v_{DS}}{V_p}\right)^2\right]$$

$$= 8 \times 10^{-3}\left[2\left(1 - \frac{1}{-2}\right)\frac{v_{DS}}{2} - \left(\frac{v_{DS}}{-2}\right)^2\right] = 12 \times 10^{-3}v_{DS} - 2 \times 10^{-3}v_{DS}^2$$

Using this expression and Ohm's law, we get

$$i_D = 12 \times 10^{-3}v_{DS} - 2 \times 10^{-3}v_{DS}^2 = \frac{16 - v_{DS}}{750}$$

This yields the quadratic equation

$$3v_{DS}^2 - 20v_{DS} + 32 = 0$$

the solutions of which are $v_{DS_1} = 4$ and $v_{DS_2} = \frac{8}{3}$. Since $v_{DS_1} = 4 > v_{GS} - V_p = 3$ implies active-region operation, and $v_{DS_2} = \frac{8}{3} < v_{GS} - V_p = 3$ implies ohmic-region operation, we have that $v_{DS} = \frac{8}{3}$ V. The corresponding drain current, therefore, is

$$i_D = \frac{16 - \frac{8}{3}}{750} = 17.78 \text{ mA}$$

The Enhancement MOSFET

Consider again the MOSFET depicted in Fig. 8.13. Let us remove the *n*-type material that constitutes the channel between the two n^+ regions. The resulting device is shown in Fig. 8.17, where again the substrate is connected to the source. Suppose that $v_{GS} = 0$ V. To have a (positive) current go from drain to source, we require that the drain voltage be greater than the source voltage. But this means that the *p-n* junction formed by the drain n^+ region and the *p*-type substrate is reverse biased. Consequently, there is no drain current.

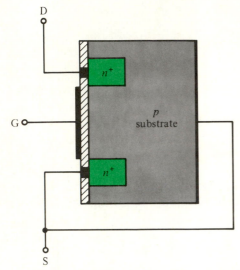

Fig. 8.17 An *n*-channel enhancement MOSFET.

Next, let us consider the case that v_{GS} is made positive. This will induce negative charges near the *p*-type semiconductor (substrate) surface beneath the gate. The more positive v_{GS} is, the more induced negative charges there are. When v_{GS} is made sufficiently large, an "inversion layer" of electrons results and this, in effect, forms an *n*-type channel. The voltage required for this to happen is called the **gate-source threshold voltage** and is designated by $V_{GS(th)}$, or more simply V_T. (*Note:* This is not the same quantity as the volt-equivalent of temperature $V_T = T/11,586$ discussed in Chapter 6.) Typically, V_T ranges between 1 and 3 V.

As v_{GS} is increased beyond V_T, the channel is widened or enhanced. Once a channel has been established (i.e., for $v_{GS} > V_T$), the behavior of such a device, called an **enhancement MOSFET,** is similar to that discussed previously. The drain characteristics of a typical *n*-channel enhancement MOSFET are shown in Fig. 8.18. For this enhancement NMOS transistor, the (gate-source) threshold voltage is $V_T = 2$ V. Thus, if $v_{GS} \leq 2$ V, then the gate-source voltage is not sufficient to produce a channel, and the MOSFET is cut off. If $v_{GS} > 2$ V, then the gate-source voltage is large enough to establish a channel. Under this circumstance,

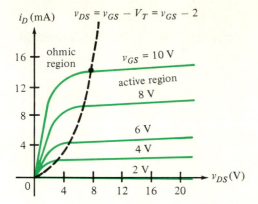

Fig. 8.18 Drain characteristics of a typical _n_-channel enhancement MOSFET.

however, the value of v_{DS} will determine whether or not the channel is pinched off, and, therefore, whether the MOSFET is in the active or ohmic region.

   More than one circuit symbol is also used for an n-channel enhancement MOSFET. Two of the more common symbols are shown in Fig. 8.19. The typical situation that the substrate is connected to the source is depicted in Fig. 8.19(a). Note that the simplified circuit symbol shown in Fig. 8.19(b) is identical to the n-channel depletion MOSFET circuit symbol given in Fig. 8.15(b). Therefore, when these simplified circuit symbols are employed, the relevant discussions will indicate whether the MOSFETs are the depletion type or the enhancement type. Again, by reversing the direction of the arrow for the circuit symbol of an n-channel MOSFET, we obtain the circuit symbol for a p-channel MOSFET.

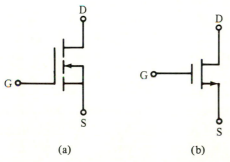

(a)   (b)

Fig. 8.19 Circuit symbols for an _n_-channel enhancement MOSFET.

   Suppose that a channel has been established ($v_{GS} > V_T$) in an enhancement NMOS transistor. Because of the symmetry of the device, this channel will be pinched off when $v_{GD} \le V_T$. Since $v_{GD} = v_{GS} - v_{DS}$, pinchoff occurs when

$$v_{GS} - v_{DS} \le V_T$$

or when

$$v_{DS} \geq v_{GS} - V_T$$

Under this condition, therefore, the MOSFET is in the active region. Of course, this means that it is in the ohmic region when $v_{DS} < v_{GS} - V_T$. For the drain characteristics given in Fig. 8.18, the dashed curve described by

$$v_{DS} = v_{GS} - V_T$$

therefore, is the border between the active and the ohmic regions.

Unlike the JFET and the depletion MOSFET, an enhancement MOSFET in the active region satisfies the equation

$$i_D = K(v_{GS} - V_T)^2 \tag{8.6}$$

where the value of the constant K depends on the construction parameters of the device. A plot of the curve described by this equation is the transfer characteristic of the MOSFET. Such a typical i_D-versus-v_{GS} plot is shown in Fig. 8.20. If we know V_T and the values of current and voltage for any point on this curve (for $v_{GS} > V_T$), we can determine K easily by using Equation (8.6). As an example of an alternative method for finding K, note that from the drain characteristics given in Fig. 8.18, when the MOSFET is in the active region and $v_{GS} = 6$ V, then $i_D \approx 5$ mA. Thus, from Equation (8.6), we have the approximation

$$K = \frac{i_D}{(v_{GS} - V_T)^2} = \frac{5 \times 10^{-3}}{(6 - 2)^2} = 3.125 \times 10^{-4} \text{ A/V}^2$$

Ideally, this value of K is the same throughout the active region. However, if we consider the case that $v_{GS} = 8$ V, then $i_D \approx 10$ mA, and from Equation (8.6) we get the approximation

$$K = \frac{10 \times 10^{-3}}{(8 - 2)^2} = 2.78 \times 10^{-4} \text{ A/V}^2$$

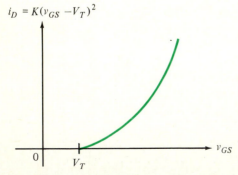

Fig. 8.20 Transfer characteristic of an enhancement MOSFET.

Therefore, for the sake of computation, it is reasonable to use the approximate value $K = 3 \times 10^{-4}$ A/V^2.

Previously we mentioned that the dashed curve in Fig. 8.18, which separates the active and ohmic regions, corresponds to $v_{DS} = v_{GS} - V_T$. In substituting this fact into Equation (8.6), we get the equation of this dashed curve as

$$i_D = K v_{DS}^2$$

and, therefore, this curve is a parabola.

For the case that an enhancement MOSFET is in the ohmic region, it can be shown that

$$i_D = K[2(v_{GS} - V_T)v_{DS} - v_{DS}^2] \qquad (8.7)$$

where the constant K (ideally) has the same value as for active-region operation. For the case that v_{DS} is small, we can approximate Equation (8.7) by

$$i_D \approx K[2(v_{GS} - V_T)v_{DS}]$$

or

$$\frac{i_D}{v_{DS}} \approx 2K(v_{GS} - V_T)$$

Thus, for small values of v_{DS}, when the MOSFET behaves as a linear resistor, the approximate value of drain-source resistance r_{DS} is given by

$$r_{DS} = \frac{v_{DS}}{i_D} \approx \frac{1}{2K(v_{GS} - V_T)} \qquad (8.8)$$

For example, for the n-channel enhancement MOSFET discussed above, from Equation (8.8), when $v_{GS} = 8$ V, we have that for small values of v_{DS}

$$r_{DS} \approx \frac{1}{2(3 \times 10^{-4})(8 - 2)} = 278 \ \Omega$$

while when $v_{GS} = 6$ V,

$$r_{DS} \approx \frac{1}{2(3 \times 10^{-4})(6 - 2)} = 417 \ \Omega$$

Example 8.4

Consider the circuit shown in Fig. 8.21. From the characteristics of the n-channel enhancement MOSFET, suppose it is determined that, in the active region, when $v_{GS} = 8$ V, then $i_D \approx 9$ mA. Furthermore, it is given that $V_T = 2$ V.

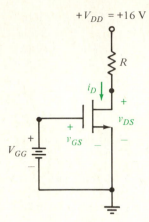

Fig. 8.21 Enhancement NMOS circuit.

From Equation (8.6), we have, therefore, that the approximate value of K is

$$K = \frac{i_D}{(v_{GS} - V_T)^2} = \frac{9 \times 10^{-3}}{(8 - 2)^2} = 2.5 \times 10^{-4} \text{ A/V}^2$$

For the case that $R = 250 \ \Omega$, and $V_{GG} = 10$ V, let us assume active-region operation. Since $v_{GS} = V_{GG} = 10$ V, from Equation (8.6) we have

$$i_D = K(v_{GS} - V_T)^2 = 2.5 \times 10^{-4}(10 - 2)^2 = 16 \text{ mA}$$

Thus,

$$v_{DS} = -250(16 \times 10^{-3}) + 16 = 12 \text{ V}$$

Since $v_{DS} = 12 > v_{GS} - V_T = 10 - 2 = 8$, active-region operation is confirmed.

Now suppose that only the resistance is changed to $R = 1$ kΩ. Again assuming active-region operation, as above, $i_D = 16$ mA. Thus,

$$v_{DS} = -1000(16 \times 10^{-3}) + 16 = 0 \text{ V}$$

But since $v_{DS} = 0 < v_{GS} - V_T = 8$, this is a contradiction to operation in the active region. Therefore, let us assume ohmic-region operation. From Equation (8.7),

$$i_D = K[2(v_{GS} - V_T)v_{DS} - v_{DS}^2] = 2.5 \times 10^{-4}[2(10 - 2)v_{DS} - v_{DS}^2]$$

Furthermore, by Ohm's law,

$$i_D = 2.5 \times 10^{-4}[16v_{DS} - v_{DS}^2] = \frac{16 - v_{DS}}{1000}$$

Simplifying this expression, we get the quadratic equation

$$v_{DS}^2 - 20v_{DS} + 64 = 0$$

the solutions of which are $v_{DS_1} = 4$ and $v_{DS_2} = 16$. However, as stated above, for ohmic-region operation we require that $v_{DS} < 8$ V. Thus, we choose the solution $v_{DS} = 4$ V. As a result of this, we obtain

$$i_D = \frac{16 - 4}{1000} = 12 \text{ mA}$$

8.3 MOSFET INVERTERS

Let us now concentrate on the graphical approach to the analysis of MOSFET circuits. To be specific, consider the simple circuit shown in Fig. 8.22, where the enhancement NMOS transistor has $V_T = 2$ V and its drain (output) characteristics are as given in Fig. 8.23. For the given circuit,

$$i_D = \frac{V_{DD} - v_{DS}}{R} = \frac{12 - v_{DS}}{1000} = -10^{-3}v_{DS} + 12 \times 10^{-3}$$

The associated load line is plotted on the drain characteristics as indicated.

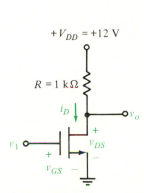

Fig. 8.22 Simple *n*-channel enhancement MOSFET circuit.

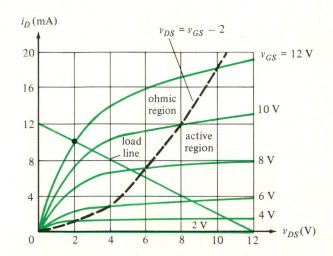

Fig. 8.23 Enhancement MOSFET drain characteristics.

We are now in a position to determine graphically, quite easily, the output voltage $v_o = v_{DS}$ that results when an input voltage $v_1 = v_{GS}$ is applied. For example, if the input is a high voltage of $v_1 = 12$ V, then from the intersection of the load line and the curve labeled $v_{GS} = 12$ V, we have that $i_D = 10$ mA and $v_{DS} = 2$ V. Thus, when the input is $v_1 = 12$ V, the output is $v_o = v_{DS} = 2$ V. Conversely, if the input is a low voltage of $v_1 = 2$ V, then from the intersection of the load line and the curve labeled $v_{GS} = 2$ V (which coincides with the horizontal axis), we have that $i_D = 0$ and $v_{DS} = 12$ V. Thus, when the input is $v_1 = 2$ V, the output is

$v_o = v_{DS} = 12$ V. Hence, choosing the low and high voltages to be 2 V and 12 V, respectively, the circuit given in Fig. 8.22 is an example of an inverter (NOT gate). Note that when the input is low, the MOSFET is in cutoff (OFF), and when the input is high, the MOSFET is in the ohmic region (ON).

Active-Region Load

In the inverter circuit just discussed, the load is a resistor R. In an IC, however, a resistor can occupy 20 times the space that a MOSFET does. Therefore, let us use a MOSFET as a resistance, albeit nonlinear, in place of the linear resistance R. To see how this is done, consider the n-channel enhancement MOSFET Q_1 shown in Fig. 8.24, where the gate is short-circuited to the drain. Suppose that $V_T = 2$ V and the drain characteristics of this NMOS transistor are as given in Fig. 8.25.

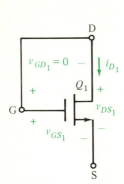

Fig. 8.24 An *n*-channel enhancement MOSFET.

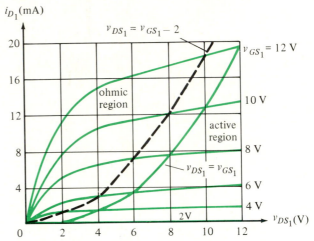

Fig. 8.25 MOSFET drain characteristics.

Since $v_{DS_1} = -v_{GD_1} + v_{GS_1} = v_{GS_1}$, the MOSFET will operate only at some point on the curve labeled $v_{DS_1} = v_{GS_1}$ indicated in Fig. 8.25. Note, therefore, that this NMOS transistor will always be in the active region. By inspection, we see that the characteristic curve $v_{DS_1} = v_{GS_1}$ indicates that there is a nonlinear relationship between i_{D_1} and v_{DS_1}. In other words, between the drain and the source, the MOSFET has a nonlinear resistance.

As a consequence of this discussion, let us now consider the (resistorless) MOSFET circuit shown in Fig. 8.26, where the enhancement NMOS transistor Q_1 is the **load,** and the enhancement NMOS transistor Q_2 is known as the **driver.** Let us suppose that Q_1 and Q_2 have $V_T = 2$ V and their drain characteristics are identical to those given in Fig. 8.25. Note that due to the connection of Q_1 and Q_2, we have $i_{D_1} = i_{D_2} = i_D$. For the case that the load is a (linear) resistor R (see Fig. 8.22), the

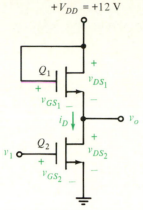

$+V_{DD} = +12$ V

Fig. 8.26 Enhancement NMOS inverter with active-region load.

plot of the relationship between i_D and v_{DS} is the load line. However, for the circuit given in Fig. 8.26, the load is a MOSFET (nonlinear resistor), and the corresponding plot is called the **load curve.** In other words, the load curve is the plot of i_D versus v_{DS_2}.

To obtain the load curve, we use the fact that

$$v_{DS_2} = -v_{DS_1} + V_{DD} = -v_{DS_1} + 12 \qquad (8.9)$$

From the plot of $i_D = i_{D_1}$ versus v_{DS_1} for the load MOSFET Q_1 and Equation (8.9), we can compute the following values.

$v_{GS_1} = v_{DS_1}$, V	i_D, mA	$v_{DS_2} = -v_{DS_1} + 12$, V
2	0	10
4	1.4	8
6	3.4	6
8	7.8	4
10	13	2
12	19	0

The first two columns of this table were obtained graphically from Fig. 8.25, and the third column resulted from Equation (8.9). Using these calculated values, we can now sketch the load curve, that is, a plot of $i_D = i_{D_2}$ versus v_{DS_2}. This is shown in Fig. 8.27 as the solid trace.

We may now plot the output voltage $v_o = v_{DS_2}$ versus the input voltage $v_1 = v_{GS_2}$ by using the load curve to determine various points graphically. In particular, from Fig. 8.27, we obtain the following values.

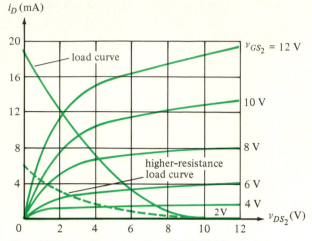

$v_1 = v_{GS_2}$, V	$v_o = v_{DS_2}$, V
12	2.3
10	3.0
8	4.3
6	6.0
4	7.4
2	10.0
0	10.0

Fig. 8.27 Load curves for MOSFET inverter.

The resulting plot, known as the **transfer curve** of the circuit, is shown in Fig. 8.28 as the solid trace. From Fig. 8.28, we see that when the input voltage v_1 is low, the output voltage v_o is high, and vice versa. Thus, the MOSFET circuit given in Fig. 8.26 behaves as an inverter. However, this behavior is by no means ideal. If the input is between 0 and 2 V, then Q_2 is OFF and the output is 10 V. Conversely, if the input is 10 V, then Q_2 is ON and the output is 3 V. A lower output voltage is obtained when the input is increased to 12 V—the output then is 2.3 V. The ideal (positive-logic) situation is to have the low voltage 0 V and the high voltage $V_{DD} = 12$ V.

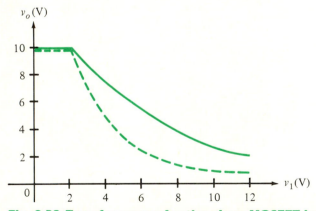

Fig. 8.28 Transfer curves for the given MOSFET inverter.

To get a behavior closer to ideal, let us replace the load Q_1 with a MOSFET that has a larger (nonlinear) resistance. Such a device is fabricated by increasing the channel length and decreasing its width. Doing so, suppose that the resulting load curve is given by the dashed trace in Fig. 8.27. The voltages corresponding to this load curve are the following.

$v_1 = v_{GS_2}$, V	$v_o = v_{DS_2}$, V
12	0.7
10	1.0
8	1.5
6	2.3
4	4.5
2	10.0
0	10.0

The resulting transfer curve, shown in Fig. 8.28 as the dashed trace, indicates a behavior closer to ideal than that for the original load.

Ohmic-Region Load

For the inverter just discussed, the high voltage is $V_{DD} - V_T = 12 - 2 = 10$ V. We can increase the high voltage, however, by modifying the circuit as shown in Fig. 8.29. Suppose that Q_1 and Q_2 are n-channel enhancement MOSFETs with $V_T = 2$ V and the drain characteristics shown in Fig. 8.30. If $V_{GG} = V_{DD}$, then this circuit is identical to the one in Fig. 8.26. However, if $V_{GG} = V_{DD} + V_T + a$, where $a > 0$, then

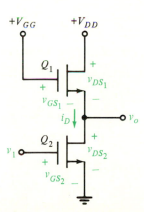

Fig. 8.29 Enhancement NMOS inverter with ohmic-region load.

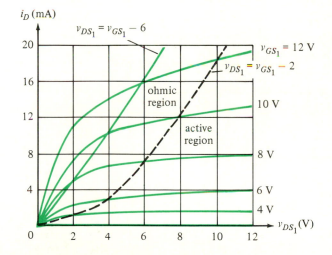

Fig. 8.30 MOSFET drain characteristics.

$$v_{DS_1} = V_{DD} - V_{GG} + v_{GS_1} \tag{8.10}$$

$$= V_{DD} - (V_{DD} + V_T + a) + v_{GS_1} = v_{GS_1} - V_T - a$$

and

$$v_{DS_1} < v_{GS_1} - V_T$$

which means that Q_1 operates in the ohmic region.

Suppose that, as before, $V_{DD} = 12$ V. Let us choose $a = 4$. Then

$$V_{GG} = V_{DD} + V_T + a = 12 + 2 + 4 = 18 \text{ V}$$

From Equation (8.10),

$$v_{DS_1} = V_{DD} - V_{GG} + v_{GS_1} = 12 - 18 + v_{GS_1} = v_{GS_1} - 6$$

A plot of this relationship is shown on the drain characteristics given in Fig. 8.30. The resulting curve, which is nearly a straight line, also describes the relationship between $i_D = i_{D_1}$ and v_{DS_1}. We can use this nearly linear relationship, however, to sketch a plot of $i_D = i_{D_2}$ versus v_{DS_2}—this is the load curve. This is done by using the fact that $v_{DS_2} = -v_{DS_1} + 12$ and graphically obtaining points from the i_D-v_{DS_1} characteristic shown in Fig. 8.30. Doing so, we get the following values.

v_{GS_1}, V	$v_{DS_1} = v_{GS_1} - 6$, V	i_D, mA	$v_{DS_2} = -v_{DS_1} + 12$, V
6	0	0	12
8	2	5.0	10
10	4	10.5	8
12	6	16.0	6

Using the values in the third and fourth columns, we can sketch the load curve. This plot of $i_D = i_{D_2}$ versus v_{DS_2} is shown in Fig. 8.31 as the solid trace. From this load

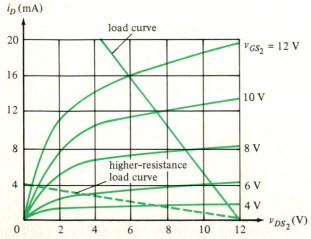

Fig. 8.31 Load curves for MOSFET inverter.

curve, we can graphically determine the output voltages for different input voltages. By inspection, we obtain the following values.

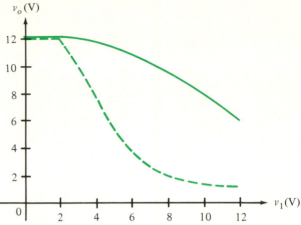

$v_1 = v_{GS_2}$, V	$v_o = v_{DS_2}$, V
12	6.0
10	7.5
8	9.0
6	10.5
4	11.5
2	12.0
0	12.0

Fig. 8.32 Transfer curves for MOSFET inverter.

The resulting transfer curve is shown in Fig. 8.32 as the solid trace.

Again, as with the case of the MOSFET inverter with the active-region load (Fig. 8.26), a more ideal behavior can be obtained if the load MOSFET Q_1 has a higher resistance. In particular, if Q_1 is fabricated such that the result is the higher-resistance load curve indicated by the dashed trace in Fig. 8.31, then the dashed trace in Fig. 8.32 is the subsequent transfer curve. Note that the highest output voltage for the inverter with the active-region load (Fig. 8.26) is $V_{DD} - V_T = 12 - 2 = 10$ V, while for the inverter with the ohmic-region load (Fig. 8.29) it is $V_{DD} = 12$ V. The latter property, however, has to be paid for with the use of an additional supply voltage V_{GG}.

Depletion-MOSFET Load

For the two types of inverters just discussed, one type had an active-region load and the other an ohmic-region load. In both cases, however, the load MOSFETs were enhancement NMOS transistors (as were the drivers). Now let us consider the circuit shown in Fig. 8.33, where Q_1 is a depletion MOSFET and Q_2 is again an enhancement MOSFET. Suppose that the drain characteristics of the enhancement NMOS driver transistor Q_2 are given by Fig. 8.31, and the drain characteristic of Q_1 for the case $v_{GS_1} = 0$ is given by the solid trace in Fig. 8.34. Only this characteristic is needed since the gate of Q_1 is short-circuited to its source, and, thus, $v_{GS_1} = 0$ V.

From Fig. 8.34, using the solid curve labeled $v_{GS_1} = 0$, we can graphically determine the drain current $i_{D_1} = i_D$ for different values of v_{DS_1}. Then, using the fact that $v_{DS_2} = -v_{DS_1} + V_{DD} = -v_{DS_1} + 12$, we can calculate the corresponding values of v_{DS_2}. The numbers obtained are given below.

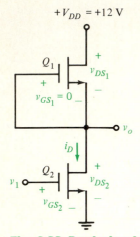

Fig. 8.33 Depletion-load MOSFET inverter.

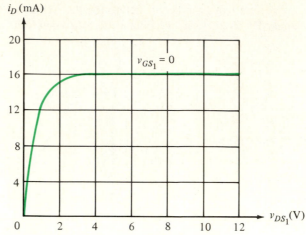

Fig. 8.34 Drain characteristics for depletion-load MOSFET.

v_{DS_1}, V	i_D, mA	$v_{DS_2} = -v_{DS_1} + 12$, V
0	0	12
1	13	11
2	15	10
3	16	9
4	16	8
6	16	6
8	16	4
10	16	2
12	16	0

From these values, we can sketch a plot of $i_D = i_{D_2}$ versus v_{DS_2} on the drain characteristics for Q_2. The resulting load curve is indicated by the solid trace in Fig. 8.35. By using this load curve, we graphically obtain the following values for the input and output voltages.

$v_1 = v_{GS_2}$, V	$v_o = v_{DS_2}$, V
12	6.0
10	10.8
8	11.4
6	11.6
4	11.8
2	12.0
0	12.0

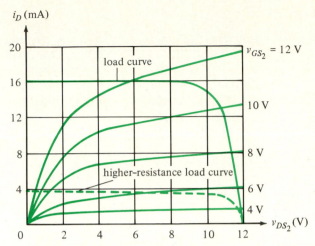

Fig. 8.35 Load curves for depletion-load MOSFET inverter.

The resulting transfer curve is shown in Fig. 8.36 as the solid trace. This behavior, of course, is less than ideal. To obtain a more ideal behavior, let us use a higher-resistance MOSFET for the load Q_1. Specifically, if this higher-resistance MOSFET yields the load curve indicated by the dashed trace shown in Fig. 8.35, then the corresponding transfer curve is as given by the dashed trace in Fig. 8.36. Thus, as for the case of the enhancement-load inverter given in Fig. 8.29, the depletion-load inverter given in Fig. 8.33 can have nearly ideal behavior. As we have seen, however, the former inverter requires two supply voltages V_{DD} and V_{GG}, whereas the latter needs only the single supply voltage V_{DD}.

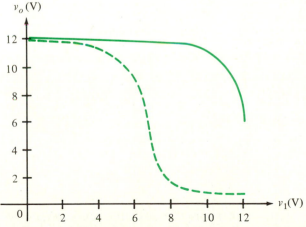

Fig. 8.36 Transfer curves for depletion-load MOSFET inverter.

8.4 MOSFET GATES AND SWITCHES

Consider the MOSFET circuit shown in Fig. 8.37, where Q_1, Q_2, and Q_3 are enhancement NMOS transistors having threshold voltage V_T. Suppose that $V_{DD} \gg V_T$ and that both driver Q_1 and driver Q_2 have drain characteristics such as those shown in Fig. 8.38, while the load Q_3 is a MOSFET with higher resistance than Q_1 and Q_2. Let us assume the ideal case that the low voltage is 0 V and the high voltage is V_{DD}.

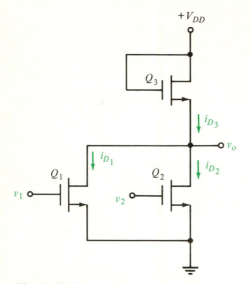

Fig. 8.37 Enhancement MOSFET NOR gate.

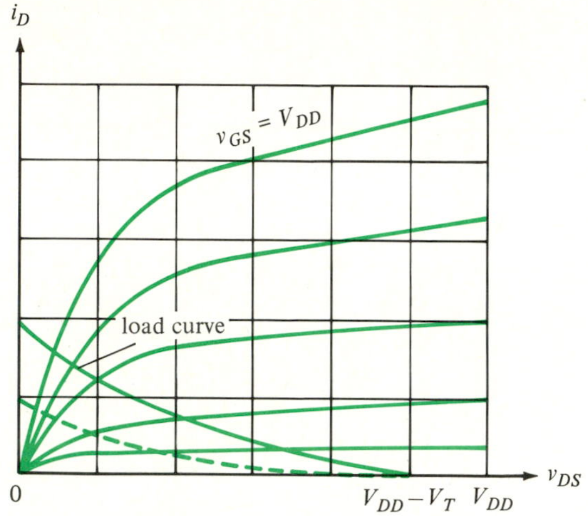

Fig. 8.38 Drain characteristics for Q_1 and Q_2.

If the input voltages are $v_1 = v_{GS_1} = 0$ and $v_2 = v_{GS_2} = 0$, since $v_{GS_1} = v_{GS_2} < V_T$, then Q_1 and Q_2 are cut off (OFF). Under these circumstances, $i_{D_1} = i_{D_2} = 0$, and, therefore, $i_{D_3} = i_{D_1} + i_{D_2} = 0$. If $v_{DS_3} = v_{GS_3} > V_T$, then $i_{D_3} > 0$—a contradiction. Thus, $v_{DS_3} = v_{GS_3} \le V_T$ and Q_3 is OFF. Since $v_o = -v_{GS_3} + V_{DD}$ and $V_{DD} \gg V_T$, the approximate output voltage is $v_o = V_{DD}$.

If $v_1 = 0$ and $v_2 = V_{DD}$, then Q_1 is OFF and $i_{D_1} = 0$. However, $v_2 = v_{GS_2} = V_{DD} > V_T$ means that Q_2 can conduct current ($i_{D_2} = i_{D_3} \ge 0$), so it is ON. The intersection of the curve $v_{GS_2} = V_{DD}$ and the load curve that results when Q_1 is OFF yields the values of the drain current and output voltage v_o. From Fig. 8.38, we see that the output voltage is small—let us approximate it by $v_o = 0$.

If $v_1 = V_{DD}$ and $v_2 = 0$, using reasoning as for the case $v_1 = 0$ and $v_2 = V_{DD}$, we deduce that Q_1 is ON, Q_2 is OFF, and the output voltage is approximately $v_o = 0$.

Finally, if $v_1 = v_2 = V_{DD}$, then Q_1 and Q_2 are ON. Since Q_1 and Q_2 have the same characteristics, $i_{D_1} = i_{D_2} = i_{D_3}/2$. This means that Q_1 and Q_2 each effectively has half of the load that it had when the other transistor was OFF. The corresponding load curve is given by the dashed trace in Fig. 8.38. The intersection of this and the curve $v_{GS} = V_{DD}$ demonstrates that again the output voltage is small—ideally, $v_o = 0$.

A summary of the output voltages v_o that result when the input voltages v_1 and v_2 take on the values 0 and V_{DD} is given by the following table.

v_1	v_2	v_o
0	0	V_{DD}
0	V_{DD}	0
V_{DD}	0	0
V_{DD}	V_{DD}	0

Hence, we see that the enhancement MOSFET circuit given in Fig. 8.37 is a (positive-logic) NOR gate.

Next consider the MOSFET circuit shown in Fig. 8.39. Again, Q_1, Q_2, and Q_3 are enhancement NMOS transistors with threshold voltage V_T, and we assume that $V_{DD} \gg V_T$. Furthermore, the load MOSFET Q_3 has a higher resistance than Q_1 and Q_2.

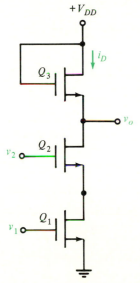

Fig. 8.39 Enhancement MOSFET NAND gate.

If $v_1 = 0$, then Q_1 is OFF and $i_{D_1} = i_{D_2} = i_{D_3} = i_D = 0$. If $v_{DS_3} = v_{GS_3} > V_T$, then $i_{D_3} = i_D > 0$—a contradiction. Thus, $v_{DS_3} = v_{GS_3} \leq V_T$ and Q_3 is OFF. Since $v_o = -v_{GS_3} + V_{DD}$ and $V_{DD} \gg V_T$, the approximate output voltage is $v_o = V_{DD}$.

If $v_1 = V_{DD}$ and $v_2 = 0$, then $v_2 = v_{GS_2} + v_{DS_1} = 0$. Thus, $v_{GS_2} =$

$-v_{DS_1} < V_T$. Therefore, Q_2 is OFF and $i_{D_2} = i_D = 0$. As above, this implies that $v_o = V_{DD}$.

Finally, if $v_1 = v_2 = V_{DD}$, then $v_{GS_1} = V_{DD} > V_T$ and Q_1 is ON. Assuming some load curve, as illustrated in Fig. 8.38, then $v_{DS_1} \approx 0$. Furthermore, $v_2 = v_{GS_2} + v_{DS_1} = V_{DD} \approx v_{GS_2}$. Since $v_{GS_2} \approx V_{DD} > V_T$, then Q_2 is ON and $v_{DS_2} \approx 0$. Hence, $v_o = v_{DS_1} + v_{DS_2} \approx 0 + 0 = 0$.

The resulting table for this circuit is given below.

V_1	V_2	V_o
0	0	V_{DD}
0	V_{DD}	V_{DD}
V_{DD}	0	V_{DD}
V_{DD}	V_{DD}	0

Hence, we see that the enhancement MOSFET circuit given in Fig. 8.39 is a (positive-logic) NAND gate.

The NMOS circuits in Figs. 8.37 and 8.39 are DCTL gates. Fortunately, though, MOSFET DCTL does not have any of the disadvantages that were listed for BJT DCTL.

A MOSFET Switch

Having now seen some MOSFET applications to digital circuits, specifically logic gates, let us now present an example of the use of MOSFETs for analog purposes. Specifically, consider the simple n-channel enhancement MOSFET circuit shown in Fig. 8.40. We will demonstrate that in this context the MOSFET Q can be used as a switch.

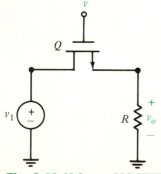

Fig. 8.40 Using a MOSFET as a switch.

Suppose that the applied voltage v_1 varies between 0 and V. Depending upon the condition of the MOSFET, the output voltage v_o will be some value in the interval $0 \leq v_o \leq V$. For example, if Q is OFF, then $i_D = 0$, and as a consequence, $v_o = Ri_D = 0$. Conversely, if Q is sufficiently ON and the result is $v_{DS} \approx 0$, then $v_o \approx v_1$.

Since Q is OFF when $v_{GS} \leq V_T$, then Q is OFF when $v - v_o \leq V_T$ or $v \leq V_T + v_o$. Since the minimum value of the output voltage is 0, then to guarantee that Q is OFF, we should use $v \leq V_T$. For this condition, $v_o = 0$ and Q behaves as an open circuit (an open switch).

To get Q to behave as a short circuit (a closed switch), we should make v large enough such that $v_{DS} \approx 0$. This can be done by making $v_{GS} \gg V_T$, that is, $v - v_o \gg V_T$ or $v \gg V_T + v_o$. Since the maximum value of the output voltage is V, we want to have $v \gg V_T + V$. Although a large value of v would guarantee a small value of v_{DS}, a smaller value of v can suffice if the value of R is sufficiently large (just recall a load line plotted on the drain characteristics of a MOSFET).

As a numerical example, suppose that Q has the drain characteristics shown in Fig. 8.41, where $V_T = 2$ V. Suppose that $R = 5$ kΩ and the applied voltage is in the range $0 \leq v_1 \leq 5$ V. We can turn Q OFF by setting $v = 0$ (since $0 < V_T = 2$). Conversely, if Q is ON, we wish to have $v_{DS} \approx 0$. When $v_1 = 5$ V (the maximum applied voltage), we see that the intersection of the corresponding load line and the curve $v_{GS} = 12$ V will yield $v_{DS} \approx 0$. Thus, we can achieve reasonable accuracy by having $v_{GS} = v - v_o = 12$, or $v = 12 + v_o = 12 + 5 = 17$ V. In summary, therefore, for the circuit given in Fig. 8.40, the MOSFET Q (with the drain characteristics given in Fig. 8.41) can be used as a switch (for applied voltages in the range $0 \leq v_1 \leq 5$ V), which is open when $v = 0$ V and closed when $v = 17$ V.

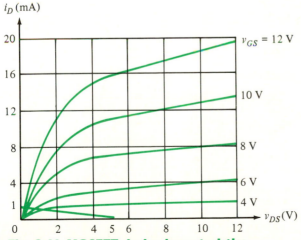

Fig. 8.41 MOSFET drain characteristics.

Complementary MOSFETs (CMOS)

If two enhancement MOSFETs, one NMOS and one PMOS, have identical characteristics (differing only by a minus sign), we say that they possess **complementary symmetry.** This property can be obtained by constructing the MOSFETs on the same IC chip. The resulting devices are known as **complementary** MOSFETs, or CMOS for short. (They are also referred to as COSMOS by RCA, and as McMOS by Motorola.*)

Figure 8.42 illustrates the construction of individual *n*-channel and *p*-channel enhancement MOSFETs (ignoring the substrate terminal). Figure 8.43 shows an example of CMOS, where the drains are connected together, as are the gates. Note that to fabricate an NMOS transistor, a *p*-type region, or "well," must be formed in the *n*-type substrate.

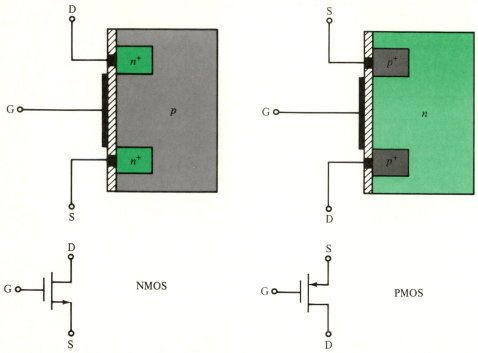

Fig. 8.42 NMOS and PMOS transistors.

Let us use the device depicted in Fig. 8.43 to form the simple CMOS circuit shown in Fig. 8.44. If the threshold voltage for the NMOS transistor is V_T, then for the PMOS transistor it is $-V_T$. Typical drain characteristics for Q_1 and Q_2 are shown in Figs. 8.45(a) and (b) respectively. Let us assume that $V_{DD} > V_T$.

---

*And by McDonald's.

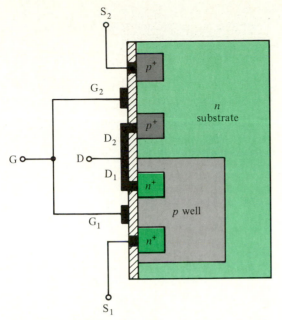

Fig. 8.43 Complementary MOSFETs (CMOS).

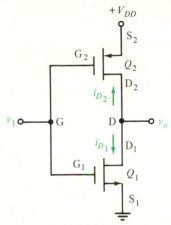

Fig. 8.44 A CMOS inverter.

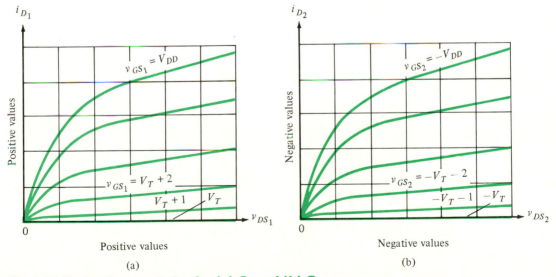

(a)   (b)

Fig 8.45 Drain characteristics for (a) Q_1 and (b) Q_2.

If $v_1 = 0$, then $v_{GS_1} = v_1 = 0 < V_T$. Thus, Q_1 is OFF and $i_{D_1} = 0$. But $v_{GS_2} = v_1 - V_{DD} = -V_{DD} < -V_T$. Therefore, Q_2 is ON and tends to conduct current. Specifically, the operation of Q_2 is described by the curve $v_{GS_2} = -V_{DD}$ given in Fig. 8.45(b). However, the only point on the curve $v_{GS_2} = -V_{DD}$ that satisfies the condition that $i_{D_2} = -i_{D_1} = 0$ is the origin. Thus, $v_{DS_2} = 0$, and, hence, $v_o = v_{DS_2} + V_{DD} = V_{DD}$.

If $v_1 = V_{DD}$, then $v_{GS_1} = v_1 = V_{DD} > V_T$. Thus, Q_1 is ON and tends to conduct current. Specifically, the operation of Q_1 is described by the curve $v_{GS_1} = V_{DD}$ given in Fig. 8.45(a). But $v_{GS_2} = v_1 - V_{DD} = 0 > -V_T$. Therefore, Q_2 is OFF and $i_{D_2} = 0$. However, the only point on the curve $v_{GS_1} = V_{DD}$ that satisfies the condition that $i_{D_1} = -i_{D_2} = 0$ is the origin. Thus, $v_{DS_1} = 0$, and, hence, $v_o = v_{DS_1} = 0$.

This discussion confirms that the CMOS circuit given in Fig. 8.44 is an inverter. Note, too, that when the input voltage is either high or low, the drain current is essentially zero. Even though the drain current is nonzero when the inverter changes state, on the whole such a device consumes a relatively small amount of power.

Having discussed the CMOS NOT gate given in Fig. 8.44, let us look at a more complicated CMOS logic circuit. Consider the CMOS circuit shown in Fig. 8.46. In essence, the driver (n-channel) enhancement MOSFETs Q_1 and Q_2 are connected in series, while the load (p-channel) enhancement MOSFETs Q_3 and Q_4 are connected in parallel. Suppose that Q_1 and Q_2 have a threshold voltage of V_T, and Q_3 and Q_4 have a threshold voltage of $-V_T$. Further, assume that $V_{DD} > V_T$.

If $v_1 = 0$, then $v_{GS_1} = v_1 = 0 < V_T$. Thus, Q_1 is OFF and $i_{D_1} = 0$. In addition, $v_{GS_3} = v_1 - V_{DD} = -V_{DD} < -V_T$, so Q_3 is ON. But $i_{D_2} = i_{D_1} = 0$ regardless

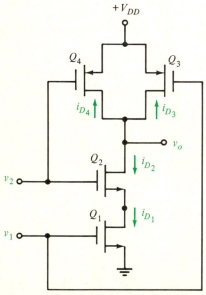

Fig. 8.46 A CMOS NAND gate.

of the value of v_2, and, therefore, $i_{D_3} + i_{D_4} = -i_{D_2} = 0$. Since $i_{D_3} \leq 0$ and $i_{D_4} \leq 0$, we deduce that $i_{D_3} = i_{D_4} = 0$. However, as was discussed for the CMOS inverter, if Q_3 is ON and $i_{D_3} = 0$, then we must have that $v_{DS_3} = 0$. Hence, $v_o = v_{DS_3} + V_{DD} = V_{DD}$.

If $v_1 = V_{DD}$ and $v_2 = 0$, then $v_2 = v_{GS_2} + v_{DS_1} = 0$. Thus, $v_{GS_2} = -v_{DS_1} < V_T$. This means that Q_2 is OFF and $i_{D_2} = 0$. In addition, $v_{GS_4} = v_2 - V_{DD} = -V_{DD} < -V_T$, so Q_4 is ON. Hence, using reasoning as above, we deduce that $v_o = V_{DD}$.

If $v_1 = v_2 = V_{DD}$, then $v_{GS_3} = v_1 - V_{DD} = 0 > -V_T$. Thus, Q_3 is OFF and $i_{D_3} = 0$. Similarly, Q_4 is OFF and $i_{D_4} = 0$. Therefore, $i_{D_1} = i_{D_2} = -i_{D_3} - i_{D_4} = 0$. However, $v_1 = v_{GS_1} = V_{DD} > V_T$ means that Q_1 is ON. But since $i_{D_1} = 0$, then, as discussed before, $v_{DS_1} = 0$. Furthermore, $v_2 = v_{GS_2} + v_{DS_1} = v_{GS_2} = V_{DD} > V_T$ means that Q_2 is ON. But since $i_{D_2} = 0$, then $v_{DS_2} = 0$. Hence, $v_o = v_{DS_1} + v_{DS_2} = 0$.

Using the above results, we can construct the following table.

v_1	v_2	v_o
0	0	V_{DD}
0	V_{DD}	V_{DD}
V_{DD}	0	V_{DD}
V_{DD}	V_{DD}	0

Therefore, we see that the CMOS circuit given in Fig. 8.46 is a NAND gate.

Another CMOS logic circuit is shown in Fig. 8.47. Conversely to the NAND

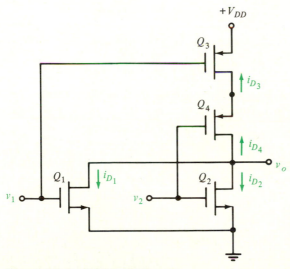

Fig. 8.47 A CMOS NOR gate.

gate given in Fig. 8.46, the driver MOSFETs Q_1 and Q_2 are in parallel, while the load MOSFETs Q_3 and Q_4 are in series. Again suppose that Q_1 and Q_2 have a threshold voltage of V_T, and Q_3 and Q_4 have a threshold voltage of $-V_T$. Also, assume that $V_{DD} > V_T$.

If $v_1 = v_2 = 0$, then $v_{GS_1} = v_{GS_2} = 0$. Thus, Q_1 and Q_2 are OFF, and $i_{D_1} = i_{D_2} = 0$. It follows that $i_{D_3} = i_{D_4} = -i_{D_1} - i_{D_2} = 0$. Since $v_{GS_3} = v_1 - V_{DD} = -V_{DD} < -V_T$, then Q_3 is ON. But since $i_{D_3} = 0$, then $v_{DS_3} = 0$. Therefore, $v_{GS_4} = v_2 - (v_{DS_3} + V_{DD}) = -V_{DD} < -V_T$, so Q_4 is ON. Again, since $i_{D_4} = 0$, then $v_{DS_4} = 0$. Hence, $v_o = v_{DS_4} + v_{DS_3} + V_{DD} = V_{DD}$.

If $v_1 = 0$ and $v_2 = V_{DD}$, then $v_{GS_1} = 0$. Thus, Q_1 is OFF and $i_{D_1} = 0$. Also, $v_{GS_2} = V_{DD}$, and, hence, Q_2 is ON. Since $v_{GS_3} = v_1 - V_{DD} = -V_{DD} < -V_T$, then Q_3 is ON as well. But since $v_{DS_3} \leq 0$, then $v_{GS_4} = v_2 - (v_{DS_3} + V_{DD}) = V_{DD} - v_{DS_3} - V_{DD} = -v_{DS_3} \geq 0 > -V_T$. Thus, Q_4 is OFF and $i_{D_4} = 0$. Therefore, $i_{D_2} = -i_{D_1} - i_{D_4} = 0$, and since Q_2 is ON, we must have that $v_{DS_2} = 0$. Hence, $v_o = v_{DS_2} = 0$.

If $v_1 = V_{DD}$ and $v_2 = 0$, then Q_1 is ON, Q_2 is OFF, and $i_{D_2} = 0$. Also, $v_{GS_3} = v_1 - V_{DD} = 0$, so Q_3 is OFF and $i_{D_3} = 0$. Thus, $i_{D_1} = -i_{D_2} - i_{D_3} = 0$, and since Q_1 is ON, then $v_{DS_1} = 0$. Hence, $v_o = v_{DS_1} = 0$.

Finally, if $v_1 = v_2 = V_{DD}$, then Q_1 and Q_2 are ON. However, $v_{GS_3} = v_1 - V_{DD} = 0$, so Q_3 is OFF and $i_{D_3} = 0$. Since $i_{D_1} \geq 0$, $i_{D_2} \geq 0$, and $i_{D_1} + i_{D_2} = -i_{D_3} = 0$, then we deduce that $i_{D_1} = i_{D_2} = 0$. But Q_1 and Q_2 are ON. Thus, $v_{DS_1} = v_{DS_2} = 0$, and, hence, $v_o = 0$.

The results determined above give us the table below.

V_1	V_2	V_o
0	0	V_{DD}
0	V_{DD}	0
V_{DD}	0	0
V_{DD}	V_{DD}	0

Therefore, we see that the CMOS circuit given in Fig. 8.47 is a NOR gate.

The Transmission Gate

In the chapter on BJTs, we saw how a BJT can be used as an electronic switch that is turned ON (saturated) and OFF (cut off) by base current. We have also seen that a MOSFET can be used as a switch that is turned ON (ohmic-region operation) and OFF (cut off) by an applied gate voltage. (A JFET may be used as a switch, too.) However, a more sophisticated device, known as a **transmission gate,** can be realized by the CMOS circuit shown in Fig. 8.48. For this circuit, there is a low voltage, say 0, and a high voltage, say $V > V_T$, where V_T and $-V_T$ are the threshold voltages of Q_1 and Q_2, respectively. To operate this gate, we apply the high voltage to the gate of Q_1 and the low voltage to the gate of Q_2, or vice versa.

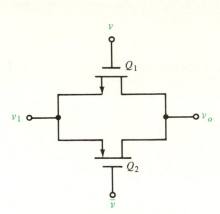

Fig. 8.48 Transmission gate.

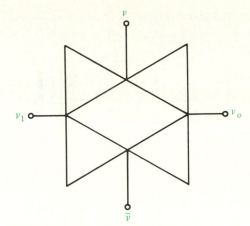

Fig. 8.49 Transmission-gate circuit symbol.

First consider the case that $v = V$ and $\bar{v} = 0$. If $v_1 = 0$, then $v_{GS_1} = v - v_1 = V - 0 = V > V_T$. Thus, Q_1 is ON. Furthermore, $v_{GS_2} = \bar{v} - v_1 = 0 - 0 = 0 > -V_T$, so Q_2 is OFF and $i_{D_2} = 0$. With no load connected to the output, we must have that $i_{D_1} = 0$. But since Q_1 is ON, then $v_{DS_1} = 0$ and $v_o = v_1 = 0$. Conversely, if $v_1 = V$, then $v_{GS_1} = v - v_1 = V - V = 0 < V_T$. Thus, Q_1 is OFF and $i_{D_1} = 0$. In addition, $v_{GS_2} = \bar{v} - v_1 = 0 - V = -V < -V_T$, so Q_2 is ON. But since (with no load) $i_{D_2} = 0$, then $v_{DS_2} = 0$ and $v_o = v_1 = V$. In summary, therefore, when $v = V$ and $\bar{v} = 0$, the gate is "closed" ($v_o = v_1$) for either input $v_1 = 0$ or $v_1 = V$.

Now consider the case that $v = 0$ and $\bar{v} = V$. If $v_1 = 0$, then $v_{GS_1} = v - v_1 = 0 - 0 = 0 < V_T$ and Q_1 is OFF. Furthermore, $v_{GS_2} = \bar{v} - v_1 = V - 0 = V > -V_T$ and Q_2 is OFF. Thus, the device behaves as an open circuit between output and input. If $v_1 = V$, then $v_{GS_1} = v - v_1 = 0 - V = -V < V_T$ and Q_1 is OFF. In addition, $v_{GS_2} = \bar{v} - v_1 = V - V = 0 > -V_T$ and Q_2 is OFF. Therefore, effectively there is again an open circuit between output and input. In summary, when $v = 0$ and $\bar{v} = V$, for either input $v_1 = 0$ or $v_1 = V$, the gate is "open."

We have just seen that the transmission gate given in Fig. 8.48 is closed when $v = V$ and $\bar{v} = 0$, while it is open when $v = 0$ and $\bar{v} = V$. Even though the statements were verified for the cases that $v_1 = 0$ and $v_1 = V$, it can be shown that they also hold for values of v_1 in the interval $0 \le v_1 \le V$ (see Problem 8.35). In other words, a transmission gate can be used for analog as well as digital applications. The circuit symbol for the transmission gate given in Fig. 8.48 is shown in Fig. 8.49.

SUMMARY

1. The terminals of a field-effect transistor (FET) are the gate (G), the drain (D), and the source (S). Unlike the BJT, the FET is a unipolar device—majority charge carriers go from source to drain.

2. The junction field-effect transistor (JFET) operates only in the depletion mode.

3. The voltage required to close a channel is the pinchoff voltage.

4. The drain characteristics of an FET are plots of drain current versus drain-source voltage for different values of gate-source voltage.

5. The gate current of an FET is zero.

6. A JFET is in cutoff if its drain current is zero.

7. A JFET is in the active region when its drain current is nonzero and its channel is pinched off.

8. A JFET is in the ohmic region when its drain current is nonzero and its channel is not pinched off.

9. For small values of drain-source voltage, an FET behaves as a linear resistor.

10. The transfer curve of an FET is a plot of drain current versus gate-source voltage.

11. A depletion metal-oxide-semiconductor FET (MOSFET) has a layer of insulation between the gate and the channel. A depletion MOSFET can operate in the enhancement mode as well as the depletion mode.

12. An enhancement MOSFET is fabricated with no channel, but a channel results from an applied gate-source voltage. The minimum (absolute) value of gate-source voltage required to induce a channel is the threshold voltage.

13. The same circuit symbol is often used for depletion and enhancement MOSFETs.

14. An IC MOSFET inverter uses MOSFETs as both the driver and the load. Although the former is an enhancement type, the latter can be either an enhancement or a depletion MOSFET. An enhancement load can operate entirely in either the active region or the ohmic region.

15. Enhancement MOSFETs can be used to fabricate DCTL gates and switches.

16. Complementary MOSFETs (CMOS) are well suited for IC logic.

Problems for Section 8.1

8.1 An n-channel JFET has $I_{DSS} = 10$ mA and $V_p = -2$ V. For small values of v_{DS}, determine the drain-source resistance r_{DS} when (a) $v_{GS} = 0$ V and (b) $v_{GS} = -1$ V.

8.2 Given the JFET circuit shown in Fig. P8.2, suppose that $I_{DSS} = 12$ mA and $V_p = -4$ V. Given that $R = 3$ kΩ and $V_{GG} = 2$ V, find (a) i_D and (b) v_{DS}.

8.3 For the JFET circuit given in Fig. P8.2 ($I_{DSS} = 12$ mA, $V_p = -4$ V), find the value of R for which $v_{DS} = 0.1$ V when $V_{GG} = 0$ V.

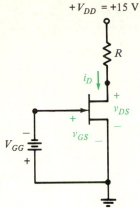

$+V_{DD} = +15$ V

R

i_D

$+$
v_{DS}
$-$

$+$
v_{GS}
$-$

V_{GG}

Fig. P8.2

8.4 Repeat Problem 8.2 for the case that $R = 1$ kΩ and $V_{GG} = 0$ V.

8.5 For the circuit given in Fig. P8.2, let $R = 5$ kΩ and $V_{DD} = 25$ V. Suppose that the drain characteristics of the JFET are as given by Fig. 8.7. Graphically determine i_D, v_{DS}, and the region of operation when (a) $V_{GG} = 0$ V, (b) $V_{GG} = 2$ V, and (c) $V_{GG} = 4$ V.

8.6 For the circuit given in Fig. 8.10, change the 1-kΩ resistor to 500 Ω. Assuming that $I_{DSS} = 16$ mA and $V_p = -4$ V, determine (a) v_{GS}, (b) i_D, and (c) v_{DS}.

8.7 Given the circuit shown in Fig. P8.7, the JFET has $V_p = -2$ V.

(**a**) Assuming active-region operation, and that $I_{DSS} = 10$ mA, determine the value of R for which $v_{GS} = 0$ V.

(**b**) Assuming active-region operation, find I_{DSS} when $R = 400$ Ω.

(**c**) If I_{DSS}, R, and the region of operation are not known, what is v_{GS} when $v_{DS} = 3$ V?

(**d**) If R is not known, what is the region of operation when $v_{DS} = 2$ V and $v_{GS} = -1.5$ V?

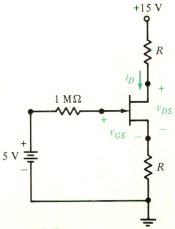

+15 V

R

i_D

$+$
v_{DS}
$-$

1 MΩ

$+$
v_{GS}
$-$

5 V

R

Fig. P8.7

8.8 For the circuit shown in Fig. P8.8, the JFET has $I_{DSS} = 9$ mA and $V_p = -3$ V. Assuming active-region operation, determine (a) v_{GS}, (b) i_D, and (c) v_{DS}. (*Hint:* Use Thévenin's theorem.)

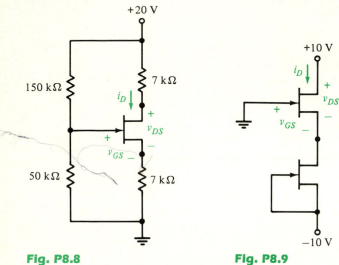

Fig. P8.8   **Fig. P8.9**

8.9 Both JFETs in the circuit shown in Fig. P8.9 have $I_{DSS} = 4$ mA and $V_p = -2$ V. Find (a) i_D, (b) v_{GS}, and (c) v_{DS}.

8.10 Repeat Problem 8.9 for the circuit shown in Fig. P8.10.

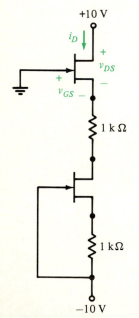

Fig. P8.10

Problems for Section 8.2

8.11 An n-channel depletion MOSFET has $I_{DSS} = 10$ mA and $V_p = -2$ V. For small values of v_{DS}, determine the drain-source resistance r_{DS} when (a) $v_{GS} = 1$ V and (b) $v_{GS} = 2$ V.

8.12 For the circuit given in Fig. 8.16 ($I_{DSS} = 8$ mA, $V_p = -2$ V), determine the value of R for which the depletion MOSFET will operate on the border between the active and the ohmic regions.

8.13 Given the MOSFET circuit shown in Fig. P8.13, the depletion NMOS transistor has $I_{DSS} = 8$ mA and $V_p = -2$ V. When $R_D = R_S = 250$ Ω, find (a) v_{GS}, (b) i_D, and (c) v_{DS}.

8.14 Consider the circuit given in Problem 8.13:

(a) If $R_D = R_S = R$, then what value of R results in active-region operation and $v_{GS} = 1$ V?

(b) Find R_D and R_S such that $v_{DS} = v_{GS} = 1$ V.

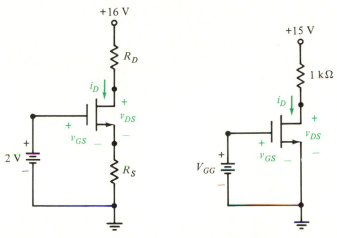

Fig. P8.13   **Fig. P8.15**

8.15 Given the circuit shown in Fig. P8.15, the depletion MOSFET has $I_{DSS} = 10$ mA and $V_p = -4$ V. Find v_{DS} when (a) $V_{GG} = 0$ V and (b) $V_{GG} = 5$ V.

8.16 An n-channel enhancement MOSFET has $K = 2.5 \times 10^{-4}$ A/V^2 and $V_T = 2$ V. For small values of v_{DS}, determine the drain-source resistance r_{DS} when (a) $v_{GS} = 4$ V, (b) $v_{GS} = 6$ V, and (c) $v_{GS} = 10$ V.

8.17 For the circuit given in Fig. 8.21 ($V_T = 2$ V, $K = 2.5 \times 10^{-4}$ A/V^2) determine the value of R for which the enhancement MOSFET will operate on the border between the active and the ohmic regions when (a) $V_{GG} = 4$ V, and (b) $V_{GG} = 10$ V.

8.18 Given the circuit shown in Fig. 8.21 with $K = 2.5 \times 10^{-4}$ A/V^2, $R = 1$ kΩ, $V_{GG} = 4$ V, and $V_T = 2$ V, find (a) i_D, and (b) v_{DS}.

8.19 For the circuit shown in Fig. P8.19, $V_{DD} = 20$ V. The enhancement NMOS

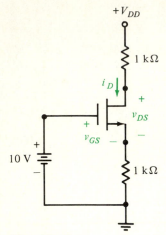

Fig. P8.19

transistor has $K = 2.5 \times 10^{-4}$ A/V^2 and $V_T = 2$ V. Find (a) v_{GS}, (b) i_D, and (c) v_{DS}. In what region does the MOSFET operate?

8.20 Repeat Problem 8.19 for the case that $V_{DD} = 9$ V.

Problems for Section 8.3

8.21 For the simple MOSFET circuit given in Fig. 8.22, the enhancement NMOS transistor has the drain characteristics shown in Fig. 8.23. For the case that $R = 625$ Ω and $V_{DD} = 10$ V, graphically determine v_{DS} when (a) $v_{GS} = 6$ V, (b) $v_{GS} = 8$ V, and (c) $v_{GS} = 11$ V.

8.22 For the simple MOSFET circuit given in Fig. 8.22, the enhancement NMOS transistor has the drain characteristics shown in Fig. 8.23. Given that $v_{DS} = 4$ V when $v_{GS} = 8$ V and $v_{DS} = 6$ V when $v_{GS} = 6$ V, determine (a) i_D when $v_{GS} = 8$ V, (b) i_D when $v_{GS} = 6$ V, (c) V_{DD}, and (d) R.

8.23 For the simple MOSFET circuit given in Fig. 8.22, the enhancement NMOS transistor has the drain characteristics shown in Fig. 8.23. Assuming that $V_{DD} = 12$ V, find the value of R for which $v_{DS} = 10$ V when $v_{GS} = 6$ V.

8.24 Change the supply voltage of the MOSFET inverter given in Fig. 8.26 to $V_{DD} = 10$ V. Given that both Q_1 and Q_2 have the drain characteristics shown in Fig. 8.27, graphically determine v_o for $v_1 = 0, 2, 4, \ldots , 12$ V.

8.25 Change the supply voltages of the MOSFET inverter given in Fig. 8.29 to $V_{DD} = 10$ V and $V_{GG} = 16$ V. Given that both Q_1 and Q_2 have the drain characteristics shown in Fig. 8.30, graphically determine v_o for $v_1 = 0, 2, 4, \ldots , 12$ V.

8.26 Change the supply voltage of the inverter given in Fig. 8.33 to $V_{DD} = 10$ V. Given that the depletion MOSFET Q_1 and the enhancement MOSFET Q_2 have the drain characteristics shown in Fig. 8.34 and Fig. 8.35, respectively, graphically determine v_o for $v_1 = 0, 2, 4, \ldots , 12$ V.

Problems for Section 8.4

8.27 For the circuit shown in Fig. P8.27, the enhancement MOSFET loads have a high resistance compared with the enhancement MOSFET drivers. Assume that all MOSFETs have threshold voltage V_T and that $V_{DD} \gg V_T$. What type of logic gate is this NMOS circuit?

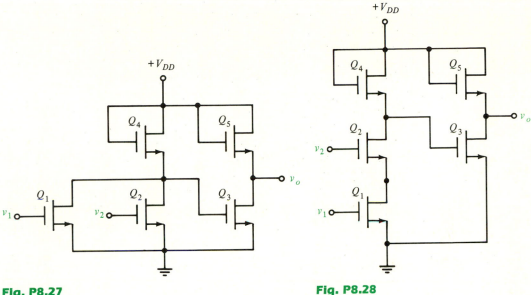

Fig. P8.27　　　　　　　　　　　　　　　　　　　**Fig. P8.28**

8.28 Repeat Problem 8.27 for the circuit shown in Fig. P8.28.
8.29 Repeat Problem 8.27 for the circuit shown in Fig. P8.29.

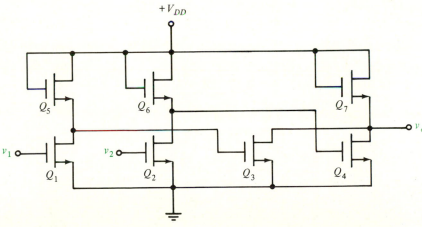

Fig. P8.29

8.30 Repeat Problem 8.27 for the circuit shown in Fig. P8.30.

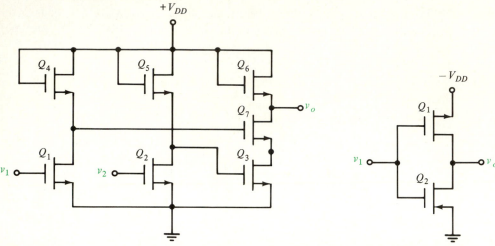

Fig. P8.30

Fig. P8.31

8.31 For the simple CMOS circuit shown in Fig. P8.31, Q_1 and Q_2 are enhancement MOSFETs with threshold voltages V_T and $-V_T$ respectively. Assuming that $V_{DD} > V_T$, determine the output voltage v_o when the input voltage is (a) $v_1 = 0$ and (b) $v_1 = -V_{DD}$.

8.32 Given the CMOS circuit shown in Fig. P8.32, the enhancement MOSFETs Q_1 and Q_2 have threshold voltage V_T, while Q_3 and Q_4 have threshold voltage $-V_T$. Suppose that $V_{DD} > V_T$, and the low voltage $-V_{DD}$ is logical 1 while the high voltage 0 V is logical 0. (This is negative logic.) What type of logic gate is this circuit?

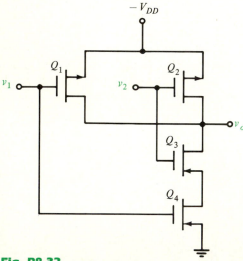

Fig. P8.32

8.33 The CMOS circuit shown in Fig. P8.33 is a "NO-DAMN-GOOD" gate. The enhancement MOSFETs Q_1 and Q_2 have threshold voltage V_T, while Q_3 and Q_4 have threshold voltage $-V_T$. Assume that, except for a difference in sign, all MOSFETs have the same characteristics. Then, assuming that $V_{DD} > V_T$, determine v_o when (a) $v_1 = v_2 = 0$, (b) $v_1 = v_2 = V_{DD}$, and (c) $v_1 = V_{DD}$, $v_2 = 0$.

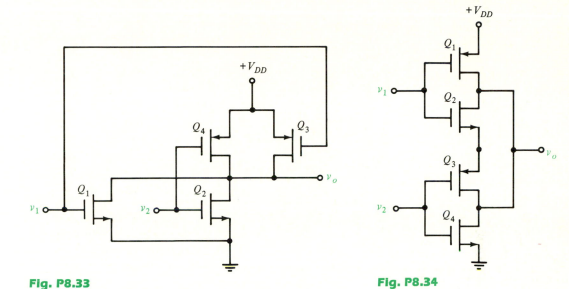

Fig. P8.33   **Fig. P8.34**

8.34 Repeat Problem 8.33 for another "NO-DAMN-GOOD" gate shown in Fig. P8.34, where $V_{DD}/2 > V_T$.

8.35 For the transmission gate given in Fig. 8.48, let $V = 5$ V and $V_T = 2$ V. Verify the transmission-gate operation for (a) $v_1 = 1$ V, (b) $v_1 = 2.5$ V, and (c) $v_1 = 4$ V.

8.36 For the transmission gate given in Fig. 8.48, $V_T = 2$ V, the low voltage is -5 V and the high voltage is 5 V. Given that $v = 5$ V, $\bar{v} = -5$ V, and the input voltage v_1 varies between -5 and 5 V, determine the range of values of v_1 for which (a) Q_1 is ON, Q_2 is OFF; (b) Q_1 is ON, Q_2 is ON; and (c) Q_1 is OFF, Q_2 is ON.

Chapter 9

Transistor Amplifiers

INTRODUCTION

Amplification is an important, yet fundamental, signal-processing task. In many situations, important signals arise, but are much too small in magnitude to be directly useful. For us to utilize their information content, such signals must be enlarged or amplified—hence, the need for amplifiers. Transistor amplifiers can be designed to result in voltage, current, and/or power gain of signals.

When desired signals are weak, typically, they are first strengthened with small-signal amplifiers. Such devices can be constructed using bipolar junction transistors (BJTs), field-effect transistors (FETs), or a combination of both types of components. These amplifiers can be constructed in either discrete or integrated-circuit (IC) form.

The range of signal frequencies that can be amplified depends upon whether or not capacitors are employed in the design of the amplifiers, and upon the inherent capacitances of the transistors themselves. By avoiding the use of capacitors for the coupling of stages of amplification, signals with frequency components extending down to direct current can be amplified. The basic device for accomplishing this is the differential amplifier.

The differential amplifier is a direct-coupled (dc) device that is typically the input stage of an operational amplifier (op amp). Because it requires no capacitors for coupling its stages, the op amp lends itself nicely to IC fabrication. It can be produced small in size and high in reliability. Operational amplifiers are useful for numerous linear applications, and nonlinear applications as well. They are versatile and predictable—they are the basic analog IC.

After small ac signals have been sufficiently amplified to produce large ac signals, they may be amplified further by power amplifiers (or large-signal amplifiers) so that there is sufficient power to drive the final load, for example, a loudspeaker, a motor, a recording instrument, or a cathode-ray tube.

9.1 BJT AMPLIFIERS

In digital-circuit applications, transistors typically operate either in cutoff or in saturation. (An exception is nonsaturating logic such as ECL where transistors are either in cutoff or in the active region close to saturation.) However, if we operate a transistor strictly in the active region, also known as the linear region, then it can exhibit amplification properties. To see how this happens, consider the collector characteristics shown in Fig. 9.1. Included with these characteristics is the load line corresponding to some associated circuit. As an example, suppose that the base current of the BJT is 40 μA. The intersection of curve $i_B = 40$ μA and the load line is the point labeled Q. This point is called the **quiescent operating point,** or **operating point,** or Q **point** for short. From Fig. 9.1, therefore, we see that the

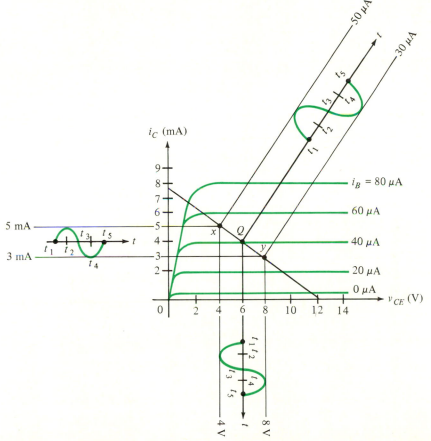

Fig. 9.1 Collector characteristics and load line of a BJT biased in the active region.

resulting collector circuit is $i_C = I_{CQ} = 4$ mA and the resulting collector–emitter voltage is $v_{CE} = V_{CEQ} = 6$ V. In other words, the transistor is **biased** (clearly in the active region) at $i_B = I_{BQ} = 40$ μA, $i_C = I_{CQ} = 4$ mA, and $v_{CE} = V_{CEQ} = 6$ V.

Now suppose that we add an ac component—in particular, a sinusoid—to the base current so that it varies between 30 and 50 μA as illustrated in Fig. 9.1. Specifically, if $t_1 = 0$, then the total base current is $i_B = I_{BQ} + i_b = 40 \times 10^{-6} + 10 \times 10^{-6} \sin \omega t$. (A lowercase subscript indicates an ac quantity.) Since the corresponding operation of the transistor is along the load line between points x and y, the resulting total collector current (ideally) is $i_C = I_{CQ} + i_c = 4 \times 10^{-3} + 1 \times 10^{-3} \sin \omega t$ and the total collector–emitter voltage (ideally) is $v_{CE} = V_{CEQ} + v_{ce} = 6 - 2 \sin \omega t$. Note that although the ac component of the base current has an amplitude of 10 μA, the ac component of the resulting collector current has an amplitude of 1 mA. In other words, the ac collector current is 100 times as large as the ac base current. Thus, the given transistor amplifies alternating current by a factor of 100—or we say that it has an ac gain of 100.

In general, we define the **ac current gain** h_{fe} (also called the **ac beta** β_{ac}) of a BJT to be the ratio of the ac collector current i_c (or variation in collector current Δi_C) to the ac base current i_b (or variation in base current Δi_B). That is,

$$h_{fe} = \frac{i_c}{i_b} = \frac{\Delta i_C}{\Delta i_B}$$

From Fig. 9.1, for example, we have that

$$h_{fe} = \frac{i_{Cx} - i_{Cy}}{i_{Bx} - i_{By}} = \frac{5 \times 10^{-3} - 3 \times 10^{-3}}{50 \times 10^{-6} - 30 \times 10^{-6}} = 100$$

Compare this with the dc beta h_{FE} (capital-letter subscripts), which is the ratio of dc (quiescent) collector current to dc (quiescent) base current. In other words, the dc beta is

$$h_{FE} = \frac{I_{CQ}}{I_{BQ}} = \frac{4 \times 10^{-3}}{40 \times 10^{-6}} = 100$$

so in this case we have that $h_{FE} = h_{fe}$. In general, however, h_{FE} and h_{fe} need not be identical.

Note that if the Q point is located too close either to cutoff or to saturation, the ac component of the base current will, respectively, either cut off or saturate the transistor. Even when the Q point is centrally located between cutoff and saturation, if this ac component becomes too large, then the transistor can be driven into cutoff or saturation. Regardless of the cause, if the transistor is cut off or saturated, the undesirable result will be that the output voltage v_{CE} and output current i_C will be distorted ("clipped").

Biasing the BJT

We have just seen that a BJT can be used as an amplifier when it is biased in the active region. Therefore, let us now concentrate on the biasing of BJTs. Shortly, we will see how to add an ac component that can be amplified.

Consider the simple dc *n-p-n* CE transistor circuit shown in Fig. 9.2. For this circuit, we have that

$$i_B = \frac{V_{CC} - v_{BE}}{R_B} \tag{9.1}$$

If the transistor is in the active region, then $v_{BE} \approx 0.7$ V for silicon (0.2 V for germanium). Once R_B and V_{CC} have been selected, therefore, the base current is, in essence, constant or fixed. For this reason, Fig. 9.2 is an example of a **fixed-bias circuit.**

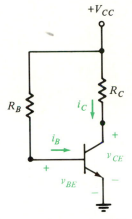

Fig. 9.2 Simple BJT circuit with fixed bias.

Figure 9.3 shows a BJT circuit with a **voltage-divider bias.** For such a situation, we can employ Thévenin's theorem. Doing so, we can erase the voltage-divider resistors R_1 and R_2, and then place a resistor $R_B = R_1 \parallel R_2 = R_1 R_2/(R_1 + R_2)$ and a voltage source $V_{BB} = R_2 V_{CC}/(R_1 + R_2)$ in series with the base as shown in Fig. 9.4. It is now apparent that the voltage-divider bias in Fig. 9.3 is really just another form of fixed bias.

For the case of fixed bias, when the transistor is in the active region, we have that

$$i_C = h_{FE} i_B$$

and

$$v_{CE} = -R_C i_C + V_{CC} = -R_C h_{FE} i_B + V_{CC}$$

Thus, we see that the operating point Q is dependent on the value of h_{FE}. If a

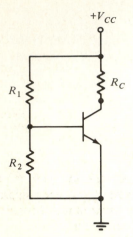

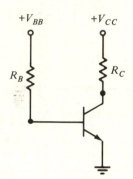

Fig. 9.3 A BJT circuit with a voltage-divider bias.

Fig. 9.4 Voltage-divider bias after application of Thévenin's theorem.

fixed-bias circuit is duplicated using another transistor of the same type as the original, the value of h_{FE} can differ significantly. Even a temperature change will alter the value of h_{FE}. In either case, the Q point can wind up at an undesirable location. For these reasons, we conclude that a fixed-bias circuit does not have good stability.

There is another important reason why a fixed-bias circuit may result in unstable operation. Recall that the reverse saturation current I_{CBO} approximately doubles for every 10°C rise in temperature. A current i_C going across the collector junction will cause a rise in its temperature. This, in turn, increases I_{CBO} (and beta, too). From Equation (7.13), we see that there will be an increase in collector current, and, thus, an even higher junction temperature. The result can be a cumulative process, known as **thermal runaway,** that will produce excessive heat and destroy the transistor. Even if such an extreme situation does not occur, it may be that the Q point moves close to or actually goes into saturation when the temperature increases.

A common way to improve bias stability is to put a resistor R_E in series with the emitter of the transistor as shown in Fig. 9.5. For analysis purposes, it is convenient to replace the voltage-division portion of the circuit by its Thévenin equivalent. Doing so, we obtain the equivalent circuit shown in Fig. 9.6, where $R_B = R_1 \parallel R_2 = R_1 R_2/(R_1 + R_2)$ and $V_{BB} = R_2 V_{CC}/(R_1 + R_2)$. By KVL, we have that

$$V_{BB} = R_B i_B + v_{BE} + R_E(i_B + i_C) \tag{9.2}$$

For active-region operation, $i_C = h_{FE}i_B$ and $v_{BE} = 0.7$ V (for silicon). Thus, Equation (9.2) becomes

$$R_B i_B + R_E(i_B + h_{FE}i_B) = V_{BB} - v_{BE}$$

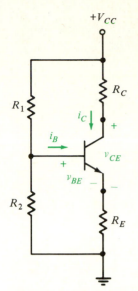

Fig. 9.5 A BJT self-biasing circuit.

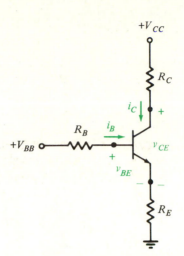

Fig. 9.6 Thévenin equivalent of self-biasing circuit.

from which

$$i_B = \frac{V_{BB} - v_{BE}}{R_B + (1 + h_{FE})R_E} \tag{9.3}$$

Unlike the case of fixed bias, from Equation (9.3), we see that the base current is dependent on h_{FE}. However, this is actually beneficial. Suppose that I_{CBO} and h_{FE} increase due to a rise in temperature. As mentioned above, this tends to increase i_C. But from Equation (9.3) we see that an increase in h_{FE} means a decrease in i_B, and this, in turn, tends to decrease i_C. Hence, the presence of R_E means that the overall increase in i_C is less than it would be if R_E were not present in the circuit. This results in more stable operation. By inserting the resistance R_E in series with the emitter of the transistor as shown in Fig. 9.5, we obtain what is known as a **self-biasing circuit**.

Example 9.1

For the self-biasing circuit given in Fig. 9.5, suppose that $R_1 = 20$ kΩ, $R_2 = 12$ kΩ, $R_C = R_E = 1$ kΩ, $V_{CC} = 10$ V, and $h_{FE} = 100$. Then, for the equivalent circuit given in Fig. 9.6, we have that

$$R_B = \frac{R_1 R_2}{R_1 + R_2} = \frac{(20 \times 10^3)(12 \times 10^3)}{20 \times 10^3 + 12 \times 10^3} = 7.5 \text{ kΩ}$$

and

$$V_{BB} = \frac{R_2}{R_1 + R_2} V_{CC} = \frac{12 \times 10^3}{20 \times 10^3 + 12 \times 10^3}(10) = 3.75 \text{ V}$$

Therefore, for active-region operation, Equation (9.3) yields

$$i_B = \frac{3.75 - 0.7}{7500 + (1 + 100)(1000)} = 28.1 \ \mu\text{A}$$

Thus,

$$i_C = h_{FE}i_B = 100(28.1 \times 10^{-6}) = 2.81 \text{ mA} \tag{9.4}$$

and

$$v_{CE} = -R_C i_C + V_{CC} - R_E(i_B + i_C)$$
$$= -1000(2.81 \times 10^{-3}) + 10 - 1000(28.1 \times 10^{-6} + 28.1 \times 10^{-3})$$
$$= 4.35 \text{ V}$$

Since v_{CE} is well above the saturation voltage of 0.2 V, this confirms that the transistor is biased in the active region, and the coordinates of the operating point are $I_{CQ} = 2.81$ mA and $V_{CEQ} = 4.35$ V.

Small-Signal ac Models

Given a transistor biased in the active region, let us see how ac components can be added to the existing quiescent operating (dc) conditions. One common way to do this is capacitively to couple an ac source (an ideal voltage source v_s in series with a resistance R_s) to the base of a transistor by using a capacitor, called a **coupling capacitor,** as seen for the CE amplifier shown in Fig. 9.7. Since the capacitor C is an open circuit to direct current, the addition of the ac source will have no effect on the Q point (dc conditions) of the transistor. However, let us assume that the capacitance C is large enough that, in essence, it behaves as a short circuit for the ac signals produced by voltage source v_s. In this circumstance, therefore, the ac source will add ac components to the transistor's voltages and currents. One way to describe the relationships between these ac components is to do so graphically as illustrated by Fig. 9.1. But this is not the only approach to the ac analysis of transistor amplifiers.

When the transistor in the circuit given in Fig. 9.7 is biased in the active (linear) region, we can apply the principle of superposition to the circuit. For the case that $v_s = 0$ and $V_{CC} \neq 0$ (the dc case), the ac source is replaced by a short circuit and the capacitor is replaced by an open circuit. The analysis then proceeds as was discussed in the preceding section ("Biasing the BJT"). On the other hand, for the case that $v_s \neq 0$ and $V_{CC} = 0$, the dc supply (implicit by the notation $+V_{CC}$) is replaced by a short circuit—and, as mentioned above, so too is the capacitor. The

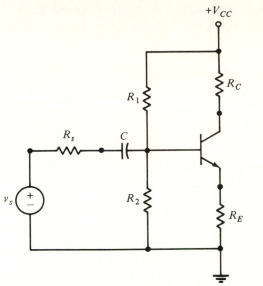

Fig. 9.7 A BJT capacitively coupled to an ac source.

resulting **ac equivalent** of the circuit given in Fig. 9.7 is shown in Fig. 9.8. Note that since a short circuit is placed between the node labeled $+V_{CC}$ and the reference, in the ac-equivalent circuit, R_1 and R_2 are connected in parallel. For the ac-equivalent circuit given in Fig. 9.8, the transistor's alternating currents and voltages are denoted by using lowercase letter subscripts (e.g., i_b, i_c, v_{be}, v_{ce}), and these are the ac components of the corresponding total currents and voltages.

Now, however, the question arises as to how we can analyze an ac-equivalent transistor circuit such as the one given in Fig. 9.8. That is, other than a graphical approach such as illustrated by Fig. 9.1, is there some description of a transistor that relates its alternating currents and voltages? The answer is yes. We can model the ac behavior of a transistor in a number of ways. We will begin, however, by

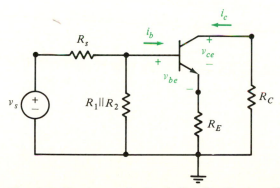

Fig. 9.8 The ac-equivalent circuit of a transistor amplifier.

considering simple models. It is always our assumption, though, that the ac components under consideration are small enough that the transistor will operate in the linear region at all times.

Another assumption will be made as well. For the sake of simplified numerical calculations, the collector characteristics given in Fig. 9.1 were so drawn that the curves appear parallel and equally spaced. In actuality, however, this is not the case. As a consequence, a sinusoidal variation in the base current will produce somewhat distorted sinusoids as the collector current and the collector–emitter voltage. Even an applied sinusoidal base–emitter voltage will produce a somewhat distorted sinusoidal base current because of the nonlinear characteristic of the base–emitter junction of a BJT (see Fig. 7.11). However, if the ac signals are small, these nonlinear distortions become insignificant, and we can model the transistor's ac behavior with a linear circuit. It is because of this that we refer to such linear circuits as **small-signal (ac) models.**

Given the transistor shown in Fig. 9.9 with some of its associated alternating currents and voltages, the first simple small-signal model we present is shown in Fig. 9.10. The ac base resistance r_π is also designated by h_{ie}. Since $i_b = v_{be}/r_\pi$, the current-dependent current source can be relabeled as

$$h_{fe}i_b = \frac{h_{fe}}{r_\pi}v_{be} = g_m v_{be}$$

where $g_m = h_{fe}/r_\pi$ is known as the **transconductance.** With this definition of transconductance, the model given in Fig. 9.10 is equivalent to the circuit shown in Fig. 9.11. For the latter case, though, the current source is dependent upon a voltage (v_{be}) instead of a current (i_b). Still another simple small-signal model of a BJT is shown in Fig. 9.12. This model makes use of the emitter resistance r_e instead of r_π. Note that

$$r_e = \frac{v_{be}}{i_b + h_{fe}i_b} = \frac{v_{be}}{(1 + h_{fe})i_b} = \frac{r_\pi}{1 + h_{fe}} \approx \frac{r_\pi}{h_{fe}} = \frac{1}{g_m} \tag{9.5}$$

from which

$$r_\pi = (1 + h_{fe})r_e \approx h_{fe}r_e \tag{9.6}$$

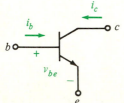

Fig. 9.9 A BJT.

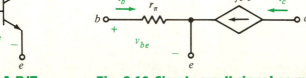

Fig. 9.10 Simple small-signal ac model of a BJT.

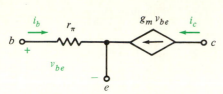

Fig. 9.11 Second small-signal model.

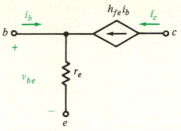

Fig. 9.12 Another small-signal model.

It can be shown that if an *n-p-n* transistor is biased in the active region with operating point coordinates $i_C = I_{CQ}$ and $v_{CE} = V_{CEQ}$, then the transconductance approximately is given by

$$g_m = \frac{I_{CQ}}{V_T}$$

where $V_T = T/11{,}586$ [Equation (6.14)] and T is the temperature in kelvins. At 25°C (298 K), we have that $V_T \approx 26$ mV, so a good approximation for transconductance is

$$g_m = \frac{I_{CQ}}{0.026} \tag{9.7}$$

From Equation (9.5), therefore, we have the approximations

$$r_e = \frac{1}{g_m} = \frac{0.026}{I_{CQ}} \tag{9.8}$$

and

$$r_\pi = h_{fe} r_e = \frac{0.026 h_{fe}}{I_{CQ}} \tag{9.9}$$

For the case of a *p-n-p* transistor, replace I_{CQ} by $|I_{CQ}|$ in Equations (9.7), (9.8), and (9.9). It should also be pointed out that a given small-signal ac model can be used for either an *n-p-n* transistor, or a *p-n-p* transistor—there is no need to be concerned with signs as in the dc case.

Example 9.2

To demonstrate the use of a small-signal model, let us replace the transistor in the ac circuit shown in Fig. 9.8 by the model given in Fig. 9.12. The result is shown in Fig. 9.13, where $R_p = R_1 \parallel R_2 = R_1 R_2/(R_1 + R_2)$. Of course, this circuit can be analyzed by nodal or mesh analysis. However, let us take an alternative approach, thereby defining some new terms along the way.

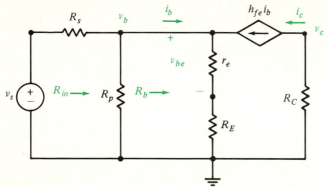

Fig. 9.13 Transistor amplifier ac-equivalent circuit incorporating small-signal model.

The resistance between the base and the reference, looking into the base of the transistor, is $R_b = v_b/i_b$, where v_b is the base voltage (with respect to the reference). Since

$$v_b = r_e(i_b + h_{fe} i_b) + R_E(i_b + h_{fe} i_b)$$

then

$$R_b = \frac{v_b}{i_b} = (1 + h_{fe})(r_e + R_E) \approx h_{fe}(r_e + R_E) \qquad \textbf{(9.10)}$$

The resistance between the base and the reference, looking into the amplifier, is called the **input resistance** R_{in} of the amplifier. We have, therefore, that

$$R_{in} = R_p \parallel R_b = \frac{R_p R_b}{R_p + R_b} \qquad \textbf{(9.11)}$$

The **voltage gain** A_v from base (amplifier input) to collector (amplifier output) is the ratio of collector voltage v_c to base voltage v_b, that is, $A_v = v_c/v_b$. For the circuit given in Fig. 9.13, the voltage gain is

$$A_v = \frac{v_c}{v_b} = \frac{-R_C h_{fe} i_b}{R_b i_b} = \frac{-h_{fe} R_C}{(1 + h_{fe})(r_e + R_E)} \approx \frac{-R_C}{r_e + R_E} \qquad \textbf{(9.12)}$$

Since $v_c = A_v v_b$ and A_v is a negative number, then when the source voltage v_s is a sinusoid, the input v_b is a sinusoid, and the output voltage v_c is a sinusoid that is 180 degrees out of phase with the input v_b.

From Equation (9.12), we see that if we decrease the denominator $r_e + R_E$, we increase the voltage gain A_v. The resistance r_e is determined by the transistor's Q point [Equation (9.8)]. However, R_E is external to the transistor. Although setting R_E to zero will increase the gain, it will also change the operating point and cause a decrease in the stability of operation. Suppose, therefore, that we place in parallel with R_E a capacitor whose impedance to the ac signals produced by v_s is negligible compared with R_E. The addition of such a capacitor will not affect the dc conditions (quiescent operating point) of the circuit, so r_e remains unchanged. However, since for ac signals R_E is short-circuited or bypassed by the capacitor, in the ac circuit we replace R_E by a short circuit. Thus, by adding such an element, called a **bypass capacitor,** to the amplifier, we get a voltage gain of

$$A_v = \frac{-h_{fe}R_C}{(1 + h_{fe})r_e} \approx -\frac{R_C}{r_e} \qquad (9.13)$$

The voltage gain A_s from the ac source to the amplifier output (collector) is the ratio of collector voltage v_c to source voltage v_s, that is, $A_{vs} = v_c/v_s$. For the circuit given in Fig. 9.13, by voltage division

$$v_b = \frac{R_{in}}{R_{in} + R_s} v_s$$

Thus,

$$A_{vs} = \frac{v_c}{v_s} = \frac{v_c}{v_b} \frac{v_b}{v_s} = \frac{A_v R_{in}}{R_{in} + R_s}$$

where R_{in} is given by Equation (9.11) and A_v is given by Equation (9.12).

■

Additional Amplifier Principles

Although the CE amplifier given in Fig. 9.7 is not loaded, in general, it may have some load resistance R_L. Such a situation, in which R_L is capacitively coupled to the amplifier, is depicted in Fig. 9.14. Note that for this circuit, in addition to the coupling capacitors C_1 and C_2, there is also a bypass capacitor C_E.

Figure 9.15 shows typical collector characteristics for the transistor. Also shown are the quiescent operating point Q and the load line obtained for the dc case (C_1, C_2, and C_E behave as open circuits). This line, which is described by the equation

$$i_C = -\frac{v_{CE}}{R_C + R_E} + V_{CC}$$

is, for obvious reasons, called the **dc load line.** It passes through the Q point and

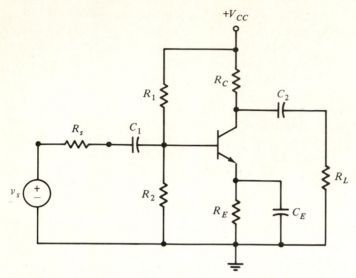

Fig. 9.14 A CE amplifier with a capacitively coupled external load.

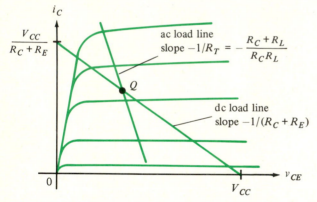

Fig. 9.15 Collector characteristics with ac and dc load lines.

has a slope of $-1/(R_C + R_E)$. For the ac case, we again assume that C_1, C_2, and C_E behave as short circuits. The resulting ac-equivalent circuit is shown in Fig. 9.16, where $R_p = R_1 \parallel R_2 = R_1 R_2/(R_1 + R_2)$ and $R_T = R_C \parallel R_L = R_C R_L/(R_C + R_L)$. In this case, however, the transistor operates on a line that passes through the Q point but has a slope of $-1/R_T$. This line, known as the **ac load line,** is indicated in Fig. 9.15. [If the bypass capacitor C_E is omitted, the slope of the ac load line is $-1/(R_T + R_E)$.]

If we replace the transistor in the ac-equivalent circuit of Fig. 9.16 by the simple model given in Fig. 9.12, we get the circuit shown in Fig. 9.17. As for the case of the circuit in Fig. 9.13, we may determine that $R_b = (1 + h_{fe})r_e \approx h_{fe}r_e$, $R_{in} = R_p \parallel R_b = R_p R_b/(R_p + R_b)$, and $A_v = -h_{fe}R_T/(1 + h_{fe})r_e \approx -R_T/r_e$. Be-

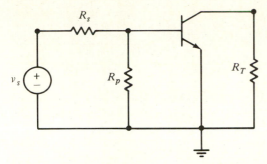

Fig. 9.16 The ac-equivalent circuit of the given transistor amplifier.

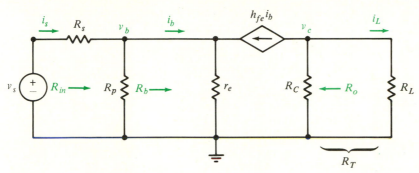

Fig. 9.17 Transistor amplifier incorporating small-signal model.

cause there is an external load R_L (that is why R_T is explicitly shown as the parallel combination of R_C and R_L), another important parameter of the amplifier is its **output** (Thévenin-equivalent) **resistance** R_o. As we know from our study of electric circuits, this is the resistance between the collector and the reference (with the load removed) when the input v_s is set to zero. It is a routine matter of circuit analysis to show that $R_o = R_C$ (see Problem 9.13).

Another network function is the **current gain.** For the given transistor amplifier, the (ac) current gain A_i from the base to the load is $A_i = i_L/i_b$. From the circuit given in Fig. 9.17, by current division we immediately have

$$A_i = \frac{i_L}{i_b} = \frac{-h_{fe}R_C}{R_C + R_L} \tag{9.14}$$

The current gain A_{is} from the source to the load is defined to be $A_{is} = i_L/i_s$. For the given amplifier, by current division, we know that $i_b = R_p i_s/(R_p + R_b)$. Thus, we have that

$$A_{is} = \frac{i_L}{i_s} = \frac{i_L}{i_b}\frac{i_b}{i_s} = \frac{A_i R_p}{R_p + R_b}$$

where $R_b = (1 + h_{fe})r_e$ and A_i is given by Equation (9.14).

Example 9.3

As another example of the use of a small-signal model, consider the BJT circuit shown in Fig. 9.18, where the resistor values are such that the transistor is biased in the active region. Figure 9.19 shows the ac-equivalent circuit for the case that the transistor is replaced by the model given in Fig. 9.10, and where $R_p = R_1R_2/(R_1 + R_2)$. Note that since R_L is the load, the emitter voltage v_e is the output voltage and the collector is common to both input and output. Therefore, such a circuit is known as a **common-collector amplifier.**

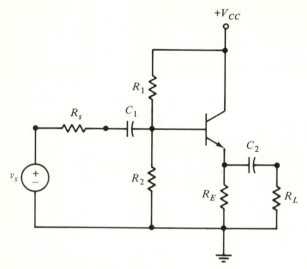

Fig. 9.18 Emitter follower.

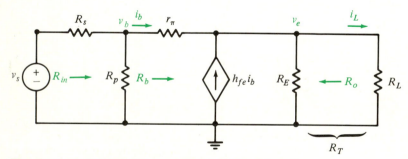

Fig. 9.19 Emitter-follower ac-equivalent circuit.

Let us designate the parallel connection of R_E and R_L by R_T. Thus,

$$v_e = R_T(i_b + h_{fe}i_b) = (1 + h_{fe})R_Ti_b$$

and

$$v_b = r_\pi i_b + v_e = [r_\pi + (1 + h_{fe})R_T]i_b \qquad (9.15)$$

Hence, the voltage gain A_v from base to emitter is

$$A_v = \frac{v_e}{v_b} = \frac{(1 + h_{fe})R_T}{r_\pi + (1 + h_{fe})R_T} \approx \frac{h_{fe}R_T}{r_\pi + h_{fe}R_T} \qquad (9.16)$$

For the case that $h_{fe}R_T \gg r_\pi$, we have that $A_v \approx 1$. In other words, $v_e \approx v_b$ and the emitter voltage "follows" the base voltage. It is for this reason that the common-collector circuit is usually called an **emitter follower.**

The fact that the output (emitter) voltage approximates the base voltage is not the only important property of an emitter follower. The resistance between the base and the reference looking into the base is, from Equation (9.15),

$$R_b = \frac{v_b}{i_b} = r_\pi + (1 + h_{fe})R_T \approx r_\pi + h_{fe}R_T \qquad (9.17)$$

and for large values of h_{fe} and R_T, R_b can be made very large. The input resistance of the emitter follower, therefore, is

$$R_{in} = R_p \parallel R_b = \frac{R_p R_b}{R_p + R_b}$$

where $R_p = R_1 R_2/(R_1 + R_2)$. If R_1 and R_2 are selected to be large, the input resistance can be made large.

The output resistance R_o of of the emitter follower can be determined from the circuit shown in Fig. 9.20. Since $R_o = R_E \parallel R'$, let us first find R'. For the sake of simplicity, denote the parallel connection of R_s and R_p by R. By KCL,

$$\frac{v_b}{R} + \frac{v_b - v_o}{r_\pi} = 0$$

from which

$$v_b = \frac{Rv_o}{r_\pi + R}$$

Substituting this expression for v_b into

$$i' = -(1 + h_{fe})i_b = -(1 + h_{fe})\frac{v_b - v_o}{r_\pi}$$

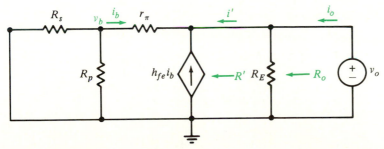

Fig. 9.20 Circuit for determining R_o for an emitter follower.

we eventually get

$$R' = \frac{v_o}{i'} = \frac{r_\pi + R}{1 + h_{fe}} \approx \frac{r_\pi + R}{h_{fe}} \qquad\qquad (9.18)$$

where $R = R_s R_p / (R_s + R_p)$, and the output resistance is

$$R_o = R_E \parallel R'$$

Dividing the value $r_\pi + R$ by $1 + h_{fe}$ as indicated in Equation (9.18) means that R' can be made relatively small and, consequently, so can the output resistance R_o.

The properties of approximately unity gain, a high input resistance, and a low output resistance allow the emitter follower to be used as a buffer or isolation amplifier for connecting a low-resistance load to a high-resistance source.

Having considered some simple small-signal ac models of a BJT biased in the active region, let us now discuss a model that yields greater accuracy. Shown in Fig. 9.21 is a BJT and its (CE) **hybrid-parameter** or **h-parameter** model. The four h parameters used in this model can be determined as follows:

(1) If we place a short circuit between the collector and emitter ($v_{ce} = 0$), then from Fig. 9.21(b) we see that $v_{be} = h_{ie} i_b$. Thus,

$$h_{ie} = \left. \frac{v_{be}}{i_b} \right|_{v_{ce}=0} \qquad \text{(ohms)}$$

and h_{ie} is called the **short-circuit input resistance.**

(2) If the base is open-circuited ($i_b = 0$), then $v_{be} = h_{re} v_{ce}$. Thus,

$$h_{re} = \left. \frac{v_{be}}{v_{ce}} \right|_{i_b=0} \qquad \text{(dimensionless)}$$

and h_{re} is called the **open-circuit reverse voltage gain.**

(3) If $v_{ce} = 0$, then there is no current through h_{oe} and $i_c = h_{fe} i_b$. Thus,

$$h_{fe} = \left. \frac{i_c}{i_b} \right|_{v_{ce}=0} \qquad \text{(dimensionless)}$$

and h_{fe} is called the **short-circuit forward current gain.**

(4) If $i_b = 0$, then $v_{ce} = i_c / h_{oe}$. Thus,

$$h_{oe} = \left. \frac{i_c}{v_{ce}} \right|_{i_b=0} \qquad \text{(mhos or siemens)}$$

and h_{oe} is called the **open-circuit output conductance.**

Typical numerical values for h parameters are:

$$h_{ie} = 2\text{ k}\Omega, \qquad h_{re} = 10^{-4}, \qquad h_{fe} = 100, \qquad h_{oe} = 10^{-5}\ \mho$$

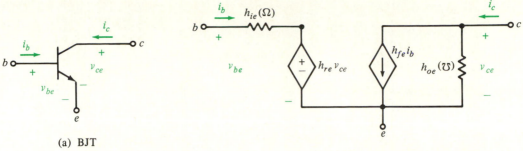

(a) BJT

(b) h-parameter model of a BJT

Fig. 9.21 A BJT and its h-parameter model.

For the case that we ignore (set to zero) h_{re} and h_{oe}, the h-parameter model given in Fig. 9.21(b) is reduced to the simple small-signal model shown in Fig. 9.10.

If we replace the BJT in the ac circuit given in Fig. 9.16 by the h-parameter model given in Fig. 9.21(b), for example, we get the ac circuit shown in Fig. 9.22. Note that, in this case, since $v_e = 0$, then $v_{be} = v_b$ and $v_{ce} = v_c$.

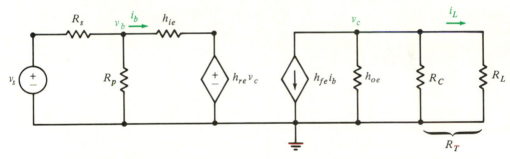

Fig. 9.22 Transistor amplifier incorporating the h-parameter BJT model.

9.2 FET AMPLIFIERS

Just as a BJT biased in the active region can be used for amplification purposes, so too can an FET biased in the active region be employed as a small-signal ac amplifier. We begin our discussion of small-signal FET amplifiers by considering the case of depletion-type FETs, that is, JFETs and depletion MOSFETs.

Figure 9.23 shows an n-channel JFET (depletion-type FET) circuit, and Fig. 9.24 shows the transfer characteristic of the JFET. Since $i_G = 0$, then $v_{GS} = -V_{GG}$. Thus, the intersection of the vertical line, whose equation is $v_{GS} = -V_{GG}$, and the transfer characteristic curve at point Q graphically determines the quiescent drain current I_{DQ}. Analytically, the equation $i_D = I_{DSS}(1 - v_{GS}/V_p)^2$ can be used. From a realistic point of view, however, a JFET's transfer characteristic varies with temperature and from device to device. Suppose that such a variation range corresponds to the shaded region indicated in Fig. 9.24. In other

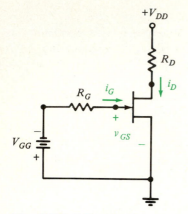

Fig. 9.23 A fixed-bias JFET circuit.

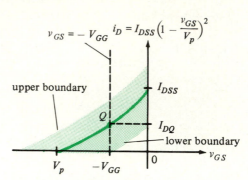

Fig. 9.24 JFET transfer characteristic and its range of variation.

words, let us allow the possibility that the transfer characteristic of the JFET could be anywhere within the shaded region. For the extreme case, for instance, that the transfer characteristic corresponds to the lower boundary, since v_{GS} is "fixed" at $v_{GS} = -V_{GG}$, the JFET actually will be cut off ($i_D = 0$). Because of such a possibility, this type of bias, called **fixed bias,** is undesirable.

An improvement on fixed bias is the arrangement shown in Fig. 9.25. In this case (since $i_G = 0$), we have that $i_D = -v_{GS}/R_S$ and the line described by this equation is plotted with the JFET's transfer characteristic (and range of variation) in Fig. 9. 26. Note, however, that even for the extreme case the transfer characteristic is at the lower boundary, the JFET will not be cut off. Thus, this type of bias

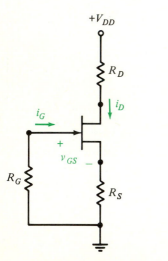

Fig. 9.25 Self-bias JFET circuit.

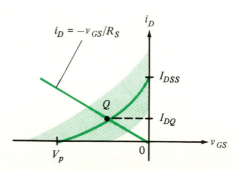

Fig. 9.26 JFET transfer characteristic and its range of variation.

is superior to fixed bias. Since $v_{GS} = -R_S i_D$, the value of v_{GS} depends upon one of the JFET's own parameters (i_D), and, therefore, this is an example of what is known as **self-bias.** Note that if the transfer characteristic curve changes position, then, in general, so will the value of the quiescent drain current I_{DQ}. The possible variation in I_{DQ} can be reduced by making the line $i_D = -v_{GS}/R_S$ more nearly horizontal (i.e., by increasing R_S). Doing this, though, will put the Q point closer to cutoff. There is, however, an alternative way to decrease the dependence of I_{DQ} on the variation of the JFET's characteristics.

To accomplish this, we combine self-bias and fixed bias as shown in Fig. 9.27. In this case, we have that $i_D = (-v_{GS} + V_{GG})/R_S = -v_{GS}/R_S + V_{GG}/R_S$, and the line described by this equation is shown in Fig. 9.28. Of course, a separate second supply voltage V_{GG} need not be used as the series connection of V_{GG} and R_G can represent the Thévenin equivalent of a voltage divider connected to the supply voltage V_{DD}.

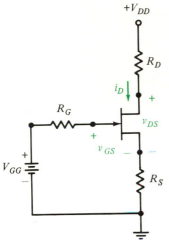

Fig. 9.27 Combination of fixed and self-bias.

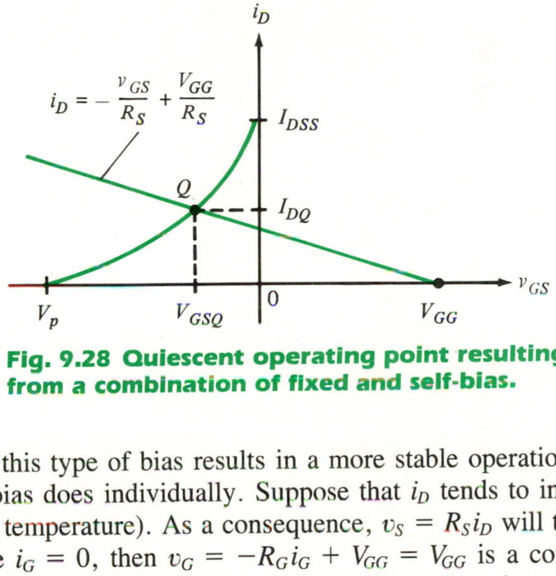

Fig. 9.28 Quiescent operating point resulting from a combination of fixed and self-bias.

Let us now see why this type of bias results in a more stable operation than either fixed bias or self-bias does individually. Suppose that i_D tends to increase (e.g., due to a change in temperature). As a consequence, $v_S = R_S i_D$ will tend to increase. However, since $i_G = 0$, then $v_G = -R_G i_G + V_{GG} = V_{GG}$ is a constant, and $v_{GS} = v_G - v_S$ will tend to decrease. This, in turn, will tend to decrease i_D, thereby opposing the original tendency for an increase in i_D.

Example 9.4

For the amplifier circuit given in Fig. 9.29, the JFET has $I_{DSS} = 10$ mA and $V_p = -4$ V. Given $R_1 = 500$ kΩ, $R_2 = 250$ kΩ, and $V_{DD} = 15$ V, for the dc-

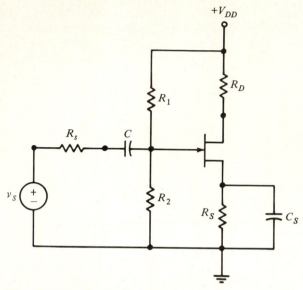

Fig. 9.29 A JFET amplifier.

equivalent circuit shown in Fig. 9.27, by Thévenin's theorem we have that $R_G = R_1 \parallel R_2 = 167$ kΩ and $V_{GG} = R_2 V_{DD}/(R_1 + R_2) = 5$ V. Suppose that we desire to have $v_{GS} = -2$ V $= V_{GSQ}$. Then

$$i_D = I_{DSS}\left(1 - \frac{v_{GS}}{V_p}\right)^2 = 10 \times 10^{-3}\left(1 - \frac{-2}{-4}\right)^2 = 2.5 \text{ mA} = I_{DQ}$$

Since the slope of the straight line in Fig. 9.28 is $-1/R_S$, then

$$R_S = \frac{V_{GG} + |V_{GSQ}|}{I_{DQ}} = \frac{5 + 2}{2.5 \times 10^{-3}} = 2.8 \text{ kΩ}$$

Of course, to confirm active-region operation, the value of R_D must be chosen such that $v_{DS} \geq v_{GS} - V_p = -2 + 4 = 2$ V. Specifically, for the case that $R_D = 1$ kΩ, then

$$v_{DS} = -R_D i_D + V_{DD} - R_S i_D = -(R_D + R_S)i_D + V_{DD}$$

$$= -(1000 + 2800)(2.5 \times 10^{-3}) + 15 = 5.5 \text{ V}$$

and $v_{DS} > v_{GS} - V_p$.

∎

Biasing Enhancement MOSFETs

Let us now consider the case of an enhancement-type FET. Figure 9.30 shows an *n*-channel enhancement MOSFET circuit, and Fig. 9.31 shows the transfer characteristic curve for the MOSFET. Again, by Thévenin's theorem, we can replace the

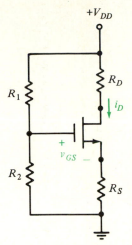

Fig. 9.30 Bias configuration for an enhancement MOSFET.

voltage divider consisting of R_1 and R_2 by a voltage source $V_{GG} = R_2 V_{DD}/(R_1 + R_2)$ and resistance $R_G = R_1 \parallel R_2 = R_1 R_2/(R_1 + R_2)$ in series with the gate of the MOSFET (see Fig. 9.27). For the circumstance that $R_S = 0$, then $v_{GS} = V_{GG}$. However, as for the fixed-bias case of a JFET, a variation in the MOSFET's characteristics can result in a significant change in the quiescent drain current. For this reason, we should have $R_S > 0$.

In general, $i_D = (-v_{GS} + V_{GG})/R_S = -v_{GS}/R_S + V_{GG}/R_S$, and the intersection of the line described by this equation and the MOSFET transfer characteristic determines the Q point. Such a situation is illustrated in Fig. 9.31. As was the case for the JFET, the resistance R_S produces a biasing stabilizing effect.

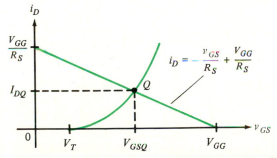

Fig. 9.31 Enhancement MOSFET transfer characteristic and resulting Q point.

Example 9.5

For the amplifier circuit shown in Fig. 9.32, the enhancement MOSFET has $V_T = 2$ V and $K = 2.5 \times 10^{-4}$ A/V^2. Given $R_1 = 1$ MΩ, $R_2 = 2$ MΩ, and

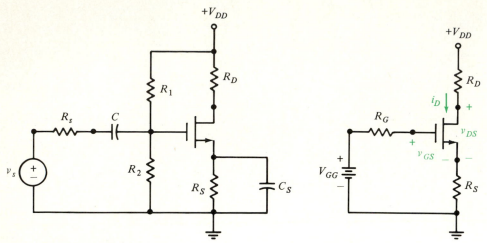

Fig. 9.32 A MOSFET amplifier.   **Fig. 9.33 The dc-equivalent circuit.**

$V_{DD} = 12$ V, for the dc-equivalent circuit shown in Fig. 9.33, $R_G = R_1 \parallel R_2 = 667$ kΩ and $V_{GG} = R_2 V_{DD}/(R_1 + R_2) = 8$ V. Suppose that we wish to have $i_D = 4$ mA $= I_{DQ}$. Then from

$$i_D = K(v_{GS} - V_T)^2$$

we have that

$$v_{GS} = V_T + \sqrt{i_D/K} = 2 + \sqrt{4 \times 10^{-3}/2.5 \times 10^{-4}} = 6 \text{ V} = V_{GSQ}$$

Since the slope of the straight line in Fig. 9.31 is $-1/R_S$, then

$$R_S = \frac{V_{GG} - V_{GSQ}}{I_{DQ}} = \frac{8 - 6}{4 \times 10^{-3}} = 500 \text{ }\Omega$$

Again, in order to confirm active-region operation, the value of R_D must be chosen such that $v_{DS} \geq v_{GS} - V_T = 6 - 2 = 4$ V. Specifically, for the case that $R_D = 1$ kΩ,

$$v_{DS} = -R_D i_D + V_{DD} - R_S i_D = -(R_D + R_S)i_D + V_{DD}$$

$$= -(1000 + 500)(4 \times 10^{-3}) + 12 = 6 \text{ V}$$

and $v_{DS} > v_{GS} - V_T$.

■

An alternative popular biasing scheme for an enhancement MOSFET is shown in Fig. 9.34. In this case, there is a "feedback" resistance R_F connected between the drain and the gate of the MOSFET. The MOSFET is in the active region when $v_{DS} \geq v_{GS} - V_T$. Since there is no gate current, $v_{GS} = v_{DS}$. Thus, $v_{DS} = v_{GS} > v_{GS} - V_T$ guarantees active-region operation.

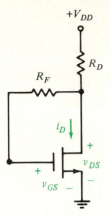

Fig. 9.34. **Biasing an enhancement MOSFET with the use of a feedback resistance.**

Suppose that for this type of bias, i_D tends to increase. As a consequence of this, $v_{GS} = v_{DS} = -R_D i_D + V_{DD}$ tends to decrease. However, a tendency of v_{GS} to decrease means that i_D has a tendency to decrease—and this opposes the original tendency for i_D to increase. Thus, we see that the feedback-bias arrangement shown in Fig. 9.34 possesses operational stability.

Example 9.6

The MOSFET in the circuit given in Fig. 9.34 has $K = 2.5 \times 10^{-4}$ A/V^2 and $V_T = 2$ V. Suppose that $R_D = 1$ kΩ, $R_F = 1$ MΩ, and $V_{DD} = 10$ V. Since

$$i_D = K(v_{GS} - V_T)^2$$

and

$$i_D = \frac{V_{DD} - v_{DS}}{R_D} = \frac{V_{DD} - v_{GS}}{R_D}$$

then

$$K(v_{GS} - V_T)^2 = \frac{V_{DD} - v_{GS}}{R_D}$$

$$2.5 \times 10^{-4}(v_{GS} - 2)^2 = \frac{10 - v_{GS}}{1000}$$

Solving this equation results in $v_{GS} = 6$ V $= V_{GSQ}$, and by using this value, we get $i_D = 4$ mA $= I_{DQ}$.

Small-Signal ac Models

Just as we modeled the small-signal ac behavior of a BJT, soo too we can use small-signal (ac) models for FETs. Figure 9.35 depicts a generic circuit symbol of an FET, that is, it can represent either a JFET, a depletion MOSFET, or an enhancement MOSFET. Some of the FET's associated alternating voltages and currents are indicated. A simple small-signal model of the given FET is presented in Fig. 9.36(a), where the ac resistance between gate and source is $r = \infty$. (Practically speaking, $r > 10^{10}\ \Omega$.) Instead of using the circuit symbol for a resistance, we can indicate the infinite resistance by an open circuit, and, consequently, the model in Fig. 9.36(a) can be redrawn in the equivalent form shown in Fig. 9.36(b). (Compare the FET small-signal model given in Fig. 9.36 with the BJT model given in Fig. 9.11.)

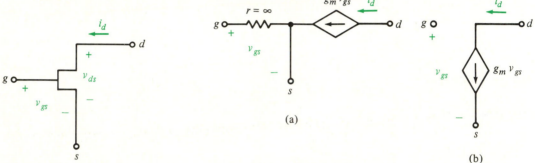

(a)

(b)

Fig. 9.35 A JFET, a depletion MOSFET, or an enhancement MOSFET.

Fig. 9.36 Simple small-signal ac model of a FET.

As can be seen from the drain characteristics of an FET, the drain current is dependent upon the gate–source voltage. For the small-signal (linear) model of the FET given in Fig. 9.36, the ac drain current is directly proportional to the ac gate–source voltage, that is, $i_d = g_m v_{gs}$. The proportionality constant g_m is the **transconductance** of the FET, and, as we can see, g_m is given by

$$g_m = \frac{i_d}{v_{gs}}$$

Since g_m is the ratio of the ac drain current i_d (that is, the variation in i_D) to the ac gate–source voltage v_{gs} (that is, the variation in v_{GS}), we can also write for fixed v_{DS} that

$$g_m = \frac{di_D}{dv_{GS}}$$

For the case of a (*n*-channel) JFET or depletion MOSFET, recall that

$$i_D = I_{DSS}\left(1 - \frac{v_{GS}}{V_p}\right)^2 \tag{9.19}$$

Thus,

$$g_m = \frac{di_D}{dv_{GS}} = -\frac{2I_{DSS}}{V_p}\left(1 - \frac{v_{GS}}{V_p}\right) \tag{9.20}$$

From Equation (9.19), we have that

$$1 - v_{GS}/V_p = \sqrt{i_D/I_{DSS}}$$

Substituting this fact into Equation (9.20) yields the alternative expression for transconductance

$$g_m = -\frac{2I_{DSS}}{V_p}\sqrt{\frac{i_D}{I_{DSS}}} = -\frac{2}{V_p}\sqrt{i_D I_{DSS}} \tag{9.21}$$

For the special case that $v_{GS} = 0$, the transconductance is designated g_{m0}. From Equation (9.20), setting $v_{GS} = 0$, we get

$$g_{m0} = -\frac{2I_{DSS}}{V_p} \tag{9.22}$$

Practically speaking, for a FET, g_{m0} and I_{DSS} can be measured much more accurately than can be V_p. Thus, by measuring g_{m0} and I_{DSS}, Equation (9.22) can be utilized to accurately determine the pinchoff voltage V_p. Furthermore, by using Equation (9.22), we have the additional formulas for the transconductance

$$g_m = g_{m0}\left(1 - \frac{v_{GS}}{V_p}\right) = g_{m0}\sqrt{\frac{i_D}{I_{DSS}}}$$

The values of g_m typically range between 10^{-4} and $2 \times 10^{-2}\,\mho$.

A more accurate small-signal model of a FET is shown in Fig. 9.37, where r_d is called the **drain** or **output resistance** of the FET. Since

$$i_d = g_m v_{gs} + v_{ds}/r_d$$

then when $v_{ds} = 0$, we have that $g_m = i_d/v_{gs}$ as before. For the case that $v_{gs} = 0$, though, we see that

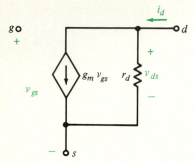

Fig. 9.37 More accurate small-signal model of a FET.

$$r_d = \frac{v_{ds}}{i_d}$$

This equation relating ac quantities is tantamount to saying for fixed v_{GS} that

$$r_d = \frac{dv_{DS}}{di_D}$$

A typical value of r_d for a JFET is $100\,\text{k}\Omega$, while for a MOSFET it is $10\,\text{k}\Omega$.

Example 9.7

For the JFET circuit given in Fig. 9.29, suppose that the capacitors C and C_S are large enough that they behave as short circuits for the ac signals produced by v_s. The corresponding ac-equivalent circuit is shown in Fig. 9.38, where $R_p = R_1 \parallel R_2 = R_1 R_2 / (R_1 + R_2)$.

In replacing the JFET by the simple small-signal model given in Fig. 9.36, we get the circuit shown in Fig. 9.39.

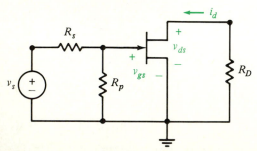

Fig. 9.38 An ac-equivalent circuit of the JFET amplifier given in Fig. 9.29.

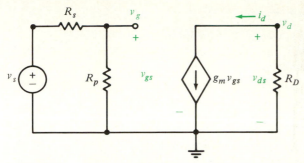

Fig. 9.39 A JFET amplifier ac-equivalent circuit incorporating small-signal model.

Since

$$v_{ds} = -R_D i_d = -R_D g_m v_{gs}$$

we have that the voltage gain A_v from gate to drain is

$$A_v = \frac{v_d}{v_g} = \frac{v_{ds}}{v_{gs}} = -g_m R_D \qquad\qquad (9.23)$$

If, instead of using the small-signal model given in Fig. 9.36, we had used the model of Fig. 9.37, then in the expression given by Equation (9.23) we would replace R_D by $r_d \parallel R_D = r_d R_D / (r_d + R_D)$.

In Example 9.4 (see Fig. 9.29), for the case that $I_{DSS} = 10$ mA, $V_p = -4$ V, $R_1 = 500$ kΩ, $R_2 = 250$ kΩ, $R_S = 2.8$ kΩ, and $V_{DD} = 15$ V, we have that $v_{GS} = -2$ V and $i_D = 2.5$ mA. Under these circumstances, from Equation (9.20), we have that the transconductance is

$$g_m = -\frac{2I_{DSS}}{V_p}\left(1 - \frac{v_{GS}}{V_p}\right)$$

$$= -\frac{2(10 \times 10^{-3})}{-4}\left(1 - \frac{-2}{-4}\right) = 2.5 \times 10^{-3} \, \mho$$

Thus, for the case that $R_D = 1$ kΩ, the voltage gain is

$$A_v = -g_m R_D = -2.5 \times 10^{-3}(1000) = -2.5$$

Note, however, that the (instantaneous) power absorbed by the input of the amplifier is $p_g = v_g^2 / R_p$, while the power absorbed by the output resistance R_D is $p_d = v_d^2 / R_D$. Therefore, the power gain A_p from gate to drain is

$$A_p = \frac{p_d}{p_g} = \frac{v_d^2/R_D}{v_g^2/R_p} = \left(\frac{v_d}{v_g}\right)^2 \frac{R_p}{R_D} = A_v^2 \frac{R_p}{R_D}$$

$$= (-2.5)^2 \frac{167 \times 10^3}{1000} = 1044$$

Amplifiers with High Input Resistance

When discussing BJT amplifiers, we saw how an emitter follower (see Fig. 9.18) can be used as an isolation amplifier. There is also a FET amplifier that corresponds to the BJT emitter follower.

Example 9.8

The JFET amplifier shown in Fig. 9.40 is known as a **source follower.** Replacing the JFET by the model given in Fig. 9.36, we obtain the ac-equivalent circuit shown in Fig. 9.41. In this case, the amplifier input is the gate voltage v_g and the output is the source voltage v_s. Note, too, that in ac terms the drain is common to both the input and the output. It is because of this that the source follower is also known as a **common-drain amplifier.**

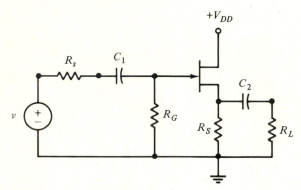

Fig. 9.40 Source follower.

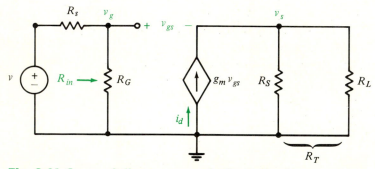

Fig. 9.41 Source-follower ac-equivalent circuit.

To determine the voltage gain $A_v = v_s/v_g$, let us use the fact that

$$v_s = R_T g_m v_{gs} \tag{9.24}$$

where $R_T = R_S \parallel R_L = R_S R_L/(R_S + R_L)$. Since $v_{gs} = v_g - v_s$, Equation (9.24) becomes

$$v_s = R_T g_m (v_g - v_s)$$

from which

$$A_v = \frac{v_s}{v_g} = \frac{g_m R_T}{1 + g_m R_T} = \frac{R_T}{1/g_m + R_T} \qquad (9.25)$$

For the case that $R_T \gg 1/g_m$, we have that $A_v \approx 1$. In other words, $v_s \approx v_g$ and the source voltage (output) "follows" the gate voltage (input)—hence, the name "source follower."

Since the resistance between the gate and the reference looking into the gate is infinite, then clearly the input resistance of the source follower is

$$R_{in} = R_G$$

Thus, if R_G is selected to be large, then so is the amplifier input resistance R_{in}.

The output resistance R_o of the source follower is equal to the ratio v_o/i_o for the circuit shown in Fig. 9.42. Since the current into the parallel combination of R_s and R_G must be zero (just apply KCL at the node labeled v_g), then by Ohm's law, $v_g = 0$. Therefore, since $v_s = v_o$, then

$$v_{gs} = v_g - v_s = -v_s = -v_o$$

Also, by KCL,

$$i_o = \frac{v_s}{R_S} - g_m v_{gs} = \frac{v_o}{R_S} - g_m(-v_o) = \left(\frac{1}{R_S} + g_m \right) v_o$$

Thus, the output resistance is

$$R_o = \frac{v_o}{i_o} = \frac{1}{1/R_S + g_m} = \frac{R_S}{1 + g_m R_S} = \frac{(1/g_m)R_S}{1/g_m + R_S} = \frac{1}{g_m} \parallel R_S$$

and, typically, this is a low resistance.

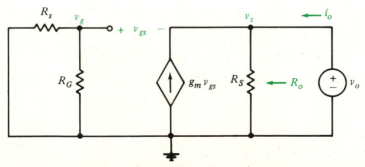

Fig. 9.42 Circuit for determining the output resistance of the source follower.

For the source follower given in Fig. 9.40, if we use the model shown in Fig. 9.37 in place of the model in Fig. 9.36, we only need replace the term R_S by $r_d \parallel R_S = r_d R_S/(r_d + R_S)$ in the above results.

In summary, because a source follower has approximately unity gain, high input resistance, and low output resistance, as for the emitter follower, it sees application as an isolation amplifier.

For the case of an (n-channel) enhancement MOSFET, recall that

$$i_D = K(v_{GS} - V_T)^2$$

Thus, for such a FET, the transconductance is given by

$$g_m = \frac{di_D}{dv_{GS}} = 2K(v_{GS} - V_T)$$

Example 9.9

Figure 9.43(a) shows an enhancement MOSFET amplifier, and Fig. 9.43(b) shows its ac-equivalent circuit after the MOSFET has been replaced by the model given in Fig. 9.36. Defining $R_T = R_D \parallel R_L = R_D R_L/(R_D + R_L)$, and since $v_g = v_{gs}$, then for the circuit given in Fig. 9.43(b), by KCL,

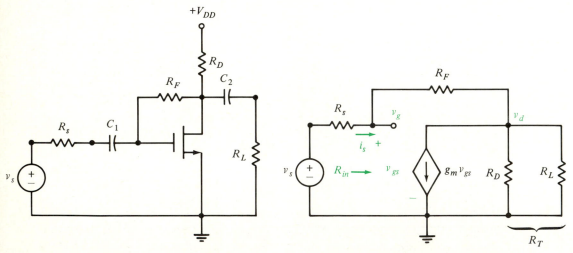

(a) An enhancement MOSFET amplifier

(b) The ac-equivalent circuit of the enhancement MOSFET amplifier

Fig. 9.43

$$g_m v_g + \frac{v_d}{R_T} + \frac{v_d - v_g}{R_F} = 0$$

Solving this equation for the voltage gain $A_v = v_d/v_g$, we get

$$A_v = \frac{v_d}{v_g} = \frac{R_T(1 - g_m R_F)}{R_F + R_T} = \frac{R_T[(1/g_m) - R_F]}{(1/g_m)(R_F + R_T)} \qquad (9.26)$$

The input resistance of the amplifier is $R_{in} = v_{gs}/i_s$. We can use Equation (9.26) to express v_d in terms of v_g and then substitute this into the fact that $i_s = (v_g - v_d)/R_F$. Thus, we obtain

$$i_s = \frac{v_g - [R_T(1 - g_m R_F)/(R_F + R_T)]v_g}{R_F}$$

and simplifying this expression results in

$$R_{in} = \frac{R_F + R_T}{1 + g_m R_T} = \frac{(1/g_m)(R_F + R_T)}{(1/g_m) + R_T} \qquad (9.27)$$

A typical value of g_m for a FET is 5 m℧, while for a BJT it is 40 m℧. This means that for small-signal applications, when it comes to gain, the BJT is superior to the FET. In addition to this fact, the transfer characteristic of a BJT is much closer to being linear than that of a FET. Consequently, nonlinear distortion for a BJT is less than for a FET. With regard to input resistance, however, the FET is superior to the BJT in that the FET has a much higher resistance than the BJT—the FET draws negligible current. As a consequence of these facts, the BJT has dominated in analog IC design.

9.3 FREQUENCY EFFECTS

In the preceding discussions of BJT and FET small-signal amplifiers, we assumed that the coupling and bypass capacitors behaved as short circuits to the ac signals that were being amplified. Since the impedance $\mathbf{Z}_C$ of a capacitor C is $\mathbf{Z}_C = 1/j\omega C$, when the frequency of an ac signal becomes sufficiently low, the impedances of coupling and bypass capacitors no longer can be ignored. Therefore, we will investigate the effects that coupling and bypass capacitors have on the frequency responses of amplifiers. We first consider the case of a BJT amplifier, and then study a typical FET amplifier.

Before we proceed, by way of a review, consider the simple RC circuit shown in Fig. 9.44. It is a routine exercise to show that if the input voltage v_1 is a sinusoid, then the output sinusoid v_2 will lead v_1. Consequently, such a circuit is referred to as a lead network. It should be obvious that for the circuit shown in Fig. 9.45, v_2 again leads v_1. Since v_3 is directly proportional to v_2 (by voltage division), then v_3 leads v_1. Therefore, the circuit shown in Fig. 9.45 is also a lead network. The lead

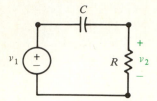

Fig. 9.44 Simple lead network.

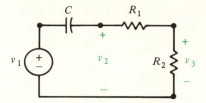

Fig. 9.45 Another lead network.

network shown in Fig. 9.44 is, in fact, the high-pass filter given in Fig. 5.5, and as discussed in Section 5.1, it has a cutoff frequency of $\omega_c = 1/RC$. So, too, the high-pass filter shown in Fig. 9.45 has a cutoff frequency of $\omega_c = 1/RC$, where $R = R_1 + R_2$.

Consider now a BJT amplifier such as the one given in Fig. 9.14. To begin with, let us concentrate on the effects of the coupling capacitors C_1 and C_2. Treating the bypass capacitor C_E as a short circuit, but not ignoring C_1 and C_2, the ac-equivalent circuit of the amplifier is as shown in Fig. 9.46, where $R_p = R_1 \parallel R_2$. To find the effect due solely to C_1, let us ignore C_2. As was pointed out in the discussion in conjunction with Fig. 9.14, the amplifier input resistance R_{in} is independent of R_C and R_L, and is given by

$$R_{in} = R_p \parallel R_b$$

where $R_p = R_1 R_2/(R_1 + R_2)$, $R_b = (1 + h_{fe})r_e$, and $r_e = I_{CQ}/0.026$. Consequently, the ac input voltage v_b to the amplifier can be determined from the equivalent circuit shown in Fig. 9.47. This base-equivalent lead network, therefore, has a cutoff frequency of

$$\omega_1 = \frac{1}{(R_s + R_{in})C_1} \qquad (9.28)$$

To determine the frequency effect on the amplifier due solely to C_2, we ignore C_1 and replace the portion of the circuit (given in Fig. 9.46) to the left of R_C by its

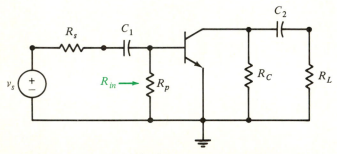

Fig. 9.46 Coupling capacitors included in the ac-equivalent circuit of the amplifier given in Fig. 9.14.

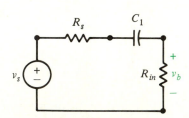

Fig. 9.47 Base-equivalent lead network.

Thévenin equivalent. The result is given by Fig. 9.48. It can be shown (see Problem 9.13) that $R_o = R_C$. It should be clear that the output voltage v_L will not be affected if the circuit given in Fig. 9.48 is cast into the form of Fig. 9.45. Thus, the collector-equivalent lead network shown in Fig. 9.48 has a cutoff frequency of

$$\omega_2 = \frac{1}{(R_C + R_L)C_2} \qquad (9.29)$$

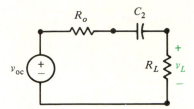

Fig. 9.48 Collector-equivalent lead network.

Having determined the cutoff frequencies due to the coupling capacitors C_1 and C_2, let us now deal with the bypass capacitor C_E. For the amplifier given in Fig. 9.14, treating C_1 and C_2 as short circuits, we get the ac-equivalent circuit shown in Fig. 9.49, where $R_p = R_1 \| R_2$ and $R_T = R_C \| R_L$. To calculate the effective resistance R_o that is in parallel with C_E, we replace the transistor by its small-signal model and set v_s to zero as shown in Fig. 9.50. From the resulting circuit, we can find $R_o = v_o/i_o$.

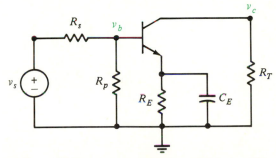

Fig. 9.49 Bypass capacitor included in the ac-equivalent circuit of the amplifier given in Fig. 9.14.

Let us define $R_q = R_s \| R_p$. Applying KCL to node v_b, we get

$$\frac{v_b}{R_q} + \frac{v_b - v_o}{r_e} = h_{fe}i_b \qquad (9.30)$$

Since $i_b + h_{fe}i_b = (v_b - v_o)/r_e$, then $i_b = (v_b - v_o)/(1 + h_{fe})r_e$. Substituting this fact into Equation (9.30), we obtain

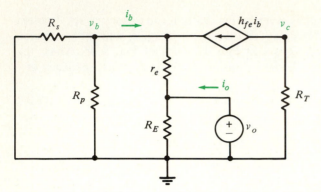

Fig. 9.50 Circuit for determining the effective resistance R_o in parallel with C_E.

$$\frac{v_b}{R_q} + \frac{v_b - v_o}{r_e} = \frac{h_{fe}(v_b - v_o)}{(1 + h_{fe})r_e}$$

Manipulating this equation results in

$$v_b = \frac{R_q/(1 + h_{fe})}{r_e + R_q/(1 + h_{fe})} v_o$$

Substituting this expression into the fact that $i_o = v_o/R_E + (v_o - v_b)/r_e$, we obtain

$$\frac{1}{R_o} = \frac{i_o}{v_o} = \frac{1}{r_e + R_q/(1 + h_{fe})} + \frac{1}{R_E}$$

In other words,

$$R_o = R_E \parallel \left(r_e + \frac{R_q}{1 + h_{fe}} \right) \qquad (9.31)$$

For the ac-equivalent amplifier circuit shown in Fig. 9.49, for the case that R_E is unbypassed (C_E is an open circuit), just as for the derivation of Equation (9.12), we have that

$$A_v = \frac{v_c}{v_b} \approx -\frac{R_T}{r_e + R_E}$$

However, if R_E is bypassed (C_E is a short circuit), then

$$A_v \approx -\frac{R_T}{r_e}$$

In other words, as frequency decreases and C_E behaves less and less as a short circuit, the voltage gain decreases. For the coupling capacitors C_1 and C_2, the critical frequencies are given by Equations (9.28) and (9.29), respectively. For the bypass capacitor C_E, though, the critical frequency is

$$\omega_3 = \frac{1}{R_o C_E} \tag{9.32}$$

where R_o is given by Equation (9.31).

Assuming that the values of the frequencies given in Equations (9.28), (9.29), and (9.32) are sufficiently different, the lower cutoff frequency for the BJT amplifier given in Fig. 9.14 is equal to the highest of those three values. In practice, quite often ω_3 has the highest value, and, therefore, it determines the lower cutoff frequency.

For a FET amplifier, such as the one shown in Fig. 9.51, the corresponding ac-equivalent circuit, including the coupling and bypass capacitors, is shown in Fig. 9.52, where $R_p = R_1 \parallel R_2$. Similar to the case of the BJT amplifier just discussed, the cutoff frequency due to C_1 is given by Equation (9.28). In this case,

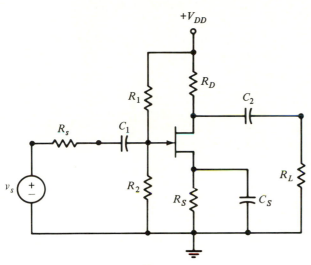

Fig. 9.51 A FET amplifier.

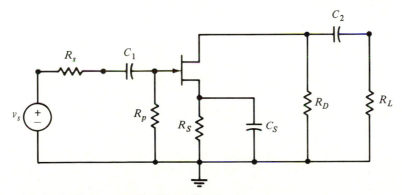

Fig. 9.52 The ac-equivalent circuit.

however, we have that $R_{in} = R_p$. Therefore, the gate-equivalent lead network has a cutoff frequency of

$$\omega_1 = \frac{1}{(R_s + R_p)C_1} \tag{9.33}$$

Again similar to the BJT amplifier discussed above, the cutoff frequency of the drain-equivalent lead network is

$$\omega_2 = \frac{1}{(R_D + R_L)C_2} \tag{9.34}$$

Finally, to determine the effect of the bypass capacitor C_S, we replace C_1, C_2, and v_s by short circuits and determine $R_o = v_o/i_o$ from the circuit shown in Fig. 9.53, where $R_T = R_D \| R_L$.

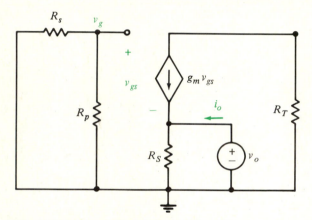

Fig. 9.53 Circuit for determining R_o.

Since, by KCL, there is no current going into the parallel combination of R_p and R_s, then $v_g = 0$. Consequently, by KVL, $v_{gs} = -v_o$. Since

$$i_o = \frac{v_o}{R_S} - g_m v_{gs} = \frac{v_o}{R_S} + g_m v_o$$

then

$$\frac{1}{R_o} = \frac{i_o}{v_o} = \frac{1}{R_S} + g_m$$

In other words,

$$R_o = R_S \| \frac{1}{g_m}$$

Thus, we have

$$\omega_3 = \frac{1}{R_o C_S} \tag{9.35}$$

Again, as was the situation for the BJT amplifier discussed above, for the case of sufficiently different values, the lower cutoff frequency for the FET amplifier shown in Fig. 9.51 is given by the highest of the values determined by Equations (9.33), (9.34), and (9.35).

Miller's Theorem

Having discussed how to determine the lower cutoff frequency for some transistor amplifiers, we now consider the problem of finding the upper cutoff frequency. To simplify our analysis, we use an important result known as "Miller's theorem."

We begin by considering the situation depicted in Fig. 9.54. Here, the box represents an amplifier whose input is v_1 and whose output is v_2. There is a feedback element, an impedance $\mathbf{Z}_f$, connected between the output and the input. With the feedback impedance $\mathbf{Z}_f$ in place, the gain of the amplifier is $v_2/v_1 = -A$. Further, the input impedance of the amplifier is $v_1/i_a = \mathbf{Z}_a$. By KCL, we have that*

$$i_1 = i_a + i_f = \frac{v_1}{\mathbf{Z}_a} + \frac{v_1 - v_2}{\mathbf{Z}_f} = \frac{v_1}{\mathbf{Z}_a} + \frac{(1 + A)v_1}{\mathbf{Z}_f} \tag{9.36}$$

and

$$\frac{i_1}{v_1} = \frac{1}{\mathbf{Z}_a} + \frac{1}{\mathbf{Z}_f/(1 + A)}$$

Thus, the input impedance $\mathbf{Z}_1$ to the overall feedback amplifier is

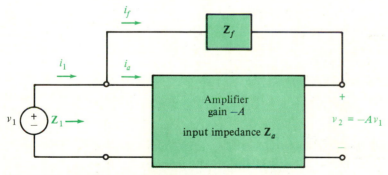

Fig. 9.54 Amplifier with feedback.

---

*Strictly speaking, we should use current and voltage phasor notation in the following equations.

$$\mathbf{Z}_1 = \frac{v_1}{i_1} = \mathbf{Z}_a \parallel \frac{\mathbf{Z}_f}{1 + A} \tag{9.37}$$

We also have that

$$i_f = \frac{v_1 - v_2}{\mathbf{Z}_f} = \frac{-v_2/A - v_2}{\mathbf{Z}_f} = \frac{-v_2 - Av_2}{A\mathbf{Z}_f} = \frac{-v_2}{A\mathbf{Z}_f/(1 + A)} \tag{9.38}$$

Now note that Equations (9.36), (9.37), and (9.38) are also satisfied for the configuration shown in Fig. 9.55. In other words, the effect on the amplifier in the box is the same as that for the circuit in Fig. 9.54, and the impedance $\mathbf{Z}_1$ is also the same. This result is known as **Miller's theorem,** and the equivalent manner in which the feedback element $\mathbf{Z}_f$ is reflected across the input as the impedance $\mathbf{Z}_f/(1 + A)$ and across the output as $A\mathbf{Z}_f/(1 + A)$ is referred to as the **Miller effect.**

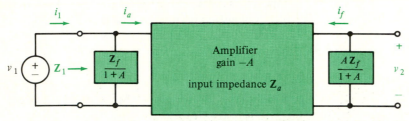

Fig. 9.55 Miller equivalent of the given amplifier with feedback.

Before we see how Miller's theorem is used to determine how high frequencies affect transistor amplifiers, again by way of a review, we mention that the circuit shown in Fig. 9.56 is an example of a **lag network.** In other words, if the input voltage v_1 is a sinusoid, then the output sinusoid v_2 will lag v_1. The lag network shown in Fig. 9.56 is, in fact, the low-pass filter given in Fig. 5.8, and it has a cutoff frequency of $\omega_c = 1/RC$.

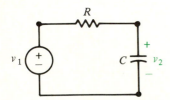

Fig. 9.56 Simple lag network.

The small-signal ac models of BJTs and FETs encountered previously are valid for low and middle frequencies. However, for high frequencies, junction capacitances become significant factors. Redrawing the small-signal BJT model given in Fig. 9.11, and including the base–emitter capacitance C_e (also denoted C_π) and the collector–emitter capacitance C_c (also denoted C_μ), we get the high-frequency BJT small-signal model shown in Fig. 9.57. In amplification applications, the

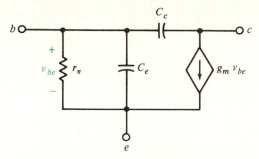

Fig. 9.57 High-frequency BJT model.

base–emitter junction of a BJT is forward biased and the collector–emitter junction is reverse biased. Consequently, C_e is comprised of a transition capacitance and a diffusion capacitance (see Section 6.7), while C_c is strictly a transition capacitance. A typical value for C_e is 100 pF, while a typical value for C_c is 5 pF.

Again consider the BJT amplifier given in Fig. 9.14 and its ac-equivalent circuit shown in Fig. 9.16, where $R_p = R_1 \parallel R_2$ and $R_T = R_C \parallel R_L$. Replacing the BJT in Fig. 9.16 by the model given in Fig. 9.57, we get the circuit shown in Fig. 9.58. For low and middle frequencies when C_c and C_e can be ignored (treated as open circuits), we have already seen in the discussion associated with Fig. 9.16 that the corresponding voltage gain is $A_v = v_c/v_b \approx -R_T/r_e$. However, from Equation (9.5), $r_e = 1/g_m$. Thus, $A_v \approx -g_m R_T = -A$.

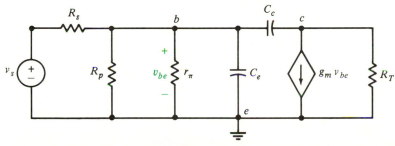

Fig. 9.58 Amplifier ac-equivalent circuit using high-frequency BJT model.

Now, by Miller's theorem, the capacitance C_c can be reflected across the input (base–emitter) and the output (collector–emitter) by appropriately valued capacitances. Specifically, since $\mathbf{Z}_f = 1/j\omega C_c$, then the reflected input impedance is

$$\frac{\mathbf{Z}_f}{1 + A} = \frac{1}{j\omega C_c(1 + A)} \tag{9.39}$$

which is the impedance of a capacitance of value $(1 + A)C_c$, while the reflected output impedance is

$$\frac{AZ_f}{1 + A} = \frac{1}{j\omega C_c(1 + A)/A} \tag{9.40}$$

which is the impedance of a capacitance of value $(1 + A)C_c/A$. The use of Miller's theorem, therefore, yields the equivalent circuit shown in Fig. 9.59.

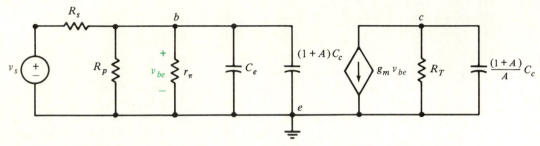

Fig. 9.59 Amplifier ac-equivalent circuit resulting from Miller's theorem.

Since C_e and $(1 + A)C_c$ are connected in parallel, they can be combined into a single capacitance $C_g = C_e + (1 + A)C_c$. Furthermore, we can apply Thévenin's theorem to the remainder of the base portion of the circuit and represent it as a single voltage source v_{oc} in series with a single resistance $R_g = R_s \parallel R_p \parallel r_\pi$. In addition, the current source $g_m v_{be}$ in parallel with R_T is equivalent to a voltage source $R_T g_m v_{be}$ in series with R_T (see Problem 9.45). Therefore, the circuit given in Fig. 9.59 can be cast into the form shown in Fig. 9.60, which explicitly exhibits base and collector lag networks. We deduce that the cutoff frequency resulting from the base lag network is

$$\omega_g = \frac{1}{R_g C_g} \tag{9.41}$$

where $R_g = R_s \parallel R_p \parallel r_\pi$ and $C_g = C_e + (1 + A)C_c$. The cutoff frequency due to the collector lag network is

$$\omega_T = \frac{1}{R_T C_T} \tag{9.42}$$

where $R_T = R_C \parallel R_L$ and $C_T = (1 + A)C_c/A$.

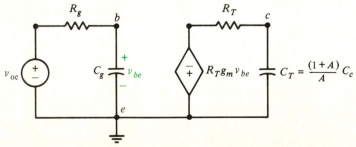

Fig. 9.60 Equivalent circuit showing base and collector lag networks.

Assuming that ω_g and ω_T have sufficiently different values, the upper cutoff frequency for the BJT amplifier given in Fig. 9.14 is equal to the smaller of these two values.

Having determined the upper cutoff frequency ω_u of the amplifier as just described and the lower cutoff frequency ω_l as previously discussed, we can define the **bandwidth** BW of the amplifier by

$$BW = \omega_u - \omega_l \tag{9.43}$$

The capacitance C_c, which typically lies in the range 1–20 pF, is often designated on BJT specification sheets as C_{ob}. On the other hand, C_e, which typically lies in the range 25–500 pF, does not have a similar designation. Instead, C_e can be determined from another parameter called the **current-gain/bandwidth product** f_T. The parameter f_T is the frequency (in hertz) for which the effect of the capacitances has become so significant that the short-circuit ($v_{ce} = 0$) current gain (i_c/i_b) equals unity. The range of f_T typically goes from 100 to 1000 MHz. It can be shown that

$$f_T \approx \frac{g_m}{2\pi(C_c + C_e)}$$

from which

$$C_c + C_e \approx \frac{g_m}{2\pi f_T}$$

Since typically $C_e \gg C_c$, then

$$C_e \approx \frac{g_m}{2\pi f_T} \tag{9.44}$$

From the small-signal model of an FET shown in Fig. 9.36 (or 9.37), by adding capacitances C_{gs}, C_{gd}, and C_{ds} we may obtain the high-frequency FET small-signal model shown in Fig. 9.61. This model is valid for JFETs and MOSFETs—both depletion and enhancement types. Capacitances C_{gs} and C_{gd} typically range between 1 and 3 pF, while C_{ds} is usually 1 pF or less. Similar to the case of the BJT, these capacitances cause equivalent lag networks (low-pass filters) that establish upper cutoff frequencies for FET amplifiers.

In particular, for the FET amplifier given in Fig. 9.51 at higher frequencies when the coupling and bypass capacitors behave as short circuits, using the model given in Fig. 9.61, we get the ac-equivalent circuit shown in Fig. 9.62, where $R_p = R_1 \| R_2$, and $R_T = R_D \| R_L$. For low and middle frequencies when C_{gs}, C_{gd}, and C_{ds} can be treated as open circuits, just as in the derivation of Equation (9.23), we have that $A_v = -g_m R_T = -A$. By the Miller effect, C_{gd} is reflected between the

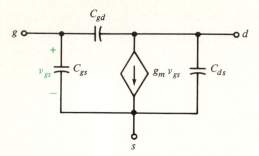

Fig. 9.61 High-frequency FET model.

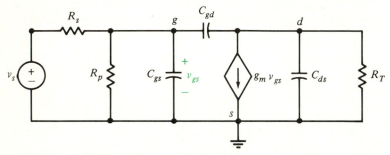

Fig. 9.62 Amplifier ac-equivalent circuit using high-frequency FET model.

gate and the source as $(1 + A)C_{gd}/A$. Combining (adding) capacitances in parallel and using Thévenin's theorem, we obtain the equivalent circuit shown in Fig. 9.63, where $C_g = C_{gs} + (1 + A)C_{gd}$, $C_T = C_{ds} + (1 + A)C_{gd}/A$, and $R_g = R_s \parallel R_p$. It is from this circuit that we notice that the gate lag network has a cutoff frequency given by Equation (9.41) and the drain lag network has a cutoff frequency given by Equation (9.42). If sufficiently different, the smaller of these two values is the upper cutoff frequency of the FET amplifier. As with the case of the BJT amplifier, if the upper cutoff frequency of the FET amplifier is ω_u and the lower cutoff frequency is ω_l, then the bandwith is $BW = \omega_u - \omega_l$.

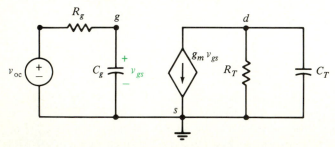

Fig. 9.63 Equivalent circuit showing gate and drain lag networks.

9.4 IC AMPLIFIERS

In the preceding section, we saw how the coupling and bypass capacitors determine the lower cutoff frequency of an amplifier. If such capacitors can be eliminated, then an amplifier's frequency response can go down to dc (zero frequency). Because its frequency response cannot go all the way down to dc, an amplifier that employs coupling capacitors is known as an **ac amplifier.** In contrast to this, if coupling is accomplished directly without the use of capacitors, we say that the amplifier is **directly coupled** and refer to it as a **direct-coupled (dc) amplifier.** Note that the two ways in which the term "dc" is used are, in essence, synonymous: Provided that a dc (direct-coupled) amplifier has no bypass capacitors, its frequency response will go down to dc (zero frequency).

The elimination of coupling and bypass capacitors from amplifier design has an extremely important consequence—amplifiers can be produced on IC chips. As a result, amplifiers—in particular, operational amplifiers—can be made small in size and economical as to price.

Differential Amplifiers

We begin our study of IC amplifiers by considering the BJT dc amplifier shown in Fig. 9.64. Here we assume that Q_1 and Q_2 have been fabricated with identical characteristics—a typical assumption for two transistors on the same IC chip. This configuration, known as a **differential pair,** has two ac input voltages, v_1 and v_2. For the corresponding dc-equivalent circuit, shown in Fig. 9.65, by symmetry Q_1 and Q_2 have identical currents and voltages. By KVL,

$$R_s i_B = v_{BE} + 2(i_B + i_C)R_E - V_{EE} = 0 \qquad\qquad (9.45)$$

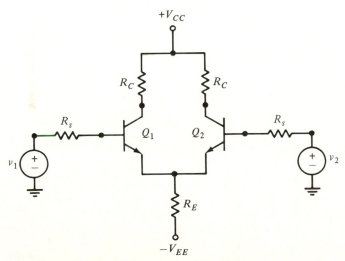

Fig. 9.64 Differential-pair configuration.

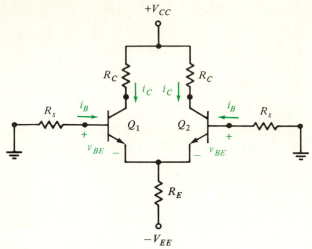

Fig. 9.65 The dc-equivalent circuit.

Since $i_C = h_{FE}i_B$ and $i_B + i_C = i_B + h_{FE}i_B = (1 + h_{FE})i_B \approx h_{FE}i_B$, Equation (9.45) becomes

$$R_s \frac{i_C}{h_{FE}} + v_{BE} + 2R_E i_C - V_{EE} \approx 0$$

from which

$$i_C \approx \frac{V_{EE} - v_{BE}}{(R_s/h_{FE}) + 2R_E}$$

There is, however, an alternative way to establish bias currents in Q_1 and Q_2. In place of R_E, another transistor Q_3 and its associated resistors R_1, R_2, and R_3 are used as illustrated in Fig. 9.66. The reason why this approach is taken in practice will be discussed shortly. First, though, let us concentrate on Q_3 and its associated resistors. By Thévenin's theorem, we can replace the voltage divider consisting of R_1 and R_2 by a voltage source $v_{oc} = -R_1 V_{EE}/(R_1 + R_2) = -KV_{EE}$ in series with a resistance $R_p = R_1 \parallel R_2 = R_1 R_2/(R_1 + R_2)$ as shown in Fig. 9.67. Applying KVL, we get

$$-KV_{EE} = R_p i_{B_3} + v_{BE_3} + (i_{B_3} + i_{C_3})R_3 - V_{EE}$$

Since $i_{C_3} = h_{FE_3}i_{B_3} \approx i_{B_3} + i_{C_3}$, then

$$V_{EE} - KV_{EE} - v_{BE_3} \approx R_p \frac{i_{C_3}}{h_{FE_3}} + R_3 i_{C_3}$$

from which

$$i_{C_3} \approx \frac{(1 - K)V_{EE} - v_{BE_3}}{(R_p/h_{FE_3}) + R_3} = I_o \qquad \textbf{(9.46)}$$

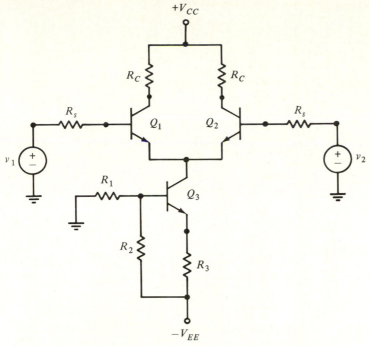

Fig. 9.66 Alternative differential-pair configuration.

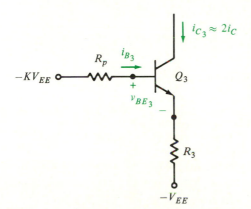

Fig. 9.67 Result of applying Thévenin's theorem to the base circuit of Q_3.

where $K = R_1/(R_1 + R_2)$ and $R_p = R_1 R_2/(R_1 + R_2)$. However, since $i_{C_3} \approx 2i_C$, then $i_C \approx i_{C_3}/2 = I_o/2$. In other words, the collector currents of Q_1 and Q_2 are determined by Q_3 and its associated components. Thus, R_1, R_2, R_3, Q_3, and $-V_{EE}$ combine to act as a constant-current (dc) source I_o. For this reason, the differential pair given in Fig. 9.66 is often drawn as shown in Fig. 9.68.

To perform an ac analysis of the differential pair given in Fig. 9.68, we first consider its ac-equivalent circuit shown in Fig. 9.69. Note that because I_o is a dc

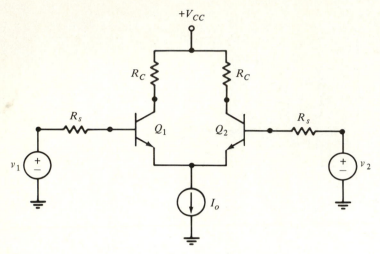

Fig. 9.68 Differential pair with current-source biasing.

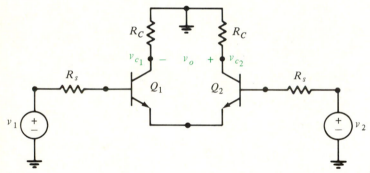

Fig. 9.69 The ac-equivalent circuit of the differential pair with current-source biasing.

(constant-current) source, in ac terms it is a current source of value zero (i.e., an open circuit). We proceed by using the principle of superposition.

To find the response due solely to v_1, set v_2 equal to zero, that is, replace voltage source v_2 with a short circuit. Also, replace Q_1 and Q_2 by the small-signal BJT model given in Fig. 9.12. The resulting circuit is shown in Fig. 9.70. By KVL, we have

$$v_1 = R_s i_{b_1} + r_e(1 + h_{fe})i_{b_1} + r_e(1 + h_{fe})i_{b_1} - R_s i_{b_2} \tag{9.47}$$

Applying KCL at the node where both resistances labeled r_e meet, we get $i_{b_1} + h_{fe}i_{b_1} + i_{b_2} + h_{fe}i_{b_2} = 0$, from which we deduce that $i_{b_2} = -i_{b_1}$. Substituting this fact into Equation (9.47), we obtain

$$v_1 = 2R_s i_{b_1} + 2r_e(1 + h_{fe})i_{b_1} = 2[R_s + (1 + h_{fe})r_e]i_{b_1} \tag{9.48}$$

However, $v'_{c_1} = -R_C h_{fe}i_{b_1}$. Thus, $i_{b_1} = -v'_{c_1}/h_{fe}R_C$ and substituting this into

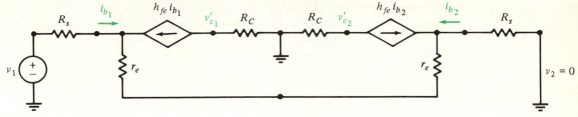

Fig. 9.70 Differential-pair ac-equivalent circuit including small-signal BJT model.

Equation (9.48) yields

$$v'_{c_1} = \frac{-h_{fe}R_C}{2[R_s + (1 + h_{fe})r_e]} v_1 \approx \frac{-h_{fe}R_C}{2(R_s + h_{fe}r_e)} v_1$$

Since $v'_{c_2} = -R_C h_{fe} i_{b_2}$, then $i_{b_1} = -i_{b_2} = v'_{c_2}/h_{fe}R_C$. Substituting this fact into Equation (9.48) results in

$$v'_{c_2} \approx \frac{h_{fe}R_C}{2(R_s + h_{fe}r_e)} v_1$$

Next consider the case that $v_1 = 0$ and $v_2 \neq 0$. By the symmetry of the differential pair, we deduce that the resulting ac collector voltages are

$$v''_{c_1} \approx \frac{h_{fe}R_C}{2(R_s + h_{fe}r_e)} v_2 \quad \text{and} \quad v''_{c_2} \approx \frac{-h_{fe}R_C}{2(R_s + h_{fe}r_e)} v_2$$

Hence, by the principle of superposition,

$$v_{c_1} = v'_{c_1} + v''_{c_1} \approx \frac{h_{fe}R_C}{2(R_s + h_{fe}r_e)}(v_2 - v_1) = \frac{A}{2}(v_2 - v_1)$$

and

$$v_{c_2} = v'_{c_2} + v''_{c_2} \approx \frac{h_{fe}R_C}{2(R_s + h_{fe}r_e)}(v_1 - v_2) = \frac{A}{2}(v_1 - v_2) \approx -v_{c_1}$$

where $A = h_{fe}R_C/(R_s + h_{fe}r_e)$.

If the output v_o of the amplifier is taken between the collectors of Q_1 and Q_2 as indicated in Fig. 9.69—this is called a **differential output**—then

$$v_o = v_{c_2} - v_{c_1} = \frac{A}{2}(v_1 - v_2) - \frac{A}{2}(v_2 - v_1) = A(v_1 - v_2) \qquad \textbf{(9.49)}$$

where $A = h_{fe}R_C/(R_s + h_{fe}r_e)$ is called the **differential gain.** Even if the output is chosen to be one of the collector voltages (with respect to the reference), called a **single-ended output,** we see that the output is directly proportional to the difference of the input voltages v_1 and v_2. Because of this, such an amplifier is referred to as a **differential amplifier** or a **difference amplifier.** The input stage of an operational amplifier is typically a differential amplifier.

Note that the differential output of a differential amplifier results in twice as much gain as that for a single-ended output. Also note that a differential amplifier can be used for one single-ended input—we need only connect the other input terminal (base) to the reference, thereby making the second input equal to zero. The output is then directly proportional to the single nonzero input. Specifically, for the case that $v_1 \neq 0$ and $v_2 = 0$, from Equation (9.49), we have that $v_o = Av_1$. Furthermore, from Equation (9.48), the resistance R_1 seen by the source v_1 is

$$R_1 = \frac{v_1}{i_{b_1}} = 2[R_s + (1 + h_{fe})r_e] \approx 2[R_s + h_{fe}r_e] \qquad (9.50)$$

For the case that the signal source resistance $R_s = 0$, the input resistance R_{in} of the differential amplifier is

$$R_{in} = 2(1 + h_{fe})r_e \approx 2h_{fe}r_e \qquad (9.51)$$

In addition to applying single-ended inputs (either one or two at a time) to a differential amplifier, we can also apply a single **differential input** v_d as shown in Fig. 9.71. Since $v_d = v_1 - v_2$, then from Equation (9.49), we have that $v_o = Av_d$, where $A = h_{fe}R_C/(R_s + h_{fe}r_e)$. Furthermore, the ac-equivalent circuit of Fig. 9.71 can be obtained from Fig. 9.70 by changing the source v_1 to v_d, erasing the reference symbols at the lower left- and right-hand portions of the circuit and joining the resulting two loose ends. Proceeding as for the circuit given in Fig. 9.70, we again find that the resistance seen by v_d is given by Equation (9.50), and the input resistance of the differential amplifier (with $R_s = 0$) is given by Equation (9.51).

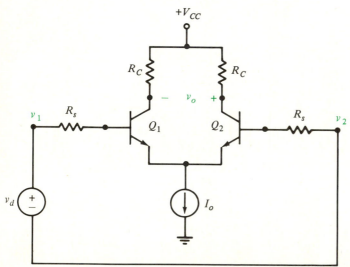

Fig. 9.71 Differential amplifier with a differential-input voltage.

Common-Mode Rejection Ratio (CMRR)

When both inputs of a differential amplifier have the same input $v_1 = v_2 = v_{cm}$, called a **common-mode input,** as shown in Fig. 9.72, from Equation (9.49), the differential-mode output is $v_o = A(v_1 - v_2) = A(v_{cm} - v_{cm}) = 0$. This property of a differential amplifier is of great practical importance. Because of environmental or technological factors, quite often a desired signal is subject to interference. Such unwanted signals, referred to as **noise,** often appear in the common mode. As an example, superimposed on the differential output of a differential amplifier can be some ripple (see Section 6.7) due to the amplifier's dc power supply. If the output of the amplifier is connected to the differential input of a second differential amplifier, the ripple will be a common-mode input. Ideally, the second differential amplifier will not amplify the ripple, that is, the second amplifier will reject the ripple.

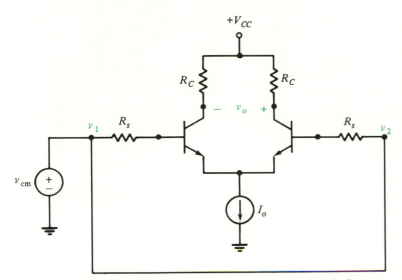

Fig. 9.72 Differential amplifier with a common-mode input.

In practice, a differential amplifier is not perfectly symmetrical. Consequently, when a common-mode input is applied, the output will be nonzero—the smaller the better. The ratio of the gain A_d for a differential-input signal to the gain A_{cm} for a common-mode input is called the **common-mode rejection ratio** (CMRR) of the amplifier. That is,

$$\text{CMRR} = \left| \frac{A_d}{A_{cm}} \right| \qquad \text{(9.52)}$$

so the larger the CMRR the better. Normally, the CMRR is expressed in decibels:

$$\text{CMRR}_{\text{dB}} = 20 \log_{10} \left| \frac{A_d}{A_{\text{cm}}} \right| \tag{9.53}$$

Previously we mentioned that, in practice, the current-source biasing scheme for a differential amplifier (see Figs. 9.66 and 9.68) was preferred over the use of an emitter resistance R_E (see Fig. 9.64). The main reason for this preference, as we now see, is that the former approach yields a higher CMRR than does the latter. Specifically, consider the differential amplifier given in Fig. 9.64, where $R_s = 0$. Let us apply a common-mode input v_{cm}. Since the resulting differential output is ideally zero, let us consider the single-ended output v_{c_2}. It can be shown (see Problem 9.56) when $R_E \gg r_e$ that $v_{c_2}/v_{\text{cm}} = -R_C/2R_E$. In addition, if a differential input v_d is applied to the circuit given in Fig. 9.64 with $R_s = 0$, then it also can be shown (see Problem 9.55) that $v_{c_2}/v_d \approx R_C/2r_e$. Thus, we deduce that for the differential amplifier given in Fig. 9.64,

$$\text{CMRR} \approx \left| \frac{R_C/2r_e}{-R_C/2R_E} \right| = \frac{R_E}{r_e} \tag{9.54}$$

Therefore, to get a large CMRR, we should try to make R_E large. This, however, reduces the quiescent current and increases r_e.

On the other hand, consider the current-source biasing scheme for a differential amplifier as shown in Fig. 9.66. Although Q_3 and its associated resistors do not form an ideal current source as shown in Fig. 9.68, an accurate model is an ideal current source in parallel with a resistance R_o. Proceeding for the resulting differential amplifier as was done for the amplifier in Fig. 9.64, we deduce that the CMRR for a differential amplifier with current source biasing is

$$\text{CMRR} \approx R_o/r_e \tag{9.55}$$

where R_o is the output (Norton-equivalent) resistance of the nonideal current source. Since R_o is typically much higher than R_E would have to be to produce the same quiescent current and, hence, the same value of r_e, the CMRR for current-source biasing is generally superior.

Example 9.10

The two outputs of a differential amplifier measure 8 mV rms and 10 mV rms when a common-mode input of 1 V rms is applied. When a differential input of 1 mV rms is applied to the amplifier, the resulting differential output is 75 mV rms. Let us determine the CMRR of this differential amplifier.

The differential output for the common-mode input of 1 V is 10 mV $-$ 8 mV $=$ 2 mV. Thus, the common-mode gain has absolute value

$$|A_{cm}| = \frac{2 \times 10^{-3}}{1} = 2 \times 10^{-3}$$

Furthermore, the absolute value of the differential gain is

$$|A_d| = \frac{75 \times 10^{-3}}{1 \times 10^{-3}} = 75$$

Thus,

$$\text{CMRR} = \left| \frac{A_d}{A_{cm}} \right| = \frac{|A_d|}{|A_{cm}|} = \frac{75}{2 \times 10^{-3}} = 37{,}500$$

and

$$\text{CMRR}_{dB} = 20 \log_{10}(37{,}500) = 91.5 \text{ dB}$$

Although our discussion has been limited to BJTs, differential amplifiers can also be constructed using FETs (see Problems 9.59–9.63). An important property of a FET differential amplifier is its high input resistance.

Operational Amplifiers

Perhaps the most important analog IC is the operational amplifier (op amp). As mentioned previously, the first stage of this dc amplifier is typically a differential amplifier. The output of an op amp, however, is usually single ended. By way of a review, the circuit symbol of an op amp is shown in Fig. 9.73. The input marked with a minus sign (v_1) is the **inverting input,** and the one marked with a plus sign (v_2) is the **noninverting input.** The single-ended output is labeled v_o. Although voltages v_1, v_2, and v_o are measured with respect to the reference so indicated, often the reference is not shown, and, therefore, is implicit. The simplified model of an op amp is shown in Fig. 9.74, where A is the gain, R_{in} is the input resistance, and R_o is the output resistance. For an **ideal operational amplifier,** $A = \infty$, $R_{in} = \infty$, and $R_o = 0 \, \Omega$.

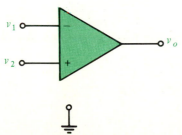

Fig. 9.73 Operational-amplifier circuit symbol.

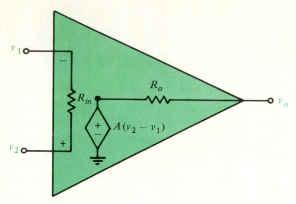

Fig. 9.74 Operational-amplifier model.

An actual operational amplifier can be described by the block diagram shown in Fig. 9.75. As already indicated, the first stage of an op amp is a differential amplifier. Since the input resistance of a simple differential amplifier is not necessarily high [see Equation (9.51)], modified versions of the basic differential amplifier are used, or FETs are used in place of BJTs. Typically, what results is an input resistance of 1 MΩ or more. Following the first stage is a high-gain amplifier that supplies additional gain. The overall gain of an op amp is typically 100,000 (100 dB) or more. After the high-gain amplifier comes a buffer, which is usually an emitter follower, and a level shifter. The level shifter is used so that the output voltage is zero when the input is zero. The last stage of the op amp, the driver, is a large-signal amplifier (this subject will be discussed in the next section) with a low output resistance—typically 100 Ω. The driver supplies the output current and voltage.

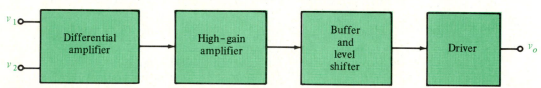

Fig. 9.75 Block diagram of operational amplifier.

Just as for the case of a differential amplifier, the CMRR of an op amp is the ratio of the differential gain to the common-mode gain [see Equations (9.52) and (9.53)]. The CMRR of an op amp typically ranges between 80 and 100 dB.

The most popular operational amplifier and the industry standard is the 741 op amp shown in Fig. 9.76. This analog (also called "linear") IC, which is produced by a number of manufacturers and is available in different types of packages, is comprised of 20 BJTs, 11 resistors, and one capacitor. Power is supplied to this device by supply voltages $+V_{CC}$ and $-V_{EE}$. Typically, $V_{CC} = V_{EE} = 15$ V. With such supply voltages, the range of the output voltage v_o is -10 V $\leq v_o \leq 10$ V,

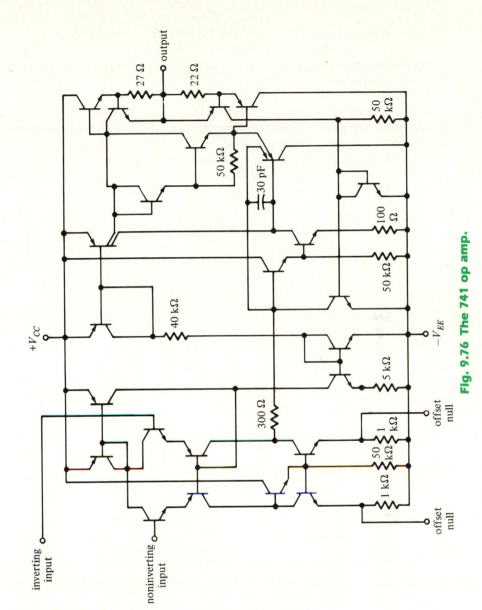

Fig. 9.76 The 741 op amp.

or wider. Also, the output of a 741 op amp is **short-circuit protected.** That is, if the output is accidentally "grounded" (i.e., connected to the reference), then the resulting short-circuit output current is limited to 25 mA, and the device will not be destroyed. Typical parameters for the 741 op amp are gain $A = 200,000$, input resistance $R_{in} = 2$ MΩ, and output resistance $R_o = 75$ Ω.

In addition to gain, input resistance, and output resistance, op amps have other important characteristics. We discuss some of these, citing typical values for the 741 op amp.

If both input terminals of an op amp are grounded, then $v_1 = v_2 = 0$, and ideally, $v_o = 0$. However, since an actual op amp is not perfectly symmetrical (or balanced), the output voltage, in general, will be nonzero. This is known as the **output offset voltage** of the op amp. By applying an appropriate voltage V_{os} to the input of the op amp, the output voltage can be made zero. The value $|V_{os}|$ is called the **input offset voltage** of the op amp. For a 741 op amp, the input offset voltage is typically 1.0 mV. Because of its definition, an op amp with a nonzero output offset voltage can be modeled by an op amp with no offset voltage and a dc voltage source V_{os} (which can be positive or negative) as shown in Fig. 9.77. Note that in the schematic diagram of the 741 op amp (Fig. 9.76), there are two terminals (pins) labeled "offset null." These can be used to yield $v_o = 0$ when $v_1 = v_2 = 0$. This is done by connecting a **potentiometer** (an adjustable resistor that is tapped) as shown in Fig. 9.78. The potentiometer (pot for short)—typically 10 kΩ—is adjusted so that $v_o = 0$ when $v_1 = v_2 = 0$.

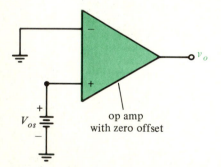

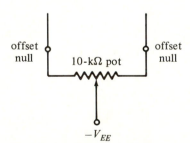

Fig. 9.77 Model of an op amp with nonzero offset.

Fig. 9.78 Offset null adjustment.

Since the input stage of an op amp is a differential amplifier, an op amp's inputs are typically the bases of two BJTs. For an op amp to function properly, the input BJTs must be appropriately biased. Thus, there must be a dc path between each base and the reference (ground). (Recall that an ac source acts as a short circuit to dc.) Consequently, in general, there are input (base) bias currents, albeit small currents, in the two input BJTs. The average value of these currents, when the output is zero due to an applied V_{os}, is called the **input bias current** of the op amp. For a 741 op amp, the input bias current is typically 80 nA. The absolute value of the difference between the two input bias currents is known as the **input offset current,** and for a 741 op amp this current is typically 20 nA.

The **slew rate** of an op amp is the maximum rate of change of the output voltages when the input is a large-signal step of voltage. For a 741 op amp, the slew rate is typically 0.5 V/μs. In other words, a change of 20 V in the output voltage due to an ideal step input typically would require 20 V ÷ 0.5 V/μs = 40 μs.

Example 9.11

If the input voltage for an inverting amplifier (see Fig. 2.13) or a noninverting amplifier [see Fig. 2.14(a)] is set to zero, what results is the op-amp circuit shown in Fig. 9.79. Let us determine the output voltage v_o of the overall circuit that is the result of the op amp's input offset voltage $|V_{os}|$.

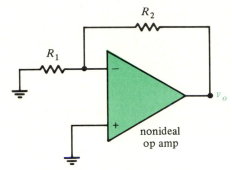

Fig. 9.79 Op-amp circuit with zero input.

As indicated in Fig. 9.77, let us model the nonideal op amp in Fig. 9.79 by an ideal op amp and a voltage source V_{os} as shown in Fig. 9.80. (An alternative model is given in Problem 9.64.) By KCL, we have that

$$\frac{v}{R_1} + \frac{v_1 - v_o}{R_2} = 0$$

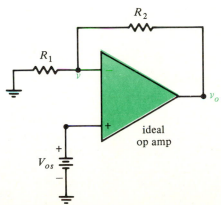

Fig. 9.80 Circuit used to determine the effect of the input offset voltage.

Solving this equation for v_o and using the fact that $v = V_{os}$, we get

$$v_o = \frac{R_1 + R_2}{R_1}v = \frac{R_1 + R_2}{R_1}V_{os} = \left(1 + \frac{R_2}{R_1}\right)V_{os} \qquad (9.56)$$

Next let us determine the effect of the input bias currents i_{B_1} and i_{B_2} on the output voltage v_o. To do this, let us model the op amp given in Fig. 9.79 as shown in Fig. 9.81. Since, in essence, $v = 0$, there is no current through R_1. This means that the current through R_2 is i_{B_1} and

$$v_o = R_2 i_{B_1} \qquad (9.57)$$

This output voltage, however, can be reduced by connecting a resistor R_3 to the noninverting input as shown in Fig. 9.82. We proceed by using the principle of superposition.

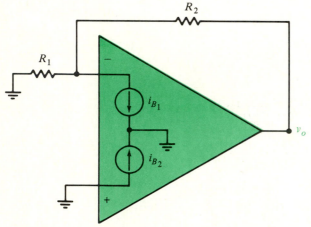

Fig. 9.81 Circuit used to determine the effect of the dc base currents.

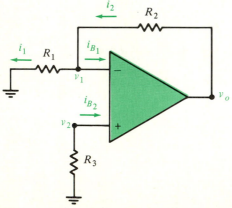

Fig. 9.82 Addition of R_3 reduces the effects of the input bias currents.

First, let $i_{B_1} \neq 0$ and $i_{B_2} = 0$. Then $v_2 = -R_3 i_{B_2} = 0$. This implies that $v_1 = 0$. Thus, $i_1 = v_1/R_1 = 0$, so $i_2 = i_{B_1}$. Therefore, the resulting output voltage is $v_{o1} = R_2 i_2 = R_2 i_{B_1}$.

Next, let $i_{B_1} = 0$ and $i_{B_2} \neq 0$. Then $v_2 = -R_3 i_{B_2} = v_1$, so $i_1 = v_1/R_1 = -R_3 i_{B_2}/R_1$. Thus, we have that the output voltage v_{o2} is

$$v_{o2} = R_2 i_2 + v_1 = R_2 i_1 + v_1 = -\frac{R_2 R_3}{R_1} i_{B_2} - R_3 i_{B_2} = -\left(\frac{R_2}{R_1} + 1\right) R_3 i_{B_2}$$

Hence, the output voltage due to the input bias currents is

$$v_o = v_{o1} + v_{o2} = R_2 i_{B_1} - \left(\frac{R_2}{R_1} + 1\right) R_3 i_{B_2} \tag{9.58}$$

Since i_{B_1} and i_{B_2} are typically close in value, v_o can be made small by making

$$R_2 = \left(\frac{R_2}{R_1} + 1\right) R_3 \tag{9.59}$$

From this equation, we easily determine that

$$R_3 = \frac{R_1 R_2}{R_1 + R_2} = R_1 \parallel R_2$$

In other words, to minimize the effects of the input bias currents on the output, we select $R_3 = R_1 \parallel R_2$. If we do so, then, from Equations (9.58) and (9.59), the resulting output voltage is

$$v_o = R_2 i_{B_1} - R_2 i_{B_2} = R_2(i_{B_1} - i_{B_2}) = R_2 I_{os} \tag{9.60}$$

where $|I_{os}|$ is the input offset current of the op amp.

Note that even if a resistor R_3 is placed in series with the voltage source V_{os} in Fig. 9.80, Equation (9.56) will still be valid. Thus, from Equations (9.56) and (9.60), the output voltage v_o that results from both the input offset voltage and input offset current of the op amp is (when R_3 is used)

$$v_o = \left(1 + \frac{R_2}{R_1}\right) V_{os} + R_2 I_{os} \tag{9.61}$$

As V_{os} and I_{os} each can be positive or negative, the worst-case output offset voltage has an absolute value of

$$|v_o| = \left(1 + \frac{R_2}{R_1}\right)|V_{os}| + R_2|I_{os}| \tag{9.62}$$

We have already seen a number of applications for op amps in preceding chapters. They are used to make inverting and noninverting amplifiers (also called scalers), buffers (voltage followers), adders, difference amplifiers, differentiators, integrators, and, hence, analog computers, and so on. In the next chapter, we see additional applications of operational amplifiers.

9.5 POWER AMPLIFIERS

In our discussion of transistor amplifiers thus far, we have been preoccupied with small-signal analysis. In many applications, small ac signals are amplified sufficiently to produce large ac signals. The resulting ac signals are then amplified by **large-signal amplifiers** or **power amplifiers,** which, in turn, provide sufficient power to drive devices such as loudspeakers, motors, recording instruments, and cathode-ray tubes.

In small-signal applications, transistors dissipate a fraction of watt, whereas when used for power amplification, they may have to be capable of dissipating greater amounts of power. It is because of this that a special class of transistors, referred to as **power transistors,** is required. Since such devices deal with large ac signals, it is desirable that their characteristics be more linear than was the case for small-signal transistors. The more linear a transistor's characteristics, the less nonlinear distortion will result.

Class-A Amplifiers

Class-A operation is the term that is used to indicate that a transistor operates entirely in the active (linear) region. In our previous discussions of small-signal amplifiers, transistors were always biased in the active region, and it was assumed that the ac signals were sufficiently small always to have class-A operation.

Let us now consider a typical ac (capacitively coupled) BJT amplifier as shown in Fig. 9.83. To obtain more general results, we have not used a bypass capacitor for R_E. Suppose that the BJT is biased in the active region, the quiescent collector current is I_{CQ}, and the quiescent collector–emitter voltage is V_{CEQ}. Figure 9.84

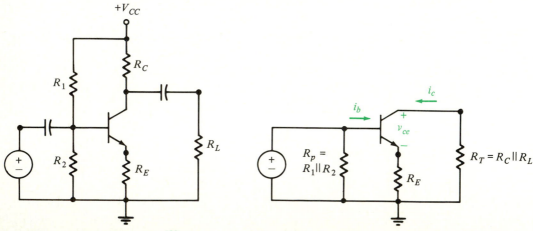

Fig. 9.83 A typical ac amplifier.   **Fig. 9.84 Amplifier ac-equivalent circuit.**

shows the ac-equivalent circuit of the amplifier, where $R_p = R_1 \parallel R_2$ and $R_T = R_C \parallel R_L$. The voltage and currents indicated in Fig. 9.84 are ac quantities, and, therefore, lowercase subscripts are used. Thus, the total collector current i_C is given by $i_C = I_{CQ} + i_c$ and the total collector–emitter voltage v_{CE} by $v_{CE} = V_{CEQ} + v_{ce}$.

From Fig. 9.84, we have

$$v_{ce} = -i_c R_T - (i_b + i_c)R_E \approx -i_c R_T - i_c R_E = -i_c(R_T + R_E)$$

from which we get the approximation

$$i_c = \frac{-v_{ce}}{R_T + R_E}$$

Thus, the (total) collector current is

$$i_C = I_{CQ} + i_c = I_{CQ} - \frac{v_{ce}}{R_T + R_E} \tag{9.63}$$

But $v_{CE} = V_{CEQ} + v_{ce}$ implies that $v_{ce} = v_{CE} - V_{CEQ}$. Substituting this fact into Equation (9.63) yields

$$i_C = I_{CQ} - \frac{v_{CE} - V_{CEQ}}{R_T + R_E} = -\frac{v_{CE}}{R_T + R_E} + \left(I_{CQ} + \frac{V_{CEQ}}{R_T + R_E}\right) \tag{9.64}$$

This is the equation of a straight line, in the i_C-v_{CE} plane, that has slope $-1/(R_T + R_E)$ and vertical-axis intercept $I_{CQ} + V_{CEQ}/(R_T + R_E)$. Setting $v_{CE} = V_{CEQ}$, we get $i_C = I_{CQ}$. Thus, the line goes through the Q point $i_C = I_{CQ}$, $v_{CE} = V_{CEQ}$. Furthermore, setting $i_C = 0$, we obtain the horizontal-axis intercept $V_{CEQ} + (R_T + R_E)I_{CQ}$. The line described by Equation (9.64) is depicted in Fig. 9.85 and is precisely the amplifier's ac load line (see Fig. 9.15 and the associated discussion). For a silicon BJT, we know that the cutoff current is $i_{C(\text{off})} \approx 0$ and the saturation voltage is $v_{CE(\text{sat})} \approx 0.2$ V. However, for the sake of simplicity, let us use $i_{C(\text{off})} = 0$ and $v_{CE(\text{sat})} = 0$. Then the saturation current is $i_{C(\text{sat})} = I_{CQ} + V_{CEQ}/(R_T + R_E)$ and the cutoff voltage is $v_{CE(\text{off})} = V_{CEQ} + I_{CQ}/(R_T + R_E)$.

For class-A operation, the transistor must not go into saturation or cutoff. Since the transistor operates along the load line, to allow for the largest possible ac signal,

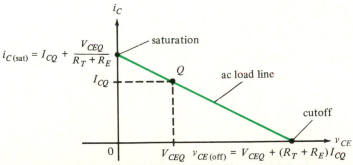

Fig. 9.85 The ac load line for the given amplifier.

we should have the Q point at the center of the load line. When the Q point is centered on the load line, then $i_{C(\text{sat})} = 2I_{CQ}$. Under this circumstance, therefore,

$$i_{C(\text{sat})} = I_{CQ} + \frac{V_{CEQ}}{R_T + R_E} = 2I_{CQ}$$

From this equation,

$$\boxed{\frac{V_{CEQ}}{I_{CQ}} = R_T + R_E} \qquad (9.65)$$

and this is the condition for a centered Q point.

Example 9.12

For the amplifier given in Fig. 9.83, suppose that $h_{FE} = 30$, $R_1 = 600 \ \Omega$, $R_2 = 300 \ \Omega$, $R_C = R_E = 100 \ \Omega$, $R_L = 80 \ \Omega$, and $V_{CC} = 30$ V. By routine dc BJT analysis, we can find that $I_{CQ} = 87.2$ mA and $v_{CEQ} = 12.56$ V. Therefore,

$$\frac{V_{CEQ}}{I_{CQ}} = \frac{12.56}{87.2 \times 10^{-3}} = 144 \ \Omega$$

Since

$$R_T = R_C \parallel R_L = \frac{R_C R_L}{R_C + R_L} = \frac{(100)(80)}{100 + 80} = 44.4 \ \Omega$$

then $R_T + R_E = 44.4 + 100 = 144.4 \ \Omega \approx V_{CEQ}/I_{CQ}$, and, hence, the Q point is centered on the load line.

Power Terminology

Returning to the class-A amplifier given in Fig. 9.83 and its ac-equivalent circuit shown in Fig. 9.84, suppose that the Q point is centered on the load line. It follows that the largest possible sinusoidal collector current has amplitude I_{CQ}. The corresponding rms value is $I_{CQ}/\sqrt{2}$. The resulting average signal (ac) power absorbed by R_T is

$$P_T = \left(\frac{I_{CQ}}{\sqrt{2}}\right)^2 R_T = \frac{1}{2}I_{CQ}^2 R_T$$

Since $i_e \approx i_c$, the (average) signal power absorbed by R_E is

$$P_E = \left(\frac{I_{CQ}}{\sqrt{2}}\right)^2 R_E = \frac{1}{2}I_{CQ}^2 R_E$$

Thus, the total signal power (or output power) P_o absorbed by R_T and R_E is

$$P_o = P_T + P_E = \tfrac{1}{2}I_{CQ}^2(R_T + R_E) \tag{9.66}$$

and this is the maximum output signal power since it was assumed that the sinusoidal collector current had maximum amplitude. Using Equation (9.65)—the condition for a centered Q point—we can substitute $R_T + R_E = V_{CEQ}/I_{CQ}$ into Equation (9.66) to get the maximum output signal power

$$\boxed{P_o = \tfrac{1}{2}V_{CEQ}I_{CQ}} \tag{9.67}$$

For instance, the amplifier described in Example 9.12 has a maximum output signal power of

$$P_o = \tfrac{1}{2}(12.56)(87.2 \times 10^{-3}) = 547.6 \text{ mW} \approx 0.55 \text{ W}$$

Since the dc collector current of a BJT is typically much greater than the dc base current, we define the **instantaneous power dissipation** p of a BJT to be

$$\boxed{p = v_{CE}i_C} \tag{9.68}$$

We know that a BJT operates somewhere on its load line. But what point on the load line corresponds to maximum instantaneous power dissipation? With reference to Fig. 9.85, for the sake of simplicity, let us define the symbols $a = v_{CE(\text{off})}$ and $b = i_{C(\text{sat})}$. Then the equation of the load line can be written as

$$i_C = -\frac{b}{a}v_{CE} + b \tag{9.69}$$

Substituting this into Equation (9.68) yields

$$p = v_{CE}\left(-\frac{b}{a}v_{CE} + b\right) = -\frac{b}{a}v_{CE}^2 + bv_{CE}$$

To determine what value of v_{CE} results in a maximum value of p, let us take the derivative of p with respect to v_{CE} and set the resulting expression to zero. Then we get

$$\frac{dp}{dv_{CE}} = -\frac{2b}{a}v_{CE} + b = 0$$

from which

$$v_{CE} = \frac{a}{2} = \frac{v_{CE(\text{off})}}{2}$$

Substituting this into Equation (9.69) yields

$$i_C = -\frac{b}{a}\left(\frac{a}{2}\right) + b = \frac{b}{2} = \frac{i_{C(sat)}}{2}$$

Since

$$\frac{d^2p}{dv_{CE}^2} = -\frac{2b}{a} < 0$$

the point $i_C = i_{C(sat)}/2$, $v_{CE} = v_{CE(off)}/2$ gives the maximum value of p. In other words, the instantaneous power dissipation of a BJT is maximum when the BJT is operating at the center of the load line.

The **quiescent power dissipation** P_{DQ} of a BJT is defined to be

$$\boxed{P_{DQ} = V_{CEQ}I_{CQ}} \tag{9.70}$$

This is the power dissipated by the transistor when no ac signal is present. According to the discussion above, P_{DQ} is maximum when the Q point is centered on the load line. A centered Q point, however, is required for maximum ac output power. Under this circumstance, though, the average power P_D dissipated by the transistor will be less than P_{DQ}. Thus, given a centered Q point, the value of the quiescent power dissipation $P_{DQ} = V_{CEQ}I_{CQ}$ is the upper limit for the average power that is dissipated by the transistor.

The maximum average power that can be safely dissipated by a BJT is denoted $P_{D(max)}$ and is found on a transistor's specification sheet. From the preceding discussion, $P_{D(max)}$ should be chosen to be greater than or equal to P_{DQ} corresponding to a centered Q point. That is,

$$\boxed{P_{D(max)} \geq P_{DQ} = V_{CEQ}I_{CQ}} \tag{9.71}$$

for a centered Q point. For instance, for the BJT used in the amplifier described in Example 9.12, $P_{DQ} = V_{CEQ}I_{CQ} = (12.56)(87.2 \times 10^{-3}) = 1.096$ W. Therefore, the transistor should have $P_{D(max)} \geq 1.096$ W.

From Equations (9.67) and (9.70), note that the maximum output signal power of a class-A amplifier is only half of the maximum quiescent power dissipation of the BJT. Thus, for example, if a class-A amplifier is to have a maximum output signal power of 5 W, then the transistor used must be rated at $P_{D(max)} \geq 10$ W.

For a BJT, typically the collector current is much greater than the base current. For this reason, let us define the dc power P_{dc} delivered by the supply V_{CC} to be $P_{dc} = V_{CC}I_{CQ}$. In addition, let us define the **efficiency** η of an amplifier to be the ratio of the output signal power P_o to the dc power P_{dc} delivered by the supply, that is,

$$\boxed{\eta = P_o/P_{dc}}$$

It can be shown (see Problem 9.73) that the maximum efficiency for a class-A amplifier of the type shown in Fig. 9.83 is $\eta = 1/4$.

The power absorbed by a transistor is dissipated as heat. For a BJT, typically $P_{D(max)}$ is rated at 25°C. If the ambient (surrounding) temperature is greater than 25°C, the value of $P_{D(max)}$ must be **derated** (reduced) appropriately. This is done by using the **power derating factor** given on the transistor's specification sheet.

Example 9.13

A 2N1711 silicon power BJT is rated at $P_{D(max)} = 3$ W at 25°C. Its power derating factor is 17 mW/°C. Therefore, at 75°C, $P_{D(max)}$ must be derated by

$$(75 - 25)(17 \times 10^{-3}) = 0.85 \text{ W}$$

In other words, at 75°C,

$$P_{D(max)} = 3 - 0.85 = 2.15 \text{ W}$$

The value of $P_{D(max)}$ at which a BJT is rated can be increased by connecting the transistor's case to a relatively large metallic surface with good heat-radiating properties. Such a structure, called a **heat sink,** is often used with power transistors.

Class-B Amplifiers

Although class-A operation is both simple and stable, the $P_{D(max)}$ rating of a transistor must be at least twice the maximum ac output power. Also, for the case of no ac signal, a centered Q point means that the current drain on the voltage supply is $i_{C(sat)}/2$. For low-power applications, these properties may not be serious drawbacks. However, there are situations in which the use of class-A amplifiers yields results that are too inefficient. Fortunately, though, there are efficient alternatives to class-A operation. Such an alternative is the subject of this section.

Class-B operation is the term that is used to indicate that a transistor is biased at cutoff, and for an applied ac signal, the transistor remains cut off for half of the cycle and operates in the active region for the other half. Since a transistor is biased at cutoff for class-B operation, then necessarily $I_{CQ} = 0$.

Suppose that the circuit shown in Fig. 9.86 is the ac-equivalent circuit of an emitter follower that is biased at cutoff. For the time being, let us assume the ideal case that when the base–emitter junction is forward biased, the voltage is $v_{BE} = 0$. Further, let us assume that the cut-in voltage is $V_\gamma = 0$. Thus, when the ac signal

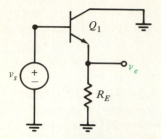

Fig. 9.86 Emitter follower biased at cutoff.

voltage $v_s > 0$, the base–emitter junction will be forward biased, and, therefore, $v_e = v_s$. Conversely, when $v_s \leq 0$, the base–emitter junction remains reverse biased, and, therefore, $v_e = 0$.

Figure 9.87(a) shows the ac-equivalent circuit of an emitter follower that uses a *p-n-p* transistor, which is again biased at cutoff. This circuit is redrawn in Fig. 9.87(b). Proceeding as was done for the case of the emitter follower given in Fig. 9.87, we deduce that when $v_s < 0$, then $v_e = v_s$; when $v_s \geq 0$, then $v_e = 0$.

If we combine the emitter followers given in Figs. 9.86 and 9.87 as shown in Fig. 9.88, then when $v_s > 0$, Q_1 conducts and Q_2 is cut off, and $v_e = v_s$. Similarly, when $v_s < 0$, Q_2 conducts and Q_1 is cut off, and $v_e = v_s$. (Of course, when

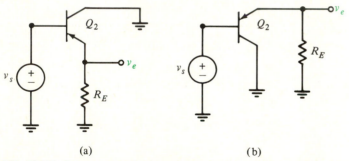

(a)   (b)

Fig. 9.87 A *p-n-p* emitter follower biased at cutoff.

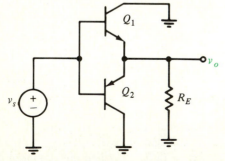

Fig. 9.88 Push–pull emitter follower.

$v_s = 0$, then $v_e = 0$.) Thus, for an ac signal v_s (which is not large enough to saturate either Q_1 or Q_2), we have that $v_e = v_s$. The circuit given in Fig. 9.88 is an example of a **push–pull circuit**—specifically, it is a push–pull emitter follower. Although the output voltage v_o (ideally) equals the input voltage v_s, an emitter follower can produce a significant power gain (e.g., see Problem 9.17).

In the discussion above, we assumed the ideal situation that a forward-biased base–emitter junction had a voltage of $v_{BE} = 0$. Of course, we know that, in actuality, for a silicon BJT such a voltage is about 0.7 V. Furthermore, for the circuit given in Fig. 9.88, even when $v_s > 0$, there will be essentially no current in Q_1 until v_s exceeds the cut-in voltage $V_\gamma = 0.5$ V. Thus, $v_o \approx 0$ for $0 < v_s < 0.5$ V. By symmetry, we deduce that $v_o \approx 0$ for $-0.5 < v_s < 0$ V. Consequently, if v_s is a sinusoid, for example, the resulting output voltage v_o will be nonsinusoidal and will be a function like the one shown in Fig. 9.89. The resulting distortion is known as **crossover distortion.**

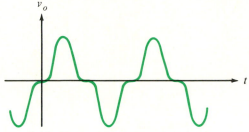

Fig. 9.89 Crossover distortion.

To minimize crossover distortion, we slightly forward bias the base–emitter junctions of the transistors. This is referred to as a **trickle bias.** As a consequence of having a trickle bias, I_{CQ} is nonzero, and has a value that typically is between 1% and 5% of

$$i_{C(\text{sat})} = I_{CQ} + V_{CEQ}/(R_T + R_E) \approx V_{CEQ}/R_E$$

When $I_{CQ} > 0$, strictly speaking, we no longer have class-B operation. Technically, operation between class A and class B is called **class-AB operation.** However, since I_{CQ} is a relatively small quantity for a trickle bias, we still say that we have class-B operation.

An example of an ac-coupled, class-B push–pull emitter follower is shown in Fig. 9.90(a). The transistors Q_1 and Q_2 are assumed to be **complementary**—that is, they have identical characteristics that differ only in sign (polarity) since one is an *n-p-n* BJT and the other is *p-n-p*. The dc-equivalent circuit of the emitter follower is shown in Fig. 9.90(b) where the quiescent currents and voltages are indicated.

By symmetry, we have that $V_{CEQ} = V_{CC}/2$, as is the voltage across R_1 and R_2. For a trickle bias, I_{CQ} is small, as is I_{BQ}. For the ideal case that $I_{BQ} = 0$, by voltage division we have

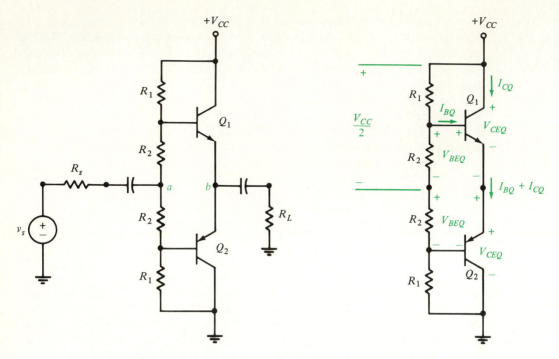

(a) Class-B push–pull emitter follower   (b) The dc-equivalent circuit

Fig. 9.90 A class-B push–pull emitter follower: (a) complete circuit, (b) dc-equivalent circuit.

$$V_{BEQ} = \frac{R_2}{R_1 + R_2}\left(\frac{V_{CC}}{2}\right)$$

(9.72)

Since $I_{BQ} \neq 0$, Equation (9.72) is an approximation, but it is a good approximation if the current through R_2 is much greater than I_{BQ}. In other words, Equation (9.72) is accurate when $I_{BQ} \ll 0.7/R_2$, that is, when $R_2 \ll 0.7/I_{BQ}$.

Example 9.14

For the circuit given in Fig. 9.90, suppose $h_{FE} = 25$, $R_2 = 200\ \Omega$, $R_L = 100\ \Omega$, and $V_{CC} = 20$ V. Let us find the value of R_1 required to obtain a trickle bias of $I_{CQ} = 0.02i_{C(\text{sat})}$.

Since $i_{C(\text{sat})} = I_{CQ} + V_{CEQ}/(R_T + R_E)$, $R_T = R_L$, and $R_E = 0$, because I_{CQ} is small, we have the approximation

$$i_{C(\text{sat})} = \frac{V_{CEQ}}{R_L} = \frac{10}{100} = 100\ \text{mA}$$

Thus,

$$I_{CQ} = 0.02 i_{C(\text{sat})} = 0.02(100 \times 10^{-3}) = 2 \text{ mA}$$

and

$$I_{BQ} = \frac{I_{CQ}}{h_{FE}} = \frac{2 \times 10^{-3}}{25} = 80 \ \mu\text{A}$$

Suppose that the BJT's transfer characteristics indicate that $v_{BE} = 0.67$ V when $i_C = 2$ mA. Then, since $R_2 = 200 \ \Omega \ll 0.7/I_{BQ} = 8750 \ \Omega$, from Equation (9.72) we have

$$V_{BEQ} = \frac{200}{R_1 + 200}\left(\frac{20}{2}\right) = 0.67 \text{ V}$$

Solving this equation for R_1, we get $R_1 = 2785 \ \Omega$.

■

The need for the coupling capacitors shown in Fig. 9.90(a) can be eliminated if the point where the collector of Q_2 and the lower resistor R_1 meet is connected to a supply voltage of $-V_{CC}$ instead of ground. Under such a circumstance, due to the symmetry of the circuit, the dc potentials at nodes a and b would be 0 V. As a consequence of no dc voltage across R_L and R_s (assuming that v_s is purely an ac voltage source), the direct current through them is zero, so no coupling capacitors are required. The resulting dc, class-B push–pull emitter follower is shown in Fig. 9.91. For such a circuit, Equation (9.72) becomes

$$V_{BEQ} = \frac{R_2}{R_1 + R_2} V_{CC} \qquad (9.73)$$

and, as was the case for Equation (9.72), this equation is accurate when $R_2 \ll 0.7/I_{BQ}$.

Push–pull emitter followers are indeed important—the output stage of most IC amplifiers is a push–pull emitter follower.

Class-B push–pull amplifiers exist in other than emitter-follower configurations. For example, the circuit shown in Fig. 9.92 is a CE push–pull amplifier. In addition to power gain, an amplifier such as this can provide a voltage gain greater than unity. Because of the nonlinear characteristic of the base–emitter junction of a BJT, emitter resistors labeled R_E are included as part of the amplifier to reduce nonlinear distortion. Unfortunately, these resistors also reduce the voltage gain.

Now consider the case that a class-B push–pull emitter follower or amplifier is driven with a sinusoidal signal such that the maximum output results. Specifically, Fig. 9.93 shows the load line for the *n-p-n* transistor along with the associated waveforms for i_C and v_{CE}. For the sake of notational simplicity, let us define

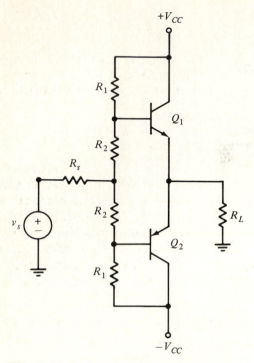

Fig. 9.91 A dc, class-B push–pull emitter follower.

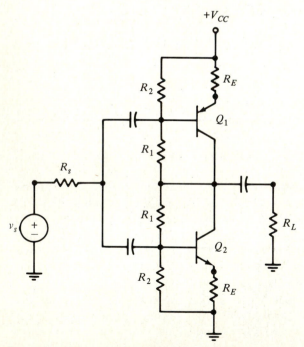

Fig. 9.92 Class-B push–pull CE amplifier.

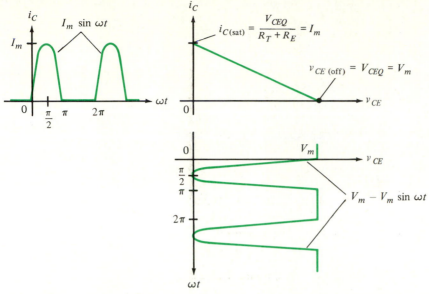

Fig. 9.93 Load line and waveforms of i_C and v_{CE} for *n-p-n* transistor when maximum output is achieved.

$I_m = i_{C(\text{sat})} = V_{CEQ}/(R_T + R_E)$ and $V_m = v_{CE(\text{off})} = V_{CEQ}$. The waveforms shown in Fig. 9.93 are the current and the voltage associated with one (the *n-p-n* BJT) of the two complementary transistors. The ac output current and voltage are sinusoids with amplitudes I_m and V_m, respectively. As a consequence, the maximum output signal power P_o is

$$P_o = \left(\frac{V_m}{\sqrt{2}}\right)\left(\frac{I_m}{\sqrt{2}}\right) = \frac{1}{2}V_m I_m$$

or

$$P_o = \tfrac{1}{2} V_{CEQ} i_{C(\text{sat})} \qquad (9.74)$$

Since class-B operation means that transistors are biased at cutoff, then the quiescent power dissipation P_{DQ} of the transistors is (ideally) zero, as is the dc power P_{dc} delivered by the supply (also called the no-signal power drain). However, the average power P_D dissipated by each transistor is not zero, and with a little bit of effort, it can be calculated.

The instantaneous power dissipated by a BJT is, from Equation (9.68), $p = v_{CE} i_C$. With reference to Fig. 9.93, for the maximum ac-signal case, the average power P_D dissipated by the corresponding transistor is

$$P_D = \frac{1}{2\pi}\int_0^{2\pi} p \, d(\omega t) = \frac{1}{2\pi}\int_0^{\pi}(V_m - V_m \sin \omega t)(I_m \sin \omega t) \, d(\omega t) \qquad (9.75)$$

It can be shown (see Problem 9.82) that from this integral we get

$$P_D = \frac{4 - \pi}{4\pi} V_m I_m = 0.0683 \, V_{CEQ} i_{C(\text{sat})} \tag{9.76}$$

This, however, is not the maximum average power that is dissipated by a transistor. In particular, if the ac signal is less than maximum and the entire load line is not utilized, then a larger value of P_D can be obtained. Specifically, suppose that the sinusoidal portions of i_{CE} and v_{CE} have amplitudes kI_m and kV_m, where $0 < k < 1$. Then the average power dissipated by each transistor is

$$P_D = \frac{1}{2\pi} \int_0^\pi (V_m - kV_m \sin \omega t)(kI_m \sin \omega t) \, d(\omega t) \tag{9.77}$$

It can be shown (see Problem 9.83) that from this integral we get

$$P_D = \frac{V_m I_m}{2\pi} \left(2k - \frac{\pi}{2} k^2 \right) \tag{9.78}$$

To find the value of k that yields the maximum value of P_D, let us take the derivative of P_D with respect to k and set the result equal to zero. Doing so, we get

$$\frac{dP_D}{dk} = \frac{V_m I_m}{2\pi} (2 - \pi k) = 0$$

from which we deduce that P_D is maximum when $k = 2/\pi$. Substituting this result into Equation (9.78), we find that the maximum average power P_{DM} dissipated by each transistor is

$$P_{DM} = \frac{V_m I_m}{2\pi} \left(\frac{4}{\pi} - \frac{2}{\pi} \right) = \frac{V_m I_m}{\pi^2}$$

or

$$\boxed{P_{DM} = 0.1 V_{CEQ} i_{C(\text{sat})}} \tag{9.79}$$

Thus, for each BJT we require that

$$\boxed{P_{D(\text{max})} \geq 0.1 V_{CEQ} i_{C(\text{sat})}} \tag{9.80}$$

From Equation (9.74), we know that the maximum output signal power is $P_o = 0.5 V_{CEQ} i_{C(\text{sat})}$. Thus, the maximum output signal power of a class-B amplifier is five times as large as the maximum power dissipated by each BJT. For example, if a class-B amplifier is to have a maximum output signal power of 5 W, then each transistor used must be rated at $P_{D(\text{max})} \geq 1$ W. Recall that for the corresponding class-A amplifier, $P_{D(\text{max})} \geq 10$ W. From an alternative point of view, as another example, given BJTs with $P_{D(\text{max})} = 4$ W, a class-A amplifier will safely yield an output signal power of at most 2 W, while a class-B amplifier can produce an output signal power of up to 20 W. This much higher efficiency is a major reason for the selection of class-B amplifiers over class-A amplifiers.

With regard to efficiency, reconsider the class-B push–pull emitter follower given in Fig. 9.91. Since the transistors are (ideally) biased at cutoff, then the dc power P_{dc} delivered by the supply is zero. However, there is a current going through the supply. For the case of maximum output signal power, this current is (approximately) equal to the collector current i_C shown in Fig. 9.93. Thus, while Q_1 is conducting, the instantaneous power delivered by the supply is $V_{CC} I_m \sin \omega t$. Therefore, the average power P_d delivered by the supply while Q_1 is conducting is

$$P_d = \frac{1}{\pi} \int_0^{\pi} V_{CC} I_m \sin \omega t \, d(\omega t) = \frac{2}{\pi} V_{CC} I_m = \frac{2}{\pi} V_m I_m$$

since $V_m = V_{CEQ} = V_{CC}$ for the given circuit. By symmetry, this is also the average power delivered while Q_2 is conducting. From Equation (9.74), the maximum output signal power is $V_m I_m / 2$. Hence, the maximum efficiency for the push–pull emitter follower under consideration is

$$\eta = \frac{(1/2) V_m I_m}{(2/\pi) V_m I_m} = \frac{\pi}{4}$$

The fact that class-B amplifiers have a maximum efficiency of 78.54% ($\eta = \pi/4$), while class-A amplifiers have a maximum efficiency of 25% ($\eta = 1/4$), affirms the practicality of class-B amplifiers.

An important concern with class-B amplifiers is their sensitivity to temperature changes. As we have seen, a push–pull configuration is trickle biased by using a voltage divider consisting of resistors R_1 and R_2 (e.g., see Fig. 9.91). The voltage across R_2 establishes V_{BEQ}, which, in turn, determines the trickle-bias collector current I_{CQ}. This current should be between 1% and 5% of $i_{C(sat)}$. However, for a given value of V_{BEQ}, an increase in temperature can result in an excessive amount of current (see Fig. 6.15). We can compensate for this by replacing the resistors labeled R_2 with diodes that are directed downward (see Problem 9.85). We should select diodes whose characteristics closely match those of the base–emitter junctions of the BJTs. This is not easy to do for discrete components, but it is readily accomplished on the same IC chip. Since $V_{BEQ} \approx 0.7$ V, then typically $R_1 \gg R_2$, and the current through R_2 is determined by the value of R_1. When R_2 is replaced by a diode, the current through the diode is still determined by the value of R_1. If the temperature increases, as the diode current remains the same, the diode voltage will decrease (see Fig. 6.15). This, in turn, means that V_{BEQ} decreases, and when the characteristics of the diode and the base–emitter junction are matched, I_{CQ} remains unchanged.

Power FETs

Although our discussion of power amplifiers has concentrated on BJTs, there are FETs that are fabricated specifically for power applications. Of particular importance is the n-channel enhancement MOSFET illustrated in Fig. 9.94(a). Com-

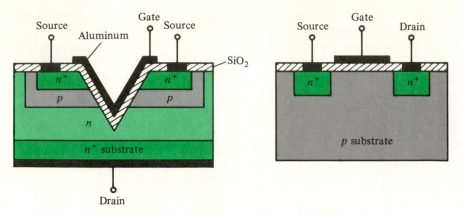

(a) Cross-sectional view of a VMOS   (b) Ordinary *n*-channel enhancement MOSFET

Fig. 9.94 Two types of *n*-channel enhancement MOSFETs.

pare this with the ordinary *n*-channel enhancement MOSFET shown in Fig. 9.94 (b). For the latter device, current going from drain to source travels horizontally, whereas for the former, it travels vertically. For this reason, as well as because of the shape of its gate, the "vertical" or "V-shaped" FET depicted in Fig. 9.94(a) is designated **VMOS**. In previous discussions, it was mentioned that ordinary FETs are symmetrical devices in that source and drain can be interchanged. This, however, is not the case with the VMOS.

Not only is the VMOS capable of handling much more power than an ordinary MOSFET, but VMOS characteristics are much closer to being linear. This property is quite desirable for large-signal applications. Whereas the peak current for an ordinary MOSFET is typically 50 mA, it is not uncommon for a VMOS to have a peak current of 2 A. Although power ratings of ordinary MOSFETs are limited to a few watts, the power ratings for some VMOS are in the tens of watts.

SUMMARY

1. For use as small-signal amplifiers, transistors are biased in the active region.

2. For BJTs and FETs, bias stability is obtained by using a combination of fixed and self-biasing.

3. The ac analysis of transistor amplifiers is simplified by using small-signal models.

4. Analysis with the use of the *h*-parameter model of a BJT results in a high degree of accuracy.

5. The coupling and bypass capacitors of an amplifier determine its lower cutoff frequency.

6. A transistor's junction capacitances determine an amplifier's upper cutoff frequency.

7. Direct-coupled (dc) amplifiers are easily fabricated on IC chips.

8. Differential amplifiers are dc amplifiers that are typically the first stages in operational amplifiers (op amps).

9. Differential amplifiers, and, hence, op amps, have the ability to reject common-mode signals (noise). This ability is expressed in terms of the common-mode rejection ratio (CMRR).

10. An ideal operational amplifier has infinite gain, infinite input resistance, and zero output resistance.

11. An op amp typically has a differential input and a single-ended output.

12. The industry standard is the 741 op amp.

13. To obtain maximum power output from a class-A amplifier, a transistor's Q point is centered on the ac load line. The maximum power dissipated by the transistor is twice the maximum output signal power.

14. The maximum efficiency of a (transformerless) class-A amplifier is 25%.

15. The transistors of a class-B push–pull amplifier are biased at cutoff. The maximum power dissipated by each transistor is only one fifth of the maximum output signal power.

16. The maximum efficiency of a class-B amplifier is 78.54%.

Problems for Section 9.1

9.1 For the circuit given in Fig. 9.2, suppose that $h_{FE} = 100$, $R_B = 330$ kΩ, $R_C = 2.7$ kΩ, and $V_{CC} = 10$ V. Find the quiescent operating point of the BJT.

9.2 For the circuit given in Fig. 9.3, suppose that $h_{FE} = 100$, $R_1 = 90$ kΩ, $R_2 = 10$ kΩ, $R_C = 1$ kΩ, and $V_{CC} = 16$ V. Find the Q point of the transistor.

9.3 For the circuit given in Fig. 9.5, suppose that $h_{FE} = 100$, $R_1 = 90$ kΩ, $R_2 = 10$ kΩ, $R_C = 5$ kΩ, $R_E = 1$ kΩ, and $V_{CC} = 20$ V. Find the operating point of the BJT.

9.4 For the circuit given in Fig. 9.5, suppose that $h_{FE} = 100$, $R_E = 100$ Ω, and $V_{CC} = 20$ V. Given that the parallel combination of R_1 and R_2 is 2.7 kΩ, find R_1, R_2, and R_C such that the operating-point conditions are $I_{CQ} = 10.2$ mA and $V_{CEQ} = 10.2$ V.

9.5 For the circuit given in Fig. 9.5, suppose that $h_{FE} = 100$, $R_1 = 6$ kΩ, $R_2 = 3$ kΩ, and $V_{CC} = 12$ V. Determine R_E and R_C such that the Q point is at $I_{CQ} = 10$ mA and $V_{CEQ} = 6$ V.

9.6 The transistor in the circuit shown in Fig. P9.6 has $h_{FE} = 100$. Determine the Q point of this BJT.

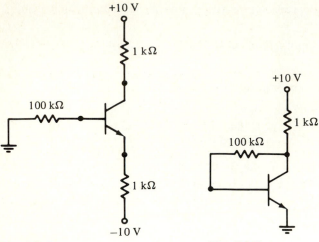

Fig. P9.6

Fig. P9.7

9.7 The transistor circuit shown in Fig. P9.7 is an example of **collector-feedback bias.** Find the operating point of the transistor given that $h_{FE} = 100$.

9.8 Both transistors in the circuit shown in Fig. P9.8 have $h_{FE} = 100$. Find the operating points for Q_1 and Q_2.

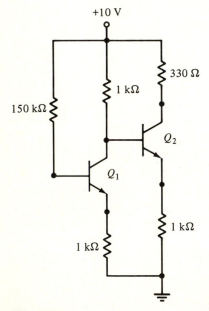

Fig. P9.8

9.9 The small-signal transistor amplifier given in Fig. 9.7 has $h_{fe} = h_{FE} = 100$,

$R_1 = 90$ kΩ, $R_2 = 10$ kΩ, $R_C = 1$ kΩ, $R_E = 0$ Ω, and $V_{CC} = 16$ V. Find (a) g_m, (b) r_e, (c) R_{in}, and (d) A_v.

9.10 Repeat Problem 9.9 after making the following changes: $R_C = 5$ kΩ, $R_E = 1$ kΩ, and $V_{CC} = 20$ V.

9.11 For the BJT amplifier given in Fig. 9.7, use the small-signal ac model shown in Fig. 9.10 to derive formulas for (a) R_b and (b) A_v.

9.12 The transistor amplifier given in Fig. 9.14 has $h_{fe} = h_{FE} = 100$, $R_1 = 90$ kΩ, $R_2 = 10$ kΩ, $R_C = 5$ kΩ, $R_E = 1$ kΩ, $R_L = 5$ kΩ, and $V_{CC} = 20$ V. Find (a) R_{in}, (b) A_v, and (c) A_i.

9.13 For the ac-equivalent circuit given in Fig. 9.17, show that the output (Thévenin-equivalent) resistance is $R_o = R_C$.

9.14 For the **feedback amplifier** given in Fig. P9.14 (see Problem 9.7), $h_{fe} = h_{FE} = 100$. Find (a) $A_v = v_c/v_b$, (b) $R_b = v_b/i_b$, (c) $R_{in} = v_b/i_s$, (d) $A_i = i_L/i_b$, and (e) $A_{is} = i_L/i_s$.

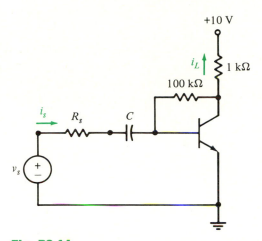

+10 V

i_L ↑ 1 kΩ

100 kΩ

i_s R_s C

v_s

Fig. P9.14

9.15 The two-stage amplifier shown in Fig. P9.15 has $h_{fe} = h_{FE} = 100$ for both transistors Q_1 and Q_2. Find (a) $A_v = v_{c_2}/v_{b_1}$ and (b) $A_i = i_{c_2}/i_{b_1}$.

9.16 For the emitter follower given in Fig. 9.18, $h_{fe} = h_{FE} = 100$, $R_1 = R_2 = 400$ kΩ, $R_E = R_s = 1$ kΩ, $R_L = 9$ kΩ, and $V_{CC} = 20$ V. Determine (a) A_v, (b) R_{in}, (c) R_o, (d) $A_i = i_L/i_b$, and (e) the power gain $A_p = p_L/p_b$, where $p_L = v_e^2/R_L$ and $p_b = v_b^2/R_b$. Is there a relationship between A_p and the product $A_v A_i$?

9.17 For the circuit given in Fig. 9.22, $R_C = R_L = 5$ kΩ, $R_p = 9$ kΩ, and the h parameters are $h_{ie} = 2.2$ kΩ, $h_{re} = 10^{-4}$, $h_{fe} = 100$, and $h_{oe} = 10^{-5}$. Find (a) $A_i = i_L/i_b$, (b) $A_v = v_c/v_b$, and (c) $R_{in} = R_p \parallel R_b$, where $R_b = v_b/i_b$. Compare these results with those obtained for Problem 9.12.

9.18 Given the emitter follower shown in Fig. 9.18 with $R_1 = R_2 = 400$ kΩ, $R_E = 1$ kΩ, and $R_L = 9$ kΩ, supposing the BJT has the h parameters $h_{ie} = 840$ Ω, $h_{re} = 10^{-4}$, $h_{fe} = 100$, and $h_{oe} = 10^{-5}$ ℧, use the small-signal

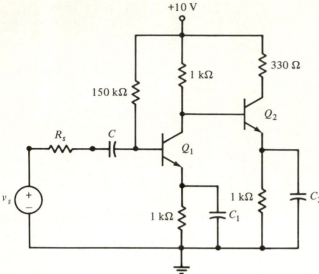

Fig. P9.15

h-parameter model given in Fig. 9.21(b) to determine (a) $A_v = v_e/v_b$, (b) $A_i = i_L/i_b$, and (c) $R_{in} = R_p \parallel R_b$, where $R_b = v_b/i_b$. Compare these results with those obtained for Problem 9.16.

9.19 Figure P9.19 shows the ac small-signal **hybrid-π model** of a BJT. Use the typical hybrid-π parameters of $r_b = 100\ \Omega$, $r_\pi = 2\ \text{k}\Omega$, $r_\mu = 20\ \text{M}\Omega$, $r_o = 200\ \text{k}\Omega$, and $g_m = 45\ \text{m}\mho$ to find the voltage gain $A_v = v_c/v_b$ for the amplifier given in Fig. 9.7, where $R_1 = 90\ \text{k}\Omega$, $R_2 = 10\ \text{k}\Omega$, $R_C = 5\ \text{k}\Omega$, and $R_E = 1\ \text{k}\Omega$.

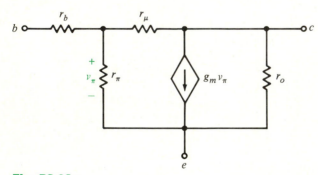

Fig. P9.19

Problems for Section 9.2

9.20 For the circuit given in Fig. 9.23, suppose that $I_{DSS} = 12\ \text{mA}$, $V_p = -4\ \text{V}$, $R_D = 1\ \text{k}\Omega$, $R_G = 1\ \text{M}\Omega$, $V_{DD} = 12\ \text{V}$, and $V_{GG} = 2\ \text{V}$. Find (a) v_{GS}, (b) i_D, and (c) v_{DS}.

9.21 For the circuit given in Fig. 9.25, suppose that $I_{DSS} = 16$ mA, $V_p = -4$ V, $R_D = 1$ kΩ, $R_G = 1$ MΩ, $R_S = 500$ Ω, and $V_{DD} = 10$ V. Find (a) v_{GS}, (b) i_D, and (c) v_{DS}.

9.22 For the circuit given in Fig. 9.27, $I_{DSS} = 12.5$ mA, $V_p = -2$ V, $R_D = R_S = 400$ Ω, $R_G = 1$ MΩ, $V_{GG} = 5$ V, and $V_{DD} = 15$ V. Find (a) v_{GS}, (b) i_D, and (c) v_{DS}.

9.23 For the circuit given in Fig. 9.27, $I_{DSS} = 12$ mA, $V_p = -2$ V, $R_D = 1$ kΩ, $R_G = 1$ MΩ, $V_{GG} = 5$ V, and $V_{DD} = 15$ V. For the case that $v_{GS} = -1$ V, find (a) i_D, (b) R_S, and (c) v_{DS}.

9.24 For the circuit given in Fig. 9.27, $I_{DSS} = 16$ mA, $V_p = -2$ V, $R_D = 1.5$ kΩ, $R_G = 1$ MΩ, $V_{GG} = 5$ V, and $V_{DD} = 15$ V. For the case that $i_D = 1$ mA, find (a) v_{GS}, (b) R_S, and (c) v_{DS}.

9.25 For the circuit given in Fig. 9.27, $I_{DSS} = 16$ mA, $V_p = -4$ V, $R_G = 1$ MΩ, $V_{GG} = 1$ V, and $V_{DD} = 15$ V. For the case that $i_D = 1$ mA and $v_{DS} = 7$ V, find (a) v_{GS}, (b) R_S, and (c) R_D.

9.26 Given the circuit shown in Fig. 9.33, suppose that $K = 2.5 \times 10^{-4}$ A/V^2, $V_T = 2$ V, $R_D = R_S = 1$ kΩ, $R_G = 1$ MΩ, $V_{DD} = 20$ V, and $V_{GG} = 10$ V. Find (a) v_{GS}, (b) i_D, and (c) v_{DS}.

9.27 Given the circuit shown in Fig. 9.33, suppose that $K = 2.5 \times 10^{-4}$ A/V^2, $V_T = 2$ V, $R_D = 1$ kΩ, $R_G = 1$ MΩ, $V_{DD} = 12$ V, and $V_{GG} = 6$ V. For the case that $v_{GS} = 4$ V, find (a) i_D, (b) R_S, and (c) v_{DS}.

9.28 Given the circuit shown in Fig. 9.33, suppose that $K = 2.5 \times 10^{-4}$ A/V^2, $V_T = 2$ V, $R_G = 1$ MΩ, $V_{DD} = 24$ V, and $V_{GG} = 12.5$ V. For the case that $i_D = 9$ mA and $v_{DS} = 6$ V, find (a) v_{GS}, (b) R_S, and (c) R_D.

9.29 For the circuit given in Fig. 9.34, $K = 2.5 \times 10^{-4}$ A/V^2, $V_T = 2$ V, $R_D = 4$ kΩ, $R_F = 1$ MΩ, and $V_{DD} = 22$ V. Find (a) v_{GS}, (b) i_D, and (c) v_{DS}.

9.30 For the circuit given in Fig. 9.34, $K = 2.5 \times 10^{-4}$ A/V^2, $V_T = 2$ V, $R_F = 1$ MΩ, and $V_{DD} = 9$ V. Find the value of R_D for which $v_{GS} = v_{DS} = 4$ V.

9.31 For the circuit given in Fig. 9.34, $K = 2.5 \times 10^{-4}$ A/V^2, $V_T = 2$ V, $R_F = 1$ MΩ, and $V_{DD} = 12$ V. Find the value of R_D for which $i_D = 4$ mA.

9.32 For the enhancement MOSFET circuit shown in Fig. P9.32, $v_{DS} \neq v_{GS}$. Given that $K = 2.5 \times 10^{-4}$ A/V^2, $V_T = 2$ V, $R_D = 4$ kΩ, $R_F = R_G = 1$ MΩ, and $V_{DD} = 12$ V, find (a) v_{GS}, (b) i_D, and (c) v_{DS}.

9.33 For the circuit given in Fig. P9.32, suppose that $K = 2.5 \times 10^{-4}$ A/V^2, $V_T = 2$ V, $R_F = 1$ MΩ, $R_G = 2$ MΩ, and $V_{DD} = 15$ V. For the case that $v_{GS} = 6$ V, find (a) i_D, (b) v_{DS}, and (c) R_D.

9.34 For the circuit given in Fig. 9.29, remove the capacitor C_S. When $I_{DSS} = 10$ mA, $V_p = -4$ V, $R_1 = 500$ kΩ, $R_2 = 250$ kΩ, $R_D = 1$ kΩ, $R_S = 2.8$ kΩ, and $V_{DD} = 15$ V, then $v_{GS} = -2$ V and $i_D = 2.5$ mA. Use the small-signal model given in Fig. 9.36 to determine (a) the voltage gain $A_v = v_d/v_g$, and (b) the power gain $A_p = p_d/p_g$.

9.35 For the circuit given in Fig. 9.29, remove the capacitor C_S. Use the small-signal model given in Fig. 9.36 to find an expression for (a) the voltage gain $A_v = v_d/v_g$, and (b) the power gain $A_p = p_d/p_g$.

9.36 For the source follower given in Fig. 9.40, suppose that $I_{DSS} = 9$ mA,

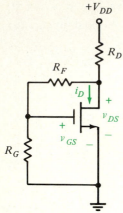

Fig. P9.32

$V_p = -3$ V, $R_G = 1$ MΩ, $R_L = 6$ kΩ, $R_S = 2$ kΩ, and $V_{DD} = 10$ V. Find (a) $A_v = v_s/v_g$, (b) R_{in}, (c) R_o, and (d) $A_p = p_s/p_g$.

9.37 The JFET in the source follower shown in Fig. P9.37 has $I_{DSS} = 9$ mA and $V_p = -3$ V. Find (a) $A_v = v_s/v_g$, (b) R_{in}, (c) R_o, and (d) $A_p = p_s/p_g$.

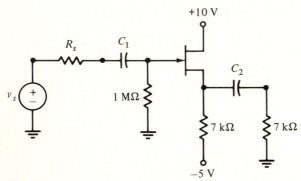

Fig. P9.37

9.38 For the enhancement MOSFET amplifier given in Fig. 9.43, let $K = 2.5 \times 10^{-4}$ A/V^2, $V_T = 2$ V, $R_D = 4$ kΩ, $R_L = 6$ kΩ, $R_F = 10$ MΩ, and $V_{DD} = 14$ V. Find (a) $A_v = v_d/v_g$, (b) R_{in}, and (c) $A_p = p_d/p_g$.

9.39 Repeat Problem 9.38 using the small-signal model given in Fig. 9.37, where $r_d = 21.6$ kΩ.

9.40 Show that the output resistance R_o for the enhancement MOSFET amplifier of Fig. 9.43 is given by

$$\frac{1}{R_o} = \frac{1}{R_D} + \frac{1 + g_m R_S}{R_F + R_S}$$

Problems for Section 9.3

9.41 For the common-emitter (CE) BJT amplifier given in Fig. 9.14, suppose that $C_1 = C_2 = 1\ \mu\text{F}$, $C_E = 47\ \mu\text{F}$, $C_c = 5\ \text{pF}$, $f_T = 150\ \text{MHz}$, $h_{fe} = 100$, $R_1 = 90\ \text{k}\Omega$, $R_2 = 10\ \text{k}\Omega$, $R_C = 5\ \text{k}\Omega$, $R_E = R_s = 1\ \text{k}\Omega$, $R_L = 5\ \text{k}\Omega$, and $r_e = 22\ \Omega$. Find (a) the lower cutoff frequency, (b) the upper cutoff frequency, and (c) the bandwidth.

9.42 The CE amplifier given in Fig. 9.14 has $C_1 = C_2 = 1\ \mu\text{F}$, $C_E = 10\ \mu\text{F}$, $C_c = 5\ \text{pF}$, $f_T = 120\ \text{MHz}$, $h_{fe} = 100$, $R_1 = 60\ \text{k}\Omega$, $R_2 = 30\ \text{k}\Omega$, $R_C = R_E = 10\ \text{k}\Omega$, $R_L = 30\ \text{k}\Omega$, $R_s = 1\ \text{k}\Omega$, and $r_e = 26\ \Omega$. Find (a) the lower cutoff frequency and (b) the upper cutoff frequency.

9.43 The common-source JFET amplifier given in Fig. 9.51 has $C_1 = 0.1\ \mu\text{F}$, $C_2 = 1\ \mu\text{F}$, $C_S = 5\ \mu\text{F}$, $C_{gs} = 2\ \text{pF}$, $C_{gd} = 3\ \text{pF}$, $C_{ds} = 1\ \text{pF}$, $g_m = 2.5 \times 10^{-3}\ \mho$, $R_1 = 500\ \text{k}\Omega$, $R_2 = 250\ \text{k}\Omega$, $R_D = R_s = 1\ \text{k}\Omega$, $R_L = 4\ \text{k}\Omega$, and $R_S = 2.8\ \text{k}\Omega$. Find (a) the lower cutoff frequency, (b) the upper cutoff frequency, and (c) the bandwidth.

9.44 For the JFET amplifier given in Fig. 9.51, suppose that $C_1 = 0.01\ \mu\text{F}$, $C_2 = 0.1\ \mu\text{F}$, $C_S = 10\ \mu\text{F}$, $C_{gs} = 4\ \text{pF}$, $C_{gd} = 2\ \text{pF}$, $C_{ds} = 1\ \text{pF}$, $g_m = 2 \times 10^{-3}\ \mho$, $R_1 = R_2 = 10\ \text{M}\Omega$, $R_D = R_L = 4\ \text{k}\Omega$, and $R_s = R_S = 1\ \text{k}\Omega$. Find (a) the lower cutoff frequency, and (b) the upper cutoff frequency.

9.45 Show that the circuits given in Fig. P9.45(a) and (b) are equivalent by writing an expression relating i and v for both circuits.

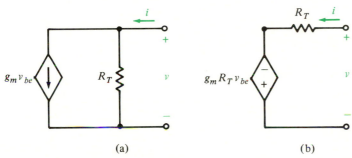

(a)   (b)

Fig. P9.45

9.46 For the emitter follower (common-collector BJT amplifier) shown in Fig. P9.46(a), use the small-signal BJT model given in Fig. 9.57 to obtain the ac-equivalent circuit shown in Fig. P9.46(b).

9.47 For the emitter-follower ac-equivalent circuit given in Fig. P9.46(b), show for low and middle frequencies that the voltage gain from base to emitter is

$$A_v = \frac{v_e}{v_b} = \frac{R_E}{R_E + R}$$

where $R = r_e \parallel r_\pi$.

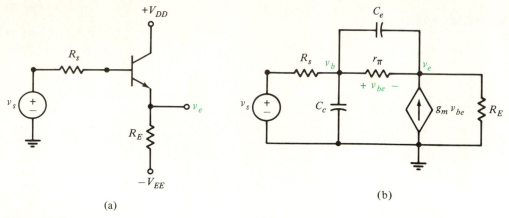

Fig. P9.46

9.48 For the emitter-follower ac-equivalent circuit shown in Fig. P9.46(b), given that $A_v = v_e/v_b = -A$, use Miller's theorem to find an equivalent ac circuit that explicitly exhibits the base and emitter lag networks.

9.49 For the emitter-follower ac-equivalent circuit given in Fig. P9.46(b), suppose that $C_c = 2$ pF, $f_T = 400$ MHz, $g_m = 38.46$ m℧, $R_E = R_s = 1$ kΩ, and $r_\pi = 2.6$ kΩ. Determine the upper cutoff frequency.

9.50 A CB amplifier typically has a higher upper cutoff frequency than a CE amplifier. Given the CB amplifier shown in Fig. P9.50, find the upper cutoff frequency for the case that $C_c = 5$ pF, $C_e = 50$ pF, and $h_{fe} = h_{FE} = 100$.

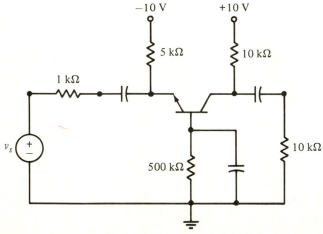

Fig. P9.50

Problems for Section 9.4

9.51 For the differential amplifier given in Fig. 9.68, suppose that $h_{FE} = 100$, $I_o = 2$ mA, $R_C = 1$ kΩ, and $R_s = 100$ Ω. Find the output voltage v_o.

9.52 Given the differential amplifier shown in Fig. 9.64, let $R_s = 0$. Show when $R_E \gg r_e$ that $v_o = v_{c_2} - v_{c_1} \approx A(v_1 - v_2)$, where $A = R_C/r_e$.

9.53 Given the differential amplifier shown in Fig. 9.64, let $R_s = 0$ and $v_2 = 0$. Show when $R_E \gg r_e$ that the input resistance R_{in} of the differential amplifier is $R_{in} = v_1/i_{b_1} \approx 2h_{FE}r_e$.

9.54 For the differential amplifier with a differential input voltage given in Fig. 9.71, set $R_s = 0$ and replace the current source I_o by a resistor R_E in series with a supply voltage $-V_{EE}$ (see Fig. 9.64). Show that the input resistance R_{in} of the resulting differential amplifier is $R_{in} = v_d/i_{b_1} \approx 2h_{FE}r_e$.

9.55 For the differential amplifier described in Problem 9.54, show that the gain $v_{c_2}/v_d \approx R_C/2r_e$.

9.56 For the differential amplifier given in Fig. 9.64, let $R_s = 0$. Apply a common-mode input v_{cm} and show when $R_E \gg r_e$ that (a) the gain $v_{c_2}/v_{cm} \approx -R_C/2R_E$, and (b) the input resistance is $v_{cm}/(i_{b_1} + i_{b_2}) \approx h_{FE}R_E$.

9.57 Given the differential amplifier shown in Fig. 9.64, where $h_{FE} = 100$, $R_C = R_E = 4.7$ kΩ, $R_s = 0$, and $V_{CC} = V_{EE} = 10$ V, find CMRR$_{dB}$.

9.58 Given the differential amplifier shown in Fig. 9.66, where $h_{FE} = 100$, $R_1 = 47$ kΩ, $R_2 = 100$ kΩ, $R_3 = 2.7$ kΩ, $R_C = 4.7$ kΩ, $R_s = 0$, and $V_{CC} = V_{EE} = 10$ V, find CMRR$_{dB}$ assuming that $R_o = 650$ kΩ.

9.59 In addition to BJTs, differential amplifiers can be fabricated using FETs. The differential amplifier shown in Fig. P9.59 has identical JFETs with $I_{DSS} = 12$ mA and $V_p = -6$ V. Find the quiescent value of (a) v_{GS}, (b) i_D, and (c) v_{DS}.

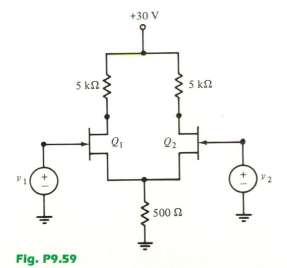

Fig. P9.59

9.60 For the JFET differential amplifier given in Fig. P9.59, suppose that $g_m = 2$ m℧. Find (a) the gain v_{d_2}/v_1 given that $v_1 \neq 0$, $v_2 = 0$; and (b) the gain v_{d_2}/v_1 given that $v_1 = v_2$.

9.61 The differential amplifier shown in Fig. P9.61 has $R_D = 5$ kΩ, $I_{DSS} = 12$ mA, and $V_p = -6$ V for identical JFETs Q_1 and Q_2, while $I_{DSS} = 6$ mA for Q_3. Assume that Q_3 behaves as an ideal current source, and find the gain v_o/v_d.

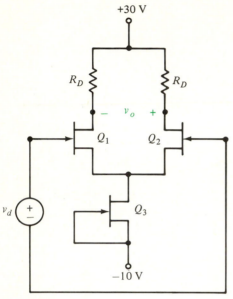

Fig. P9.61

9.62 For the circuit shown in Fig. P9.61, JFETs Q_1 and Q_2 are identical and have $g_m = 2$ m℧. Suppose that Q_3 behaves as an ideal current source in parallel with $R_o = 50$ kΩ. If $R_D = 5$ kΩ, what is CMRR$_{dB}$ for this differential amplifier?

9.63 Find a formula for the CMRR of the differential amplifier shown in Fig. P9.61 given that the identical JFETs Q_1 and Q_2 have transconductance g_m, and Q_3 has output resistance R_o.

9.64 An alternative circuit for determining the effect of an op amp's input offset voltage is shown in Fig. P9.64. Derive an expression for v_o, and compare it with Equation (9.56).

9.65 For the inverting amplifier shown in Fig. P9.65, $R_1 = 100$ kΩ, $R_2 = 900$ kΩ, and the op amp has an input offset voltage of 1 mV, an input offset current of 20 nA, and an input bias current of 80 nA. Find the worst-case output offset voltage of the overall amplifier when R_3 is chosen to minimize the output offset voltage.

9.66 For the inverting amplifier described in Problem 9.65, let $R_3 = 0$. Assume that $i_{B_1} \approx i_{B_2}$ and determine the worst-case output offset voltage for the input offset voltage and current given in Problem 9.65.

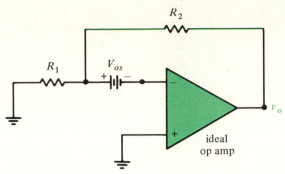

Fig. P9.64

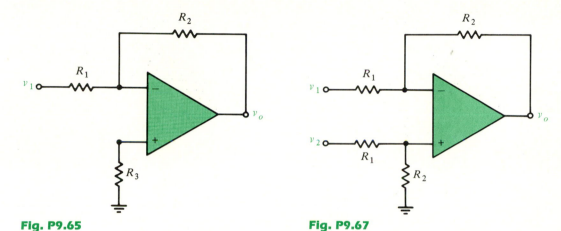

Fig. P9.65   **Fig. P9.67**

9.67 Given the differential amplifier shown in Fig. P9.67, where $R_1 = 100$ kΩ and $R_2 = 900$ kΩ, suppose that the op amp has an input offset voltage of 1 mV, an input offset current of 20 nA, and an input bias current of 80 nA. Find the worst-case output offset voltage of the overall amplifier.

9.68 Assume that the op amp in the differential amplifier shown in Fig. P9.67 is ideal. If $R_1 = 100$ kΩ and $R_2 = 900$ kΩ, then the differential gain of the overall amplifier has absolute value $|A_d| = 9$. If the top resistor labeled R_2 changes to 900,100 Ω, then the differential gain is still $|A_d| \approx 9$. Determine the CMRR$_{dB}$ for the resulting slightly unbalanced amplifier.

Problems for Section 9.5

9.69 For the circuit given in Fig. 9.83, suppose that $h_{FE} = 20$, $R_1 = 300$ Ω, $R_2 = 200$ Ω, $R_C = 50$ Ω, $R_E = 75$ Ω, and $V_{CC} = 20$ V. Find the value of R_L that results in a centered Q point.

9.70 Given the class-A amplifier shown in Fig. P9.70, determine the value of R_L that results in a centered Q point when $h_{FE} = 20$.

9.71 For the class-A amplifier given in Fig. P9.71, $h_{FE} = 20$, $R_1 = 300$ Ω, $R_2 =$

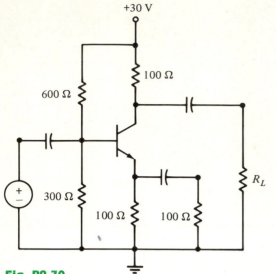

Fig. P9.70

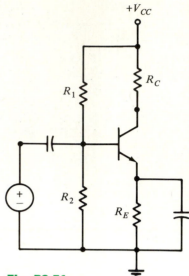

Fig. P9.71

200 Ω, $R_E = 75$ Ω, and $V_{CC} = 20$ V. Find the value of R_C that results in a centered Q point.

9.72 Find the maximum output signal power for the amplifier described in (a) Problem 9.69, (b) Problem 9.70, and (c) Problem 9.71.

9.73 Given the class-A amplifier shown in Fig. P9.73, show that $\eta = \frac{1}{4}$ when the output signal power is maximum.

9.74 Find the minimum value of $P_{D(\text{max})}$ for the transistor in the amplifier described in (a) Problem 9.69, (b) Problem 9.70, and (c) Problem 9.71.

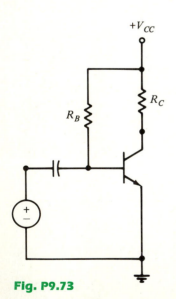

Fig. P9.73

9.75 A silicon *n-p-n* transistor has $P_{D(max)} = 1.8$ W at 25°C and a derating factor of 12 mW/°C. Find $P_{D(max)}$ at (a) 50°C, (b) 100°C, and (c) 150°C.

9.76 The circuit shown in Fig. P9.76(a) is an example of a "transformer-coupled" class-A amplifier. Figure P9.76(b) shows the dc-equivalent circuit of the amplifier, and Fig. P9.76(c) gives the ac-equivalent circuit, where $R_p = R_1 \| R_2$ and N is the "turns ratio" of the transformer. Given that $h_{FE} = 20$, $R_1 = R_2 = 200$ Ω, $R_E = 175$ Ω, and $R_L = 8$ Ω, determine the value of the turns ratio N that results in a centered Q point.

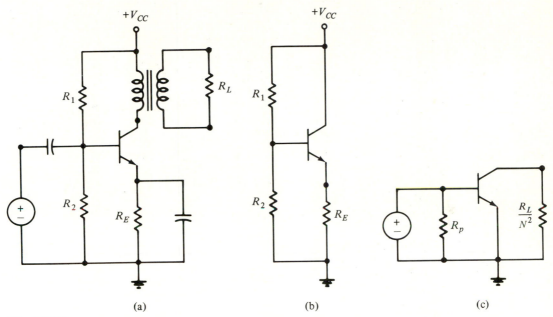

(a)   (b)   (c)

Fig. P9.76

9.77 For the transformer-coupled class-A amplifier given in Fig. P9.76(a), remove the bypass capacitor and set $R_E = 0$. Show that the resulting amplifier has efficiency $\eta = \frac{1}{2}$ when the output signal power is maximum.

9.78 Given the class-B push–pull emitter follower shown in Fig. 9.90, suppose that $h_{FE} = 20$, $R_1 = 2$ kΩ, and $V_{CC} = 30$ V.
(a) Choose R_L such that $i_{C(sat)} = 200$ mA.
(b) Determine the value of R_2 that will trickle bias the BJTs at $I_{CQ} = 2$ mA, $V_{BEQ} = 0.67$ V.
(c) Justify the use of Equation (9.71).

9.79 For the dc, class-B push–pull emitter follower given in Fig. 9.91, suppose that $h_{FE} = 25$, $R_2 = 100$ Ω, and $V_{CC} = 10$ V.
(a) Choose R_L such that $i_{C(sat)} = 100$ mA.
(b) Determine the value of R_1 that will trickle bias the BJTs at $I_{CQ} = 2$ mA, $V_{BEQ} = 0.67$ V.
(c) Justify the use of Equation (9.72).

9.80 Given the dc, class-B push–pull emitter follower shown in Fig. 9.91, suppose that $h_{FE} = 25$, $R_1 = 1.2$ kΩ, and $V_{CC} = 12$ V.
(a) Choose R_L such that $i_{C(sat)} = 100$ mA.
(b) Determine the value of R_2 that will trickle bias the BJTs at $I_{CQ} = 2$ mA, $V_{BEQ} = 0.68$ V.
(c) Justify the use of Equation (9.72).

9.81 For the class-B push–pull CE amplifier given in Fig. 9.92, suppose that $h_{FE} = 25$, $R_1 = 12$ kΩ, $R_E = 50$ Ω, and $V_{CC} = 20$ V.
(a) Choose R_L such that $i_{C(sat)} = 100$ mA.
(b) Determine the value of R_2 that will trickle bias the BJTs at $I_{CQ} = 2$ mA, $V_{BEQ} = 0.66$ V.

9.82 Show that the integral given in Equation (9.75) yields Equation (9.76). (*Hint:* $\sin^2 \theta = \frac{1}{2} - \frac{1}{2} \cos 2\theta$.)

9.83 Show that the integral given in Equation (9.77) yields Equation (9.78). (*Hint:* $\sin^2 \theta = \frac{1}{2} - \frac{1}{2} \cos 2\theta$.)

9.84 Find the maximum output signal power and the $P_{D(max)}$ rating of the transistors for the class-B amplifier described in (a) Problem 9.78, (b) Problem 9.79, (c) Problem 9.80, and (d) Problem 9.81.

9.85 Given the diode-compensated push–pull emitter follower shown in Fig. P9.85, suppose that, for this circuit, the trickle bias is $I_{CQ} = 0.02 i_{C(sat)}$.
(a) Ignore the dc base currents and determine the dc diode current I_{DQ}.
(b) Find I_{CQ}.
(c) Determine the no-signal power drain P_{dc}.
(d) Find the maximum output signal power P_o.
(e) Determine the $P_{D(max)}$ rating of the transistors.

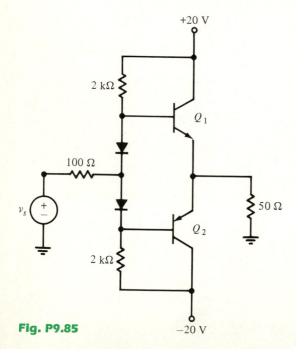

Fig. P9.85

Chapter 10

Electronic Circuits and Applications

INTRODUCTION

Feedback can be produced in a circuit or system by returning a portion of the output to the input. Because the incorporation of feedback generally will modify the performance and characteristics of a circuit, feedback principles are quite useful in electronic circuit design. In this chapter, we will see how negative feedback can be used to improve amplifier operation, and how positive feedback can be used to produce oscillators.

Oscillators are devices with no external inputs, but which generate repetitive or periodic waveforms such as sinusoids, square waves, and triangular waves. Although an oscillator may seem to be just another electronic device, its importance should not be underestimated. Specifically, even if an electronic system does not contain some type of oscillator, it almost assuredly will be used in conjunction with a system that does. Among the major systems that require oscillators are computers, digital wristwatches, oscilloscopes, radio receivers, television sets, and various types of measuring instruments.

A major aspect of the field of communications is the concept of modulation. By changing or modulating a characteristic of a waveform, it is possible to transmit information efficiently and economically from one place to another. Although the most familiar methods of communication, such as radio and television, typically utilize amplitude modulation (AM) and frequency modulation (FM), there are numerous modulation schemes and a wide variety of applications.

563

10.1 FEEDBACK

When a portion of the output of a circuit or system is returned to the input, the result is known as **feedback.** The consequences of feedback are both useful and important. In this section, we study some of them.

Let us begin by considering the situation depicted in Fig. 10.1. Shown is an amplifier that produces an output voltage v_o across a load R_L. The output voltage is also applied to the input of some feedback network (circuit), and the output of the feedback network is returned to the input portion of the amplifier. If the amplifier alone has voltage gain $v_o/v_1 = A$, then the voltage gain v_o/v_{in} of the overall circuit, in general, will be some value other than A—what it is depends upon the feedback network.

The overall circuit shown in Fig. 10.1 is an example of a **feedback amplifier,** and it illustrates one of the four basic forms that feedback amplifiers take. In particular, since the output of the feedback network is in series with the amplifier input, while the input of the feedback network is in parallel with amplifier output, this is an example of **series-parallel feedback.** (It is also called **voltage-series feedback.**) Since this type of feedback offers the most benefits for voltage-amplification applications (as we shall see), the other forms of feedback will be relegated to the problem section (see Problems 10.3–10.8).

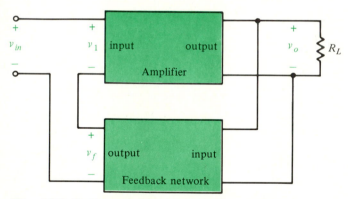

Fig. 10.1 Series-parallel feedback amplifier.

An example of a series-parallel feedback amplifier is shown in Fig. 10.2. The amplifier here is an op amp and the feedback network is a voltage divider consisting of resistors R_1 and R_2. The voltage that is being amplified by the op amp is $v_1 = v_{in} - v_f$. Since the feedback voltage v_f is subtracted from the overall input voltage v_{in}, this is an example of **negative feedback.**

To find the gain $A_F = v_o/v_{in}$ of the (overall) feedback amplifier, we use the fact that for an op amp, $v_1 = v_{in} - v_f = 0$, or $v_{in} = v_f$. Since the inputs of an (ideal) op amp draw no current, then by voltage division we have

$$v_{in} = v_f = \frac{R_1}{R_1 + R_2} v_o \qquad \textbf{(10.1)}$$

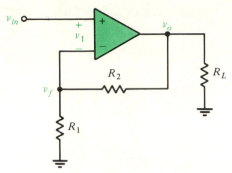

Fig. 10.2 Example of a series-parallel feedback amplifier.

from which

$$A_F = \frac{v_o}{v_{in}} = \frac{R_1 + R_2}{R_1} = 1 + \frac{R_2}{R_1} \tag{10.2}$$

Of course, the same result is obtained by a routine application of nodal analysis. Note that the gain B of the feedback network, which we define as $B = v_f/v_o$, is, by voltage division,

$$B = \frac{v_f}{v_o} = \frac{R_1}{R_1 + R_2} = \frac{1}{A_F} \tag{10.3}$$

Since the op amp shown in Fig. 10.2 is ideal, the noninverting input draws no current. Therefore, the feedback amplifier of which the op amp is part also draws no current, and, hence, the feedback amplifier also has an infinite input resistance. The output resistance R_{of} of the feedback amplifier given in Fig. 10.2 can be determined from the circuit shown in Fig. 10.3. For an ideal op amp, $v_f = v_{in} = 0$. Thus, $i_1 = v_f/R_1 = 0$, and since the inverting input of the op amp draws no current, then $i_2 = i_1 = 0$. Consequently, $v_o = R_1 i_1 + R_2 i_2 = 0$, and, hence, $R_{of} = v_o/i_o = 0\ \Omega$. Therefore, for this example, the feedback network has not altered the input and output resistances.

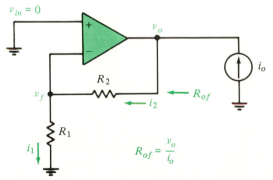

Fig. 10.3 Circuit for determining the output resistance of the feedback amplifier.

Now suppose that the op amp shown in Fig. 10.2 is no longer ideal in that its gain is a finite value A. We will still assume that its input resistance is infinite and its output resistance is zero. Since $v_f = Bv_o$, where $B = R_1/(R_1 + R_2)$ is the gain of the feedback network, then

$$v_o = A(v_{in} - v_f) = Av_{in} - Av_f = Av_{in} - ABv_o$$

Thus,

$$v_o + ABv_o = Av_{in}$$

from which the overall gain $A_F = v_o/v_{in}$ of the given feedback amplifier is

$$A_F = \frac{v_o}{v_{in}} = \frac{A}{1 + AB} \tag{10.4}$$

For very large values of A, we have the approximation

$$A_F = \frac{A}{1 + AB} \approx \frac{A}{AB} = \frac{1}{B} = \frac{R_1 + R_2}{R_1}$$

and the gain A_F becomes independent of the op-amp gain. In the limit as $A \to \infty$, this becomes an equality—which is Equation (10.2), as it should be.

Example 10.1

With the use of Equation (10.4), we may now demonstrate how negative feedback produces a more stable amplifier gain. For the feedback amplifier given in Fig. 10.2, let $A = 200,000$, $R_1 = 1 \text{ k}\Omega$, and $R_2 = 100 \text{ k}\Omega$. We then have that

$$B = \frac{R_1}{R_1 + R_2} = \frac{1000}{1000 + 100,000} = 0.0099$$

From Equation (10.4), the overall gain of the feedback amplifier is

$$A_F = \frac{200,000}{1 + (0.0099)(200,000)} = 100.959$$

Now suppose that, because of a change in temperature, the amplifier gain increases 10% to $A = 220,000$. The result of this will be that the feedback amplifier gain A_F changes to

$$A_F = \frac{220,000}{1 + (0.0099)(220,000)} = 100.964$$

which is an increase of only 0.005% in the gain of the feedback amplifier. The price that we pay for this more stable operation is a reduction of voltage gain from around 200,000 to about 100. Thus, in applications where more gain is needed, further stages of amplification would be required.

■

Next, let us investigate what happens to the input resistance R_{if} of the feedback amplifier, given that the input resistance R_{in} of the op amp is finite. In replacing the op amp in Fig. 10.2 by its model (still assuming that its output resistance is $R_o = 0$), we get the circuit shown in Fig. 10.4. By definition $R_{if} = v_{in}/i_1$. However, since $i_1 = v_1/R_{in}$, then

$$R_{if} = \frac{v_{in}}{i_1} = \frac{v_{in}}{v_1}R_{in} \qquad (10.5)$$

From the fact that $v_o = Av_1$, we have that $v_1 = v_o/A$. Substituting this into Equation (10.5) yields

$$R_{if} = \frac{v_{in}}{v_o}AR_{in} \qquad (10.6)$$

For the (typical) case that $R_{in} \gg R_1$, from Equation (10.4), we have the approximate formula $v_o/v_{in} = A/(1 + AB)$. Substituting this into Equation (10.6) results in

$$R_{if} = (1 + AB)R_{in} \qquad (10.7)$$

Thus, we see that the input resistance R_{if} of the feedback amplifier is the product of the input resistance R_{in} of the op amp and the factor $(1 + AB)$. This, therefore, shows how negative feedback can be used to increase the input resistance of an amplifier.

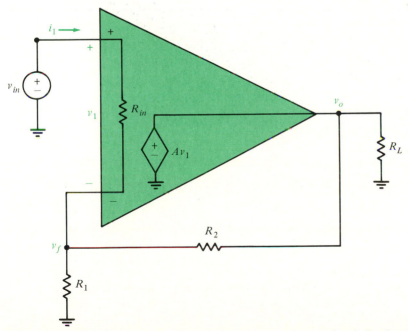

Fig. 10.4 Circuit for determining the input resistance of the feedback amplifier.

Now let us take the output resistance R_o of an op amp into consideration, and see how this affects the output resistance R_{of} of the feedback amplifier. To determine R_{of} for the feedback amplifier given in Fig. 10.2, we will replace the op amp by its model and set the input voltage to zero. The resulting circuit is shown in Fig. 10.5, where $R_{of} = v_o/i_o$. Let us assume that $R_{in} \gg R_1$, which is a typical situation. Then, by voltage division, we get the approximation

$$v_f = \frac{R_1}{R_1 + R_2} v_o = Bv_o$$

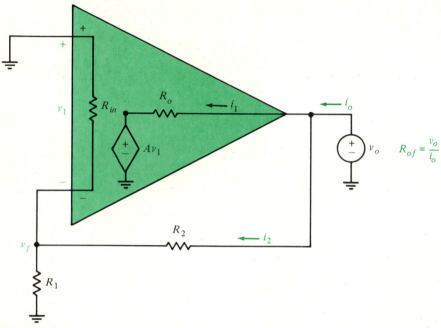

Fig. 10.5 Circuit for determining the output resistance of the feedback amplifier.

Since $v_f = -v_1$, then

$$i_o = i_1 + i_2 = \frac{v_o - Av_1}{R_o} + \frac{v_o - v_f}{R_2} = \frac{v_o + Av_f}{R_o} + \frac{v_o - v_f}{R_2}$$

$$= \frac{v_o + ABv_o}{R_o} + \frac{v_o - [R_1/(R_1 + R_2)]v_o}{R_2} = \frac{1 + AB}{R_o} v_o + \frac{1}{R_1 + R_2} v_o$$

Thus,

$$\frac{1}{R_{of}} = \frac{i_o}{v_o} = \frac{1 + AB}{R_o} + \frac{1}{R_1 + R_2}$$

In other words, R_{of} is equal to the parallel combination of $R_o/(1 + AB)$ and

$R_1 + R_2$. That is,

$$R_{of} = \frac{R_o}{1 + AB} \parallel (R_1 + R_2) \approx \frac{R_o}{1 + AB} \qquad (10.8)$$

since, typically, $R_1 + R_2 \gg R_o/(1 + AB)$. Whereas negative feedback caused the input resistance of an op amp to be multiplied by the factor $(1 + AB)$, we now see that it causes the output resistance to be divided by that same factor.

Example 10.2

For the feedback amplifier given in Fig. 10.2, suppose that the op amp has $A = 200{,}000$, $R_{in} = 2$ MΩ, and $R_o = 75$ Ω. When $R_1 = 1$ kΩ and $R_2 = 100$ kΩ, then $R_{in} \gg R_1$, and

$$B = \frac{R_1}{R_1 + R_2} = \frac{1000}{1000 + 100{,}000} = 0.0099$$

From Equation (10.7), the input resistance of the feedback amplifier is

$$R_{if} = (1 + AB)R_{in} = [1 + (200{,}000)(0.0099)](2 \times 10^6) = 3.962 \times 10^9$$

$$= 3962 \text{ M}\Omega$$

while from Equation (10.8), the output resistance of the feedback amplifier is

$$R_{of} = \frac{R_o}{1 + AB} = \frac{75}{1 + (200{,}000)(0.0099)} = 0.038 \text{ }\Omega$$

■

As mentioned earlier, other forms of feedback are considered in Problems 10.3 through 10.8.

Frequency Effects

In our discussion of operational amplifiers prior to this point, we have assumed that a nonideal op amp's gain is some constant A for all frequencies. In actuality, the gain A of an op amp is constant only over a limited range of frequencies. Since it is a dc amplifier, the lower cutoff frequency of an op amp is zero. As frequency increases, the internal capacitances of transistors become more significant. However, the upper cutoff frequency of an op amp can be determined by an appropriately added capacitance. (An example is the 30-pF capacitance shown in the 741 op-amp schematic given in Fig. 9.76.) Furthermore, such a capacitance can produce an op-amp frequency response that is essentially equivalent to a simple lag network. Specifically, for such an **internally compensated** op amp, the gain, referred to as the **open-loop gain,** is a function of frequency. Let us denote this function by **A**. Then **A** has the form

$$\mathbf{A} = \frac{A}{1 + j\omega/\omega_u} \tag{10.9}$$

where ω_u is the upper cutoff frequency.

Now suppose that the op amp in the feedback amplifier shown in Fig. 10.2 has an open-loop gain of the form of Equation (10.9). Thus, from Equation (10.4), the gain $\mathbf{A}_F$ of the feedback amplifier is also a function of frequency, which is given by

$$\mathbf{A}_F = \frac{\mathbf{A}}{1 + \mathbf{A}B} = \frac{\left[\dfrac{A}{1 + j\omega/\omega_u}\right]}{1 + \left[\dfrac{A}{1 + j\omega/\omega_u}\right]B} \tag{10.10}$$

where $B = R_1/(R_1 + R_2)$. Simplifying Equation (10.10), we get

$$\mathbf{A}_F = \frac{A}{(1 + AB) + j\omega/\omega_u} = \frac{A}{(1 + AB)\left(1 + j\dfrac{\omega}{(1 + AB)\omega_u}\right)}$$

$$= \frac{A}{1 + AB} \frac{1}{1 + j\left[\dfrac{\omega}{(1 + AB)\omega_u}\right]} = \frac{A_F}{1 + j\omega/\omega_{uF}}$$

where $A_F = A/(1 + AB)$ is the feedback-amplifier gain for low and middle frequencies (called the **dc gain**), and $\omega_{uF} = (1 + AB)\omega_u$ is the upper cutoff frequency for the feedback amplifier. Therefore, as was indicated in Equation (10.4), when placed in the feedback configuration shown in Fig. 10.2, the dc gain A of the op amp is divided by the factor $1 + AB$. However, the upper cutoff frequency ω_u (and, hence, the bandwidth) of the op amp is multiplied by the factor $1 + AB$. Thus, the use of an op amp as part of a feedback amplifier drastically can increase an amplifier's bandwidth.

Example 10.3

The **gain-bandwidth product** f_T of an op amp is just the product of its dc gain A and its upper cutoff frequency (bandwidth) f_u given in hertz. That is,

$$f_T = Af_u \tag{10.11}$$

Using the 741 op amp typical values of $A = 200{,}000$ and $f_T = 1$ MHz, we obtain

$$f_u = \frac{f_T}{A} = \frac{1 \times 10^6}{2 \times 10^5} = 5 \text{ Hz}$$

and

$$\omega_u = 2\pi f_u = 10\pi = 31.4 \text{ rad/s}$$

If such an op amp is part of the feedback amplifier shown in Fig. 10.2, where $R_1 = 1$ kΩ and $R_2 = 100$ kΩ, then

$$B = \frac{R_1}{R_1 + R_2} = \frac{1000}{1000 + 100,000} = 0.0099$$

and the resulting upper cutoff frequency (bandwidth) is

$$\omega_{uF} = (1 + AB)\omega_u = [1 + (200,000)(0.0099)](10\pi) = 6.22 \times 10^4 \text{ rad/s}$$

or

$$f_{uF} = \frac{\omega_{uF}}{2\pi} = 9.9 \times 10^3 = 9.9 \text{ kHz}$$

By changing the values of R_1 and R_2, we can change the value of the feedback gain $B = R_1/(R_1 + R_2)$, and, hence, the feedback-amplifier dc gain $A_F = A/(1 + AB)$. Although this means that the feedback-amplifier upper cutoff frequency (bandwidth) $\omega_{uF} = (1 + AB)\omega_u = 2\pi f_{uF} = 2\pi(1 + AB)f_u$ will change, the product of the gain and bandwidth of the feedback amplifier is

$$A_F f_{uF} = \frac{A}{1 + AB}(1 + AB)f_u = Af_u = f_T \qquad \textbf{(10.12)}$$

In other words, the gain-bandwidth product of the feedback amplifier is constant and is equal to the gain-bandwidth product of the op amp.

10.2 SINUSOIDAL OSCILLATORS

In the preceding section, we saw an example of a feedback amplifier (Fig. 10.2) whose gain A_F was given by

$$A_F = \frac{A}{1 + AB}$$

where A is the op-amp gain and B is the feedback-network gain. For the case that the feedback network is a resistive circuit, such as a voltage divider, then $B > 0$ and, hence, $AB > 0$. Therefore, $1 + AB > 1$ and $A_F < A$. When $A_F < A$, we say that the feedback is **negative** (or **degenerative**). On the other hand, when $A_F > A$ due to some other type of feedback network, we call the feedback **positive** (or **regenerative**).

For the feedback amplifier shown in Fig. 10.2, we have seen how negative feedback produced more stability in voltage gain, a higher input resistance, and a lower output resistance. In addition to the usefulness of negative feedback, though, positive feedback can produce desirable results in areas other than voltage amplification.

Consider again the general series-parallel feedback amplifier depicted in Fig. 10.1, where A is the gain of the amplifier and B is the gain of the feedback network. Before an input voltage is applied, the output voltage is $v_o = 0$, and, hence, the output of the feedback network is $v_f = Bv_o = 0$. Therefore, when an input voltage v_{in} is applied, initially, $v_o = A(v_{in} - v_f) = Av_{in}$. This output voltage then gives rise to a nonzero voltage at the output of the feedback network, in particular, $v_f = Bv_o = BAv_{in}$. However, now that v_f is no longer zero, let us suppose that we can immediately make the original input voltage zero. This means that the resulting input to the gain-A amplifier will be $v_1 = v_{in} - v_f = 0 - BAv_{in} = -BAv_{in}$. Therefore, under the condition that $BA = -1$, the gain-A amplifier input will be $v_1 = -BAv_{in} = -(-1)v_{in} = v_{in}$, which is equal to the original input. In other words, the feedback network allows the maintenance of a voltage equal to the original voltage v_{in} at the input of the gain-A amplifier even after the original input voltage is set to zero. Thus, also, the same output voltage v_o is maintained in spite of the fact that an external input is no longer applied.

A feedback configuration such as the one just described, which produces a nonzero output when no external input is applied, is called an **oscillator.** We define the **loop gain** of a feedback configuration with no external input to be the product of the individual gains around the loop. In particular, for the configuration given in Fig. 10.1 with the external input v_{in} set to zero, the amplifier gain is $v_o/v_1 = A$ and the feedback-network gain is $v_f/v_o = B$. Note, too, the additional gain $v_1/v_f = -1$. Thus, the loop gain for this case is $-BA$. From the discussion above, we see that the condition for oscillation is $-BA = 1$, that is, unity loop gain. This condition is known as the **Barkhausen criterion.** Substitution of $BA = -1$ into the feedback-gain formula $A_F = A/(1 + AB)$ indicates that the feedback gain is infinite. This, in turn, justifies the fact that we can have a nonzero output when there is no external input.

Phase-Shift Oscillators

If the amplifier gain A is a constant (not frequency dependent) for the configuration shown in Fig. 10.1, one way to get a unity loop gain is to use a frequency-dependent feedback network having a transfer function **B,** where the magnitude of **B** is $|\mathbf{B}| = 1/A$ and the phase angle (phase shift) of **B** is $\theta = 180$ degrees. If the phase shift of the feedback network is 180 degrees for just a single frequency, then the loop gain will be unity for only one frequency and the oscillator output will be a sinusoid having that particular frequency.

Since a simple RC lead or lag network (which consists of one R and one C) will produce a phase shift of less than 90 degrees for finite frequencies, a minimum of three simple RC stages is required to obtain an 180-degree phase shift. In particular, consider the RC phase-shift network shown in Fig. 10.6. By KCL, using phasor notation, we have

$$j\omega C(\mathbf{V}_3 - \mathbf{V}_o) + j\omega C(\mathbf{V}_3 - \mathbf{V}_2) + \frac{\mathbf{V}_3}{R} = 0$$

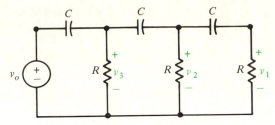

Fig. 10.6 An *RC* phase-shift network.

as well as

$$j\omega C(\mathbf{V_2} - \mathbf{V_3}) + j\omega C(\mathbf{V_2} - \mathbf{V_1}) + \frac{\mathbf{V_2}}{R} = 0$$

and

$$j\omega C(\mathbf{V_1} - \mathbf{V_2}) + \frac{\mathbf{V_1}}{R} = 0$$

Manipulating these equations eventually results in the voltage transfer function

$$\mathbf{B} = \mathbf{B}(j\omega) = \frac{\mathbf{V_1}}{\mathbf{V_o}} = \frac{\omega^3 R^3 C^3}{\omega RC(\omega^2 R^2 C^2 - 5) + j(1 - 6\omega^2 R^2 C^2)} \qquad \textbf{(10.13)}$$

The angle of this transfer function is 180 degrees when $1 - 6\omega^2 R^2 C^2 = 0 \Rightarrow \omega = 1/\sqrt{6}RC$, and substituting the frequency $\omega_o = 1/\sqrt{6}RC$ into Equation (10.13), we get

$$\mathbf{B}(j\omega_o) = -\tfrac{1}{29} = \tfrac{1}{29}\underline{/180°} = |\mathbf{B}|\underline{/\theta}$$

In other words, the gain (the magnitude of the transfer function) of this feedback network is $|\mathbf{B}| = \frac{1}{29}$ and the phase shift (angle) is $\theta = 180$ degrees when $\omega = 1/\sqrt{6}RC = \omega_o$.

  We now can see how the circuit given in Fig. 10.6 can be used as a feedback network to produce an oscillator. Specifically, consider the op-amp circuit shown in Fig. 10.7. Assuming that the op amp is ideal, the gain v_o/v_1 due to the op amp and resistors R and R_F is $A = -R_F/R$ (see Fig. 2.13). Since the inverting input of the op amp must be at 0 V, then the resistance between node v_1 and ground is equal to R. In other words, the *RC* feedback network is loaded with a resistance of value R. Thus, the transfer function of the feedback network is given by Equation (10.13). Hence, when

$$\boxed{\omega = \omega_o = \frac{1}{RC\sqrt{6}}}$$

the loop gain is

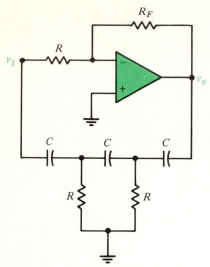

Fig. 10.7 An op-amp phase-shift oscillator.

$$\mathbf{B}A = -\frac{1}{29}\left(-\frac{R_F}{R}\right) = \frac{R_F}{29R}$$

which is unity when $R_F = 29R$ (i.e., $A = -29$). This, therefore, is the condition for oscillation, and ω_o is the corresponding frequency of oscillation.

If the loop gain is less than unity, from the discussion above, we know that oscillation will not occur. However, if the loop gain is greater than unity, not only will oscillation occur, but because of positive feedback, the oscillation amplitude will increase with time. Since an actual amplifier is a physical device, its output cannot grow without bound, and eventually the operation of the amplifier will become nonlinear. This, in turn, will limit the oscillation amplitude.

Because of the variations of transistor characteristics due to such things as temperature, voltage, and age, the theoretical requirement of unity loop gain cannot be met practically. Furthermore, an oscillator such as the one shown in Fig. 10.7 at no time has an explicit external input. However, small noise voltages are present that are produced by thermally agitated electrons in resistors. When the loop gain is greater than unity, the component of the noise voltage at the oscillation frequency grows in amplitude so as to produce the desired oscillation. For these reasons, for a practical oscillator, the loop gain is chosen to be greater than unity—typically by about 5 percent.

In the discussion above, we assumed that the op amp shown in Fig. 10.7 was ideal. In particular, we ignored its upper cutoff frequency, and, hence, the upper cutoff frequency of the amplifier comprised of the op amp and resistors R and R_F. If the latter cutoff frequency is much greater than the frequency of oscillation, then the same approach can be taken for the case of a nonideal op amp as was taken for an ideal op amp. Yet, higher oscillation frequencies can be obtained if, instead of an op amp, we use a single transistor.

The phase-shift oscillator shown in Fig. 10.8 uses the RC phase-shift feedback network described earlier (see Fig. 10.6). The amplifier portion of the oscillator consists of a JFET and its associated components. Assuming that $R \gg R_D$ and ignoring the ac output resistance r_d of the JFET, the mid-band gain of the amplifier is $A = -g_m R_D$. (To take r_d into consideration, replace R_D by $R_L = r_d \parallel R_D$.) Therefore, at the oscillation frequency

$$\omega_o = \frac{1}{RC\sqrt{6}}$$

setting the loop gain to unity we have

$$\left(-\frac{1}{29}\right)(-g_m R_D) = 1 \quad \Rightarrow \quad g_m = 29/R_D$$

Thus, to get oscillation, we must have $g_m \geq 29/R_D$.

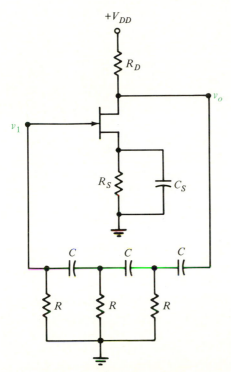

Fig. 10.8 A JFET phase-shift oscillator.

In addition to utilizing a FET, a phase-shift oscillator can be constructed by using a BJT as shown in Fig. 10.9. Unlike a FET, a BJT does not have an (ideally) infinite input resistance. Since the feedback network is to be loaded with a resistance equal to R, we should choose R_s such that the resistance R_{in} between node v_1 and the reference looking into the amplifier is equal to R. In other words, we want

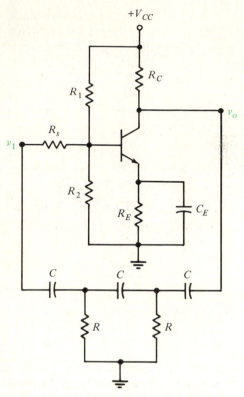

Fig. 10.9 A BJT phase-shift oscillator.

to satisfy the equation

$$R_{in} \approx R_s + R_1 \parallel R_2 \parallel h_{fe}r_e = R$$

and since $g_m = 1/r_e$ for a BJT, we wish to have

$$R_s = R - \left(R_1 \parallel R_2 \parallel \frac{h_{fe}}{g_m} \right)$$

If the collector resistor R_C is not ignored, then after a considerable amount of mathematical manipulation, it is found that the loop gain is unity at the oscillation frequency

$$\omega_o = \frac{1}{RC\sqrt{6 + 4R_C/R}}$$

when

$$h_{fe} = 23 + \frac{29R_C}{R} + \frac{4R}{R_C}$$

The Wien-Bridge Oscillator

A practical op-amp oscillator circuit that sees frequent application is the Wien-bridge oscillator shown in Fig. 10.10. Assuming an ideal op amp, the gain of the amplifier that consists of the op amp and resistors R_1 and R_2, by Equation (10.2), is

$$A = \frac{v_o}{v_1} = \frac{R_1 + R_2}{R_1} = 1 + \frac{R_2}{R_1}$$

For the RC feedback network, let us designate the parallel RC connection by $\mathbf{Z}_1$ and the series RC connection by $\mathbf{Z}_2$. Then

$$\mathbf{Z}_1 = \mathbf{Z}_R \parallel \mathbf{Z}_C = \frac{(R)(1/j\omega C)}{R + (1/j\omega C)} = \frac{R}{1 + j\omega RC}$$

and

$$\mathbf{Z}_2 = \mathbf{Z}_R + \mathbf{Z}_C = R + \frac{1}{j\omega C} = \frac{1 + j\omega RC}{j\omega C}$$

Therefore, by voltage division, the voltage transfer function of the feedback network is

$$\mathbf{B} = \mathbf{B}(j\omega) = \frac{\mathbf{V}_1}{\mathbf{V}_o} = \frac{\mathbf{Z}_1}{\mathbf{Z}_1 + \mathbf{Z}_2} = \frac{\dfrac{R}{1 + j\omega RC}}{\dfrac{R}{1 + j\omega RC} + \dfrac{1 + j\omega RC}{j\omega C}}$$

After a few steps of algebraic manipulation, we get

$$\mathbf{B} = \mathbf{B}(j\omega) = \frac{j\omega RC}{1 - \omega^2 R^2 C^2 + j3\omega RC}$$

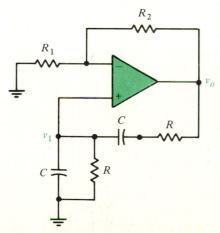

Fig. 10.10 Wien-bridge oscillator.

Note, however, that when $\omega = \omega_o = 1/RC$, then $\mathbf{B}(j\omega_o) = \frac{1}{3}$. Thus, for $\omega = \omega_o = 1/RC$, the loop gain is unity when

$$\mathbf{B}A = \frac{1}{3}A = \frac{1}{3}\left(1 + \frac{R_2}{R_1}\right) = 1$$

that is, when $R_2 = 2R_1$. In summary, the circuit given in Fig. 10.10 oscillates when $R_2 = 2R_1$ (or $R_2 > 2R_1$), and the frequency of oscillation is

$$\boxed{\omega_o = 1/RC}$$

The circuit given in Fig. 10.10 can be drawn in the equivalent form shown in Fig. 10.11. From this figure, it is explicitly seen that the resistors and capacitors constitute a configuration known as a **Wien bridge**—hence, the name of the oscillator. Because of the op amp, we must have that $v_1 = v_2$, and for this condition we say that the bridge is **balanced.**

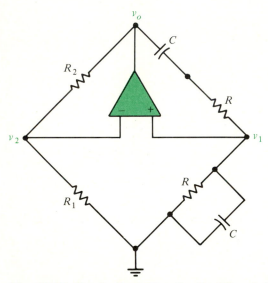

Fig. 10.11 Equivalent circuit of a Wien-bridge oscillator.

Colpitts and Hartley Oscillators

We may develop other types of oscillators by investigating the general feedback circuit given in Fig. 10.12. This circuit shows an inverting amplifier with input resistance R_{in}, output resistance R_o, and gain A_1. For the sake of simplified calculations, let us assume that the input resistance is infinite—a reasonable assumption when the amplifier is realized with a FET or an op amp.

To determine an expression for the feedback-network transfer function

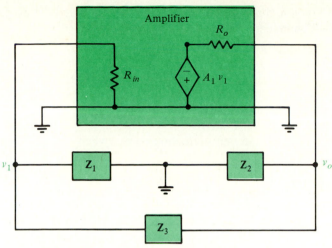

Fig. 10.12 General feedback circuit.

$\mathbf{B} = \mathbf{V}_1/\mathbf{V}_o$, we apply KCL at node v_1. The result is

$$\frac{\mathbf{V}_1}{\mathbf{Z}_1} + \frac{\mathbf{V}_1 - \mathbf{V}_o}{\mathbf{Z}_3} = 0$$

From this equation, we get (the voltage-division formula)

$$\mathbf{B} = \frac{\mathbf{V}_1}{\mathbf{V}_o} = \frac{\mathbf{Z}_1}{\mathbf{Z}_1 + \mathbf{Z}_3}$$

Applying KCL at node v_o and using the fact that $\mathbf{V}_1 = \mathbf{B}\mathbf{V}_o$, we get

$$\frac{-A_1\mathbf{V}_1 - \mathbf{V}_o}{R_o} = \frac{\mathbf{V}_o}{\mathbf{Z}_2} + \frac{\mathbf{V}_o - \mathbf{B}\mathbf{V}_o}{\mathbf{Z}_3}$$

After substituting $\mathbf{B} = \mathbf{Z}_1/(\mathbf{Z}_1 + \mathbf{Z}_3)$ into this equation and doing some manipulation, we obtain the amplifier transfer function

$$\mathbf{A} = \frac{-A_1\mathbf{Z}_2(\mathbf{Z}_1 + \mathbf{Z}_3)}{R_o(\mathbf{Z}_1 + \mathbf{Z}_2 + \mathbf{Z}_3) + \mathbf{Z}_2(\mathbf{Z}_1 + \mathbf{Z}_3)}$$

Hence, the loop gain is

$$\mathbf{AB} = \frac{-A_1\mathbf{Z}_1\mathbf{Z}_2}{R_o(\mathbf{Z}_1 + \mathbf{Z}_2 + \mathbf{Z}_3) + \mathbf{Z}_2(\mathbf{Z}_1 + \mathbf{Z}_3)} \qquad \textbf{(10.14)}$$

Suppose that we let each impedance $\mathbf{Z}$ be a pure reactance, that is, $\mathbf{Z} = jX$, where $X = \omega L$ for an L-henry inductor or $X = -1/\omega C$ for a C-farad capacitor. Then Equation (10.14) becomes

$$\mathbf{AB} = \frac{A_1 X_1 X_2}{jR_o(X_1 + X_2 + X_3) - X_2(X_1 + X_3)} \qquad \textbf{(10.15)}$$

To obtain oscillation, we require that the loop gain be unity (or greater). To make the imaginary part of the denominator of **AB** equal to zero, we set

$$X_1 + X_2 + X_3 = 0 \qquad (10.16)$$

From this equation, $X_1 + X_3 = -X_2$, and, therefore, the loop gain is

$$\mathbf{AB} = \frac{A_1 X_1 X_2}{-X_2(X_1 + X_3)} = -\frac{A_1 X_1}{X_1 + X_3} = \frac{A_1 X_1}{X_2} \qquad (10.17)$$

Setting the loop gain to unity, we find that oscillation occurs when

$$A_1 = X_2/X_1$$

(or more).

For Equation (10.16) to hold, the reactances can be neither all inductive nor all capacitive. Furthermore, since the loop gain must be positive, and it was assumed implicitly that $A_1 > 0$, then from Equation (10.17), both X_1 and X_2 must be either capacitive or inductive.

First consider the case that X_1 and X_2 are both capacitive, for example, $\mathbf{Z}_1 = 1/j\omega C_1$ and $\mathbf{Z}_2 = 1/j\omega C_2$. To satisfy Equation (10.16), let us make X_3 inductive, for example, $\mathbf{Z}_3 = j\omega L$. Thus, we have

$$X_1 = -\frac{1}{\omega C_1}, \qquad X_2 = -\frac{1}{\omega C_2}, \qquad X_3 = \omega L$$

Therefore, Equation (10.16) is satisfied when

$$-\frac{1}{\omega C_1} - \frac{1}{\omega C_2} + \omega L = 0$$

or when

$$\omega L = \frac{1}{\omega C_1} + \frac{1}{\omega C_2} = \frac{1}{\omega}\left(\frac{1}{C_1} + \frac{1}{C_2}\right) = \frac{1}{\omega C} \qquad (10.18)$$

where we define C by

$$\frac{1}{C} = \frac{1}{C_1} + \frac{1}{C_2} \qquad \Rightarrow \qquad C = \frac{C_1 C_2}{C_1 + C_2}$$

Let us denote the frequency ω that satisfies Equation (10.18) by ω_o. Then from

$$\omega_o L = \frac{1}{\omega_o C}$$

we have that

$$\boxed{\omega_o = \frac{1}{\sqrt{LC}}}$$

where $C = C_1 C_2 / (C_1 + C_2)$, and ω_o is the frequency of oscillation. Note that C is, in effect, the series combination of C_1 and C_2, and ω_o is the resonance frequency of the parallel combination of L and C. For this case, the circuit given in Fig. 10.12 is known as a **Colpitts oscillator.** An example of a JFET Colpitts oscillator is shown in Fig. 10.13. Here, the amplifier indicated in Fig. 10.12 consists of a JFET, a gate resistor R_G, a drain resistor R_D, a coupling capacitor C_D, and a power supply voltage V_{DD}.

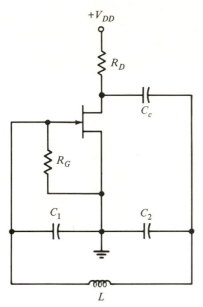

Fig. 10.13 A JFET Colpitts oscillator.

Suppose that instead of an inductor L, the impedance $\mathbf{Z}_3$ consists of an inductor L in series with a capacitor C_3. Then

$$X_3 = \omega L - \frac{1}{\omega C_3}$$

Substituting this into Equation (10.16) along with the fact that $X_1 = -1/\omega C_1$ and $X_2 = -1/\omega C_2$, we find that the oscillation frequency is

$$\omega_o = \frac{1}{\sqrt{LC}}$$

where

$$\frac{1}{C} = \frac{1}{C_1} + \frac{1}{C_2} + \frac{1}{C_3}$$

The resulting oscillator is referred to as a **Clapp oscillator.** If $C_1 \gg C_3$ and $C_2 \gg C_3$, then $C \approx C_3$, and the frequency of oscillation is $\omega_o = 1/\sqrt{LC_3}$. Therefore, by selecting C_1 and C_2 to be much greater than C_3, we can make ω_o essentially independent of C_1 and C_2. This may be desirable since junction capacitances of transistors change with temperature, and thereby can affect the values of C_1 and C_2. Thus, by eliminating the dependence of ω_o on C_1 and C_2, greater oscillation-frequency stability is achieved.

In the analysis of the general feedback configuration given in Fig. 10.12, we assumed that the input resistance of the amplifier was infinite. This, however, is not a good approximation for a single BJT amplifier. But a similar approach can be taken for the case of a BJT, and the result is that oscillation will occur for sufficient amplifier gain, and the oscillation frequency will be the same as the FET case. An example of a BJT Colpitts oscillator is shown in Fig. 10.14. Again the oscillation frequency is $\omega_o = 1/\sqrt{LC}$, where $C = C_1 C_2/(C_1 + C_2)$.

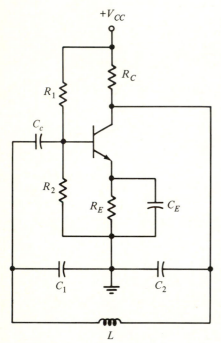

Fig. 10.14 A BJT Colpitts oscillator.

Next let us consider the case that X_1 and X_2 are both inductive, for example, $\mathbf{Z}_1 = j\omega L_1$ and $\mathbf{Z}_2 = j\omega L_2$. Proceeding in a dual manner to the preceding (Colpitts oscillator) approach, we make X_3 capacitive (e.g., $\mathbf{Z}_3 = 1/j\omega C$), and deduce that the frequency of oscillation is

$$\omega_o = \frac{1}{\sqrt{LC}}$$

where $L = L_1 + L_2$. This type of oscillator is an example of a **Hartley oscillator.**
Figures 10.15 and 10.16 show JFET and BJT Hartley oscillators respectively.

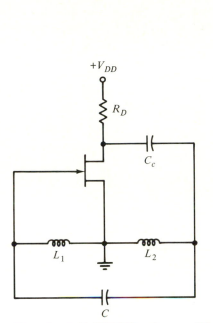

Fig. 10.15 A JFET Hartley oscillator.

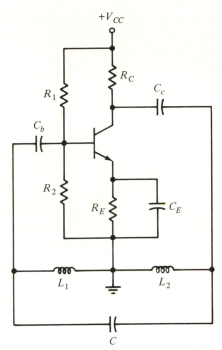

Fig. 10.16 A BJT Hartley oscillator.

Crystal Oscillators

While RC oscillators can achieve frequency stabilities of about 0.1% (e.g.,
1000 Hz $\pm$ 1 Hz), stabilities of 0.01% are more typical of LC oscillators. Much
greater oscillation-frequency stabilities, however, are obtained with oscillators that
employ crystals—typically quartz. Such **crystal oscillators** can achieve stabilities
of 0.001% (also denoted as 10 parts per million), and even 0.0001% (one part per
million). As a consequence, such oscillators are used in digital wristwatches.

Quartz crystals have the property that if they are mechanically vibrated, they
produce an ac voltage. Conversely, if an ac voltage is applied across them, they will
vibrate at the frequency of the applied voltage. This phenomenon is known as the
piezoelectric effect, and quartz is an example of a **piezoelectric crystal.**

Although a quartz crystal will vibrate at the frequency of an applied ac voltage, the magnitude of the vibration is a function of frequency. Specifically, the crystal's internal friction, mass, and compliance give rise to a series *RLC* electric equivalent circuit. Further, since a quartz crystal is mounted between two metal plates, what results is a **mounting capacitance** that is in parallel with the series *RLC* connection. Shown in Fig. 10.17(a) is the schematic symbol for a quartz crystal (designated XTAL), and Fig. 10.17(b) gives its electric equivalent circuit.

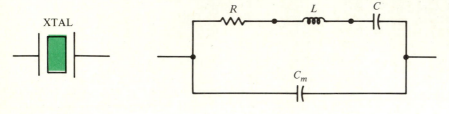

(a) Quartz-crystal symbol   (b) Equivalent circuit

Fig. 10.17 Quartz crystal and its equivalent circuit.

For the circuit shown in Fig. 10.17(b), the series *LC* combination behaves as a short circuit at its resonance frequency $\omega_s = 1/\sqrt{LC}$. There is, however, an additional resonance frequency due to the parallel connection of C_m. Since R typically is relatively small, let us ignore it in determining this second resonance frequency. Neglecting R, the admittance of the equivalent circuit shown in Fig. 10.17(b) is

$$
Y = j\omega C_m + \cfrac{1}{j\omega L + \cfrac{1}{j\omega C}} = \cfrac{(j\omega C_m)(j\omega L) + \cfrac{j\omega C_m}{j\omega C} + 1}{j\omega L + \cfrac{1}{j\omega C}}
$$

$$
= \cfrac{-\omega^2 L C_m + \cfrac{C_m}{C} + 1}{j\left(\omega L - \cfrac{1}{\omega C}\right)} = \cfrac{j\left(\omega^2 L C_m - \cfrac{C_m}{C} - 1\right)}{\omega L - \cfrac{1}{\omega C}}
$$

At the resonance frequency ω_p, we must have that

$$
\omega_p^2 L C_m - \frac{C_m}{C} - 1 = 0
$$

from which

$$
\boxed{\omega_p = \sqrt{\frac{1}{L}\left(\frac{1}{C} + \frac{1}{C_m}\right)}}
$$

Typically, however, for a quartz crystal $C_m \gg C$, and consequently

$$\frac{1}{C} + \frac{1}{C_m} \approx \frac{1}{C}$$

so that the parallel resonance frequency is

$$\omega_p \approx \sqrt{\frac{1}{LC}} = \frac{1}{\sqrt{LC}} = \omega_s$$

In employing a quartz crystal for an oscillator, we can utilize either its series resonance ω_s or its parallel resonance ω_p—they are approximately equal. In the former case, the crystal behaves as a small (ideally zero) impedance, and in the latter it behaves as a large (ideally infinite) impedance.

One way to employ the series resonance of a quartz crystal is to use it in place of the inductor L in the Colpitts oscillator shown in either Fig. 10.13 or Fig. 10.14. Another way is indicated by Fig. 10.18, which shows a JFET **Pierce oscillator.** On the other hand, the modified BJT Colpitts oscillator shown in Fig. 10.19 utilizes the parallel resonance of a quartz crystal.

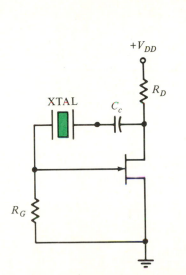

Fig. 10.18 Pierce oscillator.

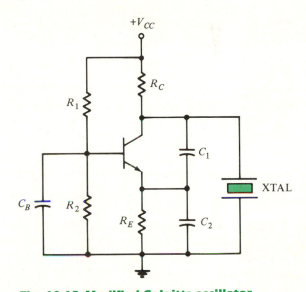

Fig. 10.19 Modified Colpitts oscillator.

By electric-circuit standards, the quality factor Q of a quartz crystal is extremely high. For discrete electric-circuit components, a Q of 100 is considered high. However, a quartz crystal with a Q of 10,000 or higher is not uncommon. Even Q values of up to 1,000,000 are possible. These high-Q values for quartz crystals, along with the fact that quartz characteristics are quite stable with respect to temperature and time, mean that crystal oscillators have excellent oscillation-frequency stability.

Quartz crystals with resonance frequencies in the range of 10 kHz to 10 MHz are commonly available. The equivalent circuit, given in Fig. 10.17, of a crystal is actually a simplification of a more complicated model that contains additional series *RLC* combinations. These give rise to additional resonances, called **over-tones,** which are integer multiples of ω_s, that is, $2\omega_s$, $3\omega_s$, and so on. By using crystals designed for overtone-mode operation, it is possible to obtain crystal oscillators with oscillation frequencies that go beyond 200 MHz.

10.3 THE SCHMITT TRIGGER

In our previous discussions, some type of feedback was always associated with the use of operational amplifiers. There is, however, an important application—albeit nonlinear—for an operational amplifier that does not employ feedback. To demonstrate this, consider the op amp shown in Fig. 10.20, where the supply voltages $+V_{CC}$ and $-V_{CC}$ are explicitly indicated. Let us assume that the gain A of the op amp is large. (Recall that a typical gain for a 741 op amp is $A = 200,000$.) Let us suppose that one of the inputs, the inverting input v_2, is constrained to be some reference voltage $V_r > 0$, that is, suppose $v_2 = V_r$. Although the output of the op amp tends to be $v_o = A(v_1 - v_2) = A(v_1 - V_r)$, because of the large gain, even if v_1 is just slightly larger than V_r, the op-amp output will saturate and (ideally) we get that $v_o = +V_{CC}$. Similarly, if v_1 is less than V_r, then $v_o = -V_{CC}$. These results are summarized by the transfer characteristic shown in Fig. 10.21. A nonlinear device such as this is an example of a **comparator.** Specifically, for the simple comparator shown in Fig. 10.20, the input voltage v_1 is compared with the reference voltage $v_2 = V_r$; and if $v_1 > V_r$, then $v_o = +V_{CC}$, while if $v_1 < V_r$, then $v_o = -V_{CC}$.

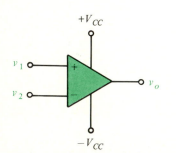

Fig. 10.20 Simple comparator.

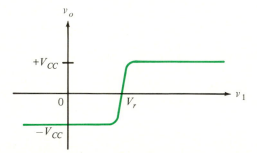

Fig. 10.21 Comparator transfer characteristic.

For the case that the reference voltage is nonzero ($V_r \neq 0$), a comparator is called a **level detector;** if the reference voltage is zero ($V_r = 0$), it is referred to as a **zero-crossing detector.** Figure 10.22(a) shows a zero-crossing detector, and Fig. 10.22(b) shows a sinusoidal input voltage v_1 and the corresponding output voltage

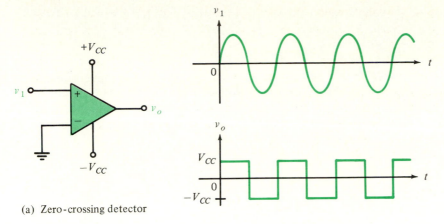

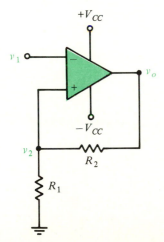

(a) Zero-crossing detector

(b) Input and output of zero-crossing detector

Fig. 10.22

v_o. Thus, this is an example of a device that converts a sine wave into a square wave.

Although a general-purpose high-gain op amp can be used as a comparator, some op amps are specially designed for such use and are so designated. (These devices have higher speeds than ordinary op amps.) In addition to a single comparator, IC chips that contain two or four independent comparators are on the market.

Now suppose that we apply positive feedback to an op amp (comparator) as shown in Fig. 10.23. Such a regenerative comparator is known as a **Schmitt trigger.** When the loop gain is unity or greater, since the feedback is positive (regenerative), the output voltage will saturate at either $+V_{CC}$ or $-V_{CC}$. Which of these two values is taken on depends not only on the value of the input, but also on the value of the output. Let us see why.

Fig. 10.23 Schmitt trigger.

Assume that the output voltage is $v_o = +V_{CC}$. Then, by voltage division,

$$v_2 = \frac{R_1}{R_1 + R_2} v_o = BV_{CC}$$

where $B = R_1/(R_1 + R_2)$. If $v_1 < BV_{CC}$, then the op-amp input voltage is $v_2 - v_1 = BV_{CC} - v_1 > 0$ and the output voltage remains at $+V_{CC}$. Conversely, if $v_1 > BV_{CC}$, then the op-amp input is $v_2 - v_1 = BV_{CC} - v_1 < 0$, and the output becomes $-V_{CC}$. The transfer characteristic of the Schmitt trigger for an increasing input voltage v_1 is shown in Fig. 10.24(a).

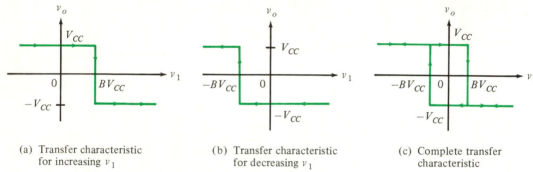

(a) Transfer characteristic for increasing v_1

(b) Transfer characteristic for decreasing v_1

(c) Complete transfer characteristic

Fig. 10.24 Schmitt-trigger transfer characteristic.

Next assume that the output voltage is $v_o = -V_{CC}$. Then

$$v_2 = \frac{R_1}{R_1 + R_2} v_o = -BV_{CC}$$

If $v_1 > -BV_{CC}$, then the op-amp input is $v_2 - v_1 = -BV_{CC} - v_1 < 0$ and the output remains at $-V_{CC}$. Conversely, if $v_1 < -BV_{CC}$, then the op-amp input is $v_2 - v_1 = -BV_{CC} - v_1 > 0$ and the output becomes $+V_{CC}$. The transfer characteristic of the Schmitt trigger for decreasing v_1 is shown in Fig. 10.24(b).

Combining the Schmitt-trigger transfer characteristics shown in Figs. 10.24(a) and (b), we get the complete transfer characteristic shown in Fig. 10.24(c). Note that an output voltage of $+V_{CC}$ does not change to $-V_{CC}$ until the input voltage increases beyond BV_{CC}, and an output of $-V_{CC}$ does not change to $+V_{CC}$ until the input decreases below $-BV_{CC}$. This phenomenon is known as **hysteresis.** The **width** of the hysteresis is the difference between the two **threshold** (or **triggering**) **voltages** BV_{CC} and $-BV_{CC}$. Thus, in this case, the hysteresis width is $2BV_{CC}$.

The Schmitt trigger is a **bistable** device in the sense that it has two stable states. For one of the stable states, the output is $v_o = +V_{CC}$, while for the other stable state, it is $v_o = -V_{CC}$. The circuit will remain in either of its two stable states unless and until it is triggered into its other stable state by the appropriate input voltage—$v_1 > BV_{CC}$ for v_o to change to $-V_{CC}$, and $v_1 < -BV_{CC}$ for v_o to change to $+V_{CC}$.

Like a zero-crossing detector, a Schmitt trigger can convert a sinusoid (or some other periodic function with evenly spaced zero crossings) into a square wave. In addition to having the ability to "square" a function, a Schmitt trigger has another important use. Specifically, consider the input voltage v_1, which is a sinusoid that is corrupted by noise, given by Fig. 10.25(a). If this voltage were applied to a zero-crossing detector's input, the output would become $+V_{CC}$ at time $t = 0$. However, as a result of the noise, soon thereafter ($0 < t < t_1$) the output would become $-V_{CC}$ and then $+V_{CC}$ again. The same situation could recur later. A Schmitt trigger, on the other hand, offers some noise immunity. Suppose that the voltage v_1 is applied to a Schmitt trigger having an upper threshold of V_1 and a lower threshold of V_2 as indicated in Fig. 10.25(a). If the Schmitt-trigger output is $+V_{CC}$, the output will not change to $-V_{CC}$ until v_1 exceeds V_1. As shown, this occurs when $t = t_1$. However, if the output is $-V_{CC}$, it will not change back to $+V_{CC}$ until v_1 becomes less than V_2. Therefore, even though v_1 falls slightly below V_1 just after $t = t_1$, that will in no way cause the output to change from $-V_{CC}$. The output will not change to $+V_{CC}$ until $t = t_2$, at which time v_1 falls below V_2. The Schmitt-trigger output corresponding to the input given in Fig. 10.25(a) is shown in Fig. 10.25(b).

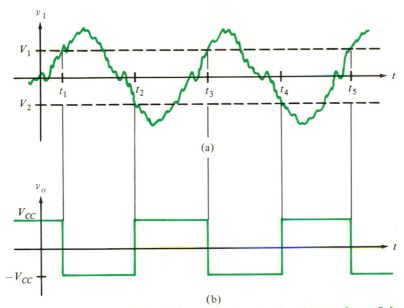

(a)

(b)

Fig. 10.25 Noise-corrupted input and output voltages for a Schmitt trigger.

The Schmitt trigger shown in Fig. 10.26 is a generalization of the one shown in Fig. 10.23. The latter circuit can be obtained from the former by setting $V_r = 0$. Since the voltage between the output v_o and the reference V_r is $v_o - V_r$, then by voltage division we have

$$v_2 - V_r = \frac{R_1}{R_1 + R_2}(v_o - V_r) = B(v_o - V_r)$$

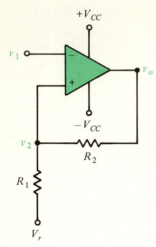

Fig. 10.26 More general Schmitt trigger.

where $B = R_1/(R_1 + R_2)$. Thus,

$$v_2 = V_r + B(v_o - V_r)$$

Assume that the output voltage is $+V_{CC}$. Then it will remain at $+V_{CC}$ as long as $v_2 - v_1 > 0$, that is, as long as

$$v_2 - v_1 = V_r + B(V_{CC} - V_r) - v_1 > 0$$

or

$$v_1 < V_r + B(V_{CC} - V_r) = V_1$$

However, the output will change to $-V_{CC}$ when $v_2 - v_1 < 0$, that is, when

$$v_2 - v_1 = V_r + B(V_{CC} - V_r) - v_1 < 0$$

or

$$v_1 > V_r + B(V_{CC} - V_r) = V_1$$

Assuming that $V_r < V_{CC}$, the transfer characteristic describing this behavior is as shown in Fig. 10.27(a).

Now assume that the output voltage is $-V_{CC}$. Then it will remain at $-V_{CC}$ as long as $v_2 - v_1 < 0$, that is, as long as

$$v_2 - v_1 = V_r + B(-V_{CC} - V_r) - v_1 < 0$$

or

$$v_1 > V_r - B(V_{CC} + V_r) = V_2$$

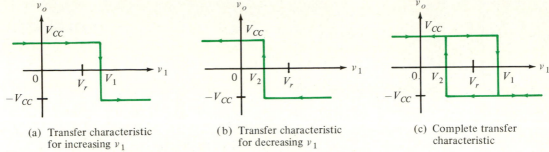

(a) Transfer characteristic for increasing v_1

(b) Transfer characteristic for decreasing v_1

(c) Complete transfer characteristic

Fig. 10.27 Schmitt-trigger transfer characteristic.

However, the output will change to $+V_{CC}$ when $v_2 - v_1 > 0$, that is, when

$$v_2 - v_1 = V_r + B(-V_{CC} - V_r) - v_1 > 0$$

or

$$v_1 < V_r - B(V_{CC} + V_r) = V_2$$

Note that $V_2 < V_r$ and $V_2 < V_1$. The transfer characteristic describing this behavior is shown in Fig. 10.27(b).

The complete transfer characteristic for the given Schmitt trigger is shown in Fig. 10.27(c), where the hysteresis width is

$$V_1 - V_2 = V_r + B(V_{CC} - V_r) - V_r + B(V_{CC} + V_r) = 2BV_{CC}$$

where $B = R_1/(R_1 + R_2)$.

Example 10.4

For the Schmitt trigger given in Fig. 10.26, suppose that $R_1 = 1$ kΩ, $R_2 = 9$ kΩ, $V_{CC} = 10$ V, and $V_r = 5$ V. Then

$$B = \frac{R_1}{R_1 + R_2} = \frac{1000}{1000 + 9000} = 0.1$$

$$V_1 = V_r + B(V_{CC} - V_r) = 5 + 0.1(10 - 5) = 5.5 \text{ V}$$

$$V_2 = V_r - B(V_{CC} + V_r) = 5 - 0.1(10 + 5) = 3.5 \text{ V}$$

and, as a check,

$$V_1 - V_2 = 2BV_{CC} = 2(0.1)(10) = 2 \text{ V}$$

If the input voltage to the Schmitt trigger is $v_1 = 5 + 2 \sin \omega t$ as depicted in Fig. 10.28(a), then the corresponding output voltage v_o is as shown in Fig. 10.28(b).

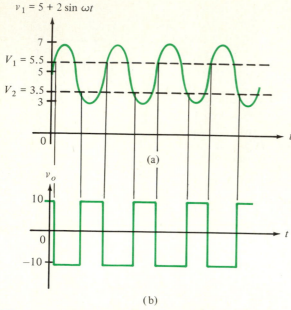

$v_1 = 5 + 2 \sin \omega t$

(a)

(b)

Fig. 10.28 Schmitt-trigger voltages: (a) input, (b) output.

10.4 NONSINUSOIDAL OSCILLATORS

To the Schmitt trigger in Fig. 10.23, let us add a resistor R and a capacitor C as shown in Fig. 10.29. Assume that $v_2 > v_1$ and $v_o = +V_{CC}$. By voltage division, we also have that

$$v_2 = \frac{R_1}{R_1 + R_2} v_o = BV_{CC} < V_{CC}$$

where $B = R_1/(R_1 + R_2)$. However, the fact that $v_o = V_{CC}$ means that C will tend to charge up to $+V_{CC}$ with a time constant of $\tau = RC$. Therefore, eventually the voltage v_1 across the capacitor will become larger than $v_2 = BV_{CC}$. When $v_1 > v_2$, then $v_o = -V_{CC}$ and $v_2 = -BV_{CC}$. This, in turn, means that C will then tend to charge to $-V_{CC}$, again with a time constant of $\tau = RC$. Hence, eventually v_1 will become more negative than v_2 (i.e., $v_2 > v_1$), and the resulting output will be $v_o = V_{CC}$, putting the circuit back into the originally assumed state. The process repeats, and what results are the waveforms shown in Fig. 10.30. The device shown in Fig. 10.29 is a nonsinusoidal oscillator whose operation is based on the charging and discharging of a capacitor. We call such a device a **relaxation oscillator.** These types of oscillators are particularly useful at lower frequencies--from 10 Hz to 10 kHz. At higher frequencies, the square-wave output is far less than ideal due to the slew rate of the op amp (comparator).

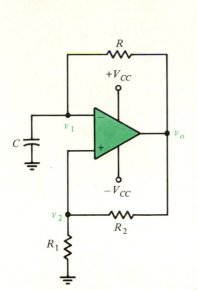

Fig. 10.29 Relaxation oscillator.

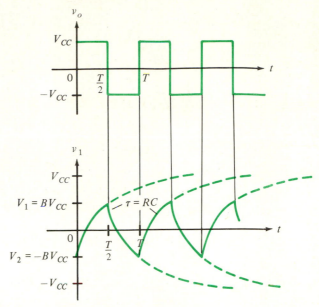

Fig. 10.30 Associated waveforms.

As shown in Fig. 10.30, the output voltage of the relaxation oscillator given in Fig. 10.29 is (ideally) a square wave, while the capacitor voltage is comprised of rising and decaying exponentials. It is from the capacitor voltage that we may determine an expression for the frequency of the oscillator. Specifically, let us consider the half period from $t = 0$ to $t = T/2$ during which the capacitor charges from $V_2 = -BV_{CC}$ to $V_1 = BV_{CC}$ with a time constant of $\tau = RC$. This corresponds to the application of the voltage V_{CC} to a series RC circuit, where the initial capacitor voltage is $-BV_{CC}$. From Equation (3.18), the capacitor voltage is

$$v_1(t) = V_{CC} + Ae^{-t/RC} \tag{10.19}$$

Setting $t = 0$, we have

$$v_1(0) = V_{CC} + A = -BV_{CC}$$

from which

$$A = -BV_{CC} - V_{CC} = -(B + 1)V_{CC}$$

Substituting this value of A into Equation (10.19), we get

$$v_1(t) = V_{CC} - (B + 1)V_{CC}e^{-t/RC} = V_{CC}[1 - (B + 1)e^{-t/RC}]$$

From Fig. 10.30, we see that $v_1(T/2) = BV_{CC}$. Thus,

$$v_1(T/2) = V_{CC}[1 + (B + 1)e^{-T/2RC}] = BV_{CC}$$

from which

$$e^{-T/2RC} = \frac{B - 1}{B + 1}$$

Solving this equation for the period T, we obtain

$$T = 2RC \ln \frac{B + 1}{B - 1} \tag{10.20}$$

and since $B = R_1/(R_1 + R_2)$, Equation (10.20) can be cast into the form

$$T = 2RC \ln\left(\frac{2R_1}{R_2} + 1\right) \tag{10.21}$$

Of course, the frequency (in hertz) of the oscillator is, therefore, $f = 1/T$.

Since the output voltage v_o of the relaxation oscillator is temporarily in one of two stable states ($v_o = +V_{CC}$ and $v_o = -V_{CC}$) for half a period and then jumps to the other stable state for the other half period, we say that the device is **astable.** As we have seen, the relaxation oscillator given in Fig. 10.29, which is also called an **astable multivibrator,** produces a square-wave output voltage and a repetitive exponentially increasing and decreasing capacitor voltage. However, the beginning portion of such an exponential curve approximates a straight-line segment. Thus, by selecting $B = R_1/(R_1 + R_2)$ to be a small quantity, the threshold voltages $V_1 = BV_{CC}$ and $V_2 = -BV_{CC}$ will be relatively small. This, in turn, means that the capacitor voltage will be approximately a triangular waveform. Thus, the relaxation oscillator given in Fig. 10.29 can produce an (approximate) triangular wave, as well as a square wave.

There is, however, a way to produce more accurate triangular waveforms. Since the input voltage is applied to the inverting input, the circuit given in Fig. 10.23 is an **inverting Schmitt trigger**. In the circuit shown in Fig. 10.31, the op amp on the left, together with resistors R_1 and R_2, is an example of a **noninverting Schmitt trigger**. Connected to this device is an integrator whose output is fed back to the input of the noninverting Schmitt trigger. (For the sake of

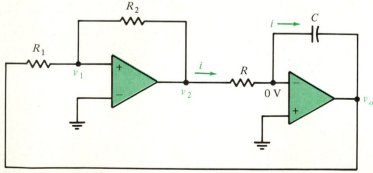

Fig. 10.31 Triangular-wave relaxation oscillator.

simplicity, the power supply voltages $+V_{CC}$ and $-V_{CC}$ are not shown.) Let us see how the entire circuit behaves.

If $v_1 < 0$, then $v_2 = -V_{CC}$. Therefore, $i = v_2/R = -V_{CC}/R$, and

$$v_o = -\frac{1}{C}\int i\,dt = -\frac{1}{C}\int -\frac{V_{CC}}{R}\,dt = \frac{V_{CC}}{RC}t + K_1 \qquad (10.22)$$

Thus, we see that the output voltage increases linearly with time. In addition, we have that

$$\frac{v_o - v_1}{R_1} + \frac{v_2 - v_1}{R_2} = 0$$

from which

$$v_1 = \frac{R_2 v_o}{R_1 + R_2} + \frac{R_1 v_2}{R_1 + R_2}$$

Although $v_2 = -V_{CC}$, since v_o is increasing with time, eventually v_1 will become positive, that is, we will get

$$v_1 = \frac{R_2 v_o}{R_1 + R_2} - \frac{R_1 V_{CC}}{R_1 + R_2} > 0$$

From this inequality, we obtain

$$v_o > \frac{R_1}{R_2} V_{CC}$$

In other words, when $v_o > R_1 V_{CC}/R_2 = V_1$ (the upper threshold voltage), then $v_1 > 0$ and the Schmitt-trigger output becomes $v_2 = +V_{CC}$. Proceeding as above, the resulting output voltage becomes

$$v_o = -\frac{V_{CC}}{RC}t + K_2$$

which indicates that the output voltage decreases linearly with time. When $v_o < -R_1 V_{CC}/R_2 = V_2$ (the lower threshold voltage), then $v_1 < 0$ and the process is repeated. The Schmitt-trigger output voltage v_2 and the integrator output voltage v_o are shown in Fig. 10.32.

Having determined the waveforms associated with the oscillator given in Fig. 10.31, it is a simple matter to determine the frequency of oscillation. For the time interval $0 \le t \le T/2$, by Equation (10.22) the (integrator) output voltage is

$$v_o(t) = \frac{V_{CC}}{RC}t + V_2 = \frac{V_{CC}}{RC}t - \frac{R_1 V_{CC}}{R_2}$$

Setting $t = T/2$, we get that

$$v_o(T/2) = \frac{V_{CC}}{RC}\left(\frac{T}{2}\right) - \frac{R_1 V_{CC}}{R_2} = V_1 = \frac{R_1 V_{CC}}{R_2}$$

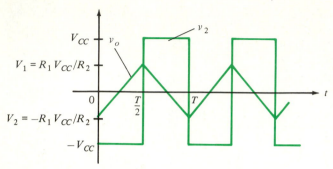

Fig. 10.32 Schmitt-trigger and integrator outputs for triangular-wave oscillator.

Solving this equation for the period T results in

$$T = \frac{4R_1 RC}{R_2}$$

so that the frequency of oscillation (in hertz) is

$$f = \frac{1}{T} = \frac{R_2}{4R_1 RC}$$

The 555 IC Timer

Although astable multivibrators (relaxation oscillators) can be constructed using op amps, a popular method for forming such devices is with the use of an IC chip known as a **timer.** The pin connections for the popular 555 timer are shown in Fig. 10.33. This IC, which contains two comparators, resistors, transistors, and a digital circuit called a "flip-flop," operates as described in the following.

The supply voltage v_8 (indicating pin 8) is made $v_8 = V_{CC}$. When an applied **threshold** voltage is $v_6 \geq \frac{2}{3}V_{CC}$, then the **output** voltage v_3 is low ($v_3 \approx 0$), and, in addition, a switch (transistor) that is internally connected between **discharge** (pin 7) and ground (pin 1) is turned ON. Conversely, when an applied **trigger** voltage is $v_2 \leq \frac{1}{3}V_{CC}$, then v_3 is high ($v_3 \approx V_{CC}$), and, in addition, the discharge transistor is turned OFF. When the **reset** voltage v_4 is low, the discharge transistor is ON. However, the reset is disabled when the reset pin is "tied" (directly connected) to the supply voltage.

We now demonstrate why the circuit shown in Fig. 10.34 is an astable multivibrator. Let us assume that the capacitor C is initially uncharged. Then, since the trigger voltage is $v_2 < \frac{1}{3}V_{CC}$, the output voltage is $v_3 \approx V_{CC}$, and the discharge transistor is OFF. Because the trigger and threshold inputs draw no current (they are connected to comparators), the capacitor C begins to charge with a time constant

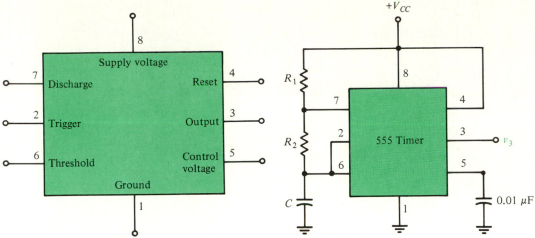

Fig. 10.33 The 555 IC timer.   **Fig. 10.34 Astable multivibrator.**

of $\tau_1 = (R_1 + R_2)C$. When the capacitor voltage reaches the threshold voltage $v_6 = \frac{2}{3}V_{CC}$, then the output voltage becomes $v_3 \approx 0$ and the discharge transistor turns ON. Thus, C can discharge through R_2 with a time constant of $\tau_2 = R_2C < \tau_1$. After the capacitor discharges to $v_2 = v_6 = \frac{1}{3}V_{CC}$, the output again becomes $v_3 \approx V_{CC}$ and the discharge transistor again goes OFF, allowing the capacitor to recharge. The process repeats and the waveforms shown in Fig. 10.35 result. For the sake of mathematical convenience, as we shall now see, time $t = 0$ is chosen to be the time when the capacitor initially charges to $\frac{2}{3}V_{CC}$.

In the time interval $0 \leq t \leq t_1$, from Equation (3.11), the capacitor voltage $v_C = v_2 = v_6$ is

$$v_C(t) = \tfrac{2}{3}V_{CC}e^{-t/R_2C}$$

At time $t = t_1$, therefore,

$$v_C(t_1) = \tfrac{2}{3}V_{CC}e^{-t_1/R_2C} = \tfrac{1}{3}V_{CC}$$

Solving this equation for t_1, we get

$$t_1 = R_2C \ln 2$$

However, for $t_1 \leq t \leq t_2$, as for Equation (3.18),

$$v_C(t) = V_{CC} + Ae^{-(t-t_1)/(R_1 + R_2)C}$$

But since

$$v_C(t_1) = V_{CC} + A = \tfrac{1}{3}V_{CC}$$

then $A = -\frac{2}{3}V_{CC}$, and

$$v_C(t) = V_{CC} - \tfrac{2}{3}V_{CC}e^{-(t-t_1)/(R_1 + R_2)C}$$

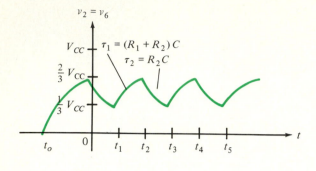

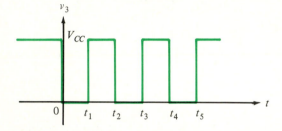

Fig. 10.35 Astable-multivibrator capacitor and output voltages.

At time $t = t_2$, therefore,

$$v_C(t_2) = V_{CC} - \tfrac{2}{3}V_{CC}e^{-(t_2 - t_1)/(R_1 + R_2)C} = \tfrac{2}{3}V_{CC}$$

Solving this equation for t_2, we obtain

$$t_2 = (R_1 + R_2)C \ln 2 + t_1 = (R_1 + R_2)C \ln 2 + R_2C \ln 2$$
$$= (R_1 + 2R_2)C \ln 2 = 0.693(R_1 + 2R_2)C$$

Since the oscillation period is $T = t_2$, then the frequency of oscillation is

$$f = \frac{1}{T} = \frac{1}{0.693(R_1 + 2R_2)C} = \frac{1.443}{(R_1 + 2R_2)C}$$

The supply voltage for a 555 timer is typically in the range 5 V $\leq V_{CC} \leq$ 16 V, and the frequency of oscillation usually is between 0.1 Hz and 100 kHz. A 556 IC chip consists of two 555 timers on the same 14-pin package.

Another useful application of the 555 timer is for a circuit known as a **one-shot** or **monostable multivibrator.** This type of device has one normal (stable) state and one temporary state that is triggered by an appropriate input. After the device leaves its temporary state, it returns to its normal state.

Figure 10.36 shows an example of a monostable multivibrator that employs a 555 timer. The trigger voltage v_2 is the input of the one-shot and v_3 is the output. Given that the input voltage is normally high ($v_2 \approx V_{CC}$) and the output is normally

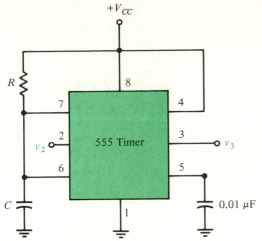

Fig. 10.36 Monostable multivibrator.

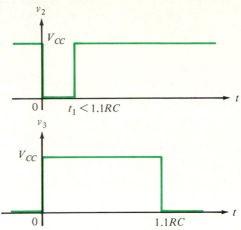

Fig. 10.37 Input and output waveforms.

low ($v_2 \approx 0$), suppose that the input becomes low for a short period of time. As soon as $v_2 < \frac{1}{3}V_{CC}$, the output goes high ($v_3 \approx V_{CC}$) and the discharge transistor turns OFF. Then C charges with a time constant of $\tau = RC$. If $t = 0$ is chosen to coincide with the time that the input is triggered, then from Equation (3.18) the capacitor voltage is given by

$$v_6(t) = V_{CC} - V_{CC}e^{-t/RC}$$

However, when $v_6 > \frac{2}{3}V_{CC}$, provided that the input is no longer low (otherwise there would be a contradiction), the output goes low, the discharge transistor turns ON, and the capacitor discharges quickly. This means that the output is in its high (temporary) state from time $t = 0$ until the time when

$$V_{CC} - V_{CC}e^{-t/RC} = \frac{2}{3}V_{CC}$$

Solving this equation for t, we get

$$t = RC \ln 3 = 1.099RC \approx 1.1\ RC$$

Hence, the time that the one-shot is in its temporary state is $1.1RC$. Input (v_2) and output (v_3) waveforms for the given monostable multivibrator are shown in Fig. 10.37.

A Voltage-Controlled Oscillator

In addition to a 555 timer, another IC chip—the 566—is often used as an oscillator. This device produces both a square-wave and a triangular-wave output voltage whose frequency depends on the values of an external resistor R and capacitor C.

The pin connections of a 566 IC are shown in Fig. 10.38. Pin 1 is connected to ground, while pin 8 is connected to the supply voltage $+V_{CC}$. There is no connection (NC) for pin 2, pin 3 is the square-wave output, and pin 4 is the triangular-wave output. The external resistor R is connected between pin 6 and the supply voltage $+V_{CC}$, and the external capacitor C is connected between pin 7 and ground. Although the frequency of oscillation depends on the values of R and C, it also depends on the voltage v_c, called the **control voltage,** which is applied to pin 5. Specifically, the oscillation frequency f in hertz is given by

$$f = \frac{2}{RC}\left(1 - \frac{v_c}{V_{CC}}\right)$$

(10.23)

where $v_c < V_{CC}$. Since the frequency of such a device is determined by an applied voltage (in addition to the values of R and C), we refer to it as a **voltage-controlled oscillator** (VCO).

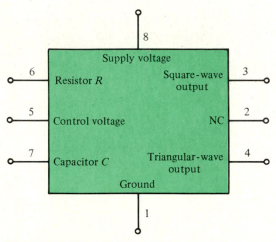

Fig. 10.38 The 566 IC chip.

For the oscillator shown in Fig. 10.39, assuming that the control-voltage input draws no current, by voltage division, the control voltage is

$$v_c = \frac{R_2}{R_1 + R_2} V_{CC}$$

and the frequency of oscillation, therefore, is

$$f = \frac{2}{RC}\left(1 - \frac{R_2}{R_1 + R_2}\right) = \frac{2}{RC}\left(\frac{R_1}{R_1 + R_2}\right)$$

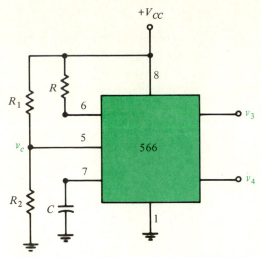

Fig. 10.39 Oscillator using a 566 IC chip.

Not only can f be changed by changing either R or C; but by varying either R_1 or R_2, the control voltage, and, hence, the frequency, can be varied.

In practical terms, the supply voltage for a 566 IC should be in the range $10 \text{ V} \leq V_{CC} \leq 24 \text{ V}$, and the control voltage should be $0.75V_{CC} \leq v_c < V_{CC}$. The range of the resistance R should be $2 \text{ k}\Omega \leq R \leq 20 \text{ k}\Omega$, and the oscillation frequency should be kept below 1 MHz.

Another way of varying the control voltage v_c is to add an ac component to it by connecting an ac signal source to pin 5 via a coupling capacitor. In this way, the frequency of the oscillator will change in accordance with the ac signal. This is an example of a process known as **frequency modulation** (FM). The subject of modulation is discussed in the next section.

10.5 INTRODUCTION TO COMMUNICATIONS

The transmission of information is an extremely important and vital process in a modern technological society. Because of the nature of electromagnetic radiation, however, information in the **audio band** or **spectrum** (20–20,000 Hz) cannot be directly transmitted efficiently through the atmosphere or space. Conversely, though, **radio-frequency** (RF) signals (above 20 kHz), can be propagated through space both efficiently and economically. Therefore, a common method for transmitting audio information (and higher-frequency information, too) is to transmit a high-frequency signal, called a **carrier,** which is altered in some way corresponding to the information to be sent. We refer to this as **modulation.** After such a signal is received, the original information is extracted by a process known as **demodulation** or **detection.**

Amplitude Modulation (AM)

Let us be specific by considering a particular type of modulation that is simple, but still very useful. We begin by selecting the carrier to be a high-frequency sinusoid $v_c(t) = V_c \cos \omega_c t$. For the sake of mathematical convenience, suppose that the audio information to be transmitted is simply a sinusoid $v_m(t) = V_m \cos \omega_m t$, called the **modulating signal,** where we require that the carrier frequency ω_c be much higher than the modulating-signal frequency ω_m (i.e., $\omega_c \gg \omega_m$). Suppose, now, that we make the amplitude of the carrier $v_c(t)$ change according to the modulating signal $v_m(t)$. This is known as **amplitude modulation** (AM).

Since the amplitude of a sinusoid is a positive quantity, we cannot simply let the amplitude of the carrier be equal to $V_m \cos \omega_m t$. However, we can let the amplitude of the carrier be

$$V_c + kV_m \cos \omega_m t$$

where k is a positive constant such that $V_c \geq kV_m$. In other words, an AM signal can be written in the form

$$v(t) = (V_c + kV_m \cos \omega_m t) \cos \omega_c t$$

$$= V_c\left(1 + \frac{kV_m}{V_c} \cos \omega_m t\right) \cos \omega_c t$$

$$= V_c(1 + m \cos \omega_m t) \cos \omega_c t \tag{10.24}$$

where $m = kV_m/V_c$ is called the **modulation index.** Note that when there is no modulating signal $[v_m(t) = V_m \cos \omega_m t = 0]$, then $V_m = 0$ and, hence $m = 0$. Under this circumstance (zero modulation), the AM signal is simply equal to the carrier, that is, $v(t) = V_c \cos \omega_c t = v_c(t)$. Conversely, since $V_c \geq kV_m$, then $1 \geq kV_m/V_c = m$. Thus, the modulation index is bounded by $m \leq 1$. We refer to the case that $m = 1$ as 100% modulation. Shown in Fig. 10.40 is an example of a carrier $v_c(t) = V_c \cos \omega_c t$, a modulating signal $v_m(t) = V_m \cos \omega_m t$ (where $\omega_m \ll \omega_c$), and an AM signal $v(t) = V_c(1 + m \cos \omega_m t) \cos \omega_c t$ for the case of

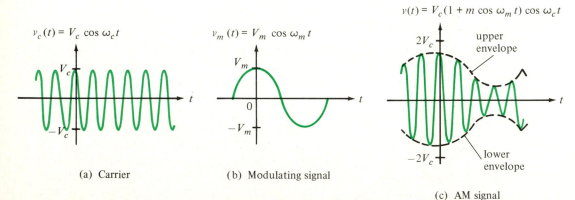

(a) Carrier (b) Modulating signal

(c) AM signal

Fig. 10.40 Carrier, modulating-signal, and AM waveforms.

roughly 50% modulation ($m = 0.5$). The upper boundary for the AM signal is described by $V_c(1 + m \cos \omega_m t)$ and is referred to as the **upper envelope.** Similarly, $-V_c(1 + m \cos \omega_m t)$ is known as the **lower envelope.** The upper and lower envelopes are indicated in Fig. 10.40(c) by dashed curves.

Although the carrier $v_c(t)$ is an ordinary sinusoid, as is the modulating signal $v_m(t)$, the AM signal $v(t)$ is a sinusoid whose amplitude varies. Yet we can express $v(t)$ in an alternative form as follows. From Equation (10.24), we have

$$v(t) = V_c \cos \omega_c t + mV_c \cos \omega_c t \cos \omega_m t$$

Using the trigonometric identity $\cos A \cos B = \frac{1}{2} \cos(A + B) + \frac{1}{2} \cos(A - B)$, we get that

$$v(t) = V_c \cos \omega_c t + \tfrac{1}{2} mV_c \cos(\omega_c + \omega_m)t + \tfrac{1}{2} mV_c \cos(\omega_c - \omega_m)t \qquad \textbf{(10.25)}$$

Therefore, the AM signal $v(t)$ given by Equation (10.24) is actually the sum of three (ordinary) sinusoids as demonstrated by Equation (10.25). One sinusoid is just the carrier $v_c(t) = V_c \cos \omega_c t$ whose amplitude is V_c. Although the other two sinusoids both have amplitudes of $\frac{1}{2} mV_c = \frac{1}{2} kV_m$, one has a frequency of $\omega_c + \omega_m$, and the other a frequency of $\omega_c - \omega_m$. A plot of the amplitudes of the frequency components (versus frequency) of the AM signal $v(t)$ is shown in Fig. 10.41. We refer to such a plot as the **frequency spectrum,** or simply **spectrum,** of $v(t)$. The sinusoid whose frequency is $\omega_c + \omega_m$ is known as the **upper sideband,** and the one whose frequency is $\omega_c - \omega_m$ is the **lower sideband.**

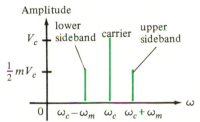

Fig. 10.41 Spectrum of AM signal.

In the preceding discussion, we considered the case that the modulating signal is sinusoidal, that is, $v_m(t) = V_m \cos \omega_m t$. However, now suppose that the modulating signal $v_m(t)$ is nonsinusoidal and has a spectrum that is limited to frequencies between 0 and ω_u as illustrated in Fig. 10.42(a), where V_M is the maximum amplitude of the frequency components comprising $v_m(t)$. Then, as in the discussion above, the spectrum of the AM signal is as depicted in Fig. 10.42(b).

For standard AM radio, the bandwidth of the modulating signal is limited to $f_u = \omega_u/2\pi = 5$ kHz. In audio terms, by no means would this be considered "high fidelity." However, because of such a small bandwidth, the carrier frequencies for AM radio stations (which range between 540 and 1600 kHz) are spaced 10 kHz apart.

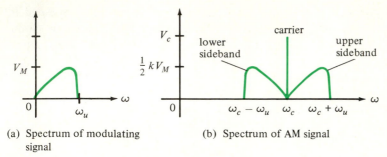

(a) Spectrum of modulating
    signal

(b) Spectrum of AM signal

Fig. 10.42

     An example of a simple BJT amplitude modulator is shown in Fig. 10.43. This circuit is, in essence, an amplifier whose input is the relatively high-frequency carrier $v_c(t)$, and whose gain depends upon the relatively low-frequency modulating signal $v_m(t)$. The coupling and bypass capacitors are chosen so that they behave as small impedances (ideally zero) for the carrier frequency and as large impedances (ideally infinite) for the modulating-signal frequencies. Practically speaking, therefore, ω_c should be at least 100 times the value of ω_u. Placing a large modulating signal in series with the emitter will vary r_e, and, hence, the gain of the amplifier as $v_m(t)$ varies. The result is an output voltage $v_o(t)$ that is a sinusoid having the carrier frequency ω_c and an amplitude that varies with $v_m(t)$—in other words, an AM signal.

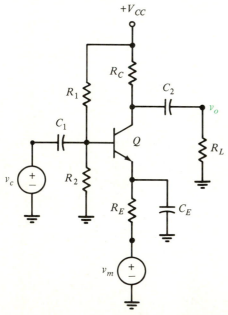

Fig. 10.43 A BJT amplitude modulator.

The problem of dealing with amplifier coupling and bypass capacitors, such as the ones shown in Fig. 10.43, can be eliminated by using the dc-amplifier modulator shown in Fig. 10.44. Again, the modulating signal will cause the gain of the differential pair to vary accordingly. Although a portion of the modulating signal appears at the collectors of Q_1 and Q_2, because it is in the common mode, this voltage is not present when the output v_o is taken between the two collectors as indicated. The result, as was the case for the circuit given in Fig. 10.43, is an AM signal.

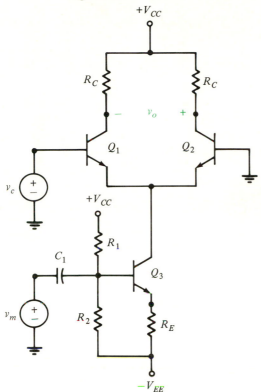

Fig. 10.44 Differential-amplifier modulator.

AM Detection

After an AM signal is transmitted and then received, the original information it contains must be extracted. The process that accomplishes this is called **amplitude demodulation** or **AM detection.** Detecting an AM signal is even less involved than generating it. Specifically, the peak detector given in Fig. 6.55 and repeated in Fig. 10.45 is an example of an AM detector, also called an **envelope detector.** If the input to this circuit is the AM signal $v(t)$ shown in Fig. 10.40(c), assuming an ideal diode, when $v(t)$ reaches its first peak, the capacitor will charge up to this value. During the time it takes for $v(t)$ to go from its first peak to its second peak, the

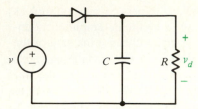

Fig. 10.45 Envelope detector.

capacitor voltage will discharge with a time constant of $\tau = RC$. This time constant should be large compared with the period T_c of the carrier. In other words,

$$RC \gg T_c = \frac{1}{f_c} = \frac{2\pi}{\omega_c}$$

However, the time constant $\tau = RC$ should not be too large; otherwise, the capacitor (output) voltage v_d will not be able to follow the successive peaks of the AM signal. In particular, the time constant $\tau = RC$ should be small compared with the period T_m of the modulating signal. In other words,

$$RC \ll T_m = \frac{1}{f_m} = \frac{2\pi}{\omega_m}$$

[In the general case where the modulating signal $v_m(t)$ is nonsinusoidal, the time constant should be small compared with the period $T_u = 2\pi/\omega_u$ of the highest frequency component of the modulating signal.] Combining the preceding two inequalities, we have that the time constant $\tau = RC$ is chosen such that

$$\frac{2\pi}{\omega_c} \ll RC \ll \frac{2\pi}{\omega_m} \tag{10.26}$$

From this constraint, we see that we must have that $\omega_c \gg \omega_m$—which was an original condition for amplitude modulation.

When the time constant $\tau = RC$ is chosen according to inequality (10.26), the output voltage $v_d(t)$ of the peak detector given in Fig. 10.45 is as shown in Fig. 10.46 and will be essentially equal to the upper envelope of the applied AM signal. That is,

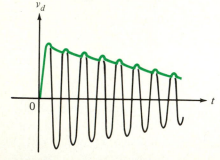

Fig. 10.46 Output voltage of envelope detector.

$$v_d(t) = V_c(1 + m \cos \omega_m t) = V_c + mV_c \cos \omega_m t$$

$$= V_c + \frac{kV_m}{V_c} V_c \cos \omega_m t = V_c + kV_m \cos \omega_m t$$

Since the output $v_d(t)$ contains the original information $v_m(t) = V_m \cos \omega_m t$, this means that the AM signal has been detected. Connecting the output of the peak detector to a simple RC low-pass filter will eliminate the constant V_c and produce an output that is directly proportional to $V_m \cos \omega_m t$—if such an output is desired.

Let us again consider a band-limited modulating signal $v_m(t)$ having a spectrum, referred to as the **baseband,** as depicted in Fig. 10.47(a). If this signal amplitude modulates a carrier that has the frequency $\omega_c \gg \omega_u$, then the spectrum of the resulting AM signal is as shown in Fig. 10.47(b), where the upper and lower sidebands are symmetrical with respect to the carrier frequency.

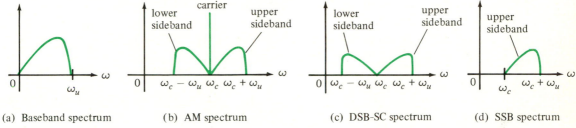

(a) Baseband spectrum   (b) AM spectrum   (c) DSB-SC spectrum   (d) SSB spectrum

Fig. 10.47 Some types of amplitude modulation.

If the carrier is removed (or suppressed) from the AM signal, the resulting signal is called a **double-sideband suppressed-carrier** (DSB-SC) signal, and it has the spectrum illustrated in Fig. 10.47(c). Since the carrier is not sent along, transmission of a DSB-SC signal requires less power than is needed for the corresponding AM signal. There is a price, however, for this saving of power. To detect a DSB-SC signal, an oscillator is required at the receiver so that the carrier can be reinserted. Hence, a DSB-SC detector is more complicated than an ordinary AM detector. This type of modulation is used in conjunction with FM stereo broadcasting.

Since the lower sideband of an AM signal is the mirror image of the upper sideband, it is redundant. By eliminating one of the sidebands, say the lower, and the carrier, we obtain a signal whose spectrum is as shown in Fig. 10.47(d). Such a signal, referred to as a **single-sideband** (SSB) signal, requires less power than is needed for either AM or DSB-SC, and has half of the bandwidth. However, because of the missing carrier and sideband, a more complicated detector is needed. Owing to its high efficiency, this type of modulation is popular with radio amateurs.

Yet another form of amplitude modulation is where the carrier, one complete sideband, and only a part or "vestige" of the other sideband are transmitted. The resulting signal is known as a **vestigial-sideband** (VSB) signal. This type of modulation is used for transmitting television video signals. A standard video signal has a maximum frequency component of about $f_u = 4$ MHz

$[\omega_u = 2\pi f_u = 2\pi(4 \times 10^6) \approx 25 \text{ Mrad/s}]$. If such a signal were transmitted by ordinary AM, a bandwidth of 8 MHz would be required. However, since television stations are allocated a bandwidth of only 6 MHz, VSB modulation is used. (Although a smaller bandwidth is needed for SSB, it is much more complicated to detect than is VSB.)

Figure 10.48(a) shows the baseband for a video signal, where the horizontal axis is labeled in megahertz. If this signal is to be broadcast on channel 3, whose carrier frequency is $f_c = 61.25$ MHz, then the spectrum for channel 3 is as shown in Fig. 10.48(b). The entire carrier is transmitted along with the upper sideband. However, only 1.25 MHz of the lower sideband is broadcast. Note, though, that the $4 + 1.25 = 5.25$ MHz required for the VSB video signal does not occupy the allotted 6 MHz band. This is so because part of channel 3's spectrum, which extends from 60 to 66 MHz, is required for the audio signal. Specifically, at a frequency of 4.5 MHz above the video carrier (i.e., at $61.25 + 4.5 = 65.75$ MHz) is the audio carrier. But the audio signal is not AM—it is FM.

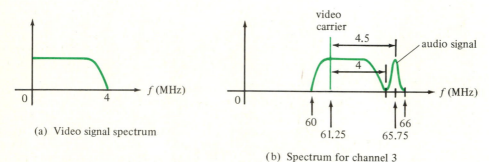

(a) Video signal spectrum

(b) Spectrum for channel 3

Fig. 10.48 Video baseband and spectrum for a TV signal.

Frequency Modulation (FM)

Amplitude modulation (AM) is the process by which the amplitude of a waveform—quite often a sinusoid—is varied in accordance with some modulating signal. If the frequency of the waveform (carrier) changes, instead of the amplitude, then the result is known as **frequency modulation** (FM). In particular, if ω_c is the carrier frequency and $v_m(t)$ is the modulating signal, then the expression

$$v(t) = A \cos [\omega_c + k_1 v_m(t)]t$$

where k_1 is a constant, is an example of an FM signal.

Figure 10.49 shows waveforms of an FM carrier $v_c(t) = A \cos \omega_c t$, a modulating signal $v_m(t) = V_m \sin \omega_m t$, where $\omega_m \ll \omega_c$, and the resulting FM signal $v(t)$. Note that the change in amplitude of the modulating signal causes the change in the frequency of the FM signal.

There is another type of modulation, called **phase modulation** (PM), that is closely related to FM. In this case, the phase angle ϕ is dependent upon the modulating signal $v_m(t)$. Specifically, for a sinusoidal carrier, the expression

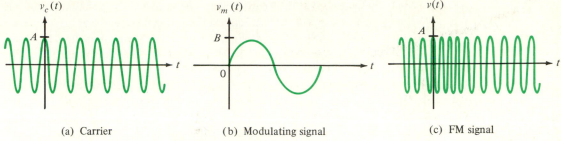

(a) Carrier (b) Modulating signal (c) FM signal

Fig. 10.49 Frequency-modulation waveforms.

$$v(t) = A \cos \left[\omega_c t + \phi(t) \right] = A \cos \left[\omega_c t + k_2 v_m(t) \right]$$

where k_2 is a constant, is an example of a PM signal. Both FM and PM are generically termed **angle modulation**.

FM signals can be produced in numerous ways. One approach is to use the VCO given in Fig. 10.39. The frequency of this device depends on the control voltage v_c [see Equation (10.23)]. By using a coupling capacitor to connect a modulating-signal source to the control-voltage input (pin 5), the frequency of oscillation will vary in accordance with the modulating signal. Although the output of this VCO is triangular or square in shape, by using the appropriate circuits (e.g., a narrow bandpass filter), a frequency-modulated sinusoid can be produced.

The bandwidth of an AM signal is determined strictly by the highest frequency component f_u of the modulating signal. In particular, for ordinary AM, the bandwidth is $BW_{AM} = 2f_u$. For an FM signal, however, the bandwidth is determined not only by f_u, but by the amplitude of the modulating signal as well. It is the amplitude of the modulating signal that causes the deviation in the frequency of the carrier, and it is the frequency of the modulating signal that determines the rate of change of this deviation. Although a precise discussion of the bandwidth BW_{FM} of an FM signal is quite involved, an approximate formula for it is given by

$$BW_{FM} \approx 2(\Delta_f + f_u) \tag{10.27}$$

where Δ_f is the maximum frequency deviation of the carrier and f_u is the highest frequency component of the modulating signal. We refer to the case $\Delta_f > f_u$ as **wideband** FM, and $\Delta_f < f_u$ as **narrowband** FM.

The commercial FM broadcasting band extends from 88 to 108 MHz (compare this with 0.54–1.6 MHz for AM). Carrier frequencies, which are spaced 200 kHz (0.2 MHz) apart, begin at 88.1 MHz and end at 107.9 MHz. The highest frequency component of an audio modulating signal is 15 kHz (compared with 5 kHz for AM). Thus, we see that FM allows for much greater fidelity than does AM. The maximum frequency deviation allowed for broadcast FM is $\Delta_f = 75$ kHz. From Equation (10.27), we see that the bandwidth of an FM signal with maximum frequency deviation is

$$BW_{FM} \approx 2(75,000 + 15,000) = 180 \text{ kHz}$$

Therefore, we see that the bandwidth of such a wideband FM signal falls within the 200 kHz allocated to a station by the Federal Communications Commission.

As mentioned in Chapter 5, an FM signal can be demodulated or detected with the use of a phase-locked loop (PLL). The PLL block diagram shown in Fig. 5.36 is repeated in Fig. 10.50. The free-running (no input voltage) frequency of the VCO is set equal to the FM carrier frequency. The input of the PLL is the FM signal. If the modulating signal is zero, then the input and the VCO will have the same frequency, and the output essentially will be zero. If the input frequency changes, the (low-pass filter) output voltage will change so as to match the frequency of the VCO with that of the input. The result of this is that the output voltage changes, as does the frequency of the input (FM signal). In other words, the output is the detected FM signal. A popular IC chip for commercial broadcast FM detection is the 560 PLL.

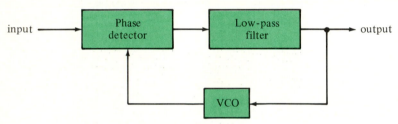

Fig. 10.50 Block diagram of a PLL.

As mentioned previously, the audio signal for broadcast TV is FM. The audio carrier is located 4.5 MHz above the video carrier [see Fig. 10.48(b) for example]. The maximum frequency deviation of the audio carrier that is allowed is 25 kHz, and the bandwidth allocated for TV audio is 50 kHz.

In addition to FM radio and TV audio, FM has numerous other applications. Among these are narrowband voice transmissions used by aircraft, fire and police departments, and weather services. Bandwidth requirements for such uses typically are in the range from 10 to 30 kHz.

Electric noise arising from numerous causes (e.g., lightning, automobile ignitions, static electricity discharges) is generally additive to electric signals. As a result of this phenomenon, the amplitude of a signal is affected more by noise than is its frequency. Thus, AM is much more susceptible to noise than is FM. Also, the greater the frequency deviation (and, hence, the bandwidth), the greater are the noise-immunity properties of FM. Therefore, in terms of noise, wideband FM is superior to narrowband FM.

FM Stereo

In the preceding discussion of FM, it was assumed that there was a single modulating signal (monaural FM). We are, however, familiar with the concept of FM stereo in which there are two modulating signals—a left and a right. Let us see how FM stereo is produced.

We begin with two signals v_L and v_R, one for the left and one for the right respectively. We then form the sum $v_L + v_R$ and the difference $v_L - v_R$ of the two signals. To achieve high fidelity, the maximum frequency components of $v_L + v_R$ and $v_L - v_R$ are allowed to be 15 kHz. However, since the spectra of both $v_L + v_R$ and $v_L - v_R$ cover the same frequency band (0–15 kHz) as indicated in Figs. 10.51(a) and (b), we will modulate one of the signals. In particular, for $v_L - v_R$, we use DSB-SC modulation employing a carrier frequency of 38 kHz as indicated in Fig. 10.51(c). We now combine (add) the signal for $v_L + v_R$ and the DSB-SC modulated signal of $v_L - v_R$ along with a 19-kHz sinusoid called a **pilot subcarrier.** The spectrum of the resulting signal is shown in Fig. 10.52. (Also shown in Fig. 10.52 is a frequency band labeled "SCA"—ignore this for the time being.) It is this signal that modulates the FM carrier; that is, the frequency spectrum shown in Fig. 10.52 is the baseband for the FM stereo signal. The procedure just described is an example of **frequency multiplexing.**

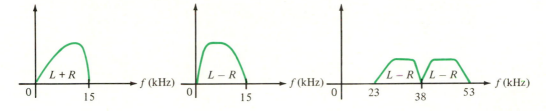

(a) Spectrum of $v_L + v_R$ (b) Spectrum of $v_L - v_R$ (c) DSB-SC spectrum of $v_L - v_R$

Fig. 10.51 Frequency spectra for $v_L + v_R$, $v_L - v_R$, and modulated form of $v_L - v_R$.

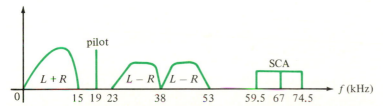

Fig. 10.52 Spectrum of an FM stereo modulating signal.

An ordinary (monaural) FM receiver will detect the $v_L + v_R$ portion of a stereo FM signal, so stereo FM is compatible with ordinary FM. A stereo FM receiver not only detects the $v_L + v_R$ signal, but it also detects the DSB-SC signal and the 19-kHz pilot.* The frequency of the pilot is doubled to produce a 38-kHz carrier for the detection of the DSB-SC signal, and DSB-SC demodulation yields the $v_L - v_R$ signal. Taking the sum of $v_L + v_R$ and $v_L - v_R$ gives $2v_L$; taking the

---

*It is the detection of the pilot subcarrier that causes a receiver's stereo indicator light to go on.

difference gives $2v_R$. In this way, the left-channel and the right-channel signals emerge, and stereophonic sound results.

Once an FM stereo signal is detected—for example, by a PLL, as mentioned previously—the job of demodulating the DSB-SC signal and combining the $v_L + v_R$ and $v_L - v_R$ signals appropriately can be accomplished with an IC chip such as the CA3090 produced by RCA.

Since the highest frequency component of an FM stereo signal is much higher than a monophonic signal, to be within the allocated bandwidth, a smaller maximum frequency deviation is required. The result is that FM stereo is narrowband FM, and, therefore, does not have the superior noise characteristics of monophonic FM.

Some FM stations also broadcast background music for department stores, elevators, grocery stores, and the like. This type of transmission, known as **subsidiary communication authorization** (SCA), is accomplished with a narrowband FM signal having a carrier* frequency of 67 kHz and a bandwidth of 15 kHz. The spectrum of an SCA signal is included as part of the FM stereo modulating-signal spectrum given in Fig. 10.52. After such a modulating signal is detected, the SCA signal can then be demodulated. In particular, the circuit shown in Fig. 10.53 will detect SCA signals. The IC chip used in this circuit is a 565 PLL. Connected to the phase-detector input terminals (pins 2 and 3) is a high-pass filter consisting of 500-pF capacitors and 5-kΩ resistors. The resistors also are part of a voltage divider that establishes the appropriate dc bias voltage. The frequency of the VCO square-wave output (pin 4) is compared with the input frequency at pin 5. The free-running frequency of the VCO is determined by the resistance connected to pin 8 and the capacitance connected to pin 9. The cutoff frequency of the PLL's low-pass filter is determined by the capacitor which is connected between pins 7 and 8. Finally, connected to the PLL output (pin 7) is a low-pass filter that reduces the high-frequency noise that often occurs with SCA transmissions.

Subscription television (STV) services encode or "scramble" their audio and video signals so that they cannot be received properly unless an appropriate decoder is used. Scrambling the video consists of moving the video signal's sync pulses to positions that will not be correctly identified by a television set's sync section. A number of different schemes are in use to accomplish this. On the other hand, there are basically but two methods for scrambling (actually, hiding) the audio. One scheme is the frequency multiplexing that is used for FM stereo, and the other is the SCA transmission technique. In fact, the SCA detector given in Fig. 10.53 will decode the audio signal for some STV systems.

Stereo TV sound is accomplished as is FM stereo. In stereo TV, however, the pilot frequency is the same as the television's horizontal sweep rate, that is, 15.75 kHz. However, instead of an SCA signal, the TV audio signal may include a **second audio program** (SAP) channel. The SAP channel can be used to supply

---

*Since this FM signal is part of an overall FM signal, the term **subcarrier** actually is more accurate.

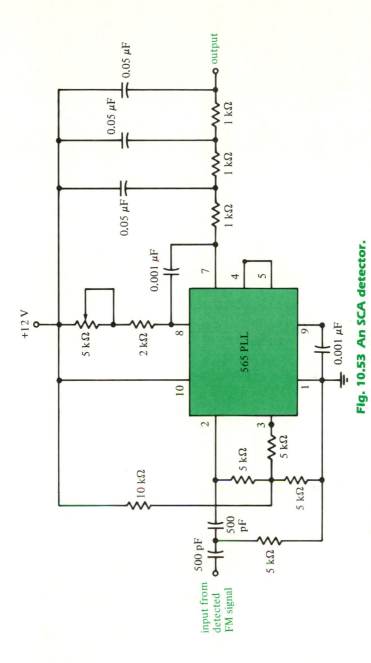

Fig. 10.53 An SCA detector.

input from detected FM signal

output

+12 V

565 PLL

500 pF
500 pF
10 kΩ
5 kΩ
5 kΩ
5 kΩ
5 kΩ
5 kΩ
0.001 μF
2 kΩ
5 kΩ
0.001 μF
1 kΩ
1 kΩ
1 kΩ
0.05 μF
0.05 μF
0.05 μF

sound in a second language simultaneously. The capability of stereo TV sound and a SAP channel is known as **multichannel television sound** (MTS).

Previously, we stated that a commercial TV broadcast signal consists of an AM (VSB) video signal and a FM audio signal, where the bandwidth of a channel is 6 MHz. In addition to broadcast TV, television signals are transmitted by networks to their affiliates, and pay-TV services send their programming to cable companies throughout the country. A very popular way of program distribution is with the use of communication satellites. Although such satellites have numerous channels, called **transponders,** their bandwidths are not 6 MHz, as is the case for standard broadcast TV. The reason is that satellite-TV transmissions use FM for video as well as audio—a transponder bandwidth of 30 MHz being typical.

SUMMARY

1. Negative feedback can be used to improve amplifier stability. It also can be used to improve a voltage amplifier's input and output characteristics. The cost is a reduction of gain.

2. The gain-bandwidth product of a feedback amplifier is a constant.

3. Positive feedback can be used to produce oscillators.

4. The loop gain of an oscillator must be unity (or greater).

5. Phase-shift and Wien-bridge oscillators are examples of RC sinusoidal oscillators.

6. Colpitts, Hartley, and Clapp oscillators are examples of LC sinusoidal oscillators.

7. Crystal oscillators are used when extreme stability is required.

8. An operational amplifier can be used as a comparator. Comparators can be used as level detectors and zero-crossing detectors.

9. The Schmitt trigger is a bistable device that exhibits hysteresis.

10. Comparators and Schmitt triggers can be used to convert sine waves into square waves.

11. A nonsinusoidal oscillator that is based on the charging and discharging of a capacitor is an example of a relaxation oscillator or astable multivibrator.

12. The 555 IC timer is frequently used to construct astable and monostable (one-shot) multivibrators.

13. A voltage-controlled oscillator (VCO), such as the 566 IC chip, can also be used to produce nonsinusoidal, periodic waveforms.

14. Information can be transmitted efficiently with the use of modulation.

15. Amplitude modulation (AM) consists of varying the amplitude of a carrier in accordance with some information source (modulating signal); frequency modulation (FM) consists of varying the frequency of the carrier.

16. Demodulation or detection is the process of extracting the original information from a modulated waveform.

17. Amplitude modulation can take the form of ordinary AM, double-sideband suppressed-carrier (DSB-SC), single-sideband (SSB), and vestigial-sideband (VSB) modulation.

18. Frequency modulation can be either narrowband FM or wideband FM.

19. Commercial broadcast television uses amplitude modulation (VSB) for the video information and FM for the audio information.

20. DSB-SC modulation is utilized as part of FM stereo broadcasting.

Problems for Section 10.1

10.1 The op amp in Fig. 10.2 has $A = 200{,}000$, $R_{in} = 2$ MΩ, and $R_o = 75$ Ω. For the given feedback amplifier, $R_1 = 1$ kΩ and $R_2 = 9$ kΩ. Find (a) A_F, (b) R_{if}, and (c) R_{of}.

10.2 Repeat Problem 10.1 for the case that R_2 is changed to 1 MΩ.

10.3 Figure P10.3(a) illustrates **parallel-parallel feedback** (also called **voltage-shunt feedback**). Verify that Fig. P10.3(b) shows a specific example. Assume that the op amp in Fig. P10.3(b) has input resistance $R_{in} = \infty$, output resistance $R_o = 0$, and finite gain A. For the feedback amplifier given in Fig. P10.3(b), find the input resistance $R_{if} = v_{in}/i_1$, and show that

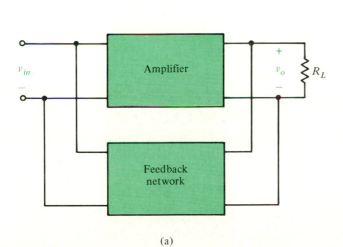

(a)

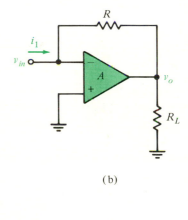

(b)

Fig. P10.3

$$v_o = -ARi_1/(1 + A) \approx -Ri_1$$

for $A \gg 1$.

10.4 Figure P10.4(a) illustrates **series-series feedback** (also called **current-series feedback**). Verify that Fig. P10.4(b) shows a specific example. Assume that the op amp in Fig. P10.4(b) has input resistance $R_{in} = \infty$, output resistance $R_o = 0$, and finite gain A. For the feedback amplifier given in Fig. P10.4(b), show that

$$i_o = \frac{A}{(1 + A)R + R_L} v_{in}$$

so that when $A \gg 1 + R_L/R$, then $i_o \approx v_{in}/R$.

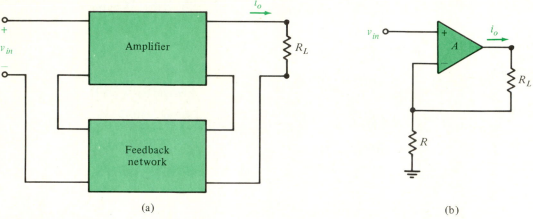

(a) (b)

Fig. P10.4

10.5 For the feedback amplifier shown in Fig. P10.4(b), the op amp has gain A and input resistance R_{in}. Assuming that $R_{in} \gg R$ and the output resistance of the op amp is $R_o = 0$, show that the feedback-amplifier input resistance is

$$R_{if} = \left(1 + \frac{AR}{R + R_L}\right)R_{in}$$

10.6 For the feedback amplifier shown in Fig. P10.4(b), the op amp has gain A, input resistance R_{in}, and output resistance R_o. Assume that $R_{in} \gg R$ and find the feedback-amplifier output resistance R_{of}.

10.7 Figure P10.7(a) illustrates **parallel-series feedback** (also called **current-shunt feedback**). Verify that Fig. P10.7(b) shows a specific example. Assume that the op amp in Fig. P10.7(b) has input resistance $R_{in} = \infty$, output resistance $R_o = 0$, and finite gain A. For the feedback amplifier given in Fig. P10.7(b), show that

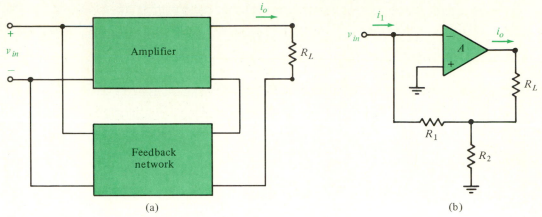

(a)

(b)

Fig. P10.7

$$i_o = -\frac{AR_1 + (1 + A)R_2}{R_L + (1 + A)R_2} i_1 \approx \left(1 + \left[\frac{A}{1 + A}\right]\frac{R_1}{R_2}\right)i_1$$

when $A \gg R_L/R_2$, and, thus, when $A \gg 1$, then $i_o \approx -(1 + R_1/R_2)i_1$.

10.8 For the feedback amplifier shown in Fig. P10.7(b), assume that the op amp has input resistance $R_{in} = \infty$, output resistance $R_o = 0$, and finite gain A. Show that the feedback-amplifier input resistance is

$$R_{if} = \frac{R_1}{\left(\dfrac{R_L}{R_2} + 1 + A\right)\left(\dfrac{R_T}{R_L + R_T}\right)}$$

where $R_T = R_1 \parallel R_2 = R_1R_2/(R_1 + R_2)$, so that when $A \gg R_L/R_2$, then $R_{if} \approx R_1(R_L + R_T)/(1 + A)R_T$.

10.9 For the feedback amplifier given in Fig. 10.2, the op amp has a gain of 200,000 and a gain-bandwidth product of 1 MHz. Find the bandwidth of the feedback amplifier for the case that $R_1 = 1$ kΩ and (a) $R_2 = 9$ kΩ, (b) $R_2 = 1$ MΩ.

10.10 Given the feedback amplifier shown in Fig. 10.2, the op amp has a gain of 200,000 and a gain-bandwidth product of 1 MHz. Assuming that $R_1 = 5$ kΩ, determine the value of R_2 that will result in a bandwidth of 200 kHz for the feedback amplifier.

10.11 By adding a resistance to the feedback amplifier given in Fig. P10.3(b), we obtain the feedback amplifier shown in Fig. P10.11. Assume that the op amp has input resistance $R_{in} = \infty$, output resistance $R_o = 0$, and finite gain A. Show that the overall voltage gain is

$$A_F = \frac{v_o}{v_{in}} = \frac{-R_2}{R_1 + (R_1 + R_2)/A}$$

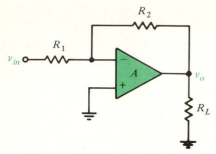

Fig. P10.11

10.12 For the feedback amplifier shown in Fig. P10.11, the op amp has gain A and input resistance R_{in}. Assume that the output resistance of the op amp is $R_o = 0$, and find an expression for the input resistance of the feedback amplifier.

10.13 For the feedback amplifier shown in Fig. P10.11, the op amp has gain A and output resistance R_o. Assume that the input resistance of the op amp is $R_{in} = \infty$, and find an expression for the output resistance of the feedback amplifier.

10.14 The voltage gain of the feedback amplifier shown in Fig. P10.11 is given in Problem 10.11, where A is the gain of the op amp. If ω_u is the bandwidth (upper cutoff frequency) of the op amp, show that the bandwidth of the feedback amplifier is

$$\omega_{uF} = \left(\frac{R_1 A}{R_1 + R_2} + 1 \right) \omega_u$$

10.15 For the feedback amplifier shown in Fig. P10.11, the op amp has a gain of 200,000 and a gain-bandwidth product of 1 MHz. Use the result given in Problem 10.14 to determine the bandwidth of the feedback amplifier for the case that $R_1 = 1$ kΩ and $R_2 = 9$ kΩ.

Problems for Section 10.2

10.16 For the op-amp phase-shift oscillator given in Fig. 10.7, determine R and the minimum value of R_F required to obtain an oscillation frequency of 10 kHz when $C = 0.001$ μF.

10.17 Given the op-amp phase-shift oscillator shown in Fig. 10.7, assuming $R = 1.5$ kΩ, determine C and the minimum value of R_F required for an oscillation frequency of 1.2 kHz.

10.18 Given the JFET phase-shift oscillator shown in Fig. 10.8, assuming $g_m = 5$ m$\mho$ and $R = 100$ kΩ, determine C and the minimum value of R_D required for an oscillation frequency of 1 kHz.

10.19 For the JFET phase-shift oscillator given in Fig. 10.8, determine R and the minimum value of R_D required to obtain an oscillation frequency of 1 kHz when $C = 0.001~\mu F$ and $g_m = 10~m℧$.

10.20 Given the BJT phase-shift oscillator shown in Fig. 10.9, suppose that $R_1 = 27~k\Omega$, $R_2 = 3~k\Omega$, $R_C = 860~\Omega$, and $R = 1~k\Omega$. Verify that $h_{fe} = 100$ is sufficient to produce oscillation, and determine R_s and C for an oscillation frequency of 1 kHz given that $r_e = 2.16~\Omega$.

10.21 For the BJT phase-shift oscillator described in Problem 10.20, find the value for R_C that results in an oscillation frequency of 1 kHz when $C = 0.047~\mu F$. Determine the minimum value of h_{fe} required for oscillation.

10.22 Given the Wien-bridge oscillator shown in Fig. 10.10, find C and the minimum value of R_2 that results in an oscillation frequency of 10 kHz when $R = R_1 = 10~k\Omega$.

10.23 For the Wien-bridge oscillator given in Fig. 10.10, $C = 0.02~\mu F$ and $R_2 = 10~k\Omega$. Determine R such that the frequency of oscillation is 5 kHz, and find the maximum value of R_1.

10.24 Consider the oscillator shown in Fig. P10.24.

(a) Show that the voltage transfer function of the feedback network is

$$\mathbf{B} = \frac{\mathbf{V_1}}{\mathbf{V_0}} = \frac{1}{3 + j(\omega RC - 1/\omega RC)}$$

(b) Find the oscillation frequency ω_o and the relationship between R_1 and R_2 for which oscillation occurs.

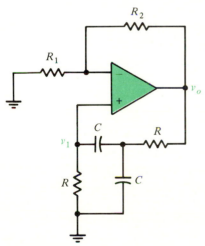

Fig. P10.24

10.25 For the JFET Colpitts oscillator shown in Fig. 10.13, $C_1 = 1200~pF$, $C_2 = 2400~pF$, $L = 50~\mu H$, $R_D = 1~k\Omega$, and $R_G = 10~M\Omega$.

(a) Find the oscillation frequency ω_o.

(b) Determine the minimum value of g_m required for oscillation.

10.26 For the JFET Hartley oscillator shown in Fig. 10.15, $C = 200$ pF, $L_1 = L_2 = 100$ μH, and $R_D = 1$ kΩ.
  (a) Find the oscillation frequency ω_o.
  (b) Determine the minimum value of g_m required for oscillation.

10.27 Given the oscillator shown in Fig. P10.27, and assuming that the op amp is ideal, find (a) the frequency of oscillation and (b) the minimum value of R for which oscillation occurs.

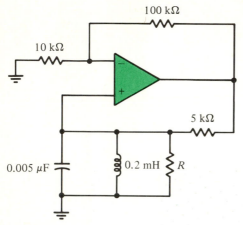

Fig. P10.27

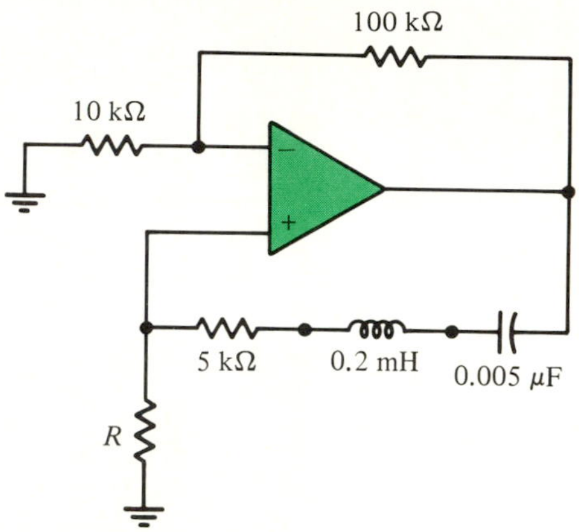

Fig. P10.28

10.28 Repeat Problem 10.27 for the oscillator shown in Fig. P10.28.

10.29 The electric equivalent circuit of a quartz crystal has the values $R = 1$ kΩ, $L = 5$ H, $C = 0.04$ pF, and $C_m = 12$ pF. Find (a) the series resonance ω_s, (b) the parallel resonance ω_p, and (c) the quality factor Q at the series resonance frequency.

Problems for Section 10.3

10.30 Given the simple comparator shown in Fig. 10.20, where $V_{CC} = 10$ V, assuming the input voltage is $v_1 = 6 \sin \omega t$, sketch the output voltage v_o for the case that (a) $v_2 = V_r = 5$ V and (b) $v_2 = V_r = -5$ V.

10.31 For the simple comparator given in Fig. 10.20, let $v_1 = V_r > 0$ be the reference voltage and let v_2 be the input voltage.
  (a) Sketch the transfer characteristic v_o versus v_2.
  (b) Sketch the output voltage v_o when the input voltage is $v_2 = V_r + V_r \sin \omega t$.
  (c) Sketch the output voltage v_o when the input voltage is $v_2 = 2V_r \sin \omega t$.

10.32 Given the Schmitt trigger shown in Fig. 10.23, suppose that $R_1 = R_2 = 10$ kΩ and $V_{CC} = 10$ V. Sketch the output voltage v_o that corresponds to the input voltage $v_1 = 10 \sin \omega t$.

10.33 For the Schmitt trigger given in Fig. 10.26, suppose that $R_1 = 1\ \mathrm{k\Omega}$, $R_2 = 9\ \mathrm{k\Omega}$, and $V_{CC} = 10\ \mathrm{V}$. Sketch the transfer characteristic for the case that (a) $V_r = 5\ \mathrm{V}$, (b) $V_r = 10\ \mathrm{V}$, and (c) $V_r = 20\ \mathrm{V}$.

10.34 Given the Schmitt trigger shown in Fig. 10.26, suppose that $R_1 = 1\ \mathrm{k\Omega}$, $R_2 = 9\ \mathrm{k\Omega}$, $V_{CC} = 10\ \mathrm{V}$, and $V_r = 5\ \mathrm{V}$. Sketch the output voltage v_o for the case that $v_1 = 6\sin \omega t$.

10.35 Since the input is applied to the inverting input, the circuit given in Fig. 10.23 is an inverting Schmitt trigger. For the noninverting Schmitt trigger shown in Fig. P10.35, verify that

$$v = \frac{R_1}{R_1 + R_2}\, v_o + \frac{R_2}{R_1 + R_2}\, v_1$$

and sketch the transfer characteristic for this device.

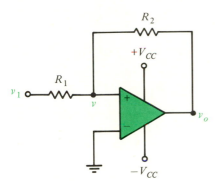

Fig. P10.35

Problems for Section 10.4

10.36 Given the oscillator shown in Fig. 10.29, where $C = 0.05\ \mu\mathrm{F}$, $R_1 = R_2 = 10\ \mathrm{k\Omega}$, and $R = 91\ \mathrm{k\Omega}$, find the frequency of oscillation.

10.37 For the oscillator given in Fig. 10.29, suppose that $C = 0.1\ \mu\mathrm{F}$, $R = 20\ \mathrm{k\Omega}$, and $R_2 = 10\ \mathrm{k\Omega}$. Determine the value of R_1 that yields an oscillation frequency of 100 Hz.

10.38 Given the oscillator shown in Fig. 10.29, where $R = R_1 = R_2 = 22\ \mathrm{k\Omega}$, find the value of C that results in a frequency of oscillation of 100 Hz.

10.39 For the oscillator given in Fig. 10.29, suppose that $C = 0.1\ \mu\mathrm{F}$, $R = 20\ \mathrm{k\Omega}$, and $R_1 = 10\ \mathrm{k\Omega}$. Determine the value of R_2 that yields an oscillation frequency of 100 Hz.

10.40 Given the oscillator shown in Fig. 10.31, where $C = 0.05\ \mu\mathrm{F}$ and $R = R_1 = R_2 = 10\ \mathrm{k\Omega}$, find the frequency of oscillation.

10.41 For the oscillator given in Fig. 10.31, suppose that $C = 0.1\ \mu\mathrm{F}$ and $R = R_1 = 10\ \mathrm{k\Omega}$. Determine the value of R_2 that yields an oscillation frequency of 300 Hz.

10.42 Given the oscillator shown in Fig. 10.31, where $R = R_1 = R_2 = 10$ kΩ, find the value of C that results in a frequency of oscillation of 250 Hz.

10.43 For the oscillator given in Fig. 10.31, suppose that $C = 0.1$ μF and $R = R_1 = 10$ kΩ. Determine the value of R_1 that results in an oscillation frequency of 100 Hz.

10.44 Given the astable multivibrator shown in Fig. 10.34, where $C = 0.1$ μF and $R_1 = R_2 = 5$ kΩ, find the frequency of oscillation.

10.45 For the astable multivibrator given in Fig. 10.34, suppose that $R_1 = R_2 = 5$ kΩ. Find the value of C that results in a frequency of oscillation of 1600 Hz.

10.46 Given the astable multivibrator shown in Fig. 10.34, where $C = 0.005$ μF, and supposing the frequency of oscillation is to be 10 kHz, determine (a) R_1 when $R_2 = 10$ kΩ and (b) R_2 when $R_1 = 10$ kΩ.

10.47 For the oscillator given in Fig. 10.39, find the frequency of oscillation when $R_1 = 1$ kΩ, $R_2 = 9$ kΩ, $R = 10$ kΩ, and $C = 0.001$ μF.

10.48 Given the oscillator shown in Fig. 10.39, suppose that $C = 0.001$ μF and $R = 10$ kΩ. To obtain an oscillation frequency of 40 kHz, find (a) R_1 when $R_2 = 20$ kΩ and (b) R_2 when $R_1 = 3$ kΩ.

10.49 For the oscillator shown in Fig. 10.39, given that $R_1 = 2$ kΩ, $R_2 = 18$ kΩ, and $R = 10$ kΩ, find the value of C that results in an oscillation frequency of 100 kHz.

Problems for Section 10.5

10.50 Determine the number of distinct carrier frequencies (i.e., the maximum number of stations for a given location) in the AM broadcast band.

10.51 The envelope detector given in Fig. 10.45 is to be used to demodulate an AM signal whose carrier frequency is 540 kHz, where the highest frequency component of the modulating signal is 5 kHz.

(a) Determine the time constant $\tau = RC$ that is geometrically centered for inequality (10.26), where an inequality $a < x < b$ is geometrically centered when $b/x = x/a$.

(b) Given that $R = 1$ kΩ, find the value of C that meets the condition given in part (a).

10.52 Find the frequencies of the video and audio carriers for the following television channels: (a) channel 4 (66–72 MHz), (b) channel 5 (76–82 MHz), (c) channel 7 (174–180 MHz).

10.53 Determine the number of distinct carrier frequencies (i.e., the maximum number of stations for a given location) in the FM broadcast band.

10.54 For an FM signal, the ratio of the frequency deviation to the modulating frequency is called the **modulation index.** The ratio of the maximum frequency deviation Δ_f to the maximum modulating frequency f_u is often called the **deviation ratio.** Assuming maximum frequency deviation (referred to as 100% modulation), find the deviation ratio for monophonic FM radio.

10.55 An FM radio station transmits a stereo signal but no SCA signal. Assuming that the station fully occupies its 200-kHz bandwidth, determine the signal's deviation ratio (see Problem 10.54). Is this narrowband or wideband FM?

10.56 Repeat Problem 10.55 for the case that an SCA signal is also transmitted.

PART III
Digital Systems

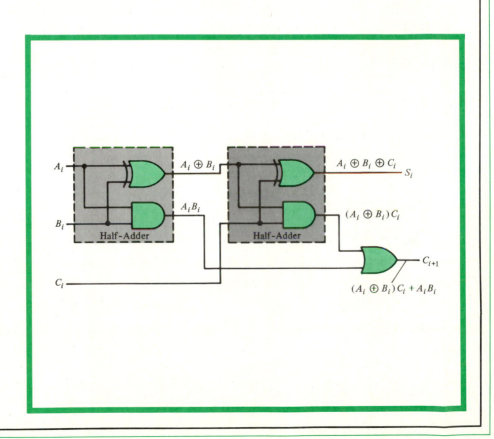

Chapter 11

Digital Logic

INTRODUCTION

Almost everyone is familiar with digital clocks, digital wristwatches, pocket calculators, and video games. Less familiar, perhaps, are general-purpose digital computers, microcomputers, and microprocessors. Yet these systems, as well as uncountable others, have a common characteristic—they manipulate information that is in a discrete form. Since this information can be represented by digits, such systems have come to be called **digital systems.**

In our studies in preceding chapters, our preoccupation was with analog signals—that is, signals that could assume any one of an infinite number of values. (For example, at some instant of time the sinusoid $v(t) = V \cos \omega t$ can take on any value between $-V$ and V.) Now, however, we turn our attention to discrete or digital signals and the systems with which they are used.

A mechanically operated switch is a device that is either open or closed. Thus, we can associate two discrete values, say 0 and 1, with a switch. Because it can be in either of two possible states, a switch is a **binary** device. Since a transistor can be used as a switch (when it is either OFF or ON), a transistor also can be employed as a binary device. Such binary devices are used to construct digital systems, from the simplest to the most complicated. Before the transistor was invented, electromechanically operated switches (relays) were often used in digital-system design. Although relays were superseded by discrete-component transistor electronics, it was the advances in integrated-circuit (IC) technology that led to the present-day proliferation of digital systems.

Perhaps the most important digital system is the digital computer. As shown in the simplified block diagram in Fig. 11.1, a digital computer has five basic

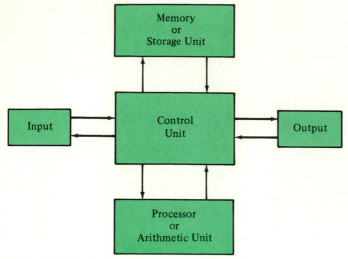

Fig. 11.1 Block diagram of a digital computer.

components. A set of instructions, called a **program,** which will be followed, is stored in the computer's memory. Also stored in memory are the data on which the program will be executed. The data and program enter the computer via the **input.** Among the numerous types of input devices are magnetic-tape readers, paper-tape readers, punched-card readers, typewriters, display scopes, and various types of sensors. The flow of information from the input to the memory is supervised by the **control unit,** as is the information flow between the other units of the computer. The **processor** or **arithmetic unit** manipulates the data, under the direction of the control unit, as specified by the program's instructions. These data manipulations can be arithmetic operations (e.g., addition, subtraction, multiplication, division) or other data-processing tasks. The results of the computations are available to the user at the computer's **output.** Very often the output is printed on paper or appears in graphical form on either plotters or cathode-ray tube (CRT) displays. Outputs printed on a CRT screen are also quite popular.

We now begin our study of the fundamentals of digital computers, as well as of digital systems in general. Since such systems are binary, our first concern is with the representation of information in binary form.

11.1 BINARY NUMBERS

It is said that the number system we ordinarily use is based on the fact that we have ten fingers. Such a system employs the digits 0, 1, 2, 3, 4, 5, 6, 7, 8, 9 and is said to be **decimal.** For example, the decimal number

1983.64

represents the following sum of the powers of 10:

$$1 \times 10^3 + 9 \times 10^2 + 8 \times 10^1 + 3 \times 10^0 + 6 \times 10^{-1} + 4 \times 10^{-2}$$

In general, the decimal number

$$a_n a_{n-1} \ldots a_3 a_2 a_1 a_0 . a_{-1} a_{-2} a_{-3} \ldots a_{-m}$$

represents

$$a_n \times 10^n + a_{n-1} \times 10^{n-1} + \cdots + a_3 \times 10^3 + a_2 \times 10^2 + a_1 \times 10^1$$
$$+ a_0 \times 10^0 + a_{-1} \times 10^{-1} + a_{-2} \times 10^{-2} + a_{-3} \times 10^{-3}$$
$$+ \cdots + a_{-m} \times 10^{-m}$$

where each coefficient a_i is one of the digits 0, 1, 2, 3, 4, 5, 6, 7, 8, 9.

Since it uses powers of 10, we say that the decimal number system is **base** 10, or **radix** 10. By choosing a different base, however, we get a different number system. Specifically, numbers in the base-r system also have the form

$$a_n a_{n-1} \ldots a_3 a_2 a_1 a_0 . a_{-1} a_{-2} a_{-3} \ldots a_{-m}$$

which represents

$$a_n \times r^n + a_{n-1} \times r^{n-1} + \cdots + a_3 \times r^3 + a_2 \times r^2 + a_1 \times r^1 + a_0 \times r^0$$
$$+ a_{-1} \times r^{-1} + a_{-2} \times r^{-2} + a_{-3} \times r^{-3} + \cdots + a_{-m} \times r^{-m}$$

where each coefficient a_i is one of the elements in the set 0, 1, 2, . . . , $r - 1$.

For example, the **binary** (base-2) number

$$1\ 1\ 0\ 1\ 0\ 1\ .\ 1\ 0\ 1$$

is

$$1 \times 2^5 + 1 \times 2^4 + 0 \times 2^3 + 1 \times 2^2 + 0 \times 2^1 + 1 \times 2^0$$
$$+ 1 \times 2^{-1} + 0 \times 2^{-2} + 1 \times 2^{-3}$$

and this is the decimal number 53.625. To be explicit as to the number system, we will place parentheses around a number and use a subscript to indicate the base. For instance, from the discussion above, we have

$$(1\ 1\ 0\ 1\ 0\ 1\ .\ 1\ 0\ 1)_2 = (53.625)_{10}$$

Because of our training in decimal-number arithmetic, converting a binary number into a decimal number—as was done above—is a straightforward, routine procedure. Now, however, let us see how we can convert a decimal number to binary in a simple, systematic manner. To begin with, let the decimal number I be an integer. Expressing I in terms of powers of 2, we have that

$$I = a_n \times 2^n + a_{n-1} \times 2^{n-1} + \cdots + a_3 \times 2^3 + a_2 \times 2^2 + a_1 \times 2^1 + a_0 \times 2^0$$
$$= 2(a_n \times 2^{n-1} + a_{n-1} \times 2^{n-2} + \cdots + a_3 \times 2^2 + a_2 \times 2^1 + a_1 \times 2^0) + a_0$$

If we divide I by 2, then the remainder will be a_0, where a_0 is 0 or 1, and the quotient is

$$a_n \times 2^{n-1} + a_{n-1} \times 2^{n-2} + \cdots + a_3 \times 2^2 + a_2 \times 2^1 + a_1 \times 2^0$$

If this is divided by 2, then the remainder will be a_1, where a_1 is 0 or 1. The process can be repeated until the quotient is zero, and the successive remainders will be $a_2, a_3, \ldots, a_{n-1}, a_n$.

Example 11.1

Let us determine the binary form of the decimal number 25. By successive divisions by 2, we get the following table.

Decimal Number	Division By 2	Quotient	Remainder
25	$\frac{25}{2}$	12	$1 = a_0$
12	$\frac{12}{2}$	6	$0 = a_1$
6	$\frac{6}{2}$	3	$0 = a_2$
3	$\frac{3}{2}$	1	$1 = a_3$
1	$\frac{1}{2}$	0	$1 = a_4$

Thus, we conclude that

$$(25)_{10} = (a_4 a_3 a_2 a_1 a_0)_2 = (11001)_2$$

Instead of filling out a table as shown above, a more convenient way of converting a decimal integer into binary is by performing repeated divisions from right to left as shown below.

	Third remainder	Second remainder	First remainder		
	a_2	a_1	a_0		
$- - - -$	Third quotient	Second quotient	First quotient	I	2

Specifically, for the decimal integer 25, we get

a_4	a_3	a_2	a_1	a_0		
1	1	0	0	1		
0	1	3	6	12	25	2

Now suppose that the decimal number F is a fraction that is less than one. Expressing F in terms of powers of 2, we have that

$$F = a_{-1} \times 2^{-1} + a_{-2} \times 2^{-2} + a_{-3} \times 2^{-3} + a_{-4} \times 2^{-4} + \cdots$$
$$+ a_{-m} \times 2^{-m} + \cdots$$

If we multiply F by 2, then we get

$$2F = a_{-1} + a_{-2} \times 2^{-1} + a_{-3} \times 2^{-2} + a_{-4} \times 2^{-3} + \cdots$$
$$+ a_{-m} \times 2^{-m+1} + \cdots$$

which is a number whose integer part is equal to a_{-1}, where a_{-1} is 0 or 1. We then can form

$$2F = a_{-1} + a_{-2} \times 2^{-1} + a_{-3} \times 2^{-2} + a_{-4} \times 2^{-3} + \cdots$$
$$+ a_{-m} \times 2^{-m+1} + \cdots$$

Multiplying this by 2 yields a number whose integer part is equal to a_{-2}, where a_{-2} is 0 or 1. By repeating this procedure, we can obtain the remaining coefficients a_{-3}, $a_{-4}, \ldots, a_{-m}, \ldots$.

Example 11.2

Let us determine the binary form of the decimal number 0.684. By successive multiplications by 2, we get the following table.

Decimal number	Multiplication by 2	Integer part
0.684	1.368	$1 = a_{-1}$
0.368	0.736	$0 = a_{-2}$
0.736	1.472	$1 = a_{-3}$
0.472	0.944	$0 = a_{-4}$
0.944	1.888	$1 = a_{-5}$
0.888	1.776	$1 = a_{-6}$
0.776	1.552	$1 = a_{-7}$
0.552	1.104	$1 = a_{-8}$
etc.		

Thus, we conclude that

$$(0.684)_{10} = (0.a_{-1}a_{-2}a_{-3}a_{-4}a_{-5}a_{-6}a_{-7}a_{-8} \ldots)_2 = (0.10101111 \ldots)_2$$

Combining the results of the preceding two examples, we have

$$(25.684)_{10} = (11001.10101111 \ldots)_2$$

Other Number Systems

Two other number systems, which are closely related to the binary number system, are sometimes utilized. One of these, the **octal** (base-8) number system, employs the digits 0, 1, 2, 3, 4, 5, 6, 7.

Example 11.3

An example of a number conversion from octal to decimal is

$$(261.64)_8 = 2 \times 8^2 + 6 \times 8^1 + 1 \times 8^0 + 6 \times 8^{-1} + 4 \times 8^{-2}$$

$$= (146.8125)_{10}$$

Converting a decimal number to an octal number can be accomplished as was done for the decimal-to-binary conversion techniques described earlier, with the exception that division or multiplication by 8 is used in place of 2. For instance, for the decimal number $(937)_{10}$, successive divisions by 8 yield

$$
\begin{array}{ccccc}
1 & 6 & 5 & 1 & \\
\overline{(\ 0} & (\ 1 & (\ 14 & (\ 117 & (\ 937 \quad (8 \\
\end{array}
$$

Thus,

$$(937)_{10} = (1651)_8$$

Furthermore, for the decimal number $(0.194)_{10}$, successive multiplications by 8 yield

$$0.194 \times 8 = 1.552 \quad \Rightarrow \quad a_{-1} = 1$$
$$0.552 \times 8 = 4.416 \quad \Rightarrow \quad a_{-2} = 4$$
$$0.416 \times 8 = 3.328 \quad \Rightarrow \quad a_{-3} = 3$$
$$0.328 \times 8 = 2.624 \quad \Rightarrow \quad a_{-4} = 2$$
$$0.624 \times 8 = 4.992 \quad \Rightarrow \quad a_{-5} = 4$$
$$0.992 \times 8 = 7.936 \quad \Rightarrow \quad a_{-6} = 7$$

etc.

Thus,

$$(0.194)_{10} = (0.143247 \ldots)_8$$

The conversion of numbers from binary to octal, and vice versa, is quite simple. Since $2^3 = 8$, the digits of a binary number can be partitioned into groups of three, moving to the left and to the right from the binary point. (If there are insufficient digits to complete a group of three, then a zero or zeros can be added.) Replacing each group of three binary digits by its octal equivalent yields the octal form of the number. For example, with the addition of some zeros, the digits of the binary number 10101011000.1110011 can be partitioned into groups of three as follows:

```
Additional              Additional
   zero                    zeros
   ⌒                         ⌒
010 101 011 000.111 001 100
 ⌣   ⌣   ⌣   ⌣   ⌣   ⌣   ⌣
  2   5   3   0   7   1   4
```

Substituting the octal equivalent for each group of three binary digits as indicated results in the fact that

$$(10101011000.1110011)_2 = (2530.714)_8$$

Conversely, given an octal number, by replacing each digit by its three-digit binary equivalent, we get the binary form of the number. For example, for the octal number 1704.536, we make the following associations:

```
 1   7   0   4  .  5   3   6

001 111 000 100   101 011 110
```

This means that

$$(1704.536)_8 = (1111000100.10101111)_2$$

The other number system that was mentioned as being closely related to the binary number system is the **hexadecimal** (base-16) system. Since such a system requires 16 symbols, the digits 0 through 9 do not suffice. For this reason, the first six letters of the alphabet are also used. In other words, the digits for the hexadecimal number system are 0, 1, 2, 3, 4, 5, 6, 7, 8, 9, A, B, C, D, E, F. Converting numbers from binary to hexadecimal, and vice versa, can be accomplished as between binary and octal—the difference being that groups of four binary digits are used instead of groups of three. For example, for the binary number 10101011000.1110011, we have the grouping

```
0101 0101 1000 . 1110 0110
  ⌣    ⌣    ⌣      ⌣    ⌣
  5    5    8      E    6
```

so that

$$(10101011000.1110011)_2 = (558.E6)_{16}$$

Furthermore, for the hexadecimal number 6A9.0C, we have

$$6 \quad A \quad 9 \;.\; 0 \quad C$$
$$\underbrace{0110}\ \underbrace{1010}\ \underbrace{1001}\ \underbrace{0000}\ \underbrace{1100}$$

so that

$$(6A9.0C)_{16} = (11010101001.000011)_2$$

A partial summary of number systems is given by Table 11.1, which shows the integers zero through fifteen in four different bases.

Table 11.1

Decimal (base 10)	Binary (base 2)	Octal (base 8)	Hexadecimal (base 16)
0	0	0	0
1	1	1	1
2	10	2	2
3	11	3	3
4	100	4	4
5	101	5	5
6	110	6	6
7	111	7	7
8	1000	10	8
9	1001	11	9
10	1010	12	A
11	1011	13	B
12	1100	14	C
13	1101	15	D
14	1110	16	E
15	1111	17	F

11.2 BINARY ARITHMETIC

Just as we can perform decimal arithmetic operations, so too can we add, subtract, multiply, and divide binary numbers.

To perform addition, we use the following rules for adding the binary digits 0 and 1:

$$
\begin{array}{cccc}
0 & 0 & 1 & 1 \\
+\,0 & +\,1 & +\,0 & +\,1 \\
\hline
0 & 1 & 1 & 1\;0
\end{array}
$$

Note that for the first three additions, the sum is a single binary digit. However, for the fourth addition—which is the binary version of "one plus one equals two"—the sum consists of two binary digits. This means, just as in ordinary decimal addition, that we can have "carries" when we add binary numbers. For example, the binary equivalent of the decimal sum $60 + 42 = 102$ is the following.

```
carries → 1   1
        1  1  1  1  0  0
      + 1  0  1  0  1  0
        1  1  0  0  1  1  0
```

Subtraction of binary numbers is accomplished as is decimal subtraction, where it may be necessary to "borrow" from a higher number position. For example, the binary equivalent of the decimal difference $60 - 42 = 18$ is the following.

```
              borrow
        1  1  1  1  0  0
      - 1  0  1  0  1  0
        0  1  0  0  1  0
```

To subtract a positive number from a smaller positive number, simply subtract the smaller number from the larger and place a negative sign in front of the difference.

Multiplication of binary numbers can be performed by using the following product rules for 0 and 1:

```
   0        0        1        1
 × 0      × 1      × 0      × 1
   0        0        0        1
```

We then proceed as we do for decimal numbers. For example, the binary equivalent of the decimal product $23 \times 5 = 115$ can be obtained as follows:

```
        1  0  1  1  1
            ×  1  0  1
        1  0  1  1  1
     0  0  0  0  0
  1  0  1  1  1
  1  1  1  0  0  1  1
```

Division of binary numbers can be accomplished as ordinary long division is for decimal numbers. For example, the decimal division $87 \div 6 = 14.5$ when performed with binary numbers is as follows:

```
                    1  1  1  0 . 1
        1  1  0 )1  0  1  0  1  1  1 . 0
                 1  1  0
                 1  0  0  1
                    1  1  0
                       1  1  1
                       1  1  0
                          1  1  0
                          1  1  0
                                0
```

Subtraction using "carries" may be well suited for hand calculations using pencil and paper, but other techniques are more appropriate for digital-computer calculations. In this vein, we say that the **1's complement** of a binary number is obtained by changing the 0s and 1s and the 1s to 0s. (For a fraction less than unity, the 0 to the left of the binary point is left as is.) For example,

1's complement of 1 0 1 1 0 0 1 0 = 0 1 0 0 1 1 0 1
1's complement of 0 . 1 1 1 0 1 0 0 = 0 . 0 0 0 1 0 1 1
1's complement of 1 0 1 1 . 0 0 1   = 0 1 0 0 . 1 1 0

The difference between two positive binary numbers n and m (i.e., $n - m$) can be obtained using the 1's complement as follows:

1. To n add the 1's complement of m.

2. If the sum has a carry at the left end, add a 1 to the least significant digit (this is called an **end-around carry**) of the sum. The result is $n - m$.

3. If the sum has no carry at the left end, take the 1's complement of the sum and place a negative sign in front of it. The result is $n - m$.

Example 11.4

Given $n = 1101001$ and $m = 0101110$, to obtain $n - m$, we proceed as follows:

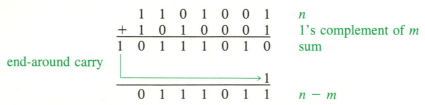

	1	1	0	1	0	0	1	n
+	1	0	1	0	0	0	1	1's complement of m
1	0	1	1	1	0	1	0	sum

end-around carry → 1

| 0 | 1 | 1 | 1 | 0 | 1 | 1 | $n - m$ |

Now to form $m - n$, we have the following:

	0	1	0	1	1	1	0	m
+	0	0	1	0	1	1	0	1's complement of n
	1	0	0	0	1	0	0	sum
−	0	1	1	1	0	1	1	$m - n$

∎

The **10's complement** of a positive decimal number whose integer part has p digits is obtained by subtracting that number from 10^p. For instance, the 10's complement of 47 is $10^2 - 47 = 53$, while the 10's complement of 125.84 is $10^3 - 125.84 = 874.16$. The 10's complement of 0 is defined to be 0.

The **2's complement** of a positive binary number whose integer part has p digits is obtained by subtracting that number from 2^p. For instance, the 2's complement of 11010 is $2^5 - 11010 = 100000 - 11010 = 00110$, while the 2's complement of 1011.001 is $2^4 - 1011.001 = 10000 - 1011.001 = 0100.111$. The 2's complement of 0 is defined to be 0.

The 2's complement of a positive binary number can also be obtained by adding a 1 to the rightmost digit of its 1's complement.

Example 11.5

1's complement of 1 0 1 1 0 0 1 0 = 0 1 0 0 1 1 0 1
$$\phantom{\text{2's complement of 1 0 1 1 0 0 1 0 = 0 1 0 0 1 1}}+ 1$$
2's complement of 1 0 1 1 0 0 1 0 = 0 1 0 0 1 1 1 0

1's complement of 0 . 1 1 1 0 1 0 0 = 0 . 0 0 0 1 0 1 1
$$\phantom{\text{2's complement of 0 . 1 1 1 0 1 0 0 = 0 . 0 0 0 1 0}}+ 1$$
2's complement of 0 . 1 1 1 0 1 0 0 = 0 . 0 0 0 1 1 0 0

1's complement of 1 0 1 1 . 0 0 1 = 0 1 0 0 . 1 1 0
$$\phantom{\text{2's complement of 1 0 1 1 . 0 0 1 = 0 1 0 0 . 1}}+ 1$$
2's complement of 1 0 1 1 . 0 0 1 = 0 1 0 0 . 1 1 1

■

We are all familiar with ordinary decimal subtraction. However, there is an alternative technique for performing this type of operation. Specifically, instead of subtracting a number, we may add its 10's complement and ignore any carry that results. For example, the alternative to $8 - 3 = 5$ is to take the 10's complement of 3, which is 7 (because $10 - 3 = 7$), and form the sum $8 + 7 = 15$. Ignoring the carry, we obtain 5—the desired result. Similarly, the alternative to $27 - 40 = -13$ is to take the 10's complement of 40, which is $10^2 - 40 = 60$, and form the sum $27 + 60 = 87$. Since there is no carry, we take the 10's complement of 87 (which is $10^2 - 87 = 13$) and insert a negative sign. The result is -13—as it should be.

The difference between two positive binary numbers n and m (i.e., $n - m$) can be obtained by using the 2's complement as follows.

1. To n add the 2's complement of m.

2. If the sum has a carry at the left end, remove it. The result is $n - m$.

3. If the sum has no carry at the left end, take the 2's complement of the sum and insert a negative sign. The result is $n - m$.

Example 11.6

Given $n = 1101001$ and $m = 0101110$, to obtain $n - m$, we proceed as follows:

$$
\begin{array}{cccccccl}
1 & 0 & 1 & 0 & 0 & 0 & 1 & \text{1's complement of } m \\
  &   &   &   &   & + & 1 & \\
\hline
1 & 0 & 1 & 0 & 0 & 1 & 0 & \text{2's complement of } m \\
+1 & 1 & 0 & 1 & 0 & 0 & 1 & n \\
\hline
\text{remove} \longrightarrow 1\ 0 & 1 & 1 & 1 & 0 & 1 & 1 & \text{sum} \\
0 & 1 & 1 & 1 & 0 & 1 & 1 & n - m
\end{array}
$$

Now to form $m - n$, we have the following:

$$
\begin{array}{cccccccl}
0 & 0 & 1 & 0 & 1 & 1 & 0 & \text{1's complement of } n \\
  &   &   &   &   & + & 1 & \\
\hline
0 & 0 & 1 & 0 & 1 & 1 & 1 & \text{2's complement of } n \\
+0 & 1 & 0 & 1 & 1 & 1 & 0 & m \\
\hline
1 & 0 & 0 & 0 & 1 & 0 & 1 & \text{sum} \\
0 & 1 & 1 & 1 & 0 & 1 & 0 & \text{1's complement of sum} \\
\text{insert} &   &   &   &   & + & 1 & \\
\hline
\text{negative sign} - 0 & 1 & 1 & 1 & 0 & 1 & 1 & m - n
\end{array}
$$

■

Binary Codes

We have already seen how we can express numerical information in binary (base-2) form. However, other types of information, such as letters of the alphabet and various kinds of symbols, are also very important. To be handled by digital systems, such as digital computers, information must be in some type of binary form. This can be accomplished by "encoding" information with the use of "binary codes."

The binary **encoding** of information consists of representing various pieces of information by sequences of the binary digits, called **bits,** 0 and 1. A sequence of 8 bits is referred to as a **byte.** The resulting set of binary sequences is known as a **binary code,** and each sequence is a **code word.** For example, suppose that we had four pieces of information W, X, Y, and Z that we would like to encode in binary form. Since there are four distinct binary pairs (binary sequences with 2 bits), those being 00, 01, 10, and 11, one possibility for a binary encoding of W, X, Y, and Z is to make the associations shown below:

$$W \leftrightarrow 0\ 0$$

$$X \leftrightarrow 0\ 1$$

$$Y \leftrightarrow 1\ 0$$

$$Z \leftrightarrow 1\ 1$$

The resulting set of binary sequences (00, 01, 10, 11) is the binary code.

When dealing with computers, users prefer to use the decimal representations of numbers rather than the binary versions. Because of this, even though real numbers can be expressed in binary (base-2) form, often it is desirable to encode decimal numbers with a binary code. The encoded decimal numbers can then be converted into binary numbers for purposes of arithmetic operations, or alternatively, the arithmetic can be performed directly on the encoded decimal numbers.

Since there are ten decimal digits (0, 1, 2, 3, 4, 5, 6, 7, 8, 9), a minimum of 4 bits is required to encode these digits into binary. (This is a consequence of the fact that there are only eight distinct binary triples: 000, 001, 010, 011, 100, 101, 110, 111.) A popular scheme for encoding the decimal digits is to use binary quadruples (sequences of 4 bits), where each quadruple is the binary (base-2) representation of the corresponding decimal digit. The resulting code, referred to as **binary-coded decimal** (BCD), is given by Table 11.2. For example, the BCD version of the decimal number 1983 is 0001 1001 1000 0011.

Let us now suppose that eight pieces of information are encoded into the eight distinct binary triples 000, 001, 010, 011, 100, 101, 110, and 111. During the transmission of information, it is possible that electrical noise can cause an error in a code word—that is, either a 0 is changed to 1, or a 1 is changed to 0. If this happens, when the triple is decoded, the wrong piece of information will result. To help prevent such an occurrence, an additional bit—called a **parity bit**—can be added to the triples so that a new code consisting of binary quadruples will be produced. One possible rule for adding the parity bit is to make each new code word have an even number of 1s. Table 11.3 indicates what parity bit should be chosen and the resulting new code. Since any word in the new code differs from any other word in at least two positions, a single error cannot change one code word into another. If a single error is made in a code word, then the result will be an odd number of 1s in that word. Thus, by counting the number of 1s in a code word, a single error can be detected. However, two errors in a code word will cause an even number of 1s, and, therefore, two errors cannot be detected. The same statement

Table 11.2

Decimal Digit	BCD			
0	0	0	0	0
1	0	0	0	1
2	0	0	1	0
3	0	0	1	1
4	0	1	0	0
5	0	1	0	1
6	0	1	1	0
7	0	1	1	1
8	1	0	0	0
9	1	0	0	1

Table 11.3

Coded Information	Parity Bit	New Code
0 0 0	0	0 0 0 0
0 0 1	1	0 0 1 1
0 1 0	1	0 1 0 1
0 1 1	0	0 1 1 0
1 0 0	1	1 0 0 1
1 0 1	0	1 0 1 0
1 1 0	0	1 1 0 0
1 1 1	1	1 1 1 0

can be made about four errors. Since three errors in a code word also produce an odd number of 1s, three errors can be detected as well.

Table 11.3 shows an error-detecting code that is obtained by adding a parity bit such that each resulting code word has an even number of 1s. Another error-detecting code can be formed by making the number of 1s in each code word odd (see Problem 11.18). These are normally referred to as **even** and **odd parity** respectively. Although such codes have error-detection capabilities, they do not have error-correction capabilities. In other words, if a single error is made in one of the code words of the new code given in Table 11.3, the result is a binary quadruple that is not a code word. However, there is now a way of knowing for sure what the original code word was. To construct codes with the ability to correct errors, additional parity bits are required. The subject of such **error-correcting codes** is beyond the scope of this book.

Having seen how information can be expressed or represented in binary form by using binary numbers or an appropriate binary encoding scheme, let us turn our attention to the binary operation of electric and electronic circuits.

11.3 DIGITAL LOGIC CIRCUITS

Consider the simple circuit consisting of a battery V, a switch S, and a light bulb L shown in Fig. 11.2. Suppose that we indicate that the switch is open by $S = 0$ and is closed by $S = 1$. In other words, the variable S designates the state of the switch (open or closed) and S takes on one of two values (0 or 1). Since S can take on only two values, we say that it is a **binary variable.** When the switch is open ($S = 0$), there is an open circuit between a and b (the terminals of the switch), and, therefore, the light bulb is off. However, when the switch is closed ($S = 1$), there is a short circuit between a and b, and the light bulb is off and $L = 1$ means that the light bulb is on.

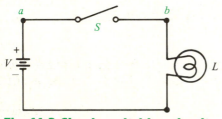

Fig. 11.2 Simple switching circuit.

Suppose that we also designate the state of the light bulb by $L = 0$ when the light bulb is off, and $L = 1$ when the light bulb is on. Doing this, we see that L is also a binary variable, and $S = 0 \Rightarrow L = 0$, while $S = 1 \Rightarrow L = 1$. In other words, $L = S$, and a table indicating the relationship between S and L is given by Table 11.4(a). Such a table is called a **truth table.** If the light bulb being off is designated by $L = 1$ and being on is designated by $L = 0$, then we get the truth

Table 11.4 Truth Tables for (a) $L = S$ and (b) $L = \bar{S}$.

	(a)		(b)
S	$L = S$	S	$L = \bar{S}$
0	0	0	1
1	1	1	0

table given by Table 11.4(b). Under this circumstance, we say that L is the **complement** of S and express this mathematically as $L = \bar{S}$. However, unless indicated otherwise, $L = 0$ denotes that the light bulb is off and $L = 1$ means that the light bulb is on.

Switching Circuits

A circuit such as the one shown in Fig. 11.2, where a light bulb L is either off or on depending upon whether a switch (or switches) is open or closed, is referred to as a **switching circuit.**

Let us consider the parallel connection of switches A and B shown in Fig. 11.3, where although the battery V and the light bulb L of the corresponding switching circuit are not shown, they are assumed to be connected as depicted in Fig. 11.2. When either switch (or both) is closed, there is a short circuit between a and b, and, therefore, the light bulb is on. In other words, when either $A = 1$ or $B = 1$, then $L = 1$. When both switches are open ($A = 0$ and $B = 0$), then the light bulb is off ($L = 0$). The behavior of the two switches connected in parallel is summarized by Table 11.5. This truth table describes the **OR operation** for binary variables. By specifying A and B to be input variables and L to be the output variable, then the output $L = 1$ when either input $A = 1$ OR $B = 1$ (or both)—otherwise $L = 0$. Instead of writing $L = A$ OR B, we use the mathematical notation $L = A + B$.

Table 11.5 Truth Table for the OR Operation.

A	B	$L = A + B$
0	0	0
0	1	1
1	0	1
1	1	1

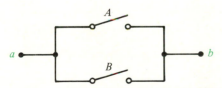

Fig. 11.3 Two switches connected in parallel.

Figure 11.4 shows the series connection of switches A and B. In this case, the only situation for which $L = 1$ (i.e., there is a short circuit between a and b) is when both $A = 1$ AND $B = 1$ (i.e., both switches are closed). The resulting truth

Table 11.6 Truth Table for the AND Operation.

A	B	L = AB
0	0	0
0	1	0
1	0	0
1	1	1

Fig. 11.4 Two switches connected in series.

table, given by Table 11.6, describes the **AND operation.** The mathematical notation for this operation is $L = A \cdot B$, or, more simply, $L = AB$.

For the connection of three switches shown in Fig. 11.5, the resulting truth table is given by Table 11.7. In this case, the truth table requires eight rows instead of four so that all of the possible combinations of open and closed switches are included. Since switches B and C are connected in parallel, that connection can be described by $B + C$. This parallel combination, however, is connected in series with switch A. Thus, we may write that $L = A(B + C)$. But does the distributive law hold for binary variables? In other words, does $A(B + C) = AB + AC$? The answer to this question can be found by solving Problem 11.19.

Table 11.7 Truth Table for Three-Switch Connection.

A	B	C	L = A(B + C)
0	0	0	0
0	0	1	0
0	1	0	0
0	1	1	0
1	0	0	0
1	0	1	1
1	1	0	1
1	1	1	1

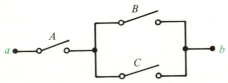

Fig. 11.5 Connection of three switches.

Given a switch A and a switch B, if B is closed when A is open (if $B = 1$ when $A = 0$), and if B is open when A is closed (if $B = 0$ when $A = 1$), then from the truth table corresponding to these conditions (Table 11.8), we see that $B = \bar{A}$.

We can use this concept of a complement in the design of an arbitrary switching circuit—that is, we can connect switches together to realize a given truth table. For example, consider Table 11.9. (This truth table describes the **exclusive-OR operation.**) Since $L = 1$ when $A = 0$ ($\bar{A} = 1$) AND $B = 1$, OR when $A = 1$ AND $B = 0$ ($\bar{B} = 1$), then we have that $L = \bar{A}B + A\bar{B}$. A connection of switches that realizes this expression is shown in Fig. 11.6.

Table 11.9 Truth Table for the Exclusive-OR Operation.

A	B	$L = \bar{A}B + A\bar{B}$
0	0	0
0	1	1
1	0	1
1	1	0

Table 11.8

A	$B = \bar{A}$
0	1
1	0

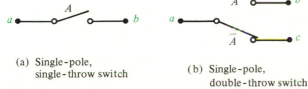

Fig. 11.6 Realization of exclusive-OR operation.

Binary operations such as AND, OR, exclusive-OR, and complement are referred to as **logic operations,** and circuits that perform these operations are called **logic gates.**

Up to this point we have been dealing with switches of the type shown in Fig. 11.7(a). Such a switch, called a **single-pole, single-throw** (SPST) switch, either connects a and b or it does not. Another type of switch, a **single-pole, double-throw** (SPDT) switch, is shown in Fig. 11.7(b). For this three-terminal device, either a and b are connected or a and c are connected—there is no other switch condition. This means that if a and b are considered to be the terminals of switch A, then terminals a and c comprise switch $\bar{A}$. Of course, by not using either terminal b or terminal c, a SPDT switch can be used as a SPST switch.

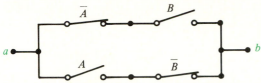

(a) Single-pole, single-throw switch

(b) Single-pole, double-throw switch

Fig. 11.7

By using SPDT switches, it is a simple matter to realize the exclusive-OR operation given by Table 11.9. Specifically, the connection of two SPDT switches as shown in Fig. 11.8 yields $L = \bar{A}B + A\bar{B}$.

The realization of the exclusive-OR operation with switches has a very practical application—a two-way light switch. To see why this is true, again consider Table 11.9. Suppose that a room has two doors, where switch A is next to one door and switch B is next to the other door. We begin with both switches open ($A = B = 0$) and the room light off ($L = 0$). Upon entering door A, the switch is

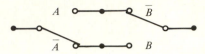

Fig. 11.8 Alternative realization of exclusive-OR operation.

thrown ($A = 1$, $B = 0$) and the light goes on ($L = 1$). Leaving by the same door results in the initial conditions. However, leaving by door B instead and throwing that switch ($A = 1$, $B = 1$) also turns the light off ($L = 0$). If, after leaving by door B, door A is entered and the switch is thrown, the result is $A = 0$, $B = 1$, and $L = 1$—that is, the light goes on. Leaving by either door and throwing the corresponding switch again will turn off the light.

The concept of a two-way light switch can be extended to a three-way switch, and so on. However, any further discussion of this topic is left as an exercise for the reader (see Problems 11.24 and 11.25).

Logic Gates

We have seen how we can construct circuits corresponding to binary expressions (or functions) by using mechanically actuated switches. However, because transistors can be used as electronic switches, manually operated switches and relays have become passé for purposes of performing logical operations. Instead, as discussed in the electronics portion of this book, various types of electronic circuits are used to realize logic gates. At this point, however, we will not be concerned with the electronic details of how such gates are constructed, but instead will introduce digital-circuit symbols to signify the various types of logic gates.

A digital logic gate whose input is a binary variable, say A, and whose output is also A, is referred to as a **buffer.** The circuit symbol for such a gate is shown in Fig. 11.9. A logic circuit, whose output is $\overline{A}$ (the complement of A) when the input is A, is called a **NOT gate** or **inverter.** The circuit symbol for this gate is obtained from the buffer symbol by placing a small circle at the buffer's output as shown in Fig. 11.10. The truth table for a NOT gate is given by Table 11.8.

Fig. 11.9 Buffer.

Fig. 11.10 A NOT gate.

Another logic circuit is the **OR gate.** Figure 11.11 shows a two-input OR gate whose inputs are A and B, and whose output is $A + B$ (see Table 11.5). Although a NOT gate has only one input, an OR gate can have two or more inputs. (Very shortly we will discuss the situation of more than two inputs in greater detail.)

Figure 11.12 shows the circuit symbol of a two-input **AND gate,** that is, a gate that has an output of AB (see Table 11.6) when its inputs are A and B.

Fig. 11.11 Two-input OR gate.

Fig. 11.12 Two-input AND gate.

By appropriately connecting AND, OR, and NOT gates, it is possible to construct a digital logic circuit that realizes any binary function (expression). As an example of this fact, recall the exclusive-OR operation mentioned previously. We use the symbol $\oplus$ to denote this important operation. From Table 11.9, which is repeated by Table 11.10, we have that $A \oplus B = \overline{A}B + A\overline{B}$. The logic circuit shown in Fig. 11.13 utilizes two AND gates, two NOT gates, and one OR gate to realize this operation. Thus, we have used AND, OR, and NOT gates to construct an **exclusive-OR gate.** However, since exclusive-OR gates are used frequently, such a gate has its own circuit symbol—it is shown in Fig. 11.14.

Table 11.10 Truth Table for the Exclusive-OR Operation.

A	B	$A \oplus B = \overline{A}B + A\overline{B}$
0	0	0
0	1	1
1	0	1
1	1	0

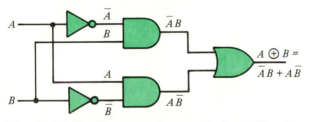

Fig. 11.13 Realization of an exclusive-OR gate.

Fig. 11.14 Two-input exclusive-OR gate.

Another very important logic operation is obtained by taking the complement after performing the AND operation. The truth table that describes this situation is given by Table 11.11, and we refer to this operation as **NAND**—a contraction of NOT–AND.* Although this operation can be realized with an AND gate followed by a NOT gate as shown in Fig. 11.15(a), we give the **NAND gate** its own circuit symbol. This is shown in Fig. 11.15(b).

---

*The contraction may be misleading because the AND operation is performed prior to the NOT operation.

Table 11.11 Truth Table for the NAND Operation.

A	B	AB	$\overline{AB}$
0	0	0	1
0	1	0	1
1	0	0	1
1	1	1	0

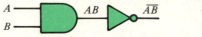

(a) AND-gate and NOT-gate combination

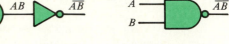

(b) Circuit symbol of a NAND gate

Fig. 11.15 Two-input NAND gates.

By following an OR gate with a NOT gate as shown in Fig. 11.16(a), we obtain a **NOR gate** (NOR is a contraction of NOT–OR). As with the NAND gate, the NOR gate has its own circuit symbol—it is shown in Fig. 11.16(b). The truth table for the NOR operation is given by Table 11.12.

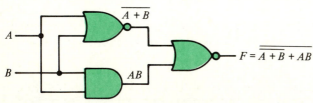

(a) OR-gate and NOT-gate combination.

(b) Circuit symbol of a NOR gate.

Fig. 11.16 Two-input NOR gates.

Table 11.12 Truth Table for the NOR Operation.

A	B	$A + B$	$\overline{A + B}$
0	0	0	1
0	1	1	0
1	0	1	0
1	1	1	0

Example 11.7

For the logic circuit shown in Fig. 11.17, clearly the output is characterized by the binary function $F = \overline{\overline{A + B} + AB}$. By substituting the values of 0 and 1 for the variables A and B, we can obtain the corresponding truth table. This approach is illustrated by Table 11.13.

$$A + B$$

$$F = \overline{\overline{A + B} + AB}$$

$$AB$$

Fig. 11.17 Exclusive-OR gate.

Table 11.13 Truth Table for Example.

A	B	$A + B$	$\overline{A + B}$	AB	$\overline{A + B} + AB$	$\overline{\overline{A + B} + AB}$
0	0	0	1	0	1	0
0	1	1	0	0	0	1
1	0	1	0	0	0	1
1	1	1	0	1	1	0

■

From Example 11.7 we see that the truth table for $F = \overline{\overline{A + B} + AB}$ is identical to that for the exclusive-OR operation. Thus, the logic circuit given in Fig. 11.17 realizes an exclusive-OR gate. Furthermore, we must have that

$$F = \overline{\overline{A + B} + AB} = \overline{A}B + A\overline{B} = A \oplus B \tag{11.1}$$

Trying to simplify a binary function by constructing its corresponding truth table is a very tedious procedure. Fortunately, though, there is an analytical means for simplifying binary expressions. This is the next topic of discussion.

11.4 BOOLEAN ALGEBRA

We have seen that by connecting logic gates together, we can construct a logic circuit whose output is described by some binary function (or expression) F. As indicated by Equation (11.1), for example, such an expression can have more than one form. We can, however, obtain one form from another by using mathematical manipulations that do not require the construction of truth tables. Let us see how this is accomplished.

We are all familiar with the mathematical system consisting of the set of real numbers together with the operations of multiplication and addition.* Another important mathematical system consists of the set of all complex numbers together with the operations of complex-number multiplication and complex-number addition. Fortunately, there is also a structure to the binary elements 0 and 1 along with the operations of AND, OR, and NOT (complement). Such a mathematical system is referred to as **switching algebra** (because of the use of switches in the early days) or **Boolean**[†] **algebra.** We will now describe the rules for this algebra.

First, to recapitulate, suppose that A and B are binary variables. Then the truth tables for the operations AND (denoted $A \cdot B$ or AB) and OR (denoted $A + B$) are given by Tables 11.14 and 11.15 respectively. Furthermore, for the binary variable A, the truth table for the operation NOT, or complement (denoted $\overline{A}$), is given by

---

*Dividing by a number is equivalent to multiplying by the reciprocal of that number. Subtracting a number is equivalent to adding the negative of that number.

[†]Named for the English mathematician George Boole (1815–1864).

Table 11.14 Truth Table for A AND B.

A	B	AB
0	0	0
0	1	0
1	0	0
1	1	1

Table 11.15 Truth Table for A OR B.

A	B	A + B
0	0	0
0	1	1
1	0	1
1	1	1

Table 11.16 Truth Table for NOT A (the complement of A).

A	$\overline{A}$
0	1
1	0

Table 11.16. For these operations, it is simply a matter of filling out the corresponding truth tables to show the following equalities:

$$A \cdot 1 = A \qquad A + 0 = A$$

$$A \cdot A = A \qquad A + A = A$$

$$A \cdot 0 = 0 \qquad A + 1 = 1$$

$$A \cdot \overline{A} = 0 \qquad A + \overline{A} = 1$$

A number of other results, which can be confirmed with the use of truth tables, are listed in the following.

Involution:	$\overline{\overline{A}} = A$
Commutative law for AND:	$AB = BA$
Commutative law for OR:	$A + B = B + A$
Associative law for AND:	$A(BC) = (AB)C$
Associative law for OR:	$A + (B + C) = (A + B) + C$
Distributive law for AND over OR:	$A(B + C) = AB + AC$
Distributive law for OR over AND:	$A + BC = (A + B)(B + C)$

Implicit for the above distributive laws is the fact that the AND operation has precedence over the OR operation. Specifically, for the distributive law $A + BC =$

$(A + B)(A + C)$, the AND operation BC is performed prior to the OR operation in the term $A + BC$. Since this distributive law may seem strange (when compared with the distribution of real numbers), let us verify the expression with a truth table. Because the columns labeled $A + BC$ and $(A + B)(A + C)$ are identical, Table 11.17 confirms the distributive law for OR over AND.

Table 11.17 Truth Table for $A + BC = (A + B)(A + C)$

A	B	C	BC	A + BC	A + B	A + C	(A + B)(A + C)
0	0	0	0	0	0	0	0
0	0	1	0	0	0	1	0
0	1	0	0	0	1	0	0
0	1	1	1	1	1	1	1
1	0	0	0	1	1	1	1
1	0	1	0	1	1	1	1
1	1	0	0	1	1	1	1
1	1	1	1	1	1	1	1

Two absorption laws for Boolean algebra are the following:

$$A + AB = A \qquad \text{and} \qquad A(A + B) = A \qquad \text{absorption laws}$$

Instead of verifying these equalities with truth tables, let us use algebraic manipulations. For example,

$$A + AB = A \cdot 1 + AB \qquad \text{since } A \cdot 1 = A$$
$$= A(1 + B) \qquad \text{distributive law}$$
$$= A \cdot 1 \qquad \text{since } 1 + B = B + 1 \quad \text{and} \quad B + 1 = 1$$
$$= A \qquad \text{since } A \cdot 1 = A$$

Also,

$$A(A + B) = AA + AB \qquad \text{distributive law}$$
$$= A + AB \qquad \text{since } AA = A$$
$$= A \qquad \text{preceding absorption law}$$

Two additional types of absorption laws are obtained as follows:

$$A + \overline{A}B = (A + \overline{A})(A + B) \qquad \text{distributive law}$$
$$= 1(A + B) \qquad \text{since } A + \overline{A} = 1$$
$$= A + B \qquad \text{since } 1 \cdot A = A$$

and

$$A(\bar{A} + B) = A\bar{A} + AB \qquad \text{distributive law}$$

$$= 0 + AB \qquad \text{since } A\bar{A} = 0$$

$$= AB \qquad \text{since } 0 + A = A + 0 \quad \text{and} \quad A + 0 = A$$

In summary, therefore,

$$\boxed{A + \bar{A}B = A + B \qquad \text{and} \qquad A(\bar{A} + B) = AB}$$

Two very important results, known as **De Morgan's theorems**,* are

$$\boxed{\overline{A + B} = \bar{A}\bar{B} \qquad \text{and} \qquad \overline{AB} = \bar{A} + \bar{B}}$$

The simplest way of establishing these facts is with the use of truth tables. This is done by Tables 11.18 and 11.19 respectively.

Table 11.18 Proof of $\overline{A + B} = \bar{A}\bar{B}$.

A	B	$A + B$	$\overline{A + B}$	$\bar{A}$	$\bar{B}$	$\bar{A}\bar{B}$
0	0	0	1	1	1	1
0	1	1	0	1	0	0
1	0	1	0	0	1	0
1	1	1	0	0	0	0

Table 11.19 Proof of $\overline{AB} = \bar{A} + \bar{B}$.

A	B	AB	$\overline{AB}$	$\bar{A}$	$\bar{B}$	$\bar{A} + \bar{B}$
0	0	0	1	1	1	1
0	1	0	1	1	0	1
1	0	0	1	0	1	1
1	1	1	0	0	0	0

Example 11.8

Let us verify algebraically that $\overline{\overline{A + B} + AB} = \bar{A}B + A\bar{B}$ [Equation (11.1)].

---

*Named for the English mathematician Augustus De Morgan (1806–1871).

$$\overline{A + B} + AB = (\overline{A + B})(\overline{AB}) \qquad \text{since } \overline{A + B} = \overline{A}\,\overline{B}$$

$$= (A + B)(\overline{AB}) \qquad \text{involution}$$

$$= (A + B)(\overline{A} + \overline{B}) \qquad \text{since } \overline{AB} = \overline{A} + \overline{B}$$

$$= (A + B)\overline{A} + (A + B)\overline{B} \qquad \text{distributive law}$$

$$= \overline{A}(A + B) + \overline{B}(B + A) \qquad \text{commutative laws}$$

$$= \overline{A}B + \overline{B}A = \overline{A}B + A\overline{B} \qquad \text{since } A(\overline{A} + B) = AB \text{ and } \overline{B}A = A\overline{B}$$

■

A frequently used property of Boolean algebra is the principle of **duality.** Two expressions are said to be **duals** if all the AND and OR operations are interchanged, and all of the 0s and 1s are interchanged. (The variable names and complements remain unchanged.) The following pairs are some examples of duals:

AB	and	$A + B$	
$A + 0 = A$	and	$A \cdot 1 = A$	
$B \cdot 0 = 0$	and	$B + 1 = 1$	
$A(BC) = (AB)C$	and	$A + (B + C) = (A + B) + C$	associative laws
$A(B + C) = AB + AC$	and	$A + BC = (A + B)(A + C)$	distributive laws
$A + AB = A$	and	$A(A + B) = A$	absorption laws
$A + \overline{A}B = A + B$	and	$A(\overline{A} + B) = AB$	absorption laws
$\overline{A + B} = \overline{A}\,\overline{B}$	and	$\overline{AB} = \overline{A} + \overline{B}$	De Morgan's theorems

Once an equality has been demonstrated, its dual is automatically valid, and, therefore, does not need to be verified. Thus, we get two proofs for the price of one.

Just because two expressions are duals, that does not mean that they are equivalent—in general, they are not equivalent.

Example 11.9

The dual of the exclusive-OR operation

$$A \oplus B = \overline{A}B + A\overline{B}$$

is called the **exclusive-NOR** (or **equivalence**) operation, and is denoted A ⊙ B. In other words,

$$A \odot B = (\overline{A} + B)(A + \overline{B})$$

The truth table for this operation is given by Table 11.20, and the circuit symbol for an **exclusive-NOR gate** is shown in Fig. 11.18.

Table 11.20 Truth Table for Exclusive-NOR Operation.

A	B	$\bar{A}$	$\bar{B}$	$\bar{A} + B$	$A + \bar{B}$	$A \odot B$
0	0	1	1	1	1	1
0	1	1	0	1	0	0
1	0	0	1	0	1	0
1	1	0	0	1	1	1

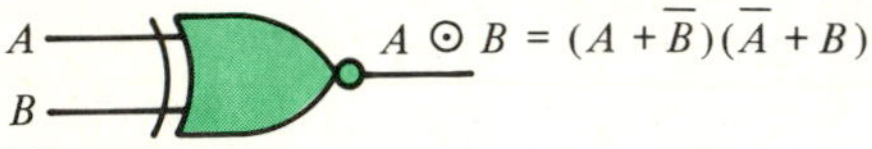

$A \odot B = (A + \bar{B})(\bar{A} + B)$

Fig. 11.18 Exclusive-NOR gate.

Multiple-Input Gates

The associativity law for the OR operation states that $A + (B + C) = (A + B) + C$. This property, along with the commutative law for the OR operation, $A + B = B + A$, means that the order in which the OR operations are performed on three binary variables does not matter—the results are the same. Therefore, we can write $A + B + C$, and its meaning is unambiguous. A three-input OR gate is a logic gate whose output is $A + B + C$ when its inputs are A, B, and C. The circuit symbol for such a logic gate is shown in Fig. 11.19.

For the case of the AND operation, just as was deduced in the discussion of the OR operation, it is unambiguous to write ABC. The circuit symbol for a three-input AND gate is shown in Fig. 11.20.

$A + B + C$

Fig. 11.19 Three-input OR gate.

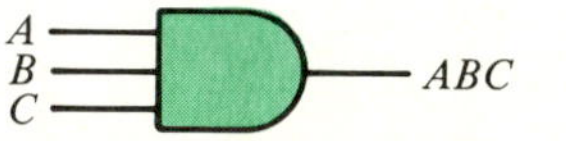

ABC

Fig. 11.20 Three-input AND gate.

For the reasons just discussed, the number of inputs for AND gates and OR gates can be extended to four or more.

Now consider the NOR operation. Let us denote this operation by the symbol $\downarrow$. Thus, $A \downarrow B = \overline{A + B}$. Clearly, commutivity holds for this operation— $A \downarrow B = B \downarrow A$—because the OR operation is commutative. However, associativity does not hold, that is,

$$A \downarrow (B \downarrow C) \neq (A \downarrow B) \downarrow C$$

because

$$A \downarrow (B \downarrow C) = \overline{A + \overline{B + C}} = \overline{A + \overline{B}\overline{C}} = \overline{A}(\overline{\overline{B}\overline{C}}) = \overline{A}(B + C)$$

$$(A \downarrow B) \downarrow C = \overline{\overline{A + B} + C} = \overline{\overline{A}\overline{B} + C} = (\overline{\overline{A}\overline{B}})\overline{C} = (A + B)\overline{C}$$

Therefore, we shall define a three-input NOR gate to be a logic gate whose output is $\overline{A + B + C}$ when its inputs are A, B, and C. The circuit symbol for such a logic gate is shown in Fig. 11.21.

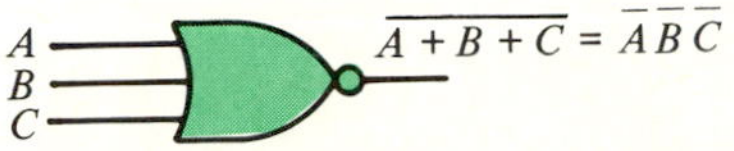

A
B ———▷ $\overline{A + B + C} = \overline{A}\,\overline{B}\,\overline{C}$
C

Fig. 11.21 Three-input NOR gate.

Let us denote the NAND operation by the symbol $\uparrow$. Thus, $A \uparrow B = \overline{AB}$. Similar to the discussion above, the NAND operation is commutative but not associative (see Problem 11.41). Therefore, we define the output of a three-input NAND gate to be $\overline{ABC}$ when the inputs are A, B, and C. The circuit symbol for such a logic gate is shown in Fig. 11.22.

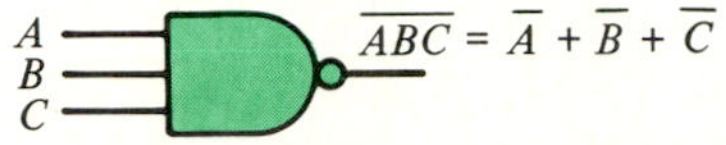

A
B ———▷ $\overline{ABC} = \overline{A} + \overline{B} + \overline{C}$
C

Fig. 11.22 Three-input NAND gate.

Note that

$$\overline{A + B + C} = \overline{A + (B + C)} = \overline{A}(\overline{B + C}) = \overline{A}(\overline{B}\overline{C}) = \overline{A}\overline{B}\overline{C}$$

and

$$\overline{ABC} = \overline{A(BC)} = \overline{A} + \overline{BC} = \overline{A} + (\overline{B} + \overline{C}) = \overline{A} + \overline{B} + \overline{C}$$

These are just the three-variable versions of De Morgan's theorems, and, therefore, we have alternative expressions for the outputs of the NOR and NAND gates shown in Figs. 11.23 and 11.24 respectively.

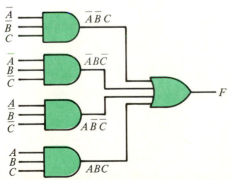

Fig. 11.23 Realization of the Boolean function
$F = \overline{A}BC + \overline{A}B\overline{C} + AB\overline{C} + ABC.$

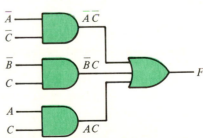

Fig. 11.24 Implementation of
$F = \overline{A}C + \overline{B}C + AC.$

Clearly, NOR and NAND gates can have more than three inputs, and De Morgan's theorems can be extended as well. For example, for the case of the four inputs A, B, C, and D, the output of a four-input NOR gate is

$$\overline{A + B + C + D} = \overline{A}\overline{B}\overline{C}\overline{D}$$

and the output of a four-input NAND gate is

$$\overline{ABCD} = \overline{A} + \overline{B} + \overline{C} + \overline{D}$$

11.5 STANDARD FORMS OF BOOLEAN FUNCTIONS

A binary function comprised of binary variables, along with AND, OR, and NOT operations, is referred to as a **Boolean function.** An example of a Boolean function of the two variables A and B is

$$F_1 = \overline{A}B + A\overline{B}$$

As we have already seen, this function describes the exclusive-OR operation, as well as a two-way light switch (see Fig. 11.9). In addition to an algebraic expression, a Boolean function can also be specified in terms of a truth table. The truth table that corresponds to $F_1 = \overline{A}B + A\overline{B}$ is given by Table 11.21.

Another example of a Boolean function is $F_2 = (A + B)(\overline{AB})$. The truth table for this function (along with the columns for $A + B$, AB, and $\overline{AB}$, which aid computation) is given by Table 11.22. Since the columns for A, B, and F_1 in Table 11.21 are identical to those for A, B, and F_2 in Table 11.22, respectively, then F_1 and F_2 have the same truth tables. This means that $F_1 = F_2$.

Table 11.21 Truth Table for $F_1 = \overline{A}B + A\overline{B}$.

A	B	F_1
0	0	0
0	1	1
1	0	1
1	1	0

Table 11.22 Truth Table for $F_2 = (A + B)(\overline{AB})$.

A	B	$A + B$	AB	$\overline{AB}$	F_2
0	0	0	0	1	0
0	1	1	0	1	1
1	0	1	0	1	1
1	1	1	1	0	0

Even though $F_1 = \overline{A}B + A\overline{B} = (A + B)(\overline{AB}) = F_2$, it is the expression $F_1 = \overline{A}B + A\overline{B}$ that is evident from the truth table. That is so because $F_1 = 1$ when $A = 0$ ($\overline{A} = 1$) AND $B = 1$, OR $A = 1$ AND $B = 0$ ($\overline{B} = 1$).

The Sum of Products

In general, given a function F of n variables, the corresponding truth table has 2^n rows. Typically, the rows of the first n columns consist of the binary numbers ordered from 0 to $2^n - 1$, where the first row is the number 0 and the last row is the number $2^n - 1$. For example, a function F of three variables is described by Table 11.23. In this case, we can proceed to write an expression for F as was suggested above. Specifically,

$$F = \overline{A}\overline{B}C + \overline{A}B\overline{C} + A\overline{B}\overline{C} + ABC$$

Note that F consists of four terms ($\overline{A}\overline{B}C$, $\overline{A}B\overline{C}$, $A\overline{B}\overline{C}$, and ABC)—called **minterms**—which are combined by the OR operation. Because the OR operation is represented as addition is, F is said to be a **sum** of the four minterms. Every minterm consists of the three variables, each being complemented or uncomplemented, which are combined by the AND operation. Because the AND operation is represented as multiplication is, each of the four minterms is said to be a **product** of the three variables. For these reasons, the expression for the Boolean function F given above is said to be in the form of a **sum of products.**

Table 11.23 Truth Table for a Boolean Function F.

A	B	C	F
0	0	0	0
0	0	1	1
0	1	0	1
0	1	1	0
1	0	0	1
1	0	1	0
1	1	0	0
1	1	1	1

Table 11.24 Minterms for the Case of Three Variables A, B, C.

Binary Number			Minterms for Variables A, B, C
0	0	0	$m_0 = \overline{A}\overline{B}\overline{C}$
0	0	1	$m_1 = \overline{A}\overline{B}C$
0	1	0	$m_2 = \overline{A}B\overline{C}$
0	1	1	$m_3 = \overline{A}BC$
1	0	0	$m_4 = A\overline{B}\overline{C}$
1	0	1	$m_5 = A\overline{B}C$
1	1	0	$m_6 = AB\overline{C}$
1	1	1	$m_7 = ABC$

For a Boolean function of three variables, there are $2^3 = 8$ minterms—one for each row of the truth table. In other words, there is one minterm for each number between 0 and $2^3 - 1 = 7$. These minterms are designated m_0, m_1, m_2, m_3, m_4, m_5, m_6, m_7. (For the case of n variables, there are 2^n minterms, and these are designated m_0, m_1, m_2, m_3, . . . , m_{2^n-1}.) Every minterm consists of a product of the three variables, where each variable is either uncomplemented or complemented. Specifically, if the first bit of the binary number i is 0 (or 1), then the first variable in the minterm m_i is complemented (or uncomplemented), and so on. Table 11.24 shows the binary numbers ranging from 0 to 7 (the case of three variables) and the corresponding minterms.

Given the truth table, an expression for a Boolean function F is obtained by summing the minterms that correspond to the rows in which there is a 1 in the column labeled F. For Table 11.23, we have that

$$F = m_1 + m_2 + m_4 + m_7 = \overline{A}\overline{B}C + \overline{A}B\overline{C} + A\overline{B}\overline{C} + ABC$$

A logic circuit whose output is described by this Boolean function is shown in

Fig. 11.23. For this circuit, which consists of four three-input AND gates and one four-input OR gate, it is assumed that each variable and its complement are available. Furthermore, to avoid the pictorial cluttering that results from wires crossing, inputs on different gates are labeled with the same (uncomplemented or complemented) variable.

Given an expression for a Boolean function, it is not necessary to construct a truth table in order to write the function as a sum of minterms. For example, consider the function

$$F = \overline{A}\overline{C} + (A + \overline{B})C$$

This can be written as

$$F = \overline{A}\overline{C} + \overline{B}C + AC$$

To obtain minterms, we can multiply each product in the sum by 1 in an appropriate form. Specifically, since $\overline{A} + A = 1$, then

$$F = \overline{A}(\overline{B} + B)\overline{C} + (\overline{A} + A)\overline{B}C + A(\overline{B} + B)C$$

from which

$$F = \overline{A}\overline{B}\overline{C} + \overline{A}B\overline{C} + \overline{A}\overline{B}C + A\overline{B}C + A\overline{B}C + ABC$$

$$= \overline{A}\overline{B}\overline{C} + \overline{A}B\overline{C} + \overline{A}\overline{B}C + A\overline{B}C + ABC$$

$$= m_0 + m_1 + m_2 + m_5 + m_7$$

As exemplified in Fig. 11.23, a logic circuit whose output is characterized by a Boolean function, which is expressed as a sum of products, can be constructed by using two columns of gates. The first column is composed of AND gates, and the second column is a single multi-input OR gate. Such a circuit is an example of what is known as **two-level logic.**

We have just demonstrated that the Boolean function $F = m_0 + m_1 + m_2 + m_5 + m_7$ is equivalent to the sum of products

$$F = \overline{A}\overline{C} + \overline{B}C + AC$$

A two-level logic circuit that realizes, or implements, this function is shown in Fig. 11.24. Although this realization employs AND gates, an OR gate, and (implicit) NOT gates, it is possible to implement the given function using only NAND gates. Let us see how.

Suppose that the same input A is applied to both inputs of a two-input NAND gate as shown in Fig. 11.25(a). Since the output of the gate is $\overline{AA} = \overline{A}$, the gate is an inverter (NOT gate). The equivalent situation, a one-input NAND gate, is symbolized in Fig. 11.25(b).

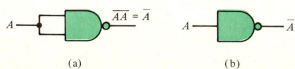

(a)   (b)

Fig. 11.25 NAND-gate realization of an inverter (NOT gate).

We will now show how NAND gates can be connected to form an OR gate and how they can be connected to form an AND gate. By De Morgan's theorem,

$$A + B = \overline{\overline{A + B}} = \overline{\overline{A}\,\overline{B}}$$

and this expression can be realized by the logic circuit, which, therefore, is an OR gate, shown in Fig. 11.26. Furthermore, since

$$AB = \overline{\overline{AB}}$$

then the NAND-gate circuit shown in Fig. 11.27 realizes an AND gate.

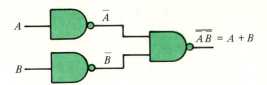

Fig. 11.26 NAND-gate realization of an OR gate.

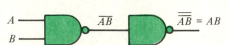

Fig. 11.27 NAND-gate realization of an AND gate.

Since any Boolean function can be implemented with AND, OR, and NOT gates, it can also be implemented using only NAND gates. This is so because, as we have seen, AND, OR, and NOT gates can be constructed from NAND gates. However, there is a more direct method for obtaining a NAND-gate realization of a Boolean function.

Let us see how this is done by looking at a specific example. We begin with a sum of products such as $F = \overline{A}\overline{C} + \overline{B}C + AC$. Using De Morgan's theorem, we can write

$$F = \overline{A}\overline{C} + \overline{B}C + AC = \overline{\overline{\overline{A}\overline{C} + \overline{B}C + AC}} = \overline{(\overline{\overline{A}\overline{C}})(\overline{\overline{B}C})(\overline{AC})}$$

and this expression can be implemented by the two-level NAND circuit shown in Fig. 11.28. This implementation, therefore, is an alternative to the one given in Fig. 11.24.

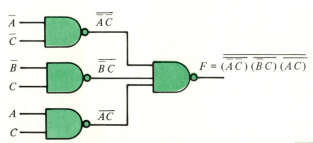

Fig. 11.28 NAND-gate implementation of $F = \overline{A}\overline{C} + \overline{B}C + AC$.

Up to this point, we have used an AND gate symbol followed by a small circle (to denote inversion) to signify a NAND gate. This logic-circuit symbol is repeated in Fig. 11.29(a) for the case of three input variables. However, we may also use

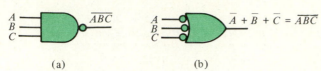

Fig. 11.29 Alternative circuit symbols for the NAND gate.

the circuit symbol shown in Fig. 11.29(b) to indicate a NAND gate. In this case, the small circles preceding the OR gate cause the input variables to be complemented before the OR operation is performed. The resulting output, therefore, is, by De Morgan's theorem,

$$\bar{A} + \bar{B} + \bar{C} = \overline{ABC}$$

and this confirms the NAND operation.

Since an arbitrary Boolean function can be implemented by a logic circuit consisting only of NAND gates, we say that the NAND gate is an **universal gate.**

A sum of products is one type of **standard form** of a Boolean function. There is one other type of standard form—a product sums. We now look at this functional form.

The Product of Sums

Given the truth table for a Boolean function, we have seen how we can express that function as a sum of minterms. Let us describe a procedure for obtaining an alternative, equivalent expression of the function from its truth table. For demonstration purposes, reconsider Table 11.23, which is repeated as Table 11.25. Here a column corresponding to the function $\bar{F}$, the complement of F, has been attached. As described previously, we can express $\bar{F}$ as

$$\bar{F} = m_0 + m_3 + m_5 + m_6 = \bar{A}\bar{B}\bar{C} + \bar{A}BC + A\bar{B}C + AB\bar{C}$$

Taking the complement of both sides of this equation, we get

$$F = \overline{\bar{A}\bar{B}\bar{C} + \bar{A}BC + A\bar{B}C + AB\bar{C}}$$

By De Morgan's theorems, we have that

$$F = (\overline{\bar{A}\bar{B}\bar{C}})(\overline{\bar{A}BC})(\overline{A\bar{B}C})(\overline{AB\bar{C}})$$
$$= (A + B + C)(A + \bar{B} + \bar{C})(\bar{A} + B + \bar{C})(\bar{A} + \bar{B} + C)$$

which is a **product of sums.** Each such sum is called a **maxterm.** For a Boolean function of three variables, there are $2^3 = 8$ maxterms—one for each row of the truth table. The maxterms are designated $M_0, M_1, M_2, M_3, M_4, M_5, M_6, M_7$. Each maxterm consists of the sum of the three variables, where each variable is either uncomplemented or complemented. Specifically, if the first bit of the binary number i is 0 (or 1), then the first variable in the maxterm M_i is uncomplemented (or complemented), and so on. Table 11.26 shows the binary numbers ranging from 0 to 7 and the corresponding maxterms.

Table 11.25 Truth Table for F and $\overline{F}$.

A	B	C	F	$\overline{F}$
0	0	0	0	1
0	0	1	1	0
0	1	0	1	0
0	1	1	0	1
1	0	0	1	0
1	0	1	0	1
1	1	0	0	1
1	1	1	1	0

Table 11.26 Maxterms for the Case of Three Variables A, B, C.

Binary Number	Maxterms for Variables A, B, C
0 0 0	$M_0 = A + B + C$
0 0 1	$M_1 = A + B + \overline{C}$
0 1 0	$M_2 = A + \overline{B} + C$
0 1 1	$M_3 = A + \overline{B} + \overline{C}$
1 0 0	$M_4 = \overline{A} + B + C$
1 0 1	$M_5 = \overline{A} + B + \overline{C}$
1 1 0	$M_6 = \overline{A} + \overline{B} + C$
1 1 1	$M_7 = \overline{A} + \overline{B} + \overline{C}$

Given the truth table for a Boolean function F, an alternative expression to the sum of minterms is obtained by taking the product of the maxterms that correspond to the rows of the table in which there is a 0 in the column labeled F. For Table 11.25, we have that

$$F = M_0 M_3 M_5 M_6 = (A + B + C)(A + \overline{B} + \overline{C})(\overline{A} + B + \overline{C})(\overline{A} + \overline{B} + C)$$

A two-level logic circuit whose output is described by this Boolean function is shown in Fig. 11.30. This circuit is composed of four three-input OR gates and one four-input AND gate—again it is assumed that each variable and its complement are available.

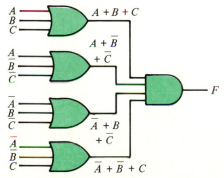

Fig. 11.30 Realization of the Boolean function
$$F = (A + B + C)(A + \overline{B} + \overline{C})(\overline{A} + B + \overline{C})(\overline{A} + \overline{B} + C).$$

Given an expression for a Boolean function, it is not necessary to construct a truth table in order to write the function as a product of maxterms. For example, consider the function

$$F = \overline{A}\,\overline{C} + (A + \overline{B})C$$

Using the distributive law $A + BC = (A + B)(A + C)$, we can write

$$F = (\overline{A}\overline{C} + A + \overline{B})(\overline{A}\overline{C} + C) = (A + \overline{A}\overline{C} + \overline{B})(C + \overline{C}\overline{A})$$

and since $A + \overline{A}B = A + B$, we have the product of sums

$$F = (A + \overline{C} + \overline{B})(C + \overline{A}) = (A + \overline{B} + \overline{C})(\overline{A} + C)$$

Although $A + \overline{B} + \overline{C}$ is a maxterm, $\overline{A} + C$ is not. To obtain maxterms, to each sum that is not a maxterm we can add 0 in an appropriate form. Specifically, since $B\overline{B} = 0$, then

$$F = (A + \overline{B} + \overline{C})(B\overline{B} + \overline{A} + C) = (A + \overline{B} + \overline{C})(B + \overline{A} + C)(\overline{B} + \overline{A} + C)$$

$$= (A + \overline{B} + \overline{C})(\overline{A} + B + C)(\overline{A} + \overline{B} + C)$$

$$= M_3 M_4 M_6$$

Instead of implementing F in this form, let us use the simpler product-of-sums expression $F = (A + \overline{B} + C)(\overline{A} + C)$ obtained above. A two-level logic circuit that realizes this function is shown in Fig. 11.31. Although this circuit employs an AND gate, NOR gates, and (implicit) NOT gates, it is possible to implement the given function using only NOR gates. Let us see how.

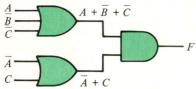

Fig. 11.31 Implementation of $F = (A + \overline{B} + \overline{C})(\overline{A} + C)$.

As was the case for NAND gates, we only need show that AND, OR, and NOT gates can be constructed from NOR gates. By connecting both inputs of a NOR gate together as shown in Fig. 11.32(a), we obtain a NOT gate. This is a consequence of the fact that $\overline{A + A} = \overline{A}$. The equivalent situation of the gate given in Fig. 11.32(a) is the one-input NOR gate symbolized in Fig. 11.32(b).

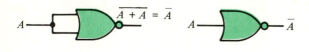

(a)   (b)

Fig. 11.32 NOR-gate realization of an inverter.

Since

$$A + B = \overline{\overline{A + B}}$$

then the connection of NOR gates shown in Fig. 11.33 realizes an OR gate.

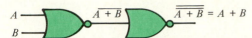

Fig. 11.33 NOR-gate realization of an OR gate.

Furthermore, since

$$AB = \overline{\overline{AB}} = \overline{\overline{A} + \overline{B}}$$

then the NOR-gate circuit shown in Fig. 11.34 realizes an AND gate. This, therefore, demonstrates the fact that an arbitrary Boolean function can be implemented using only NOR gates. Like the NAND gate, the NOR gate is a universal gate.

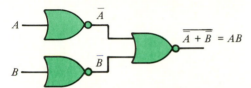

Fig. 11.34 NOR-gate realization of an AND gate.

Instead of constructing AND, OR, and NOT gates from NOR gates, a more direct approach can be taken. To do this, we use De Morgan's theorem on a Boolean function that is expressed as a product of sums. For example, for the function $F = (A + \overline{B} + \overline{C})(\overline{A} + C)$ encountered above, we have that

$$F = (A + \overline{B} + \overline{C})(\overline{A} + C) = \overline{\overline{(A + \overline{B} + \overline{C})(\overline{A} + C)}}$$

$$= \overline{\overline{(A + \overline{B} + \overline{C})} + \overline{(\overline{A} + C)}}$$

and this expression can be implemented by the two-level NOR circuit shown in Fig. 11.35. Such a logic circuit is an alternative to those given in Figs. 11.24, 11.28, and 11.31.

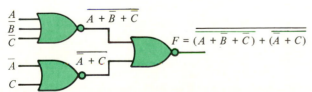

Fig. 11.35 NOR-gate implementation of $F = (A + \overline{B} + \overline{C})(\overline{A} + C)$.

As indicated above, a NOR gate is signified by the symbol for an OR gate followed by a small circle. This logic-circuit symbol is repeated in Fig. 11.36(a) for the case of three input variables. However, a NOR gate can also be represented by the circuit symbol shown in Fig. 11.36(b). In this case, the small circles

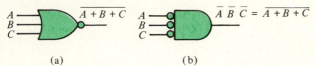

Fig. 11.36 Alternative circuit symbols for the NOR gate.

preceding the AND gate indicate the inversion of the input variables prior to the performing of the AND operation. By De Morgan's theorem, the resulting output is given by

$$\overline{A}\,\overline{B}\,\overline{C} = \overline{A + B + C}$$

and this confirms NOR operation.

The realization of Boolean functions using NAND gates or NOR gates is not just an academic exercise. Because they are more easily fabricated, these types of gates see greater use than do AND gates and OR gates. NAND and NOR gates are the basic gates of IC logic.

11.6 SIMPLIFICATION OF BOOLEAN FUNCTIONS

When a Boolean function is expressed either as a sum of minterms or as a product of maxterms, we say that such a function is in **canonical form.** We refer to an uncomplemented variable or a complemented variable as a **literal.** When the output of a logic circuit is characterized by a Boolean function in either canonical form, each literal in the expression corresponds to an input of a gate, and each term (product or sum) in the expression corresponds to a logic gate. If a Boolean function can be simplified so as to consist of fewer literals and terms, the resulting realization of such a function will require less hardware than is needed for the unsimplified form of the function.

One way to simplify Boolean functions is to use algebraic manipulations that employ the relationships given in Section 11.4. For example, consider the following Boolean function expressed as a sum of minterms:

$$F = \overline{A}BC + A\overline{B}\,\overline{C} + A\overline{B}C + AB\overline{C} + ABC$$

It follows that

$$F = \overline{A}BC + A\overline{B}(\overline{C} + C) + AB(\overline{C} + C)$$
$$= \overline{A}BC + A\overline{B} + AB$$
$$= \overline{A}BC + A(\overline{B} + B)$$
$$= \overline{A}BC + A$$
$$= A + BC$$

and the function cannot be simplified further.

Karnaugh Maps

Instead of taking an algebraic approach, Boolean functions can be simplified graphically using maps referred to as **Karnaugh* maps.** Specifically, we will discuss only the cases of either three or four variables. We shall avoid the complexity that arises when dealing with more than four variables.

For the case of three variables, we form a map consisting of $2^3 = 8$ boxes as shown in Fig. 11.37. Here the eight boxes form two rows and four columns. The rows, labeled 0 and 1, are associated with variable A, while the columns are associated with variables B and C. Note that the columns, going from left to right, are labeled 00, 01, 11, and 10. The labeling is done in this way so that in going from left to right the binary pairs change in only one bit.

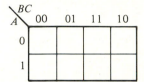

Fig. 11.37 Three-variable Karnaugh map.

Each box corresponds to a distinct binary triple. The first bit of the triple is the box's row designation, while the second and third bits are indicated by the box's column designation. For example, the box in the upper left-hand corner corresponds to 000, while the box in the lower right-hand corner corresponds to 110. Therefore, with each box we can identify one of the minterms m_0, m_1, m_2, m_3, m_4, m_5, m_6, m_7, where m_i is associated with the box corresponding to the binary number i. The minterms associated with the boxes of a three-variable Karnaugh map are as shown in Fig. 11.38.

BC / A	00	01	11	10
0	m_0	m_1	m_3	m_2
1	m_4	m_5	m_7	m_6

Fig. 11.38 Minterms associated with a three-variable Karnaugh map.

Given a Boolean function in the form of a sum of minterms, place a 1 in each of the boxes that corresponds to a minterm in the sum. Place a 0 in each remaining box, or, to avoid the cluttering of symbols, leave the remaining boxes empty. For example, consider the function

$$F = m_0 + m_1 + m_2 + m_6 = \overline{A}\,\overline{B}\,\overline{C} + \overline{A}\,\overline{B}C + \overline{A}B\overline{C} + AB\overline{C}$$

---

*Named for the American electrical engineer Maurice Karnaugh (1924–).

The corresponding Karnaugh map is shown in Fig. 11.39. The two boxes in the upper left with 1s in them represent the sum $m_0 + m_1 = \overline{A}\overline{B}\overline{C} + \overline{A}\overline{B}C$. Algebraically, we know that this reduces to $\overline{A}\overline{B}(\overline{C} + C) = \overline{A}\overline{B}$. The graphical interpretation is the following: Group the adjacent 1s corresponding to the minterms m_0 and m_1 together as shown in Fig. 11.40. Both 1s are identified with $A = 0$ and $B = 0$. However, $C = 0$ for one box, while $C = 1$ for the other. Thus, the term that corresponds to the two 1s in the upper left portion of the map is $\overline{A}\overline{B}$.

Using this interpretation, when the 1s on the right are grouped together as shown in Fig. 11.40, since $A = 0$ for one box and $A = 1$ for the other, while $B = 1$ and $C = 0$ for both boxes, the corresponding term is $B\overline{C}$.

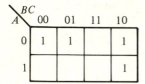

Fig. 11.39 Karnaugh map for
$F = m_0 + m_1 + m_2 + m_6.$

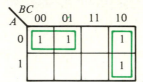

Fig. 11.40 Grouping of minterms.

The result of the two groupings shown is the expression

$$F = \overline{A}\overline{B} + B\overline{C}$$

Although this expression is also a sum of products, its form is simpler than the original sum of minterms.

A pair of 1s in a Karnaugh map can be adjacent even though they do not appear to be at first glance. For example, for the map shown in Fig. 11.41, the two ones indicated are in adjacent boxes, and, therefore, are grouped as depicted. In this case, for both boxes $A = 1$ and $C = 0$, while $B = 0$ for one box and $B = 1$ for the other. Thus, the term corresponding to this pair of 1s is $A\overline{C}$.

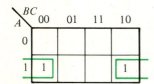

Fig. 11.41 Karnaugh map with adjacent 1s.

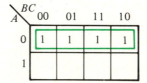

Fig. 11.42 Karnaugh map with four adjacent 1s.

A greater reduction in literals can be made when there are 1s in four adjacent boxes. An example is shown in Fig. 11.42. In this case, each 1 in the group corresponds to $A = 0$, while B takes on the values 0 and 1, as does C. Thus, as can be proved algebraically, the term corresponding to this group of four 1s is simply $\overline{A}$.

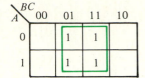

Fig. 11.43 Another group of four adjacent 1s.

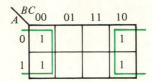

Fig. 11.44 Karnaugh map with a group of four adjacent 1s.

Another group of four 1s is shown in Fig. 11.43. Here the only condition corresponding to all four boxes is $C = 1$. Thus, the term that results is simply C. For the Karnaugh map shown in Fig. 11.44, the group of four adjacent 1s has the common condition $C = 0$. Thus, the corresponding term is $\overline{C}$.

Having considered some examples of grouping the 1s in a Karnaugh map, let us now list the steps to take for function simplification.

Boolean-Function Simplification with a Karnaugh Map

1. Given a Boolean function F, express it as a sum of minterms, and place 1s in the corresponding boxes of a Karnaugh map.

2. Group adjacent 1s so as to form the largest possible groups. A particular 1 can be placed in more than one group.

3. A 1 that is not adjacent to any other 1 forms its own group.

4. After all the 1s have been placed in the largest possible groups, without forming more groups than required, take the sum of the terms corresponding to each group—this is the simplified expression for the Boolean function.

Following this procedure results in a sum of products with the minimum number of literals. However, if alternative groupings are possible, then other expressions, also with a minimum number of literals, may be obtained.

Example 11.10

Consider a Boolean function F_1 whose Karnaugh map is as shown in Fig. 11.45. In this case, the bottom four 1s form a group that corresponds to the term A. The two 1s on the lower right can be used again with the two on the upper right to form another group of four 1s. This group corresponds to the term B. Finally, grouping the two 1s on the extreme left with the two 1s on the extreme right, we

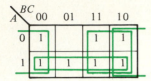

Fig. 11.45 Karnaugh map with groups of four adjacent 1s.

get the term $\overline{C}$. Thus, the function has the simplified form

$$F_1 = A + B + \overline{C}$$

Next consider a Boolean function F_2 having the Karnaugh map shown in Fig. 11.46. In this case, there are no groups consisting of four adjacent 1s. However, as indicated, there are three groups that are comprised of a pair of 1s. The corresponding expression for F_2 is

$$F_2 = AC + \overline{A}\,\overline{C} + \overline{B}C$$

If the 1s are grouped by the alternative scheme shown in Fig. 11.47, then the corresponding equivalent expression is

$$F_2 = AC + \overline{A}\,\overline{C} + \overline{A}\,\overline{B}$$

Even though both of these expressions for F_2 have the minimum number of literals (six) for a sum of products, an expression not in the form of a sum of products may contain less literals. Specifically, from the expressions obtained above, we have that

$$F_2 = (A + \overline{B})C + \overline{A}\,\overline{C} = AC + \overline{A}(B + \overline{C})$$

which consists of only five literals.

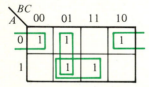

Fig. 11.46 Karnaugh map with groups of two adjacent 1s.

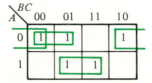

Fig. 11.47 Karnaugh map with alternative grouping.

As still another example, consider the Boolean function F_3 whose Karnaugh map is as shown in Fig. 11.48. In this case, again there is no group consisting of four adjacent 1s. But, as illustrated, there are two groups, each consisting of a pair of 1s. There is also a 1 that is not adjacent to any other 1. This single 1 corresponds to the minterm ABC, and the Boolean function in simplified sum-of-product form is

$$F_3 = \overline{A}\,\overline{B} + \overline{B}\,\overline{C} + ABC$$

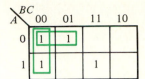

Fig. 11.48 Karnaugh map with a group consisting of a single 1.

∎

Four-Variable Maps

A Karnaugh map for a Boolean function of the four variables A, B, C, and D is shown in Fig. 11.49. In this case, there are 16 boxes that correspond to the 16 minterms m_0, m_1, m_2, . . . , m_{15}. Each of the map's boxes is labeled with its associated minterm.

$\frac{CD}{AB}$	00	01	11	10
00	m_0	m_1	m_3	m_2
01	m_4	m_5	m_7	m_6
11	m_{12}	m_{13}	m_{15}	m_{14}
10	m_8	m_9	m_{11}	m_{10}

Fig. 11.49 Four-variable map displaying the corresponding minterms.

The simplification procedure is essentially the same as that described for the case of three variables. For four variables, however, it may be possible to form groups of eight adjacent 1s—for three variables, the largest group of adjacent 1s consists of four 1s. (We are ignoring the trivial case that every box has a 1—this corresponds to the Boolean function $F = 1$.)

Example 11.11

The map for a Boolean function F_1 is shown in Fig. 11.50. The group of two 1s located in approximately the center of the map corresponds to the term $B\overline{C}D$. Since the pair of 1s in the upper right-hand corner is adjacent to the 1s in the lower right-hand corner, these 1s form a group of four. The corresponding term is $\overline{B}C$. Since the remaining 1 in the lower left-hand corner can be grouped with the 1 in the lower right-hand corner, the corresponding term is $A\overline{B}\overline{D}$. Therefore, the simplified sum-of-products form for the given function is

$$F_1 = B\overline{C}D + \overline{B}C + A\overline{B}\overline{D}$$

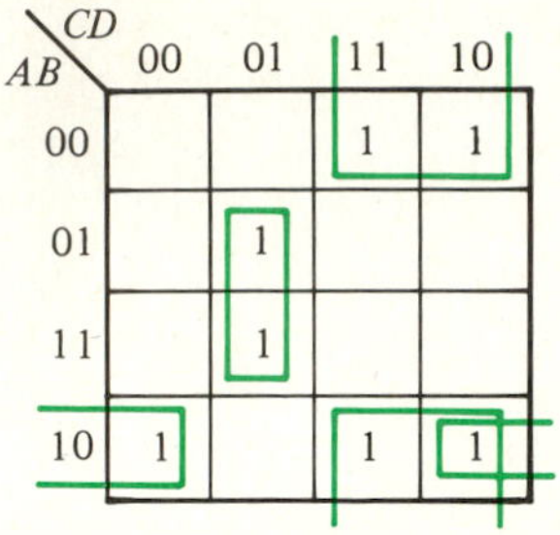

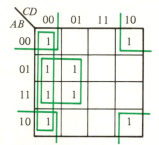

Fig. 11.50 Example of a map for a function with four variables.

Another Boolean function F_2 has the Karnaugh map shown in Fig. 11.51. In this case, two groups of four 1s should be immediately obvious. One group consists of the four 1s comprising the column on the far left. The corresponding term is $\overline{C}\overline{D}$. Another group consists of the four 1s representing minterms m_4, m_5, m_{12}, and m_{13}. This group corresponds to the term $B\overline{C}$. What remains is the pair of 1s in the upper right-hand corner and the lower right-hand corner of the map. Even though these 1s are adjacent and form a group of two, note that they are also adjacent to the 1s in the upper and lower left-hand corners of the map. Therefore, the 1s in the corners of the map form a group of four 1s as illustrated—the corresponding term being $\overline{B}\overline{D}$. Hence, the simplified form of F_2 is

$$F_2 = \overline{C}\overline{D} + B\overline{C} + \overline{B}\overline{D}$$

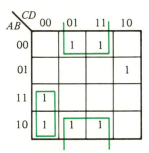

Fig. 11.51 Map with three groups of four 1s.

Fig. 11.52 Map with a group consisting of a single 1.

For the Karnaugh map shown in Fig. 11.52, the pair of 1s on top can be grouped with the pair of 1s directly under them on the bottom as illustrated. The corresponding term is $\overline{B}D$. The pair of 1s on the extreme left form a group of two 1s whose corresponding term is $A\overline{C}\overline{D}$. The 1 on the extreme right cannot be grouped with any other 1. Since this corresponds to the minterm $m_6 = \overline{A}BC\overline{D}$, the simplified form of the function F_3 associated with the map is

$$F_3 = \overline{B}D + A\overline{C}\overline{D} + \overline{A}BC\overline{D}$$

Instead of associating the boxes of a Karnaugh map with minterms, we can associate them with maxterms. Specifically, maxterm M_i is associated with the box corresponding to the binary number i. If the first bit of the binary number i is 0 (or 1), then the first variable in the maxterm M_i is uncomplemented (or complemented), and so on. The grouping of the 0s that correspond to maxterms is performed just as was the grouping of the 1s that correspond to minterms. When writing the simplified expression in the form of a product of sums, for the bits designating a group of 0s, a 0 signifies an uncomplemented variable while a 1 signifies a complemented variable.

Example 11.12

The Boolean function F_1 whose Karnaugh map is given in Fig. 11.50 is repeated in Fig. 11.53, where the 0s are shown explicitly and the 1s are implicit. The four 0s in the center form a group that corresponds to the sum $\bar{B} + \bar{D}$. The term that corresponds to the two 0s in the upper left-hand portion of the map is $A + C + D$. Finally, the two 0s in the upper and lower right-hand corners form a group that corresponds to $B + \bar{C} + D$. Thus, an alternative expression for F_1 is the simplified product of sums

$$F_1 = (A + C + D)(B + \bar{C} + D)(\bar{B} + \bar{D})$$

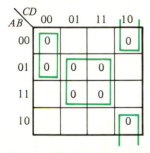

Fig. 11.53 Karnaugh map with 0s shown explicitly.

The Karnaugh map for the Boolean function F_2 described by Fig. 11.51 is repeated in Fig. 11.54 with the 0s shown explicitly. There are two groups, each containing four 0s. The terms that correspond to these groups are $\bar{B} + \bar{C}$ and $\bar{C} + \bar{D}$. The term corresponding to the group consisting of the two 0s, as illustrated, is $B + C + \bar{D}$. Thus, we have that

$$F_2 = (\bar{B} + \bar{C})(\bar{C} + \bar{D})(B + C + \bar{D})$$

The Karnaugh map given in Fig. 11.52 is redrawn in Fig. 11.55. In this case, there is only one group with four 0s, and it corresponds to $\bar{B} + \bar{D}$. The group of two 0s in the upper left corresponds to $A + C + D$, while the group in the lower right corresponds to $\bar{A} + \bar{C} + D$. The 0 in the upper right-hand corner can be

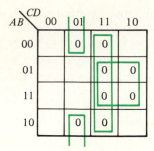

Fig. 11.54 Map with two groups of four 0s and one group of two 0s.

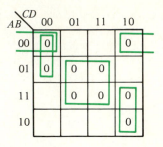

Fig. 11.55 Karnaugh map for the Boolean function F₃.

grouped with either the 0 in the upper left-hand corner or the 0 in the lower right-hand corner. Arbitrarily selecting the former, we get $A + B + D$. The result of these groupings is the simplified form

$$F_3 = (\bar{B} + \bar{D})(A + C + D)(\bar{A} + \bar{C} + D)(A + B + D)$$

◼

Even though we demonstrated how to group 0s and write the corresponding product-of-sums functional forms for four-variable Karnaugh maps, it should be obvious that essentially the same technique can be employed for three-variable maps. The details are left as exercises in the following problem set.

SUMMARY

1. The representation of a number depends upon the base of the number system selected.

2. Arithmetic can be performed in the binary number system as well as the decimal number system.

3. Nonnumerical information can be put in binary form by using codes.

4. The position of a switch (i.e., open, closed) in a circuit can be characterized by the binary digits (bits) 0 and 1.

5. AND, OR, NOT (complement), exclusive-OR, NAND, and NOR operations are examples of logic operations.

6. Logic gates are circuits that perform logic operations.

7. Switching or logic circuits are characterized by Boolean functions.

8. Boolean functions can be manipulated by using the rules of Boolean algebra.

9. Boolean functions in standard form can be implemented with two-level logic.

10. NAND and NOR gates are universal gates—either type can be used exclusively to realize arbitrary Boolean functions.

11. Boolean functions can be simplified graphically by using Karnaugh maps.

12. Grouping the 1s on a Karnaugh map produces a simplified sum-of-products expression, while grouping the 0s yields a simplified product-of-sums functional form.

Problems for Section 11.1

11.1 Make a list of the binary numbers from zero to fifteen.

11.2 Express the following binary numbers in decimal form: (a) 11011, (b) 101011, (c) 0.11011, (d) 0.10101, (e) 10100.011, (f) 10011.101.

11.3 Express the following decimal numbers in binary form: (a) 43, (b) 27, (c) 0.84375, (d) 0.65625, (e) 19.625, (f) 20.375.

11.4 Express the following octal numbers in decimal form: (a) 33, (b) 53, (c) 0.66, (d) 0.52, (e) 24.3, (f) 23.5.

11.5 Express the following decimal numbers in octal form: (a) 43, (b) 27, (c) 0.84375, (d) 0.65625, (e) 19.625, (f) 20.375.

11.6 Express the following binary numbers in octal form: (a) 110010110.1, (b) 101100101.01, (c) 11101001.111.

11.7 Express the following octal numbers in binary form: (a) 20.375, (b) 413.702, (c) 2610.35.

11.8 Repeat Problem 11.6 converting from binary to hexadecimal numbers.

11.9 Express the following hexadecimal numbers in binary form: (a) 1983.605, (b) 2B2.A07, (c) C3B0.82D2.

Problems for Section 11.2

11.10 Find the sum $n + m$ for the following binary numbers: (a) $n = 101100$, $m = 10100$; (b) $n = 1110.01$, $m = 1001.11$; (c) $n = 101.011$, $m = 1101.01$; (d) $n = 11001$, $m = 101110$.

11.11 Find the difference $n - m$ for the binary numbers given in Problem 11.10. (Do not use complements.)

11.12 Find the product $n \times m$ for the following binary numbers: (a) $n = 110.01$, $m = 11000$; (b) $n = 101.11$, $m = 10100$; (c) $n = 11001.1$, $m = 101$; (d) $n = 10010.01$, $m = 110$.

11.13 Divide m by n, that is, find $m \div n$, for the following binary numbers: (a) $m = 10010110$, $n = 11000$; (b) $m = 1110011$, $n = 10100$; (c) $m = 11111111$, $n = 110011$; (d) $m = 11011011$, $n = 110$.

11.14 Use 1's complements to find the difference $n - m$ for the binary numbers given in Problem 11.10.

11.15 Use 2's complements to find the difference $n - m$ for the binary numbers given in Problem 11.10.

11.16 Express the following decimal numbers in BCD: (a) 1127, (b) 1940, (c) 261.64, (d) 607.83.

11.17 Express the following BCD numbers in decimal form: (a) 100101000011, (b) 0011010100100011, (c) 10011000.0110.

11.18 List the words of the binary code obtained by adding a parity bit to the eight distinct binary triples such that each resulting code word has an odd number of 1s.

Problems for Section 11.3

11.19 Determine the truth table for the connection of switches shown in Fig. P11.19. (The two distinct switches labeled A are both open when $A = 0$ and both closed when $A = 1$.) Write a mathematical expression for the output variable L in terms of the input variables A, B, and C. How is this expression related to $A(B + C)$?

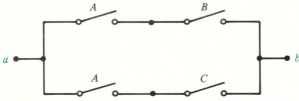

Fig. P11.19

11.20 Determine the truth table for the connection of switches shown in Fig. P11.20. Write a mathematical expression for L in terms of A, B, and C.

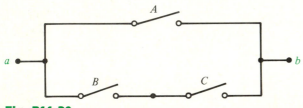

Fig. P11.20

11.21 Repeat Problem 11.19 for the connection of switches shown in Fig. P11.21. How is this expression for L related to the one obtained in Problem 11.20?

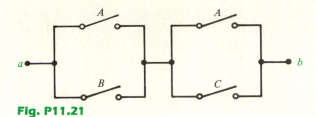

Fig. P11.21

11.22 Determine the truth table for the connection of switches shown in Fig. P11.22. Write a mathematical expression for L in terms of A, $\bar{A}$, and B. Is this expression for L related either to the OR operation $A + B$ or the AND operation AB?

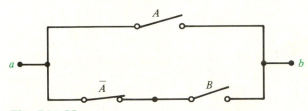

Fig. P11.22

11.23 Repeat Problem 11.22 for the connection of switches shown in Fig. P11.23.

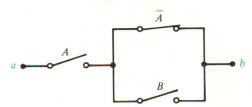

Fig. P11.23

11.24 A room has three doors and three associated switches A, B, and C. By entering any door and throwing the associated switch, the room light is to turn on. Upon leaving through any door and throwing the associated switch, the room light is to turn off. Assume that the light is off when $A = B = C = 0$.

(a) Determine the corresponding truth table.

(b) Show that

$$L = (AB + \bar{A}\bar{B})C + (A\bar{B} + \bar{A}B)\bar{C}$$

11.25 Figure P11.25 shows a **double-pole, double-throw** (DPDT) switch. This device is, in essence, two distinct SPDT switches that are operated together. Specifically, when terminals a and b are connected together, then

a' and b' are connected together. In addition, when a and c are connected together, so are a' and c'. Use one SPDT switch and two DPDT switches to construct the three-way light switch described in Problem 11.24.

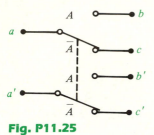

Fig. P11.25

Table P11.26

A B	A ⊙ B
0 0	1
0 1	0
1 0	0
1 1	1

11.26 The truth table given by Table P11.26 describes the exclusive-NOR (or equivalence) operation. This operation is denoted by the symbol ⊙. Express $A \odot B$ as a function of the binary variables A and B, and use this expression to realize an exclusive-NOR gate using AND, OR, and NOT gates.

11.27 Determine the function F that characterizes the output of the logic circuit shown in Fig. P11.27. Construct the corresponding truth table. Does this table describe any operation that was studied previously?

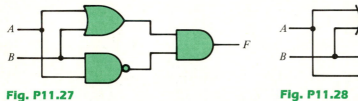

Fig. P11.27

Fig. P11.28

11.28 Repeat Problem 11.27 for the logic circuit shown in Fig. P11.28.
11.29 Repeat Problem 11.27 for the logic circuit shown in Fig. P11.29.

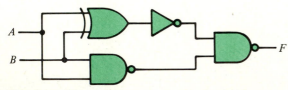

Fig. P11.29

11.30 Repeat Problem 11.27 for the logic circuit shown in Fig. P11.30.

Fig. P11.30

11.31 Repeat Problem 11.27 for the logic circuits shown in Fig. P11.31.

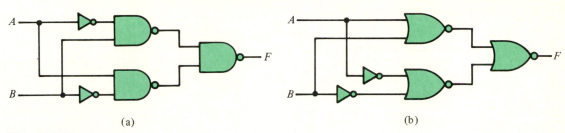

(a)   (b)

Fig. P11.31

Problems for Section 11.4

11.32 Verify the distributive law for AND over OR by constructing a truth table.

11.33 Use algebraic manipulations to simplify the following:

$$\overline{(A\bar{B} + \bar{A}B)}\,(\bar{A}\bar{B})$$

11.34 Use algebraic manipulations to simplify the following:

$$(A\bar{B} + \bar{A}B) + \overline{(A + B)}$$

11.35 Use algebraic manipulations to show that the exclusive-NOR operation has the equivalent forms

$$A \odot B = (A + \bar{B})(\bar{A} + B) = AB + \bar{A}\bar{B}$$

11.36 Use algebraic manipulations to show that the exclusive-OR operation has the equivalent forms

$$A \oplus B = A\bar{B} + \bar{A}B = (A + B)(\overline{AB})$$

11.37 Use algebraic manipulations to show that

$$AB + \bar{A}\bar{C} = (\bar{A} + B)(A + \bar{C})$$

11.38 Find the dual of the expression given in (a) Problem 11.33, (b) Problem 11.34, (c) Problem 11.35, (d) Problem 11.36, and (e) Problem 11.37.

11.39 Use algebraic manipulations to show that the exclusive-OR operation $A \oplus B = A\bar{B} + \bar{A}B$ is (a) commutative, that is, $A \oplus B = B \oplus A$; and (b) associative, that is, $A \oplus (B \oplus C) = (A \oplus B) \oplus C$.

11.40 Repeat Problem 11.39 for the exclusive-NOR operation $A \odot B = AB + \bar{A}\bar{B}$.

11.41 Show that (a) $A \uparrow B = B \uparrow A$, and (b) $A \uparrow (B \uparrow C) \neq (A \uparrow B) \uparrow C$.

Problems for Section 11.5

11.42 Construct a truth table for $F_1 = A \uparrow (B \uparrow C)$, where $\uparrow$ denotes the NAND operation. Express F_1 as (a) a sum of minterms and (b) a product of maxterms. Repeat for $F_2 = (A \uparrow B) \uparrow C$.

11.43 Realize an exclusive-OR gate using only (a) NAND gates, and (b) NOR gates. Repeat for an exclusive-NOR gate.

11.44 A **majority gate** is a logic circuit whose output is 1 when a majority (more than half) of its inputs are 1. Construct the truth table for a three-input majority logic gate. Express the Boolean function F that describes the output of the gate as (a) a sum of minterms and (b) a product of maxterms. Realize both of these expressions using AND gates and OR gates.

11.45 Show that the Boolean function $F = AB + BC + AC$ describes the output of a three-input majority gate (see Problem 11.44) by constructing the corresponding truth table. Implement this function with a two-level logic circuit employing only NAND gates.

11.46 Show that the Boolean function $F = (A + B)(B + C)(A + C)$ describes the output of a three-input majority gate (see Problem 11.44) by constructing the corresponding truth table. Implement this function with a two-level logic circuit employing only NOR gates.

11.47 Express the Boolean function $F = A + BC$ as (a) a sum of minterms and (b) a product of maxterms. Realize the expression obtained in part (a), the expression obtained in part (b), and the expression $F = A + BC$ by using AND gates and OR gates. Which realization requires the fewest logic gates?

11.48 Implement the Boolean function $F = A + BC = (A + B)(A + C)$ with a two-level logic circuit employing only (a) NAND gates and (b) NOR gates.

11.49 Use algebraic manipulations to express the Boolean function $F = AB + BC + AC$ as (a) a sum of minterms and (b) a product of maxterms.

11.50 Repeat Problem 11.49 for the function $F = \bar{A}\bar{B} + B\bar{C}$.

Problems for Section 11.6

11.51 Find the simplified sum-of-products form of the Boolean function (a) $\bar{A}\bar{B}C + \bar{A}B\bar{C} + A\bar{B}\bar{C} + ABC$; (b) $\bar{A}\bar{B}C + \bar{A}B\bar{C} + \bar{A}B C + A\bar{B}C + ABC$; (c) $\bar{A}BC + A\bar{B}C + AB\bar{C} + ABC$; and (d) $\bar{A}B\bar{C} + \bar{A}BC + \bar{A}B\bar{C} + AB\bar{C}$.

11.52 Find the simplified product-of-sums forms of the Boolean functions given in Problem 11.51.

11.53 Find the simplified sum-of-products forms of the Boolean functions that correspond to the Karnaugh maps shown in Fig. P11.53.

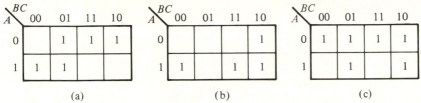

Fig. P11.53

11.54 Find the simplified product-of-sums forms of the Boolean functions corresponding to the Karnaugh maps given in Fig. P11.53.

11.55 Find the simplified product-of-sums form of the Boolean functions:

(a) $(A + B + C)(A + B + \overline{C})(A + \overline{B} + \overline{C})(\overline{A} + B + \overline{C})(\overline{A} + \overline{B} + \overline{C})$

(b) $(A + B + \overline{C})(A + \overline{B} + \overline{C})(A + \overline{B} + C)$

(c) $(A + B + \overline{C}) \, (\overline{A} + B + \overline{C})(\overline{A} + \overline{B} + \overline{C})(\overline{A} + \overline{B} + C)$

11.56 Find the simplified sum-of-products forms of the Boolean functions given in Problem 11.55.

11.57 Find the simplified sum-of-products forms of the Boolean functions that correspond to the Karnaugh maps shown in Fig. P11.57.

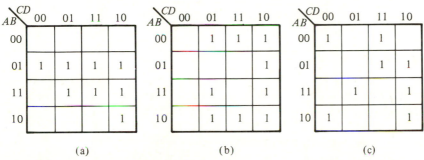

Fig. P11.57

11.58 Find the simplified product-of-sums forms of the Boolean functions corresponding to the Karnaugh maps given in Fig. P11.57.

11.59 Find the simplified sum-of-products form of the Boolean function (a) $\overline{A}(\overline{B} + \overline{C}) + \overline{A}B + ABC$; (b) $A(B + \overline{D}) + \overline{B}(C + AD)$; and (c) $\overline{B}D + \overline{A}B\overline{C} + A(\overline{B}C + B\overline{C})$.

11.60 Find the simplified product-of-sums forms of the Boolean functions given in Problem 11.59.

Chapter 12

Logic Design

INTRODUCTION

In the preceding chapter, we had the opportunity to analyze and design some relatively simple logic circuits. In this chapter, we consider the analysis and design of more complicated logic circuits—with the emphasis on design. In addition to the classical approach of designing logic circuits with individual logic gates that are available as small-scale-integration (SSI) devices, we will also study the design of logic circuits by using medium-scale-integration (MSI) and large-scale-integration (LSI) packages. Examples of these are binary and BCD adders, magnitude comparators, encoders, decoders, multiplexers, demultiplexers, read-only memories (ROMs), and programmable logic arrays (PLAs). Devices such as these, that see application in numerous types of digital systems, especially digital computers, will be studied in this chapter.

In addition to the devices mentioned above, the logic circuits encountered in the preceding chapter are examples of combinational logic. For such a circuit, the output at a particular time is determined directly from the combination of inputs applied at that time. Although important, a combinational logic circuit is not the only type of logic circuit. In this chapter, we will also discuss sequential logic circuits (or sequential circuits for short). Such a circuit contains storage or memory elements, typically in the form of flip-flops, and the output of the circuit at a particular time depends on the contents of the memory as well as the combination of inputs applied at that time. The chapter concludes with a discussion of flip-flops and the analysis and design of sequential circuits.

12.1 COMBINATIONAL LOGIC

A logic circuit whose output (or outputs) at any given time is determined directly from the combination of inputs applied at that time is referred to as a **combinational logic circuit** or **combinational logic.** All the logic circuits we have encountered so far are examples of combinational logic. Later we will study noncombinational logic.

Given a combinational logic circuit, the process of **analysis** consists of determining a Boolean function (or functions), constructing a truth table, or formulating a verbal description that characterizes the circuit. As an example of the analysis of a combinational logic circuit, consider the two-level NAND-gate circuit shown in Fig. 12.1. As indicated, the outputs of the two-input NAND gates are, from top to bottom, $\overline{AB}$, $\overline{BC}$, and $\overline{AC}$ respectively. Thus, the Boolean function F describing the output of the circuit is

$$F = \overline{(\overline{AB})(\overline{BC})(\overline{AC})} \tag{12.1}$$

From this expression, we can construct a truth table such as Table 12.1. From this table, we have alternative expressions for F—one being the sum of minterms

$$F = m_3 + m_5 + m_6 + m_7 = \overline{A}BC + A\overline{B}C + AB\overline{C} + ABC \tag{12.2}$$

and one being the product of maxterms

$$F = M_0 M_1 M_2 M_4 = (A + B + C)(A + B + \overline{C})(A + \overline{B} + C)(\overline{A} + B + C) \tag{12.3}$$

The corresponding Karnaugh maps are given in Figs. 12.2(a) and (b) respectively. For the former, the simplified expression for F is

$$F = AB + BC + AC \tag{12.3}$$

while for the latter it is

$$F = (A + B)(B + C)(A + C) \tag{12.4}$$

For this function, however, constructing a truth table and Karnaugh maps can be avoided by using the De Morgan theorem $\overline{ABC} = \overline{A} + \overline{B} + \overline{C}$. Applying this to Equation (12.1) yields

$$F = \overline{(\overline{AB})(\overline{BC})(\overline{AC})} = \overline{\overline{AB}} + \overline{\overline{BC}} + \overline{\overline{AC}} = AB + BC + AC$$

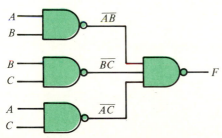

Fig. 12.1 Two-level NAND-gate logic circuit.

Table 12.1 Truth Table for $F = \overline{(\overline{AB})(\overline{BC})(\overline{AC})}$.

A	B	C	AB	$\overline{AB}$	BC	$\overline{BC}$	AC	$\overline{AC}$	$(\overline{AB})(\overline{BC})(\overline{AC})$	F
0	0	0	0	1	0	1	0	1	1	0
0	0	1	0	1	0	1	0	1	1	0
0	1	0	0	1	0	1	0	1	1	0
0	1	1	0	1	1	0	0	1	0	1
1	0	0	0	1	0	1	0	1	1	0
1	0	1	0	1	0	1	1	0	0	1
1	1	0	1	0	0	1	0	1	0	1
1	1	1	1	0	1	0	1	0	0	1

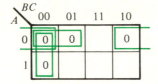

(a) Karnaugh map for
$F = m_3 + m_5 + m_6 + m_7$

(b) Karnaugh map for
$F = M_0\, M_1\, M_2\, M_4$

Fig. 12.2

Furthermore,

$AB + BC + AC$

$= (AB + BC + A)(AB + BC + C)$	since $A + BC = (A + B)(A + C)$
$= (A + BC)(AB + C)$	since $A + AB = A$
$= (A + B)(A + C)(A + C)(B + C)$	since $A + BC = (A + B)(A + C)$
$= (A + B)(B + C)(A + C)$	since $AA = A$

where commutivity was also employed even though it was not explicitly stated. Therefore, algebraic manipulation is the simplest type of analysis in this case.

Adders

The process of going from a verbal description, a truth table, or a Boolean function (or functions) describing a logic circuit to a circuit diagram is known as **synthesis** or **design.** Analysis is unique in the sense that a given circuit behaves in a particular way. Conversely, synthesis is not unique because, in general, more than one logic circuit can produce a desired behavior. As an example of synthesis, let us synthesize a combinational logic circuit that will add binary numbers, that is, let us design an adder.

One way to describe the addition of two n-bit binary numbers

$A = A_{n-1}A_{n-2} \cdots A_1A_0$ and $B = B_{n-1}B_{n-2} \cdots B_1B_0$ is with a truth table. As a consequence, there are $2n$ variables $(A_0, A_1, A_2, \cdots, A_{n-2}A_{n-1}, B_0, B_1, B_2, \cdots, B_{n-1})$ associated with the table. In other words, such a table has 2^{2n} rows. Just to be able to add numbers between 0 and 500 requires that $n = 9$ (because $2^9 = 512$), and, therefore, such a truth table has $2^{2n} = 2^{18} = 262{,}144$ rows. Hence, we should take a different approach.

What we will do, instead, is approach the problem from a bit-by-bit point of view. Specifically, recall the rules for binary arithmetic for single bits. They are

$$
\begin{array}{cccc}
0 & 0 & 1 & 1 \\
+\,0 & +\,1 & +\,0 & +\,1 \\
\hline
0 & 1 & 1 & 10
\end{array}
$$

For the first three cases, addition results in a sum—but no carry—while in the fourth case $(1 + 1)$, there is a sum (0) and a carry (1). A truth table describing this process of addition is given by Table 12.2, where A and B are the binary variables to be added, S is the sum, and C is the carry. From this table, we see that the resulting Boolean functions for S and C are

$$S = \bar{A}B + A\bar{B} = A \oplus B$$
$$C = AB$$

We now recognize that the expression for S describes the exclusive-OR operation, while C indicates the AND operation. Therefore, these functions can be implemented simply with an exclusive-OR gate and an AND gate, respectively, as shown in Fig. 12.3(a). This logic circuit with two inputs A and B, and two outputs S and C, is known as a **half-adder.** A block diagram for a half-adder is shown in Fig. 12.3(b).

Table 12.2 Truth Table for the Addition of 2 Bits.

A	B	S	C
0	0	0	0
0	1	1	0
1	0	1	0
1	1	0	1

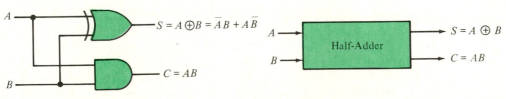

(a) Half-adder   (b) Block diagram of half-adder

Fig. 12.3

Now let us see how we can extend the concept of a half-adder. First consider the example of adding $(93)_{10} = (1011101)_2$ to $(121)_{10} = (1111001)_2$ as shown below.

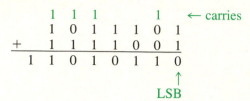

```
    1  1  1        1      ← carries
    1  0  1  1  1  0  1
 +  1  1  1  1  0  0  1
 1  1  0  1  0  1  1  0
                 ↑
                LSB
```

In determining the **least significant bit** (LSB)—the bit on the extreme right—2 bits are added together. This can be accomplished with a half-adder. However, to obtain any other bit, there must be the capability to deal with carries. In general, given two n-bit binary numbers $A_{n-1}A_{n-2} \ldots A_3A_2A_1A_0$ and $B_{n-1}B_{n-2} \ldots B_3B_2B_1B_0$, the process of addition is described as follows:

$$
\begin{array}{l}
C_nC_{n-1}C_{n-2} \ldots C_3C_2C_1 \\
\quad A_{n-1}A_{n-2} \ldots A_3A_2A_1A_0 \\
+ \quad B_{n-1}B_{n-2} \ldots B_3B_2B_1B_0 \\
\hline
S_n S_{n-1} S_{n-2} \ldots S_3 S_2 S_1 S_0
\end{array}
$$

Here S_i is the sum of A_i, B_i, and C_i for $i = 1, 2, 3, \ldots, n - 1$, and C_{i+1} is the resulting carry. This means that the sums $S_1, S_2, S_3, \ldots, S_{n-2}, S_{n-1}$ are formed as are carries $C_2, C_3, \ldots, C_{n-2}, C_{n-1}, C_n$, according to Table 12.3(a). Furthermore, Table 12.3(b) indicates how to obtain S_0 and C_1. (This is a half-adder truth table.) In addition, $S_n = C_n$. The result of adding the two given binary numbers is the binary number $S_n S_{n-1}S_{n-2} \ldots S_3S_2S_1S_0$.

Table 12.3(a) Truth Table for S_i and C_{i+1}, for $i = 1, 2, \ldots, n - 1$.

A_i	B_i	C_i	S_i	C_{i+1}
0	0	0	0	0
0	0	1	1	0
0	1	0	1	0
0	1	1	0	1
1	0	0	1	0
1	0	1	0	1
1	1	0	0	1
1	1	1	1	1

Table 12.3(b) Truth Table for S_0 and C_1.

A_0	B_0	S_0	C_1
0	0	0	0
0	1	1	0
1	0	1	0
1	1	0	1

From Table 12.3(a), we have that

$$S_i = m_1 + m_2 + m_4 + m_7 = \bar{A}_i\bar{B}_iC_i + \bar{A}_iB_i\bar{C}_i + A_i\bar{B}_i\bar{C}_i + A_iB_iC_i \qquad \textbf{(12.5)}$$

and

$$C_{i+1} = m_3 + m_5 + m_6 + m_7 = \overline{A}_i B_i C_i + A_i \overline{B}_i C_i + A_i B_i \overline{C}_i + A_i B_i C_i \quad \textbf{(12.6)}$$

The Karnaugh map for S_i is given in Fig. 12.4. From this map, we see that the expression for S_i cannot be simplified as a sum of products. On the other hand, the expression for C_{i+1} has the form of Equation (12.2) and the Karnaugh map in Fig. 12.2(a), and, therefore, can be written as is Equation (12.3). In other words, we can put C_{i+1} in the form

$$C_{i+1} = A_i B_i + B_i C_i + A_i C_i \quad \textbf{(12.7)}$$

The expressions for S_i and C_i can be implemented by using the two-level AND-OR logic circuits shown in Fig. 12.5. A logic circuit, such as the one given in Fig. 12.5, whose inputs are A_i, B_i, C_i and whose outputs are S_i and C_{i+1} as described by Equations (12.5) and (12.7), respectively, is called a **full-adder.** A block diagram for a full-adder is shown in Fig. 12.6.

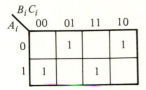

Fig. 12.4 Karnaugh map for S_i.

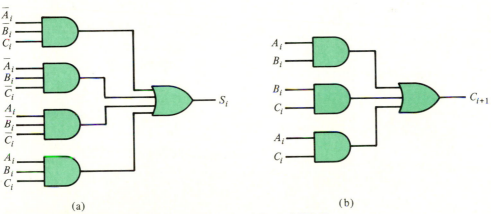

(a) (b)

Fig. 12.5 AND-OR implementation of (a) S_i and (b) C_{i+1}.

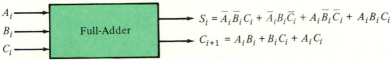

Fig. 12.6 Block diagram for a full-adder.

By connecting a half-adder and $n - 1$ full-adders as shown in Fig. 12.7, we get a combinational logic circuit that will add any two n-bit binary numbers $A_{n-1}A_{n-2} \ldots A_3A_2A_1A_0$ and $B_{n-1}B_{n-2} \ldots B_3B_2B_1B_0$.

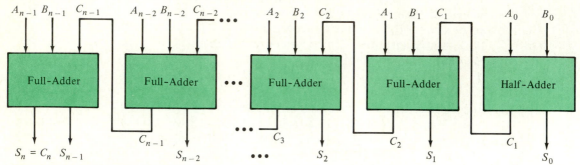

Fig. 12.7 An n-bit binary adder.

An important alternative to the full-adder implementation given in Fig. 12.5 may also be derived. Although Equation (12.5) is in simplified sum-of-products form, we can manipulate S_i algebraically as follows:

$$S_i = \overline{A}_i\overline{B}_iC_i + \overline{A}_iB_i\overline{C}_i + A_i\overline{B}_i\overline{C}_i + A_iB_iC_i$$
$$= \overline{A}_i(\overline{B}_iC_i + B_i\overline{C}_i) + A_i(\overline{B}_i\overline{C}_i + B_iC_i)$$

But note that

$$\overline{B}_i\overline{C}_i + B_iC_i = \overline{\overline{\overline{B}_i\overline{C}_i + B_iC_i}} = \overline{(\overline{\overline{B}_i\overline{C}_i})(\overline{B_iC_i})}$$
$$= \overline{(B_i + C_i)(\overline{B}_i + \overline{C}_i)} = \overline{B_i\overline{B}_i + B_i\overline{C}_i + \overline{B}_iC_i + C_i\overline{C}_i}$$
$$= \overline{B_i\overline{C}_i + \overline{B}_iC_i}$$

If we define $D_i = \overline{B}_iC_i + B_i\overline{C}_i$, then $\overline{D}_i = \overline{\overline{B}_iC_i + B_i\overline{C}_i} = \overline{B}_i\overline{C}_i + B_iC_i$ and we can write

$$S_i = \overline{A}_iD_i + A_i\overline{D}_i = A_i \oplus D_i$$

where

$$D_i = \overline{B}_iC_i + B_i\overline{C}_i = B_i \oplus C_i$$

In other words, we can write S_i as

$$S_i = A_i \oplus (B_i \oplus C_i)$$

However, since the exclusive-OR operation is associative (see Problem 11.39), it is unambiguous to write

$$S_i = A_i \oplus B_i \oplus C_i$$

In addition to this, from Equation (12.6), we have

$$C_{i+1} = \overline{A}_i B_i C_i + A_i \overline{B}_i C_i + A_i B_i \overline{C}_i + A_i B_i C_i$$

$$= (\overline{A}_i B_i + A_i \overline{B}_i)C_i + A_i B_i(\overline{C}_i + C_i) = (A_i \oplus B_i)C_i + A_i B_i$$

Using these expressions for S_i and C_{i+1}, we now see that the logic circuit shown in Fig. 12.8, which is composed of two half-adders and an OR gate, realizes a full-adder.

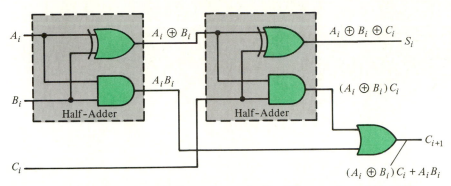

Fig. 12.8 Full-adder consisting of two half-adders and an OR gate.

Subtractions can be accomplished by taking complements and then performing addition (see Section 11.2). Because of this fact, adders can also be used for finding the difference between two binary numbers.

Don't-Care Conditions

Figure 12.9(a) shows either an LCD (liquid-crystal display) or an LED (light-emitting diode) seven-segment display that is used to indicate any of the decimal digits 0, 1, 2, 3, 4, 5, 6, 7, 8, 9. Figure 12.9(b) shows how the decimal digits can be formed by using the various segments of the display. For example, the decimal digit 1 is comprised of the segments labeled b and c, while the digit 2 consists of the segments labeled a, b, d, e, and g.

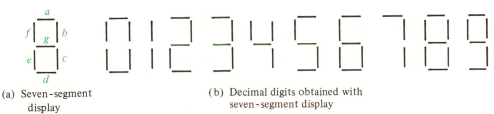

(a) Seven-segment display

(b) Decimal digits obtained with seven-segment display

Fig. 12.9

Assume that a segment of the display is visible when the bit 1 is applied to it and is invisible when a 0 is applied to it. Given the BCD representation of decimal digits, suppose that we wish to have such a digit shown on the display. This situation is described by the truth table given in Table 12.4. The input variables A, B, C, and D give the BCD representation of a digit. The seven output functions a, b, c, d, e, f, and g indicate whether or not the corresponding segment is to be visible. For example, the segment labeled a is to be visible for the digits 0, 2, 3, 5, 6, 7, 8, and 9. Thus, there are 1s in column a only in the rows corresponding to these digits. Similarly, column b has 1s in the rows corresponding to the digits 0, 1, 2, 3, 4, 7, 8, and 9.

Table 12.4 Truth Table Characterizing the Seven-Segment Display.

A	B	C	D	a	b	c	d	e	f	g
0	0	0	0	1	1	1	1	1	1	0
0	0	0	1	0	1	1	0	0	0	0
0	0	1	0	1	1	0	1	1	0	1
0	0	1	1	1	1	1	1	0	0	1
0	1	0	0	0	1	1	0	0	1	1
0	1	0	1	1	0	1	1	0	1	1
0	1	1	0	1	0	1	1	1	1	1
0	1	1	1	1	1	1	0	0	0	0
1	0	0	0	1	1	1	1	1	1	1
1	0	0	1	1	1	1	1	0	1	1
1	0	1	0	×	×	×	×	×	×	×
1	0	1	1	×	×	×	×	×	×	×
1	1	0	0	×	×	×	×	×	×	×
1	1	0	1	×	×	×	×	×	×	×
1	1	1	0	×	×	×	×	×	×	×
1	1	1	1	×	×	×	×	×	×	×

In the discussion above, we assumed that information expressed as decimal digits is to be displayed, and such digits are in BCD form. Let us also assume that a binary quadruple that is not the BCD representation of a decimal digit is not produced. Therefore, since the binary quadruples in the last six rows of Table 12.4 (under the variables A, B, C, D) cannot occur, then it does not matter whether a 0 or a 1 is indicated for each output function. Because of this, the symbol × is used, and this is referred to as a **don't-care condition.**

The Karnaugh map for the Boolean function a is shown in Fig. 12.10(a), where the don't-care conditions are included. When simplifying the expression for a, each × can be used as a 0 or as a 1—whichever yields the simplest expression. In this case, by making each × a 1, we can form two groups of eight and two groups of four, as indicated. The result is that

$$a = A + C + BD + \overline{B}\,\overline{D}$$

and a two-level logic circuit that implements this Boolean function is shown in Fig. 12.11(a)

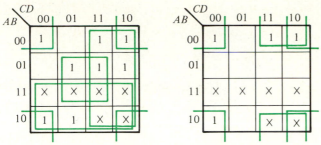

(a) Karnaugh map for function a (b) Karnaugh map for function e

Fig. 12.10

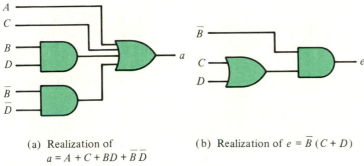

(a) Realization of
$a = A + C + BD + \bar{B}\,\bar{D}$

(b) Realization of $e = \bar{B}\,(C + D)$

Fig. 12.11

As another example, the Karnaugh map for the Boolean function e is shown in Fig. 12.10(b). In this case, two groups of four encompass all the 1s, and this is accomplished by utilizing only two of the six don't-care conditions. As a result, we have that

$$e = \bar{B}C + \bar{B}D = \bar{B}(C + D)$$

and this function can be implemented as shown in Fig. 12.11(b).

Under the circumstance that binary quadruples that do not correspond to the BCD representations of decimal digits are allowed to appear, all of the don't-care conditions indicated by an $\times$ in Table 12.4 should be changed to 0s. Doing this will result in an invisible (blank) display when an invalid (non-BCD) quadruple arises.

Integrated-Circuit Gates

Ordinarily, logic gates in integrated-circuit (IC) form are used in logic design. Such ICs typically come in two types of packages—the **dual-in-line package** (DIP) shown in Fig. 12.12(a) and the **flat package** shown in Fig. 12.12(b). Because it is easily installed in circuit boards or sockets, and is low in price, the former IC package is much more popular than the latter. The number of pins on a package can

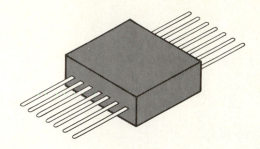

(a) Dual-in-line package (DIP)   (b) Flat package

Fig. 12.12 IC packages.

be as few as eight and as many as 64—the 14-pin and 16-pin packages being quite common. The dimensions of a typical 14-pin DIP are $20 \times 8 \times 5$ millimeters, while a 16-pin flat package can be $20 \times 10 \times 3$ millimeters.

The number of gates on an IC chip indicates what type of device that IC is. If an IC consists of 30 or fewer gates, it is called a **small-scale-integration** (SSI) device. A **medium-scale-integration** (MSI) device contains between 30 and 300 gates, while a **large-scale-integration** (LSI) device has between 300 and 3000 gates. An IC with more than 3000 gates is referred to as a **very-large-scale-integration** (VLSI) device.

Even though there are basically two types of digital logic IC packages, the IC itself can be a member of one of a number of logic-circuit families. The most popular logic families are **transistor-transistor logic** (TTL), **emitter-coupled logic** (ECL), **metal-oxide semiconductor** (MOS), **complementary metal-oxide semiconductor** (CMOS), and **integrated-injection logic** (I^2L).

The labeling on a package indicates to which family that IC belongs. ICs belonging to the 5400 series or the 7400 series are TTL ICs. The ICs in the 5400 series operate over a wider temperature range (-55 to $125°C$) than those in the 7400 series (0 to $70°C$), and, therefore, the 5400 series is designated for military use. On the other hand, ICs in the 7400 series are suitable for industrial applications. The 7400 ICs are designated 7400, 7401, 7402, 7403, and so on. Figure 12.13 shows the functional description of a 7400 TTL IC. This device is a quad 2-input NAND gate, that is, the IC contains four distinct 2-input NAND gates as illustrated. The top view of the IC shows that pins 1 and 2 are the inputs to one NAND gate and pin 3 is the output of that gate. The pin connections for the other NAND gates are also displayed. Pin 7 is labeled GND to indicate that it is to be connected to the reference or "ground." A supply voltage of $V_{CC} = 5$ V connected to pin 14 is required for proper operation. As another example of a TTL IC and its pin connections, Fig. 12.14 shows the 7427 triple 3-input NOR gate.

The 10,000 series is a common ECL designation. An ECL 10107 triple 2-input exclusive-OR/NOR gate is shown in Fig. 12.15. For this particular IC, the typical value for the negative supply-voltage pin is $V_{EE} = 5.2$ V, while pins 1 and 16 are connected to ground. Note that this IC is in a 16-pin DIP. Furthermore, each individual gate has two outputs, the one without the small circle indicating the

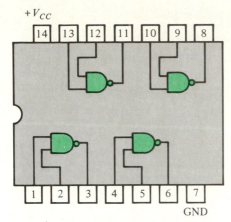

Fig. 12.13 A 7400 quad 2-input NAND gate (TTL).

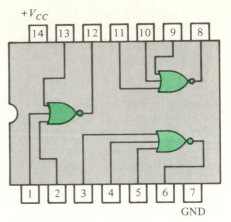

Fig. 12.14 A 7427 triple 3-input NOR gate (TTL).

exclusive-OR function and the other (with the small circle) designating the exclusive-NOR function. Figure 12.16 shows a CMOS IC in the 4000 series. This 4002 dual 4-input NOR gate utilizes only 12 of the 14 pins of the DIP. Specifically, pins 6 and 8 are labeled NC to indicate "no connection." The supply voltage V_{DD} for this IC can be any value in the range from 3 to 15 V.

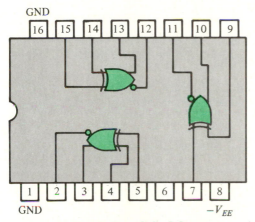

Fig. 12.15 A 10107 triple 2-input exclusive-OR/NOR gate (ECL).

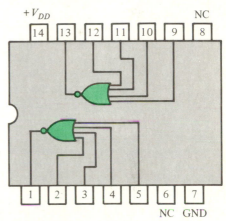

Fig. 12.16 A 4002 dual 4-input NOR gate (CMOS).

Each IC family uses different values for supply voltages, and they have different values of low- and high-voltage levels.* Table 12.5 summarizes the typical values for the TTL, ECL, and CMOS families.

---

*For positive logic, the low voltage is logical 0 and the high voltage is logical 1; conversely for negative logic.

Table 12.5 Comparison of Typical Values Used for IC Families.

Family	Basic Gate	Supply Voltage	Low-Voltage Level	High-Voltage Level
TTL	NAND	5 V	0.2 V	3.4 V
ECL	OR/NOR	−5.2 V	−1.8 V	−0.8 V
CMOS	NOT	3 to 15 V	0 V	3 to 15 V

Not only are there different IC families, but there are also different versions of the same family. In particular, there are a number of forms of TTL, such as standard, high-power standard (designated H), low-power standard (L), Schottky (S), and low-power Schottky (LS). The specific version of an IC is indicated by placing the appropriate letter designation after its series number. For example, the low-power-Schottky version of the 7400 quad 2-input NAND gate is designated 74LS00, and the high-power-standard version is 74H00. The Schottky version of the 7427 triple 3-input NOR gate is designated 74S27, and the standard-version labeling remains 7427.

Typical characteristics for the basic gate of different versions of TTL, as well as ECL and CMOS, are listed in Table 12.6. Recall from Chapter 7 that the maximum number of similar output gates that can load a given gate, and still have that gate function properly, is called the **fan-out** of the gate. Furthermore, the minimum noise voltage at the input of a gate that causes the output to deviate from its proper value is known as the **noise margin** of the gate. The **power dissipation** of a gate is the average power that is delivered to that gate by the supply voltage (power supply). The **propagation delay** is the average time delay for a voltage pulse to propagate through the gate.

The classical approach to logic design, as described earlier, is to write a Boolean function (or functions) and implement it using individual logic gates. As

Table 12.6 Characteristics of IC Logic Families.

Logic Family	Fan-out	Noise Margin, V	Power Dissipation, mW	Propagation Delay, ns
Standard TTL	10	0.4	10	10
High-power standard TTL	10	0.4	22	6
Low-power standard TTL	20	0.4	1	33
Schottky TTL	10	0.4	22	3
Low-power Schottky TTL	20	0.4	2	10
ECL	25	0.2	25	2
CMOS	50	3	0.1	25

we have just discussed, these gates are available in a wide variety of SSI packages that are fabricated from different IC families. (The MOS and I^2L families generally see LSI and VLSI applications.) There can be, however, many instances in which commercially available MSI or LSI packages can be utilized for the design process. It may be that such packages produce the exact implementations that are desired, or that such realizations can be obtained by appropriately connecting ICs together. This approach to design is our next topic of study.

12.2 MSI AND LSI DESIGN

In the preceding section, we saw how to design an n-bit binary adder (see Fig. 12.7) using a half-adder and $n - 1$ full-adders. To be able to include the case of a possible previous carry, the half-adder can be replaced by a full-adder. Figure 12.17 shows this more general **binary parallel adder** for the case that $n = 4$. Such a device is available as an MSI package (e.g., the TTL 74283 IC), as are other n-bit ($n \neq 4$) binary parallel adders. By connecting such ICs together appropriately, larger adders can be formed. For example, by connecting two 4-bit adders as shown in Fig. 12.18, we get an 8-bit binary parallel adder.

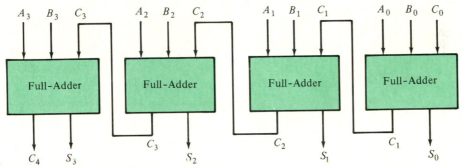

Fig. 12.17 A 4-bit binary parallel adder.

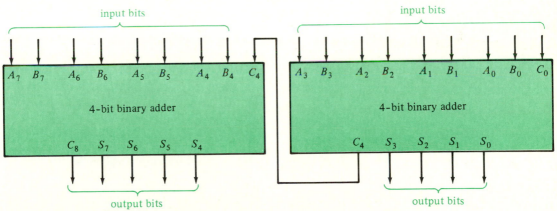

Fig. 12.18 An 8-bit binary parallel adder.

Not only can adders be used to construct larger adders, but they also can be used for other types of applications as well. An example of such a situation is the conversion from BCD to the **excess-3 code**. Table 12.7 shows the BCD representations of the decimal digits and the corresponding words of the excess-3 code. Note that the words of the excess-3 code can be obtained by adding the binary number 3 (0011) to the corresponding BCD representations of the decimal digits. Because of this property, we can use a 4-bit binary parallel adder if we make $B_0 = B_1 = 1$ and $B_2 = B_3 = C_0 = 0$. Doing this will result in the desired addition, and thus produce outputs that are the appropriate excess-3 code words. An MSI implementation of such a code converter is shown in Fig. 12.19.

Table 12.7 Excess-3 Code-Word Correspondence with BCD.

BCD				Excess-3 Code			
A_3	A_2	A_1	A_0	S_3	S_2	S_1	S_0
0	0	0	0	0	0	1	1
0	0	0	1	0	1	0	0
0	0	1	0	0	1	0	1
0	0	1	1	0	1	1	0
0	1	0	0	0	1	1	1
0	1	0	1	1	0	0	0
0	1	1	0	1	0	0	1
0	1	1	1	1	0	1	0
1	0	0	0	1	0	1	1
1	0	0	1	1	1	0	0

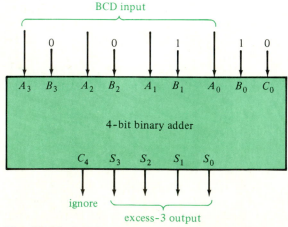

Fig. 12.19 A BCD-to-excess-3-code converter.

BCD Adders

There is an alternative to the binary addition of numbers—specifically, an alternative is decimal addition. By this, we mean the addition of decimal digits in BCD form.

Suppose that we perform binary addition on the BCD representations of two decimal digits. If the sum of the two digits is 9 or less, then performing binary addition results in the appropriate number. However, if the sum of the digits is greater than 9, then what results from the addition is the binary number representation of the sum, not the BCD representation. Table 12.8 lists the sums that result from the binary addition of decimal digits in BCD form, and it also shows the BCD versions of these sums. Even though $9 + 9 = 18$, the table extends to include sums equaling 19 so that carries from a preceding addition can be included.

Table 12.8 Sums of BCD Digits in Binary Form and BCD Form.

Sums in Binary Form					Sums in BCD Form				
S_4	S_3	S_2	S_1	S_0	D_4	D_3	D_2	D_1	D_0
0	0	0	0	0	0	0	0	0	0
0	0	0	0	1	0	0	0	0	1
0	0	0	1	0	0	0	0	1	0
0	0	0	1	1	0	0	0	1	1
0	0	1	0	0	0	0	1	0	0
0	0	1	0	1	0	0	1	0	1
0	0	1	1	0	0	0	1	1	0
0	0	1	1	1	0	0	1	1	1
0	1	0	0	0	0	1	0	0	0
0	1	0	0	1	0	1	0	0	1
0	1	0	1	0	1	0	0	0	0
0	1	0	1	1	1	0	0	0	1
0	1	1	0	0	1	0	0	1	0
0	1	1	0	1	1	0	0	1	1
0	1	1	1	0	1	0	1	0	0
0	1	1	1	1	1	0	1	0	1
1	0	0	0	0	1	0	1	1	0
1	0	0	0	1	1	0	1	1	1
1	0	0	1	0	1	1	0	0	0
1	0	0	1	1	1	1	0	0	1

By applying two decimal digits in BCD form, say $A_3A_2A_1A_0$ and $B_3B_2B_1B_0$, to a 4-bit binary adder (see Fig. 12.17), we obtain a sum $S_4S_3S_2S_1S_0$ in binary form, where $S_4 = C_4$ is the carry bit. If this binary number is 9 or less, then this sum is also in BCD form. However, if this binary number is greater than 9, we must modify the sum to get its BCD form. From Table 12.8, we see that if we add the binary number 6 (0110) to the sum in binary form, we get the sum in BCD form. But how can we state this in logical terms?

From Table 12.8, we have that the binary number 6 should be added to a sum in binary form when $S_4 = 1$, OR $S_3 = 1$ AND $S_2 = 1$, OR $S_3 = 1$ AND $S_1 = 1$. This description corresponds to the Boolean function

$$F = S_4 + S_3S_2 + S_3S_1$$

As a consequence of this discussion, we can realize a **BCD adder,** that is, a logic circuit that adds decimal digits in BCD form, by using two 4-bit binary adders as shown in Fig. 12.20. Two decimal digits in BCD form are added together in binary by the top 4-bit adder. If the resulting binary number $S_4S_3S_2S_1S_0$ is 9 or less, then $S_4S_3S_2S_1S_0 = S_3S_2S_1S_0$, $F = 0$, and 0 is added to $S_3S_2S_1S_0$ by the bottom 4-bit adder. However, if the resulting binary number $S_4S_3S_2S_1S_0$ is greater than 9, then $F = 1$ and 6 is added to $S_3S_2S_1S_0$ by the bottom 4-bit adder. Because it is a 4-bit adder and S_4 is not applied to its input, the output carry D_4 cannot be obtained for

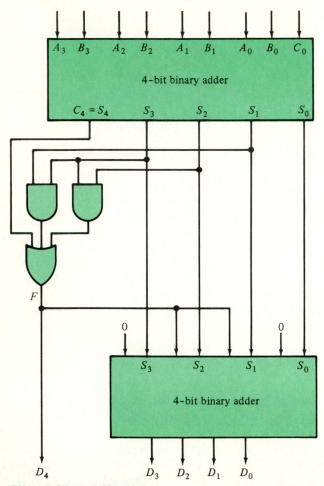

Fig. 12.20 A BCD adder.

the sums 16, 17, 18, and 19. However, this is no problem since a carry results when $F = 1$. Therefore, we can use the output of the OR gate to produce D_4. Note, too, that the bottom 4-bit adder employs no carry from a preceding stage, and, therefore, no input for such a carry is indicated.

Even though we have spent time discussing how to construct them from two 4-bit adders, BCD adders are available as MSI packages (e.g., the TTL 82S83 IC). Despite this, it is worthwhile studying how to combine MSI circuits that implement one kind of function, to produce logic circuits that realize another type of function.

Magnitude Comparators

Another very important digital logic function is performed by a **magnitude comparator,** and such a device is also available in MSI form (e.g., the TTL 7485 IC is a 4-bit magnitude comparator). As its name suggests, the function that is performed is the comparison of two numbers A and B, and the determination of whether $A < B$, $A > B$, or $A = B$.

Given two n-bit binary numbers $A = A_{n-1} \cdots A_2A_1A_0$ and $B = B_{n-1} \cdots B_2B_1B_0$, an n-bit magnitude comparator therefore will have $2n$ inputs and 3 outputs C, D, and E, which signify $A < B$, $A > B$, and $A = B$ respectively. (For example, $C = 0$, $D = 1$, $E = 0 \Rightarrow A > B$.) A truth table characterizing magnitude comparison has 2^{2n} rows—therefore, the construction of such a table will be avoided.

Specifically, let us consider the case that $n = 3$. In other words, suppose that we wish to compare the 3-bit binary numbers $A = A_2A_1A_0$ and $B = B_2B_1B_0$. The equality of a pair of bits A_i and B_i can be determined from Table 12.9. From this truth table, we see that $A_i = B_i$ is described by the Boolean expression $F_i = \bar{A}_i\bar{B}_i + A_iB_i = A_i \odot B_i$. We wish to have the output $C = 1$ when $A < B$. This inequality occurs when $A_2 = 0$ ($\bar{A}_2 = 1$) AND $B_2 = 1$, OR when $A_1 = 0$ ($\bar{A}_1 = 1$) AND $B_1 = 1$ given that (AND) $A_2 = B_2$ ($F_2 = 1$), OR when $A_0 = 0$ ($\bar{A}_0 = 1$) AND $B_0 = 1$ given that (AND) $A_2 = B_2$ AND $A_1 = B_1$ ($F_2 = 1$ AND $F_1 = 1$). This description can be characterized by the Boolean function

$$C = \bar{A}_2B_2 + \bar{A}_1B_1F_2 + \bar{A}_0B_0F_1F_2$$

where $F_1 = \bar{A}_1\bar{B}_1 + A_1B_1 = A_1 \odot B_1$ and $F_2 = \bar{A}_2\bar{B}_2 + A_2B_2 = A_2 \odot B_2$ are exclusive-NOR functions.

Table 12.9 Truth Table for the Equality of a Pair of Bits.

A_i	B_i	F_i
0	0	1
0	1	0
1	0	0
1	1	1

Proceeding in a similar manner for the case that $A > B$, we get the Boolean function

$$D = A_2\bar{B}_2 + A_1\bar{B}_1F_2 + A_0\bar{B}_0F_1F_2$$

Finally, for the case that $A = B$ (i.e., $A_2 = B_2$ AND $A_1 = B_1$ AND $A_0 = B_0$), we have that

$$E = F_2F_1F_0 = (A_2 \odot B_2)(A_1 \odot B_1)(A_0 \odot B_0)$$

A logic circuit for a 3-bit magnitude comparator is shown in Fig. 12.21.

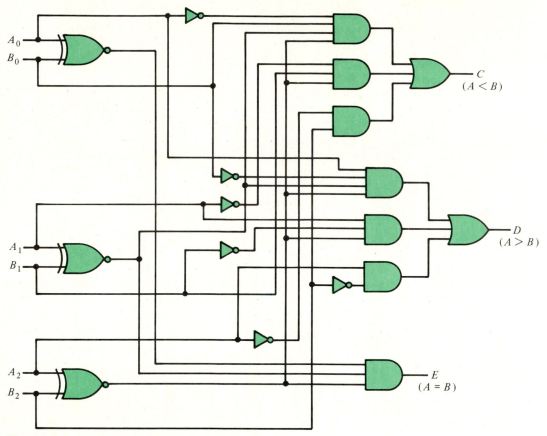

Fig. 12.21 A 3-bit magnitude comparator.

Encoders and Decoders

In Section 11.2, we mentioned that various pieces of information could be represented by a code, that is, a set of binary sequences. As a simple example, consider the four pieces of information A_1, A_2, A_3, and A_4. To have four distinct binary sequences, each code word must consist of a minimum of 2 bits. Even though we can use sequences of more than 2 bits, for the sake of brevity, let us pick code words with 2 bits as given in Table 12.10. This table indicates that the code words for A_1, A_2, A_3, and A_4 are 00, 01, 10, and 11, respectively.

Table 12.10 A Binary Encoding.

Pieces of Information				Code Words	
A_1	A_2	A_3	A_4	B_1	B_2
1	0	0	0	0	0
0	1	0	0	0	1
0	0	1	0	1	0
0	0	0	1	1	1

Suppose that we wish to construct a logic circuit to encode each piece of information. In other words, when $A_1 = 1$, we want 00 to be produced; when $A_2 = 1$, we want 01 to result, and so on. We will also assume that only one piece of information is encoded at a time (e.g., if $A_3 = 1$, then $A_1 = A_2 = A_4 = 0$). Under these conditions, from Table 12.10, we can write the Boolean expressions

$$B_1 = A_3 + A_4$$

and

$$B_2 = A_2 + A_4$$

A combinational logic circuit that implements these functions is shown in Fig. 12.22. (Note that input A_1 is not connected to anything.) Such a device is called an **encoder.** In general, an encoder for m pieces of information has m inputs (one for each piece of information) and n outputs, where $2^n \geq m$. This can be called an **m-to-n-line** (designated $m \times n$) **encoder.**

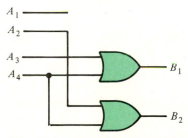

Fig. 12.22 Encoder corresponding to Table 12.15.

Now let us consider the situation that two or more inputs are equal to 1 at the same time. For such a case, an encoder's output has no meaning—two pieces of information cannot be encoded at the same time. Therefore, let us assign a priority to the information to be encoded. For the example above, let the subscript of the information indicate its priority, that is, A_1 has the highest priority, A_2 is second, A_3 is third, and A_4 is last. Thus, if more than one input is 1, the piece of information with the highest priority is encoded. These conditions are described by Table 12.11. Here the symbol $\times$ indicates a don't-care condition. For example, if $A_3 = 1$, it doesn't matter whether $A_4 = 0$ or $A_4 = 1$; since A_3 has priority over A_4, the resulting encoding is 10. From Table 12.11, we can write the Boolean expressions

$$B_1 = \bar{A}_1\bar{A}_2A_3 + \bar{A}_1\bar{A}_2\bar{A}_3A_4 = \bar{A}_1\bar{A}_2(A_3 + \bar{A}_3A_4) = \bar{A}_1\bar{A}_2(A_3 + A_4)$$

and

$$B_2 = \bar{A}_1A_2 + \bar{A}_1\bar{A}_2\bar{A}_3A_4 = \bar{A}_1(A_2 + \bar{A}_2\bar{A}_3A_4) = \bar{A}_1(A_2 + \bar{A}_3A_4)$$

A logic circuit for such an encoder, known as a **priority encoder,** is shown in Fig. 12.23.

Table 12.11 A Priority Encoding.

Pieces of Information				Code Words	
A_1	A_2	A_3	A_4	B_1	B_2
1	$\times$	$\times$	$\times$	0	0
0	1	$\times$	$\times$	0	1
0	0	1	$\times$	1	0
0	0	0	1	1	1

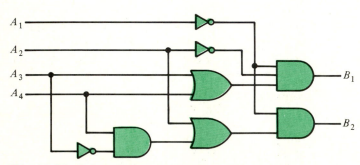

Fig. 12.23 A 4-to-2-line priority encoder.

Because of its simplicity of construction, ordinary (nonpriority) encoders do not come in the form of MSI packages. However, priority encoders are available as MSI units. For example, the 74147 IC is a 10-to-4-line decimal-to-BCD priority encoder, and the IC 74148 is an 8-to-3-line octal-to-binary priority encoder.

The converse of the process of encoding is called **decoding.** In general, an **n-to-m-line decoder** has n inputs and m outputs, where $2^n \geq m$. For each n-bit input, there is a unique output that is equal to 1—the remaining outputs being equal to 0.

An example of a 3-to-8-line decoder is a device whose input is a 3-bit binary number between 0 and 7, and whose output is the corresponding octal digit. The truth table describing this decoder is given by Table 12.12, where D_0 through D_7 designate the octal digits 0 through 7 respectively. From this table, we see that

$$D_i = m_i \qquad \text{for } i = 0, 1, 2, 3, 4, 5, 6, 7$$

Therefore, each output function is one of the minterms of the input variables. The implementations of these Boolean expressions with logic circuits should be a routine exercise by now, so they are omitted. However, do note that each output function can be implemented with one (3-input) AND gate.

Table 12.12 Truth Table for a (3-to-8-Line) Binary-to-Octal Decoder.

Inputs			Outputs							
A	B	C	D_0	D_1	D_2	D_3	D_4	D_5	D_6	D_7
0	0	0	1	0	0	0	0	0	0	0
0	0	1	0	1	0	0	0	0	0	0
0	1	0	0	0	1	0	0	0	0	0
0	1	1	0	0	0	1	0	0	0	0
1	0	0	0	0	0	0	1	0	0	0
1	0	1	0	0	0	0	0	1	0	0
1	1	0	0	0	0	0	0	0	1	0
1	1	1	1	0	0	0	0	0	0	1

Unlike (nonpriority) encoders, decoders are available in MSI form (e.g., the TTL 74138 IC is a 3-to-8-line decoder).

The 3×8 decoder characterized by Table 12.12 is an example of a decoder with n inputs and $m = 2^n$ outputs. The output functions for such a device are all of the minterms of the input variables. Since any Boolean function can be expressed as a sum of minterms, this type of decoder can be connected to an OR gate to implement any Boolean function.

Example 12.1

From Equations (12.5) and (12.6), we know that a full-adder is characterized by the Boolean functions

$$S_i = m_1 + m_2 + m_4 + m_7$$

and

$$C_{i+1} = m_3 + m_5 + m_6 + m_7$$

Therefore, we can use OR gates and a 3-to-8-line decoder such as the one described by Table 12.12 to implement a full-adder. Figure 12.24 gives the realization.

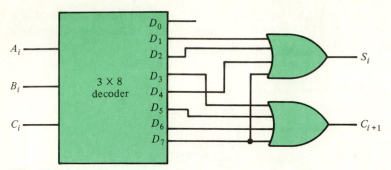

Fig. 12.24 Realization of a full-adder with a 3-to-8-line decoder.

◼

Multiplexers and Demultiplexers

A **digital multiplexer** is a logic circuit typically with 2^n inputs, one output, and n selection variables or lines. For such a device, the n selection lines are used to make the output equal to one of the inputs. Because of this, a digital multiplexer (designated MUX) is also referred to as a **data selector.**

Figure 12.25 shows a realization of a 4-to-1-line multiplexer (designated 4×1 MUX). For this case ($n = 2$), there are four input lines labeled I_0, I_1, I_2, I_3; two selection lines labeled A, B; and one output line F. Note that when $A = B = 0$, then the output of the top AND gate is $\bar{A}\bar{B}I_0 = I_0$ and the output of each remaining AND gate is 0. Thus, $F = I_0$. Similarly, when $A = 0$ and $B = 1$, then the output of the second (from the top) AND gate is $\bar{A}BI_1 = I_1$ and each of the other AND gates has an output of 0. Thus, $F = I_1$. Continuing in this manner, we see that the output is

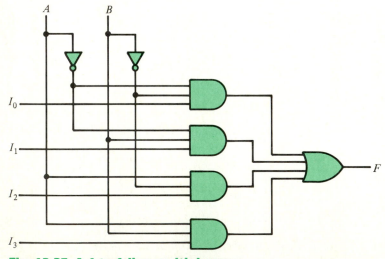

Fig. 12.25 A 4-to-1-line multiplexer.

equal to one of the inputs—which one depends on the selection-line values. These results are summarized by Table 12.13. The MSI representation of a 4-to-1-line multiplexer is shown in Fig. 12.26. Note that a Boolean expression for the output F is given by

$$F = \overline{A}\overline{B}I_0 + \overline{A}BI_1 + A\overline{B}I_2 + ABI_3 \qquad\qquad (12.8)$$

Table 12.13 Table Characterizing a 4-to-1-Line Multiplexer.

Selection Lines		Output
A	B	F
0	0	I_0
0	1	I_1
1	0	I_2
1	1	I_3

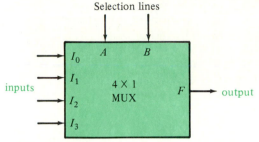

Selection lines

inputs

output

Fig. 12.26 Circuit symbol of a 4-to-1-line multiplexer.

Let us now see how we can use a multiplexer to implement Boolean functions. Suppose that we would like to realize the function F characterized by Table 12.14. From this truth table, we can write

$$F = \overline{A}BC + A\overline{B}\overline{C} + AB\overline{C} + ABC = \overline{A}BC + A\overline{B}\overline{C} + AB(\overline{C} + C)$$

$$F = \overline{A}\overline{B}(0) + \overline{A}BC + A\overline{B}\overline{C} + AB(1) \qquad\qquad (12.9)$$

Therefore, by putting the expression for F [Equation (12.9)] into the form of Equation (12.8), we can make the associations

$$I_0 = 0 \qquad I_1 = C \qquad I_2 = \overline{C} \qquad I_3 = 1$$

This means that the 4-to-1-line multiplexer shown in Fig. 12.27 implements the Boolean function F described by Table 12.14.

Table 12.14 Truth Table for a Boolean Function F.

A	B	C	F
0	0	0	0
0	0	1	0
0	1	0	0
0	1	1	1
1	0	0	1
1	0	1	0
1	1	0	1
1	1	1	1

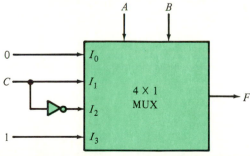

Fig. 12.27 Realization of F with a 4-to-1-line multiplexer.

The converse process of multiplexing is **demultiplexing.** Specifically, a **demultiplexer** is a device with one input, n selection lines, and 2^n outputs. For such a device, the n selection lines are used to make one of the outputs equal to the input. Figure 12.28 shows a realization of a 1-to-4-line demultiplexer. For this case ($n = 2$), there is one input line labeled E; two selection lines labeled A, B; and four output lines labeled D_0, D_1, D_2, D_3. Note that when $A = B = 0$, then the output of the top AND gate is $D_0 = \overline{A}\overline{B}E = E$ and the remaining outputs are $D_1 = D_2 = D_3 = 0$. Similarly, when $A = 0$ and $B = 1$, then $D_1 = \overline{A}BE = E$ and $D_0 = D_2 = D_3 = 0$. Continuing in this manner, we see that the input is transferred to one of the outputs—which one depends on the selection-line values. These results are summarized by Table 12.15. The MSI representation of a 1-to-4-line demultiplexer is shown in Fig. 12.29.

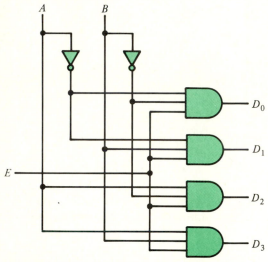

Fig. 12.28 A 1-to-4-line demultiplexer.

Table 12.15 Table Characterizing a 1-to-4-line demultiplexer.

Selection Lines		Outputs			
A	B	D_0	D_1	D_2	D_3
0	0	E	0	0	0
0	1	0	E	0	0
1	0	0	0	E	0
1	1	0	0	0	E

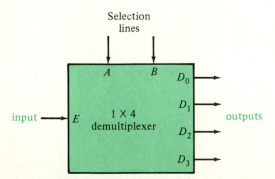

Fig. 12.29 Circuit symbol of a 1-to-4-line demultiplexer.

From Table 12.15, note that if we always have the condition that $E = 1$, then the 1-to-4-line demultiplexer becomes a 2-to-4-line decoder. On the other hand, if we always have $E = 0$, then the device does nothing—all the outputs are 0 regardless of the values of the selection lines. Thus, we can use a demultiplexer as a decoder; to do this, we just set $E = 1$ and we use the selection lines as the decoder inputs. In this case, E is referred to as the **enable** input; making $E = 1$ enables decoding, whereas $E = 0$ does not. Using the 1-to-4-line demultiplexer given in Fig. 12.29 as a 2-to-4-line decoder yields the MSI representation shown in Fig. 12.30.

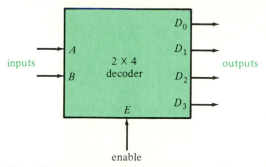

Fig. 12.30 A 2-to-4-line decoder with an enable input.

Read-Only Memories (ROMs)

Previously, we saw how decoders can be used in conjunction with OR gates to implement Boolean functions. A $2^n \times m$ **read-only memory** (ROM) is a device with an n-to-2^n-line decoder and m OR gates, each having 2^n inputs, all on the same IC. Such a device has n inputs and m outputs and is depicted in Fig. 12.31 for the

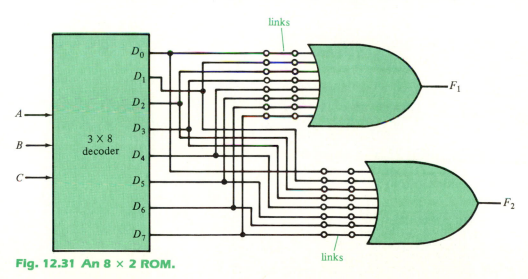

Fig. 12.31 An 8 × 2 ROM.

case that $n = 3$ and $m = 2$. In other words, Fig. 12.31 shows a $2^n \times m = 8 \times 2$ ROM. Note that at the inputs of the OR gates there are connections, referred to as **links,** that can be broken if desired. Whether or not a link is broken depends on what the Boolean functions F_1 and F_2 are to be. The decoder outputs are all of the minterms of the input variables, and any or all of these minterms can be applied to each OR gate. If a particular minterm is not to be applied to the input of an OR gate, then the corresponding link is broken. (It is assumed that an open terminal at the input of an OR gate behaves as an input of 0.)

As an example, we have already seen that the output functions of a full-adder are characterized by the Boolean expressions

$$S_i = m_1 + m_2 + m_4 + m_7$$

and

$$C_{i+1} = m_3 + m_5 + m_6 + m_7$$

Therefore, for the 8×2 ROM shown in Fig. 12.31 to realize a full-adder, if $F_1 = S_i$ and $F_2 = C_{i+1}$, then for the upper OR gate the links connected to the decoder outputs D_0, D_3, D_5, and D_6 are broken. In addition, for the lower OR gate, the links connected to D_0, D_1, D_2, and D_4 are broken.

Establishing a pattern of connections between the outputs of the decoder and the inputs of the OR gates by breaking links is referred to as **programming** the ROM. Since complicated Boolean expressions can be realized with a programmed ROM, such functions can be implemented with a single IC. This eliminates the need for interconnecting a number of ICs.

A $2^n \times m$ ROM has n inputs and m outputs. Thus, there are 2^n possible n-bit input combinations, and each one is called an **address.** Each m-bit output combination is called a **word** or **output word.** For every address, therefore, there corresponds an output word, and such a word is "read" (appears at the output) when the corresponding address (input) is applied. Consequently, it is said that a ROM stores 2^n words—hence, the use of the word "memory." Because there are 2^n m-bit words stored in the ROM, we say that it contains $2^n \times m$ bits. For instance, the ROM shown in Fig. 12.31 contains $2^3 \times 2 = 16$ bits, so it is referred to as a 16-bit ROM.

A ROM with 15 inputs and 8 outputs contains $2^{15} \times 8 = 32,768 \times 8 = 262,144$ bits, and, therefore, is a 262,144-bit ROM. Clearly, describing a ROM in this manner is awkward. However, by using the letter K to designate the number $2^{10} = 1024$, we can call such a ROM a 256K-bit ROM, or 256K ROM for short. Furthermore, since 1 byte $= 8$ bits, we may also refer to it as a 32K-byte ROM.

A ROM may be programmed by its manufacturer during its fabrication process —typically the last step. In this case, the user of the ROM is required to specify the characteristics of the particular device by giving the manufacturer the truth table for the ROM. Such a procedure is expensive, but if a large number of the same ROM are to be produced, then such a procedure is economical.

An alternative to a ROM that is programmed by its manufacturer is a **programmable** ROM, designated PROM. For a PROM, all the links are initially intact, and the device is programmed by applying current pulses (through output terminals) to break the appropriate links.

If a ROM is programmed in the manufacturing process, its programming cannot be changed by its user. In addition, once a link of a PROM has been broken, it cannot be replaced. It is in this sense that the programming of a ROM or PROM is irreversible. There is, however, another type of ROM—an **erasable PROM,** designated EPROM. For one type of such a device, the programming can be erased by placing it under a certain kind of ultraviolet light for a specified period of time. For another type, erasure can be accomplished electrically; such a device may be referred to as an **electrically alterable ROM,** or EAROM for short.

Programmable Logic Arrays (PLAs)

Suppose that 64 pieces of information are encoded by using a code with 12-bit binary sequences.* Furthermore, suppose that these code words are to be converted into 6-bit binary sequences, one for each of the original pieces of information. If a ROM is used for this purpose, the device requires 12 inputs and 6 outputs, that is, a $2^{12} \times 6$ (24K-bit) ROM is needed. Even though such a device has $2^{12} = 4096$ addresses and can accommodate 4096 words, only 64 of these words are utilized. This is considered a wasteful use of a ROM.

Another type of LSI device that is similar to a ROM is the **programmable logic array** (PLA). A PLA does not contain a decoder that generates all possible min-terms as does a ROM. Instead of a decoder, a PLA contains a group of AND gates, and the inputs of each AND gate are connected to every input variable and the complement of every variable through links. If the PLA has n inputs, then each AND gate has $2n$ inputs, and $2n$ links are associated with each AND gate. If the PLA has m outputs, then there are m OR gates. If there are p AND gates, then each OR gate has p inputs; the output of every AND gate is connected to each OR gate through links. The output of each OR gate is one output of the PLA. Such a device can be called an $n \times p \times m$ PLA.

An example of a $3 \times 4 \times 2$ PLA is shown in Fig. 12.32. To use a device such as this to realize Boolean functions F_1 and F_2, we must be able to express both F_1 and F_2 as a sum of at most four products of the input variables or their complements. These products will be the outputs of the four AND gates, and the outputs of the OR gates will be F_1 and F_2. Specifically, we can realize the Boolean functions given by Table 12.16 using this PLA. From the given truth table, we get the Karnaugh

---

*Even though there are $2^6 = 64$ distinct 6-bit binary sequences, appropriately using 12 bits for each code word gives the code error-detecting and error-correcting capabilities. This can be very desirable when transmitting information in the presence of noise.

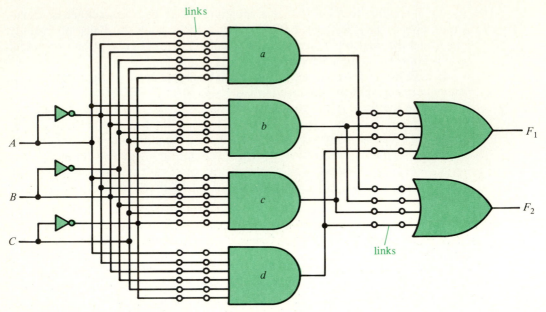

Fig. 12.32 A 3 × 4 × 2 PLA.

Table 12.16 Truth Table for Example.

A	B	C		F_1	F_2
0	0	0		1	0
0	0	1		1	1
0	1	0		0	0
0	1	1		1	1
1	0	0		0	0
1	0	1		0	1
1	1	0		1	1
1	1	1		0	0

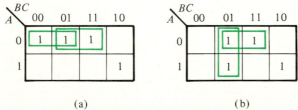

(a)   (b)

Fig. 12.33 Karnaugh maps for (a) F_1 and (b) F_2.

maps shown in Fig. 12.33. From these maps we have that

$$F_1 = \overline{A}\,\overline{B} + \overline{A}C + AB\overline{C}$$

and

$$F_2 = \overline{A}C + \overline{B}C + AB\overline{C}$$

Thus, we see that these two expressions contain four distinct products—$\overline{A}\,\overline{B}$, $\overline{A}C$, $\overline{B}C$, and $AB\overline{C}$. Therefore, to implement F_1 and F_2 with the given PLA, we need only break the appropriate links. In particular, for the AND gate labeled a, break the links associated with A, B, C, and $\overline{C}$. Then the resulting output of AND gate a is $\overline{A}\,\overline{B}$. For AND gate b, break the links A, B, $\overline{B}$, and $\overline{C}$. The resulting output is $\overline{A}C$. For AND gate c, break the links A, $\overline{A}$, B, and $\overline{C}$. The resulting output is $\overline{B}C$. Finally, for AND gate d, break the links $\overline{A}$, $\overline{B}$, and C. The resulting output is $AB\overline{C}$. (We have assumed here that an open terminal at the input of an AND gate behaves as an input of 1.) Next, for the upper OR gate, break the link connected to the output of AND gate c. Assuming that an open terminal at the input of an OR gate behaves as an input of 0, the output of the OR gate is $F_1 = \overline{A}\,\overline{B} + \overline{A}C + AB\overline{C}$. For the lower OR gate, break the link connected to the output of AND gate a. Then the output of this OR gate is $F_2 = \overline{A}C + \overline{B}C + AB\overline{C}$. Hence, we have the desired implementations.

The above example of the use of a PLA resulted in a relatively small logic circuit. From a practical point of view, implementations of such Boolean functions are better accomplished using the previously studied approaches. The same thing can be said about the example dealing with the use of a ROM. These examples were presented for purposes of demonstration. ROMs and PLAs are useful for the design of large combinational logic circuits—typically the former for the case that most or all of the addresses (possible input combinations) are utilized, and the latter for the case that only a relatively few addresses are needed. As with a ROM, a PLA can be programmed by the manufacturer or the user. For the latter situation, the device is referred to as a **field-programmable logic array** (FPLA). A typical $n \times p \times m$ PLA, such as the TTL 82S100 IC, has $n = 16$, $p = 48$, and $m = 8$.

12.3 SEQUENTIAL LOGIC

Up to this point, our concern has been focused on combinational logic. For this type of circuit, the output (or outputs) at a particular time is determined by a combination of the inputs applied at that time. A more general situation of a logic circuit is to have the output determined not only from a combination of the inputs applied at a particular time, but from the condition or "state" of the circuit at that time as well. We refer to this type of logic as **sequential logic,** and we may represent a general sequential logic circuit by the block diagram shown in Fig. 12.34. Here the information stored in the memory elements at a particular time defines the state of the sequential circuit at that time.

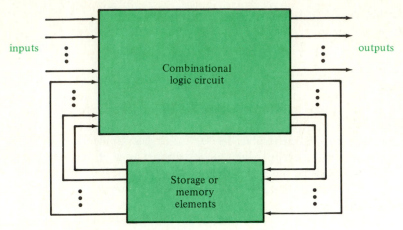

Fig. 12.34 General sequential-logic circuit.

SR Flip-Flops

Typically, sequential-circuit memory elements are **flip-flops.*** Such a device stores 1 bit of information, either a 0 or a 1, and has two outputs—one is equal to the bit stored, and the other is equal to its complement.

Consider the connection of the two NOR gates shown in Fig. 12.35. This logic circuit has two inputs labeled R and S, and two outputs labeled Q and $\overline{Q}$. The output Q indicates the state of the circuit—$Q = 1$ ($\overline{Q} = 0$) is called the **set state** and $Q = 0$ ($\overline{Q} = 1$) is called the **reset** or **clear state.** Such a device is known as a **set-reset** (SR) **flip-flop**[†] or an **SR latch.** Let us analyze this sequential circuit.

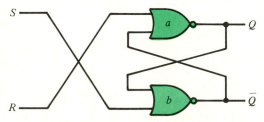

Fig. 12.35 An SR (set–reset) flip-flop.

Suppose that at some time the flip-flop is in the reset state $Q = 0$ ($\overline{Q} = 1$). If $R = S = 0$, then the output of NOR gate a is $\overline{R + \overline{Q}} = \overline{0 + 1} = 0$ and the output of NOR gate b is $\overline{Q + S} = \overline{0 + 0} = 1$. Thus, when the flip-flop is in the reset state $Q = 0$ and the input $R = S = 0$ is applied, the flip-flop remains in the reset state. Similarly, it can be shown that if the flip-flop is in the set state $Q = 1$ ($\overline{Q} = 0$), then when $R = S = 0$ is applied, the flip-flop remains in the set state.

---

*A flip-flop is a **bistable multivibrator.**
[†]Also called an RS flip-flop.

Now suppose that the flip-flop is in the reset state $Q = 0$ ($\overline{Q} = 1$). If $R = 1$ and $S = 0$, then the output of NOR gate a is $\overline{R + \overline{Q}} = \overline{1 + 1} = 0$ and the output of NOR gate b is $\overline{Q + S} = \overline{0 + 0} = 1$. Thus, the flip-flop remains in the reset state. However, if $R = 0$ and $S = 1$, then the output of NOR gate b becomes $\overline{Q + S} = \overline{0 + 1} = 0$—and 0 is the new value for $\overline{Q}$. This, in turn, causes the output of NOR gate a to be $\overline{R + \overline{Q}} = \overline{0 + 0} = 1$—and 1 is the new value for Q. Does this then affect the new output value of NOR gate b? Since $\overline{Q + S} = \overline{1 + 1} = 0$, it does not. Thus, the flip-flop changes to the set state. In summary, if the flip-flop is in the reset state ($Q = 0$), then the inputs $R = 1$ and $S = 0$ cause the flip-flop to remain in the reset state ($Q = 0$), while the inputs $R = 0$ and $S = 1$ cause the flip-flop to change to the set state ($Q = 1$).

Similarly, it can be shown that if the flip-flop is in the set state ($Q = 1$), then the inputs $R = 0$ and $S = 1$ cause the flip-flop to remain in the set state ($Q = 1$), while the inputs $R = 1$ and $S = 0$ cause the flip-flop to change to the reset state ($Q = 0$).

Finally, if $R = S = 1$, no matter what the state of the flip-flop is, both NAND gate outputs are equal to 0. This would mean that both Q and $\overline{Q}$ would take on the value of 0. However, such a condition violates the definition of a flip-flop. Therefore, to use the logic circuit shown in Fig. 12.34 as an SR flip-flop, we will insist that the condition $R = S = 1$ not be allowed to occur.

The above discussion of the operation of the SR flip-flop shown in Fig. 12.35 is summarized by Table 12.17. In this table, Q indicates the state at some time t, and R and S designate the inputs at that time. The next state that results, denoted $Q(t + 1)$, may be the same or different from the present state Q. An input of $S = 1$ sets the next state to 1, and an input of $R = 1$ resets the next state to 0.

Table 12.17 Truth Table for SR Flip-Flop.

Inputs		Present State	Next State
S	R	Q	$Q(t + 1)$
0	0	0	0
0	0	1	1
0	1	0	0
0	1	1	0
1	0	0	1
1	0	1	1
1	1	NOT ALLOWED	

Typically, the inputs of a flip-flop are connected to the outputs of some logic circuit. It may be that for some of the time these outputs are not equal to their proper values as a result of propagation delays or spurious effects. Because of this, we can modify the SR flip-flop shown in Fig. 12.35 so that the device will ignore the inputs except over some specified time interval when it is known that they will take on their correct values. This can be done by adding two AND gates to the circuit as shown in Fig. 12.36. For this sequential circuit, however, in addition to the set (S) and reset (R) inputs, there is an input for a **clock pulse,** designated C, as shown in Fig. 12.37(a). When the clock pulse is equal to 0 (i.e., $C = 0$), then the outputs of the AND gates are 0. Such values will not change the value of Q. But when $C = 1$, then the output of AND gate a is $CS = (1) S = S$ and the output of AND gate b is $CR = (1) R = R$. These values will affect the present state Q as described by Table 12.17. (It is assumed that the clock pulse is sufficiently long that the device will be allowed to reach its appropriate next state.) The sequential circuit shown in Fig. 12.36 is known as a **clocked** (or **gated**) **SR flip-flop,** and its block-diagram circuit symbol is shown in Fig. 12.37(b). A circuit such as this, in which operations are performed in conjunction with repetitive or periodic clock pulses, is called a **synchronous sequential circuit.** Typically, these clock pulses are produced by a timing device known as a **master-clock generator.**

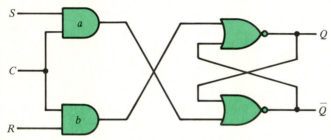

Fig. 12.36 Clocked (gated) SR flip-flop.

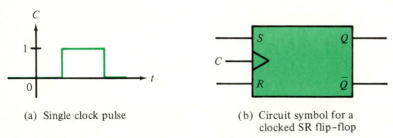

(a) Single clock pulse

(b) Circuit symbol for a clocked SR flip-flop

Fig. 12.37

When power is first supplied to a flip-flop, its state is arbitrary. By slightly modifying the SR flip-flop circuit given in Fig. 12.36, the flip-flop can be given the capability of being directly **cleared** (making $Q = 0$) or of being directly **preset** (making $Q = 1$) before a clock pulse is applied. This is accomplished by the logic circuit shown in Fig. 12.38. This circuit is the same as the one in Fig. 12.36 with the exception that there is an additional input to each NOR gate. For the upper NOR gate, this input is labeled CLR (for clear), and for the lower NOR gate it is labeled PR (for preset). When 0s are applied to both of these inputs, the circuit behaves identically to the SR flip-flop given in Fig. 12.36. In the absence of a clock pulse

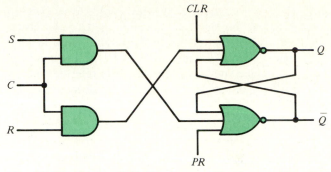

Fig. 12.38 Clocked SR flip-flop with clear and preset capability.

($C = 0$), the middle inputs of both NOR gates are 0. Making $CLR = 1$ then results in $Q = 0$ and $\overline{Q} = 1$, thereby clearing the flip-flop. On the other hand, making $PR = 1$ causes $\overline{Q} = 0$ and $Q = 1$, thereby presetting the flip-flop. Because these inputs are applied without the need for a clock pulse, they are referred to as **asynchronous** inputs. The block-diagram circuit symbol for a clocked SR flip-flop having direct clear and preset inputs is shown in Fig. 12.39.

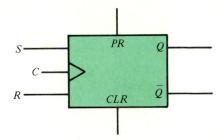

Fig. 12.39 An SR flip-flop with clear and preset inputs.

D Flip-Flops and JK Flip-Flops

Consider the synchronous sequential circuit shown in Fig. 12.40. For this circuit, the inputs to the clocked SR flip-flop are $S = D$ and $R = \overline{D}$. Note that because of this fact, we never can have the undesirable condition for an SR flip-flop that $R = S = 1$. We see from Table 12.17 for the case that the present state is $Q = 0$, when $D = S = 0$ ($R = \overline{D} = 1$) then the next state is $Q(t + 1) = 0$. For the case that the present state is $Q = 1$, when $D = S = 0$ ($R = \overline{D} = 1$), then the state is

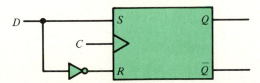

Fig. 12.40 A D flip-flop.

reset to $Q(t + 1) = 0$. Similarly, whether the present state is $Q = 0$ or $Q = 1$, when $D = S = 1$ ($R = \bar{D} = 0$), then the state is set to $Q(t + 1) = 1$. These results are summarized by Table 12.18. Therefore, for the circuit given in Fig. 12.40, an input of 0 results in a next state of 0, and an input of 1 results in a next state of 1. For these reasons, the device shown in Fig. 12.40 is referred to as a (clocked) **delay** or **data** (D) **flip-flop,** and its block diagram circuit symbol is as depicted by Fig. 12.41.

Table 12.18 Truth Table for D Flip-Flop.

Input D	Present State Q	Next State $Q(t + 1)$
0	0	0
0	1	0
1	0	1
1	1	1

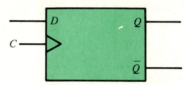

Fig. 12.41 Circuit symbol of a D flip-flop.

Let us now combine two AND gates with a clocked SR flip-flop as shown in Fig. 12.42. We designate the inputs of the resulting sequential circuit by J and K as illustrated.

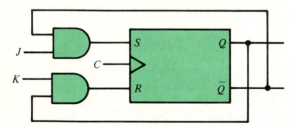

Fig. 12.42 A JK flip-flop.

For this circuit, suppose that $J = K = 0$ and $Q = 0$. Then $S = J\bar{Q} = 0(1) = 0$, $R = KQ = 0(0) = 0$, and the next state is $Q(t + 1) = 0$. When $J = K = 0$ and $Q = 1$, then $S = J\bar{Q} = 0$, $R = KQ = 0$, and the next state is $Q(t + 1) = 1$.

Next suppose that $J = 0$ and $K = 1$. When $Q = 0$, then $S = J\bar{Q} = 0(1) = 0$, $R = KQ = 1(0) = 0$, and $Q(t + 1) = 0$. When $J = 0$, $K = 1$, and $Q = 1$, then $S = J\bar{Q} = 0(0) = 0$, $R = KQ = 1(1) = 1$, and the state is reset to $Q(t + 1) = 0$.

In a similar manner, it can be seen that when $J = 1$, $K = 0$, and $Q = 0$ or $Q = 1$, then $Q(t + 1) = 1$.

The characterization thus far for the sequential circuit shown in Fig. 12.42 is given by the first six rows of Table 12.19. Since these rows are identical to the first six rows of Table 12.17, the circuit shown in Fig. 12.42 behaves as an SR flip-flop except for the case that both inputs are equal to 1. For an SR flip-flop, this condition must be avoided—however, it is allowed for the circuit given in Fig. 12.42. Specifically, suppose that $J = K = 1$. If $Q = 0$, then $S = J\overline{Q} = 1(1) = 1$, $R = KQ = 1(0) = 0$, and the state is set to $Q(t + 1) = 1$. Furthermore, if $J = K = Q = 1$, then $S = J\overline{Q} = 1(0) = 0$, $R = KQ = 1(1) = 1$, and the state is reset to $Q(t + 1) = 0$. These facts yield the last two rows in Table 12.19, and this is the truth table for the circuit shown in Fig. 12.42. Such a device is known as a **JK flip-flop,** and its block-diagram circuit symbol is given by Fig. 12.43. In summary, a JK flip-flop behaves as an SR flip-flop except that a JK flip-flop can have both inputs equal to 1. Connecting the inputs of a JK flip-flop together results in what is known as a **toggle** (T) **flip-flop** (see Problem 12.33).

Table 12.19 Truth Table for JK Flip-Flop.

Inputs		Present State	Next State
J	K	Q	$Q(t + 1)$
0	0	0	0
0	0	1	1
0	1	0	0
0	1	1	0
1	0	0	1
1	0	1	1
1	1	0	1
1	1	1	0

Fig. 12.43 Circuit symbol of a JK flip-flop.

There is a practical problem associated with the JK flip-flop just discussed. As long as the input to the flip-flop is $J = K = 1$, the state **toggles,** i.e., it changes either from 0 to 1, or from 1 to 0. Therefore, if the clock pulse is too long, there will be more than one state transition, and the final state of the flip-flop will be indeterminate. To get around this problem, instead of constructing a JK flip-flop from a single SR flip-flop as indicated in Fig. 12.42, let us consider the "master–slave" connection of two SR flip-flops shown in Fig. 12.44. Note that for this sequential circuit, when its clock input is 1 (i.e., $C_2 = 1$), the slave flip-flop simply transfers its input values to its output. This is so because, if $S_2 = 0$ and $R_2 = 1$, then when $C_2 = 1$, the flip-flop is reset to $Q_2 = 0$ and $\overline{Q}_2 = 1$. Similarly, if $S_2 = 1$ and $R_2 = 0$, then when $C_2 = 1$, the flip-flop is set to $Q_2 = 1$ and $\overline{Q}_2 = 0$.

Suppose that the circuit shown in Fig. 12.44, called a **JK master–slave flip-flop,** has inputs $J = K = 0$. Then the inputs to the master flip-flop are $S_1 = R_1 = 0$. In the absence of a clock pulse ($C = 0$), the clock input to the slave

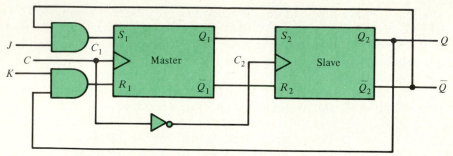

Fig. 12.44 A JK master-slave flip-flop.

is $C_2 = \overline{C}_1 = \overline{C} = 1$. Thus, the slave state is the same as the master state. When a clock pulse occurs ($C = C_1 = 1$), the master flip-flop does not change state (see Table 12.17). When the clock pulse ends ($C = 0$), then $C_2 = 1$ and the input of the slave flip-flop (output of the master flip-flop) gets transferred to the output of the slave flip-flop. Thus, the state of the slave flip-flop remains unchanged.

Now suppose that $J = 0$ and $K = 1$. When $Q = 0$, then $S_1 = J\overline{Q} = 0(1) = 0$ and $R_1 = KQ = 1(0) = 0$. This means that the master state remains unchanged, as does the slave state. However, when $Q = 1$, then $S_1 = J\overline{Q} = 0(0) = 0$ and $R_1 = KQ = 1(1) = 1$. This, in turn, means that the master flip-flop is reset to 0, and the slave flip-flop follows.

Proceeding in a similar manner, when $J = 1$ and $K = 0$, the outputs that result are $Q = 1$ and $\overline{Q} = 0$.

Finally, consider the case that $J = K = 1$. When $Q = 1$, then $R_1 = KQ = 1(1) = 1$—which means that the master and slave flip-flops are reset to 0. Conversely, when $Q = 0$, then $S_1 = J\overline{Q} = 1(1) = 1$ and both flip-flops are set to 1. This gives us the toggle operation of the JK flip-flop. However, note that the master flip-flop operates when the clock pulse is $C = 1$, while the operation of the slave flip-flop follows when $C = 0$. Because the next-state values of the flip-flop occur after the clock pulse goes to 0—and the master flip-flop is inoperative—these values cannot cause the unwanted toggling that can occur for the JK flip-flop given in Fig. 12.42. The fact that the slave flip-flop does not change state until after the clock pulse returns to 0 is why this JK master–slave flip-flop is referred to as a **trailing-edge-triggered flip-flop.**

Sequential-Circuit Analysis

As mentioned previously, the general form of a sequential circuit is as shown in Fig. 12.33, where the storage elements typically are of flip-flops. Let us consider a specific example and see what is involved in analyzing a sequential circuit.

For the sake of simplicity, the clock inputs for the two JK flip-flops in the circuit given in Fig. 12.45 are not shown. The contents (i.e., the bits stored) of the flip-flops at a particular time is said to be the **state** of the sequential circuit at that time. Therefore, we can label the state of the circuit with the binary pair Q_1Q_2. For

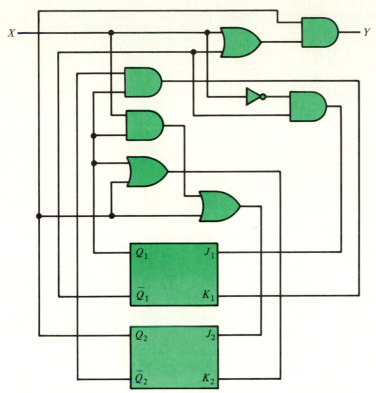

Fig. 12.45 Sequential circuit with one input, one output, and two JK flip-flops.

example, if at some time $Q_1 = 0$ and $Q_2 = 1$, then we can designate that state as 01. There is one input X to the given circuit and one output Y. Clearly, the expression relating the output Y to the input X and the flip-flop outputs Q_1 and Q_2 is

$$Y = (X + \overline{Q}_1)Q_2$$

Furthermore, we have that

$$J_1 = \overline{X}\overline{Q}_1$$

$$K_1 = Q_1\overline{Q}_2$$

$$J_2 = XQ_1 + Q_2$$

$$K_2 = Q_1 + Q_2$$

Let us now construct a table that includes all possible combinations of the input and the present state. The first three columns of Table 12.20 enumerate these combinations. Using the expressions for J_1, K_1, J_2, K_2, and Y, we can then fill in the corresponding columns in Table 12.20. Furthermore, knowing the characteristics of a JK flip-flop (Table 12.19), the columns in the table designating the next

Table 12.20 Transition Table for Given Sequential Circuit.

Input X	Present State		Flip-Flop Inputs				Next State		Output Y
	Q_1	Q_2	J_1	K_1	J_2	K_2	$Q_1(t+1)$	$Q_2(t+1)$	
0	0	0	1	0	0	0	1	0	0
0	0	1	1	0	1	1	1	0	1
0	1	0	0	1	0	1	0	0	0
0	1	1	0	0	1	1	1	0	0
1	0	0	0	0	0	0	0	0	0
1	0	1	0	0	1	1	0	0	1
1	1	0	0	1	1	1	0	1	0
1	1	1	0	0	1	1	1	0	1

state can be determined. For example, suppose that the present state of the circuit is $Q_1Q_2 = 01$ and an input of $X = 0$ is applied. This results in $J_1 = \overline{X}\overline{Q}_1 = 1(1) = 1$, $K_1 = Q_1\overline{Q}_2 = 0(0) = 0$, $J_2 = XQ_1 + Q_2 = 0(0) + 1 = 1$, $K_2 = Q_1 + Q_2 = 0 + 1 = 1$, and $Y = (X + \overline{Q}_1)Q_2 = (0 + 1)1 = 1$. The present state of the first flip-flop is $Q_1 = 0$, and an input of $J_1 = 1$, $K_1 = 0$ sets the state to $Q_1(t + 1) = 1$. The present state of the second flip-flop is $Q_2 = 1$, and an input of $J_2 = 1$, $K_2 = 1$ toggles the state to $Q_2(t + 1) = 0$. Thus, we have obtained the information for the second row of the table. Similarly, the remaining rows of the table can be determined. This table is called the **transition table** or the **state table** for the sequential circuit.

The information contained in the transition table (excluding the flip-flop inputs) can be displayed graphically by a **state diagram.** For such a diagram, each state is represented by a circle, called a **node,** that is labeled with its associated state. Thus, for the example above, there are four nodes, and they are labeled 00,01,10, and 11. If an input of X causes a state to change from Q_1Q_2 to $Q'_1Q'_2$, then there is a line, called an **edge,** directed from Q_1Q_2 to $Q'_1Q'_2$. Furthermore, if the resulting

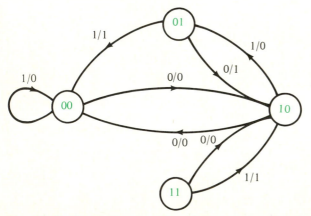

Fig. 12.46 State diagram for given sequential circuit.

output is Y, then the edge is labeled X/Y. For instance, since an input of 0 changes state 01 to 10, then there is an edge directed from node 01 to node 10, and since the resulting output is 1, then this edge is labeled $0/1$. Figure 12.46 shows the state diagram for the sequential circuit given in Fig. 12.45. From this diagram, we see that once the circuit leaves the state 11, due either to an input of 0 or of 1, there is no possible sequence of inputs that will return the circuit to that state.

A sequential circuit is completely characterized by either its transition table or its state diagram. Knowing either, the sequences of states and outputs of a sequential circuit that result from a sequence of inputs are easily obtained. Therefore, the analysis of a sequential circuit essentially consists of determining either its transition table or its state diagram.

Sequential-Circuit Design

The converse process of analysis is synthesis or design. The design of a sequential circuit consists of starting with a description of the circuit and ending with a logic-circuit diagram of the sequential circuit. Although a sequential circuit can be specified by a word description from which a state diagram may be obtained, the state diagram of a sequential circuit is often given as the initial circuit specification.

Since a sequential circuit employs flip-flops (the type of flip-flop is the choice of the designer), the behavior of different types of flip-flops is a design factor. Instead of directly using the truth tables of the flip-flops studied earlier, it is more convenient to put flip-flop descriptions in alternative forms called **excitation tables.** For example, Table 12.21(a) repeats the truth table for the SR flip-flop. The information contained in this table is used to construct the excitation table for the SR flip-flop given by Table 12.21(b). For such a table, the first column designates the present state Q and the second column designates the next state $Q(t + 1)$. The third and fourth columns, labeled S and R, indicate the inputs required to change the flip-flop's present state Q to the next state $Q(t + 1)$. Specifically, if the present state of the flip-flop is 0, to get the next state to be 0, we see from Table 12.21(a)

Table 12.21 (a) Truth Table for the SR Flip-Flop.

Inputs		Present State	Next State
S	R	Q	$Q(t + 1)$
0	0	0	0
0	0	1	1
0	1	0	0
0	1	1	0
1	0	0	1
1	0	1	1
1	1	} NOT ALLOWED	
1	1		

Table 12.21 (b) Excitation Table for the SR Flip-Flop.

Present State Q	Next State Q(t + 1)	Inputs S	R
0	0	0	×
0	1	1	0
1	0	0	1
1	1	×	0

that the inputs should be either $S = 0$, $R = 0$, or $S = 0$, $R = 1$. Therefore, the value of R is irrelevant—as long as $S = 0$ the flip-flop will go from $Q = 0$ to $Q(t + 1) = 0$. This characterization is given by the first row of Table 12.21(b). Here R is marked with the don't-care condition ×.

If an SR flip-flop is to change from the present state $Q = 0$ to the next state $Q(t + 1) = 1$, we see from Table 12.21(a) that the inputs should be $S = 1$ and $R = 0$. This condition is given by the second row of Table 12.21(b). Proceeding in this manner, the remainder of the excitation table for the SR flip-flop can be obtained. Table 12.21(b) is this excitation table.

The excitation tables for the D flip-flop and the JK flip-flop are given by Tables 12.22(b) and 12.23(b), respectively. The corresponding truth tables are reviewed by Tables 12.22(a) and 12.23(a), respectively. The confirmation of these results is left as an exercise for the reader.

Table 12.22 (a) Truth Table for the D Flip-Flop.

Input D	Present State Q	Next State Q(t + 1)
0	0	0
0	1	0
1	0	1
1	1	1

Table 12.22 (b) Excitation Table for the D Flip-Flop.

Present State Q	Next State Q(t + 1)	Input D
0	0	0
0	1	1
1	0	0
1	1	1

Table 12.23 (a) Truth Table for the JK Flip-Flop.

Inputs		Present State	Next State
J	K	Q	Q(t + 1)
0	0	0	0
0	0	1	1
0	1	0	0
0	1	1	0
1	0	0	1
1	0	1	1
1	1	0	1
1	1	1	0

Table 12.23 (b) Excitation Table for the JK Flip-Flop.

Present State Q	Next State Q(t + 1)	Inputs	
		J	K
0	0	0	×
0	1	1	×
1	0	×	1
1	1	×	0

We will now see how an excitation table is utilized by going through an example of the design of a sequential circuit. Specifically, let us design a modulo-4 binary up/down counter. Such a device has four states representing the binary numbers 0, 1, 2, and 3. An input of 1 counts up, that is, changes state from 0 to 1, or 1 to 2, or 2 to 3, or 3 to 0. An input of 0 counts down, that is, changes state from 3 to 2, or 2 to 1, or 1 to 0, or 0 to 3. The output of the counter is 0 if the next state is even (0 or 2), and it is 1 if the next state is odd (1 or 3). This description of the sequential circuit is displayed graphically by the state diagram shown in Fig. 12.47.

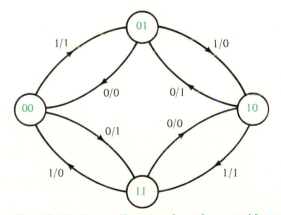

Fig. 12.47 State diagram for given up/down counter.

With the information given, we may now begin to construct the transition table for the sequential circuit. We start with columns for the input X and the present state Q_1Q_2. In these three columns, we put all possible binary triples. This is done in the first three columns of Table 12.24. The next two columns designate the next state $Q_1(t + 1) Q_2(t + 1)$. These columns, as well as the one corresponding to the output

Table 12.24 Transition Table for the Up/Down Counter.

Input X	Present State Q_1	Q_2	Next State $Q_1(t+1)$	$Q_2(t+1)$	Flip-Flop Inputs S_1	R_1	S_2	R_2	Output Y
0	0	0	1	1	1	0	1	0	1
0	0	1	0	0	0	×	0	1	0
0	1	0	0	1	0	1	1	0	1
0	1	1	1	0	×	0	0	1	0
1	0	0	0	1	0	×	1	0	1
1	0	1	1	0	1	0	0	1	0
1	1	0	1	1	×	0	1	0	1
1	1	1	0	0	0	1	0	1	0

Y, can be filled in by referring to the state diagram. For example, the third row of the table represents the condition that an input of 0 is applied when the present state is 10. From the state diagram, we see that the next state is 01 and the output is 1. These values are placed in the appropriate columns in the table.

After the next-state and output columns of the transition table have been completed, the columns for the flip-flop inputs must be filled in. The values that are entered depend on which type of flip-flop is selected. Different flip-flop types will result, in general, in different combinational logic circuits that control the flip-flops. There is no way of knowing ahead of time which type of flip-flop will produce the simplest circuits. For the given problem, let us arbitrarily select SR flip-flops. Doing so, we now employ the excitation table for the SR flip-flop [Table 12.21(b)]. For example, for the third row of the transition table, the state is to change from $Q_1Q_2 = 10$ to $Q_1(t+1)Q_2(t+1) = 01$. This means that the first flip-flop is to change its state from 1 to 0. From Table 12.21(b), the corresponding inputs are $S_1 = 0$ and $R_1 = 1$. The second flip-flop, on the other hand, is to change its state from 0 to 1. Again from Table 12.21(b), the corresponding inputs are $S_2 = 1$ and $R_2 = 0$. Hence, we have the entries in the third row in the columns for the flip-flop inputs. Proceeding in this manner, the remaining entries can be determined. These are shown in Table 12.24.

Once the columns corresponding to the flip-flop inputs have been obtained, Boolean expressions for the inputs can be found, and then the corresponding logic-circuit implementations may be realized. For the flip-flop input columns and the output column given in Table 12.24, we have the Karnaugh maps shown in Fig. 12.48. From these maps we find that

$$S_1 = \overline{X}\overline{Q_1}\overline{Q_2} + X\overline{Q_1}Q_2 = (\overline{X}\overline{Q_2} + XQ_2)\overline{Q_1}$$

$$R_1 = \overline{X}Q_1\overline{Q_2} + XQ_1Q_2 = (\overline{X}\overline{Q_2} + XQ_2)Q_1$$

$$S_2 = Y = \overline{Q_2}$$

$$R_2 = Q_2$$

Using these expressions, we obtain the sequential circuit shown in Fig. 12.49.

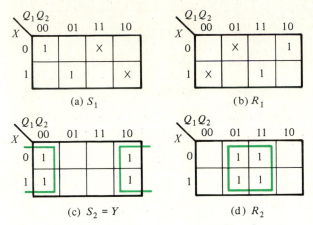

(a) S_1 (b) R_1

(c) $S_2 = Y$ (d) R_2

Fig. 12.48 Karnaugh maps for flip-flop inputs.

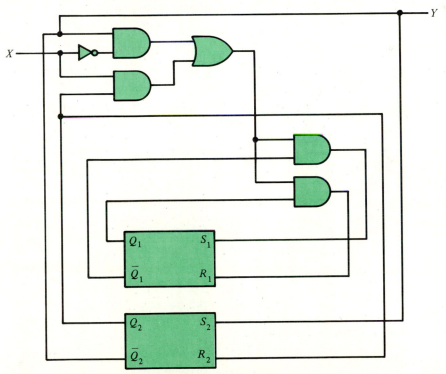

Fig. 12.49 Realization of up/down counter.

SUMMARY

1. The output of a combinational logic circuit is determined by the inputs that are applied to the circuit. The output of a sequential logic circuit depends on the state of the circuit as well as the applied inputs.

2. An n-bit binary adder can be constructed with $n - 1$ full-adders and one half-adder or with n full-adders.

3. A truth-table entry that can be either a 0 or a 1 is designated by the don't-care condition $\times$.

4. The most popular integrated-circuit (IC) package is the dual-in-line package (DIP).

5. Digital ICs are available as small-scale-integration (SSI), medium-scale-integration (MSI), large-scale-integration (LSI), and very-large-scale-integration (VLSI) devices.

6. Devices such as adders, magnitude comparators, encoders, decoders, multiplexers, demultiplexers, read-only memories (ROMs), and programmable logic arrays (PLAs) are available as MSI and LSI packages.

7. Binary adders can be combined to form larger binary adders. They also can be used to implement BCD adders.

8. A magnitude comparator determines whether one binary number is greater than, less than, or equal to another binary number.

9. An m-to-n-line ($m \times n$) encoder converts m pieces of information to binary n-tuples (sequences of n bits), where $2^n \geq m$.

10. An n-to-m-line ($n \times m$) decoder converts binary n-tuples to m pieces of information, where $2^n \geq m$.

11. A digital multiplexer (MUX) with 2^n inputs and one output employs n selection lines to make the output equal to one of the inputs.

12. A demultiplexer with one input and 2^n outputs uses n selection lines to make one of the outputs equal to the input. A demultiplexer can be used as a decoder.

13. A read-only memory (ROM) consists of a decoder and OR gates. Such devices can be programmed by the manufacturer or the user by breaking appropriate links that connect the decoder outputs to the inputs of the OR gates.

14. A PLA contains a group of AND gates that replaces the decoder in a ROM. PLAs can also be programmed by the manufacturer or the user. PLAs are preferred over ROMs when relatively few input combinations are to be utilized.

15. A flip-flop is a 1-bit memory. Common flip-flops are designated SR (set–reset), JK, D (data or delay), and T (toggle).

16. Flip-flops typically are clocked (synchronous) sequential circuits and often have inputs for direct clear and preset.

17. Flip-flop operation can be improved with a master–slave connection.

18. Flip-flops are characterized by their excitation tables as well as their truth tables.

19. The analysis of a sequential circuit amounts to finding the transition table or the state diagram of the circuit.

20. The design of a sequential circuit typically begins with a word description or a state diagram and culminates with a logic-circuit diagram.

Problems for Section 12.1

12.1 Analyze the NOR-gate logic circuit shown in Fig. P12.1 by (a) determining the Boolean function F, (b) constructing the truth table for F, (c) expressing F as a simplified sum of products, and (d) expressing F as a simplified product of sums.

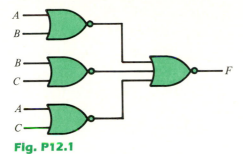

Fig. P12.1

12.2 For the NAND-gate logic circuits shown in Fig. P12.2, express F as a sum of products using only algebraic manipulations.

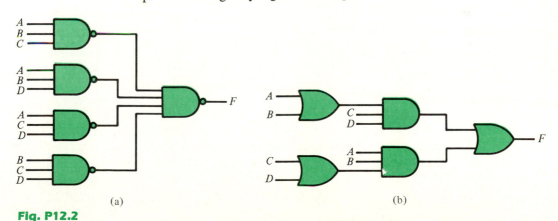

(a)   (b)

Fig. P12.2

12.3 Table P12.3 is the truth table for a **half-subtractor.** Such a device subtracts 1 bit (designated B) from another (A). The difference is $D = A - B$, where R indicates whether a borrow is required. Write expressions for D and R, and design logic circuits that realize these expressions.

Table P12.3

A	B	D	R
0	0	0	0
0	1	1	1
1	0	1	0
1	1	0	0

12.4 Table P12.4 is the truth table for a **full-subtractor.** Such a device performs subtraction and takes into consideration that there may have been a previous borrow. The difference between A_i and B_i taking a previous borrow R_i into account is designated D_i, and R_{i+1} indicates whether a borrow is required. Write expressions for D_i and R_{i+1}, and design logic circuits that realize these expressions.

Table P12.4

A_i	B_i	R_i	D_i	R_{i+1}
0	0	0	0	0
0	0	1	1	1
0	1	0	1	1
0	1	1	0	1
1	0	0	1	0
1	0	1	0	0
1	1	0	0	0
1	1	1	1	1

12.5 Show how the full-subtractor described in Problem 12.4 can be realized by using an OR gate and two of the half-subtractors (with possible slight modifications) described in Problem 12.3.

12.6 Write a simplified sum-of-products expression for the Boolean function f described in Table 12.4, and implement this function with a two-level AND-OR logic circuit. Repeat for the case that all the don't-care conditions are changed to 0s.

12.7 Repeat Problem 12.6 for the Boolean function g.

12.8 Table P12.8 shows the truth table that describes the generation of a parity bit that, when attached to its corresponding BCD quadruple, produces a binary quintuple with an even number of 1s. Write a simplified product-of-sums expression for the Boolean function P, and implement this function with a two-level AND-OR logic circuit.

Table P12.8

A	B	C	D	P
0	0	0	0	0
0	0	0	1	1
0	0	1	0	1
0	0	1	1	0
0	1	0	0	1
0	1	0	1	0
0	1	1	0	0
0	1	1	1	1
1	0	0	0	1
1	0	0	1	0
1	0	1	0	×
1	0	1	1	×
1	1	0	0	×
1	1	0	1	×
1	1	1	0	×
1	1	1	1	×

Table P12.9

BCD Input				Excess-3 Code Output			
A	B	C	D	E	F	G	H
0	0	0	0	0	0	1	1
0	0	0	1	0	1	0	0
0	0	1	0	0	1	0	1
0	0	1	1	0	1	1	0
0	1	0	0	0	1	1	1
0	1	0	1	1	0	0	0
0	1	1	0	1	0	0	1
0	1	1	1	1	0	1	0
1	0	0	0	1	0	1	1
1	0	0	1	1	1	0	0
1	0	1	0	×	×	×	×
1	0	1	1	×	×	×	×
1	1	0	0	×	×	×	×
1	1	0	1	×	×	×	×
1	1	1	0	×	×	×	×
1	1	1	1	×	×	×	×

12.9 Table P12.9 shows the truth table that describes the conversion from BCD to the excess-3 code indicated. (Note that the excess-3 code is obtained by adding the binary number 3 to each BCD quadruple.) Design logic circuits that implement the Boolean functions E, F, G, and H.

12.10 Design a logic circuit whose input is any possible binary quadruple and whose output is the 2's complement of the input.

Problems for Section 12.2

12.11 Use a 4-bit binary adder to construct an excess-3-to-BCD code converter.

12.12 Design a logic circuit whose output is 1 when two 4-bit binary numbers are equal and 0 when they are not equal.

12.13 Given two 4-bit binary numbers $A = A_3 A_2 A_1 A_0$ and $B = B_3 B_2 B_1 B_0$. A 4-bit magnitude comparator whose inputs are A and B is to have three outputs C, D, and E such that (a) $C = 1$, $D = E = 0$ when $A < B$; (b) $C = 0$, $D = 1$, $E = 0$ when $A > B$; and (c) $C = D = 0$, $E = 1$ when $A = B$. Write Boolean expressions for C, D, and E.

12.14 Design an 8 × 3 (nonpriority) encoder.

12.15 Write the Boolean expressions for the outputs B_1, B_2, and B_3 of an

8-to-3-line priority encoder having inputs A_1, A_2, A_3, A_4, A_5, A_6, A_7, and A_8, where A_i has priority over A_j if $i < j$.

12.16 (a) The truth table for a half-subtractor is given by Table P12.4. Use a 3-to-8-line decoder to realize a half-subtractor.

(b) A majority gate is a logic circuit whose output is 1 when a majority (more than half) of its inputs are 1. Use a 3-to-8-line decoder to implement a 3-input majority gate.

12.17 Table P12.17 shows the truth table that describes the conversion from binary to the **reflected** (or **Gray**) **code** indicated. Note that adjacent words in the reflected code differ by only 1 bit. Design a binary-to-Gray code converter by using a 3-to-8-line decoder.

Table P12.17

Binary			Reflected Code		
A	B	C	D	E	F
0	0	0	0	0	0
0	0	1	0	0	1
0	1	0	0	1	1
0	1	1	0	1	0
1	0	0	1	1	0
1	0	1	1	1	1
1	1	0	1	0	1
1	1	1	1	0	0

12.18 Design a BCD-to-excess-3-code converter by employing a 4-to-16-line decoder.

12.19 Realize a 3-input majority gate [see Problem 12.16(b)] using a 4-to-1-line multiplexer.

12.20 Implement the "even-parity-bit" or "three-way-light-switch" truth table given by Table P12.20 with a 4-to-1-line multiplexer.

Table P12.20

A	B	C	F
0	0	0	0
0	0	1	1
0	1	0	1
0	1	1	0
1	0	0	1
1	0	1	0
1	1	0	0
1	1	1	1

12.21 Implement a full-adder with two 4-to-1-line multiplexers.

12.22 Find the Boolean function F that corresponds to the 4-to-1-line multiplexer shown in Fig. P12.22.

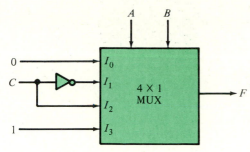

Fig. P12.22

12.23 A Boolean function F of the four variables A, B, C, and D is given by the sum of minterms

$$F = m_1 + m_2 + m_4 + m_7 + m_8$$

(See Table P12.8.) Implement this function by using an 8-to-1-line multiplexer.

12.24 Specify which links must be broken for the ROM given in Fig. 12.31 to implement the full-subtractor characterized by Table P12.24.

Table P12.24

A	B	C	F_1	F_2
0	0	0	0	0
0	0	1	1	1
0	1	0	1	1
0	1	1	0	1
1	0	0	1	0
1	0	1	0	0
1	1	0	0	0
1	1	1	1	1

12.25 For the PLA shown in Fig. 12.32, add a fifth AND gate (label it e) to form a 3 × 5 × 2 PLA. Specify which links should be broken so that this new PLA will implement the full-subtractor characterized by Table P12.24.

Problems for Section 12.3

12.26 For the logic circuit shown in Fig. P12.26, the inputs are S and R, while the ouputs are Q and $\overline{Q}$. Construct a truth table including the present state

Q and the next state $Q(t + 1)$ for this circuit. Assuming that the input condition $R = S = 1$ does not occur, what type of sequential circuit is this?

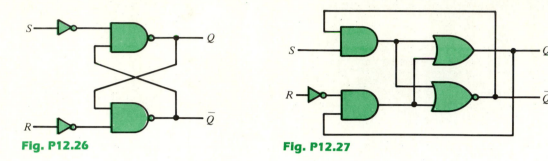

Fig. P12.26

Fig. P12.27

12.27 Repeat Problem 12.26 for the logic circuit shown in Fig. P12.27. What happens if the input condition $S = R = 1$ is allowed?

12.28 The logic circuit shown in Fig. P12.28 is an example of a **set–dominate** SR flip-flop. Construct the truth table for this device.

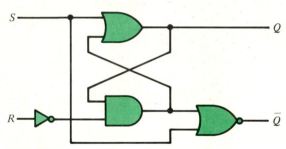

Fig. P12.28

12.29 Use the results of Problem 12.28 to design a set–dominate SR flip-flop using only (a) NAND gates and AND gates, and (b) NOR gates and OR gates.

12.30 The logic circuit shown in Fig. P12.30 is an example of a **reset–dominate** SR flip-flop. Construct the truth table for this device.

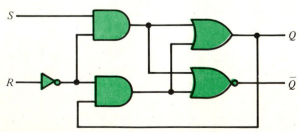

Fig. P12.30

12.31 Use the results of Problem 12.30 to design a reset–dominate SR flip-flop using only (a) NAND gates and AND gates, and (b) NOR gates and OR gates.

12.32 Repeat Problem 12.26 for the clocked sequential NAND-gate circuit shown in Fig. P12.32.

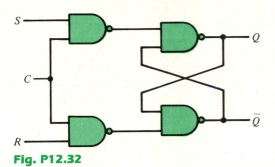

Fig. P12.32

12.33 Connecting the inputs of a JK flip-flop together as shown in Fig. P12.33(a) results in a toggle or T flip-flop, the circuit symbol of which is shown in Fig. P12.33(b). Construct the truth table for a T flip-flop.

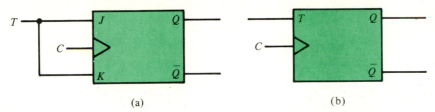

(a)   (b)

Fig. P12.33

12.34 The transition table (excluding the columns corresponding to the flip-flop inputs) for a sequential circuit is given by Table P12.34. Determine the state diagram for this sequential circuit.

Table P12.34

X	Q_1	Q_2	$Q_1(t+1)$	$Q_2(t+1)$	Y
0	0	0	0	1	1
0	0	1	1	1	0
0	1	0	1	1	0
0	1	1	1	0	1
1	0	0	0	0	0
1	0	1	0	1	1
1	1	0	0	1	0
1	1	1	0	0	0

12.35 A sequential circuit has one input X, one output Y, and two flip-flops whose states are Q_1 and Q_2. Given that the output and next-state functions are given by

$$Y = XQ_2 + \overline{Q}_1 Q_2$$

$$Q_1(t + 1) = \overline{X}\,\overline{Q}_1 + Q_1 Q_2$$

$$Q_2(t + 1) = XQ_1 \overline{Q}_2$$

determine the state diagram for the sequential circuit.

12.36 Repeat Problem 12.35 for

$$Y = \overline{X}\,\overline{Q}_2 + \overline{Q}_1$$

$$Q_1(t + 1) = XQ_2 + \overline{Q}_1 \overline{Q}_2$$

$$Q_2(t + 1) = \overline{X}\,\overline{Q}_2 + Q_1$$

12.37 Find the transition table for the sequential circuit shown in Fig. P12.37.

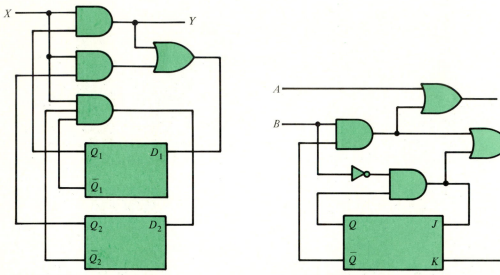

Fig. P12.37   **Fig. P12.38**

12.38 Find the transition table for the 2-input sequential circuit shown in Fig. P12.38.

12.39 Find the transition table for the sequential circuit shown in Fig. P12.39.

12.40 Construct the excitation table for the T flip-flop described in Problem 12.33.

12.41 Use JK flip-flops to realize a sequential circuit that is characterized by the state diagram given in Fig. 12.47.

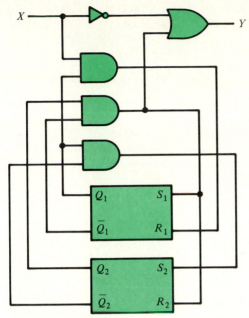

Fig. P12.39

12.42 Repeat Problem 12.41 using D flip-flops.

12.43 Use D flip-flops to realize a sequential circuit that is characterized by the state diagram given in Fig. 12.46.

12.44 Repeat Problem 12.43 using JK flip-flops. Compare the sequential circuit obtained with the one given in Fig. 12.45. Which circuit is simpler?

Chapter 13

Digital Devices

INTRODUCTION

In the preceding two chapters, we studied various combinational and sequential logic circuits that can be used in the design of digital systems. In this chapter, we turn our attention to the major components of digital computers, and discuss some of the digital devices that are integral parts of computers.

Counters, as their name suggests, are used to count events in digital systems. In addition to this, though, they can be used to generate timing sequences that control operations in computers. Both synchronous and asynchronous counters are discussed. A counter is one important application of a connection of flip-flops called a register. In this chapter, we also see the usefulness of registers in digital computers.

Information storage in digital computers is accomplished with memory devices, and different types of memories are employed for different purposes. Fast information retrieval is obtained with memories that have random-access capability. In one such memory, the read-only memory (ROM), after its initial programming, information may be retrieved or "read," but not stored or "written." A read-and-write memory (RAM)—also known as a random-access memory—has the capability, however, for both reading and writing information. Both ROMs and RAMs are commonly available as semiconductor IC packages. Although they are not as popular as semiconductor memories, random-access magnetic-core memories are still being used in some digital computers. For mass storage of information, bulk-storage devices such as magnetic tape and magnetic disks are in widespread use. Another type of semiconductor memory, the charge-coupled device (CCD), is also studied here.

This chapter includes a discussion of the digital devices that make up the typical digital computer. The five major components of a digital computer are the input, the control unit, the arithmetic-logic unit (ALU), the memory, and the output. When the control unit and the ALU are placed on one semiconductor chip, that IC is referred to as a microprocessor.

Although computers process information in digital form, analog information is inherent in many situations. The conversion of information from analog to digital form, and vice versa, is covered with the discussions on analog-to-digital converters (ADCs) and digital-to-analog converters (DACs), respectively.

13.1 COUNTERS

At the end of the preceding chapter, we saw how to design a modulo-4 up/down counter using SR flip-flops (see Fig. 12.49). Such a device also can be implemented with either D or JK flip-flops (see Problems 12.41 and 12.42). A **counter** is a sequential circuit that goes through a prescribed sequence of states when pulses are applied to its input. Most digital systems contain counters. They are an integral part of digital computers. Not only are counters used to keep track of the number of occurrences of an event, but they also are used for controlling operations in digital systems by generating timing sequences.

Binary Counters

A counter that counts in binary is called, quite obviously, a **binary counter.** An n-bit binary counter requires n flip-flops and can have as many as 2^n states. Therefore, it is possible to count from 0 to $2^n - 1$ with a binary counter. The number of states m (where $m \leq 2^n - 1$) of an n-bit binary counter is referred to as its **modulus.** A binary counter with modulus m is called a **modulo-m (or mod-m) counter.**

The input pulses to a counter, known as **count pulses,** can be either clock pulses or externally generated pulses; and they may be periodically or randomly produced. Let us design a 3-bit binary counter that counts clock pulses modulo 8. The state diagram for such a device is shown in Fig. 13.1. In this case, since the transition from state to state is the result of a clock pulse, there is no indication of an external input in the state diagram. Furthermore, since a display of the state of the counter can be considered the output, no additional external output need be indicated.

The transition table for the counter is given by Table 13.1. As shown in the table, JK flip-flops have been selected for the counter. The columns corresponding to the flip-flop inputs were obtained with the aid of the excitation table for the JK flip-flop [Table 12.23(b)]. The Karnaugh maps for the flip-flop input variables J_1, K_1, J_2, and K_2 are shown in Fig. 13.2. The resulting Boolean expressions are

$$J_1 = K_1 = Q_2Q_3$$

$$J_2 = K_2 = Q_3$$

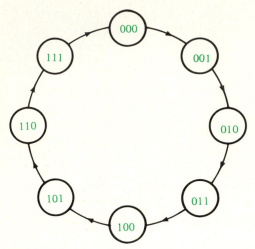

Fig. 13.1 State diagram for a 3-bit modulo-8 binary counter.

Table 13.1 Transition Table for 3-Bit Modulo-8 Binary Counter.

Present State			Next State			Flip-Flop Inputs					
Q_1	Q_2	Q_3	$Q_1(t+1)$	$Q_2(t+1)$	$Q_3(t+1)$	J_1	K_1	J_2	K_2	J_3	K_3
0	0	0	0	0	1	0	×	0	×	1	×
0	0	1	0	1	0	0	×	1	×	×	1
0	1	0	0	1	1	0	×	×	0	1	×
0	1	1	1	0	0	1	×	×	1	×	1
1	0	0	1	0	1	×	0	0	×	1	×
1	0	1	1	1	0	×	0	1	×	×	1
1	1	0	1	1	1	×	0	×	0	1	×
1	1	1	0	0	0	×	1	×	1	×	1

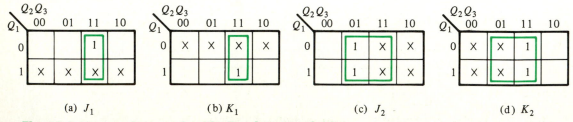

(a) J_1 (b) K_1 (c) J_2 (d) K_2

Fig. 13.2 Karnaugh maps for flip-flop input variables.

Moreover, we do not even have to draw Karnaugh maps for J_3 and K_3 since we can make all the don't-care conditions equal to 1; this results simply in

$$J_3 = K_3 = 1$$

We may now implement the modulo-8 binary counter with three JK flip-flops as shown in Fig. 13.3. The state of the sequential circuit is $Q_1 Q_2 Q_3$, and at any given time this is a binary number between 0 and 7.

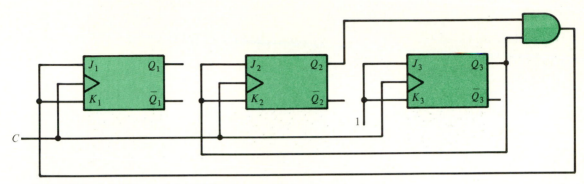

Fig. 13.3 A 3-bit modulo-8 binary counter.

The waveforms associated with the clock pulses and the flip-flop contents comprise its **timing diagram.** Figure 13.4 shows the timing diagram for the modulo-8 binary counter, where it is assumed that the counter is in the state 000 when the first clock pulse arrives. This diagram shows that the flip-flops are triggered on the leading edges of the clock pulses. As indicated by Table 13.1, the first clock pulse changes Q_3 from 0 to 1. The second clock pulse changes Q_3 from 1 to 0 and Q_2 from 0 to 1, and so on. Note that the frequency of the waveform for Q_3 is half the frequency of the waveform for the clock C. Also, the frequency of Q_2 is a fourth of the frequency of C, and the frequency of Q_1 is an eighth of the frequency of C. Because of this, we can refer to a modulo-8 counter as a divide-by-8 counter. More generally, a modulo-m counter can be called a **divide-by-m counter.**

Fig. 13.4 Timing diagram for modulo-8 binary counter.

Table 13.2 (a) Truth Table for the T Flip-Flop.

Input T	Present State Q	Next State Q(t + 1)
0	0	0
0	1	1
1	0	1
1	1	0

Table 13.2 (b) Excitation Table for the T Flip-Flop.

Present State Q	Next State Q(t + 1)	Input T
0	0	0
0	1	1
1	0	1
1	1	0

Note that for each flip-flop in Fig. 13.3, $J = K$. A JK flip-flop in which the inputs are connected together is a toggle or T flip-flop. From the truth table for a JK flip-flop (Table 12.19), we see that the truth table for a T flip-flop is given by Table 13.2(a), and the excitation table for the T flip-flop is given by Table 13.2(b). We leave it as an exercise for the reader to confirm these results (see Problems 12.33 and 12.40). Note that an input of 0 will not affect the state of the flip-flop. On the other hand, an input of 1 changes or "toggles" the state from 0 to 1, or from 1 to 0. The circuit symbol for a T flip-flop is shown in Fig. 13.5.

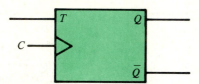

Fig. 13.5 Circuit symbol for the T flip-flop.

Ripple Counters

The binary counters discussed above are examples of **synchronous counters.** Such counters have clock pulses applied to the clock inputs of all of the flip-flops. The counter shown in Fig. 13.6 is not a synchronous counter, therefore, it is said to be **asynchronous.** The small circles at the clock-pulse inputs of the T flip-flops signify that they are trailing-edge-triggered devices.

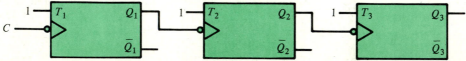

Fig. 13.6 A 3-bit binary ripple counter.

Suppose that the initial state of the counter is 000. Since the flip-flops trigger on the trailing edge of a pulse, Q_1 will not change from 0 to 1 until the first input pulse is completed as shown in Fig. 13.7. Since the transition of Q_1 changing from

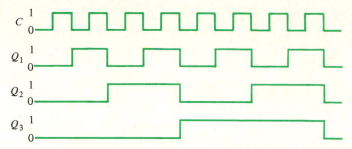

Fig. 13.7 Timing diagram for 3-bit binary ripple counter.

0 to 1 is the leading edge of a pulse that is applied to the second flip-flop, Q_2 is not affected by the change of Q_1. However, after the second input pulse, Q_1 changes from 1 to 0. This, in turn, is the trailing edge of a pulse that is applied to the second flip-flop, and, therefore, Q_2 changes from 0 to 1. But Q_2 will not change from 1 to 0 until the next time Q_1 changes from 1 to 0. In a similar manner, we see that Q_3 does not change from 0 to 1 until the first time that Q_2 changes from 1 to 0; and the second time that Q_2 changes from 1 to 0, Q_3 changes from 1 to 0. The operation of the counter is summarized by the timing diagram shown in Fig. 13.7.

For the counter just described, consider what occurs due to the trailing edge of the fourth input pulse. The value of Q_1 changes from 1 to 0. The resulting trailing edge changes Q_2 from 1 to 0. This, in turn, changes Q_3 from 0 to 1. It is because of this ripple effect that we say that the device shown in Fig. 13.6 is a **ripple counter.** Further, note from Fig. 13.7 that this device is a modulo-8 binary counter.

Next consider the ripple counter shown in Fig. 13.8. Suppose that this counter is in the state 000 when the first clock pulse arrives. Since $J_1 = K_1 = \overline{Q}_3 = 1$, $J_3 = Q_1Q_2 = 0(0) = 0$, and $K_3 = Q_3 = 0$, then after the first pulse, $Q_1 = 1$ and $Q_3 = 0$. Because the flip-flops trigger on the trailing edges of pulses, it remains that $Q_2 = 0$.

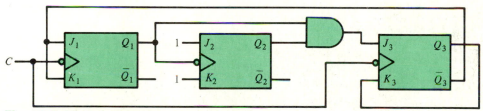

Fig. 13.8 Modulo-5 binary ripple counter.

Since $J_1 = K_1 = \overline{Q}_3 = 1$, $J_3 = Q_1Q_2 = 1(0) = 0$, and $K_3 = Q_3 = 0$, then after the second clock pulse, $Q_1 = 0$ and $Q_3 = 0$. The transition of Q_1 from 1 to 0, therefore, results in $Q_2 = 1$.

Since $J_1 = K_1 = \overline{Q}_3 = 1$, $J_3 = Q_1Q_2 = 0(1) = 0$, and $K_3 = Q_3 = 0$, then after the third clock pulse, $Q_1 = 1$ and $Q_3 = 0$. The transition of Q_1 from 0 to 1 means that $Q_2 = 1$ remains.

Since $J_1 = K_1 = \bar{Q}_3 = 1$, $J_3 = Q_1 Q_2 = 1(1) = 1$, and $K_3 = Q_3 = 0$, then after the fourth clock pulse, $Q_1 = 0$ and $Q_3 = 1$. The transition of Q_1 from 1 to 0, therefore, results in $Q_2 = 0$.

Finally, since $J_1 = K_1 = \bar{Q}_3 = 0$, $J_3 = Q_1 Q_2 = 0(0) = 0$, and $K_3 = Q_3 = 1$, then after the fifth clock pulse, $Q_1 = 0$ and $Q_3 = 0$. Because there is no change in Q_1, it remains that $Q_2 = 0$. Thus, after five clock pulses, the counter returns to its initial state 000. In other words, the device shown in Fig. 13.8 is a modulo-5 counter. The timing diagram for this sequential circuit is shown in Fig. 13.9.

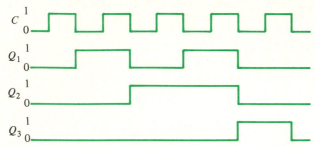

Fig. 13.9 Timing diagram for modulo-5 ripple counter.

Using an analysis similar to the one above, it can be confirmed that the ripple counter shown in Fig. 13.10 is a modulo-10 or BCD counter. The details of this are left to the reader as an exercise. An alternative way of obtaining a BCD counter is to cascade a modulo-2 counter and a modulo-5 counter (see Problem 13.6).

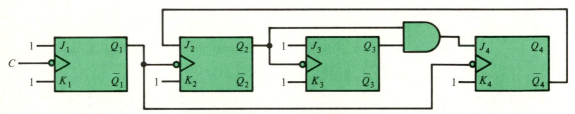

Fig. 13.10 A BCD ripple counter.

The timing diagrams displayed up to this point have been ideal in that transitions, triggered either on the leading edge or the trailing edge of a pulse, occurred instantaneously. In a physical device, there is some nonzero propagation delay associated with a flip-flop, that is, there is a time differential between when a flip-flop is triggered and when its output responds. Because of the ripple effect, however, in a ripple counter time delays are cumulative. Thus, although ripple counters generally have simpler constructions, synchronous counters are faster. Another problem associated with ripple counters is the occurrence of **glitches.** These are voltage spikes that arise as a result of temporary flip-flop states that are caused by propagation delays. This problem may be eliminated by using synchronous counters.

Ring Counters

The synchronous counter shown in Fig. 13.11 is an example of a **ring counter.** Let us assume that the D flip-flops can be directly preset and cleared (inputs not shown), and that the initial state of this counter is $Q_1 = 1$ and $Q_2 = Q_3 = 0$. Since $D_1 = Q_3 = 0$, $D_2 = Q_1 = 1$, and $D_3 = Q_2 = 0$, then after the flip-flops are triggered by the trailing edge of the first clock pulse, the state of the counter is $Q_1 = 0$, $Q_2 = 1$, and $Q_3 = 0$. Similarly, after the second clock pulse, $Q_1 = Q_2 = 0$ and $Q_3 = 1$. Finally, after the third clock pulse, $Q_1 = 1$ and $Q_2 = Q_3 = 0$, so the counter is back in its initial state. This behavior is summarized by Table 13.3. The timing diagram for this 3-bit ring counter is shown in Fig. 13.12. Note that the frequency of the waveform associated with each flip-flop is a third of the frequency of the clock-pulse waveform. Therefore, the device shown in Fig. 13.11 is a divide-by-3 counter.

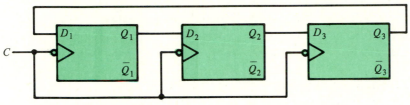

Fig. 13.11 A 3-bit ring counter.

Table 13.3 Behavior of 3-Bit Ring Counter.

Number of	State of Counter		
Applied Clock Pulses	Q_1	Q_2	Q_3
0	1	0	0
1	0	1	0
2	0	0	1
3	1	0	0

Fig. 13.12 Timing diagram for 3-bit ring counter.

A ring counter in conjunction with an appropriate encoder can be used to construct a glitch-free binary modulo-m counter—however, m flip-flops are required. For example, the first four columns of Table 13.4 indicate the states of a 4-bit ring counter (assuming that 1000 is the initial state). If $B_1 B_0$ is the binary-number representation indicated by Table 13.4, then we have

$$B_0 = \overline{Q}_1 Q_2 \overline{Q}_3 \overline{Q}_4 + \overline{Q}_1 \overline{Q}_2 \overline{Q}_3 Q_4$$

and

$$B_1 = \overline{Q}_1 \overline{Q}_2 Q_3 \overline{Q}_4 + \overline{Q}_1 \overline{Q}_2 \overline{Q}_3 Q_4$$

Therefore, the combinational logic circuit (encoder) shown in Fig. 13.13, when connected to a 4-bit ring counter, will yield a binary modulo-4 counter.

Table 13.4 Truth Table for Binary Counter Implementation.

Q_1	Q_2	Q_3	Q_4	B_1	B_0
1	0	0	0	0	0
0	1	0	0	0	1
0	0	1	0	1	0
0	0	0	1	1	1

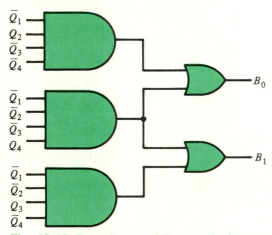

Fig. 13.13 Encoder used for producing a modulo-4 counter.

More important, a ring counter can be used to generate timing signals that control a sequence of operations in a digital system. An m-bit ring can generate m timing signals by using m flip-flops. If $m = 2^n$, then 2^n flip-flops are required. An alternative approach to generating m timing signals is to use an n-bit binary counter (which has 2^n distinct states) in conjunction with an n-to-2^n-line decoder. This, however, requires 2^n AND gates, each of which has n inputs. By using another type

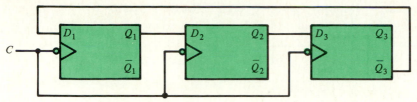

Fig. 13.14 A 3-bit twisted-ring counter.

of ring counter, we instead can use fewer flip-flops than an ordinary ring counter and not require as complicated a decoder as for a binary counter.

The synchronous counter shown in Fig. 13.14 is known as a **twisted-ring counter** (or **switchtail ring counter** or **Johnson counter**). Let us assume that the initial state of the counter is $Q_1 = Q_2 = Q_3 = 0$. Since $D_1 = \bar{Q}_3 = 1, D_2 = Q_1 = 0$, and $D_3 = Q_2 = 0$, then after the first clock pulse $Q_1 = 1$ and $Q_2 = Q_3 = 0$.

Since $D_1 = \bar{Q}_3 = 1$, $D_2 = Q_1 = 1$, and $D_3 = Q_2 = 0$, then after the second clock pulse $Q_1 = Q_2 = 1$ and $Q_3 = 0$.

Since $D_1 = \bar{Q}_3 = 1$, $D_2 = Q_1 = 1$, and $D_3 = Q_2 = 1$, then after the third clock pulse, $Q_1 = Q_2 = Q_3 = 1$.

Proceeding in this manner, the remaining sequence of states (each being a triple of the form $Q_1 Q_2 Q_3$) that results is 011, 001, and 000—which is the initial state. This behavior is summarized by Table 13.5. The corresponding timing diagram is shown in Fig. 13.15. Also included are the waveforms obtained by applying various flip-flop outputs to two-input AND gates. Specifically, the waveforms for $Q_1 \bar{Q}_2$, $Q_2 \bar{Q}_3$, $Q_1 Q_3$, $\bar{Q}_1 Q_2$, $\bar{Q}_2 Q_3$, and $\bar{Q}_1 \bar{Q}_3$ are shown. Therefore, whereas the 3-bit ring counter shown in Fig. 13.11 generates only three timing signals, we see that with the aid of six two-input AND gates, the 3-bit twisted-ring counter shown in Fig. 13.14 can produce six timing signals. Finally, note that this device is a divide-by-6 counter.

Table 13.5 Behavior of Twisted-Ring Counter.

Number of Applied Clock Pulses	State of Twisted-Ring Counter		
	Q_1	Q_2	Q_3
0	0	0	0
1	1	0	0
2	1	1	0
3	1	1	1
4	0	1	1
5	0	0	1
6	0	0	0

Various IC counters are available as MSI packages. For example, the 74160 and 74161 ICs are synchronous 4-bit counters, while the 74393 consists of two 4-bit binary ripple counters on the same IC. The 7490 and 7493 ICs are BCD ripple counters.

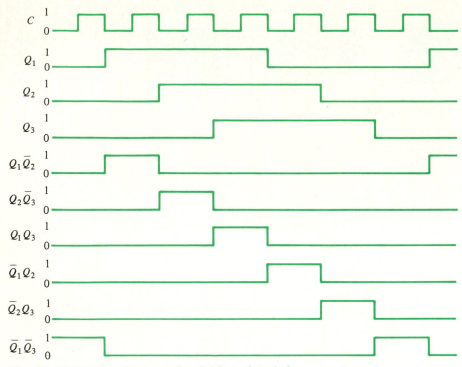

Fig. 13.15 Timing diagram for 3-bit twisted-ring counter.

13.2 REGISTERS

We have already seen that a flip-flop is a storage device that is capable of storing 1 bit of information. A **register** is a collection of flip-flops. Specifically, an n-bit register consists of n flip-flops. Combinational logic gates may or may not be part of a register. Such gates control the register's information transfer. Registers are used typically for the temporary storage of information in binary form, and they are used extensively in digital systems—especially in digital computers. A counter is a type of register whose state changes in some predetermined manner when a sequence of pulses is applied to it.

A simple register consisting of four D flip-flops is shown in Fig. 13.16. For this synchronous sequential circuit, a 4-bit word $A_1 A_2 A_3 A_4$ is applied to the inputs of the flip-flops as indicated. A clock pulse causes this word to be transferred into the register; each bit is stored in its corresponding flip-flop. Transferring information into a register is known as **loading.** The state (information content) of the register can be "read" from the output lines labeled B_1, B_2, B_3, and B_4. The contents of the register remain unchanged until the next clock pulse leads a new word (or the same word) into the register. Since all the bits enter the register at the same time and can be read at the same time, the device in Fig. 13.16 is an example of a **parallel-in, parallel-out register.**

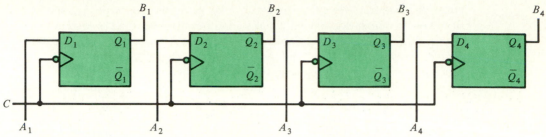

Fig. 13.16 Parallel-in, parallel-out register.

Shift Registers

There is an alternative to the parallel loading of a register as illustrated in Fig. 13.16. An example of this alternative is the simple 4-bit register shown in Fig. 13.17. For this device, information is put into the register serially. Specifically, binary information in the form of a sequence of bits is loaded into this register from left to right by applying an input A while a clock pulse occurs. For example, consider the input sequence $A_1, A_2, A_3, A_4, \ldots$. Assuming that the initial state of the register is $Q_1 = Q_2 = Q_3 = Q_4 = 0$, before the first clock pulse arrives, $D_1 = A_1$, $D_2 = Q_1$, $D_3 = Q_2$, and $D_4 = Q_3$. After the first clock pulse, the state of the register (which can be read from left to right from the output lines) is $A_1 000$. After the second clock pulse, the state of the register is $A_2 A_1 00$. After the third clock pulse, the state of the register is $A_3 A_2 A_1 0$. After the fourth clock pulse, the state of the register is $A_4 A_3 A_2 A_1$, and so on. Note how the input information shifts from left to right. Because of this property, the device shown in Fig. 13.17 is called a **shift register.** In particular, since binary information is loaded serially, and because the contents of the register can be read from the output lines simultaneously, this device is a **series-in, parallel-out shift register.**

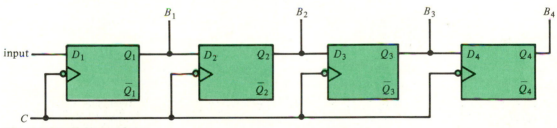

Fig. 13.17 Shift register.

Suppose that for the shift register shown in Fig. 13.17 the three output lines labeled B_1, B_2, and B_3 are removed, and only the single output line labeled B_4 remains. For the resulting shift register with an initial state of $Q_1 = Q_2 = Q_3 = Q_4 = 0$ and an input sequence of $A_1, A_2, A_3, A_4, \ldots$, as described above, the first four output bits will be 0 (the first output bit being available before the first clock pulse occurs). After the fourth clock pulse, the output is A_1; after the fifth clock pulse, the output is A_2; after the sixth clock pulse the output is A_3, and so on. Because the output is obtained serially, and because the input is loaded serially, such a device is called a **series-in, series-out shift register.** Note that the ring counter shown in Fig. 13.11 and the twisted-ring counter shown in Fig. 13.14 are series-in, series-out shift registers that have feedback connections.

Example 13.1

For the shift register shown in Fig. 13.17, suppose that the applied input sequence is 1, 1, 0, 1, 0, 0, ... Assuming that the initial state of the device is $Q_1 = Q_2 = Q_3 = Q_4 = 0$, then after the first clock pulse, the state (of the form $Q_1 Q_2 Q_3 Q_4$) of the register is 1000. After the second clock pulse, the state is 1100. The following states are 0110, 1011, 0101, 0010, 0001, 0000, ... These results are summarized by the timing diagram shown in Fig. 13.18.

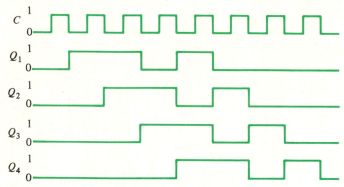

Fig. 13.18 Timing diagram of shift register for input sequence 1, 1, 0, 1, 0, 0, ...

A modification of the shift register shown in Fig. 13.17 is required to obtain a **parallel-in, series-out shift register.** One approach that can be taken is shown in Fig. 13.19. For this register, there is an enable input E. When $E = 1$, we have that $D_1 = A_1$, $D_2 = A_2$, $D_3 = A_3$, and $D_4 = A_4$. Thus, the presence of a clock pulse when $E = 1$ will result in the parallel loading of the register. Conversely, when $E = 0$, we have that $D_1 = 0$, $D_2 = Q_1$, $D_3 = Q_2$, and $D_4 = Q_3$. Under these circumstances, the presence of a clock pulse will shift the contents of the register to the right, loading a 0 into the first flip-flop. Another clock pulse will cause another shift to the right, and so on, until the loaded contents of the register can be completely put out serially.

Just as there are registers that shift to the right, there are also registers that shift to the left. (To see one, take a page with a shift-to-the-right register on it, turn the

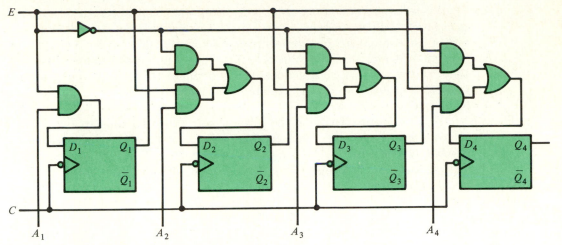

Fig. 13.19 Parallel-in, series-out shift register.

page over, and hold it up to a light.) However, there are also registers that have the capability of shifting either to the right or to the left—the direction being determined by the application of a control signal. Such a device is referred to as a **bidirectional shift register.** A bidirectional shift register that can be loaded either serially or in parallel, and also has either a serial or parallel output, is known as a **universal shift register.** As was the case for converting a series-in shift register to a parallel-loading shift register (see Fig. 13.19), bidirectional and universal shift registers can be obtained from simpler structures by adding appropriate logic gates and control signals.

A wide variety of shift registers are available as MSI packages. For example, the 7491 IC is an 8-bit series-in, series-out shift register, and the 74165 IC is an 8-bit parallel-in, series-out shift register. An example of a 4-bit universal shift register is the 74194 IC; the 74198 IC is an 8-bit universal shift register. A series-in, series-out IC shift register has only one input pin and one output pin—regardless of the number of flip-flops. Because of this, very long shift registers can be fabricated by using LSI; an example is the 2401 NMOS IC, which is a 1024-bit shift register.

A Serial Adder

A classic application of shift registers is in the construction of binary serial adders. In Chapter 12, we discussed binary parallel adders (e.g., see Figs. 12.7 and 12.17). There it was implicitly assumed that in forming the sum $S_n S_{n-1} S_{n-2} \ldots S_2 S_1 S_0$ by adding $A_{n-1} A_{n-2} \ldots A_2 A_1 A_0$ and $B_{n-1} B_{n-2} \ldots B_2 B_1 B_0$, registers for storing these numbers were required. Further, because of the nature of parallel addition, the registers storing the numbers to be added must have parallel outputs, and the register storing the sum must have a parallel loading capability. For serial addition, different types of registers are employed.

The serial addition of two n-bit binary numbers involves n 1-bit binary additions performed one after another (in series). Compare this with the parallel addition of two n-bit binary numbers where n 1-bit binary additions are performed in parallel. Although parallel addition is much faster than serial addition, parallel addition requires n full-adders. Serial addition, on the other hand, needs only one full-adder and one flip-flop. Therefore, serial addition, in general, requires less hardware.

To see how two binary numbers can be added serially, consider Fig. 13.20. Shown are two series-in, series-out shift registers that contain the n-bit binary numbers $A = A_{n-1}A_{n-2} \ldots A_2A_1A_0$ and $B = B_{n-1}B_{n-2} \ldots B_2B_1B_0$ to be added. Also shown is a full-adder (see Fig. 12.6) and a D flip-flop. After the numbers to be added are shifted into registers A and B, the least significant bit (LSB) of A (which is A_0) is added to the LSB of B (which is B_0) by the full-adder (it is assumed that the initial state of the flip-flop is 0, i.e., $C = 0$). After the next clock pulse occurs, the sum S_0 and the carry C_1, respectively, enter register S and the flip-flop. The LSB in register A then is A_1 and the LSB in register B then is B_1. The resulting outputs of the full-adder then are S_1 and C_2. The following clock pulse loads these values into register S and the flip-flop respectively. This process continues until the sum of A and B is loaded into register S. As the binary numbers A and B are shifted out of their respective registers, new binary numbers that are to be added can be shifted in.

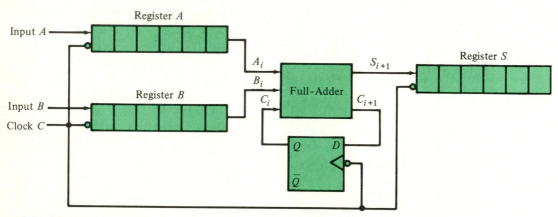

Fig. 13.20 Serial adder.

13.3 MEMORIES

The memory unit of a digital computer (see Fig. 11.1) stores instructions and information in binary form. Information put into a computer by an input device is first stored in the memory unit, and output information comes from the memory unit. The processing of data that takes place in the processor unit employs instructions and data taken from the memory unit. Also, intermediate and final data-processing results are stored there. These facts help to demonstrate the usefulness and the importance of the memory unit.

The memory unit of a computer consists of binary storage cells (e.g., flip-flops), each of which is capable of storing 1 bit of information, and associated digital circuits that are used to transfer information into and out of the cells. Typically, memory units are comprised of different types of memory devices, or **memories;** the most common are semiconductor ICs, magnetic cores, and moving-surface memories such as magnetic disks, drums, and tapes. Let us begin our discussion of memories by looking at some semiconductor LSI and VLSI memories.

Read-Only Memories (ROMs)

In our previous discussion of ROMs, we presented a logic-circuit depiction of an 8×2 ROM (see Fig. 12.31). Let us now specifically illustrate one type of ROM in which the binary cells are explicitly indicated.

Recall that a $2^n \times m$ ROM has n input lines and m output lines. Each n-bit input combination is called an **address,** and each m-bit output combination is called an **(output) word.** We can form a ROM by using an n-to-2^n-line decoder whose output lines, known as **word lines,** are connected to a matrix array of diodes. Figure 13.21 shows an example for the case that $n = 2$ and $m = 5$. Suppose that word line $D_0 = 1$ (and, therefore, $D_1 = D_2 = D_3 = D_4 = 0$). Then the diode connected between word line D_0 and output line B_4 is forward biased, resulting in $B_4 = 1$. Similarly, the diode between lines D_0 and B_2 is forward biased, as is the diode between D_0 and B_1. This results in $B_2 = 1$ and $B_1 = 1$, respectively. Because there is no diode between lines D_0 and B_3, or between lines D_0 and B_0 (and because $D_1 = D_2 = D_3 = 0$), it must be true that $B_3 = 0$ and $B_0 = 0$. Hence, the resulting output word is $B_4 B_3 B_2 B_1 B_0 = 10110$. Table 13.6 summarizes the outputs that result from inputs applied to the ROM shown in Fig. 13.21.

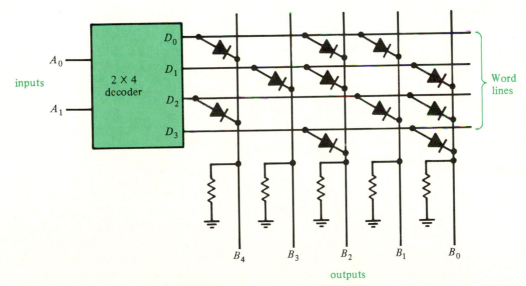

Fig. 13.21 A 4×5 (20-bit) ROM.

Table 13.6 Input–Output Description of Given ROM.

Inputs		Decoder	Outputs				
A_1	A_0	Output	B_4	B_3	B_2	B_1	B_0
0	0	D_0	1	0	1	1	0
0	1	D_1	0	1	1	0	1
1	0	D_2	1	0	0	1	1
1	1	D_3	0	0	1	0	1

For the ROM shown in Fig. 13.21, a storage cell consists of the presence or absence of a diode. Specifically, if there is a diode connected between word line D_i and output line B_j, then a 1 is stored in the i, j cell. Conversely, if there is no diode between D_i and B_j, then a 0 is stored in the corresponding cell.

A decoder output for a ROM with a diode array, such as the one shown in Fig. 13.21, is required to drive (forward bias) diodes by supplying currents to the diodes and their associated resistive loads. A design improvement is to replace the diodes with transistors. Specifically, let us connect a BJT to word line D_i and output line B_j as shown in Fig. 13.22, where the supply voltage V_{CC} is assumed to be the value of the logical 1 (high) voltage. When $D_i = 0$, the base-emitter junction of the BJT is reverse biased and the transistor is OFF. However, when $D_i = 1$, the BJT turns ON and the output line is driven by the supply voltage and $B_j = 1$.

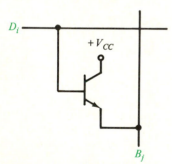

Fig. 13.22 A BJT cell.

As with diodes, we see that having a BJT connected between lines D_i and B_j, as shown in Fig. 13.22, corresponds to storing a 1 in that cell, whereas not connecting a transistor corresponds to storing a 0. Alternatively, BJTs can be placed in all of the cells initially, and then the connection (or link) between the emitter and the output line for each cell can be left alone or broken. To store a 0, the link is broken; and to store a 1, the connection is left as is. We have already mentioned that this process is called programming the ROM. Moreover, some ROMs can be programmed by their manufacturers (called **mask programming**), and some ROMs (called PROMs) can be programmed by their users.

Instead of using BJT memory cells, ROMs can be fabricated with the use of MOSFETs. In particular, each cell may have the form shown in Fig. 13.23. Instead of having a load resistance connected between each output line and ground, as shown in Fig. 13.21, a MOSFET load is connected between each output line and a supply voltage V_{DD}. For the cell shown in Fig. 13.23, when $D_i = 1$, the enhancement MOSFET is ON and $B_j = 0$. Therefore, connecting the MOSFET as shown corresponds to storing a 0. To store a 1, the MOSFET is made inoperative by fabricating it so that it has a high threshold voltage. The result is a MOSFET that is effectively absent.

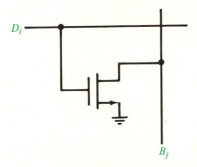

Fig. 13.23 A MOSFET cell.

For larger ROMs, the structure exemplified by Fig. 13.21 can be inefficient. For example, consider a ROM with ten inputs and four outputs, that is, a $2^{10} \times 4 = 4096 = 4\text{K-bit}$ ROM. If such a device utilizes a 10×1024 decoder as suggested, the decoder requires 1024 AND gates. Suppose, instead of using a 10×1024 decoder and a 1024×4 cellular matrix, that we use a 7×128 decoder and a 128×32 matrix as shown in Fig. 13.24. Such a matrix also stores $128 \times 32 = 4096$ bits (4K bits), but there are only 128 horizontal (word) lines. However, there are 32 vertical lines. But by using 8-to-1-line (8×1) data selectors or multiplexers (e.g., see Fig. 12.26), we can obtain the output bits, and hence, the output words, as shown. In other words, instead of determining an output bit from a column of 1024 cells, we obtain it from eight columns—each consisting of 128 cells—and an 8×1 data selector. The decoder has seven inputs (A_0, A_1, A_2, A_3, A_4, A_5, A_6), and each selector uses three additional inputs (A_7, A_8, A_9), for a total of ten inputs to the ROM—as required. Each selector can be implemented with eight AND gates (and one OR gate). Therefore, four selectors require $4 \times 8 = 32$ AND gates (and four OR gates). The 7×128 decoder uses 128 AND gates. Thus, the ROM shown in Fig. 13.24 employs $32 + 128 = 160$ AND gates. Compare this with the 1024 AND gates required for the ROM structure discussed previously. The memory shown in Fig. 13.24 is an example of a ROM with **two-dimensional addressing.**

Removal of the power supplied to a ROM does not affect the device's physical construction. Reinstating the power, therefore, does not change its programming. Because of this, we say that a ROM is an **nonvolatile memory.**

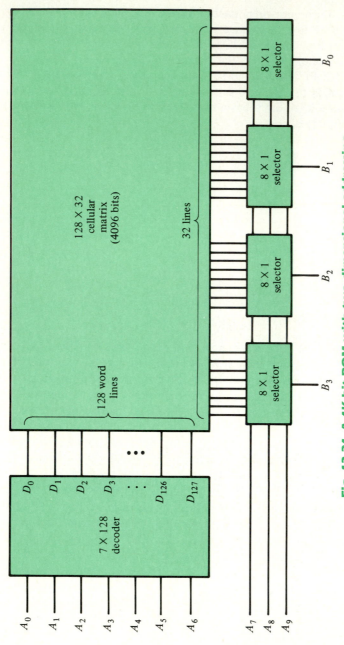

Fig. 13.24 A 4K-bit ROM with two-dimensional addressing.

In addition to memory applications, ROMs can be used for code conversions and for realizations of other combinational logic circuits.

Read-and-Write Memories (RAMs)

We know that when an input is applied to a ROM, the word at the corresponding address is read (i.e., the word appears at the output). The time required to select a word and read it is known as the **access time** of the ROM. Since the words in a ROM can be read or accessed in any order, such a device is an example of a **random-access memory.** There is another type of memory for which words not only can be read (retrieved), but can be "written" (stored) at any address in the memory as well. Such a device can be called a **read-and-write memory,** or RAM. (Although "RAM" was originally the acronym for the phrase "random-access memory," it is now used to signify a read-and-write memory.)

Typically, a ROM is used for permanent storage of data and programs that are used frequently. Examples are code converters, look-up tables, and assemblers. A RAM, on the other hand, is used for temporary storage of data and instructions. Such information can be read in accordance with a program that is to be executed, and the results then can be written in the RAM. The time required to complete a read or write operation is called the **access time** of a RAM.

A logic-equivalent circuit of a RAM cell is shown in Fig. 13.25(a). The block diagram of this cell is shown in Fig. 13.25(b). For the circuit shown in Fig. 13.25(a), there are three inputs, labeled data input (A), select (S), and read/write ($R/\overline{W}$), and one output (B).

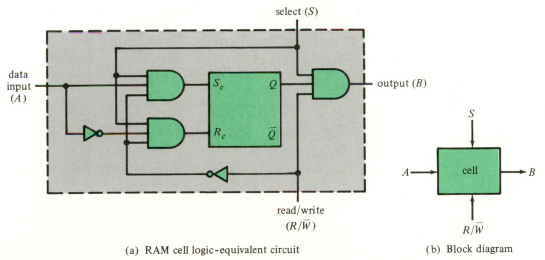

(a) RAM cell logic-equivalent circuit   (b) Block diagram

Fig. 13.25

When the select input is $S = 0$, the output of the cell is $B = 0$ regardless of the value of the data input A or the content Q of the SR flip-flop. Furthermore, $S_c = R_c = 0$, so the state of the flip-flop does not change. However, when $S = 1$, then the cell will be in either the read mode or the write mode, depending on the value of the read/write input $R/\overline{W}$. If $R/\overline{W} = 1$, then $S_c = R_c = 0$ (which does not change Q) and $B = Q$. Thus, the output reads the content of the cell. On the other hand, if $R/\overline{W} = 0$, then $S_c = A$ and $R_c = \overline{A}$ (while $B = 0$). When $A = 1$, the flip-flop is set ($Q = 1$); when $A = 0$, the flip-flop is reset ($Q = 0$). In other words, the flip-flop stores (the cell writes) A.

A RAM with n inputs and m outputs is a $2^n \times m$ RAM or $2^n \times m$-bit RAM. Figure 13.26 shows a RAM with $n = 1$ and $m = 4$, that is, an 8-bit RAM. As can be seen, the word lines are used to select a row of cells; when $D_0 = 1$, the top row is selected, and when $D_1 = 1$, the bottom row is selected. The $R/\overline{W}$ input places the selected cells in either the read mode ($R/\overline{W} = 1$) or the write mode ($R/\overline{W} = 0$). When cells are read, their contents appear at the outputs; when input data are written in cells, the outputs are 0.

Just as was the case for ROMs, increased efficiencies can be achieved for RAMs by using two-dimensional addressing. As a specific example, consider a 4096×1 (4K-bit) RAM, where, instead of having a single column of 4096 cells, the cellular array has 64 rows and 64 columns. Any cell can be selected by designating its row location and its column location. By using six inputs and a 6×64 decoder, any row can be selected. Another six inputs applied to a second 6×64 decoder can be employed to select any column. Each cell, therefore, must be able to detect signals from both the row and column decoders. Again, a cell is in the read mode when $R/\overline{W} = 1$ and in the write mode when $R/\overline{W} = 0$. For the latter case, the single data input A is written into the selected cell, and for the former case, the content of the selected cell is read and appears at the output B. Typically, amplifiers are used to write and to read or sense data. A block diagram representation of this 4K-bit RAM is shown in Fig. 13.27.

Such a RAM has 1-bit output words. Suppose that we wish to have a RAM with 4096 4-bit words (i.e., a 16K-bit RAM). We can construct such a device by combining four of the structures shown in Fig. 13.27. In this case, the 12 address inputs (A_0 through A_{11}) are applied in parallel to the four packages. The $R/\overline{W}$ inputs are connected in parallel, the data inputs are entered in parallel, and the four output bits are read in parallel.

The structure of a RAM cell depends not only on the technology to be used (i.e., BJT or MOS), but also on the mode of operation of the components. One type of cell, which contains a transistor flip-flop, maintains its state until a new state is written in (as long as the power is on). A RAM with this cell structure is known as a **static RAM.** Such RAMs usually are fabricated with either BJTs, NMOS, or CMOS.

A second type of cell may be formed with a single NMOS transistor and a capacitance as shown in Fig. 13.28. The gate of the enhancement MOSFET is connected to a word line (a line from the row decoder) and the drain is connected to a bit line (a line from the column decoder). When the capacitor voltage is high,

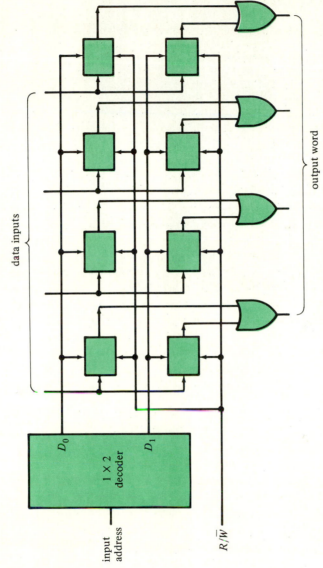

Fig. 13.26 An 8-bit RAM.

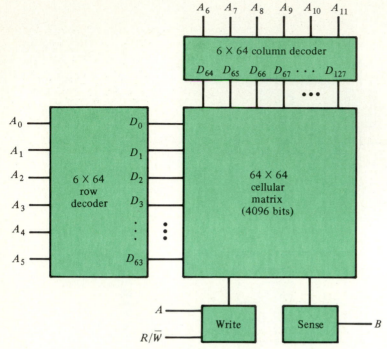

Fig. 13.27 A 4K-bit RAM with two-dimensional addressing.

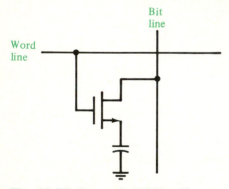

Fig. 13.28 Dynamic RAM cell.

the cell stores a 1; when the capacitor voltage is low, the cell stores a 0. Because the capacitor will lose its charge (voltage) owing to leakage, there must be a **refresh circuit** (not shown) to recharge the capacitor to its appropriate value. A RAM with cells such at this is an example of a **dynamic RAM.** Even with the added complexity of refresh circuits, because of their adaptability to VLSI, dynamic RAMs have become very popular.

Unlike the ROM, the RAM—be it static or dynamic—is a **volatile** device. That is, if the power supplied to a RAM is interrupted, the information stored will be lost.

Magnetic-Core Memories

Another type of memory with random access is a **magnetic-core memory.** Such a memory consists of a collection of **magnetic cores,** each of which is magnetic material, typically ferrite, in the shape of a doughnut. As depicted in Fig. 13.29, three conducting lines pass through the hole of each core. By putting appropriate-magnitude currents through both the word line and the bit line, a core can be magnetized in one of two directions—say negative (0) or positive (1). A current through only one line is not sufficient to magnetize the core in either direction. Once the core has been magnetized in one direction as the result of word-line and bit-line currents, it cannot be magnetized in the other direction until both word and bit reverse currents are applied. In addition to this, when the magnetization of the core changes from one direction to the other (e.g., from the positive direction to the negative direction), a voltage is induced across the sense line.

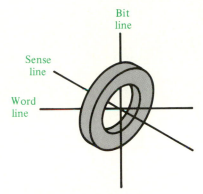

Bit
line

Sense
line

Word
line

Fig. 13.29 Magnetic core.

A simplified 16×1 (16-bit) magnetic-core memory is shown in Fig. 13.30. Let us consider a specific memory cell—the core in row 2 and column 1. To write into this cell, the row decoder selects word line D_1 and the column decoder selects bit line D_4. Then a control signal sends negative currents through the selected lines; this puts the core into the 0 state (negative magnetization). A single negative current is not sufficient to change a core to the 0 state, so only the core in row 2 and column 1 can be changed to 0. If a 0 is to be written into the cell indicated, as the core is already in the 0 state, no further action need be taken. If a 1 is to be written into the selected cell, a proper control signal sends positive currents through lines D_1 and D_4. This puts the corresponding core into the 1 state. Since a single positive current is not sufficient to change a core to the 1 state, only the core in row 2 and column 1 is changed to 1.

To read the cell in row 2 and column 1, again negative currents are sent through lines D_1 and D_4. As indicated above, this places a 0 in the core. If the initial state of the cell was 1, the change in core magnetization induces a voltage across the sense line and a 1 is entered into the memory buffer register. On the other hand, if the initial cell state was 0, the lack of a change in core magnetization means that

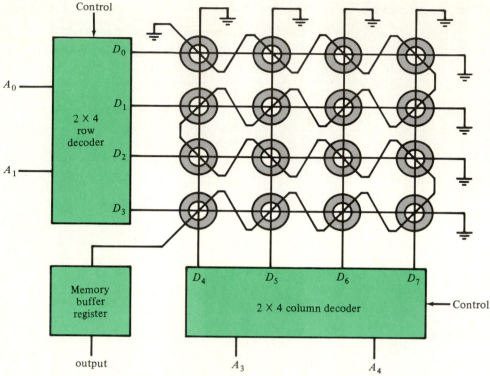

Fig. 13.30 A 16-bit magnetic-core memory.

there is no voltage induced across the sense line—so a 0 is entered into the memory buffer register. Note, however, that when a cell is read, it is put into the 0 state. This is known as **destructive readout.** Therefore, after a cell is read, its content prior to the reading must be rewritten. If the memory buffer register contains a 0, no further action is required since the cell has been put into the zero state. However, if the memory buffer register contains a 1, positive currents are passed through lines D_1 and D_4 to rewrite a 1 in the cell.

The magnetic-core memory shown in Fig. 13.30 has a 1-bit output. By combining such memories in parallel, as was indicated for the RAM given in Fig. 13.27, magnetic-core memories with n-bit outputs can be formed.

Even though a magnetic-core memory has a destructive readout and requires a rewriting cycle, if the power to a magnetic-core memory is turned off, the cores retain their magnetization. Therefore, unlike a RAM, a magnetic-core memory is a nonvolatile memory.

Although semiconductor memories are gaining greater and greater favor, magnetic-core memories are still in use.

Moving-Surface Memories

The ROM, RAM, and magnetic-core memory are examples of memories in which information can be accessed directly. These types of memories, therefore, afford rapid information access. Where high-speed information retrieval does not have high priority, other types of memories can be utilized. One nonvolatile memory medium that provides bulk or mass storage capability at low cost is magnetic tape. In essence, it is simply plastic tape with a magnetic coating. Magnetic tape is a **serial** or **sequential memory** in that information is available in the same sequence in which it was originally stored. Therefore, the access time for magnetic tape depends on where the desired information is stored and the relative location of that information with respect to the tape read-out.

For a magnetic-tape storage system, the tape moves along a **read/write head** as depicted in Fig. 13.31. The read/write head is basically an iron core (with an air gap) on which a coil of wire is wound. When the head is in the write mode, a current corresponding to the information to be stored is passed through the coil of wire. The resulting magnetic field across the gap magnetizes the magnetic surface coating of the tape as it moves across the head. When the head is in the read mode, and when a magnetized tape moves across the head, the resulting magnetic field in the core induces a voltage across the coil that corresponds to the information stored.

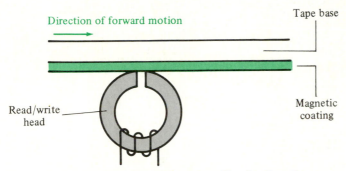

Fig. 13.31 Magnetic tape and read/write head.

Magnetic tape is an example of a **moving-surface memory.** Another moving surface memory that is used for bulk storage is the **magnetic disk.** The various types of magnetic disks differ in detail, but not in principle. A disk is shaped like a phonograph record and contains numerous concentric circles called **tracks.** As with magnetic tape, information is either read or written in a track by a read/write head.

Magnetic disks can be classified in two categories—**rigid** or **hard disks,** and **flexible disks,** called **floppy disks** or **diskettes.** The read/write head does not come into physical contact with a rigid disk, but instead is cushioned by a thin layer of air. A floppy disk does make mechanical contact with the read/write head as is done with a magnetic-tape memory. The head is movable so that is can be positioned on different tracks. Because of this, a magnetic disk is often called a **direct-access memory.** The access time consists of the time it takes to move the head to the proper track and for the desired portion of the track to reach the head. The access time for a rigid disk typically is of the order of tens of milliseconds, whereas for a floppy disk it is of the order of hundreds of milliseconds.

The rotational speed of a rigid disk is 3600 rpm (revolutions per minute), and a floppy disk rotates at 300 rpm. Although there are 77 tracks on a standard 8-inch floppy disk, and 35 tracks on a $5\frac{1}{4}$-inch floppy disk (sometimes called a **minifloppy**), a rigid disk can have 500 or more tracks. The information storage capability is measured in millions of bits for a floppy disk and tens of millions of bits for a rigid magnetic disk.

A floppy disk is placed inside an envelope similar to a jacket for a phonograph record. This protects the disk from such things as dust and scratching. The inside of the protective envelope is made so as to keep the friction between the disk and envelope low, and the disk clean. Figure 13.32 shows a floppy disk inside its protective envelope. Note that there is a slot in the envelope so that the read/write head has access to the disk. There also is a hole in the protective envelope and a hole in the disk—called an **index hole.** When these holes are aligned, a light passing through them can be used to indicate the beginning of a track.

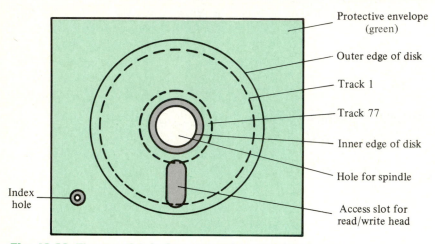

Fig. 13.32 Floppy disk in its protective envelope.

Charge-Coupled Devices (CCDs)

Another semiconductor IC memory, unlike a RAM or ROM, is a serial memory. A **charge-coupled device** (CCD) is, in effect, an enhancement MOSFET with numerous gates (see Section 8.3). Figure 13.33 illustrates such a device, where the gates are labeled ϕ_1, ϕ_2, ϕ_3, and ϕ_4. Although the number of gates is large (hundreds or more), only eight are shown. The drain and the source of the MOSFET are not shown in Fig. 13.33.

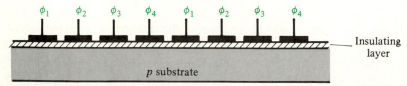

Fig. 13.33 Portion of a CCD.

If a positive voltage (with respect to the substrate) is applied to a gate, then holes (majority carriers) are repelled from a region of the p substrate directly under the gate. This region, which is called a **potential well,** is depleted of holes and is capable of attracting and storing electrons. If electrons are supplied to a potential well from the source of the MOSFET, then by applying appropriate voltages to the gates of the device, these electrons can be transferred or shifted from gate to gate along the length of the device. Therefore, a CCD is, in essence, a long shift register. The storage of bits of information in a CCD corresponds to the presence or absence of charge (electrons for the example being discussed) under the gates of the device. Quite clearly, a CCD is a serial memory.

Let us see how charge (information) can be shifted through a CCD. Suppose that a positive voltage is applied to the leftmost gate labeled ϕ_1 in Fig. 13.33. Further suppose that electrons are supplied to this well via the source of the CCD. The resulting situation is depicted by Fig. 13.34(a). If gate ϕ_3 is also made positive, then another potential well forms under that gate as shown in Fig. 13.34(b). However, electrons are absent from this newly formed well. Now suppose that in addition to ϕ_1 and ϕ_3, ϕ_2 is made positive. Assuming closely spaced gates, what results is the large potential well shown in Fig. 13.34(c), where the electrons originally stored in the well under gate ϕ_1 are distributed throughout this larger well. If the positive voltage is removed from gate ϕ_1, its associated potential well is eliminated, leaving the well shown in Fig. 13.34(d). Finally, when the positive voltage is also removed from gate ϕ_2, what results is the electron-filled well under gate ϕ_3, as shown in Fig.13.34(e). The net effect of these events is the transfer of charge from (under) gate ϕ_1 to gate ϕ_3. Because of this, a CCD is also referred to as a **charge-transfer device** (CTD).

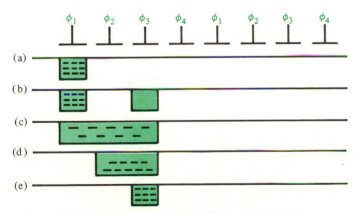

Fig. 13.34 Formation of potential wells in CCD.

A summary of the voltages that cause the transfer of charge just described is given in Fig. 13.35. For time interval T_1, the voltage on gate ϕ_1 is positive—this corresponds to the situation depicted by Fig. 13.34(a). For time interval T_2, in addition to gate ϕ_1, the voltage on gate ϕ_3 is positive—this corresponds to Fig. 13.4(b), and so on. Just as charge was transferred from gate ϕ_1 to gate ϕ_3 by using

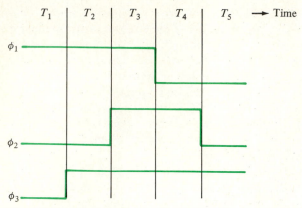

Fig. 13.35 Voltage waveforms used to transfer charge.

the clocking voltage waveforms shown in Fig. 13.35, by appropriately extending these waveforms, the charge under gate ϕ_3 can be shifted to the right along the length of the device (see Problem 13.26). However, owing to the leakage of charge, **refresh amplifiers** are inserted along the length of the structure between groups (ranging from 100 to 1000) of gates so that packets of charge can be refreshed or replenished.

13.4 DIGITAL INFORMATION PROCESSING

In the electronics section of this book, we studied the fabrication of logic gates using electronic components such as resistors, diodes, BJTs, and FETs (CMOS, NMOS, and PMOS). Logic gates are examples of small digital systems—that is, systems that operate on information in digital form. As demonstrated in Chapter 12, logic gates in SSI form can be combined to make larger digital systems. In addition, fundamental digital building blocks such as adders, magnitude comparators, encoders, decoders, multiplexers, demultiplexers, flip-flops, counters, registers, and memories are available in MSI and LSI packages. These devices are comprised of simpler digital structures, and thus are examples of more complex digital systems. And these more complicated digital devices can be combined to form digital systems that have even greater complexity. Because of the advances in integrated-circuit technology, however, more complicated structures, such as the components of digital information-processing systems (as well as entire systems such as microprocessors and some microcomputers), are available as LSI or VLSI packages. Let us now discuss the major components of some digital information-processing systems.

Digital-Computer Components

Undoubtedly, the most important digital system is the digital computer. A **digital computer** is a digital system that sequentially executes a set of instructions on data in digital form. Digital computers take many forms as to size and shape, but from a functional point of view, their behavior can be characterized by the simplified block diagram shown in Fig. 13.36.

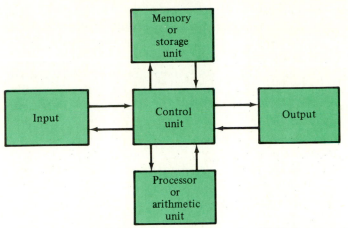

Fig. 13.36 Simplified block diagram of a digital computer.

Information in the form of instructions and data enters a computer via its input. Some of the ways in which binary information is put into computers include punched cards, magnetic tape, paper tape, magnetic disks, teletypewriters, video displays, and keyboards, In addition to the data, the set of instructions (program) to be executed is stored in a portion of the computer's memory known as **main memory.**

Main memory typically is comprised of memories with random-access capability—specifically, RAMs and ROMs. Although semiconductor IC units predominate, magnetic-core memories still see use. In addition to main memory, many computers have bulk memory devices such as magnetic disks, tapes, and drums for storing large amounts of information. Usually, information in bulk storage is not directly available for data processing, but first must be transferred into main memory.

The job of the processor or arithmetic unit is to perform all the arithmetic and logical operations. For this reason, it is also known as the **arithmetic-logic unit (ALU)** of the computer. The arithmetic and logic operations are performed by various logic circuits. Intermediate results of ALU operations may be temporarily stored in ALU registers known as **accumulators.** However, overall, the ALU receives data from the memory unit, and after the appropriate operations are performed in the ALU, the results are stored in the memory unit.

The control unit is the heart of the computer. After it reads and interprets an instruction, it sends the appropriate timing and control signals to the portions of the computer that perform the required operations. Upon the completion of these operations, the control unit proceeds to the next instruction. Control units contain counters, combinational logic circuits, and registers for the temporary storage of information. The combination of the control unit and the ALU is called the **central processing unit** (CPU). A CPU on a single IC chip is referred to as a **microprocessor.**

Information can be put out of a computer by various types of output devices. Examples are teletypewriters, printers, video (CRT) displays, and plotters.

There are basically three classifications of computers. Large computers are known as **mainframe computers** or **maxicomputers.** Intermediate-size computers are referred to as **minicomputers,** and small computers are called **microcomputers.** A computer transfers and manipulates binary information in groups of bits. The size of these groups is called the **word size** of the computer. Typically, the word size for a microcomputer is 8 bits (referred to as 1 byte). The word size for a minicomputer usually ranges from 8 to 32 bits, with 16 bits being quite common; mainframe computers have word sizes ranging from 16 to 64 bits, with 32 bits being typical.

A microcomputer can be formed from a microprocessor, a clock, a ROM, a RAM, and input/output capability. Although the CPU is usually on one chip and the other components on separate chips, it is possible to have all the major components on one LSI chip. Such a "computer on a chip," of course, has greater limitations than does a multi-chip microcomputer. However, single-chip computers can be gainfully employed in such things as electronic test equipment, microwave ovens, automobile ignition systems, supermarket scales, and video games.

The classical approach to digital-system design is to combine such devices as logic gates, flip-flops, counters, registers, and various combinational logic circuits (e.g., multiplexers, decoders) so as to implement a particular system behavior. However, an alternative method of digital-system design employs the microcomputer. Instead of interconnecting various digital devices, a microcomputer can be programmed so that a specific behavior is realized. A major advantage of this approach is that if the operating characteristics of the system must be modified, then only the programming of the microcomputer need be changed. In contrast to this, a classically designed digital system could require significant changes in its interconnections and/or components. Furthermore, having a microcomputer available on just one or a few IC chips can result in a digital system that is smaller in size and lower in cost than one that is classically designed.

Digital-to-Analog Converters (DACs)

Digital computers and microprocessors are digital information-processing systems, but information quite often is in analog form (e.g., speech, music, and video signals). To process this type of information with sophisticated digital techniques, the information first must be converted from analog to digital form. A device that accomplishes this is known as an **analog-to-digital converter** (ADC or A/D converter). Since many types of electronic equipment are inherently analog devices (e.g., stereo amplifiers, radio and television receivers), there are many occasions when it is necessary to transform digital information into analog information. This can be accomplished with a **digital-to-analog converter** (DAC or D/A converter).

The circuit shown in Fig. 13.37 is an example of a simple DAC. The input to the circuit is the 3-bit binary number $A_2A_1A_0$, where the variables A_0, A_1, and A_2 each takes on the value 0 V or 1 V. Summing currents at the inverting input of the operational amplifier (op amp) on the left yields

$$\frac{v_1}{R_F} + \frac{A_0}{R} + \frac{A_1}{R/2} + \frac{A_2}{R/4} = 0$$

from which

$$v_1 = -\frac{R_F}{R}(4A_2 + 2A_1 + A_0)$$

Since the second op-amp stage has a voltage gain of $v_2/v_1 = -R_2/R_1$, the output voltage v_2 is given by

$$v_2 = \frac{R_2R_F}{R_1R}(4A_2 + 2A_1 + A_0)$$

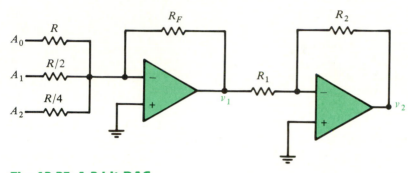

Fig. 13.37 A 3-bit DAC.

Consider the case that $R_1 = R_F$ and $R_2 = R$. Then we have

$$v_2 = 4A_2 + 2A_1 + A_0$$

Under this circumstance, when the input is $A_2A_1A_0 = 000$, the output is $v_2 = 0$ V; while when the input is $A_2A_1A_0 = 111$, the output is $v_2 = 7$ V. In general, for these choices of R_1 and R_2, the value of the output voltage is equal to value of the binary number put in. Different choices for R_1 and R_2 will result in different output voltages (see Problem 13.27). Further, if logical 0 and logical 1 correspond to voltages other than 0 V and 1 V, respectively, different output voltages will again result (see Problem 13.28).

Extending the circuit shown in Fig. 13.37 to implement a DAC with more than three input bits requires additional input resistors $R/8, R/16, R/32$, and so on (see Problem 13.29). Because of the wide variation of input-resistance values required for DACs with many input bits, a better DAC design is the one shown in Fig. 13.38. This DAC uses a connection of resistors called an **R-2R ladder,** and employs a single op amp.

Since the input of an (ideal) op amp draws no current, we can analyze the ladder network as if the op amp were absent. Let us find the voltage v_1 by using

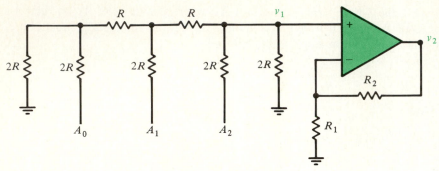

Fig. 13.38 A 3-bit DAC using an *R*-2*R* ladder.

the principle of superpostion. If $A_0 = A_1 = 0$, then all the resistors, except the $2R$ resistor connected between A_2 and v_1, can be reduced by parallel and series combinations to an R resistor. Then, by voltage division, the voltage at the noninverting input of the op amp is

$$v_{1a} = \frac{R}{R + 2R}A_2 = \tfrac{1}{3}A_2$$

Taking similar approaches, if $A_0 = A_2 = 0$, then $v_{1b} = \tfrac{1}{6}A_1$; if $A_1 = A_2 = 0$, then $v_{1c} = \tfrac{1}{12}A_0$. Therefore, by the principle of superposition,

$$v_1 = v_{1a} + v_{1b} + v_{1c} = \tfrac{1}{3}A_2 + \tfrac{1}{6}A_1 + \tfrac{1}{12}A_0 = \tfrac{1}{12}(4A_2 + 2A_1 + A_0)$$

Since the voltage gain of the op-amp stage is $v_2/v_1 = 1 + R_2/R_1$, the output voltage is

$$v_2 = \left(1 + \frac{R_2}{R_1}\right)\left(\frac{1}{12}\right)(4A_2 + 2A_1 + A_0)$$

$$= \frac{R_1 + R_2}{12R_1}(4A_2 + 2A_1 + A_0)$$

For the case that $R_2 = 11R_1$, we again have that

$$v_2 = 4A_2 + 2A_1 + A_0$$

As for the 3-bit DAC given in Fig. 13.37, the 3-bit DAC shown in Fig. 13.38 can be extended such that n-bit ($n > 3$) binary information can be converted from digital to analog form (see Problem 13.31).

Analog-to-Digital Converters (ADCs)

There are a number of approaches that can be taken for the construction of an analog-to-digital converter (ADC). One technique incorporates a DAC as shown in Fig. 13.39. For this device, an analog input voltage is applied to one of the inputs of a voltage comparator (see Section 10.3). The output of the DAC is applied to the other input of the comparator. As long as the DAC output voltage is less than the analog input voltage, the comparator output is high (logical 1) and clock pulses

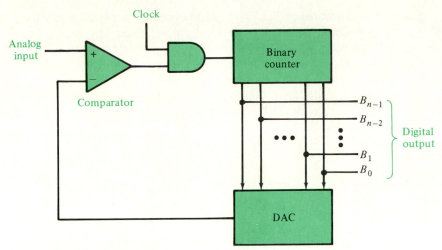

Fig. 13.39 An ADC that employs a DAC.

are applied to the binary counter. However, when the count of the binary counter is high enough such that the value of the DAC output exceeds the analog input, the comparator output is low (logical 0), so additional clock pulses are not applied to the counter. Therefore, the counter stops counting and the resulting n-bit output is $B_{n-1}B_{n-2} \ldots B_1B_0$. The counter is then reset to zero (provision for this is not indicated in the figure), and the next analog input is applied. Such a device is called a **counting ADC.**

For the ADC shown in Fig. 13.39, it was assumed that the analog input remained constant during the time interval required for the counter to reach its final state. If this is not the case, then the analog signal can be "sampled" by a switch S (which may be implemented with a FET, for example) as illustrated by Fig. 13.40. The switch S closes momentarily and the capacitor C charges to the value of the analog voltage. Since the input resistance of the voltage-follower stage is very high, with the switch open, the capacitor "holds" its charge, and, hence, holds the value of the sampled analog signal. After the ADC has had sufficient time to reach its final value, the **sample-and-hold circuit** shown in Fig. 13.40 takes the next sample of the analog signal.

For the ADC shown in Fig. 13.39, the binary counter may be required to count up to $2^n - 1$ clock pulses for some analog inputs. A faster method of analog-to-

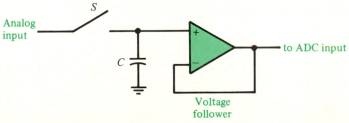

Fig. 13.40 Sample-and-hold circuit.

digital conversion is illustrated in Fig. 13.41. For this circuit, a reference (constant) voltage V_r is divided among the comparator inputs as indicated. The analog input is applied to each comparator as shown. When the analog voltage exceeds $V_r/4$, the output of comparator C_1 is 1; otherwise it is 0. Similarly, when the analog voltage exceeds $V_r/2$, the output of C_2 is 1; otherwise it is 0. And when the analog voltage exceeds $3V_r/4$, the output of C_3 is 1; otherwise it is 0. The outputs of the comparators are encoded to give the desired digital output $B_1 B_0$. The truth table for the encoder is given by Table 13.7.

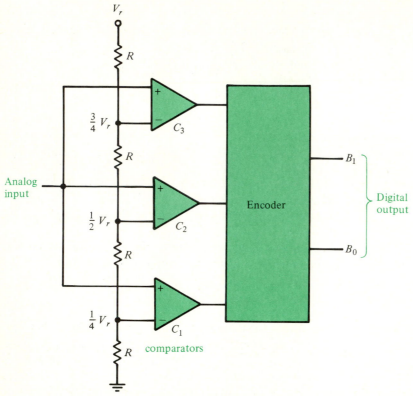

Fig. 13.41 A 2-bit parallel-comparator ADC.

Table 13.7 Truth Table for Encoder in ADC.

Encoder Inputs			Encoder Outputs	
C_3	C_2	C_1	B_1	B_0
0	0	0	0	0
0	0	1	0	1
0	1	1	1	0
1	1	1	1	1

The ADC shown in Fig. 13.41 is a parallel-comparator ADC. In particular, it is a 2-bit ADC. For a 3-bit parallel-comparator ADC, $2^3 - 1 = 7$ comparators are required. In general, $2^n - 1$ comparators are needed for an n-bit parallel-comparator ADC. This is the fastest type of ADC.

Various types of ADCs and DACs are available in IC form.

SUMMARY

1. Counters are used in digital systems to count events and to generate timing sequences.

2. Modulo-m counters are also known as divide-by-m counters.

3. Ripple counters are examples of asynchronous counters.

4. Synchronous counters are faster and do not have the noise problems associated with ripple counters.

5. Ring counters and twisted-ring counters are used to generate timing signals.

6. Registers may be loaded serially or in parallel, and they may be unloaded serially or in parallel.

7. Shift registers can be used to implement serial adders.

8. A read-only memory (ROM) is a nonvolatile memory. Two-dimensional addressing of ROMs results in more efficient device fabrications.

9. Read-and-write memories (RAMs) are either static or dynamic. In either case, a semiconductor RAM is a volatile memory.

10. Magnetic-core memories are random-access devices that are nonvolatile.

11. Magnetic-tape memories are serial memories that are nonvolatile and are used for bulk storage.

12. Magnetic disks are also nonvolatile and used for bulk storage, but they are direct-access devices.

13. A charge-coupled device (CCD) is a serial semiconductor memory. Such a device is, in essence, a MOSFET shift register.

14. The major components of a digital computer are the input, the control unit, the arithmetic-logic unit (ALU), the memory, and the output.

15. The central processing unit (CPU) is comprised of the ALU and the control unit. A microprocessor is a CPU in IC form.

16. Information an be changed from analog to digital form with an analog-to-digital converter (ADC), and from digital to analog form with a digital-to-analog converter (DAC).

Problems for Section 13.1

13.1 Design a modulo-6 binary counter using JK flip-flops.

13.2 Use four JK flip-flops to design a modulo-10 binary counter (also known as a BCD counter).

13.3 Use D flip-flops to design a synchronous counter that is characterized by the state diagram shown in Fig. P13.3.

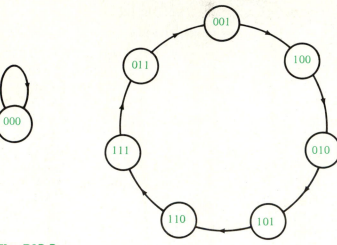

Fig. P13.3

13.4 The asynchronous counter shown in Fig. P13.4 employs three D flip-flops, each of which triggers on the leading edge of a pulse applied to its clock-pulse input. Draw the timing diagram for this counter. What type of counter is this?

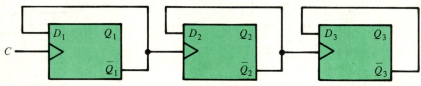

Fig. P13.4

13.5 The 3-bit ripple counter shown in Fig. P13.5 is an example of a binary count-down counter. Draw the timing diagram for this counter given that each T flip-flop triggers on the trailing edge of a pulse applied to its clock-pulse input.

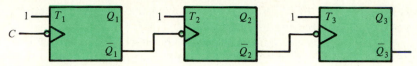

Fig. P13.5

13.6 The JK flip-flop shown in Fig. P13.6 is an example of a modulo-2 counter. Cascade this counter with the modulo-5 counter given in Fig. 13.8 by connecting the output Q of this counter to the clock-pulse input of the modulo-5 counter. Draw the circuit diagram and the timing diagram for the resulting BCD ripple counter.

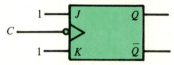

Fig. P13.6

13.7 Design a combinational logic circuit (encoder) to convert a 6-bit ring counter into a binary modulo-6 counter.

13.8 Design a combinational logic circuit to convert a 3-bit twisted-ring counter into a binary modulo-6 counter.

Problems for Section 13.2

13.9 For the shift register shown in Fig. 13.17, draw the timing diagram for the case that the input is the sequence 1, 0, 1, 1, 0, 0, . . .

13.10 The device shown in Fig. P13.10 is an example of a **feedback shift register.** Find the state diagram for this shift register.

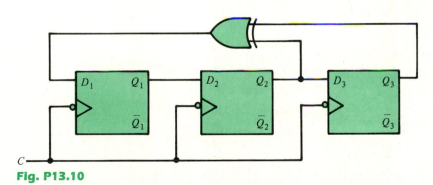

Fig. P13.10

13.11 Repeat Problem 13.10 for the case that the exclusive-OR gate is replaced by an exclusive-NOR gate.

13.12 Repeat Problem 13.10 for the case that the exclusive-OR gate is replaced by (a) an OR gate and (b) an AND gate.

13.13 The series-in, series-out shift register and the associated combinational logic circuit shown in Fig. P13.13 are to count clock pulses as indicated by Table P13.13. Such a device can be called a **serial counter.** Design an appropriate combinational logic circuit.

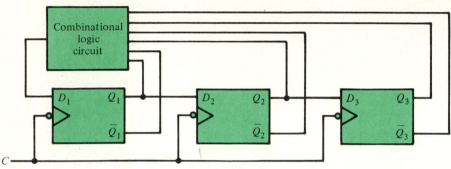

Fig. P13.13

Table P13.13

Number of Clock Pulses	State of Register		
	Q_1	Q_2	Q_3
0	0	0	0
1	1	0	0
2	0	1	0
3	1	0	1
4	1	1	0
5	1	1	1
6	0	1	1
7	0	0	1
8	0	0	0

13.14 Design an additional combinational logic circuit which, when combined with the serial counter given in Fig. P13.13, will produce a modulo-8 binary counter.

13.15 The serial counter shown in Fig. P13.13 produces a sequence of eight distinct states as indicated by Table P13.13. Use the state diagram to find a different sequence of eight distinct states, and design a corresponding combinational logic circuit.

13.16 The feedback shift register shown in Fig. P13.10 can be used to generate the words of a code. A code word is formed by taking the first seven output (Q_3) bits after a binary triple is placed in the register. Determine the eight words in the code.

Problems for Section 13.3

13.17 Draw the circuit diagram for a diode-matrix ROM that is characterized by Table P13.17.

Table P13.17

Inputs			Outputs			
A_2	A_1	A_0	B_3	B_2	B_1	B_0
0	0	0	0	1	1	0
0	0	1	0	1	0	1
0	1	0	1	0	1	0
0	1	1	0	1	1	1
1	0	0	1	0	1	1
1	0	1	1	1	0	0
1	1	0	0	0	1	1
1	1	1	1	0	0	1

13.18 Repeat Problem 13.17 for a MOSFET ROM.

13.19 A 4K-bit ROM stores 4096 bits. Find the number of output lines given that the number of input lines is (a) 10, (b) 8, (c) 6.

13.20 A 2K-bit ROM stores 2048 bits. Find the number of input lines given that the number of output lines is (a) 4, (b) 8, and (c) 16.

13.21 Given a 16 × 8 ROM with inputs A_0, A_1, A_2, A_3 and outputs B_0, B_1, B_2, B_3, B_4, B_5, B_6, B_7. Convert this ROM into a 32 × 4 ROM by adding an additional input line A_4 and using logic gates to form new output lines C_0, C_1, C_2, C_3 such that $C_0 = B_0$, $C_1 = B_1$, $C_2 = B_2$, $C_3 = B_3$ when $A_4 = 0$; while $C_0 = B_4$, $C_1 = B_5$, $C_2 = B_6$, $C_3 = B_7$ when $A_4 = 1$.

13.22 Draw the block diagram for a 256 × 8 (2K-bit) ROM with two-dimensional addressing, where a 64 × 32 cellular matrix is utilized.

13.23 Draw the block diagram for a 4 × 3 (12-bit) RAM that has the form of the RAM shown in Fig. 13.26.

13.24 Determine the number of AND gates required for the address decoder of a 4096 × 1 (4K-bit) RAM that has the form shown in (a) Fig. 13.26 and (b) Fig. 13.27. How many inputs does each AND gate have?

13.25 Each track of an 8-inch floppy disk is partitioned into 26 sectors—each containing 128 bytes (1 byte = 8 bits). Determine the information-storage content of this floppy disk. What is the information-storage content if the number of sectors is increased to 32?

13.26 Extend the waveforms ϕ_1, ϕ_2, and ϕ_3 shown in Fig. 13.35 and add the waveform for ϕ_4 such that the packet of charge under gate ϕ_3 is shifted to the right to the next gate labeled ϕ_1. (The device that produces waveforms ϕ_1, ϕ_2, ϕ_3, and ϕ_4 is known as a **four-phase clock.**)

Problems for Section 13.4

13.27 For the 3-bit DAC shown in Fig. 13.37, suppose that $R_1 = 7R_F$, and $R_2 = 10R$. Construct a table indicating the output voltages that correspond to the eight possible 3-bit binary inputs, given that a logical 0 and a logical 1 correspond to 0 V and 1 V, respectively.

13.28 For the 3-bit DAC shown in Fig. 13.37, suppose that $R_1 = R_F$ and $R_2 = R$. Construct a table indicating the output voltages that correspond to the eight possible 3-bit binary inputs, given that a logical 0 has the value 0 V and a logical 1 has the value 1.5 V.

13.29 Based on the DAC shown in Fig. 13.37, design a 4-bit DAC whose maximum output voltage is 10 V, given that a logical 0 and a logical 1 correspond to 0 V and 1 V, respectively.

13.30 For the 3-bit DAC shown in Fig. 13.38, suppose that $R_2 = 17R_1$. Construct a table indicating the output voltages that correspond to the eight possible 3-bit binary inputs, given that a logical 0 and a logical 1 correspond to 0 V and 1 V, respectively.

13.31 Repeat Problem 13.29 for the DAC shown in Fig. 13.38.

13.32 The DAC shown in Fig. P13.32 is comprised of three individual DACs. For the two DACs on the left, when the input is the binary representation of the decimal digit D, the output is D volts. Express R_0 and R_1 in terms of R_F and R such that when the input of the overall DAC is the decimal number $D_1 D_0$, the output that results is $v_2 = D_1.D_0$ volts.

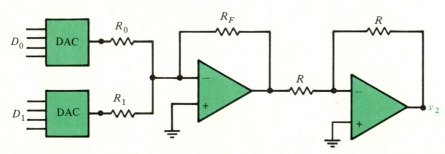

Fig. P13.32

13.33 For the 2-bit parallel-comparator ADC shown in Fig. 13.41, suppose that $V_r = 8$ V, and the analog input voltage varies between 0 and 8 V. Determine the input-voltage interval that corresponds to a digital output $B_1 B_0$ of (a) 00, (b) 01, (c) 10, and (d) 11.

13.34 Draw a 3-bit parallel-comparator ADC. Construct the truth table relating the encoder inputs and outputs.

13.35 Given an n-bit ADC whose analog input voltage varies over a range of V_r volts, the **resolution** of the ADC is defined to be $V_r/2^n$. In terms of the percentage of V_r, the resolution is $1/2^n$. Determine the resolution of an n-bit ADC for the case that (a) $n = 2$, (b) $n = 3$, (c) $n = 4$, (d) $n = 6$, and (e) $n = 8$.

PART IV
Electromechanics

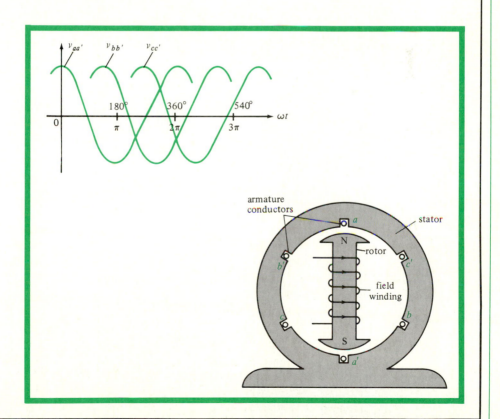

Chapter 14
Electromagnetics

INTRODUCTION

Most electric power is generated and consumed by devices that employ magnetic fields. In this chapter, we discuss the fundamentals of "magnetic circuits" and see how electromagnetics gives rise to a device known as a "transformer." Such a device, typically a four-terminal circuit element, is obtained by placing two inductors in close physical proximity.

Transformers, which are capable of efficiently transferring electric energy, come in various sizes and have a variety of uses. Relatively small transformers are used in radios and television sets to couple amplifier stages, and larger transformers are used in power-supply sections of these and numerous types of electronic equipment. Large, massive transformers are employed by electric utilities in the distribution of electric power.

14.1 MAGNETICS

We are all familiar with **permanent magnets** made of steel or iron alloys. A magnet exerts force (attraction and/or repulsion) on other magnets and on moving electric charge (i.e., current) as well. To account for this phenomenon, we visualize a **magnetic field** around a magnet. For example, Fig. 14.1 shows a bar magnet and its associated magnetic field by means of **lines of magnetic force** or **magnetic flux** (or simply **flux**). The ends of the magnet are its **poles**. The **north pole** (labeled N in Fig. 14.1) is the pole from which the flux emanates, and the **south pole** (labeled S) is the end to which the flux returns. As lines of magnetic flux always form closed loops, the lines of flux pass through the magnet going from the south pole to the

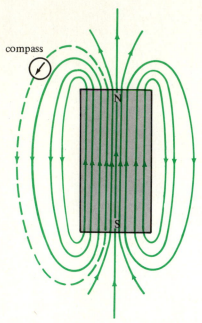

Fig. 14.1 Magnetic field around a magnet.

north pole. A compass located in the magnetic field will indicate the direction of the flux line passing through the point at which the compass is placed (see Fig. 14.1).

The total flux passing through a given area A is denoted by ϕ, where the unit of flux is the **weber***(Wb). When the flux is uniformly distributed over the area A, we say that the **magnetic flux density** B is given by

$$B = \frac{\phi}{A}$$

and the unit of flux density is the **tesla**[†](T). Consequently, 1 T = 1 weber per square meter (Wb/m²).

As flux has direction, so does flux density. The symbol B used above refers to the magnitude of the flux density, which is a vector **B.** In other words, $|\mathbf{B}| = B$. With reference to Fig. 14.1, note that the flux lines nearer the magnet are more closely spaced. This is so because the strength of the magnetic field decreases with distance; that is, the magnitude of **B** decreases with distance.

---

*Named for the German physicist Wilhelm Weber (1804–1891).
[†]Named for the American electrician and inventor Nikola Tesla (1856–1943).

As mentioned above, a magnetic field exerts a force on moving electric charge. Suppose that a charge of q coulombs has velocity **u** (a vector with magnitude and direction), where $|\mathbf{u}| = u$ is measured in meters per second (m/s). If the location of q has magnetic flux density **B**, then exerted on q is a force **f** (a vector with magnitude and direction), where the unit of $|\mathbf{f}| = f$ is the **newton***(N). The magnitude of the force is given by

$$f = quB \sin \theta \qquad\qquad \textbf{(14.1)}$$

where θ is the angle between the vectors **u** and **B**, and this angle is measured in the plane containing **u** and **B**. The vector **f** is normal to the plane containing **u** and **B**, and the direction of **f** is obtained by the **right-hand screw rule:** If the fingers on the right hand are rotated from **u** to **B** (by an amount less than 180°), then the thumb points in the direction of **f**. This concept is illustrated in Fig. 14.2, where vectors **u** and **B** are in the x-y plane. As a consequence of the right-hand screw rule, we see that the vector **f** is on the z axis and is pointed in the positive z direction. The magnitude of **f** is given by Equation (14.1).

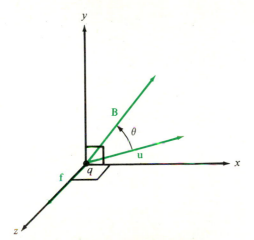

Fig. 14.2 Right-hand screw rule.

A magnetic field not only is produced by a permanent magnet, but a current going through a conductor also gives rise to a magnetic field. In particular, consider a current i going through a long wire as depicted in the three-dimensional view shown in Fig. 14.3. We see that the lines of flux form concentric circles around the conductor, and the direction of the magnetic field is given by another **right-hand rule:** If the thumb on the right hand points in the direction of the current, then the

---

*Named for the English scientist Sir Isaac Newton (1642–1727).

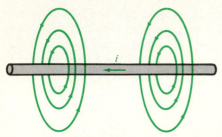

Fig. 14.3 Magnetic field around a conductor.

fingers curl in the direction of the magnetic field. Specifically, for the two-dimensional view shown in Fig. 14.4, a point whose distance is d (measured perpendicularly) from the wire has a flux density of magnitude B given by

$$B = \frac{\mu}{2\pi d}i \tag{14.2}$$

where the term μ is a property of the material that surrounds the conductor, and is called the **permeability** of the material. As 2π is a dimensionless constant, the units of μ are webers per ampere-meter (Wb/A-m). For the situation illustrated in Fig. 14.4, by the right-hand rule, the magnetic field at a point above the wire is perpendicular to and directed into the page (indicated by the cross $\times$), while below the wire, the field is perpendicular to and directed out of the page (indicated by the dot $\cdot$).

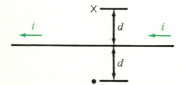

Fig. 14.4 Direction of magnetic field.

We have seen the type of magnetic field that results from a current-carrying straight wire. However, we can obtain a magnetic field similar to the type produced by the bar magnet in Fig. 14.1 by winding a current-carrying wire into a coil having N turns as shown in Fig. 14.5. When it comes to a coil, the right-hand rule can be restated for the magnetic field in the core of the coil: If the fingers on the right hand curl in the direction of the current, then the thumb points in the direction of the magnetic field inside the core. A cylindrical coil of wire, whether it has an air core, an iron core, or some other type, is called a **solenoid.** If the radius of the core is r, and the length of the cylinder is l, then when $l \gg r$, it can be shown that the flux density of the interior of the core is approximated by

$$B = \frac{\mu N}{l}i \tag{14.3}$$

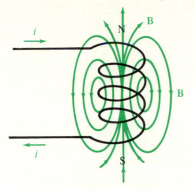

Fig. 14.5 Magnetic field for a coil.

Since the cross-sectional area of the core is $A = \pi r^2$, the total flux is given by

$$\phi = BA = \frac{\mu N \pi r^2}{l} i \qquad\qquad (14.4)$$

From Equation (14.4), we see that the flux in the core of a solenoid depends on the permeability μ of the core material. The permeability of free space is designated by μ_0 and has the value

$$\mu_0 = 4\pi \times 10^{-7} \text{ Wb/A-m}$$

We define the **relative permeability** μ_r to be the ratio

$$\mu_r = \frac{\mu}{\mu_0}$$

The relative permeability of most materials is approximately unity (examples are air and copper). However, **ferromagnetic** materials (such as iron, nickel, steel, and cobalt) have quite high relative permeabilities. A typical value for a good magnetic material is $\mu_r = 1000$ or greater. Thus, we see that the flux, and hence the flux density, of a solenoid with a ferromagnetic core may be a thousand or more times greater than for an air core.

Suppose that instead of a cylindrical core, a coil of wire is wound on a doughnut-shaped core as shown in Fig. 14.6. For this **toroid,** by the right-hand rule, for the indicated current, the direction of the magnetic flux in the core is clockwise. The expression for this flux ϕ is given by Equation (14.4), where r is the radius of the cross-sectional area and l is the average length (or mean length) of the core. Since $l = 2\pi R$, then Equation (14.4) becomes

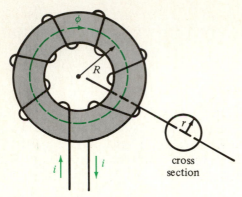

Fig. 14.6 Coil with a toroidal core.

$$\phi = \frac{\mu N r^2}{2R} i \qquad\qquad (14.5)$$

and the magnetic flux density is

$$B = \frac{\phi}{A} = \frac{\mu N}{2\pi R} i \qquad\qquad (14.6)$$

Example 14.1

The toroid shown in Fig. 14.6 consists of 500 turns wound on an iron core having relative permeability $\mu_r = 1500$, average radius $R = 0.1$ m (10 cm*), and cross-sectional radius $r = 0.02$ m (2 cm). Let us determine the current that yields a magnetic flux density of magnitude $B = 0.5$ T in the core.

Since $\mu_r = \mu/\mu_0$, the permeability of the core is

$$\mu = \mu_r \mu_0 = 1500(4 \times 10^{-7})$$

$$= 1.885 \times 10^{-3} \text{ Wb/A-m}$$

Then from Equation (14.6), we have

$$i = \frac{2\pi R B}{\mu N}$$

$$= \frac{2\pi(0.1)(0.5)}{(1.885 \times 10^{-3})(500)} = 0.333 \text{ A}$$

---

*The abbreviation for centimeter (0.01 m) is cm.

Since the cross-sectional area A of the core of a toroid need not necessarily be circular, from Equations (14.3) and the fact that $\phi = BA$, a more general version of Equation (14.4) is

$$\phi = \frac{\mu NA}{l}i \tag{14.7}$$

For this equation, we know that the cross-sectional area A and average length l depend on the geometry of the core, and its permeability μ is a property of the material from which the core is constructed. We see, therefore, that for a given core, the flux ϕ is proportional to the product of the number of coil turns N and the current i. We call this product the **magnetomotive force** (mmf), and designate it by the symbol $\mathscr{F}$. In other words,

$$\mathscr{F} = Ni \tag{14.8}$$

Although N is a dimensionless number, we shall call the units of mmf ampere-turns (A-t) to stress the typical situation of a coil with many turns.

Magnetic Circuits

Returning to Equation (14.7), we now can write it in the form

$$\phi = \frac{1}{\mathscr{R}}\mathscr{F} \tag{14.9}$$

from which

$$\mathscr{F} = \mathscr{R}\phi \tag{14.10}$$

where $\mathscr{R} = l/\mu A$ is the **reluctance** of the path of the flux through the core. From Equation (14.10), it is apparent that the units of reluctance are ampere-turns per weber (A-t/Wb).

Example 14.2

For the toroid described in Example 14.1, we determined the current required to establish $B = 0.5$ T. That is, we found the current (0.333 A) needed to produce a flux of

$$\phi = BA = (0.5)(\pi)(0.02)^2 = 6.283 \times 10^{-4} \text{ Wb}$$

The reluctance of the path of the flux is

$$\mathscr{R} = \frac{l}{\mu A} = \frac{2\pi R}{\mu \pi r^2}$$

$$= \frac{2(0.1)}{(1.885 \times 10^{-3})(0.02)^2} = 2.65 \times 10^5 \text{ A-t/Wb}$$

Thus, the mmf is

$$\mathscr{F} = \mathscr{R}\phi$$

$$= (2.65 \times 10^5)(6.283 \times 10^{-4})$$

$$= 166.5 \text{ A-t}$$

and from Equation (14.8),

$$i = \frac{\mathscr{F}}{N} = \frac{166.5}{500} = 0.333 \text{ A}$$

and this agrees with our previous calculation for the current.

Although we obtained the same results, the approach taken in Example 14.2 was slightly different from the one taken in Example 14.1. Example 14.2 employed Equation (14.10), which may remind you of Ohm's law ($v = Ri$). Drawing such an analogy, we can talk about **magnetic circuits** and employ the analogies summarized in Table 14.1.

Having established electromagnetic analogies, we can use electric circuit principles for analyzing magnetic circuits. The electric analog for the toroid given in Examples 14.1 and 14.2 is shown in Fig. 14.7.

Table 14.1 Analogies for Electric and Magnetic Circuits.

Electric	Magnetic
Current i	Flux ϕ
Voltage v	Mmf $\mathscr{F}$
Resistance R	Reluctance $\mathscr{R}$
$v = Ri$ (Ohm's law)	$\mathscr{F} = \mathscr{R}\phi$

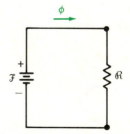

Fig. 14.7 Analogous electric circuit.

Example 14.3

Consider the rectangular iron core, which has a relative permeability of $\mu_r = 1500$, shown in Fig. 14.8. Suppose that there are $N = 200$ turns, and $i = 2$ A.

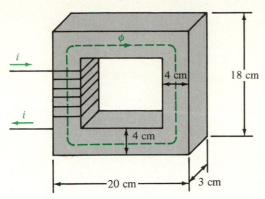

Fig. 14.8 Magnetic circuit with rectangular core.

Since the mean length of the flux path is $l = 16 + 14 + 16 + 14 = 60$ cm $= 0.6$ m, and the cross-sectional area is $A = (4$ cm$)(3$ cm$) = (0.04)(0.03) = 0.0012$ m^2, then the reluctance of the core is

$$\mathcal{R}_c = \frac{l}{\mu A}$$

$$= \frac{0.6}{(4\pi \times 10^{-7})(1500)(1.2 \times 10^{-3})} = 2.65 \times 10^5 \text{ A-t/Wb}$$

and the flux is

$$\phi = \frac{\mathcal{F}}{\mathcal{R}_c} = \frac{Ni}{\mathcal{R}_c}$$

$$= \frac{200(2)}{2.65 \times 10^5} = 1.51 \times 10^{-3} \text{ Wb}$$

The electric analog for this magnetic circuit is again as shown in Fig. 14.7.

Now suppose that a 1-cm section of the core shown in Fig. 14.8 is removed so as to produce an **air gap** as shown in Fig. 14.9. Although the flux that passes through the iron core also passes through the air gap, there is a **fringing** of the flux as illustrated in Fig. 14.9. To account for the effect of this fringing in short gaps, traditionally the dimensions of the gap's cross-sectional area are increased by the gap length l_g. In particular, to account for fringing, the effective cross-sectional area is

$$A_g = (4 + 1)(3 + 1) = 20 \text{ cm}^2$$

$$= (0.05)(0.04) = 0.002 \text{ m}^2$$

so the reluctance of the air gap is

$$\mathcal{R}_g = \frac{l_g}{\mu_0 A_g} = \frac{0.01}{(4\pi \times 10^{-7})(0.002)} = 3.98 \times 10^6 \text{ A-t/Wb}$$

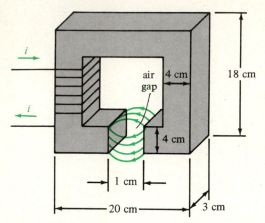

Fig. 14.9 Rectangular core with an air gap.

Since the mean length of the flux path through the iron core material is $l_c = l - 1 = 59$ cm $= 0.59$ m, then the reluctance of the core is

$$\mathcal{R}_c = \frac{l_c}{\mu A} = \frac{0.59}{(4\pi \times 10^{-7})(1500)(1.2 \times 10^{-3})} = 2.61 \times 10^5 \text{ A-t/Wb}$$

These values illustrate a common occurrence—the reluctance of the air gap is much greater than the reluctance of the iron core. Specifically, in this example, $\mathcal{R}_g$ is more than 15 times larger than $\mathcal{R}_c$.

The electric analog for the magnetic circuit is shown in Fig. 14.10. Since the iron core and the air gap are in series, the total reluctance is

$$\mathcal{R} = \mathcal{R}_c + \mathcal{R}_g = 2.61 \times 10^5 + 3.98 \times 10^6 = 4.24 \times 10^6 \text{ A-t/Wb}$$

and we see that most of the reluctance is attributable to the air gap. Furthermore,

$$\mathcal{F} = \mathcal{F}_c + \mathcal{F}_g$$

where $\mathcal{F}_c = \mathcal{R}_c \phi$ and $\mathcal{F}_g = \mathcal{R}_g \phi$.

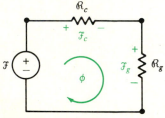

Fig. 14.10 Effect of air gap on analogous electric circuit.

To obtain the same flux as was obtained for the core without an air gap (Fig. 14.8), from

$$\mathcal{F} = Ni = \mathcal{R}\phi$$

we have that

$$i = \frac{\mathcal{R}\phi}{N} = \frac{(4.24 \times 10^6)(1.51 \times 10^{-3})}{200} = 32 \text{ A}$$

In other words, the introduction of the air gap necessitates 16 times the current required to produce the same flux that was obtained without an air gap.

We now must point out that our previous magnetic-circuit calculations were by no means exact. One of the reasons for this is the fringing of the flux as the result of an air gap. However, there are additional reasons. More of the flux is concentrated in the inner portion of a core than in the outer portion, as illustrated in the side view of a core depicted in Fig. 14.11. Some of the flux even may take a shortcut through the air and not reach the air gap. Such **leakage flux** is also shown in Fig. 14.11. Furthermore, the permeability, and hence the reluctance, of ferromagnetic core material varies with the flux through it. Because of this inherent nonlinearity, the analysis of magnetic circuits often employs graphical techniques. Let us see how.

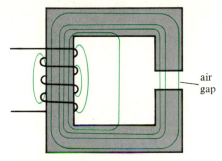

air
gap

Fig. 14.11 Flux associated with rectangular core.

Magnetization Curves

As magnetic flux density $B = \phi/A$, and because $\phi = \mathcal{F}/\mathcal{R}$, then

$$B = \frac{\mathcal{F}}{\mathcal{R}A} = \frac{\mathcal{F}}{(l/\mu A)A} = \frac{\mu\mathcal{F}}{l}$$

since reluctance $\mathcal{R} = l/\mu A$. Thus, we can write

$$\boxed{B = \mu H} \tag{14.11}$$

where

$$H = \frac{\mathcal{F}}{l} = \frac{Ni}{l}$$

(14.12)

Hence, we see that H, called the **magnetic field intensity** or **magnetization force,** has as units ampere-turns per meter (A-t/m) and is independent of core material (it depends on the number of turns, the current, and the mean length).

A plot of magnetic flux density B versus magnetic field intensity H is called a **magnetization curve** or **B-H curve,** and graphically describes the average relationship between B and H. (Such curves are typically supplied by manufacturers of magnetic materials.) Figure 14.12 shows magnetization curves for cast iron, cast steel, and sheet steel. Note that for values of B less than 1 T for steel and 0.5 T for cast iron, the curves are approximately straight lines, so that the relationship $B = \mu H$ is essentially a linear one. (Remember that the relationship for non-magnetic material, e.g., air, is given by $B = \mu_0 H$, and, therefore, is linear.) It is for these reasons that the previously discussed magnetic-circuit analysis is reasonably accurate.

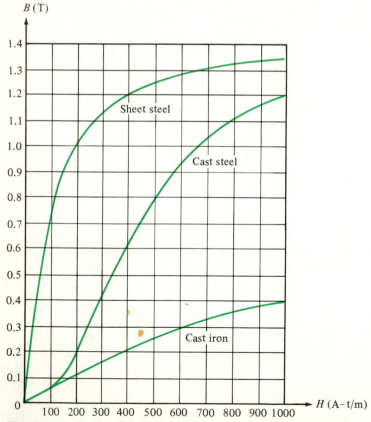

Fig. 14.12 Magnetization curves.

Example 14.4

Suppose that the toroid given in Fig. 14.6 is constructed of sheet steel, and suppose that the magnetic flux density is to be $B = 1$ T. Then from the magnetization curve given in Fig. 14.12, we have that the field intensity is $H = 200$ A-t/m. From Equation (14.11),

$$\mu = \frac{B}{H} = \frac{1}{200} = 5 \times 10^{-3} \text{ Wb/A-m}$$

and the relative permeability is

$$\mu_r = \frac{\mu}{\mu_0} = \frac{5 \times 10^{-3}}{4\pi \times 10^{-7}} = 3979$$

Suppose that $R = 0.1$ m, $r = 0.02$ m, and $N = 500$ turns. From Equation (14.12), the mmf is

$$\mathcal{F} = Hl = 200(2\pi)(0.1) = 125.66 \text{ A-t}$$

and since $\mathcal{F} = Ni$, the current required to produce the desired magnetic field intensity is

$$i = \frac{\mathcal{F}}{N} = \frac{125.66}{500} = 0.25 \text{ A}$$

Example 14.5

Suppose that the rectangular core given in Fig. 14.9 is cast iron. To establish a flux of $\phi = 2.4 \times 10^{-4}$ Wb in the iron, the magnetic flux density should be

$$B = \frac{\phi}{A} = \frac{2.4 \times 10^{-4}}{(0.03)(0.04)} = 0.2 \text{ T}$$

From the magnetization curve for cast iron (Fig. 14.12), the field intensity is $H_c = 400$ A-t/m. Thus, the mmf for the iron is

$$\mathcal{F}_c = H_c l_c = 400(0.59) = 236 \text{ A-t}$$

For the air gap, we have already calculated in Example 14.3 that the reluctance is $\mathcal{R}_g = 3.98 \times 10^6$ A-t/Wb. Thus, the mmf for the air gap is

$$\mathcal{F}_g = \mathcal{R}_g \phi = (3.98 \times 10^6)(2.4 \times 10^{-4}) = 955.2 \text{ A-t}$$

Therefore, the total mmf is

$$\mathcal{F} = \mathcal{F}_c + \mathcal{F}_g = 236 + 955.2 = 1191.2 \text{ A-t}$$

For the case of $N = 200$ turns, the current required for a flux of

$\phi = 2.4 \times 10^{-4}$ Wb is

$$i = \frac{\mathcal{F}}{N} = \frac{1191.2}{200} = 5.96 \text{ A}$$

Not only are magnetic-circuit calculations approximate for the reasons mentioned previously, but the magnetic history of magnetic materials is also a factor. Let us see why.

A **demagnetized** magnetic material is one whose magnetic flux density is zero (i.e., $B = 0$) when no external magnetizing force is applied (i.e. $H = 0$). If a magnetizing force is applied to a demagnetized magnetic material, then the relationship between the magnetic flux density B that results and the magnetizing force H is given by the magnetization curve of the material. A typical B-H curve is shown in Fig. 14.13(a). The arrowheads indicate the direction that a point on the curve moves as magnetizing force increases. For smaller values of H, the B-H curve is relatively straight. As H increases, the curve bends and then flattens out so that a further increase in H results in only a slight increase in B. Eventually, an increase in H yields no further increase in B. The material is then said to be **saturated.** If, after the material is saturated, H is reduced to zero, B will decrease. However, the original magnetization curve does not describe how B decreases. Instead, the resulting behavior is as illustrated in Fig. 14.13(b). The value of the magnetic flux density obtained when H is reduced to zero is $B = B_r$. This value is the **residual magnetism** of the material. (Permanent magnets are made from magnetic materials with very large values of residual magnetism, e.g., 1 T.) If H is increased again, the material eventually will saturate, but the path taken will not be along the one by which point B_r was reached. If H is made increasingly negative, there will be some value $H = -H_c$ for which the flux density is $B = 0$ (see Fig. 14.14). The value H_c is called the **coercive force** of the material. (Permanent magnets have a large coercive force, e.g., 10^4 A-t/m.) If H is made negative enough, the material will eventually saturate, but this time in the opposite direction. If H is then

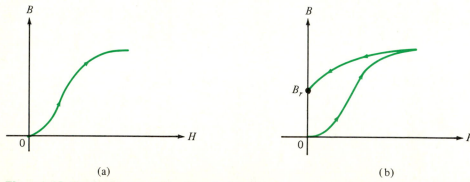

(a) (b)

Fig. 14.13 Typical magnetization behavior.

increased and made large enough, the magnetic material will again saturate in the positive direction as indicated in Fig. 14.14. The resulting loop, shown with solid lines in Fig. 14.14, is called a **hysteresis loop,** and the corresponding phenomenon is known as **hysteresis.** If the magnetizing force applied to the demagnetized material is less than that required to produce saturation, a hysteresis loop such as the dashed loop in Fig. 14.14 may be obtained.

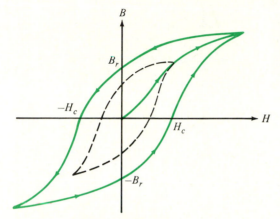

Fig. 14.14 Hysteresis loop.

Certain magnetic materials have hysteresis loops that are approximately rectangular. Such a material, called a **square-loop material,** is depicted in Fig. 14.15. Note that the slopes of the sides of this hysteresis loop are quite large. This means that a small change in H can produce a large change in B. Small cores made from such a material are used as binary memory devices in switching circuits and digital computers.

When magnetic material is periodically magnetized and demagnetized (e.g., due to an alternating current), power, and hence energy, is absorbed by the material

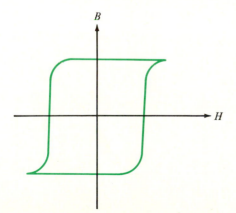

Fig. 14.15 Square hysteresis loop.

and is converted to heat. An empirical formula for this average hysteresis power loss P (in watts) was developed by Charles P. Steinmetz, and is

$$P_h = K_h f B_m^n \qquad\qquad (14.13)$$

where the constant K_h and the value of n (called the **Steinmetz exponent***) depend on the core material, B_m is the maximum flux density, and f is the frequency of the magnetic variations.

14.2 TRANSFORMERS

Electric energy can be transferred efficiently with the use of a "transformer." Although such a device, which is an example of a magnetic circuit, typically consists of two multiturn coils wound on a common core, more than two windings may also be used. Transformer applications are varied and extremely important.

One major use of transformers is in power distribution. Transformers possess the ability to "step up" and "step down" ac voltages and currents. Because of this, electric power, which, conveniently, is generated typically from 5 to 50 kV, can be stepped up in voltage and stepped down in current prior to distribution. The transmisison of power at high voltages and relatively low currents means less of an $I^2 R$ power loss in the transmission lines, and, consequently, greater efficiency. Typically, the transmission of power over long distances occurs at 200 to 400 kV, and even may be at voltages up to 1000 kV. Residential-area power lines typically range between 2 and 10 kV. After power is distributed to the load ends, transformers step down voltages typically to 120-V, 240-V levels for ordinary consumption.

In addition to power systems, transformers are commonly used in electronic, control, and communication circuits and systems. Small power transformers are employed for power supplies in various electronic devices and instruments. Transformers are often used for "impedance matching," to accomplish maximum power transfer between source and load. Since they can be employed to couple circuits (e.g., amplifier stages) without requiring a conducting electrical connection, transformers are also used to insulate one circuit from another, but at the same time they allow for the transfer of electric energy.

Magnetic Induction

It was Michael Faraday who first announced, in 1831, that magnetic flux passing through a multiturn coil would induce a voltage, called an **electromotive force** (emf), across the coil, provided that the flux was time varying. If the same flux ϕ passes through or **links** all of the N turns of a coil, then **Faraday's law** says that the emf e induced across the coil is given by

---

*A typical value of n is 1.6.

$$e = N\frac{d\phi}{dt}$$

(14.14)

where, of course, the units of e are volts. Let us designate the product of the flux ϕ linking the coil and the number of turns of the coil N by λ, that is, $\lambda = N\phi$. We call λ the number of **flux linkages,** and although N is a dimensionless number, we shall call the units of flux linkages **weber-turns** (Wb-t) to stress the typical situation of a coil with many turns. As a consequence of the definition of λ, we can rewrite Equation (14.14) as

$$e = \frac{d\lambda}{dt}$$

(14.15)

Suppose that an L-henry inductor is made with a coil of wire. We know that the voltage v across it and the current i through it are related by

$$v = L\frac{di}{dt}$$

(14.16)

Thus, for the given inductor (coil), from Equations (14.15) and (14.16), we have

$$L\frac{di}{dt} = \frac{d\lambda}{dt}$$

Therefore,

$$L\frac{di}{dt}\frac{dt}{di} = \frac{d\lambda}{dt}\frac{dt}{di}$$

and by the chain rule of calculus, we have

$$L = \frac{d\lambda}{di} = N\frac{d\phi}{di}$$

(14.17)

For the case that the core of a coil is well below saturation, we can utilize Equations (14.8) and (14.10) to deduce that

$$\phi = \frac{Ni}{\mathcal{R}} = Ki$$

(14.18)

In other words, the flux is directly proportional to the current [see Equations (14.4), (14.5), and (14.7), also]. From Equation (14.18),

$$\frac{d\phi}{di} = K$$

But, also from Equation (14.18),

$$K = \frac{\phi}{i}$$

Substituting these two facts in Equation (14.17) results in

$$L = NK = N\frac{\phi}{i}$$

or

$$L = \frac{N\phi}{i} = \frac{\lambda}{i} \tag{14.19}$$

Thus, inductance is the number of flux linkages per ampere, and we see that 1 H = 1 Wb/A. Previously, we saw that the permeability μ of a material was measured in webers/ampere-meter. Now we see that we can alternatively measure permeability in henries/meter (H/m). Actually, H/m is the SI unit for permeability.

Note also from Equation (14.18) and Equation (14.19) that

$$L = \frac{N(Ni/\mathcal{R})}{i}$$

or

$$L = \frac{N^2}{\mathcal{R}} \tag{14.20}$$

In other words, inductance is directly proportional to the square of the number of turns.

Example 14.6

A toroidal coil with 500 turns as shown in Fig. 14.6 has a flux of $\phi = 6.283 \times 10^{-4}$ Wb when $i = \frac{1}{3}$ A. The inductance of this coil, from Equation (14.19), is

$$L = \frac{N\phi}{i} = \frac{500(6.283 \times 10^{-4})}{1/3} = 0.942 \text{ H}$$

Since the energy stored in an L-henry inductor is $w = Li^2/2$, the stored energy is

$$w = \tfrac{1}{2}(0.942)(\tfrac{1}{3})^2 = 0.052 \text{ J}$$

Although the value of an induced emf can be obtained by using either Equation (14.14) or Equation (14.15), the polarity of this voltage is determined from **Lenz's law,*** which states:

> The polarity of the voltage induced by a changing flux tends to oppose the change in flux that produced the induced voltage.

To demonstrate Lenz's law, consider the core shown in Fig. 14.16. Suppose that a time-varying voltage source v is applied to the left coil and the resulting current is i as illustrated. By the right-hand rule, if at some instant of time i has a positive value, the resulting magnetic flux ϕ is clockwise as shown. If i is increasing with time, then ϕ is increasing with time since $Ni = \mathcal{R}\phi$. Under this circumstance, the coil on the right will tend to have a current in the direction shown as this would produce a counterclockwise flux that opposes the increasing flux ϕ. If nothing is connected to terminals a and b of the coil on the right, then there is no current through this coil. Whether or not there is a current through this coil, there is a voltage v_{ab} induced across it. The polarity of this voltage is easily determined by assuming that a resistor is connected to terminals a and b. The resulting current in the coil yields the polarity of v_{ab} (a positive quantity) as shown. Note that when i is increasing with time, since $v = L\, di/dt$, then v is also a positive quantity.

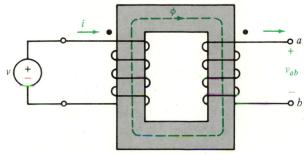

Fig. 14.16 Magnetically coupled coils.

If i is decreasing instead of increasing, then the coil on the right tends to have a current in the direction that is opposite to the one shown, and the polarity of v_{ab} is reversed. Note, also, that if i is decreasing with time, then v is a negative quantity since $v = L\, di/dt$.

The polarity of the voltage induced across the coil on the right, of course, depends upon how that coil is wound with respect to the coil on the left. One way to indicate such a relationship is with **dots.** The dots at the ends of the coils in

*Named for the German physicist Heinrich Lenz (1804–1865).

Fig. 14.16 indicate that the dotted ends are positive at the same time, and, of course, are negative at the same time. Our discussion above confirms this fact. Had we been so inclined, we could have placed both dots at the other ends of their respective coils—their significance remains the same.

Magnetic Coupling

For the rectangular core shown in Fig. 14.16, suppose that the coil on the left has N_1 turns and inductance L_1, and the coil on the right has N_2 turns and inductance L_2. Since a time-varying current going through the coil on the left induces a voltage across the coil on the right, we say that the coils are **magnetically coupled.** In particular, when there is no load connected to terminals a and b (as depicted in Fig. 14.16), since from Equation (14.18),

$$\phi = \frac{N_1 i}{\mathcal{R}}$$

then by using Equation (14.14), when there is no flux leakage, we find that the induced voltage v_{ab} is

$$v_{ab} = N_2 \frac{d\phi}{dt} = \frac{N_1 N_2}{\mathcal{R}} \frac{di}{dt} = M \frac{di}{dt} \tag{14.21}$$

where $M = N_1 N_2/\mathcal{R}$ is a positive number. Since Equation (14.21) has the form $v = L di/dt$, we deduce that M is an inductance, and we call it a **mutual inductance.** The inductances L_1 and L_2 are referred to as **self-inductances.** As for the case of self-inductance, the unit for mutual inductance is the henry.

From Equation (14.21), we see that if i is increasing with time, then v_{ab} is positive; if i is decreasing with time, then v_{ab} is negative. In accordance with our previous discussion, these are the correct values of v_{ab} for the polarity indicated in Fig. 14.16. As a consequence of these facts, we can make a general statement concerning dots: If a current i goes into the dotted end of one coil, then for the other coil, the polarity of the induced voltage $M di/dt$ is plus (+) at the dotted end. Conversely, if a current i comes out of the dotted end of one coil, then the polarity of the induced voltage $M di/dt$ is minus (−) at the dotted end for the other coil.

The pair of magnetically coupled coils shown in Fig. 14.16 is an example of a **transformer.** One of the coils, say the coil on the left, is designated as the **primary** winding or coil, and the other is the **secondary** winding or coil. A circuit symbol for a transformer is shown in Fig. 14.17(a). In general, there may be connections to both windings, so i_1 and i_2 both can be nonzero. As a result, the current i_1 going through the self-inductance L_1 produces a voltage $L_1 di_1/dt$. However, the current i_2 going through the secondary induces the voltage $M_{12} di_2/dt$ across the primary as well. By the principle of superposition, we have

$$v_1 = L_1 \frac{di_1}{dt} + M_{12} \frac{di_2}{dt} \tag{14.22}$$

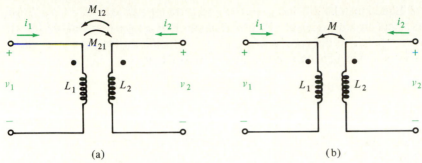

Fig. 14.17 Transformer.

where the mutual inductance that corresponds to the voltage induced across the primary winding, due to the current in the secondary winding, is M_{12}. In a similar manner, we can obtain

$$v_2 = M_{21} \frac{di_1}{dt} + L_2 \frac{di_2}{dt} \tag{14.23}$$

where the mutual inductance that corresponds to the voltage induced across the secondary, due to the current in the primary, is M_{21}.

The reasoning used to derive Equation (14.21) suggests that $M = N_1 N_2 / \mathcal{R} = M_{12} = M_{21}$. This fact can be verified by using an energy-based argument and some calculus manipulations.* Thus, we will denote the mutual inductance of a transformer by M, and we can represent a transformer as shown in Fig. 14.17(b). Furthermore, Equations (14.22) and (14.23) can be rewritten as the following general pair of equations describing a transformer:

$$v_1 = L_1 \frac{di_1}{dt} + M \frac{di_2}{dt}$$

$$v_2 = M \frac{di_1}{dt} + L_2 \frac{di_2}{dt} \tag{14.24}$$

In addition, the energy stored in the transformer can be shown to be

$$w(t) = \tfrac{1}{2} L_1 i_1^2(t) + \tfrac{1}{2} L_2 i_2^2(t) + M i_1(t) i_2(t) \tag{14.25}$$

The use of transformers is limited to non-dc applications, since for the dc case,

---

*See *Elementary Linear Circuit Analysis* by Leonard S. Bobrow, Holt, Rinehart and Winston, New York, 1981, pp. 512–514.

the pair of Equations (14.24) indicates that the primary and secondary windings behave as short circuits. Just as the time-domain expression $v = L\,di/dt$ for an inductor corresponds to the phasor relationship $\mathbf{V} = j\omega L\mathbf{I}$ for the sinusoidal case, the frequency-domain version of Equation-pair (14.24) is

$$\mathbf{V}_1 = j\omega L_1\,\mathbf{I}_1 + j\omega M\mathbf{I}_2$$
$$\mathbf{V}_2 = j\omega M\mathbf{I}_1 + j\omega L_2\,\mathbf{I}_2 \tag{14.26}$$

Example 14.7

Consider the sinusoidal circuit given in Fig. 14.18. Shown in Fig. 14.19 is the representation of this circuit in the frequency domain. For mesh $\mathbf{I}_1$, by KVL,

$$\mathbf{V} = 9\mathbf{I}_1 + j12\mathbf{I}_1 + j6\mathbf{I}_2 = (9 + j12)\mathbf{I}_1 + j6\mathbf{I}_2 \tag{14.27}$$

while for mesh $\mathbf{I}_2$, by KVL,

$$0 = j6\mathbf{I}_1 + j18\mathbf{I}_2 - j6\mathbf{I}_2 = j6\mathbf{I}_1 + j12\mathbf{I}_2 \tag{14.28}$$

From Equation (14.28), $\mathbf{I}_1 = -2\mathbf{I}_2$, and substituting this fact in Equation (14.27) yields

$$\mathbf{V} = -2(9 + j12)\mathbf{I}_2 + j6\mathbf{I}_2 = -18(1 + j)\mathbf{I}_2$$

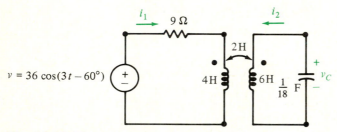

Fig. 14.18 Time-domain transformer circuit.

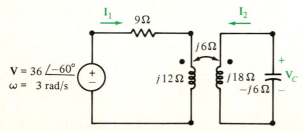

Fig. 14.19 Frequency-domain transformer circuit.

from which

$$\mathbf{I_2} = \frac{\mathbf{V}}{-18(1 + j)}$$

Thus,

$$\mathbf{V}_C = \mathbf{Z}_C(-\mathbf{I_2}) = (-j6)(-\mathbf{I_2}) = \frac{-j6\mathbf{V}}{18(1 + j)}$$

$$= \frac{(6\underline{/-90°})(36\underline{/-60°})}{(18\underline{/0°})(\sqrt{2}\underline{/45°})} = 6\sqrt{2}\underline{/-195°} = 6\sqrt{2}\underline{/165°}$$

Hence,

$$v_C(t) = 6\sqrt{2}\,\cos(3t + 165°)$$

Although a transformer is basically a pair of inductors (coils) that are magnetically coupled, when a primary terminal and a secondary terminal are connected together as shown in Fig. 14.20(a), we can model the transformer by the connection of three (uncoupled) inductors shown in Fig. 14.20(b). To verify that the circuit given in Fig. 14.20(b) is equivalent to the one in Fig. 14.20(a), simply apply KVL to the transformer in Fig. 14.20(a) and its "T-equivalent" shown in Fig. 14.20(b). The result is the Equation-pair (14.24).

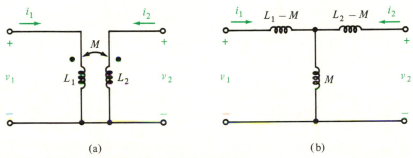

(a)   (b)

Fig. 14.20 Transformer and its model.

As with all magnetic circuits, there are certain losses in the cores of transformers. We have already discussed one type of loss—that due to hysteresis. The other type is attributable to "eddy currents." Since a magnetic core, such as iron, is a conductor, then a time-varying magnetic flux will induce a voltage that produces circulating currents, called **eddy currents,** in the core. The eddy current depicted in Fig. 14.21(a) causes an $I^2 R$ power loss that results in core heating. An expression for this eddy-current power loss is

$$P_e = K_e f^2 B_m^2$$

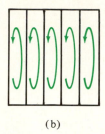

(a)   (b)

Fig. 14.21 Eddy currents.

where B_m is the maximum flux density, f is the frequency of the magnetic variations, and the constant K_e depends on the core material. To reduce eddy-current power loss, instead of using solid magnetic material, cores are constructed from thin sheets or **laminations** of the material, as illustrated in Fig. 14.21(b). The laminations,* which are typically insulated from each other with thin layers of varnish, have increased resistances and decreased induced voltages. The result is a significant reduction in eddy-current loss.

In addition to transformers, hysteresis and eddy-current losses, collectively known as **core loss,** occur in ac and dc machines. The expression for core loss P is given by

$$P = P_h + P_e = K_h f B_m^n + K_e f^2 B_m^2$$

14.3 THE IDEAL TRANSFORMER

The behavior of many iron-core transformers, especially those used in power systems, can be "idealized." This results in simplified voltage–current relationships and circuit models. These, then, can be employed to analyze circuits and obtain very accurate results. Therefore, we now proceed to develop and investigate the concept of an "ideal transformer."

Since the energy stored in a transformer is

$$w = \tfrac{1}{2}L_1 i_1^2 + \tfrac{1}{2}L_2 i_2^2 + M i_1 i_2$$
$$= \tfrac{1}{2}[L_1 i_1^2 + L_2 i_2^2] + M i_1 i_2$$

by completing the square, we get

$$w = \tfrac{1}{2}[(\sqrt{L_1}\, i_1 + \sqrt{L_2}\, i_2)^2 - 2\sqrt{L_1 L_2}\, i_1 i_2] + M i_1 i_2$$
$$= \tfrac{1}{2}(\sqrt{L_1}\, i_1 + \sqrt{L_2}\, i_2)^2 + (M - \sqrt{L_1 L_2}) i_1 i_2$$

Consider the case for which

$$\sqrt{L_1}\, i_1 + \sqrt{L_2}\, i_2 = 0$$

*A typical thickness for some types of low-frequency transformers is 0.0356 cm.

This occurs when

$$\sqrt{L_1}\, i_1 = -\sqrt{L_2}\, i_2$$

or

$$i_2 = -\sqrt{\frac{L_1}{L_2}}\, i_1$$

and the resulting energy stored is given by

$$w = (M - \sqrt{L_1 L_2})i_1\left(-\sqrt{\frac{L_1}{L_2}}\, i_1\right)$$

$$= (\sqrt{L_1 L_2} - M)\left(\sqrt{\frac{L_1}{L_2}}\right)i_1^2$$

Since the energy stored is a nonnegative quantity, we must have that

$$0 \le \sqrt{L_1 L_2} - M$$

from which

$$\boxed{M \le \sqrt{L_1 L_2}}$$

(14.29)

Thus, we have an upper bound for the mutual inductance M.

If we define the **coefficient of coupling** of a transformer, denoted by k, as

$$\boxed{k = \frac{M}{\sqrt{L_1 L_2}}}$$

then by inequality (14.29) we have that $0 \le k \le 1$.

The coupling coefficient k of a physical transformer is determined by a number of factors: the permeability of the core on which the primary and secondary coils are wound; the number of turns of each coil; the coils' relative positions and their physical dimensions. If k is close to zero, we say that the coils are **loosely coupled,** and if k is close to one, we say that they are **tightly coupled.** Air-core transformers are typically loosely coupled, and iron-core transformers are usually tightly coupled. As a matter of fact, iron-core transformers can have coupling coefficients approaching unity.

Consider the case of a transformer with **perfect coupling,** that is, $k = 1$. We represent such a device as shown in Fig. 14.22. Since

$$\mathbf{V}_1 = j\omega L_1 \mathbf{I}_1 + j\omega M \mathbf{I}_2$$

then

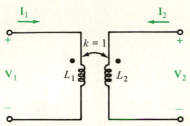

Fig. 14.22 Transformer with perfect coupling.

$$\mathbf{I}_1 = \frac{\mathbf{V}_1 - j\omega M \mathbf{I}_2}{j\omega L_1} \tag{14.30}$$

Also,

$$\mathbf{V}_2 = j\omega M \mathbf{I}_1 + j\omega L_2 \mathbf{I}_2$$

$$= j\omega M \left(\frac{\mathbf{V}_1 - j\omega M \mathbf{I}_2}{j\omega L_1} \right) + j\omega L_2 \mathbf{I}_2$$

$$= \frac{M \mathbf{V}_1}{L_1} - \frac{j\omega M^2 \mathbf{I}_2}{L_1} + j\omega L_2 \mathbf{I}_2$$

But for perfect coupling ($k = 1$), $M = \sqrt{L_1 L_2}$ so that

$$\mathbf{V}_2 = \frac{\sqrt{L_1 L_2}\, \mathbf{V}_1}{L_1} - \frac{j\omega L_1 L_2 \mathbf{I}_2}{L_1} + j\omega L_2 \mathbf{I}_2$$

$$= \sqrt{\frac{L_2}{L_1}} \mathbf{V}_1 = N \mathbf{V}_1$$

where $N = \sqrt{L_2/L_1}$.

From Equation (14.20), we know that the inductance of a coil is proportional to the square of the number of turns. Therefore, the primary and secondary self-inductances, respectively, are

$$L_1 = N_1^2/\mathcal{R} \qquad \text{and} \qquad L_2 = N_2^2/\mathcal{R}$$

Since

$$N = \sqrt{\frac{L_2}{L_1}} = \sqrt{\frac{N_2^2/\mathcal{R}}{N_1^2/\mathcal{R}}} = \frac{N_2}{N_1}$$

so we call N the **turns ratio** of the transformer.

From Equation (14.30),

$$\mathbf{I}_1 = \frac{\mathbf{V}_1}{j\omega L_1} - \frac{M \mathbf{I}_2}{L_1}$$

so that for perfect coupling

$$\mathbf{I}_1 = \frac{\mathbf{V}_1}{j\omega L_1} - \frac{\sqrt{L_1 L_2}\,\mathbf{I}_2}{L_1}$$

$$= \frac{\mathbf{V}_1}{j\omega L_1} - \sqrt{\frac{L_2}{L_1}}\,\mathbf{I}_2 \qquad\qquad \textbf{(14.31)}$$

For the case that L_1 and L_2 approach infinity such that the turns ratio remains constant (along with perfect coupling), we say that the transformer is **ideal.** For an ideal transformer, Equation (14.31) becomes

$$\mathbf{I}_1 = -N\mathbf{I}_2 \qquad \text{or} \qquad \mathbf{I}_2 = -\frac{\mathbf{I}_1}{N}$$

As good coupling is achieved by transformers with iron cores, an ideal transformer is usually represented as shown in Fig. 14.23. Since the relationships between the voltages and the currents are given by the pair

$$\mathbf{V}_2 = N\mathbf{V}_1$$
$$\mathbf{I}_2 = -\mathbf{I}_1/N$$

we can model an ideal transformer by either of the two equivalent circuits shown in Figs. 14.24 and 14.25. For the circuit given in Fig. 14.24, clearly, we have that $\mathbf{V}_2 = N\mathbf{V}_1$ and $\mathbf{I}_1 = -N\mathbf{I}_2$, while for the circuit depicted in Fig. 14.25, $\mathbf{V}_1 = \mathbf{V}_2/N$ and $\mathbf{I}_2 = -\mathbf{I}_1/N$.

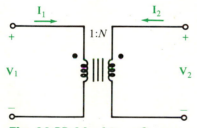

Fig. 14.23 Ideal transformer.

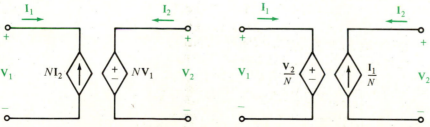

Fig. 14.24 Ideal-transformer model.   **Fig. 14.25 Alternative ideal-transformer model.**

The instantaneous power absorbed by the primary of an ideal transformer is $p_1 = v_1 i_1$, and that absorbed by the secondary is $p_2 = v_2 i_2$. Thus, the total instantaneous power absorbed by the transformer is

$$p = p_1 + p_2 = v_1 i_1 + v_2 i_2 = v_1 i_1 + (N v_1)(-i_1/N) = 0$$

Since the instantaneous power absorbed is zero, so is the average power, and, also, the energy stored must be zero. This fact can be confirmed from the formula for the energy stored in a transformer [Equation (14.25)] by using the relationships for perfect coupling ($M = \sqrt{L_1 L_2}$) and turns ratio ($N = \sqrt{L_2/L_1}$), along with the ideal transformer formulas. Consequently, we see that the ideal transformer is a lossless device.

Example 14.8

Figure 14.26 depicts the equivalent circuit of a simple "class-A" transistor power amplifier. By KVL,

$$\mathbf{V}_g = 750\mathbf{I} + 50(21\mathbf{I}) = 1800\mathbf{I}$$

from which

$$\mathbf{I} = \frac{\mathbf{V}_g}{1800}$$

Since

$$\mathbf{I}_1 = -20\mathbf{I} = -20\left(\frac{\mathbf{V}_g}{1800}\right) = -\frac{\mathbf{V}_g}{90}$$

then

$$\mathbf{I}_2 = -\frac{\mathbf{I}_1}{1/9} = -9\mathbf{I}_1 = -9\left(-\frac{\mathbf{V}_g}{90}\right) = \frac{\mathbf{V}_g}{10}$$

Thus,

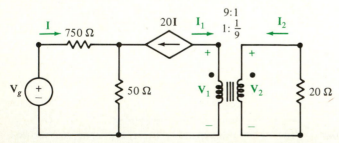

Fig. 14.26 Equivalent circuit of a power amplifier.

$$V_2 = -20I_2 = -20\left(\frac{V_g}{10}\right) = -2V_g$$

Therefore, the voltage transfer function V_2/V_g is

$$\frac{V_2}{V_g} = -2$$

When the voltage transfer function is a real number, it is often called the **voltage gain.**

The instantaneous power supplied by the independent voltage source is

$$p_g = v_g i = v_g\left(\frac{v_g}{1800}\right) = \frac{v_g^2}{1800}$$

and the power absorbed by the 20-Ω load resistor is

$$p_2 = \frac{v_2^2}{20} = \frac{(-2v_g)^2}{20} = \frac{1}{5}v_g^2$$

Thus, the **power gain** p_2/p_g is

$$\frac{p_2}{p_g} = \frac{\frac{1}{5}v_g^2}{v_g^2/1800} = 360$$

![empty box]

Impedance Matching

Consider the ideal-transformer circuit shown in Fig. 14.27. Since

$$V_g = Z_g I_1 + V_1 \tag{14.32}$$

and

$$V_2 = -Z_L I_2 \tag{14.33}$$

from Equation (14.33),

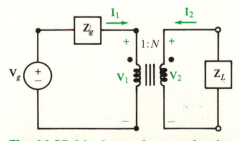

Fig. 14.27 Ideal-transformer circuit.

$$NV_1 = -Z_L\left(-\frac{I_1}{N}\right)$$

Thus,

$$V_1 = \frac{Z_L}{N^2}I_1$$

Substitution of this expression into Equation (14.32) yields

$$V_g = Z_g I_1 + \frac{Z_L}{N^2}I_1$$

so that the impedance V_g/I_1 seen by the source is

$$\frac{V_g}{I_1} = Z_g + \frac{Z_L}{N^2} \tag{14.34}$$

which is Z_g in series with the load impedance Z_L scaled by the factor $1/N^2$. The term Z_L/N^2 is often called the **reflected impedance.** We can also say that this term is the equivalent impedance **referred** to the primary side of the transformer.

From Equation (14.34),

$$I_1 = \frac{V_g}{Z_g + Z_L/N^2}$$

Therefore,

$$V_2 = -Z_L I_2 = -Z_L\left(-\frac{I_1}{N}\right) = \frac{Z_L}{N}\left(\frac{V_g}{Z_g + Z_L/N^2}\right)$$

$$V_2 = \frac{NZ_L}{N^2 Z_g + Z_L}V_g$$

As an alternative approach, we can apply Thévenin's theorem to the circuit given in Fig. 14.27. First, we determine V_{oc} from the circuit shown in Fig. 14.28. Since $I_2 = 0$, then

$$I_1 = -NI_2 = 0$$

and, by KVL,

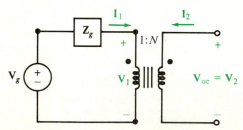

Fig. 14.28 Determination of open-circuit voltage.

$$\mathbf{V}_g = \mathbf{Z}_g\mathbf{I}_1 + \mathbf{V}_1 = \mathbf{V}_1$$

Thus,

$$\mathbf{V}_{oc} = \mathbf{V}_2 = N\mathbf{V}_1 = N\mathbf{V}_g$$

To find the Thévenin-equivalent (output) impedance, set $\mathbf{V}_g = 0$ and take the ratio of $\mathbf{V}_o$ to $\mathbf{I}_o$ for the circuit shown in Fig. 14.29. Since

$$\mathbf{V}_1 = -\mathbf{Z}_g\mathbf{I}_1 = -\mathbf{Z}_g(-N\mathbf{I}_o) = N\mathbf{Z}_g\mathbf{I}_o$$

then

$$\mathbf{V}_o = N\mathbf{V}_1 = N(N\mathbf{Z}_g\mathbf{I}_o) = N^2\mathbf{Z}_g\mathbf{I}_o$$

Hence, the output impedance is

$$\mathbf{Z}_o = \frac{\mathbf{V}_o}{\mathbf{I}_o} = N^2\mathbf{Z}_g$$

Therefore, using Thévenin's theorem, we have the circuit shown in Fig. 14.30. By voltage division,

$$\mathbf{V}_2 = \frac{\mathbf{Z}_L}{\mathbf{Z}_L + N^2\mathbf{Z}_g}(N\mathbf{V}_g) = \frac{N\mathbf{Z}_L}{N^2\mathbf{Z}_g + \mathbf{Z}_L}\mathbf{V}_g$$

as was obtained above.

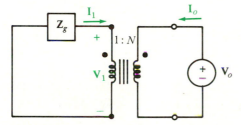

Fig. 14.29 Determination of output impedance.

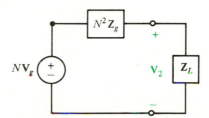

Fig. 14.30 Thévenin-equivalent circuit.

Example 14.9

For the circuit given in Fig. 14.27, suppose that $\mathbf{Z}_g = 20$ kΩ and $\mathbf{Z}_L = 8$ Ω. Let us determine the value of the turns ratio N such that the 8-Ω load resistor absorbs maximum power.

Given the Thévenin-equivalent circuit shown in Fig. 14.30, we know that the load absorbs maximum power when the load resistance $\mathbf{Z}_L = R_L = 8$ Ω is equal to the output (Thévenin-equivalent) resistance $\mathbf{Z}_o = N^2\mathbf{Z}_g = 20{,}000N^2$, that is, when

$$20{,}000N^2 = 8$$

Solving this equation, we find that the turns ratio is

$$N = 1/50$$

In this example, we have used a transformer to "match" the 8-Ω load resistor to the 20-kΩ source resistance. When a transformer is used in such an application, it can be referred to as a **matching transformer.** Typically, this type of transformer is not designated by its turns ratio N, but rather by the impedances (resistances) it matches. Specifically, the transformer in this example could be called a "20-kΩ/8-Ω matching transformer."

14.4 NONIDEAL-TRANSFORMER MODELS

There are many situations for which the assumption that an iron-core transformer is ideal yields quite good results. However, there may be times when a more accurate model of a practical iron-core transformer is desirable. One such equivalent circuit (there are others) is shown in Fig. 14.31. For this circuit, the resistor R_s results in the "copper losses" due to the resistance of the primary winding and the secondary-winding resistance referred to the primary. The reactance X_s is the result of the leakage flux in the primary and the secondary leakage reactance referred to the primary. The resistor R_m accounts for the core losses due to hysteresis and eddy currents. If there is no load connected to the secondary winding of the transformer, then $I_2 = 0$. Yet, for a physical core, even when there is no load, a voltage applied to the primary results in a primary current that will produce a flux in the core. By Equation (14.19), this indicates an inductance, called the **magnetizing inductance,** and it is accounted for by the reactance X_m. The current I_m through the magnetizing inductance is known as the **magnetizing current,** and the current I_e is called the **exciting current.** We can determine the parameters of the transformer model given in Fig. 14.31 by taking open- and short-circuit measurements.

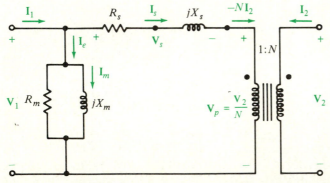

Fig. 14.31 Equivalent circuit of a practical transformer.

> **Example 14.10**

A transformer used for a power application is rated 2400:240 V, 60 Hz, 36 kVA. This rating indicates, first, that one winding, the high-voltage winding, can have a maximum rms voltage of 2400 V. The other winding, the low-voltage winding, can have a maximum rms voltage of 240 V. Although this means that the turns ratio is $N = 2400/240 = 10$ when the high-voltage winding is the secondary, either winding may be used as the secondary. The next term in the rating (60 Hz) indicates that operation at this frequency will not result in excessive core loss. Finally, the 36-kVA rating means that, whichever winding is chosen as the secondary, the product of the load's voltage and current can be maintained at this "full-load" value and excessive heating will not result. Note that we can now calculate the current ratings (maximum values) for the windings. For the high-voltage winding, the rated current is 36 kVA/2400 V = 15 A, and for the low-voltage winding, 36 kVA/240 V = 150 A is the rated current.

Suppose that the high-voltage winding is the primary. Then the turns ratio is $N = 240/2400 = 1/10$. Let us apply $|\mathbf{V}_1| = 2400$ V (rms) to the primary winding when no load is connected to the secondary. For this **open-circuit test,** suppose that we take the voltage, current, and power measurements indicated in Fig. 14.32. Assume that the voltmeter (V), ammeter (A), and wattmeter (W) are ideal, and their respective readings are 2400 V, 0.4 A, and 360 W. Since there is no load, $\mathbf{I}_2 = 0$ and there is no current through R_s (i.e., $\mathbf{I}_s = 0$). Thus, the power absorbed by R_m is 360 W, and from

$$\frac{|\mathbf{V}_1|^2}{R_m} = 360$$

we get

$$R_m = \frac{|\mathbf{V}_1|^2}{360} = \frac{(2400)^2}{360} = 16 \text{ k}\Omega$$

Furthermore, we have that $|\mathbf{I}_e| = 0.4$ A, and since

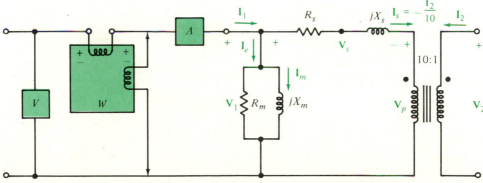

Fig. 14.32 Circuit used for open-circuit and short-circuit testing.

$$|\mathbf{V}_1|\,|\mathbf{I}_e|\cos\theta_m = 360$$

then

$$\cos\theta_m = \frac{360}{|\mathbf{V}_1||\mathbf{I}_e|} = \frac{360}{(2400)(0.4)} = 0.375$$

and

$$\theta_m = 68°$$

Since this is the angle of the impedance $\mathbf{Z}_m$ consisting of the parallel connection of R_m and X_m, then from

$$\mathbf{Z}_m = \frac{(R_m)(jX_m)}{R_m + jX_m}$$

we have that

$$\theta_m = 90° - \tan^{-1}\frac{X_m}{R_m}$$

Thus,

$$\tan^{-1}\frac{X_m}{R_m} = 90° - \theta_m = 90° - 68° = 22°$$

and

$$X_m = R_m(\tan 22°) = 16{,}000(0.404) = 6464 \ \Omega$$

Since $X_m = \omega L_m = 2\pi f L_m$, then the magnetizing inductance L_m is

$$L_m = \frac{X_m}{2\pi f} = \frac{6464}{2\pi(60)} = 17.15 \text{ H}$$

Now suppose that a short circuit is placed across the secondary winding. For this **short-circuit test,** instead of 2400 V, we apply a voltage that yields the rated current of 150 A in the secondary. Suppose that the voltmeter reads 50 V, the ammeter reads 15 A, and the wattmeter reads 315 W. The power absorbed by R_m is

$$\frac{|\mathbf{V}_1|^2}{R_m} = \frac{50^2}{16{,}000} = 0.156 \text{ W}$$

and so the power absorbed by R_s is approximately 315 W. Since the magnitude of the exciting current is

$$|\mathbf{I}_e| = \frac{|\mathbf{V}_1|}{|\mathbf{Z}_m|} = \frac{|\mathbf{V}_1|}{\dfrac{R_m X_m}{\sqrt{R_m^2 + X_m^2}}}$$

$$= \frac{50}{\dfrac{(16{,}000)(6464)}{\sqrt{16{,}000^2 + 6464^2}}} = \frac{50}{5993} = 0.008 \text{ A}$$

then $|\mathbf{I}_s| \approx 15$ A, and the secondary has the rated current of 150 A. Therefore, the power absorbed by R_s is

$$15^2 R_s = 315$$

from which

$$R_s = \frac{315}{15^2} = 1.4 \ \Omega$$

When a short circuit is placed across the secondary winding, $\mathbf{V}_2 = 0$. Under this circumstance, $\mathbf{V}_p = 0$. Thus, the voltage across $\mathbf{Z}_s$, the series connection of R_s and jX_s, is $\mathbf{V}_1$. Consequently,

$$|\mathbf{V}_1||\mathbf{I}_s| \cos \theta_s = 315$$

from which

$$\cos \theta_s = \frac{315}{|\mathbf{V}_1||\mathbf{I}_s|} = \frac{315}{(50)(15)} = 0.42$$

so

$$\theta_s = 65°$$

Since this is the angle of the impedance $\mathbf{Z}_s = R_s + jX_s$, then

$$\tan^{-1} \frac{X_s}{R_s} = 65°$$

from which we obtain the reactance

$$X_s = R_s(\tan 65°) = 1.4(2.145) = 3.0 \ \Omega$$

Since $X_s = 2\pi f L_s$, the corresponding inductance is

$$L_s = \frac{X_s}{2\pi f} = \frac{3}{2\pi(60)} = 7.96 \text{ mH}$$

In summary, the open-circuit test yields the parameters R_m and X_m, and the short-circuit test gives R_s and X_s.

Some Transformer Properties

One figure of merit of a device is its **efficiency,** which we define as the ratio (usually expressed in percent) of the output power P_o to the input power P_i. Thus, we have that the efficiency (denoted Eff) is

$$\boxed{\text{Eff} = \frac{P_o}{P_i}}$$

As some of the input power is lost in the device and the remainder reaches the output, if P_l is the power loss, we have

$$P_i = P_l + P_o \qquad \Rightarrow \qquad P_o = P_i - P_l$$

Thus, we can write

$$\boxed{\text{Eff} = \frac{P_o}{P_i} = \frac{P_o}{P_l + P_o} = \frac{P_i - P_l}{P_i} = 1 - \frac{P_l}{P_i}} \qquad (14.35)$$

In addition to the efficiency of a transformer, we can talk about its **voltage regulation** (also usually expressed in percent), which is the ratio of the change between no-load output (rms) voltage V_{NL} and full-load output (rms) voltage V_{FL} to the full-load output voltage. In other words, the voltage regulation VR is given by

$$\boxed{\text{VR} = \frac{V_{NL} - V_{FL}}{V_{FL}}} \qquad (14.36)$$

Example 14.11

Let us determine the voltage regulation and efficiency for the transformer given in Example 14.10 for the case that the transformer load has a lagging power factor (pf) of 0.866.

We know that the full-load (rated) output voltage is $V_{FL} = 240$ V. Let us take the angle of $\mathbf{V}_2$ to be zero, that is, $\mathbf{V}_2 = 240\underline{/0°}$. Then $\mathbf{V}_p = 2400\underline{/0°}$. Since the full-load output current is 150 A, then $|\mathbf{I}_2| = 150$ A. Also, since the pf is $\cos\theta = 0.866$, then $\theta = 30°$. But, as the pf is lagging, then the current lags the voltage by 30°. Thus, $-\mathbf{I}_2 = 150\underline{/-30°}$ from which $\mathbf{I}_s = -\mathbf{I}_2/10 = 15\underline{/-30°}$. Therefore,

$$\mathbf{V}_s = \mathbf{Z}_s\mathbf{I}_s = (R_s + jX_s)\mathbf{I}_s$$

$$= (1.4 + j3)(15\underline{/-30°}) = (1.4 + j3)(13 - j7.5) = 40.7 + j28.5$$

$$= 49.7\underline{/35°}$$

and

$$\mathbf{V}_1 = \mathbf{V}_s + \mathbf{V}_p = 40.7 + j28.5 + 2400 = 2440.7 + j28.5 = 2441\underline{/0.67°}$$

A phasor diagram (not drawn to scale) is shown in Fig. 14.33.

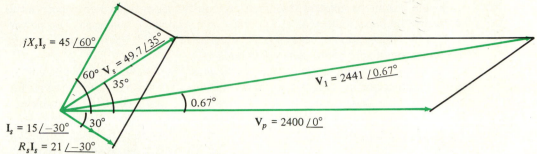

Fig. 14.33 Phasor diagram.

For the case of no load ($I_2 = 0$), we have that $I_s = 0$, and hence $V_s = 0$. Thus, $V_p = V_1$, and since $V_2 = V_p/10 = V_1/10$, then the no-load output voltage is $V_{NL} = 244.1$ V. Therefore, by Equation (14.36), the voltage regulation is

$$VR = \frac{244.1 - 240}{240} = 0.017 = 1.7\%$$

To determine the (full-load) efficiency, the copper loss of the windings is the power absorbed by the resistance R_s. This power P_s is

$$P_s = |I_s|^2 R_s = (15^2)(1.4) = 315 \text{ W}$$

which was the wattmeter reading for the short-circuit test. The core loss (due to leakage flux) is the power absorbed by the resistor R_m. This power P_m is

$$P_m = \frac{|V_1|^2}{R_m} = \frac{2441^2}{16,000} = 372 \text{ W}$$

Thus, the power loss $P_l = P_s + P_m = 315 + 372 = 687$ W.

The output power is

$$P_o = |V_2||I_2| \cos \theta$$

$$= (240)(150)(0.866) = 31,176 \text{ W}$$

and from Equation (14.35), the full-load efficiency is

$$Eff = \frac{31,176}{687 + 31,176} = 0.978 = 97.8\%$$

Normally, power transformers operate at a fixed frequency (e.g., 60 Hz), and they are designed with this in mind. However, iron-core transformers also see use in situations where a range of frequencies is encountered. One common application is in an audio system in which an "output transformer" is used to match the output of an amplifier stage to a loudspeaker load for frequencies ranging, typically, from 30 to 15,000 Hz. The behavior of such a transformer can be accurately modeled by using the circuit parameters discussed previously, along with two capacitors to represent the capacitances of the primary and secondary windings. For power-transformer applications, these capacitors can be neglected, but they become more and more significant as frequency increases.

A model of an output transformer is shown in Fig. 14.34, along with a voltage source V_g that has an internal resistance R_g and a resistive load R_L. At low frequencies, the impedance of the magnetizing inductance L_m is small. This causes a shunting effect in that a significant amount of current will go through L_m, bypassing the primary coil of the ideal transformer portion of the model. The result, for a fixed magnitude of V_g, is a small magnitude for V_2, which increases as frequency increases. After the frequency reaches around 100 Hz, the impedance of L_m becomes large enough that it can be ignored (treated as an open circuit). From 100 Hz to about 10 kHz, the magnitude of V_2 is essentially constant. However, in the vicinity of 10 kHz, the output voltage magnitude actually increases as a result of the resonance that results from C_2 and the leakage inductance L_s. For frequencies above 10 kHz, not only does the impedance of L_s get larger, but the impedances of the shunting capacitors C_1 and C_2 get smaller with increasing frequency. The result is a sharp rolloff or decrease in the magnitude of V_2 as frequency increases. The frequency dependence of the transformer modeled by the circuit given in Fig. 14.34 is summarized by the amplitude-response curve shown in Fig. 14.35.

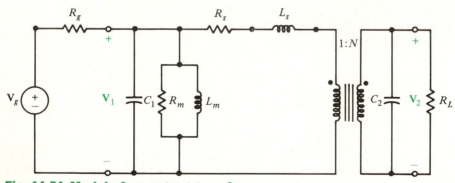

Fig. 14.34 Model of an output transformer.

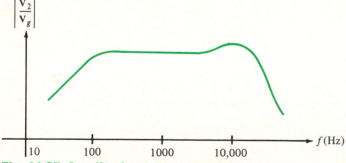

Fig. 14.35 Amplitude-response curve.

A good audio-frequency transformer is one that has a large magnetizing inductance L_m. This property can be achieved by using a core with a high permeability and appropriate geometry. In addition, a good transformer has small values for R_s and L_s. The former can be obtained by forming the primary and secondary coils from wires with large diameters, and the latter by tightly interleaving the primary and secondary windings.

SUMMARY

1. Current going through a conductor produces a magnetic field about the conductor.

2. A current going through a solenoid creates a magnetic field similar to that of a permanent magnet.

3. The relative permeability μ_r of a material is the ratio of the permeability of the material μ to the permeability of free space $\mu_0 = 4\pi \times 10^{-7}$ H/m. That is, $\mu_r = \mu/\mu_0$.

4. For a given core, magnetomotive force (mmf), designated by $\mathscr{F}$, is the product of the number of coil turns N and the current i through the coil. That is, $\mathscr{F} = Ni$.

5. The ratio of mmf $\mathscr{F}$ to magnetic flux ϕ is called reluctance and is designated by $\mathscr{R}$. That is, $\mathscr{R} = \mathscr{F}/\phi$.

6. The magnetic-circuit equation $\mathscr{F} = \mathscr{R}\phi$ is analogous to the electric-circuit equation $v = Ri$ (Ohm's law).

7. Magnetic field intensity H is the ratio of magnetic flux density B to permeability μ. That is, $H = B/\mu$.

8. A plot of magnetic flux density B versus magnetic field intensity H is called a magnetization (or B-H) curve.

9. Power losses in a core are due to hysteresis and eddy currents.

10. A changing magnetic flux passing through a coil induces an electromotive force (emf) across that coil (Faraday's law).

11. Inductance, which is the number of flux linkages per ampere, is directly proportional to the square of the number of turns.

12. The polarity of the voltage induced by a changing flux tends to oppose the change in flux that produced the induced voltage (Lenz's law).

13. The phase relationship between the windings of a transformer is indicated by dots.

14. An ideal transformer has perfect coupling and infinite self-inductances.

15. The energy stored in an ideal transformer is zero.

16. The circuit parameters for a nonideal-transformer model can be calculated from the results of open-circuit and short-circuit tests.

17. A transformer can be rated in terms of its efficiency and voltage regulation.

18. For transformers used over a range of frequencies, the capacitances of the primary and secondary windings are significant at higher frequencies.

Problems for Section 14.1

14.1 The solenoid shown in Fig. 14.5 consists of 500 turns wound on an air core of radius 0.01 m (1 cm) and length 0.2 m (20 cm). Find the current required to produce a magnetic flux density of magnitude $B = 0.5$ T in the core of the solenoid.

14.2 Repeat Problem 14.1 for the case of an iron core having relative permeability $\mu_r = 1500$.

14.3 A toroid, as shown in Fig. 14.6, consists of 500 turns wound on a nonmagnetic core ($\mu_r = 1$). Given that $R = 0.1$ m and $r = 0.02$ m, find the magnitude of the magnetic flux density in the core if the current through the coil is $i = \frac{1}{3}$ A.

14.4 A toroid, as shown in Fig. 14.6, has an iron core with relative permeability $\mu_r = 1500$, average radius $R = 0.1$ m, and cross-sectional radius $r = 0.02$ m. Assuming the current through the coil is $i = \frac{1}{3}$ A, determine the number of turns required to produce a flux of $\phi = 10^{-3}$ Wb in the core.

14.5 The rectangular iron core given in Fig. 14.8 has relative permeability $\mu_r = 1500$. Assuming the current through the coil is $i = \frac{1}{3}$ A, determine the number of turns required to produce a flux of $\phi = 10^{-3}$ Wb in the core.

14.6 Repeat Problem 14.5 for the iron core with the air gap shown in Fig. 14.9.

14.7 For the rectangular cast-iron (see Fig. 14.12) core given in Fig. 14.9, find the relative permeability when the flux in the iron is $\phi = 2.4 \times 10^{-4}$ Wb.

14.8 Use the magnetization curves given in Fig. 14.12 to determine the approximate relative permeability of (a) cast iron for $B \leq 0.4$ T, (b) cast steel for $B \leq 1$ T, and (c) sheet steel for $B \leq 1$ T.

14.9 A rectangular cast-steel core as shown in Fig. 14.8 is to have a flux of $\phi = 7.2 \times 10^{-4}$ Wb. What current is required when the coil has $N = 200$ turns?

14.10 Repeat Problem 14.9 for the case that there is a 1-cm air gap as shown in Fig. 14.9. What is the field intensity in the air gap?

14.11 A cast-iron magnetic core has an average length of $l = 0.6$ m and $N = 200$ turns, and the average hysteresis power loss is $P_h = 10$ W when the maximum current is 1.2 A. Use $n = 1.6$ to approximate P_h for the case that the maximum current is (a) 0.6 A and (b) 2.4 A.

14.12 Repeat Problem 14.11 for a cast-steel core.

Problems for Section 14.2

14.13 A toroid has a reluctance of 2.65×10^5 A-t/Wb. Find the inductance of a 500-turn coil that is wound on this core.

14.14 A rectangular core with a 200-turn coil as shown in Fig. 14.8 has a flux of 1.51×10^{-3} Wb when the current in the coil is 2 A. Find the inductance of this coil.

14.15 A rectangular core has a reluctance of 2.65×10^5 A-t/Wb. Find the inductance of a 200-turn coil that is wound on this core.

14.16 Show that the inductance of the toroidal coil shown in Fig. 14.6 is

$$L = \frac{\mu N^2 A}{l}$$

where $A = \pi r^2$ and $l = 2\pi R$.

14.17 Use the result given in Problem 14.16 to show that the energy stored in the toroid shown in Fig. 14.6 is

$$w = \frac{B^2 A l}{2\mu}$$

where $B = \phi/A$.

14.18 The cast-iron rectangular core with an air gap shown in Fig. 14.9 has a 200-turn coil and a flux of 2.4×10^{-4} Wb when the current in the coil is 5.96 A. Calculate the energy stored (a) in this inductance, (b) in the air gap, and (c) in the cast-iron core.

14.19 For the core with two coils shown in Fig. P14.19, assume that at some instant of time the current i is a positive quantity.

(a) Is the resulting magnetic flux ϕ clockwise or counterclockwise?

(b) If ϕ is increasing with time, for the coil on the right, in what direction will current tend to flow?

(c) If ϕ is increasing with time, is v_{ab} positive or negative?

(d) If ϕ is increasing with time, is v positive or negative?

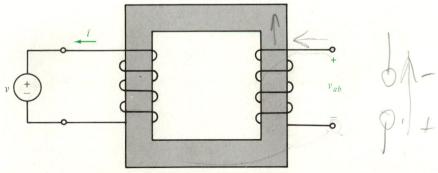

Fig. P14.19

14.20 Repeat Problem 14.19 for the case that ϕ is decreasing with time.

14.21 Place appropriate dots on the transformer shown in Fig. P14.19.

14.22 For the transformers shown in Fig. P14.22, place dots appropriately at the ends of the coils.

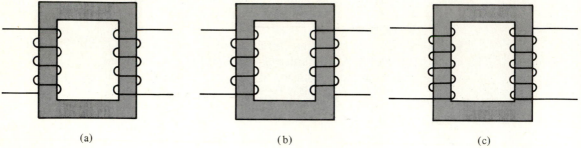

(a) (b) (c)

Fig. P14.22

14.23 For the transformer shown in Fig. P14.23, write a pair of equations in the time domain relating the voltages v_1 and v_2 to the currents i_1 and i_2. Use Equation (14.25) to show that the energy stored in this transformer is

$$w(t) = \tfrac{1}{2}L_1 i_1^2(t) + \tfrac{1}{2}L_2 i_2^2(t) - M i_1(t) i_2(t)$$

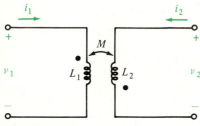

Fig. P14.23

14.24 For the circuit given in Fig. 14.18:

(a) Find the energy stored in the transformer at time $t = 0$.

(b) Find the impedance $\mathbf{V}/\mathbf{I}_1$ seen by the voltage source.

14.25 For the circuit given in Fig. 14.18, move the dot from the upper end of the secondary winding to the lower end, and find $v_C(t)$.

14.26 Find $v_2(t)$ for the circuit shown in Fig. P14.26.

14.27 The transformer given in Fig. P14.27(a) can be modeled by the T-equivalent circuit shown in Fig. P14.27(b). Use the results of Problem 14.23 to determine the values of the inductances labeled with the question marks.

14.28 A magnetic circuit with a constant ac voltage amplitude has a core loss totaling 100 W at 120 Hz and totaling 32 W at 60 Hz. Find the core loss at 50 Hz.

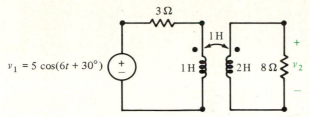

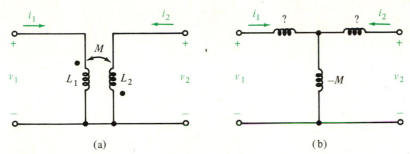

Fig. P14.26

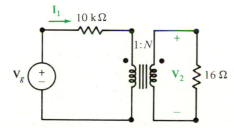

Fig. P14.27

Problems for Section 14.3

14.29 Given the ideal-transformer circuit shown in Fig. P14.29, let the turns ratio be $N = 1/10$.

(a) Find the impedance V_g/I_1 seen by the voltage source.

(b) Find the voltage gain V_2/V_g.

(c) To what value should the 16-Ω load resistance be changed so that it will absorb the maximum power?

Fig. P14.29

14.30 For the circuit given in Fig. P14.29, what value of N will result in the 16-Ω load resistor absorbing the maximum amount of power?

14.31 Repeat Problem 14.29(a) and (b) for the ideal-transformer circuit shown in Fig. P14.31.

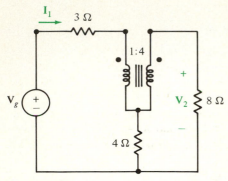

Fig. P14.31

14.32 For the circuit shown in Fig P14.32, find $v(t)$.

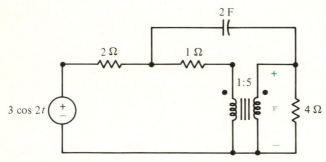

Fig. P14.32

14.33 For the power amplifier given in Fig. 14.26, use the equivalent circuit of the ideal transformer shown in Fig. 14.25 to determine the power gain p_2/p_g.

14.34 The circuit shown in Fig. P14.34 is the equivalent circuit of a simple "class-B" transistor power amplifier.

(a) Find the voltage gain V_2/V_g.

(b) Assuming p_g is the power supplied by the voltage source and p_2 is the power absorbed by the 8-Ω load resistor, find the power gain p_2/p_g.

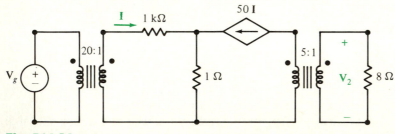

Fig. P14.34

Problems for Section 14.4

14.35 A power transformer is rated 2300:230 V, 60 Hz, 46 kVA. Performing the open-circuit test (see Fig. 14.32) results in readings of 2300 V, 0.35 A, and 400 W, and the short-circuit test results in 40 V, 20 A, and 350 W. Find the equivalent-circuit parameters R_m, X_m, R_s, and X_s.

14.36 A 2400:240-V, 60-Hz, 48-kVA transformer has equivalent-circuit parameters $R_m = 15$ kΩ, $X_m = 7$ kΩ, $R_s = 1$ Ω, and $X_s = 2$ Ω when the low-voltage winding is the secondary. Find the voltmeter, ammeter, and watt-meter readings when:

(a) The open-circuit test is performed.

(b) The short-circuit test is performed.

14.37 Repeat Problem 14.36 for the case that the high-voltage winding is the secondary.

14.38 For the transformer given in Problem 14.36, find the full-load efficiency and voltage regulation for the case that the load has a leading pf of 0.5.

14.39 For the transformer given in Problem 14.35, find the voltage regulation and efficiency for the case of a pf of unity.

14.40 Determine the lagging pf of the load for which the transformer described in Problem 14.36 has a full-load efficiency of 98%. Find the corresponding voltage regulation.

Chapter 15
Machines

INTRODUCTION

The subject of electromechanical energy conversion has been, and continues to be, a very important area of study. Because most of electric power is generated and consumed by devices that employ magnetic fields, we focus our attention on these types of machines.

The basic principles of electromechanical energy conversion are initially introduced with the use of simple translational transducers—a generator and a motor, as well as a bilateral transducer (a combination of both). In this regard, it is seen how electric and mechanical circuit elements can be used to model machines.

The concept of a transducer is extended to encompass some devices which have moving coils, and others that have moving-iron sections. The former, or dynamic transducers, find application in loudspeakers, microphones, and phonograph pickups. The latter see use as relays and solenoid plungers.

Expanding the study of transducers from translation to rotation leads to the d'Arsonval movement. This type of rotational transducer relates current and torque, and, as a consequence, has widespread application in electrical instrumentation. The same thing can be said about another rotational transducer—the electrodynamometer.

When it comes to the conversion of electromechanical energy, there is no question but that, in terms of quantity, the leader is the rotating machine. It is important as both generator and motor. In our study, for the most part, we are concerned with the steady-state behavior of rotating machines that operate under typical conditions. We will limit our discussions to include only the most common types of machines. Emphasis will be placed on simplified forms and models of machines so that any analysis can proceed in a clear and straightforward manner without becoming obscured by excessive detail.

Most rotating machines operate on the same basic principle—appropriately aligned conductors moving in a magnetic field. The different ways in which this can be accomplished result in the various types of machines. Typically, though, any machine will have a portion that rotates (the rotor) and a portion that is stationary (the stator). In a dc machine, both rotor and stator are supplied with direct current. Basically, there are two types of ac machines. In one, the induction type, alternating current is supplied to both rotor and stator. However, for the other type, the synchronous machine, rotor and stator receive different excitations—one is supplied with direct current and the other with alternating current. In this chapter, we begin our study of rotating machines with a discussion of generators, both dc and ac (or alternators), and conclude with the topic of motors. Included is material on dc motors, synchronous motors, and induction motors—both the three-phase and single-phase types.

15.1 TRANSDUCERS

In the preceding chapter, we discussed Faraday's law, which states that a changing magnetic flux ϕ that passes through an N-turn conducting coil induces an emf of $N\,d\phi/dt$ volts across that conductor. We have considered the case for which the conductor is stationary and the flux is time varying. However, if, instead, the flux is constant and the conductor moves (its position varies with time), we still achieve a changing magnetic flux relative to the conductor—and, hence, an induced emf. (Of course, if both the flux and the conductor's position are time varying, the result again will be an induced emf.) With this in mind, consider the one-turn loop formed with a movable conductor that is electrically connected to a fixed, conducting track comprised of two rails as shown in Fig. 15.1. Suppose that the length of the movable conductor is l, and a resistance is connected to the track as illustrated. Furthermore, suppose that a uniform magnetic field with a constant magnetic flux density B is directed into the page (indicated by the crosses). If the conductor moves

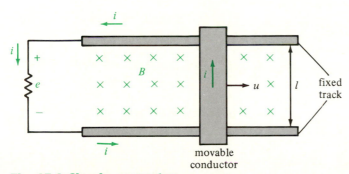

Fig. 15.1 Simple generator.

to the right at a velocity of u m/s (meters per second), then, after dt seconds, it travels $u\,dt$ meters. Since $\phi = BA$, then also after dt seconds, the flux enclosed by the single-turn loop composed of the movable conductor, the track, and the resistance increases by

$$d\phi = d(BA) = BdA = Blu\,dt$$

from which we have

$$\frac{d\phi}{dt} = Blu = e \qquad \qquad (15.1)$$

and this is the subsequent induced emf. To obtain the polarity of this voltage, we apply Lenz's law. For the current resulting from the induced emf to establish a magnetic field that opposes the increasing flux, by the right-hand rule, the current should be counterclockwise as depicted. This yields the polarity of e shown in Fig. 15.1. In the derivation of Equation (15.1), it was assumed that the movable conductor and fixed track had zero resistances, and that the magnetic field resulting from current i was negligible compared with the original magnetic field.

For the setup given in Fig. 15.1, some applied force f, measured in newtons (N), is required to move the length-l conductor with velocity u. Thus, we have that for this electromechanical system, the input is an applied force f and the response (output) is a voltage $e = Blu$. A device such as this, which converts mechanical energy into electric energy, is called a **generator.**

Let us now modify the system given in Fig. 15.1 by replacing the resistance with a current source as shown in Fig. 15.2. The result is a clockwise current i going through the loop formed by the current source, the fixed track, and the movable conductor. Suppose that u_q is the average velocity of an increment of charge dq within an incremental section dl of the movable conductor. Thus, $u_q = dl/dt$. However, since $f = quB \sin \theta$ [Equation (14.1)] and the direction of the current (charge) is normal to the magnetic field, the resulting incremental force df developed on the section dl of the movable conductor is given by

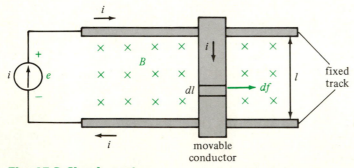

Fig. 15.2 Simple motor.

$$df = dq\ u_q B \tag{15.2}$$

and by the right-hand screw rule, this developed force is directed to the right. Since

$$i = \frac{dq}{dt} = \frac{dq}{dl}\frac{dl}{dt} = \frac{dq}{dl}u_q$$

then, by the chain rule of calculus,

$$idl = dq\ u_q$$

Substituting this result into Equation (15.2) yields

$$df = Bi\ dl$$

and summing df over the entire length l, we obtain the force f developed on the movable conductor,

$$f = Bli \tag{15.3}$$

and such a force will tend to move the conductor to the right.

Conversely to the generator given in Fig. 15.1, an applied current i is the input to the electromechanical system shown in Fig. 15.2, and the output is a force $f = Bli$ on the movable conductor. A device such as this, which converts electric energy into mechanical energy, is called a **motor.** A device that either converts electric energy to mechanical energy (e.g., a motor) or converts mechanical energy to electric energy (e.g., a generator) is called an **electromechanical transducer.**

For the setup shown in Fig. 15.2, a velocity u directed to the right due to the developed force $f = Bli$ on the movable conductor results in an induced emf $e = Blu$, just as was the case for the system given in Fig. 15.1. In this case, the electric input power supplied by the current source is

$$p_e = ei = Blui$$

and the mechanical output power is

$$p_m = fu = Bliu = p_e$$

so that energy is conserved, and we see that 1 W = 1 N-m/s. For the generator shown in Fig. 15.1, the force on the movable conductor that results from the counterclockwise current has magnitude Bli and is directed to the left. Therefore, to obtain the velocity u directed to the right, the applied force (which is directed to the right) must be $f = Bli$. Consequently, the mechanical input power is

$$p_m = fu = Bliu$$

and the electric output power (absorbed by the resistance) is

$$p_e = ei = Blui = p_m$$

as may have been anticipated.

Now consider the electromechanical system shown in Fig. 15.3, which again consists of a fixed, conducting track and a movable conductor, in addition to a series connection of a voltage source and a resistor. Since

$$i = \frac{V - e}{R}$$

for the case that the movable conductor has velocity u directed to the right, if $e = Blu < V$, then $i > 0$ and the system behaves as a motor (see Fig. 15.2). Conversely, if $e = Blu > V$, then $i < 0$ and the system behaves as a generator (see Fig. 15.1). Under this circumstance, the battery will charge (i.e., it will absorb power). Because this transducer can convert either electric energy to mechanical energy or vice versa, we say that it is **bilateral.**

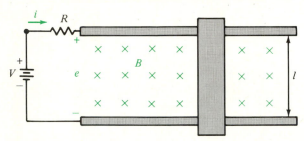

Fig. 15.3 Bilateral electromechanical transducer.

Mechanical Circuit Elements and Symbols

In our discussion of a simple generator and motor, we ignored such physical aspects as mass and friction. To describe these quantities, we introduce some mechanical circuit elements. These can be used to model electromechanical devices.

A **mass** M, whose unit is the kilogram (kg), with an applied force f and a velocity u, is depicted on a frictionless surface in Fig. 15.4(a). With the force and velocity directed to the right, we have that the force f, mass M, and velocity u are related by

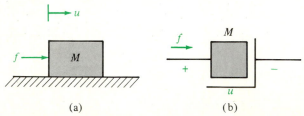

(a)   (b)

Fig. 15.4 Mass and its mechanical circuit symbol.

$$f = M\frac{du}{dt}$$ **(15.4)**

where $M\,du/dt$ is the inertia force of the mass. We can symbolize this situation by the **mechanical circuit element** representation shown in Fig. 15.4(b). Note that a mass for a mechanical circuit is analogous to a capacitor for an electric circuit when force and velocity are analogous to current and voltage respectively ($i = C\,dv/dt$).

If the surface indicated in Fig. 15.4(a) is not frictionless, then we denote by D the **coefficient of sliding friction.** If the mass has velocity u, then the frictional force will be in a direction opposite to the direction of u, and the amount of the force due to friction is Du. A mechanical element that signifies friction is the **dash pot** shown in Fig. 15.5, and the describing equation is

$$f = Du$$ **(15.5)**

Again, for the case that force and velocity are analogous to current and voltage, respectively, we see that the dash pot is the mechanical analog of conductance ($i = Gv$).

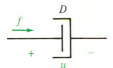

Fig. 15.5 Symbol for a dash pot.

Another mechanical circuit element is a **spring** that has a **compliance** K, which is the name for the proportionality constant arising from the property that force is proportional to displacement. Since velocity is the derivative of displacement, for a spring with compliance K, as symbolized in Fig. 15.6, the relationship between f and u is given by

$$f = \frac{1}{K}\int u\,dt \quad \Rightarrow \quad u = K\frac{df}{dt}$$ **(15.6)**

and it is interesting to note that a spring and an inductor, being analogs, have the same pictorial representation.

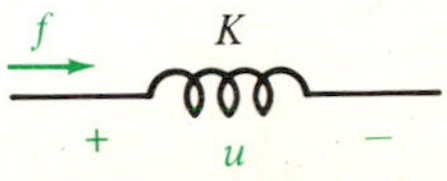

f K

$+$ u $-$

Fig. 15.6 Symbol for a spring.

Example 15.1

Figure 15.7 shows a mechanical system consisting of a spring with compliance K, a mass M on a surface with coefficient of friction D, and an applied force f. **D'Alembert's principle*** states that the inertia force and the external forces acting on a mass sum to zero. If the initial velocity and displacement are zero, then, using Equations (15.4), (15.5), and the integral form of Equation (15.6), application of D'Alembert's principle yields

$$\frac{1}{K} \int_0^t u \, dt + Du + M\frac{du}{dt} = f \tag{15.7}$$

This equation corresponds to the mechanical circuit shown in Fig. 15.8(a). The electrical analog of this mechanical circuit is shown in Fig. 15.8(b).

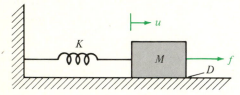

Fig. 15.7 Simple mechanical system.

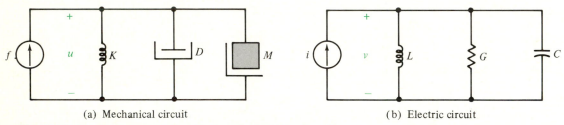

(a) Mechanical circuit   (b) Electric circuit

Fig. 15.8 Analogous mechanical and electric circuits.

The electrical analogs of mechanical circuits that we have just discussed were based on the choice of the force–current and velocity–voltage analogies. However, we could have instead selected force to be analogous to voltage and velocity to be analogous to current. Doing so would result in alternative electrical analogs (see Problems 15.5 and 15.6).

We are now in a position to generalize our previous discussion of simple motors and generators. In particular, reconsider the case of the motor shown in Fig. 15.3. Suppose that the battery is replaced by a general voltage source v, and that

---

*Named for the French mathematician Jean Le Rond D'Alembert (1717–1783).

L is the inductance of the moving conductor, and its resistance (and that of the track, too) is lumped together with R. In addition, suppose that the moving conductor has mass M, coefficient of friction D, and an external force f (directed to the left) is loading it. Then, by KVL, the equation describing the electric portion of the motor is

$$v = Ri + L\frac{di}{dt} + e = Ri + L\frac{di}{dt} + Blu \qquad \text{(15.8)}$$

Applying D'Alembert's principle to the mechanical portion of the motor,

$$f + Du + M\frac{du}{dt} = Bli$$

or

$$f = -Du - M\frac{du}{dt} + Bli \qquad \text{(15.9)}$$

Thus, we can model the motor, described by Equations (15.8) and (15.9), by the electromechanical circuit shown in Fig. 15.9.

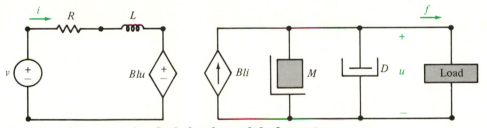

Fig. 15.9 Electromechanical circuit model of a motor.

15.2 MOVING-COIL AND MOVING-IRON DEVICES

We now have given some thought to simple motors and generators that utilize movable conductors. These kinds of elementary systems are called **translational transducers.** A **dynamic** or **moving-coil transducer** is a more practical type of translational transducer. Such a device, which can be employed in loudspeakers, microphones, and phonograph pickups, is illustrated in Fig. 15.10. The permanent magnet has an annular air gap with a radial magnetic field (only the flux at the end of the magnet is depicted). A sleeve (shown extended) with an N-turn coil fits over the south pole of the permanent magnet. The sleeve–coil combination, called the **moving coil,** is supported so that it can move axially in either direction.

For the case of a loudspeaker, the moving coil is connected to a cone or diaphragm that vibrates when alternating current passes through the coil. This results in the radiation of acoustic energy. Suppose that a voltage source v is applied to the coil whose winding resistance is R. Neglecting the inductance of the coil

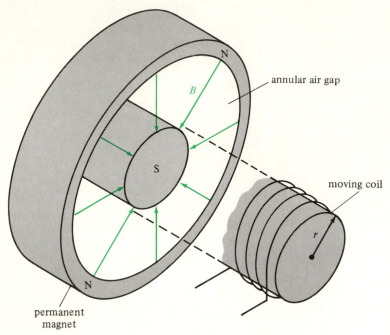

Fig. 15.10 Dynamic (moving-coil) transducer.

(i.e., $L = 0$), when a current i is directed out of the plus terminal of the voltage source, the equation describing the electric portion is, from Equation (15.8),

$$v = Ri + e = Ri + Blu \tag{15.10}$$

where the length-l coiled conductor is composed of N turns, each of which has radius r, that is, $l = 2\pi rN$. For the mechanical portion, from Equation (15.9),

$$f = -Du - M\frac{du}{dt} + Bli \tag{15.11}$$

where D is the coefficient of friction of the moving coil, M is the mass, and again $l = 2\pi rN$. This time, however, we consider that the external force f due to an acoustic load is proportional to the velocity u, that is, $f = D_L u$. Thus, we can model the loudspeaker described by Equations (15.10) and (15.11) with the electromechanical circuit shown in Fig. 15.11.

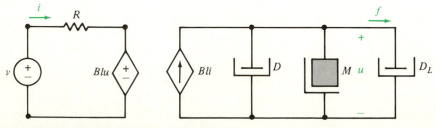

Fig. 15.11 Model of a loudspeaker.

Example 15.2

The magnetic flux density in the annular air gap of a dynamic loudspeaker (see Fig. 15.10) is $B = 0.5$ T. Suppose that the moving coil has a radius of $r = 2$ cm and a current of $i(t) = 0.1 \cos \omega t$ A. Let us find the number of turns required to develop a maximum force of $Bli = 0.25$ N.

From the fact that

$$Bli = B 2\pi r N i = 0.25$$

and since the developed force Bli is maximum when i is maximum (denoted i_m), we have that the number of turns is

$$N = \frac{0.25}{B 2\pi r i_m} = \frac{0.25}{(0.5)(2\pi)(0.02)(0.1)} \approx 40$$

Now suppose that $D = 3$ N-s/m and $D_L = 12$ N-s/m. If we ignore the mass (i.e., set $M = 0$), then using the analogy to current division (for conductances), we find that the force on the load is

$$f = \frac{D_L}{D_L + D} Bli = \frac{12}{12 + 3}(0.25) = 0.2 \text{ N}$$

■

The idea behind a dynamic microphone is the converse of the dynamic loudspeaker. Although the mechanisms are essentially the same, it is acoustic energy that applies a force f (the input) to the diaphragm of the microphone. This, in turn, translates the moving coil through the magnetic field established by the permanent magnet. The result is an induced emf $e = Blu$ that supplies energy to an electrical load (e.g., the input of an amplifier). The equations describing the behavior of a dynamic microphone are similar to Equations (15.10) and (15.11)—the difference being that some signs are reversed since the energy conversion is mechanical to electric instead of electric to mechanical. The describing equations are

$$v = -Ri + e = -Ri + Blu \tag{15.12}$$

and

$$f = Du + M\frac{du}{dt} + Bli \tag{15.13}$$

where f is the input, and i and v are the output current and output voltage respectively. These equations can be modeled in a manner similar to the case of the loudspeaker (see Problem 15.10).

Another use of the moving-coil transducer is as a **dynamic phonograph pickup,** as depicted in Fig. 15.12 by a cross-sectional view. Attached to the moving coil is a **stylus** that produces a vertical displacement $y(t)$ of the moving coil as the stylus tracks a record groove. (The variations in the groove represent the original audio signal.) The motion of the moving coil due to the displacement of

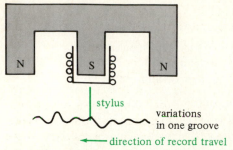

Fig. 15.12 Dynamic phonograph pickup.

the stylus induces an emf across the winding. This voltage can then be applied to an amplifier whose input resistance R_L is the load for the moving coil's winding. Since the behavior of such a phonograph pickup is that of a dynamic microphone, then from Equation (15.12) and the fact that $i = v/R_L$ (see Problem 15.10), we have

$$v = -R\left(\frac{v}{R_L}\right) + Blu$$

from which

$$v = \frac{R_L}{R_L + R}Blu = \frac{R_LBl}{R_L + R}\frac{dy}{dt}$$

since $u = dy/dt$.

For the case that a sinusoid of frequency ω rad/s and constant amplitude is recorded, the resulting vertical displacement has the form

$$y(t) = A\cos(\omega t + \theta)$$

and the resulting output voltage v is proportional to

$$\frac{dy(t)}{dt} = -A\omega\sin(\omega t + \theta) = A\omega\sin(\omega t + \theta + 180°)$$

$$= A\omega\cos(\omega t + \theta + 90°)$$

Thus, the sinusoid undergoes a phase shift of 90°, and its amplitude increases linearly with frequency as well. Because of this phenomenon, phonograph amplifiers contain **equalizing networks** to compensate for such a frequency dependence.

Moving-Iron Transducers

Not only does a magnetic field exert force on moving electric charge, but it also will exert a force on magnetic material, such as iron. (Anyone who has ever played with a permanent magnet has experienced this phenomenon.) It is this property that forms the basis for another type of transducer, called a **moving-iron transducer,** which utilizes a magnetic field to attract an initially unmagnetized, movable piece of iron.

Figure 15.13 shows a coil wound around a magnetic core that has an air gap. Such a device is often referred to as an **electromagnet.** In this particular case, part of the core is pivoted and is able to move. The movable portion, called the **armature,** is held in place by a spring. If a current passes through the coil, the resulting force opposes the restraining force of the spring and tends to move the armature so that the gap length gets smaller. This is the case regardless of the direction of the current, and, hence, regardless of the direction of the resulting flux. Also depicted in Fig. 15.13 is a rigid member extending from the armature. When the coil current is large enough, the armature will move a sufficient distance for the member to close the pair of contacts shown. (These contacts, in turn, may be used as a switch for some other system.) The entire device seen in Fig. 15.13 is a magnetically operated switch called a **relay.**

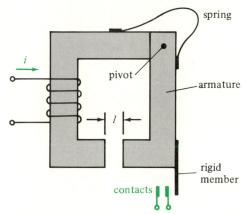

Fig. 15.13 Relay.

But how much force results from an applied current, and how much current is required to close the contacts of such a relay? To answer these questions, we proceed as follows: Suppose that, as the result of an applied current, the armature moves an infinitesimal horizontal distance dl, called a **virtual displacement,** due to the force f on it. The resulting work, called **virtual work,** done on the armature is $f\,dl$. For the sake of simplicity, let us assume that, during the duration dt of the virtual displacement, the current is adjusted so that the flux ϕ remains constant. As a consequence, the induced emf is $e = N\,d\phi/dt = 0$, and, therefore, the electric energy put into the system is $vi\,dt = Ri^2\,dt$, where R is the winding resistance of the coil. By Equation (14.20), the inductance of the device is $L = N^2/\mathcal{R}$, where the reluctance $\mathcal{R} = \mathcal{R}_c + \mathcal{R}_g$ is the sum of the core reluctance $\mathcal{R}_c$ and the gap reluctance $\mathcal{R}_g$. Assuming that the permeability of iron is so high that $\mathcal{R}_g \gg \mathcal{R}_c$, then $\mathcal{R} \approx \mathcal{R}_g$, and the energy stored in the air gap is $w_g = Li^2/2 = N^2 i^2/2\mathcal{R}$. From Equations (14.8) and (14.10), $i = \mathcal{R}\phi/N$, and, thus,

$$w_g = \frac{1}{2}\mathcal{R}\phi^2 = \frac{1}{2}\left(\frac{l}{\mu_0 A}\right)(BA)^2 = \frac{B^2 A l}{2\mu_0}$$

where A is the cross-sectional area of the gap and fringing is ignored. Since the length of the gap decreases by dl, the energy stored in the gap decreases by $B^2A\,dl/2\mu_0$. Thus, this amount of energy also goes into the system. The motion of the armature is horizontal, and so there is no change in potential energy. And, if we assume that the motion is slow enough, we can neglect kinetic energy.

As mentioned above, one result of the energy put into the system is the virtual work done on the armature. Another is the heat due to the energy absorbed by the winding resistance R—this loss is $Ri^2\,dt$. Making the further simplifying assumption of no friction, we can ignore mechanical friction loss. Finally, with a constant flux, there is no loss from hysteresis and eddy currents. Therefore, by the principle of the conservation of energy, we have

$$Ri^2\,dt + \frac{B^2A\,dl}{2\mu_0} = f\,dl + Ri^2\,dt$$

from which the force on the armature is

$$f = \frac{B^2A}{2\mu_0} \tag{15.14}$$

From Equations (14.8) and (14.10), we have that $\phi = Ni/\mathcal{R}$. Since $B = \phi/A$, then

$$B = \frac{Ni}{\mathcal{R}A} = \frac{Ni}{\left(\dfrac{l}{\mu_0 A}\right)A} = \frac{\mu_0 Ni}{l}$$

so Equation (15.14) yields a force of

$$\boxed{f = \frac{\mu_0 N^2 i^2 A}{2l^2}} \tag{15.15}$$

Example 15.3

Suppose that the relay given in Fig. 15.13 has a coil with 400 turns, a cross-sectional area of 10 cm^2, and a spring force of 0.5 N. The gap length when the contacts are open is 1 cm, and 0.4 cm when the contacts are closed. Then from Equation (15.15), the current required to close the contacts, called the **pickup** or **pull-in current**, is

$$i = \sqrt{\frac{2l^2 f}{\mu_0 N^2 A}} = \frac{l}{N}\sqrt{\frac{2f}{\mu_0 A}} \tag{15.16}$$

$$= \frac{0.01}{400}\sqrt{\frac{2(0.5)}{(4\pi \times 10^{-7})(10)(0.01)^2}} = 0.705 \text{ A}$$

The **dropout** or **hold-in current** is the minimum current required to keep the contacts closed. Some value of current which must be less than 0.705 A will result in the contacts opening. Since Equation (15.16) indicates that current is directly proportional to gap length, we deduce that the dropout current is

$$i = \frac{0.4}{1.0}(0.705) = 0.282 \text{ A}$$

Another type of moving-iron transducer is shown in Fig. 15.14. Here we have an N-turn coil (a solenoid) with a movable iron core, called a **plunger**. When a current passes through the coil, a force directed to the right (so as to center the plunger in the coil) is exerted on the plunger. The plunger may be connected to some mechanical apparatus such as a valve.

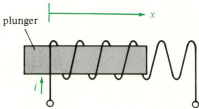

Fig. 15.14 Solenoid-plunger combination.

Suppose that, as a result of an applied current, the plunger has a virtual displacement of dx in dt seconds. Thus, the virtual work done on the plunger is $f\,dx$. Also, suppose that the resistance of the coil is zero. Then the voltage induced across the coil is $e = N\,d\phi/dt$, and the electric energy put into the system is

$$ei\,dt = N\frac{d\phi}{dt}i\,dt = Ni\,d\phi = i\,d(N\phi)$$

From Equation (14.19), $N\phi = Li$. Thus, $i\,d(N\phi) = i\,d(Li)$, and if the current is constant, then the electric energy put into the system is

$$ei\,dt = i^2\,dL$$

where dL is the change of inductance. (As the plunger moves to the right, the flux ϕ increases. This, in turn, decreases the reluctance $\mathfrak{R} = Ni/\phi$, and, hence, increases the inductance $L = N^2/\mathfrak{R}$.) Since $Li^2/2$ is the energy stored, the increase in energy stored is $i^2\,dL/2$. As for the case of the relay discussed previously, we will not be concerned with the potential and kinetic energies. Ignoring hysteresis, eddy currents, and friction, by the conservation of energy we have that

$$i^2\,dL = f\,dx + \frac{1}{2}i^2\,dL$$

from which the force exerted on the plunger is given by

$$f = \frac{1}{2} i^2 \frac{dL}{dx}$$

(15.17)

where the current i is constant.

15.3 ROTATING-COIL DEVICES

Having discussed translational transducers, let us now consider **rotational transducers,** that is, electromechanical transducers that utilize rotation. We begin with a classical example—the **d'Arsonval movement.*** The basic components of this mechanism, which employs a rotating coil, are depicted in Fig. 15.15. It consists of a stationary permanent magnet with an air gap in which a stationary iron core is placed. (The air gap between the permanent magnet and the iron core is greatly exaggerated for illustrative purposes.) Around the stationary core is an N-turn coil

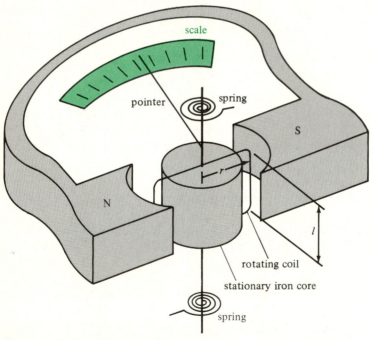

Fig. 15.15 The d'Arsonval movement.

---

*Named for the French physicist Jacques Arsène d'Arsonval (1851–1940).

that is wound on a rectangular form, and this coil is free to rotate around its vertical axis. The coil assembly, which is supported on jeweled bearings, is connected to two springs that not only supply a restraining torque, but also provide an electric connection to the rotating coil. The angular deflection γ of the coil assembly is indicated by a pointer attached to it.

A top view of the air-gap region of the d'Arsonval movement is shown in Fig. 15.16. Inserting a stationary iron core between the poles of the permanent magnet produces a uniform radial magnetic field with flux density B in the resulting air gap. Depicted in Fig. 15.16 is just one turn of the N-turn movable coil. Assuming that the coil's rotation is restricted (typically by mechanical stops) to the uniform field, when a current i goes through the coil, a force will be exerted on it. For the case indicated in which the current direction is into the page for the upper right-hand portion of the coil and out of the page for the lower left-hand portion, the resulting forces have the directions shown. Since one side of a single turn has length l, by Equation (15.3), a force of $f = Bli$ is exerted on this conductor. Consequently, the resulting torque is $fr = Blir$. Because the other side of the turn experiences the same torque, and the coil has N turns, the clockwise torque T on the movable coil is

$$T = 2NBlri \qquad\qquad (15.18)$$

In opposition to this torque is a counterclockwise torque T_0 attributable to the springs. We thus have that

$$T_0 = \frac{\gamma}{K} \qquad\qquad (15.19)$$

where K is the rotational compliance of the springs and γ is the angle of deflection of the coil. For the case of equilibrium, from Equations (15.18) and (15.19), we obtain

$$\boxed{\gamma = (2KNBlr)i} \qquad\qquad (15.20)$$

In other words, the deflection angle is directly proportional to the current in the movable coil. For this reason, the scale of the d'Arsonval movement is calibrated linearly, and such an instrument is used as a dc ammeter or dc voltmeter.

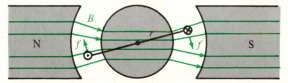

Fig. 15.16 Air-gap region of a d'Arsonval movement.

Example 15.4

A d'Arsonval movement has a moving coil with demensions $l = 2$ cm $= 0.02$ m and $r = 1$ cm $= 0.01$ m. The magnetic flux density in the air gap is $B = 0.5$ T, and the spring compliance is 2×10^8 degrees/N-m. Suppose that a full-scale deflection corresponds to $\gamma = 90°$. If a current of $i = 50$ μA is to cause a full-scale deflection, then from Equation (15.20), the number of turns required for the moving coil is

$$N = \frac{\gamma}{2KBlri}$$

$$= \frac{90}{(2)(2 \times 10^8)(0.5)(0.02)(0.01)(50 \times 10^{-6})} = 45$$

■

Clearly, the d'Arsonval movement given in the preceding example can be used as a dc ammeter to measure currents ranging from 0 to 50 μA. However, it is possible to use this 50-μA movement as a dc ammeter for other current ranges as well. For instance, suppose that the resistance of the moving coil is $R_c = 2$ kΩ. To use this movement for a 0- to 100-mA dc ammeter, we place a **shunt** resistance R_p in parallel with the moving coil as illustrated in Fig. 15.17. By current division, we have that

$$i_c = \frac{R_p}{R_p + R_c} i$$

from which

$$R_p = \frac{i_c}{i - i_c} R_c \qquad\qquad (15.21)$$

For full-scale deflection, $i_c = 50$ μA and $i = 100$ mA. Thus, from Equation

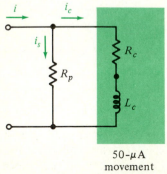

50-μA
movement

Fig. 15.17 A d'Arsonval movement ammeter.

(15.21), the value of the shunt resistance is

$$R_p = \frac{50 \times 10^{-6}}{100 \times 10^{-3} - 50 \times 10^{-6}}(2 \times 10^3) = 1 \ \Omega$$

An ideal ammeter has a resistance of zero ohms so that when it is inserted in series with an element whose current is to be measured, the resulting combined resistance remains the same—and, thus, the circuit parameters do not change. Therefore, since its resistance is 2 kΩ, the 50-μA d'Arsonval movement mentioned above is a reasonable approximation of an ideal 50-μA ammeter when it is used to measure currents through very large resistances. However, since the resistance of the 100-mA ammeter just discussed is 1 Ω, it approximates an ideal ammeter for a much wider variety of situations.

Voltmeters and Ohmmeters

We have just seen an example of how a 50-μA movement can be used as an ammeter for currents greater than 50 μA. To accomplish this, we utilized a shunt resistance R_p. However, even though a d'Arsonval movement is basically a current-measuring device, we can also use it to measure dc voltages. To do this, we place a resistance R_s in series with the movement. For instance, let us employ the 50-μA movement given in the preceding example as part of a 0- to 20-V dc voltmeter as depicted in Fig. 15.18. It is given that $R_c = 2$ kΩ, and for full-scale deflection, the current through the movement is $i = 50$ μA. Consequently, the voltage across it is $v_c = iR_c$. Since this is to correspond to the case that $v = 20$ V, from KVL and Ohm's law,

$$i = \frac{v - v_c}{R_s}$$

from which the value of the series resistance is found to be

$$R_s = \frac{v - v_c}{i} = \frac{v - iR_c}{i} = \frac{v}{i} - R_c$$

$$= \frac{20}{50 \times 10^{-6}} - 2 \times 10^3 = 3.98 \times 10^5 = 398 \ \text{k}\Omega$$

Thus, the voltmeter shown in Fig. 15.18 has a resistance of $R_s + R_c = 398,000 + 2000 = 400$ kΩ. An ideal voltmeter has infinite resistance so that, when voltage measurements are made, a circuit will not be disturbed (an ideal voltmeter draws no current). The voltmeter described above might significantly alter circuit variable values in some types of applications.

In a voltmeter such as this, consisting of a d'Arsonval movement and a series resistor R_s, instead of talking about the resistance of the voltmeter (400 kΩ in this case), typically the meter is described in terms of the number of ohms per volt of full-scale deflection. Thus, we would say that this meter is rated at

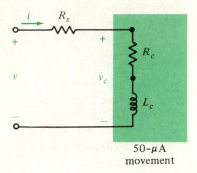

50-μA
movement

Fig. 15.18 A d'Arsonval movement voltmeter.

$$\frac{400,000 \ \Omega}{20 \ \text{V}} = 20,000 \ \Omega/\text{V} = 20 \ \text{k}\Omega/\text{V}$$

In addition to being used in ammeters and voltmeters, the d'Arsonval movement can be utilized for measuring the resistances of resistors or combinations of resistors—that is, it can be used to fabricate an ohmmeter. An elementary method for doing this is illustrated in Fig. 15.19. This ohmmeter consists of a d'Arsonval movement in series with a resistance R_s and a dc voltage source (battery) V_s. A resistance R is connected to the terminals of the ohmmeter, and its value is read on the d'Arsonval's calibrated scale.

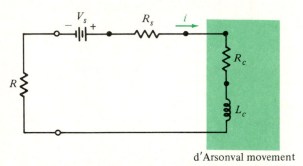

d'Arsonval movement

Fig. 15.19 A d'Arsonval movement ohmmeter.

Example 15.5

Suppose that the 50-μA movement discussed above is used for the ohmmeter shown in Fig. 15.19. Recall that $R_c = 2$ kΩ. If $V_s = 1.5$ V, then, to get full-scale deflection when the terminals of the ohmmeter are short-circuited ($R = 0 \ \Omega$), we use

$$i = \frac{V_s}{R_s + R_c}$$

to obtain

$$R_s = \frac{V_s - iR_c}{i} = \frac{V_s}{i} - R_c$$

Thus, we have

$$R_s = \frac{1.5}{50 \times 10^{-6}} - 2 \times 10^3 = 28 \text{ k}\Omega$$

Although a reading of 50 μA corresponds to the case that $R = 0\ \Omega$, for the general case of $R \neq 0$,

$$i = \frac{V_s}{R + R_s + R_c}$$

so

$$R = \frac{V_s - i(R_s + R_c)}{i} = \frac{V_s}{i} - (R_s + R_c) \qquad \text{(15.22)}$$

Therefore, for half-scale deflection ($i = 25$ μA), we have that the corresponding resistance is, from Equation (15.22),

$$R = \frac{1.5}{25 \times 10^{-6}} - (28 \times 10^3 + 2 \times 10^3) = 30 \text{ k}\Omega$$

For quarter-scale deflection ($i = 12.5$ μA), the value of the resistance is

$$R = \frac{1.5}{12.5 \times 10^{-6}} - 30 \times 10^3 = 90 \text{ k}\Omega$$

Finally, for the case that the terminals are open-circuited (R is infinite), the current is $i = 0$, so there is no meter deflection.

Unlike the ammeter and the voltmeter, which indicate values on a linear, increasing scale, as demonstrated in the preceding example, the ohmmeter displays resistance on a decreasing scale that is nonlinear. An actual ohmmeter is more complicated than the one just discussed, but the basic notion of measuring resistance is the same.

We have seen that a d'Arsonval movement can be used to measure current, voltage, and resistance. In actuality, the same d'Arsonval movement can be employed to measure a wide variety of current values. Changes in a ammeter scale are easily accomplished by switching to different resistors for the shunt. In an analogous manner, changing the series resistance of a voltmeter or an ohmmeter readily can switch the scales of the respective meters. A meter with a single d'Arsonval movement that can be used to measure multiple ranges of currents and voltages is called a **multimeter.** A common type of multimeter—a combination voltmeter, ohmmeter, and milliammeter—is abbreviated VOM.

The Electrodynamometer

Having studied the d'Arsonval movement—a permanent-magnet, moving-coil mechanism—we now will look at the **electrodynamometer.** The principle of operation of the latter mechanism is the same as for the former, but its construction is quite different. Instead of a permanent magnet, the magnetic field of the electrodynamometer is supplied by a stationary solenoid separated into two sections as illustrated in Fig. 15.20. The core of the solenoid is composed of nonmagnetic material (thus avoiding eddy-current and hysteresis losses) so that the magnetic flux density B is directly proportional to the current i_s through the solenoid [e.g., see Equation (14.3)]. In the solenoid's magnetic field is an N-turn movable coil (only one turn is depicted in Fig. 15.20) restrained by springs and equipped with a pointer (not shown), as is done for a d'Arsonval movement. Given that the length of the moving coil is l and the width is $2r$, then when a current i_c goes through it, the force (with the direction shown) on the upper right-hand portion of the coil is $f = Bli_c$. Therefore, the resulting torque is $fd = Bli_c d$. If the plane of the moving coil is at an angle ψ with the direction of B, then $d = r \cos \psi$. In combining this with the fact that the coil has N turns, the resulting clockwise torque on the movable coil is

$$T = 2NBli_c r \cos \psi$$

However, since B is directly proportional to the solenoid current i_s, that is, $B = K_s i_s$, then

$$T = (2NK_s lr \cos \psi)i_c i_s$$

Equating this torque to the spring restraining torque $T_0 = \gamma/K$, we have

$$\gamma = (2NKK_s lr \cos \psi)i_c i_s \tag{15.23}$$

Thus, we see that the angle of deflection γ depends on the product of the moving-coil current i_c and the solenoid current i_s.

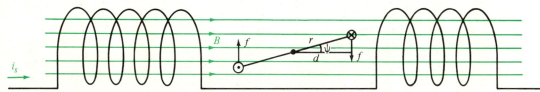

Fig. 15.20 Magnetic-field region for an electrodynamometer.

Although the electrodynamometer can be used in dc ammeters and dc voltmeters, because of a generally stronger magnetic field, the d'Arsonval movement is superior. However, the electrodynamometer is quite useful for ac measurements. When both coils carry alternating currents, say $i_c = I_c \cos(\omega t + \theta_1)$ and $i_s = I_s \cos(\omega t + \theta_2)$, the mechanical inertia of the movement causes an averaging of the angle of deflection. Since the average value of $i_c i_s$ is $\frac{1}{2}I_c I_s \cos \theta$, where $\theta = \theta_1 - \theta_2$ (see Problem 15.25), then when the instantaneous deflection-angle expression [Equation (15.23)] is averaged, we find that the angle of deflection has the form

$$\gamma = K_\psi I_c I_s \cos\theta$$

(15.24)

where the value of K_ψ varies as the angle ψ varies. For the case that the solenoid and moving coil are connected in series, the current through them will be the same. Thus, when $i_c = i_s = I\cos\omega t$, from Equation (15.24), we see that the angle of deflection is proportional to I^2. As a consequence, the scale can be graduated nonlinearly to read the rms value $(I/\sqrt{2})$ of the current. Hence, the electro-dynamometer can be used as an ac ammeter or ac voltmeter. Because of a practical limitation on the current through the moving coil, and because the reactance of the moving coil is significant in ammeter applications, the electrodynamometer's use is much more common for ac voltmeters than for ac ammeters.

Perhaps the most important use of the electrodynamometer is for the wattmeter. For the mechanism to function in this manner, the fixed coil (solenoid) is used for the load current $I\cos(\omega t + \theta_2)$, and the moving coil in series with a large resistance R_s is used for the load voltage $V\cos(\omega t + \theta_1)$. Assuming that the impedance of the moving coil is negligible compared with R_s, the amplitude of the moving-coil current is $I_c = V/R_s$. Furthermore, the amplitude of the current in the fixed coil is $I_s = I$ assuming that $i_s = i$. Thus, from Equation (15.24), we see that the angle of deflection is proportional to $I_c I_s \cos\theta$, where $\theta = \theta_1 - \theta_2$, and, therefore, it is proportional to the average power $P = \frac{1}{2}VI\cos\theta$ absorbed by the load. Hence, the scale of the electrodynamometer can be graduated to read power.

A typical connection of an electrodynamometer for use as a wattmeter is shown in Fig. 15.21. For the current directions indicated, if both i_c and i_s are positively valued, then the pointer of the meter will have an up-scale deflection. If both

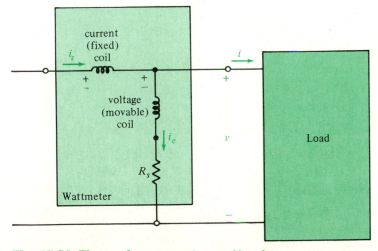

Fig. 15.21 Electrodynamometer wattmeter.

currents have a negative value, the deflection again will be up-scale. For this reason, the appropriate ends of the coils are marked with the symbol $\pm$. If one current has a positive value and the other has a negative value, then the meter will have a down-scale deflection.

In the discussion above, it was assumed that the load current is equal to the current in the fixed coil, that is $i = i_s$. However, to be more accurate, from Fig. 15.21, we see that $i = i_s - i_c$. Therefore, the load current is less than the current in the fixed coil, and, consequently, the wattmeter reading is larger than the power absorbed by the load. The power indicated by the wattmeter is the sum of the load power and the power absorbed by the series connection of the moving coil and the resistance R_s. If P_L is the power absorbed by the load and P_W is the wattmeter reading, since the amplitude of the load voltage is V, then

$$P_L = P_W - \frac{V^2}{2R_s}$$

There are wattmeters, called **compensated wattmeters,** that directly read P_L. In such a meter, the moving-coil (voltage coil) current goes through an additional fixed coil located so as to cancel the effect of i_c on the current in the fixed coil.

15.4 GENERATORS

Generators and motors are examples of electromechanical **machines.** The simple machines shown in Figs. 15.1, 15.2, and 15.3 are translational devices. However, rotating machines are inherently more practical because of their higher velocity, voltage, and power capabilities. Therefore, we will now turn our attention to the subject of rotating machines.

Let us begin with the single conducting loop imbeded in a uniform magnetic field having a constant flux density B as depicted in Fig. 15.22(a). Suppose that the loop, which is rectangular with length l and width $2r$, is forced to rotate around the z axis with a constant angular velocity of ω radians per second as indicated. At time t, the angle between the plane of the loop and the y-z plane is ωt, as shown in Fig. 15.22(a) and the end view in Fig. 15.22(b). As indicated in Fig. 15.22(a), the ends of the loop are connected to **slip rings** that rotate along with the loop. However, touching the slip rings are **brushes** that make electrical connections, but that are fixed. The area of the loop is $A = 2rl$ and the flux ϕ passing through the loop is

$$\phi = B(2r \cos \omega t)l = B2rl \cos \omega t = BA \cos \omega t$$

By Faraday's law, the emf induced across the brushes is

$$e = \frac{d\phi}{dt} = -BA\omega \sin \omega t \tag{15.25}$$

and, due to Lenz's law, the polarity of this emf is as indicated. If an N-turn coil were used in place of the single loop, the resulting induced emf would be

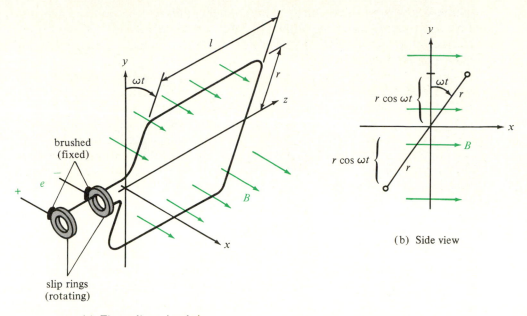

(a) Three-dimensional view

Fig. 15.22 Simple rotating machine.

$$e = -NBA\omega \sin \omega t \qquad \qquad \text{(15.26)}$$

Since the simple rotating machine shown in Fig. 15.22(a) produces a sinusoidal voltage, it is an example of an ac generator, which is referred to as an **alternator.** Note that, in addition to the number of coil turns, the magnetic field strength, and the area of the coil, the amplitude of the induced voltage also depends on the speed of rotation of the coil.

Now let us modify the simple alternator given in Fig. 15.22(a) as shown in Fig. 15.23(a). What we have done is to replace the slip rings of the alternator given in Fig. 15.22(a) by a **split-ring commutator,** a side view of which is shown in Fig. 15.23(b). Furthermore, the brushes have been repositioned on either side of the split-ring commutator. As we have already seen, the voltage induced across the loop is given by Equation (15.25), and this is the voltage v_{12} between the halves of the split-ring commutator. However, the brushes are situated such that $v = -v_{12} = -e = BA\omega \sin \omega t$ when $0 < \omega t < \pi$ and $v = v_{12} = e = -BA\omega \sin \omega t$ when $\pi < \omega t < 2\pi$. Therefore, whereas the voltage e between the brushes of the alternator given in Fig. 15.22(a) is $e = BA\omega \sin \omega t$, the voltage v between the brushes of the rotating machine given in Fig. 15.23(a) is $v = |BA\omega \sin \omega t|$. These two voltages are shown in Fig. 15.24(a) and (b) respectively. Although the voltage v shown in Fig. 15.24(b) is not constant, it never

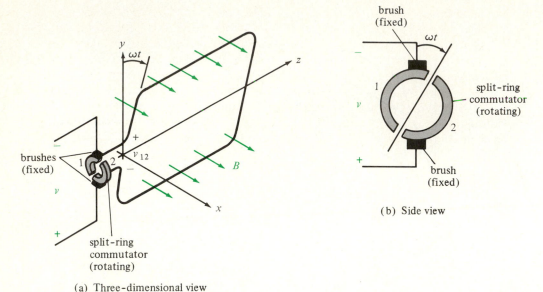

(a) Three-dimensional view

(b) Side view

Fig. 15.23 Simple rotating dc machine.

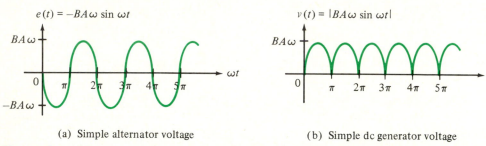

(a) Simple alternator voltage

(b) Simple dc generator voltage

Fig. 15.24 Voltages produced by simple alternator and dc generator.

takes on a negative value. Because the voltage does not change polarity, the simple machine given in Fig. 15.23(a) is an example of an elementary dc generator.

The dc Generator

As was implied in the preceding discussions of a simple alternator and a simple dc generator, a typical rotating machine is comprised of a stationary portion called the **stator** and a rotating portion called the **rotor.** In a dc generator, a magnetic field is produced when a direct current is applied to windings or coils in the stator, and so the stator is referred to as the **field,** and such coils are called **field windings.** The rotor contains the commutator and the conductors across which emfs are induced. This portion of the machine is known as the **armature.**

A cross-sectional view of a simplified dc generator is shown in Fig. 15.25. Although the stator has two poles, one labeled N and one labeled S, an even number, greater than two, of magnetic poles can be used. As indicated, situated on the poles are the field windings. The rotor consists of a cylindrical iron core that contains slots that house the armature conductors and a commutator along with the associated brushes. Shown on the armature are eight conductors that form four coils, and a commutator with four segments. The brushes are depicted as riding on the inside, but they actually make contact with the outside of the commutator. Figure 15.26 gives one possibility for connecting the coils to the commutator. Here, conductors 1 and 1' form coil 1, and so forth, but not shown is the implicit rear connections between 1 and 1', 2 and 2', 3 and 3', and 4 and 4'.

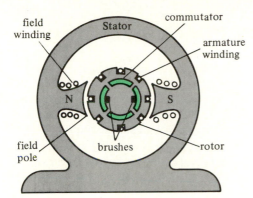

Fig. 15.25 A dc generator.

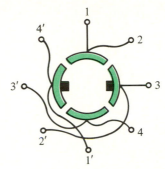

Fig. 15.26 Coil-commutator connections.

Passing (direct) current through the field windings produces essentially uniform magnetic fields in the small air gaps between the poles and the rotor. The emf e induced across a single coil is as shown in Fig. 15.27, where the maximum and minimum voltages occur when the coil is in a horizontal position. When the armature windings are connected to the commutator in an appropriate manner, such as the one indicated in Fig. 15.26, the individual induced emfs not only will be rectified, but they will be added together as well. For example, for the connection given in Fig. 15.26, the right-hand brush is connected to conductor 3, which, in turn, is (implicitly) connected to 3'. But 3' is connected to 4 (via the commutator), which, in turn, is connected to 4'. Since 4' is connected to the left-hand brush, then the series connection of coils 3 and 4 is connected between the two brushes—so their induced emfs add. Note, too, that coils 1 and 2 are also in series for the indicated brush positions. The net effect of the coil connections to the commutator

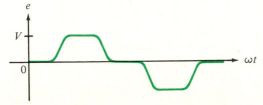

Fig. 15.27 The emf across a single coil.

is a brush voltage v such as shown in Fig. 15.28. The small notches in the waveform result from the brushes changing from the different segments of the commutator. In practice, many coils are employed and this **commutator ripple** is quite small.

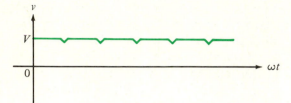

Fig. 15.28 The voltage across the brushes.

Let us now establish a formula for the voltage produced by a dc generator having p poles, each with a flux per pole of Φ, where the generator is turning at the rate of n revolutions per minute (rpm). If the rotor has a speed of n rpm, then it turns at the rate of $n/60$ revolutions per second (rps). Since there are p poles in one revolution, the rotor turns at the rate of $pn/60$ poles per second. However, each conductor experiences a flux change of Φ per pole. Therefore, the average rate of change of flux (induced emf) per conductor is $\Phi pn/60$ volts. If there are N armature conductors and a parallel paths between the brushes, then the number of conductors in series comprising each parallel path is N/a. Consequently, the generated voltage v_g is given by

$$v_g = \frac{N}{a} p \Phi \frac{n}{60} \qquad\qquad (15.27)$$

where N is the number of armature conductors, a is the number of parallel paths between the brushes, p is the number of poles, Φ is the magnetic flux per pole, and n is the speed of the rotor in rpm.

Example 15.6

A dc generator such as the one shown in Fig. 15.25 has two poles and two parallel paths between the brushes. Suppose that each pole has a surface area of $A = 0.01$ m². For the case that the air-gap magnetic field density is $B = 1$ T, the flux per hole is

$$\Phi = BA = 1(0.01) = 0.01 \text{ Wb}$$

If each of the eight armature coils has 12 turns (each turn contains two conductors), then $N = (8)(12)(2) = 192$. For a rotor speed of 1500 rpm, the generated voltage is

$$v_g = \frac{N}{a}p\Phi\frac{n}{60} = \frac{192}{2}(2)(0.01)\frac{1500}{60} = 48 \text{ V}$$

In the preceding discussion, the generated emf v_g is the voltage across the armature terminals (the brushes) when no load is connected to those terminals. If, however, there is some electrical load connected to the armature, the resulting load voltage, in general, will be different from the no-load voltage. This conclusion is a consequence of the fact that there is a resistance R_a associated with the armature as a result of such things as the resistances of the windings and the brush contacts. The following example demonstrates the effect of the armature resistance R_a of a dc generator.

Example 15.7

An 8-kW, 200-V dc generator has a full-load current of 40 A at 1200 rpm. Given that the armature resistance is $R_a = 0.5 \ \Omega$, we can model the armature as shown in Fig. 15.29, where R_L is a resistive load.

From the fact that

$$v_g = R_a i_L + v_L$$

under full-load conditions (at $n = 1200$ rpm), this equation becomes

$$v_g = (0.5)(40) + 200 = 220 \text{ V}$$

and this is the no-load voltage.

From Equation (15.27), we see that if the flux per pole Φ is kept constant, then the generated voltage is directly proportional to the armature speed n. Thus, if the field current is kept constant (and, hence, the flux per pole is kept constant), then at 900 rpm the no-load voltage is

$$v_g = \frac{900}{1200}(200) = 165 \text{ V}$$

Therefore, under the full-load condition of 40 A, at 900 rpm the full-load voltage is, by KVL,

$$v_L = -R_a i_L + v_g$$
$$= -0.5(40) + 165 = 145 \text{ V}$$

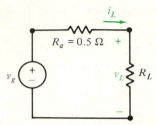

Fig. 15.29 Electric-equivalent circuit of given armature and associated load.

We can represent a dc generator symbolically as is shown in Fig. 15.30, where i_f indicates the field current and i_a denotes the armature current. Sometimes the symbol for a generator will include an indication of something, called a **prime mover,** that causes the armature to turn. However, we shall consider the prime mover to be implicit.

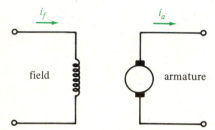

Fig. 15.30 Symbol for a dc generator.

From Equation (15.27), we know that the rotor of a dc generator turning at n rpm induces an emf (the no-load armature voltage) of

$$v_g = K\Phi n \tag{15.28}$$

where $K = Np/60a$ is a constant determined by the construction of the generator. Furthermore, the flux per pole Φ is determined by the field current i_f [see Equations (14.8) and (14.9)]. In general, the relationship between Φ and i_f is nonlinear. For a given speed n, therefore, let us plot the no-load voltage v_g versus the field current i_f. Such curves, which are shown in Fig. 15.31, can be obtained experimentally from a physical generator by holding the speed constant and measuring the value of v_g that results when different values of i_f are applied. Since v_g is proportional to Φ [Equation (15.28)], which, in turn, is proportional to B, and since i_f is proportional to $\mathscr{F}$ [Equation (14.12)], which, in turn, is proportional to H, a plot of v_g versus i_f has the same shape as a B-H curve. Thus, the curves shown in Fig. 15.31 are known as **magnetization curves** of the generator. Sometimes, in addition to being labeled in terms of field current, the horizontal axis is also labeled in terms

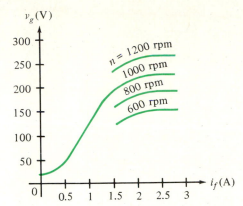

Fig. 15.31 Magnetization curves for a dc generator.

of mmf (ampere-turns) per pole. Once one magnetization curve has been determined (or measured), with the use of Equation (15.28), it is a simple matter to obtain the magnetization curve corresponding to a different speed. Note that a magnetization curve, in particular, the curve corresponding to $n = 1000$ rpm in Fig. 15.31, does not pass through the origin. This is so because, even when the field current is zero, residual pole magnetism will result in a small generated emf.

Generator Field Excitation

There are four basic ways of establishing the field current for a dc generator. The most obvious approach is to connect its own individual source to the field* as indicated in Fig. 15.32. This is an example of a **separately excited generator.** However, this requires the use of an additional source V_f. Such a requirement can be eliminated by using the generator's own voltage to produce the field current. One way this can be done is by connecting the field in parallel or in shunt with the armature, as shown in Fig. 15.33(a). The resulting **shunt-connected generator** is an example of a **self-excited generator.** In a shunt-connected generator, the field

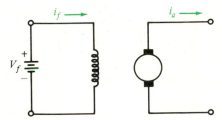

Fig. 15.32 Separately excited generator.

---

*Although the field is represented by the symbol for an ideal inductor, there is an implicit field resistance R_f that is not shown.

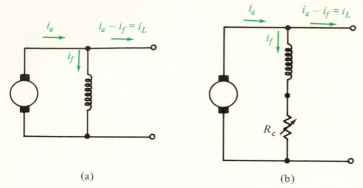

(a)   (b)

Fig. 15.33 Shunt-connected generator.

winding typically has many turns, and the relatively high field resistance results in a field current that is only a small fraction of the armature current. The field current can be further adjusted by placing a variable resistance R_c, called a **control rheo-stat,** in series with the field as shown in Fig. 15.33(b).

Another type of self-excitation is obtained for the **series-connected generator** shown in Fig. 15.34(a). Since the field current is equal to the armature current, the field winding has few turns and a relatively low resistance. Even so, under load conditions, a large armature current can cause a significant difference between the generated and load voltages. The relative effect of the series field resistance can be reduced by connecting a low resistance R_d in parallel with the field as shown in Fig. 15.34(b). Since such a resistance diverts some of the armature current from the field, R_d is called a **diverter resistance.** Despite the possibility of using diverter resistances, series-connected generators are rarely used.

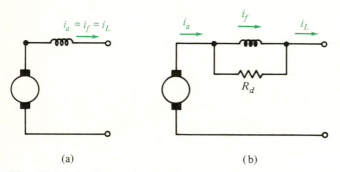

(a)   (b)

Fig. 15.34 Series-connected generator.

By combining the concepts of a shunt connection and a series connection, another type of self-excitation can be obtained. In particular, Fig. 15.35(a) shows a **compound-connected generator.** Again, a control rheostat and a diverter resistance can be employed as shown in Fig. 15.35(b). The series portion of the field can be connected so that its mmf either can be added to or subtracted from the mmf

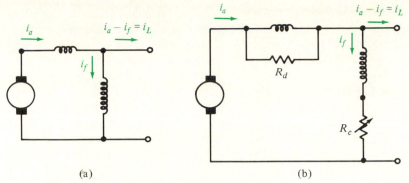

(a) (b)

Fig. 15.35 Compound-connected generator.

of the shunt portion of the field. Typically, it is the former connection, called **cumulative compounding,** that is used, rather than the latter connection, which is known as **differential compounding.**

For a separately excited generator turning at n rpm, the value of the field current (which can be adjusted by adjusting V_f as in Fig. 15.32) determines the value of the generated voltage—the magnetization curve can be used to find the voltage value. However, let us consider the shunt-connected generator shown in Fig. 15.36, where a control rheostat R_c is included, as is a field switch S. Suppose that the magnetization curve for the generator is as given in Fig. 15.37 for the case of a prime mover operating at n rpm. Even before the field switch S is closed, when $i_f = 0$, there will be a small generated emf due to residual magnetism. When the switch is closed, there will be a voltage applied across the field in series with the control rheostat. This, in turn, produces a nonzero field current, and, hence, increases the flux, thereby resulting in an increase in the generated emf. This

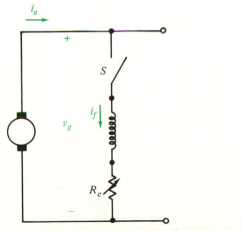

Fig. 15.36 Shunt-connected generator.

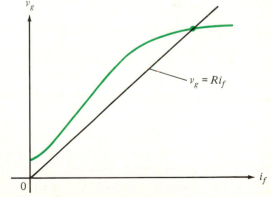

Fig. 15.37 Magnetization curve.

repetitive process of increasing generated voltage and field current is known as generator **buildup.** But how much do the voltage and current build up?

A magnetization curve is a plot of the no-load generated voltage versus the field current. For the case of no load, however, the armature resistance R_a is typically much smaller than R, the sum of the field resistance R_f and the resistance R_c of the control rheostat, that is, $R_a \ll R = R_f + R_c$. Thus, in essence, the terminal voltage equals the generated emf under no-load conditions, and, therefore, $v_g = R i_f$. A plot of this simple Ohm's law equation is also given in Fig. 15.37. Since the voltage and current values for generator buildup correspond to points on the generator's magnetization curve, and since the equation $v_g = R i_f$ also must be satisfied, the intersection of the magnetization curve and the line $v_g = R i_f$ gives the point where generator buildup ceases. At this point, the applied field voltage produces the field current necessary to generate that voltage.

Note that by decreasing the value of $R = R_f + R_c$, the intersection of the curve and the line will be to the right of the point shown. That would indicate a greater value for the generated voltage. Conversely, increasing the value of R would decrease the generated voltage. Specifically, if the value of R were increased so much that the line $v_g = R i_f$ were, for the most part, on the left side of the magnetization curve, there would be very little generator buildup and the generated voltage would be quite small. Even when the value of R is reduced so that the line $v_g = R i_f$ approximately coincides with the linear portion of the magnetization curve, the generated voltage is not uniquely determined, and it might change from one value to another. Therefore, for the most stable operation, the value of R should be chosen such that the line $v_g = R i_f$ intersects the nonlinear (saturated) portion of the magnetization curve as illustrated in Fig. 15.37.

Suppose that, for a dc generator, the load (or terminal) voltage is denoted by v_L and the load current by i_L. As depicted in Fig. 15.29, for a separately excited generator, the load voltage is related to the load current by

$$v_L = -R_a i_L + v_g \tag{15.29}$$

where v_g is the generated emf. Thus, as the load current increases, the load voltage decreases. A plot of the relationship between the load voltage and the load current (v_L versus i_L) is known as the **external characteristic** of the generator. Although Equation (15.29) indicates that the external characteristic for a separately excited generator is a straight line, in actuality it is slightly curved, as shown in Fig. 15.38. (The reason for this will be discussed shortly.) Here, the values V_r and I_r are the generator's full-load (rated) voltage and current respectively.

Now consider the case of a shunt-connected generator. As mentioned previously, the field current is small compared with the armature current, so we will assume that the load current is equal to the armature current. Just as was the case for the separately excited generator, therefore, due to the armature resistance, the load voltage will decrease as the load current increases. However, in addition, when the load (terminal) voltage decreases, the field current decreases, thereby reducing the generated voltage even further. As a consequence of this, the load voltage decreases more rapidly with increasing load current than does the separately excited generator. This situation is also depicted in Fig. 15.38.

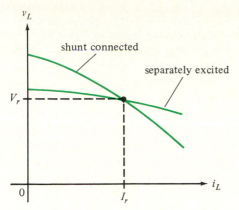

Fig. 15.38 External characteristics for separately excited and shunt-connected generators.

For some uses, it may be undesirable to have the terminal voltage of a generator decrease with increasing load current, as is indicated by Fig. 15.38. In a cumulatively compounded generator, the flux from the series-field winding reinforces the flux from the shunt-field winding. By having the series and shunt windings properly proportioned, the generator will produce a full-load (rated) voltage that is equal to the no-load voltage. The external characteristic for such a **flat-compounded generator** is shown in Fig. 15.39. When the series field reinforces the shunt field excessively, the result is an **overcompounded generator;** insufficient series field reinforcement produces an **undercompounded generator.** The external characteristics for these two cases are also depicted in Fig. 15.39.

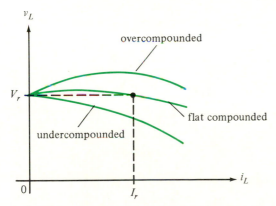

Fig. 15.39 External characteristics for a compound-connected generator.

Previously, we mentioned that an external generator characteristic was not the straight line it was predicted to be. The cause of this discrepancy is known as **armature reaction.** Current going through an armature will establish a magnetic field that can distort, especially for loaded conditions, the flux distribution of the field. In preceding discussions, we ignored this effect. However, such a distortion can adversely affect the proper commutation for the generator. To help compensate for this condition, **compensating windings,** which are connected in series with the armature, are placed in slots in the pole faces. The result is an opposition to the armature mmf, and therefore, a reduction in the distortion of the flux distribution. Compensating windings are also called **pole-face windings.**

In addition to compensating windings, many dc machines have windings, which are also connected in series with the armature, that produce small auxiliary fields that are used to aid current reversals in conductors undergoing commutation so as to alleviate sparking at the brushes. These windings are placed around narrow poles placed midway between the main poles of a machine. These additional poles are known as **commutating poles,** and the resulting fields are referred to as **commutating** or **interpole fields.**

The Alternator

We have seen that for a dc generator, typically the field (the stator) is stationary and the armature (the rotor) rotates. Conversely, for an ac generator, or alternator, usually the field windings are on the rotor while the armature conductors are embedded in slots in the stator. The reason for this is that the armature handles much more power than the field does, and by supplying direct current to the rotating field through slip rings and brushes, there are fewer mechanical and electrical problems than if the armature were to rotate.

A cross-sectional view of a simplified alternator is shown in Fig. 15.40. A constant current is applied to the field winding through stationary brushes riding on slip rings (not shown). This produces an electromagnetic field, and when the current is in the direction indicated, the resulting north (N) and south (S) poles are as shown. Embedded in slots in the stator are conductors a and a', which are connected in the rear (also not shown) so as to form a loop. As the field rotates, the changing magnetic flux induces an emf across such a coil. By shaping the poles

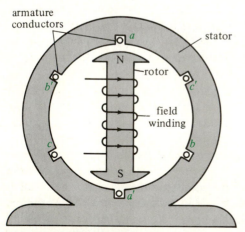

Fig. 15.40 Alternator.

appropriately, such an induced emf can be made sinusoidal. The rotor position shown in Fig. 15.40 corresponds to the case that the induced emf is maximum. Such a voltage completes one cycle for every revolution of the rotor. Therefore, if the rotor speed is n rpm, or equivalently $n/60$ rps, then the induced emf completes $n/60$ cycles per second; that is, the frequency of this voltage is $n/60$ Hz. We now see that the alternator's electrical frequency is synchronized with its mechanical speed, and, therefore, the given alternator is known as a **synchronous machine.**

In addition to conductors a and a', also placed in slots in the stator are other conductors. As with a and a', conductors b and b' are connected in the rear, as are c and c'. Note that the axis of the coil formed by b and b' differs by 120° from the axis determined by a and a'. Also, the axis for c and c' differs by 120° from the axis for b and b', and, hence, by 240° from a and a'. Thus, if the rotor turns clockwise at ω rad/s, then the emfs induced across the three coils are as depicted in Fig. 15.41. By making either a wye (Y) connection or a delta (Δ) connection of the three coils, the given alternator will realize a three-phase voltage source.

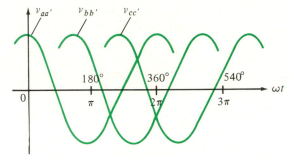

Fig. 15.41 Voltages induced across the armature coils.

The two-pole alternator (also called a "synchronous generator") shown in Fig. 15.40 has poles that are prominent or "salient," and, therefore, it is an example of what is known as a **salient-pole machine.** For such an alternator to produce a 60-Hz sinusoid, according to the discussion above, we must have that $n/60$ rps = 60 Hz, and, hence, the speed of the rotor must be $n = 3600$ rpm. However, the alternator speed can be reduced by using a rotor that has more than two poles. As an example, consider the four-pole alternator given in Fig. 15.42. For this salient-pole machine with the field winding shown, when direct current is applied in the direction depicted (the slip rings and brushes are not shown), the resulting pole polarities are as indicated. In this case, the armature conductors a_1 and a_1' are connected in the rear (not shown) so as to form a coil, as do a_2 and a_2'. By connecting a_1' and a_2 in front (also not shown), the two coils then become connected in series. As a consequence of such a construction, one complete revolution of the rotor will produce two complete cycles of the sinusoid induced across the winding consisting of coils a_1-a_1' in series with a_2-a_2'. The conductors labeled b and c are similarly connected, and again, by making either a Y or Δ connection, a three-phase alternator results.

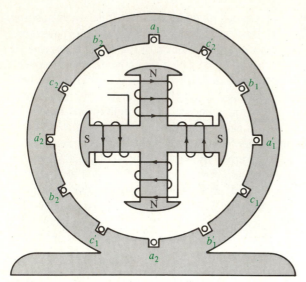

Fig. 15.42 Four-pole, three-phase alternator.

For the salient-pole alternator given in Fig. 15.42, a generated sinusoid completes two cycles for every revolution of the rotor. Thus, a rotor speed of n rpm ($n/60$ rps) corresponds to a sinusoidal frequency of $2n/60 = n/30$ Hz. In general, for an alternator with p poles (where p is even), the frequency f (in hertz) of the generated voltage is

$$f = \frac{p}{2}\frac{n}{60} = \frac{pn}{120} \qquad (15.30)$$

where n is the rotor speed in rpm.

From Equation (15.30), we see that for a fixed frequency, the rotor speed decreases as the number of poles increases. Typically, salient-pole machines have six or more poles. If either two or four poles are employed, then the rotor speed required for a frequency of 60 Hz is 3600 or 1800 rpm respectively. At such high speeds, however, a salient-pole rotor is more subject to mechanical stresses, windage losses, and noisy operation. To avoid these complications, a **cylindrical** or **nonsalient rotor** is utilized. For such a machine, the field winding is distributed around the rotor and embedded in slots. Alternators driven by steam turbines, also called **turboalternators,** typically operate at 1800 or 3600 rpm and are cylindrical (nonsalient) -rotor machines. Lower-speed alternators, such as those driven by waterwheels, are usually salient-pole machines.

Just as we modeled the armature of a dc generator with an equivalent electric circuit, we can do the same for an alternator. However, because an alternator generates alternating current instead of direct current, such a model will be slightly more complex. In particular, we shall use the equivalent circuit shown in Fig. 15.43

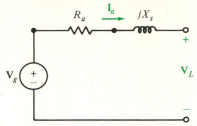

Fig. 15.43 Electric-equivalent circuit of the armature of an alternator.

to model a single phase of the voltage generated by an alternator. As for the case of a dc generator, the armature resistance is designated by R_a. The armature resistance is in series with an inductive reactance X_s, known as the **synchronous reactance,** which represents the effects that are a consequence of the armature winding. Typically, the armature resistance is much smaller than the synchronous reactance, and, therefore, quite often R_a is ignored. For a dc generator, the generated voltage is proportional to the field flux and the rotor speed [see Equation (15.28)]. This is also the case for an alternator, and, consequently, its steady-state operation can also be described in terms of a magnetization curve. The shape of such a curve, of course, is determined by the construction of the machine. Because an alternator generates ac voltages, the vertical axis of the magnetization curve indicates the rms value of the generated voltage. The field current, however, is dc. The voltage source given in Fig. 15.43 is labeled $\mathbf{V}_g$, which is the phasor representation of the generated voltage and $|\mathbf{V}_g|$ represents the rms value of the generated voltage. The armature (stator) current $\mathbf{I}_a$ is also an rms phasor.

Example 15.8

A 6-pole, three-phase 12-kVA alternator is Y connected and has a rated per-phase voltage of 125 V (rms) for a load with a lagging pf of 0.6 at 60 Hz. The synchronous reactance is 2 Ω, and the armature resistance is assumed to be negligible.

From Equation (15.30), the rotor speed of the alternator is

$$n = \frac{120f}{p} = \frac{(120)(60)}{6} = 1200 \text{ rpm}$$

Ignoring the armature resistance, the corresponding electric-equivalent per-phase armature circuit is as shown in Fig. 15.44, where $V_L = |\mathbf{V}_L| = 125$ V and $I_a = |\mathbf{I}_a|$ is to be determined. Since the load operates at a lagging pf of 0.6, then the load current $\mathbf{I}_a$ lags the load voltage $\mathbf{V}_L$ by θ, where $\cos\theta = 0.6$. Thus, the pf angle is $\theta = \cos^{-1} 0.6 = 53.13°$. Arbitrarily selecting the angle of $\mathbf{V}_L$ to be $0°$ (ang $\mathbf{V}_L = 0°$), we must have that ang $\mathbf{I}_a = -53.13°$. Since the alternator is rated at 12 kVA, the per-phase load current has an rms value of

$$I_a = \frac{(1/3)(12,000)}{V_L} = \frac{4000}{125} = 32 \text{ A}$$

From the circuit given in Fig. 15.44, by KVL the generated voltage $\mathbf{V}_g$ is found to be

$$\mathbf{V}_g = jX_s\mathbf{I}_a + \mathbf{V}_L = (j2)(32\underline{/-53.13°}) + 125\underline{/0°}$$

$$= (2\underline{/90°})(32\underline{/-53.13°}) + 125\underline{/0°}$$

$$= 64\underline{/36.87°} + 125\underline{/0°} = 51.2 + j38.4 + 125 = 176.2 + j38.4$$

$$= 180.34\underline{/12.3°}$$

The corresponding phasor diagram is shown in Fig. 15.45.

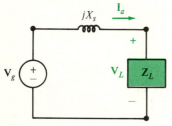

Fig. 15.44 Per-phase alternator-equivalent circuit.

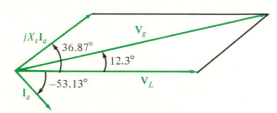

Fig. 15.45 Phasor diagram.

Suppose that the per-phase voltage of this alternator is described by the magnetization curve given in Fig. P15.31. Then from this curve, the (rms) generated voltage of 180.34 V corresponds to a (dc) field current of approximately 0.8 A.

In the preceding example, as is typical for the case of a dc generator, the generated voltage has a larger magnitude than the load voltage. However, a change in the pf of the load can result in the converse situation (see Problem 15.41). In such a case, armature reaction results in a reinforcement of the field mmf.

15.5 MOTORS

Let us return to the simple rotating machine given in Fig. 15.22. Instead of mechanically forcing this conducting loop to rotate, however, suppose that a current i is applied, as indicated in Fig. 15.46(a). (For the sake of simplicity, the slip rings and the brushes have been omitted.) The end view of the loop is shown in Fig. 15.46(b), where the current in the upper conductor is into the page (indicated by the cross $\times$) and the current in the lower conductor is out of the page (indicated by the dot $\cdot$). By the right-hand screw rule, the resulting force on the upper conductor is $f = Bli$ in the downward direction indicated. (This value of f is also

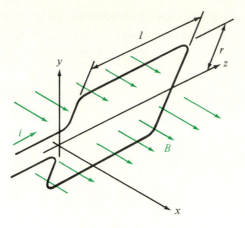

(a) Three-dimensional view

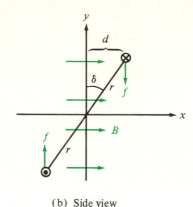

(b) Side view

Fig. 15.46 Simple rotating machine.

the force on the lower conductor.) The torque developed by the upper conductor is $fd = Blir \sin \delta$, where the angle δ made by the y-z plane and the plane of the loop is called the **torque angle** or **power angle.** If, instead of a single-turn loop, an N-turn coil is employed, then because such a coil is comprised of $2N$ conductors, the torque developed by the coil is

$$T = 2NBlir \sin \delta = NBAi \sin \delta \qquad (15.31)$$

where $A = 2rl$ is the area of the coil. Note that the developed torque is maximum when $\sin \delta = 1$ ($\delta = 90°$). Since the forces on the conductors are as shown in Fig. 15.46(b), the developed torque tends to rotate the loop clockwise, and the simple rotating machine behaves as a motor.

For the simple generator given in Fig. 15.22, at the instant of time shown, a resistive load placed across the brushes has a voltage across it with the polarity indicated. This, in turn, means that the resulting loop current is in the direction opposite to that shown in Fig. 15.46. Consequently, as in the discussion above, the developed torque tends to rotate the loop in Fig. 15.22 counterclockwise—that is, the developed torque is in opposition to that which forces the loop to rotate clockwise. Note, too, that for the simple motor given in Fig. 15.46, at the instant of time shown, the induced emf due to the loop rotating clockwise tends to produce a current that opposes the applied current.

The dc Motor

A dc motor is constructed basically as a dc generator is (e.g., see Fig. 15.25). Provided that the field current direction is maintained along with the direction of rotation, when the motor is in use, however, the direction of its armature current is opposite to that in the generator. Consequently, we can model the armature portion of a motor by the equivalent electric circuit shown in Fig. 15.47. (Note the direction of the armature current i_a as compared with that in a dc generator.) Again, R_a is the armature resistance, and by Equation (15.28), $v_g = K\Phi n$, where K is a constant dependent on the construction of the motor, Φ is the flux per pole, and n is the rotor (armature) speed in rpm. In this case, v_L indicates the applied or line voltage of the motor. As was true for the dc generator, the dc motor can be separately or self-excited as depicted in Figs. 15.32–15.35, where the directions of i_a and i_L should be reversed.

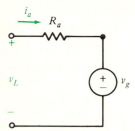

Fig. 15.47 Equivalent circuit of armature portion of a dc motor.

For a dc motor, from the circuit given in Fig. 15.47, by KVL, we have that

$$v_L = R_a i_a + v_g \tag{15.32}$$

and since $v_g = K\Phi n$, then

$$v_L = R_a i_a + K\Phi n \tag{15.33}$$

From this equation, we have that the speed (in rpm) of the motor is

$$n = \frac{v_L - R_a i_a}{K\Phi} \tag{15.34}$$

and this is known as the **speed equation** of a dc motor.

As indicated by Equation (15.31), the torque developed by a dc motor is determined by the armature current and the magnetic field density, and, hence, the field flux (flux per pole). In mathematical terms, we can write that the developed torque T is given by

$$T = K'\Phi i_a \tag{15.35}$$

where K' is a constant that is determined by the construction of the motor.

Specifically, let us consider the case of the shunt-connected motor, called a **shunt motor** for short, shown in Fig. 15.48. Suppose we are given that the line

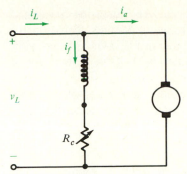

Fig. 15.48 Shunt motor.

voltage v_L is constant. For a fixed setting of the control rheostat R_c, the field flux is constant (ignoring armature reaction). Even though i_a can vary with different loading conditions, since R_a typically is small and $R_a i_a$ can be considered negligible compared with v_L, under these conditions, the shunt motor essentially has a constant speed. However, for a given load and constant line voltage, from Equation (15.34), we see that we can vary the motor speed by changing Φ. Since Φ depends on i_f and $i_f = v_L/R_c$, by adjusting R_c, we can vary the speed of the motor.

By combining Equations (15.34) and (15.35), we can derive an expression relating speed and torque. In particular, from Equation (15.35), we have that $i_a = T/K'\Phi$, and substituting this into Equation (15.34), we get

$$n = \frac{v_L}{K\Phi} - \frac{R_a}{K\Phi}\left(\frac{T}{K'\Phi}\right) = \frac{v_L}{K\Phi} - \frac{R_a}{KK'\Phi^2}T = mT + b$$

where $m = -R_a/KK'\Phi^2$ and $b = v_L/K\Phi$. Therefore, since v_L and (once R_c is fixed) Φ are constant for a shunt motor, a plot of speed n versus torque T is a straight* line ($n = mT + b$) having slope m and vertical-axis intercept b. Such a plot, known as the **speed–torque characteristic** of the motor, is shown in Fig. 15.49, where n_r is the rated speed and T_r is the rated torque at full load. Note that

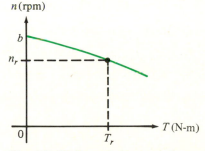

Fig. 15.49 Speed–torque characteristic of a shunt motor.

---

*Due to armature reaction, the line is slightly curved.

the speed decreases as the developed torque increases. An increasing developed torque is required for an increasing load (i.e., a resisting torque). Thus, an increasing load means a decreasing speed. From Equation (15.34), a decrease in speed means an increase in armature current i_a. Since $v_g = K\Phi n$, a decrease in speed n also means a decrease in v_g.

Example 15.9

A 220-V dc shunt motor has a speed of 1200 rpm and an armature current of 6 A at no load. Given that the armature resistance is 0.5 Ω, let us determine the motor speed when the armature current is 40 A at full load.

From Equation (15.34),

$$K\Phi = \frac{v_L - R_a i_a}{n}$$

Thus, at no load,

$$K\Phi = \frac{220 - (0.5)(6)}{1200} = 0.1808$$

Hence, at full load,

$$n = \frac{v_L - R_a i_a}{K\Phi} = \frac{220 - (0.5)(40)}{0.1808} \approx 1106 \text{ rpm}$$

As indicated in Example 15.9, the motor speed at full load, denoted n_{FL}, differs from the speed at no load, n_{NL}. A measure of the change in speed is the **speed regulation** SR of the motor, which is defined to be

$$SR = \frac{n_{NL} - n_{FL}}{n_{FL}}$$

For the preceding example, we have that

$$SR = \frac{1200 - 1106}{1106} = 0.085 = 8.5\%$$

The speed regulation for a typical shunt motor usually ranges between 5 and 12%.

As mentioned above, the speed of a shunt motor can be varied by adjusting the control rheostat R_c. If the armature and field voltages are supplied by separate sources, however, motor speed can be varied by changing the armature voltage while keeping the field voltage (and, hence, current and flux) fixed. This conclusion should be apparent from Equation (15.34). Such a technique offers a wide range of speed control, and is frequently used in industrial applications.

If the field of a dc motor is series connected as shown in Fig. 15.50, then the result is a **series motor.** Since the field current is also the armature current, and this current can be large, the field winding typically is comprised of a few turns of heavy wire. Unlike the shunt motor, where we considered Φ to be constant once the control rheostat R_c was set, for the series motor Φ is dependent on the armature current. This is so because the armature current is the field current, and the field flux Φ depends on the field current i_f.

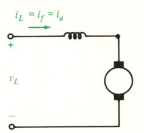

Fig. 15.50 Series motor.

For the sake of analytical convenience, let us assume that a series motor operates on the linear portion of its magnetization curve and

$$\Phi = K_f i_f = K_f i_a$$

where K_f is a constant. Since $v_g = K\Phi n$ and $T = K'\Phi i_a$, we have that

$$v_g = K(K_f i_a)n = KK_f n i_a \qquad (15.36)$$

and

$$T = K'(K_f i_a)i_a = K'K_f i_a^2 \qquad (15.37)$$

Thus, we see that the developed torque for a series motor is proportional to the square of the armature current, whereas for a shunt motor it is proportional to the armature current.

Suppose that for a series motor, R denotes the series combination of the armature resistance R_a and the resistance R_f of the field winding, that is, $R = R_a + R_f$. Then, by KVL and Equation (15.36), we get

$$v_L = Ri_a + v_g = Ri_a + KK_f n i_a = (R + KK_f n)i_a$$

from which

$$i_a = \frac{v_L}{R + KK_f n}$$

Substituting this expression for the armature current into Equation (15.37) yields

$$T = \frac{K'K_f v_L{}^2}{(R + KK_f n)^2} \tag{15.38}$$

The plot of n versus T for this expression gives us the speed–torque characteristic for a series motor as shown in Fig. 15.51. From this curve, note how rapidly the motor speed increases as the torque gets small. As a result of this fact, to avoid excessive motor speeds, and its undesirable consequences, typically a series motor is always mechanically coupled to its load. The nonconstant speed characteristic of a series motor makes it useful in certain types of variable-speed applications such as cranes, hoists, and winches.

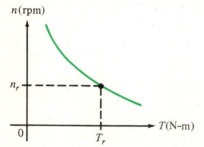

Fig. 15.51 Speed–torque characteristic for a series motor.

From Equation (15.37) we see that regardless of the direction of the current, the torque T has the same direction. Therefore, for either an ac or a dc line voltage, the developed torque for a series motor is unidirectional. Series motors designed to operate on either ac or dc are known as **universal motors.** Clearly, for ac operation the developed torque is not constant, but rather pulsating; and the mechanical output power for ac operation is less than that for dc operation. Still, universal motors are commonly used in hand power tools such as electric drills and small appliances such as vacuum cleaners.

A compound-connected motor, called a **compound motor** for short, almost always is cummulatively compounded. The speed–torque characteristic for such a machine lies between those for a shunt motor and for a series motor as illustrated in Fig. 15.52. The relative position of the characteristic for the compound motor with respect to the shunt and series motors depends on the relative strengths of the shunt and series fields.

In the preceding discussions, it was assumed that motor operation was in the steady state. Initially, however, the speed of a motor is zero. As a consequence, from Equation (15.28), the initial emf of the motor is $v_g = K\Phi n = 0$. In the case of a shunt motor, for example, this means that the entire line voltage v_L is initially across the armature resistance. If this voltage were steadily increased from zero to its final value, the motor would proceed smoothly to its rated speed. Because the

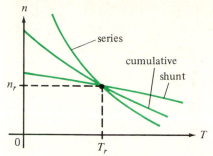

Fig. 15.52 Speed–torque dc motor characteristics.

line voltage is fixed at its given value, however, placing it across the typically small armature resistance can result in a quite excessive armature current. For this reason, when starting a large dc motor, a resistance, which is adjustable continuously or in steps, is placed in series with the armature. Typically, this resistance limits the starting current to a value between 150 and 200 percent of the rated current, and it is removed (short-circuited) when the motor reaches its appropriate (steady-state) speed. A device that will perform this duty is known as a **starting box.**

The Synchronous Motor

Previously, we mentioned that a dc motor is constructed as a dc generator is. So, too, one type of ac motor—a **synchronous motor**—is constructed as is an alternator (e.g., see Fig. 15.40). To describe this type of synchronous machine, as well as the "induction motor" that will be studied in the following section, we need to introduce the concept of a rotating stator field or flux.

To see how this phenomenon occurs, consider the simple three-phase, two-pole machine cross section given in Fig. 15.53, where, for the sake of simplicity, the rotor is not shown. Instead, depicted in the center of the machine are the flux

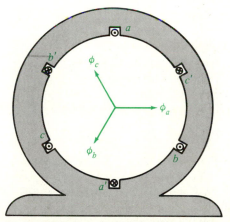

Fig. 15.53 Stator field components of a synchronous motor.

components due to the currents in each of the three stator coils a-a', b-b', and c-c'. The current i_a in coil a-a' is coming out of the page in conductor a and going into the page in conductor a' (designated by a dot · and a cross ×, respectively). By the right-hand rule, the direction of the resulting flux ϕ_a is to the right, as illustrated in Fig. 15.53. Similarly, the currents i_b and i_c in coils b-b' and c-c', respectively, yield ϕ_b and ϕ_c. By spacing the coils 120° apart, the directions of ϕ_a, ϕ_b, and ϕ_c are spaced by 120°. But what is the net flux?

To simplify our analysis, let us assume that flux is directly proportional to current, that is, $\phi = ki$. Since the flux produced by each coil has direction, as well as magnitude, let us define the complex number $\boldsymbol{\Phi} = \phi e^{j\theta}$, where ϕ is flux and θ is the angle between the direction of the flux and the horizontal. Thus,

$$\boldsymbol{\Phi}_a = ki_a e^{j0°} = ki_a$$

$$\boldsymbol{\Phi}_b = ki_b e^{-j120°}$$

$$\boldsymbol{\Phi}_c = ki_c e^{j120°}$$

If the three stator currents are equal $(i_a = i_b = i_c)$, since $\boldsymbol{\Phi} = \boldsymbol{\Phi}_a + \boldsymbol{\Phi}_b + \boldsymbol{\Phi}_c = 0$, the net flux is zero. However, let us consider the case that i_a, i_b, and i_c are the balanced set of currents

$$i_a = I \cos \omega t$$

$$i_b = I \cos(\omega t - 120°)$$

$$i_c = I \cos(\omega t - 240°) = I \cos(\omega t + 120°)$$

which are depicted in Fig. 15.54. Then we get the following set of complex numbers with time-varying magnitudes:

$$\boldsymbol{\Phi}_a = (kI \cos \omega t)e^{j0°} = kI \cos \omega t$$

$$\boldsymbol{\Phi}_b = [kI \cos(\omega t - 120°)]e^{-j120°}$$

$$\boldsymbol{\Phi}_c = [kI \cos(\omega t + 120°)]e^{j120°}$$

Using the fact that $e^{j\theta} = \cos\theta + j \sin\theta$ (Euler's formula), we have that $e^{-j120°} = -1/2 - j\sqrt{3}/2$ and $e^{j120} = -1/2 + j\sqrt{3}/2$. Thus,

$$\boldsymbol{\Phi}_a = kI \cos \omega t$$

$$\boldsymbol{\Phi}_b = -\frac{1}{2}kI \cos(\omega t - 120°) - j\frac{\sqrt{3}}{2}kI \cos(\omega t - 120°)$$

$$\boldsymbol{\Phi}_c = -\frac{1}{2}kI \cos(\omega t + 120°) + j\frac{\sqrt{3}}{2}kI \cos(\omega t + 120°)$$

and their sum is

$$\boldsymbol{\Phi} = \boldsymbol{\Phi}_a + \boldsymbol{\Phi}_b + \boldsymbol{\Phi}_c = kI\left[\cos \omega t - \frac{1}{2}\cos(\omega t - 120°) - \frac{1}{2}\cos(\omega t + 120°)\right]$$

$$+ j\frac{\sqrt{3}}{2} kI[\cos(\omega t + 120°) - \cos(\omega t - 120°)]$$

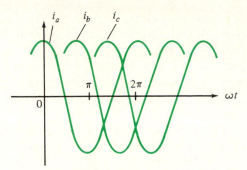

Fig. 15.54 Stator currents.

However,

$$\cos(\omega t \pm 120°) = (\cos \omega t)(\cos 120°) \mp (\sin \omega t)(\sin 120°)$$

$$= -\frac{1}{2} \cos \omega t \mp \frac{\sqrt{3}}{2} \sin \omega t$$

Therefore,

$$\mathbf{\Phi} = \mathbf{\Phi}_a + \mathbf{\Phi}_b + \mathbf{\Phi}_c$$

$$= kI\left[\cos \omega t - \frac{1}{2}\left(-\frac{1}{2} \cos \omega t + \frac{\sqrt{3}}{2} \sin \omega t \right) - \frac{1}{2}\left(-\frac{1}{2} \cos \omega t - \frac{\sqrt{3}}{2} \sin \omega t \right) \right]$$

$$+ j\frac{\sqrt{3}}{2} kI \left[\left(-\frac{1}{2} \cos \omega t - \frac{\sqrt{3}}{2} \sin \omega t \right) - \left(-\frac{1}{2} \cos \omega t + \frac{\sqrt{3}}{2} \sin \omega t \right) \right]$$

$$= \frac{3}{2} kI \cos \omega t - j\frac{3}{2} kI \sin \omega t = \frac{3}{2} kI (\cos \omega t - j \sin \omega t)$$

$$= \frac{3}{2} kI e^{-j\omega t}$$

Since $\mathbf{\Phi} = \mathbf{\Phi}_a + \mathbf{\Phi}_b + \mathbf{\Phi}_c$ has a constant magnitude $\frac{3}{2} kI$, but an angle $-\omega t$ that varies with time, this demonstrates that the flux rotates clockwise at a frequency of ω rad/s ($60\omega/2\pi = 30\omega/\pi$ rpm) when the currents i_a, i_b, and i_c given above are applied to coils a-a', b-b', and c-c' respectively.

Now let us look at the two-pole synchronous machine given in Fig. 15.53, where the rotor is explicitly depicted as in Fig. 15.55. Just as in an alternator, the rotor field is caused by a direct current applied through brushes and slip rings. At some time (e.g., for $t = 0$), the stator flux will be as indicated. Suppose that the position of the rotor is as illustrated. As the stator field rotates clockwise, the magnetic attraction between the stator and rotor poles will cause the rotor to turn at the same speed as the stator flux, that is, the rotor and stator fields will be synchronized. For the case of no load on the rotor, the torque angle δ will be small since the developed torque only needs to overcome rotational losses due to friction and windage. [As suggested by Equation (15.31), if δ is small, the developed torque is small.] Increasing the mechanical load will result in an increase in δ. If

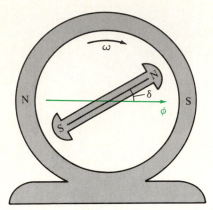

Fig. 15.55 Stator and rotor poles for a synchronous motor.

the load torque on the rotor becomes greater than can be overcome by the developed torque, then the rotor will slow down and stop. Furthermore, if the rotor- and stator-field axes coincide, that is, if $\delta = 0$, then no torque will be developed on the rotor. For motor operation, the rotor field must lag the stator field.

Because the construction of a dc motor is the same as for a dc generator, we obtained the circuit model for the former from the latter simply by changing the direction of the armature current. Again, because of the similar construction of synchronous motors and generators (alternators), this approach can also be taken for synchronous machines. Specifically, with reference to the alternator equivalent circuit given in Fig. 15.43, we have that an equivalent circuit of a synchronous motor is as shown in Fig. 15.56. Again, as in the case of an alternator, the armature resistance R_a is often considered negligible compared with the synchronous reactance X_s.

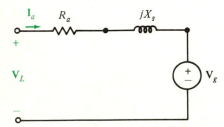

Fig. 15.56 Electric-equivalent circuit of a synchronous motor.

In an alternator (synchronous generator), the generated voltage $\mathbf{V}_g$ leads the phase voltage $\mathbf{V}_L$ (e.g., see Fig. 15.45). Conversely, for a synchronous motor, $\mathbf{V}_g$ lags $\mathbf{V}_L$, and since the rotor lags the rotating stator field by the torque angle δ, then $\mathbf{V}_g$ lags $\mathbf{V}_L$ by δ. Ignoring the armature resistance, for the circuit given in Fig. 15.56, we have

$$\mathbf{V}_L = jX_s\mathbf{I}_a + \mathbf{V}_g \tag{15.39}$$

Assuming that the angle of $\mathbf{V}_L$ is zero, an example of a phasor diagram for a synchronous motor is as shown in Fig. 15.57. For this diagram, let us construct a line labeled l and extend $\mathbf{I}_a$ as shown in Fig. 15.58. If $|\mathbf{I}_a| = I_a$, $|\mathbf{V}_g| = V_g$, and $|\mathbf{V}_L| = V_L$, we have

$$\frac{l}{V_g} = \sin \delta \qquad \text{and} \qquad \frac{l}{X_s I_a} = \cos \theta$$

from which

$$l = V_g \sin \delta = X_s I_a \cos \theta \qquad\qquad (15.40)$$

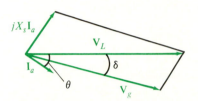

Fig. 15.57 Phasor diagram for a synchronous motor.

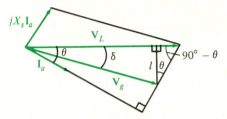

Fig. 15.58 Modified phasor diagram.

But the input power (per phase) of a synchronous motor (see Fig. 15.56) is

$$P = V_L I_a \cos \theta$$

From Equation (15.40), then

$$P = V_L I_a \frac{V_g \sin \delta}{X_s I_a} = \frac{V_L V_g}{X_s} \sin \delta \qquad\qquad (15.41)$$

and we see why the torque angle is also called the power angle. Since an increase in load torque requires an increase in power for a constant speed, the torque angle must increase when the load increases.

Example 15.10

A Y-connected, three-phase synchronous motor has a per-phase voltage of 125 V and a synchronous reactance of 5 Ω. Suppose that the motor input power is 6 kW and the torque angle is 27°.

Since the per-phase input power is

$$P = 6000/3 = 2000 \text{ W}$$

from Equation (15.41) we have

$$V_g = \frac{P X_s}{V_L \sin \delta} = \frac{(2000)(5)}{(125)(\sin 27°)} = 176.2 \text{ V}$$

If the motor were characterized by the magnetization curve given in Fig. P15.31, then the corresponding field current would be approximately 0.78 A. Furthermore, by Equation (15.39), we have

$$\mathbf{I}_a = \frac{\mathbf{V}_L - \mathbf{V}_g}{jX_s} = \frac{125\underline{/0°} - 176.2\underline{/-27°}}{5\underline{/90°}}$$

$$= \frac{125 - (157 - j80)}{5\underline{/90°}} = \frac{86.16\underline{/111.8°}}{5\underline{/90°}} = 17.23\underline{/21.8°}$$

and since the current $\mathbf{I}_a$ leads the voltage $\mathbf{V}_L$, the motor operates at a power factor of

$$\text{pf} = \cos 21.8° = 0.928 \text{ leading}$$

The phasor diagram for the motor is shown in Fig. 15.59. In this case, $V_g > V_L$, and this is an example of an **overexcited** motor. For the situation that $V_g < V_L$, we say that the motor is **underexcited.**

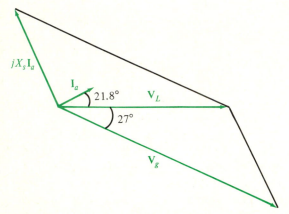

Fig. 15.59 Phasor diagram for example motor.

Consider a synchronous motor with a lagging pf, where the associated phasor diagram is as given in Fig. 15.60(a). Suppose that the load on the motor remains constant, and the field (rotor) current is increased. This will strengthen (increase) the field flux and produce a larger generated voltage magnitude V_g. The result is a greater magnetic attraction between the rotating (stator) field poles and the rotor poles, and, thus, the torque angle δ is reduced. The effect on the phasor diagram for the motor when the field current is increased is shown in Fig. 15.60(b). Hence, we see that by increasing the field current sufficiently, the pf of a motor can be changed from lagging to leading (or from lagging to less lagging). It is this property that enables a synchronous motor to be used for power-factor correction, that is, for changing the pf. Typically, a synchronous motor used for this purpose is unloaded,

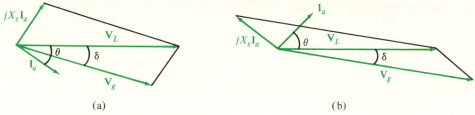

Fig. 15.60 Effect of increasing field current on phasor diagram of motor.

so the only input power required is for the motor losses. Since a capacitor has the effect of making a pf more leading, a synchronous motor that is utilized for pf correction is known as a **synchronous condenser.***

Even though our previous discussions explicitly characterized two-pole motors, just as for the case of a synchronous generator (alternator), the relationship between machine speed, ac voltage, frequency, and the number of poles of a synchronous motor is as given by Equation (15.30). From this equation, we have that the **synchronous speed** n_s of a motor (in rpm) is

$$n_s = \frac{120f}{p} \qquad\qquad (15.42)$$

For the common situation that $f = 60$ Hz, the maximum synchronous speed is obtained when $p = 2$, and under this circumstance $n_s = 3600$ rpm.

Since the speed of a synchronous motor is exactly the synchronous speed n_s, such a motor has a constant speed—even under different load conditions. However, if the load torque is made too great, the motor will lose synchronism and stop. The torque that causes this is known as the **pull-out torque.**

A variation of a synchronous motor is a motor that does not rotate continuously, but rather turns in steps as the result of an applied digital signal. A simple example of such a **stepper** (or **stepping**) **motor** is shown in Fig. 15.61. For this machine, the rotor is a permanent magnet, and the stator has four poles with their associated windings. When a current goes through a pole winding in the direction indicated, that pole is north (N); when the current has the opposite direction, that pole is south (S). When $i_1 > 0$, $i_2 < 0$, $i_3 < 0$, and $i_4 > 0$, the corresponding poles are N, S, S, and N respectively. For the situation shown, the resulting rotor angle is $\theta = 0$. When $i_1 > 0$, $i_2 > 0$, $i_3 < 0$, and $i_4 < 0$, the result is $\theta = 45°$. By applying the appropriate currents to the pole windings, the rotor can be positioned in increments of $45°$. In order to get smaller steps, more stator poles are required. For a typical motor with 24 poles, the rotor turns $7.5°$ per step. Stepper motors can be used for positioning purposes in various types of systems.

---

*A capacitor was formerly called a condenser.

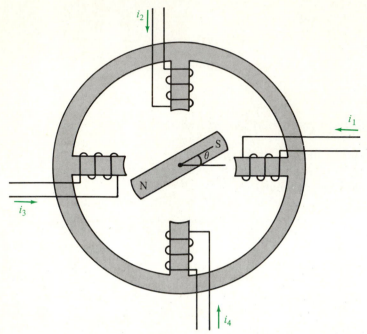

Fig. 15.61 Stepper motor.

In previous discussions, it was assumed that a synchronous motor operates at synchronous speed, for only at this speed does the motor develop a constant torque in the direction of rotation. By itself, a synchronous motor at rest will produce no starting torque. This is a consequence of the fact that a flux rotating around a motionless rotor results in a zero average torque. One way to start a synchronous motor is to modify the rotor in such a way that a starting torque can be obtained. We now discuss the principles that will achieve this end.

The Induction Motor

For a synchronous machine (either a generator or motor), a dc voltage is applied to the rotor (field) and an ac voltage is applied to the stator (armature). However, the requirement for a dc rotor excitation can be eliminated by constructing a machine that makes use of magnetic induction. **Induction motors,** which operate on this principle, are the most common type of ac motors. Conversely, an induction generator is inferior to a synchronous generator and is considered of little importance.

As in a synchronous motor, when alternating current is applied to the stator of an induction motor, a rotating magnetic flux is produced. For an induction motor, however, the stator is the field and the rotor is the armature. A common type of rotor contains conductors that form a cylinder, and the ends of the conductors are short-circuited together by conducting rings as shown in Fig. 15.62. Since such a

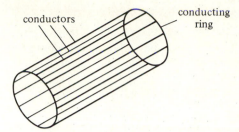

Fig. 15.62 Squirrel-cage configuration.

configuration resembles a cage that is used to exercise rodents such as hampsters and squirrels, this type of rotor is referred to as a **squirrel-cage rotor.** Unlike a synchronous motor, though, a squirrel-cage induction motor does not require brushes and slip rings for the rotor.

A cross-sectional view of a simple two-pole, three-phase induction motor is shown in Fig. 15.63. For this machine, a rotating stator flux is established as described previously and turns clockwise at an angular velocity of ω rad/s. Suppose that the rotor is initially at rest. As the field rotates, an emf will be induced in coil 1—the coil consisting of conductors 1 and 1′. With reference to Fig. 15.22, the clockwise rotating coil and fixed field are equivalent to a fixed coil and a counter-clockwise rotating field. The result is an induced emf that produces a current out of the upper conductor. Thus, for a fixed coil and a clockwise rotating field, the resulting current is into the upper conductor. This means that the current produced in coil 1 goes into conductor 1 (i.e., into the page) and comes out of conductor 1′. As illustrated in Fig. 15.46, the torque that is developed is clockwise, and so the rotor tends to turn in the direction of the rotating field. The other coils formed by the rotor conductors experience the same type of effect and contribute to the total torque of the rotor. If this starting torque is sufficient to overcome the load, the rotor will begin to turn and the motor will come up to its steady-state speed of n rpm.

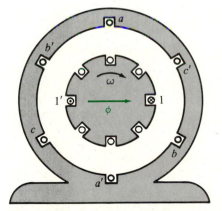

Fig. 15.63 Induction motor.

The operating speed n of an induction motor, however, cannot be equal to the synchronous speed n_s. This is so because, if the two speeds were equal, there would be no time rate of change in flux linkages. Then, by Faraday's law, there would be no induced emf, and, therefore, no induced current. The result would be no torque. Consequently, the speed n of an induction motor is less than synchronous speed n_s, since this will cause a time rate of change in flux linkages and, hence, will allow motor behavior.

As mentioned in the last section, by itself a synchronous motor at rest will produce no starting torque. But by adding some short-circuited conductors to the rotor of a synchronous motor, as in the case of an induction motor, a starting torque can be obtained. When the machine operates at synchronous speed, however, such squirrel-cage windings will make no contribution to the rotor torque. These windings, called **damping windings,** also serve to damp the tendency for the rotor to "hunt," or oscillate around its final position, when the load on the motor changes suddenly. This is a consequence of the induction-motor torque that is developed by a damping winding when the rotor momentarily operates at a nonsynchronous speed.

The **slip** s of an induction motor having speed n is defined as

$$s = \frac{n_s - n}{n_s} \qquad (15.43)$$

where $n_s = 120f/p$ is synchronous speed. From this definition, we can write that the motor speed is

$$n = n_s - sn_s = (1 - s)n_s \qquad (15.44)$$

A squirrel-cage induction motor at full load typically has a slip that is in the range of 2 to 6%.

The difference between synchronous speed n_s and rotor speed n is called the **slip speed.** From Equation (15.43), we have that the slip speed is

$$n_s - n = sn_s \qquad (15.45)$$

Since this is the relative speed of the rotating stator field with respect to the rotor, from Equation (15.30), the frequency of the induced emf is

$$\frac{psn_s}{120} = s\left(\frac{pn_s}{120}\right) = sf$$

where f is the frequency of the stator voltage. Such a voltage produces a rotor field that rotates at a speed relative to the rotor of

$$\frac{120(sf)}{p} = s\frac{120f}{p} = sn_s$$

Since the speed of the rotor is n rpm, the rotor field has an absolute speed of

$$n + sn_s = n_s$$

Thus, the rotor field has the same speed as the stator field, and, as with a synchronous motor, the developed torque for a given load is constant—despite the fact that the rotor speed n is less than the synchronous speed n_s.

Example 15.11

Suppose we are given that the speed of a three-phase induction motor is 864 rpm at 60 Hz. Since the synchronous speed is

$$n_s = \frac{120f}{p} = \frac{(120)(60)}{p} = \frac{7200}{p}$$

and since $n = 864$ rpm is slightly below n_s, we deduce that $p = 8$ because p must be even and this value will yield $n_s = 7200/8 = 900$ rpm. Thus, the slip of the motor is

$$s = \frac{n_s - n}{n_s} = \frac{900 - 864}{900} = 0.04 = 4\%$$

and the emf induced in the rotor has a frequency of

$$sf = (0.04)(60) = 2.4 \text{ Hz}$$

An Induction-Motor Model

Suppose that the rotor of an induction motor is constrained to be at rest $(s = 1)$. On a per-phase basis, the rotating stator field will induce an emf $\mathbf{V}_2$ having frequency f. If R_2 is the effective resistance of the rotor and X_2 is the reactance, then the resulting current $\mathbf{I}_2$ is given by

$$\mathbf{I}_2 = \frac{\mathbf{V}_2}{R_2 + jX_2} \tag{15.46}$$

Since the emf induced in the rotor is directly proportional to the relative velocity of the rotating field with respect to the rotor, for an induction motor having slip s, the induced emf is $s\mathbf{V}_2$, the frequency is sf, and the reactance is sX_2. Therefore, the resulting rotor current is

$$\mathbf{I}_2 = \frac{s\mathbf{V}_2}{R_2 + jsX_2} = \frac{\mathbf{V}_2}{(R_2/s) + jX_2} \tag{15.47}$$

at slip s. However, since we can write

$$\frac{R_2}{s} = \frac{R_2(1)}{s} = \frac{R_2(s + 1 - s)}{s} = \frac{R_2 s}{s} + \frac{R_2(1 - s)}{s} = R_2 + \frac{1 - s}{s}R_2$$

Equation (15.47) becomes

$$\mathbf{I}_2 = \frac{\mathbf{V}_2}{R_2 + jX_2 + (1 - s)R_2/s} \tag{15.48}$$

From this expression, we get

$$\mathbf{V}_2 = \left(R_2 + jX_2 + \frac{1 - s}{s}R_2\right)\mathbf{I}_2 = R_2\mathbf{I}_2 + jX_2\mathbf{I}_2 + \frac{1 - s}{s}R_2\mathbf{I}_2$$

which can be modeled by the equivalent circuit shown in Fig. 15.64.

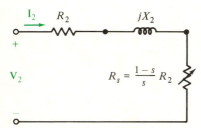

Fig. 15.64 Rotor-equivalent circuit for an induction motor.

The power absorbed by the rotor is $V_2 I_2 \cos\theta$, where $V_2 = |\mathbf{V}_2|$, $I_2 = |\mathbf{I}_2|$, and θ is the difference between the angles of $\mathbf{V}_2$ and $\mathbf{I}_2$. Since R_2 is the effective resistance of the rotor, the rotor's ohmic loss $I_2^2 R_2$ is dissipated as heat. The remainder of the power delivered to the rotor, that is, the power absorbed by the speed-dependent resistance $R_s = (1 - s)R_2/s$, represents the electric-to-mechanical power conversion. This quantity is given by $I_2^2 R_s$.

An induction motor is basically a transformer with a fixed primary (the stator) and a rotating secondary (the rotor). Consequently, one possibility for modeling an induction motor is to use the simple equivalent circuit shown in Fig. 15.65, where R_1 and X_1 are the effective resistance and reactance, respectively, of the primary. By KVL, we can write

$$\mathbf{V} = (R_1 + jX_1)\mathbf{I}_1 + \mathbf{V}_1 \tag{15.49}$$

and

$$\mathbf{V}_2 = \left(R_2 + jX_2 + \frac{1 - s}{s}R_2\right)\mathbf{I}_2 = \left(\frac{R_2}{s} + jX_2\right)\mathbf{I}_2 \tag{15.50}$$

Using the fact that for the ideal transformer portion of the circuit $\mathbf{V}_2 = N\mathbf{V}_1$ and $\mathbf{I}_2 = \mathbf{I}_1/N$, Equation (15.49) becomes

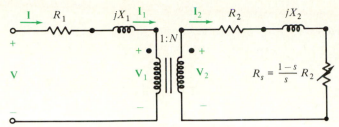

Fig. 15.65 Simplified equivalent circuit of an induction motor.

$$\mathbf{V} = (R_1 + jX_1)N\mathbf{I}_2 + \frac{\mathbf{V}_2}{N} = (R_1 + jX_1)N\mathbf{I}_2 + \frac{[(R_2/s) + jX_2]\mathbf{I}_2}{N}$$

from which

$$\mathbf{I}_2 = \frac{N\mathbf{V}}{(R_1 + jX_1)N^2 + (R_2/s) + jX_2} = \frac{\mathbf{V}/N}{R_1 + jX_1 + (R_2/sN^2) + jX_2/N^2}$$

Let us define $R_2' = R_2/N^2$ and $X = X_1 + X_2/N^2$. Then

$$\mathbf{I}_2 = \frac{\mathbf{V}/N}{R_1 + (R_2'/s) + jX} \quad \Rightarrow \quad I_2 = |\mathbf{I}_2| = \frac{V/N}{\sqrt{[R_1 + (R_2'/s)]^2 + X^2}}$$

where $V = |\mathbf{V}|$. Thus, the power absorbed by $R_s = (1 - s)R_2/s$ is

$$I_2^2 R_s = \frac{V^2/N^2}{[R_1 + (R_2'/s)]^2 + X^2} \frac{(1 - s)R_2}{s} = \frac{1 - s}{s} \frac{V^2 R_2'}{[R_1 + (R_2'/s)]^2 + X^2}$$

and this represents the per-phase electric-to-mechanical power conversion.

Torque–Speed Characteristics

Since mechanical power is the product of torque and angular velocity (i.e., $P = T\omega$), the developed torque per phase is $I_2^2 R_s/\omega$. From Equation (15.43), because angular velocity and rpm differ by a constant, it should be obvious that

$$s = \frac{n_s - n}{n_s} = \frac{\omega_s - \omega}{\omega_s}$$

This expression yields $\omega = (1 - s)\omega_s$, and, therefore, the total developed torque for a three-phase induction motor is

$$T = \frac{3I_2^2 R_s}{\omega} = \frac{3}{(1 - s)\omega_s} I_2^2 R_s = \frac{3}{(1 - s)\omega_s} \frac{1 - s}{s} \frac{V^2 R_2'}{[R_1 + (R_2'/s)]^2 + X^2}$$

or

$$T = \frac{3}{\omega_s} \frac{V^2 (R_2'/s)}{[R_1 + (R_2'/s)]^2 + X^2} \tag{15.51}$$

Since this expression relates torque T and speed, either $n = (1 - s)n_s$ or $\omega = (1 - s)\omega_s$, it is possible to plot torque versus speed. Such a typical **torque–speed curve** for a squirrel-cage induction motor is given in Fig. 15.66. Here the torque axis is labeled in terms of percent of the rated torque T_r, and the speed axis is labeled in terms of percent of synchronous speed—either ω_s or n_s. Furthermore, since slip $s = 1 - \omega/\omega_s = 1 - n/n_s$, the horizontal axis can also be labeled in terms of slip as indicated.

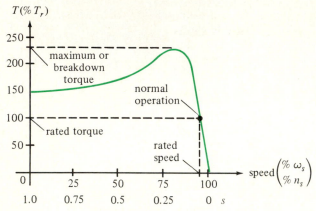

Fig. 15.66 Torque–speed curve for a typical squirrel-cage induction motor.

For normal operation, the slip s is small (between 2 and 6%), and in the range of rated load, the induction motor is, in essence, a constant-speed machine. Furthermore, for small values of s, R_1 and X are typically small compared with R_2'/s. Under this circumstance, Equation (15.51) can be approximated by

$$T \approx \frac{3sV^2}{\omega_s R_2'} \qquad \text{(normal operation)}$$

and we see that torque is directly proportional to slip. Also, torque is inversely proportional to R_2'. This means that for a given slip, a greater torque can be developed by using larger rotor conductors.

At very slow speeds ($s \approx 1$), $R_1 + R_2'/s$ is usually small compared with X and Equation (15.51) can be approximated by

$$T \approx \frac{3V^2 R_2'}{s\omega_s X^2} \qquad \text{(slow speeds)}$$

Thus, we see that, under this circumstance, torque is inversely proportional to slip s. Further, at rest (or standstill), $s = 1$ and

$$\boxed{T \approx \frac{3V^2 R_2'}{\omega_s X^2}} \qquad \text{(standstill)}$$

This, then, is the starting torque, and we see that it is directly proportional to V^2 as well as R_2'.

The **maximum** or **breakdown torque** can be determined by differentiating Equation (15.51) with respect to s and setting the result to zero. Doing so will yield the value of the slip that corresponds to the breakdown torque. We leave it to the reader (see Problem 15.59) to show that this slip is

$$s = \frac{R_2'}{\sqrt{R_1^2 + X^2}} \qquad (15.52)$$

and the maximum torque T_m is

$$T_m = \frac{3V^2}{2\omega_s(R_1 + \sqrt{R_1^2 + X^2})} \qquad (15.53)$$

Note that the breakdown torque is independent of R_2'.

Example 15.12

A three-phase induction motor has a per-phase voltage of 120 V and operates at 1152 rpm. This 60-Hz, 6-pole machine has the parameters $R_1 = 0.25\ \Omega$, $R_2' = 0.15\ \Omega$, and $X = 0.5\ \Omega$. Let us find the rated, starting, and maximum torques, as well as the rated mechanical output power.

Since $p = 6$, the synchronous speed is

$$n_s = \frac{120f}{p} = \frac{(120)(60)}{6} = 1200 \text{ rpm}$$

and thus the slip is

$$s = \frac{n_s - n}{n_s} = \frac{1200 - 1152}{1200} = 0.04 = 4\%$$

Since one revolution = 2π rad, and 1 minute = 60 s,

$$1200 \text{ rpm} = \frac{(1200)(2\pi)}{60} = 40\pi \text{ rad/s} = \omega_s$$

From Equation (15.51), the rated torque T_r is

$$T_r = \frac{3}{40\pi} \frac{(120)^2(0.15/0.04)}{[0.25 + (0.15/0.04)]^2 + (0.5)^2} = 90 \text{ N-m}$$

The starting torque T_{st} corresponds to $s = 1$. From Equation (15.51),

$$T_{st} = \frac{3}{40\pi} \frac{(120)^2(0.15)}{(0.25 + 0.15)^2 + (0.5)^2} = 125.8 \text{ N-m}$$

The maximum torque is, by Equation (15.53),

$$T_m = \frac{3(120)^2}{2(40\pi)(0.25 + \sqrt{(0.25)^2 + (0.5)^2})} = 212.5 \text{ N-m}$$

which, from Equation (15.52), occurs when

$$s = \frac{0.15}{\sqrt{(0.25)^2 + (0.5)^2}} = 0.268 = 26.8\%$$

and the corresponding motor speed is, from Equation (15.44),

$$n = (1 - s)n_s = (1 - 0.268)(1200) = 878.4 \text{ rpm}$$

Since the rated torque is $T_r = 90$ N-m, the rated mechanical output power is

$$T_r\omega = T_r(1 - s)\omega_s = 90(1 - 0.268)(40\pi) = 8278.7 \text{ W}$$

However, 1 **horsepower** (hp) = 746 W. Hence, the rated mechanical output power is $8278.7/746 \approx 11$ hp.

■

The squirrel-cage induction motor has the desirable properties of ruggedness, simplicity of construction, and efficiency. Such a machine is started easily and, for a given load, essentially has a constant speed. Another type of induction motor, however, has the capability of speed control and greater starting torque. In such a machine, referred to as a **wound-rotor motor,** the rotor has a winding similar to that on the stator. With the use of brushes and slip rings, external resistances can be connected in series with the rotor coils. In this way, the resistance of the rotor (R_2, and hence R_2') can be increased. For normal operation, an increase in R_2' means an increase in slip, and, as indicated by a torque–speed curve (e.g., Fig. 15.66), the result is a decrease in speed. Furthermore, even though the maximum developed torque [see Equation (15.53)] does not depend on the value of R_2', the starting torque does. Specifically, increasing R_2' will increase the starting torque (e.g., see Problem 15.61). The price that has to be paid for a wound-rotor induction motor is its higher cost.

Single-Phase Induction Motors

Although our discussion of induction motors thus far has been concerned with three-phase machines (which can have ratings down to a fraction of a horsepower), single-phase motors are the rule rather than the exception for low-power (under 1 hp) applications. Single-phase induction motors are similar to three-phase squirrel-cage induction motors; the main difference is that a single-phase stator winding is used in place of a three-phase winding.

In the discussion of a rotating stator field, we saw how it is produced by a three-phase stator winding. Such a rotating flux results in a unidirectional torque in an induction motor as well as a synchronous motor. For a single-phase stator winding, however, the stator field produced by a sinusoidal stator current is like one of the three components resulting from a three-phase stator winding. For example, a single-phase stator field can have a flux of the form

$$\phi = kI \cos \omega t$$

where ω is the frequency of the stator current (see the discussion associated with Figs. 15.53 and 15.54). By Euler's formula, we can rewrite this expression for ϕ as

$$\phi = \frac{kI}{2} e^{j\omega t} + \frac{kI}{2} e^{-j\omega t}$$

Therefore, we can consider ϕ to consist of two components, each having the same magnitude, but one rotating clockwise at an angular velocity of ω rad/s, and the other rotating counterclockwise at the same speed. This means that the rotor of a single-phase induction motor can turn either clockwise or counterclockwise—once it has started. However, because of the opposite rotating stator fields, the developed torque will be pulsating. Yet, on a time average, once it has been started, the torque–speed behavior of a single-phase induction motor is similar to that of a three-phase induction motor.

There is more than one way to start a single-phase induction motor. In one method, the stator has two salient poles, and a heavy short-circuited copper coil—known as a **shading coil**—is wound around half of each of these poles. A symbolic representation of such a machine is shown in Fig. 15.67. Currents induced in the shading coils delay the field in the shaded portion. This produces a small rotating-field component, which, in turn, causes a low starting torque. The direction of this torque is from the unshaded half to the shaded half of the poles. This type of single-phase machine is called a **shaded-pole motor.**

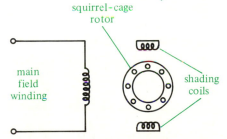

squirrel-cage
rotor

main
field
winding

shading
coils

Fig. 15.67 Shaded-pole motor.

Another technique for producing a starting torque is to have an auxiliary winding in addition to the main stator winding. Such a situation is depicted in Fig. 15.68. In this case, the main winding has a higher reactance-to-resistance ratio

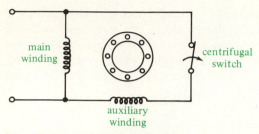

Fig. 15.68 Split-phase motor.

than the auxiliary winding. Because of this, the current in the main winding reaches its maximum value later than does the current in the auxiliary winding. This results in a rotating stator field that produces a starting torque. When such a motor, called a **split-phase motor,** approaches its operating speed, a centrifugal switch in series with the auxiliary winding opens up and that winding is disconnected.

Another type of split-phase motor is shown in Fig. 15.69. In this case, a capacitor is placed in series with the auxiliary winding to produce the phase shift between the currents in the main and auxiliary windings. Again, when the motor nears operating speed, the centrifugal switch opens and disconnects the auxiliary winding. This machine is an example of one type of **capacitor motor.** Another type can be obtained by replacing the centrifugal switch with a short circuit. Doing so can improve the running performance and pf of the motor if the capacitor value is changed, but the price paid for this is a reduction in the starting torque. However, an alternative approach can be taken. Instead of replacing the centrifugal switch with a short circuit, it is left as is and a second capacitor is placed in parallel with the original capacitor–switch series combination. In essence then, one capacitor is used for starting and one is used for running—and this type of machine may be more desirable.

Fig. 15.69 Capacitor motor.

SUMMARY

1. A conductor of length l moving with velocity u at right angles through a magnetic field of flux density B generates an emf $e = Blu$.

2. A conductor of length l carrying a current i at right angles to a magnetic field of flux density B develops a force $f = Bli$.

3. Mechanical circuits can be dealt with in a manner analogous to that for electric circuits.

4. Dynamic transducers are moving-coil devices that can be used for loud-speakers, microphones, and phonograph pickups.

5. An electromagnet is a moving-iron transducer that can be used to fabricate a relay. A solenoid plunger is also a moving-iron transducer.

6. The d'Arsonval movement is a rotational transducer that is widely used for dc ammeters, dc voltmeters, and ohmmeters.

7. The electrodynamometer is a rotational transducer that is used for ac measurements, in particular, voltmeters, and especially wattmeters.

8. A generator converts mechanical energy into electric energy; a motor converts electric energy into mechanical energy.

9. An alternator utilizes brushes and slip rings to generate alternating current. A dc generator requires the use of brushes and a split-ring commutator.

10. For a dc generator, the field is on the stator and the armature is on the rotor. For an alternator, the field is on the rotor and the armature is on the stator.

11. The voltage generated by a rotating machine is graphically displayed by its magnetization curve.

12. A dc machine can be separately excited or self-excited. Self-excited machines can be shunt connected, series connected, or compound connected.

13. The external characteristic of a dc generator graphically shows the relationship between load voltage and load current.

14. Synchronous machines can have salient-pole rotors or cylindrical (nonsalient) rotors. The former machines typically are used in lower-speed applications (less than 1800 rpm), whereas the latter often see use in higher-speed situations (1800 rpm and above).

15. Three-phase machines are usually analyzed on a per-phase basis.

16. The synchronous speed of a machine depends on the frequency of the applied excitation and the number of poles of the machine.

17. The speed of a dc motor can be easily controlled.

18. The relationship between the speed and the torque of a dc motor is graphically displayed by its speed–torque characteristic.

19. The speed regulation of a dc motor is a measure of the change in speed between no-load and full-load conditions.

20. Bringing a dc motor up to operating speed requires the use of a starting box.

21. The phenomenon of a rotating stator field is employed by an ac motor.

22. A synchronous motor operates at a constant speed over a range of load conditions. Synchronous motors can be used for power-factor correction.

23. An induction motor is basically a transformer with a fixed primary and a rotating secondary.

24. The simplest type of induction motor—the squirrel-cage motor—gets its name from its squirrel-cage rotor.

25. An induction motor typically operates at a speed slightly less than synchronous speed. For a fixed load, it is essentially a constant-speed machine.

26. The slip of an induction motor is a measure of the difference between the operating and synchronous speeds of the motor.

27. The relationship between the torque and the speed (or slip) of an induction motor is graphically given by its torque–speed curve.

28. A wound-rotor induction motor offers speed control and greater starting torque. The price for this type of motor is higher cost.

29. Single-phase induction motors are similar to three-phase squirrel-cage motors, and are used in low-power applications.

30. Shaded-pole motors, split-phase motors, and capacitor motors are all examples of single-phase induction motors.

Problems for Section 15.1

15.1 The electromechanical system shown in Fig. 15.3 is to be operated as a motor. Assuming $B = 0.4$ T and $l = 0.2$ m, find the current required to develop a force of 0.5 N directed to the right on the movable conductor. Determine the value of the resistance R that results in a velocity to the right of 30 m/s if the battery has a value of 12 V.

15.2 The electromechanical system shown in Fig. 15.3 is to be operated as a generator. Assuming $B = 0.4$ T and $l = 0.2$ m, find the current corresponding to a force of 0.5 N directed to the right on the movable conductor. Determine the velocity required to obtain this current for the case that the resistance is $1.5\ \Omega$ and the battery has a value of 12 V.

15.3 Write an equation describing the mechanical system shown in Fig. P15.3. Draw the mechanical circuit and its electric-circuit analog.

15.4 Repeat Problem 15.3 for the mechanical system shown in Fig. P15.4.

15.5 An alternative electrical analogy for a mechanical circuit is obtained when force is analogous to voltage and velocity is analogous to current.
  (a) What is the electric analog of the mass given in Fig. 15.4?
  (b) What is the electric analog of the dash pot given in Fig. 15.5?
  (c) What is the electric analog of the spring given in Fig. 15.6?

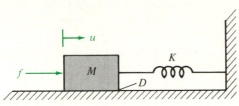

Fig. P15.3

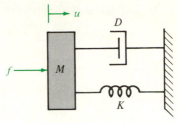

Fig. P15.4

15.6 Use the alternative electrical analogy described in Problem 15.5 to determine the analogous electric circuit of the mechanical system shown in Fig. 15.7. Is there any relationship between this electric circuit and the one shown in Fig. 15.8(b)?

15.7 The motor shown in Fig. 15.3 is described by Equations (15.8) and (15.9). Assume that the inductance $L = 0$ and the movable conductor is initially at rest. Find the velocity u for the case that a constant force f is applied at time $t = 0$.

15.8 Suppose that the electromechanical system shown in Fig. 15.3 is used as a generator when an applied force f is directed to the right. In a manner similar to that for the case of a motor, write the describing equations and model the generator with an appropriate equivalent circuit.

Problems for Section 15.2

15.9 For the dynamic loudspeaker in Example 15.2, suppose that the resistance of the moving coil is $R = 0.4\ \Omega$. The efficiency of the loudspeaker is the ratio of the output (mechanical) power $p_o = fu$ to the input (electric) power $p_{in} = vi$. Find the efficiency of the loudspeaker.

15.10 Equations (15.12) and (15.13) describe the dynamic microphone. Find an equivalent circuit that models such a device for the case that the microphone is loaded with a resistance R_L.

15.11 A force of $f = 0.25$ N is applied to the diaphragm of a dynamic microphone whose 40-turn moving-coil radius is $r = 2$ cm. Ignore the mass $(M = 0)$, assume that the coefficient of friction is $D = 3$ N-s/m, and suppose that the magnetic flux density in the annular air gap is $B = 0.5$ T. Find the efficiency of the microphone, that is, find the ratio of the output (electric) power $p_o = vi$ to the input (mechanical) power $p_{in} = fu$, for the case that the microphone has a moving-coil resistance of $R = 0.4\ \Omega$ and a load resistance of $R_L = 4\ \Omega$.

15.12 A dynamic phonograph pickup similar to that shown in Fig. 15.12 has $B = 0.3$ T, $r = 0.2$ cm, $N = 15$, and $R = 0.4\ \Omega$. Given a sinusoidal displacement of amplitude 0.002 cm, determine the rms voltage across a 4-Ω load for the case that the frequency is (a) 1000 Hz and (b) 5000 Hz.

15.13 A relay with 2000 turns has a gap length of 0.8 cm when the contacts are open and 0.2 cm when the contacts are closed. Given that the cross-sectional area is 2 cm², determine the pickup and dropout currents for the case that the spring exerts a force of 0.08 N.

15.14 For the relay shown in Fig. 15.13, initially the gap has length l_g. After the armature moves a distance x, the gap length is $l = l_g - x$. Show that the expression for the force on the armature given by Equation (15.15) has the equivalent form

$$f = \frac{1}{2} i^2 \frac{dL}{dx}$$

where $L = N^2/\mathcal{R}$, and the reluctance of the core is negligible compared with the reluctance of the air gap.

15.15 For the solenoid shown in Fig. 15.14, the length of the plunger is 10 cm, as is the length of the cyclindrical coil. Suppose that the variation of inductance with displacement is given by

$$L = \frac{1}{2} + \frac{1}{\pi} \tan^{-1}(20x - 1)$$

where x ranges between $x = 0$ and $x = 10$ cm. Determine the current required to produce a maximum force of 4 N on the plunger.
$$\left[\text{Hint: } \frac{d}{dx}(\tan^{-1}\theta) = \left(\frac{1}{1 + \theta^2}\right)\left(\frac{d\theta}{dx}\right). \right]$$

Problems for Section 15.3

15.16 A d'Arsonval movement has a 10-turn moving coil whose length is 2 cm and whose width is 2 cm. The air-gap magnetic flux density is 0.5 T and the spring compliance is 10^8 degrees/N-m. What current will yield a full-scale deflection of 100°?

15.17 A d'Arsonval movement with a moving-coil resistance of $R_c = 2$ kΩ requires 50 μA for full-scale deflection. Determine the value of the shunt resistance required to use this mechanism for a (a) 0- to 1-mA ammeter and (b) 0- to 25-mA ammeter.

15.18 For the circuit shown in Fig. P15.18, an ammeter whose resistance is

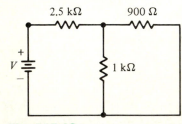

Fig. P15.18

100 Ω measures the current through the 900-Ω resistor to be 1 mA. Find the actual current through the 900-Ω resistor.

15.19 A 50-μA d'Arsonval movement with a coil resistance of $R_c = 2$ kΩ is to be used for a dc voltmeter. Determine the value of the series resistance required for a (a) 0- to 10-V voltmeter and (b) 0- to 50-V voltmeter.

15.20 With reference to Problem 15.19, what is the ohms-per-volt rating for the voltmeter in part (a) and in part (b)?

15.21 A 50-μA d'Arsonval movement with moving-coil resistance R_c is to be used for a dc voltmeter whose full-scale deflection is v volts. Show that, as long as $R_c \leq v/50 \times 10^{-6}$, the voltmeter will be rated at 20 kΩ/V.

15.22 For the circuit shown in Fig. P15.22, the voltage across the 300-kΩ resistor is measured on a 0- to 30-V voltmeter to be 8 V. If the voltmeter is rated at 20 kΩ/V, what is the actual voltage across the 300-kΩ resistor?

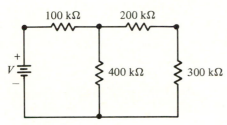

Fig. P15.22

15.23 For the ohmmeter given in Fig. 15.19, $V_s = 1.5$ V, $R_s = 28$ kΩ, $R_c = 2$ kΩ, and the d'Arsonval mechanism is a 50-μA movement. Find the value of the resistance R when the movement has a current of (a) 10 μA, (b) 20 μA, (c) 30 μA, and (d) 40 μA.

15.24 An ammeter and a voltmeter are connected to a 5-kΩ resistor. When the connection is as shown in Fig. P15.24(a), the ammeter and voltmeter read 9.98 mA and 50 V, respectively. When the connection is as depicted in Fig. P15.24(b), the respective readings are 10 mA and 49.75 V. Find the resistances of the ammeter and the voltmeter.

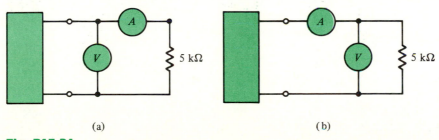

(a)   (b)

Fig. P15.24

15.25 Show that the average value of the product of $i_c = I_c \cos(\omega t + \theta_1)$ and $i_s = I_s \cos(\omega t + \theta_2)$ is $\frac{1}{2}I_c I_s \cos \theta$, where $\theta = \theta_1 - \theta_2$. Use this result to show that the average value of $I^2 \cos^2 \omega t$ is $I^2/2$.

Problems for Section 15.4

15.26 A 10-turn coil similar to the one shown in Fig. 15.22 rotates at 1200 rpm. Find the rms value of the voltage that is generated when the coil has a length of 20 cm and a width of 10 cm, and the magnetic flux density is 1 T.

15.27 Repeat Problem 15.26 for a 10-turn coil similar to the one shown in Fig. 15.23.

15.28 The poles for the dc machine shown in Fig. 15.25 have a surface area of 0.01 m². Supposing that the air-gap magnetic flux density is 1 T, and that the generated voltage is to be 80 V, determine (a) the number of turns required by each of the eight coils when the rotor speed is 1000 rpm, and (b) the rotor speed when each of the eight coils has 20 turns.

15.29 The magnetic flux per pole of a 10-kW, 4-pole, 220-V dc generator is 25 mWb. The armature has 48 coils, each having two turns, and they are connected so that there are two parallel paths between the brushes. Find the rotor speed that achieves the rated voltage at no load.

15.30 An 11-kW, 220-V dc generator has a full-load current of 50 A at 1500 rpm. Supposing the armature resistance is 0.2 Ω, find (a) the no-load voltage at 1500 rpm, (b) the no-load voltage at 1800 rpm, and (c) the full-load voltage at 1800 rpm given a full-load current of 50 A.

15.31 A separately excited dc generator has the magnetization curve shown in Fig. P15.31. Given that the field current is 1 A, find the no-load generated voltage when the generator speed is (a) 1200 rpm, (b) 1500 rpm, and (c) 1000 rpm.

15.32 A separately excited dc generator has the magnetization curve shown in Fig. P15.31. Given that the no-load voltage is 180 V, find the field current when the generator speed is (a) 1200 rpm, (b) 1500 rpm, and (c) 1000 rpm.

15.33 A separately excited 10-kW dc generator is rated at 200 V and 50 A. The magnetization curve for this machine is shown in Fig. P15.31. Given that the armature resistance is 0.4 Ω and the field resistance is 100 Ω:

(a) Find the field current for the rated conditions.

(b) Find the no-load voltage when the voltage applied to the field is 120 V.

15.34 A separately excited dc generator has the magnetization curve shown in Fig. P15.31 for the case that $n = 1200$ rpm. Suppose that the voltage applied to the field is $V_f = 200$ V and the field resistance is $R_f = 125$ Ω.

(a) Determine the no-load generated voltage assuming no control rheostat, that is, $R_c = 0$.

(b) Find the value of R_c that will result in a no-load generated voltage of 210 V.

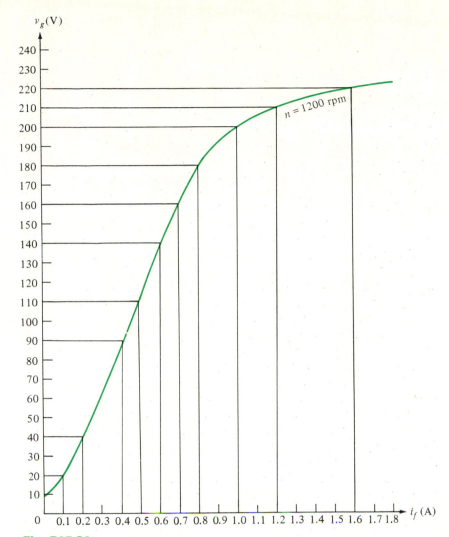

v_g(V)

i_f (A)

Fig. P15.31

15.35 A shunt-connected 8-kW, 200-V dc generator has a full-load current of 40 A at 1200 rpm. The magnetization curve for this machine is as shown in Fig. P15.31. Suppose that the armature resistance is 0.5 Ω and the field resistance is 125 Ω. Determine the value R_c of the control rheostat such that (a) the no-load voltage is 200 V, and (b) the full-load voltage is 200 V.

15.36 A 12-kW, 240-V shunt-connected dc generator has a full-load current of 50 A at 1000 rpm. Under no-load conditions, the generated voltage is 255 V. For full-load conditions, neglect the field current compared with the armature current and determine the armature resistance.

15.37 A shunt-connected dc generator has the magnetization curve shown in

Fig. P15.31. Determine the no-load voltage at 1200 rpm when the total resistance of the field circuit is $R = R_c + R_f = 175$ Ω.

15.38 A cumulatively compounded dc generator is rated at 8 kW, 200 V, and 40 A at 1200 rpm. The shunt field is separately excited, and ignoring the resistance of the series field, the no-load voltage is 200 V when the field current is 1 A, while the full-load voltage is 200 V when the field current is 1.6 A (i.e., a flat-compounded generator). Furthermore, it is given that the shunt winding has 500 turns per pole.

(a) Find the mmf in ampere-turns per pole (A-t/pole) of the shunt field at no load and at full load.

(b) Determine the mmf that must be supplied by the series field.

(c) Find the number of turns per pole for the series winding.

15.39 An alternator is to generate a 60-Hz ac voltage. Find the rotor speed for the case that the alternator has (a) 4 poles, (b) 6 poles, (c) 8 poles, and (d) 18 poles.

15.40 For the load in Example 15.8, change the pf to unity.

(a) Find the resulting generated voltage.

(b) Draw the corresponding phasor diagram.

(c) Use the magnetization curve given in Fig. P15.31 to determine the approximate field current.

15.41 Repeat Problem 15.40 for a pf of 0.8 leading.

15.42 A single-phase 5-kW alternator is rated at 110 V for a load with a pf of 0.75 lagging. If the armature resistance is 0.1 Ω, and the synchronous reactance is 0.5 Ω, what is the value of the generated voltage?

15.43 A three-phase, 10-kVA alternator is Y connected and has a rated per-phase voltage of 127 V for a load with a pf of 0.8 lagging at 60 Hz. Ignore the armature resistance and determine the synchronous reactance given that the generated voltage is 185 V and it leads the load voltage by 20°.

Problems for Section 15.5

15.44 A 10-turn coil similar to the one shown in Fig. 15.46 has a length of 20 cm and a width of 10 cm, and the magnetic flux density is 1 T.

(a) Find the maximum torque that is developed when the current is 1 A.

(b) Determine the current required to develop a maximum torque of 0.5 N-m.

15.45 A 220-V shunt motor has an armature resistance of 0.4 Ω. Given that the motor speed is 1000 rpm when the armature current is 50 A, (a) determine the speed when the armature current is 75 A, and (b) find the armature current when the speed is 1050 rpm.

15.46 A 220-V dc shunt motor has an armature current of 60 A when the speed is 1400 rpm, and an armature current of 40 A when the speed is 1500 rpm. Determine the armature resistance.

15.47 A 220-V dc shunt motor has an armature resistance of 0.5 Ω. When the speed of the motor is 1000 rpm, the armature current is 40 A. Assuming that the resisting torque of the load stays constant, (a) when a 1.5-Ω resistance is placed in series with the armature, find the resulting current and speed; and (b) determine the armature current and speed (without the 1.5-Ω resistance) when the field flux is reduced by 20%.

15.48 A 220-V dc shunt motor has an armature resistance of 0.2 Ω. When the motor speed is 1000 rpm, the armature current is 50 A and the developed torque is 100 N-m. Find the developed torque at 1100 rpm.

15.49 A 220-V dc shunt motor has an armature resistance of 0.5 Ω. At full load, the motor speed is 1000 rpm and the armature current is 40 A. At no load, the armature current is 6 A. Determine the speed regulation of the motor.

15.50 A 500-V dc series motor has an armature resistance of 0.3 Ω and a field-winding resistance of 0.2 Ω. Assuming the motor is rated at 100 A at 1000 rpm, determine the motor speed when the armature current is 46 A.

15.51 The dc series motor given in Problem 15.50 has a rated torque of 430 N-m. For the case that the motor has a speed of 2000 rpm, find (a) the armature current and (b) the torque.

15.52 A three-phase synchronous motor is Y-connected and has a per-phase voltage of 125 V. Given that the synchronous reactance is 5 Ω and the input power is 6 kW, find the torque angle for the case that the motor operates at a pf of (a) unity, (b) 0.8 lagging, and (c) 0.8 leading.

15.53 A Y-connected, three-phase synchronous motor has a per-phase voltage of 125 V. Assuming the input power is 6 kW and the torque angle is 16°, determine the synchronous reactance of the armature given that the motor operates at a pf of unity.

15.54 A Y-connected, three-phase synchronous motor has a per-phase voltage of 254 V and a synchronous reactance of 4 Ω. Suppose that the generated emf is 200 V and the torque angle is 36.87°.
(a) Find the line (armature) current.
(b) Determine the pf at which the motor operates.
(c) Find the generated emf and torque angle that result in a unity pf and the same line current.

15.55 For the motor given in Problem 15.54, suppose that the line current is 25 A and the motor operates at a pf of 0.8 leading.
(a) Find the generated emf.
(b) Determine the torque angle.

15.56 A three-phase synchronous motor operating at a pf of unity draws an armature current of 25 A.
(a) Determine the synchronous reactance given that the generated emf is 256 V and the torque angle is 20°.
(b) Find the per-phase voltage of the motor.
(c) Determine the input power of the motor.

15.57 A three-phase, 60-Hz induction motor operates at 1158 rpm.
(a) Determine the number of poles for this machine.
(b) Find the slip of the motor.
(c) Determine the induced rotor emf.

15.58 A three-phase, 60-Hz induction motor has 4 poles. Find the speed of the motor given that the emf induced in the rotor has a frequency of 3 Hz.

15.59 Show that for the developed-torque expression of an induction motor given by Equation (15.51), the breakdown torque is achieved when the slip is the value in Equation (15.52), and the breakdown torque is given by Equation (15.53).

15.60 For Example 15.12, use the approximate formulas given in the text to calculate the rated torque T_r and the starting torque T_{st}. Compare these results with those of the example and explain any discrepancies.

15.61 For Example 15.12, determine the value to which R_2' should be changed so that the maximum torque will occur at starting.

15.62 For Example 15.12, determine $I = |\mathbf{I}|$ and the power factor of the motor for the condition of (a) rated torque, (b) starting torque, and (c) maximum torque.

15.63 A three-phase induction motor has a per-phase voltage of 120 V. This 6-pole, 60-Hz machine has a rated torque of 100 N-m, and has $R_1 = 0.10 \ \Omega$, $R_2' = 0.12 \ \Omega$, and $X = 0.4 \ \Omega$. Find (a) the slip and (b) the rated speed. Justify any approximation used.

15.64 A 36-hp, 60-Hz induction motor with 4 poles has a per-phase voltage of 254 V. Given that the motor has a slip of 5% at rated speed, find the approximate value for R_2'.

15.65 A 4-pole, 60-Hz induction motor has a slip of 5% at rated speed. Given that the per-phase voltage is 254 V, and $R_1 = 0.2 \ \Omega$, $R_2' = 0.32 \ \Omega$, and $X = 0.5 \ \Omega$, find the (a) rated speed, (b) rated torque, (c) starting torque, and (d) maximum torque.

Answers to Selected Problems

1.49 5 W, $\frac{25}{12}$ W, $-\frac{25}{3}$ W, $\frac{5}{4}$ W

1.51 $\frac{15}{13}$ W, $\frac{75}{676}$ W, $-\frac{225}{169}$ W, $\frac{45}{676}$ W

1.53 (a) -18 W, 9 W, -36 W, 18 W, 27 W, (b) 18 W, 9 W, -72 W, 18 W, 27 W

Problems for Chapter 2

2.1 -1.64 A, 0.36 A, -3.36 A, 1.36 A

2.4 -3 V, 6 V, 4 V

2.6 6 V, 3 V, -3 V, -9 V

2.9 -47.76

2.11 $\frac{31}{15}\mho$

2.15 6 V, 3V, -3 V, -9 V

2.18 3 V, 0.96 V, 0.42 V

2.19 $v_o = -8v_1 + 9v_2$

2.22 $(1 + R_L/R_1)R_s$

2.24 (a) $(1 + R_2/R_L)v_s$, (b) $-R_1/R_2$

2.25 0.52

2.28 -1.8

2.29 (a) $\frac{112}{9}$ V, $\frac{29}{9}$ Ω, (b) 2 A

2.31 (a) $\frac{7}{5}$ V, $\frac{1}{5}$ Ω, (b) $\frac{7}{8}$ V, (c) $\frac{1}{5}$ Ω, (d) 2.45 W

2.33 (a) 0 V, 3 Ω, (b) 0 A, 3 Ω

2.35 $\frac{3}{2}v_1$, 3 Ω

2.36 $(1 + R_2/R_1)$ v_2, 0 Ω

2.38 $-v_s/R_1$, R_2

2.40 (a) 0 A, $\frac{4}{3}$ Ω; (b) 0 V, $\frac{4}{3}$ Ω

2.43 $\frac{7}{9}$ A

2.45 $\frac{4}{11}$ V

2.47 $-\frac{15}{8}$ A, -2A, $-\frac{31}{8}$ A

2.49 3 A, 3 V

2.51 5 A, -1 V

Problems for Chapter 3

3.1 0 for $-\infty < t < 0$, 2 for $0 \le t < 1$, 0 for $1 \le t < \infty$

0 for $-\infty < t < 0$, t^2 for $0 \le t < 1$, 1 for $1 \le t < \infty$

0 for $-\infty < t < 0$, $2t^2$ for $0 \le t < 1$, 2 for $1 \le t < \infty$

0 for $-\infty < t < 0$, 2 for $0 \le t < 1$, 2 for $1 \le t < \infty$

0 for $-\infty < t < 0$, $2t + 2$ for $0 \le t < 1$, 2 for $1 \le t < \infty$

3.3 0 for $-\infty < t < 0$, 2 for $0 \le t < 1$, 0 for $1 \le t < \infty$

0 for $-\infty < t < 0$, t^2 for $0 \le t < 1$, 1 for $1 \le t < \infty$

0 for $-\infty < t < 0$, 2 for $0 \le t < 1$, 0 for $1 \le t < \infty$

0 for $-\infty < t < 0$, 1 for $0 \le t < 1$, 0 for $1 \le t < \infty$

0 for $-\infty < t < 0$, $t + 1$ for $0 \le t < 1$, 1 for $1 \le t < \infty$

3.6 0 for $-\infty < t < 0$, $3t$ for $0 \leq t < 1$, $-3t + 6$ for $1 \leq t < 2$, 0 for $2 \leq t < \infty$
0 for $-\infty < t < 0$, 1 for $0 \leq t < 1$, -1 for $1 \leq t < 2$, 0 for $2 \leq t < \infty$
0 for $-\infty < t < 0$, $3t - 1$ for $0 \leq t < 1$, $-3t + 5$ for $1 \leq t < 2$, 0 for $2 \leq t < \infty$

3.9 0 for $-\infty < t < 0$, 2 for $0 \leq t < 1$, 0 for $1 \leq t < \infty$
0 for $-\infty < t < 0$, 1 for $0 \leq t < 1$, 0 for $1 \leq t < \infty$
0 for $-\infty < t < 0$, $-t$ for $0 \leq t < 1$, -1 for $1 \leq t < \infty$

3.12 0 for $-\infty < t < 0$, $-2t$ for $0 \leq t < 1$, -2 for $1 \leq t < \infty$
0 for $-\infty < t < 0$, $-4t$ for $0 \leq t < 1$, -4 for $1 \leq t < \infty$
0 for $-\infty < t < 0$, $-2t$ for $0 \leq t < 1$, -2 for $1 \leq t < \infty$
0 for $-\infty < t < 0$, $-3t$ for $0 \leq t < 1$, -3 for $1 \leq t < \infty$

3.15 $\frac{1}{4}t^2$ for $1 \leq t < 2$, 1 for $2 < t < \infty$

3.19 0 for $-\infty < t \leq 0$, $\frac{1}{2}t$ for $0 < t \leq 1$, $\frac{1}{2}$ for $1 < t < \infty$
0 for $-\infty < t \leq 0$, t for $0 < t \leq 1$, 1 for $1 < t < \infty$
0 for $-\infty < t \leq 0$, $t + 1$ for $0 < t \leq 1$, 1 for $1 < t < \infty$

3.23 $6 \cos 3t = \dfrac{v_1}{3} + \dfrac{1}{1} \displaystyle\int_{-\infty}^{t} (v_1 - v_2) \, dt$

$\dfrac{1}{1} \displaystyle\int_{-\infty}^{t} (v_1 - v_2) \, dt = \dfrac{v_2 - v_3}{2}$

$\dfrac{v_2 - v_3}{2} = 4 \dfrac{dv_3}{dt} + \dfrac{v_3}{5}$

3.29 10 V for $t < 0$, $10e^{-2t}$ V for $t \geq 0$
0 A for $t < 0$, $-2e^{-2t}$ A for $t \geq 0$

3.32 2 A for $t < 0$, $2e^{-2t}$ A for $t \geq 0$
0 V for $t < 0$, $-16e^{-2t}$ V for $t \geq 0$

3.35 9 V for $t < 0$, $9e^{-4t} - 18e^{-t/2}$ V for $t \geq 0$
3 A for $t < 0$, $3e^{-t/2}$ A for $t \geq 0$

3.38 $\frac{1}{2}$ V $(1 - e^{-2t/RC}) u(t)$

3.40 $\dfrac{V}{R}(1 - e^{-Rt/L})u(t)$, $Ve^{-Rt/L}u(t)$

3.43 $-\dfrac{R_2}{R_1}(1 - e^{-t/R_2 C})u(t)$

3.45 1 V for $t < 0$, $-\frac{5}{8}e^{-3t} + \frac{13}{8}e^{-t}$ V for $t \geq 0$
1 A for $t < 0$, $\frac{15}{2}e^{-3t} - \frac{13}{2}e^{-t}$ A for $t \geq 0$

3.47 0 A for $t < 0$, $e^{-4t} - e^{-2t}$ A for $t \geq 0$

3.50 1 V for $t < 0$, $e^{-t}(\cos t - \sin t) = \sqrt{2}\, e^{-t} \cos(t + \pi/4)$ V for $t \geq 0$
1 A for $t < 0$, $e^{-t}(\cos t + \sin t) = \sqrt{2}\, e^{-t} \cos(t - \pi/4)$ V for $t \geq 0$

3.51 -2 A for $t < 0$, $-(t + 2)e^{-t}$ A for $t \geq 0$
1 V for $t < 0$, $(t + 1)e^{-t}$ V for $t \geq 0$

3.53 $(1 - 2te^{-2t} - e^{-2t})u(t)$

3.55 $[\frac{1}{2} - \frac{1}{2}e^{-t}(\cos t + \sin t)]u(t) = \frac{1}{2}[1 - \sqrt{2}\, e^{-t} \cos(t - \pi/4)]u(t)$

3.57 $\frac{70}{3}(e^{-5t} - e^{-2t})u(t)$

Problems for Chapter 4

4.1 (a) $8.06e^{j60.26°}$, (c) $3.61e^{j123.7°}$, (e) 4, (g) $7e^{j90°}$

4.2 (a) $1.03 + j2.82$, (c) $2.5 - j4.33$, (e) $j6$, (g) -2

4.3 (a) $4.60 + j4.96$, (c) $1.5 - j2.60$

4.4 $12 \cos 2t$, lead

4.7 $5\sqrt{17} \cos (3t - 166°)$

4.9 (a) $0.31 + j0.017$ Ω, (b) $0.12 + j0.34$ Ω, (c) $0.041 + j0.89$ Ω

4.10 $\dfrac{4}{\sqrt{17}} \cos (4t - 164°)$

4.12 $6\sqrt{5} \cos (2t - 116.6°)$

4.14 $1.6 \cos (2t + 23.13°)$

4.16 $6.8\underline{/30°}$, $6.8\underline{/-90°}$

4.17 (a) 7.38 W, (b) 12.5 W, (c) 5.29 W

4.20 18 W

4.22 4.15 W, 0.64 W, 1.23 W

4.25 (a) 0.5 mW, 1 mW, 2 mW, (b) 0.5 mW, (c) the op-amp power is ignored

4.28 (a) 2.12, (b) 1.58, (c) 1.58, (d) 0.745

4.31 (a) $-90°$, 0; (b) leading, (c) 3.49 μF

4.32 4.58 A

4.35 (a) $13.06\underline{/-2.53°}$, (b) ≈ 1.0, (c) lagging (barely)

4.37 (a) $0.686\underline{/-31°}$ VA, (b) 0.686 VA, (c) 0.86 leading

4.41 (a) $2200\underline{/-18.2°}$ VA, (b) 2200 VA, (c) $88\underline{/-18.2°}$ Ω

4.43 $10.92\underline{/-5.9°}$ A, $15.35\underline{/-7.75°}$ A

4.44 7.66 A, 2112.4 W

4.47 $23\underline{/-53.13°}$ A, $11.5\underline{/-120°}$ A, $8.85\underline{/52.6°}$ A, 3301 W

4.51 $44.13\underline{/15.9°}$ A, $73.32\underline{/-135.7°}$ A, $40.35\underline{/75.53°}$ A, 18,833 W

4.53 3008.2 W, 0.97 leading

4.55 881.7 W, 826.6 W

4.58 7063.4 W, 11,778 W

Problems for Chapter 5

5.1 Low-pass filter

5.4 $1/\sqrt{2}RC$ rad/s, $\sqrt{2}/RC$ rad/s

5.6 $0.414/RC$ rad/s, $2.414/RC$ rad/s, bandpass filter

5.12 $1/\sqrt{LC - R^2C^2}$

5.15 1.707

5.17 (a) 10, (b) 1010 Ω, 2.02 H, (c) 10

5.19 222.7

5.21 Poles: $-0.167 - j2.23$, $-0.167 + j2.23$
Zeros: (a) 0, (b) 0, (c) 0,0, (d) no (finite) zeros

5.24 Poles: -0.34, -14.66
Zeros: (a) $-j\sqrt{5}$, $j\sqrt{5}$, (b) $-j\sqrt{5}$, $j\sqrt{5}$; (c) 0, (d) 0

5.26 (a) $\dfrac{6s + 105}{s + 15}$, (b) $\dfrac{15}{2s + 35}$, (c) $\dfrac{15V_1}{2s + 35}$, $\dfrac{90}{2s + 35}$

5.28 (a) $\dfrac{2}{(s + 1)(s + 2)}$, (b) $\dfrac{s + 2}{2s}$

5.30 (a) $\dfrac{-8}{s^2 + 2s + 4}$, (b) $\dfrac{s^2 + 2s + 4}{s^2 + s + 4}$

5.33 $\dfrac{s^2 + 1}{s^2 + 7s + 9}$

5.36 $\dfrac{1}{s + 5}$

5.39 (a) $\dfrac{s - 4}{(s + 2)(s + 8)}$, (b) $\dfrac{-4(s^2 + 7s - 9)}{s(s + 1)(s + 6)}$, (c) $\dfrac{2s + 7}{(s + 2)^2}$

5.41 (a) $(2 - 3e^{-10t} + e^{-30t})u(t)$, (b) $(10 - 6e^{-2t} - 4e^{-12t})u(t)$
5.44 $6 - 12e^{-t} + 2e^{-6t}$ for $t \geq 0$
5.46 $4e^{-2t} - 2e^{-8t}$ for $t \geq 0$, $-e^{-2t} + 2e^{-8t}$ for $t \geq 0$
5.48 $(2 + 3t)e^{-2t}$ for $t \geq 0$, $(1 - 3t)e^{-2t}$ for $t \geq 0$
5.49 $\frac{3}{4}(1 - e^{-2t} - \frac{2}{3}te^{-2t})u(t)$
5.53 (a) $-\frac{14}{3} + \frac{2}{3}e^{-3t}$ for $t \geq 0$, (b) $-4 - \sin 2t$ for $t \geq 0$

5.55 $\frac{1}{2} - \frac{1}{2}e^{-t} - \dfrac{1}{\sqrt{3}}e^{-t/2}\sin\dfrac{\sqrt{3}}{2}t$ for $t \geq 0$

5.56 (a) $\frac{1}{2}(1 - e^{-2t})u(t)$, (b) $(e^{-t} - e^{-2t})u(t)$, (c) $(\frac{1}{2} - e^{-t} + \frac{1}{2}e^{-2t})u(t)$,
(d) $te^{-2t}u(t)$

5.59 (a) $\dfrac{s + 1}{s + 2}$, (b) $\dfrac{s + 1}{s^2 + 1}$, (c) $\dfrac{s + 1}{s^2 + 2s + 2}$, (d) $\dfrac{1}{s + 1}$, (e) $\dfrac{1}{s + 2}$

Problems for Chapter 6

6.1 17.9 A
6.4 (a) 4.32×10^{-4} ℧/m, (b) 23.15 MΩ
6.7 (a) 8 Ω/m, (b) 1.25 kΩ
6.9 (a) 12.76 ℧/m, (b) 784 Ω
6.13 0.617 V
6.15 0.663 V
6.17 50
6.19 (a) 3.71 mA, (b) 0.772 V, (c) 348.5 Ω
6.21 (a) 3.91 nA, (b) 7.82 nA, (c) 11.85 nA, (d) 1.29 nA
6.24 (a) 0.275 V, (b) 0.24 V, (c) 0.36 V

6.26 303.66 K

6.28 0.394 V, -5.61 V

6.30 299 Ω

6.32 0.289 V, 1.42 A

6.34 -3 V for $v_s < -4$ V, $\frac{3}{4}v_s$ for $v_s \geq -4$ V

6.37 0 for $v_s \leq 4$ V, $v_s - 4$ for $v_s > 4$ V

6.38 $i = -I_o$ for $v \leq 0$, $v = 0$ for $i > -I_o$

6.40 $i = -I_o$ for $v \leq V_\gamma$, $v = R_f(i + I_o) + V_\gamma$ for $i > -I_o$

6.42 (a) D_1, D_3 ON; D_2, D_4 OFF; $v_o = v_s$

     (b) D_1, D_3 OFF; D_2, D_4 ON; $v_o = -v_s$

6.45 (a) 0, (b) 9 V, (c) 9.47 V

6.49 0 V

6.52 0 V

6.55 (a) 189 Ω, (b) 14 Ω, (c) 146 Ω, (d) 11.15 Ω

6.57 (a) $v_s \leq 0.2$ V, (b) $v_s > 4.6$ V, (c) 0.2 V $< v_s \leq 4.6$ V

6.61 (a) 2 V, (b) 2 V, (c) 6 V, (d) AND gate

6.63 $R < 2.82$ kΩ

6.65 (a) 10 nA, 0.036 V, -5.964 V, (b) 2.43 A, 1 V, -5 V

6.68 (a) 750 Ω, (b) 44 V, (c) 400 Ω

6.71 $v_o = v_s$ for $v_s \leq 12$ V, $v_o = 12$ V for $v_s > 12$ V

6.75 (a) $v_s + 5$, (b) $v_s + 10$, (c) $v_s + 10$

6.77 (a) $v_s - V/2$, (b) $v_s - 3V/2$

6.80 (a) 26 pF, (b) 0.577×10^{-6} m^2

Problems for Chapter 7

7.1 (a) 4.4 mA, (b) -0.6 V

7.3 (a) 4.6 V, (b) 6 V, (c) 3.4 V

7.5 (a) 8.1 V, (b) 9 V, (c) 10

7.9 (a) 3.475 kΩ, (b) 256 kΩ

7.11 (a) 26 μA, (b) 1.04 mA, (c) 13.73 V

7.14 (a) 4 V, (b) 1 V

7.16 (a) 25 μA, (b) 1.275 mA, (c) 7.45 V

7.18 (a) 40 μA, (b) 0.94 mA, (c) 6.1 V, (d) 2.48 V

7.19 (a) 19, (b) 1.07 kΩ

7.22 (a) 28.6, (b) 29.6

7.25 (a) 5 V, (b) 0.2 V, (c) 0.2 V

7.27 (a) 55.2, (b) 13.8

7.30 (a) 2.9 V, (b) 22.9, (c) 33

7.31 (a) 19.4 mW, (b) -3.5 V, 0.8 V

7.34 (a) 4.2, (b) 1

7.36 (a) 3.25 V, (b) 54.9, (c) 16

7.38 20.35 mW

7.40 (a) 2.47, ≈ 0, (b) 100

7.42 (a) 0.2 V, (b) 5 V, (c) OR gate
7.44 (a) 2 V, (b) 16, (c) −1.2 V, 0.3 V
7.48 (a) −0.35 V, (b) 0.8 V, (c) 185.5 Ω
7.50 46.8 mW
7.51 (a) AND gate, (b) NAND gate

Problems for Chapter 8

8.1 (a) 100 Ω, (b) 200 Ω
8.3 25.15 kΩ
8.4 (a) 11.67 mA, (b) 3.33 V
8.6 (a) −2 V, (b) 4 mA, (c) 14 V
8.8 (a) −2 V, (b) 1 mA, (c) 6 V
8.10 (a) 1 mA, (b) −1 V, (c) 9 V
8.12 772 Ω
8.14 (a) 55.56 Ω, (b) 1.4 kΩ, 100 Ω
8.17 (a) 14 kΩ, (b) 500 Ω
8.19 (a) 6 V, (b) 4 mA, (c) 12 V. Active region
8.21 (a) 7.9 V, (b) 5.6 V, (c) 3.2 V
8.23 545 Ω
8.25 10 V, 10 V, 9.2 V, 8.3 V, 7.0 V, 5.6 V, 4.5 V
8.27 OR gate
8.29 AND gate
8.31 (a) $-V_{DD}$, (b) 0
8.33 (a) V_{DD}, (b) 0, (c) $V_{DD}/2$
8.36 (a) $-5 \leq v_1 \leq -3$, (b) $-3 < v_1 < 3$, (c) $3 \leq v_1 \leq 5$

Problems for Chapter 9

9.1 28.2 μA, 2.82 mA, 2.386 V
9.3 11.8 μA, 1.18 mA, 12.91 V
9.5 307 Ω, 290 Ω
9.8 37 μA, 3.7 mA, 2.51 V; 54.9 μA, 5.49 mA, 2.64 V
9.10 (a) 45.4 m$\mho$, (b) 22 Ω, (c) 8.27 kΩ, (d) −4.9
9.12 (a) 1.77 kΩ, (b) −113.6, (c) −50
9.14 (a) −175, (b) 565.6 Ω, (c) 283 Ω, (d) −49.6
9.17 (a) −48.77, (b) −112.2, (c) 1.75 kΩ
9.19 −4.83
9.21 (a) −2 V, (b) 4 mA, (c) 4 V
9.23 (a) 3 mA, (b) 2 kΩ, (c) 6 V
9.26 (a) 6 V, (b) 4 mA, (c) 12 V
9.27 (a) 1 mA, (b) 2 kΩ, (c) 9 V
9.29 (a) 6 V, (b) 4 mA, (c) 6 V

9.32 (a) 4 V, (b) 1 mA, (c) 8 V
9.34 (a) −0.3125, (b) 16.3
9.36 (a) 0.75, (b) 1 MΩ, (c) 400 Ω, (d) 375
9.38 (a) −3.6, (b) 2.17 MΩ, (c) 11,720
9.41 (a) 709 rad/s, (b) 2.54 Mrad/s, (c) 2.54 Mrad/s
9.44 (a) 1250 rad/s, (b) 71.43 Mrad/s
9.49 429 Mrad/s
9.51 $37(v_1 - v_2)$
9.57 45.1 dB
9.59 (a) −3 V, (b) 3 mA, (c) 12 V
9.60 (a) 10/3, (b) −10/3
9.63 $g_m R_o + \frac{1}{2}$
9.64 Same as Equation (9.56)
9.65 28 mV
9.68 99.1 dB
9.71 111 Ω
9.75 (a) 1.5 W, (b) 0.9 W, (c) 0.3 W
9.76 0.2
9.78 (a) 75 Ω, (b) 93.5 Ω, (c) 93.5 ≪ 7000
9.81 (a) 50 Ω, (b) 987 Ω
9.85 (a) 9.65 mA, (b) 8 mA, (c) 0.706 W, (d) 4 W, (e) 0.8 W

Problems for Chapter 10

10.2 (a) 996, (b) 401.6 MΩ, (c) 0.374 Ω
10.6 $(1 + A)R + R_o$
10.9 (a) 100 kHz, (b) 1004 Hz
10.12 $R_1 + R_{in} \| R_2/(1 + A)$
10.15 100 kHz
10.17 0.036 μF, 43.5 kΩ
10.19 65 kΩ, 2.9 kΩ
10.21 1367 Ω, 65.57
10.23 1.6 kΩ, 5 kΩ
10.25 (a) 5 Mrad/s, (b) 0.5 m℧
10.27 (a) 1 Mrad/s, (b) 500 Ω
10.29 (a) 2.236 Mrad/s, (b) 2.240 Mrad/s, (c) 11,180
10.30 (a) 10 V for $v_1 > 5$ V, −10 V for $v_1 < 5$ V
         (b) 10 V for $v_1 > -5$ V, −10 V for $v_1 < -5$ V
10.34 10 V for $v_1 < 3.5$ V, −10 V for $v_1 > 5.5$ V,
         10 V if 3.5 V ≤ v_1 ≤ 5.5 V and preceding $v_o = 10$ V,
         −10 V if 3.5 V ≤ v_1 ≤ 5.5 V and preceding $v_o = -10$ V
10.37 55.9 kΩ
10.39 1.79 kΩ
10.42 0.1 μF

10.44 962 Hz
10.47 20 kHz
10.50 107
10.53 100
10.55 0.89, narrowband

Problems for Chapter 11

11.2 (a) 27, (c) 0.84375, (e) 20.375
11.3 (a) 101011, (c) 0.11011, (e) 10011.101
11.4 (a) 27, (c) 0.84375, (e) 20.375
11.5 (a) 53, (c) 0.66, (e) 23.5
11.6 (b) 545.2
11.7 (b) 100001011.11100001
11.8 (b) 165.4
11.9 (b) 1010110010.101000000111
11.10 (a) 1000000, (c) 10010.101
11.11 (a) 11000, (c) -111.111
11.12 (a) 10010110.00, (c) 1111111.1
11.13 (a) 110.01, (c) 101
11.16 (a) 0001000100100111, (c) 001001100001.01100100
11.17 (b) 3523
11.20 $A + BC$
11.22 $A + \bar{A}B$, equals the OR operation
11.26 $\bar{A}\bar{B} + AB$
11.28 $(A \oplus B) + \overline{(A + B)}$, AND operation
11.30 $(A \oplus B) \oplus (A + B)$, AND operation
11.33 $A + B$
11.38 (a) $\overline{\overline{(A + \bar{B}(\bar{A} + B)} + (A + B)}$,
         (c) $A\bar{B} + \bar{A}B = (A + B)(\bar{A} + \bar{B}) = A \oplus B$
11.42 $F_1 = m_0 + m_1 + m_2 + m_3 + m_7 = M_4M_5M_6$
         $F_2 = m_0 + m_2 + m_4 + m_6 + m_7 = M_1M_3M_5$
11.44 $m_3 + m_5 + m_6 + m_7 = M_0M_1M_2M_4$
11.47 (a) $m_3 + m_4 + m_5 + m_6 + m_7$, (b) $M_0M_1M_2$; $A + BC$ requires the fewest gates
11.49 (a) $m_3 + m_5 + m_6 + m_7$, (b) $M_0M_1M_2M_4$
11.51 (a) cannot be simplified, (c) $AC + BC + AB$
11.52 (a) $(A + B + C)(\bar{A} + B + \bar{C})(A + \bar{B} + \bar{C})(\bar{A} + \bar{B} + C)$,
         (c) $(A + B)(A + C)(B + C)$
11.53 (b) $A\bar{C} + AB + B\bar{C}$
11.54 (b) $(A + B)(A + \bar{C})(B + \bar{C})$
11.55 (b) $(A + \bar{B})(A + \bar{C})$
11.56 (b) $A + \bar{B}\bar{C}$

11.57 (b) $\overline{B}D + C\overline{D} + A\overline{C}D$

11.58 (b) $(C + D)(A + \overline{B} + D)(\overline{B} + \overline{C} + \overline{D})$

11.59 (b) $AB + \overline{B}C + A\overline{C}D$ or $AB + \overline{B}C + A\overline{B}D$

11.60 (b) $(A + \overline{B})(A + C)(B + C + \overline{D})$

Problems for Chapter 12

12.2 (a) $ABC + ABD + ACD + BCD$, (b) $ABC + ABD + ACD + BCD$

12.6 $A + B\overline{C} + B\overline{D} + \overline{C}D, A\overline{B}C + \overline{A}B\overline{C} + \overline{A}B\overline{D} + \overline{A}C\overline{D}$

12.8 $A\overline{C}\overline{D} + B\overline{C}\overline{D} + BCD + \overline{B}C\overline{D} + \overline{A}\overline{B}CD$

12.12 $(A_0 \odot B_0)(A_1 \odot B_1)(A_2 \odot B_2)(A_3 \odot B_3)$

12.14 $B_1 = A_5 + A_6 + A_7 + A_8$, $B_2 = A_3 + A_4 + A_7 + A_8$, $B_3 = A_2 + A_4 + A_6 + A_8$

12.19 $\overline{A}\overline{B}I_0 + \overline{A}BI_1 + A\overline{B}I_2 + ABI_3$; $I_0 = 0$, $I_1 = C$, $I_2 = C$, $I_3 = 1$

12.22 $AB + AC + B\overline{C}$

12.24 upper OR gate: D_0, D_3, D_5, D_6
lower OR gate: D_0, D_4, D_5, D_6

12.26 SR flip-flop

12.29 (a) $\overline{\overline{S}(\overline{\overline{R}Q})}$, (b) $\overline{\overline{S} + \overline{R}} + \overline{S + Q}$

12.32 clocked SR flip-flop

12.37

X	Q_1	Q_2	D_1	D_2	$Q_1(t + 1)$	$Q_2(t + 1)$	Y
0	0	0	0	0	0	0	0
0	0	1	0	0	0	0	0
0	1	0	0	0	0	0	0
0	1	1	0	0	0	0	0
1	0	0	0	1	0	1	0
1	0	1	1	0	1	0	0
1	1	0	1	0	1	0	1
1	1	1	1	0	1	0	1

12.41 $J_1 = K_1 = XQ_2 + \overline{X}\overline{Q}_2 = X \odot Q_2$, $J_2 = K_2 = 1$

12.43 $D_1 = \overline{X}\overline{Q}_1 + Q_1Q_2$, $D_2 = XQ_1\overline{Q}_2$, $Y = (X + \overline{Q}_1)Q_2$

Problems for Chapter 13

13.1 $J_1 = Q_2Q_3$, $K_1 = Q_3$, $J_2 = \overline{Q}_1Q_3$, $K_2 = Q_3$

13.3 $D_1 = \overline{Q}_2Q_3 + Q_2\overline{Q}_3 = Q_2 \oplus Q_3$, $D_2 = Q_1$, $D_3 = Q_2$

13.7 $m_1 + m_2, m_4 + m_8, m_1 + m_4 + m_{16}$

13.13 $D_1 = \overline{Q}_1\overline{Q}_3 + Q_2\overline{Q}_3 + Q_1\overline{Q}_2Q_3$

13.15 $D_1 = Q_1\overline{Q}_3 + \overline{Q}_2\overline{Q}_3 + \overline{Q}_1Q_2Q_3$

13.19 (b) 16

13.20 (b) 8

13.24 (a) 4096 AND gates, each with 12 inputs

13.28 0 V, 1.5 V, 3 V, 4.5 V, 6 V, 7.5 V, 9 V, 10.5 V
13.32 $R_0 = 10R_F$, $R_1 = R_F$
13.35 (a) 25%, (c) 6.25%

Problems for Chapter 14

14.1 159 A
14.3 6.67×10^{-7} Wb/m^2
14.5 795 turns
14.8 (b) 1137
14.10 15.5 A, 4.775×10^5 A-t/m
14.12 (a) 1.46 W, (b) 26.76 W
14.14 0.151 H
14.18 (a) 0.143 J, (b) 0.1146 J, (c) 0.0283 J
14.20 (a) counterclockwise, (b) to the right, (c) positive, (d) positive
14.24 (a) 0.938 J, (b) $9 + j9 = 9\sqrt{2} \underline{/45°}\ \Omega$
14.26 $2\sqrt{2} \cos(6t + 22°)$
14.28 24.17 W
14.31 (a) $\frac{23}{4}\Omega$, (b) $\frac{8}{23}\Omega$
14.34 (a) -0.095, (b) 475.7
14.36 (a) 2400 V, 0.1765 A, 384 W, (b) 44.7 V, 20 A, 400 W
14.38 96.9%, -1.0%
14.40 37.6°, 1.67%

Problems for Chapter 15

15.2 -6.25 A, 267 m/s
15.6 series RLC circuit, dual of circuit in Fig. 15.8(b)
15.7 $\dfrac{Blv - Rf}{RD + (Bl)^2}(1 - e^{-at})$ for $t \geq 0$, $a = \dfrac{D}{M} + \dfrac{(Bl)^2}{RM}$
15.9 40.8%
15.11 29.5%
15.13 100 mA, 25 mA
15.15 1.12 A
15.17 (a) 105.3 Ω, (b) 4 Ω
15.20 (a) 20 kΩ/V, (b) 20 kΩ/V
15.22 9.93 V
15.24 10.02 Ω, 995 kΩ
15.26 17.77 V
15.28 (a) 30 turns, (b) 1500 rpm
15.30 (a) 230 V, (b) 276 V, (c) 266 V
15.32 (a) 0.8 A, (b) 0.64 A, (c) 0.96 A
15.34 (a) 220 V, (b) 41.7 Ω

15.36 0.3 Ω

15.38 (a) 500 A-t/pole, 800 A-t/pole, (b) 300 A-t/pole, (c) 7.5 turns/pole

15.40 (a) 140.43 V, (c) 0.6 A

15.42 136 V

15.45 (a) 950 rpm, (b) 25 A

15.47 (a) 40 A, 350 rpm, (b) 50 A, 1218.75 rpm

15.49 8.5%

15.50 2120 rpm

15.52 (a) 32.62°, (b) 50.91°, (c) 23.39°

15.54 (a) 38.1 A, (b) 0.787 lagging, (c) 296.2 V, 30.96°

15.56 (a) 3.5 Ω, (b) 240.56 V, (c) 18,042 W

15.58 1710 rpm

15.61 0.56 Ω

15.63 (a) 3.5%, (b) 1158 rpm

15.65 (a) 1710 rpm, (b) 150 N-m, (c) 631.4 N-m, (d) 695.2 N-m

Index